Fluid Mechanics and Its Applications

Fluid Mechanics and Its Applications

Second Edition

Vijay Gupta
Vice-Chancellor
Lovely Professional University
Jalandhar

Santosh K. Gupta
L & T Chair Professor of Chemical Engineering
Indian Institute of Technology Bombay
Mumbai

PUBLISHING FOR ONE WORLD

NEW AGE INTERNATIONAL (P) LIMITED, PUBLISHERS

New Delhi • Bangalore • Chennai • Cochin • Guwahati • Hyderabad
Jalandhar • Kolkata • Lucknow • Mumbai • Ranchi

Visit us at www.newagepublishers.com

Published by New Age International (P) Ltd., Publishers
First Edition: 1984
Second Edition: 2011
Reprint: 2012

Branches:

- No. 37/10, 8th Cross (Near Hanuman Temple), Azad Nagar, Chamrajpet, **Bangalore** - 560 018.
  Tel.: (080) 26756823, Telefax: 26756820, E-mail: bangalore@newagepublishers.com
- 26, Damodaran Street, T. Nagar, **Chennai** - 600 017. Tel.: (044) 24353401, Telefax: 24351463
  E-mail: chennai@newagepublishers.com
- CC-39/1016, Carrier Station Road, Ernakulam South, **Cochin** - 682 016. Tel.: (0484) 2377004, Telefax: 4051303
  E-mail: cochin@newagepublishers.com
- Hemsen Complex, Mohd. Shah Road, Paltan Bazar, Near Starline Hotel, **Guwahati** - 781 008.Tel.: (0361) 2513881
  Telefax: 2543669, E-mail: guwahati@newagepublishers.com
- No. 105, 1st Floor, Madhiray Kaveri Tower, 3-2-19, Azam Jahi Road, Nimboliadda, **Hyderabad** - 500 027.
  Tel.: (040) 24652456, Telefax: 24652457, E-mail:hyderabad@newagepublishers.com
- RDB Chambers (Formerly Lotus Cinema)106A, 1st Floor, S.N. Banerjee Road, **Kolkata** - 700 014.
  Tel.: (033) 22273773, Telefax: 22275247, E-mail:kolkata@newagepublishers.com
- 16-A, Jopling Road, **Lucknow** - 226 001. Tel.: (0522) 2209578, 4045297, Telefax: 2204098
  E-mail: lucknow@newagepublishers.com
- 142C, Victor House, Ground Floor, N.M. Joshi Marg, Lower Parel, **Mumbai** - 400 013.Tel.: (022) 24927869
  Telefax: 24915415, E-mail: mumbai@newagepublishers.com
- 22, Golden House, Daryaganj, **New Delhi** - 110 002. Tel.: (011) 23262370, 23262368, Telefax: 43551305
  E-mail: sales@newagepublishers.com

ISBN : 978-81-224-3041-7

₹ 295.00

C-11-09-5837

Printed in India at Ajit Printers, Delhi.
Typeset at In-house.

PUBLISHING FOR ONE WORLD
NEW AGE INTERNATIONAL (P) LIMITED, PUBLISHERS
4835/24, Ansari Road, Daryaganj, New Delhi-110002
Visit us at **www.newagepublishers.com**

PREFACE TO THE SECOND EDITION

We are grateful to all the readers who have accepted this book so graciously. In this, the second edition, Chapter 9 has been re-written so as to make it easier to understand. We have placed the equation-based approach of similitude and modeling ahead of the more intuitive scale-factor approach. We found that our students appreciate this more. Other typos and errors have also been corrected.

We thank Professor Tapan K. Sengupta of the Department of Aerospace Engineering at IIT Kanpur for pointing out some errors in the older edition. In addition, we thank several instructors and tutors of the core course, Fluid Mechanics and Rate Processes, at IIT Kanpur for pointing out minor errors in that edition.

We would like to thank the Lovely Professional University, Jalandhar, Punjab, and IIT Bombay for providing us the enabling environment for revising this book.

Vijay Gupta

Santosh K. Gupta

PREFACE TO THE FIRST EDITION

This book is written as a first course in fluid mechanics for undergraduate engineering students in all disciplines. It has been our experience over a number of years of teaching that the students consider fluid mechanics to be one of the most difficult subjects to grasp. In taking up this project we were vain enough to believe that this problem was partly created by the currently available textbooks and that we could help alleviate it!

Fluid mechanics as a subject is rather tricky, requiring a large number of concepts to be invoked for solving even simple-looking problems. Also, the wide varieties of phenomena that are involved, with the flow behaviour changing suddenly with small changes in the values of parameters, make the subject frustrating to a beginner. Several popular textbooks, in their attempt to make concepts rigorous and elegant, delve to great depths into the mathematics of fluid flow using vector and even tensor calculus. But in doing so the emphasis shifts, often unintentionally, from the phenomena themselves to the mathematics. We believe that fluid-mechanical phenomena are complex enough requiring the full attention of the student without it being diverted by the demands of the newly-learnt mathematical tools which come in the way of developing a physical feel of the subject. It is because of this reason that we have kept the level of mathematics in this text to the minimum. We have used vector calculus only sparingly, and for this we have restricted most derivations to two-dimensions only, extending the results to the third dimension. We believe the elegance of vectorial or tensorial mathematics can be safely left to a second course, after the student has a firm grasp of the fundamental physical concepts.

In view of the importance we attach to the understanding of physical phenomena, we have brought forward in Chapter 1 an introduction to the various flow phenomena encountered in simple situations so that a student gets an idea of what kind of things he is to look for and becomes familiar with certain specialized vocabulary of fluid mechanics. We have found that this helps a student to become aware of the major flow phenomena and various fascinating applications of fluid mechanics and kindles an interest which sustains him through the following chapters. It should perhaps be pointed out that a beginning student should not spend too much time on Chapter 1 trying to figure out the 'whys' of the phenomena. He should only become familiar with them and come back to this chapter after the Epilogue when, hopefully, he should be able to explain.

After a survey of fluid statics in Chapter 2, we develop the basic equations of fluid flow in Chapters 3 to 7. The treatment is as rigorous as possible within the constraints that we have explicitly set for ourselves. In the initial portions, the integral approach is favoured because it helps the student get a better feel of the subject. Later, we switch to the differential approach because the detailed information is provided only by it.

Similitude and modelling has been introduced in Chapter 9 where we have made the most radical departure from current textbooks. We have found in our teaching that with the scale-factor approach that we use, it is possible to make the previously mysterious dimensional analysis more understandable. This approach, which is much more physical than the Buckingham Pi theorem route, permits the students to handle far more complex problems of modelling easily.

In Chapter 11 we have introduced a rational basis of making approximations in fluid mechanics. We find that our method of introducing boundary layers as a problem of multiple characteristic lengths finds greater favour with students than the traditional Prandtl approach or the currently-in-vogue singular-perturbation method. Our approach is more fundamental than the former but more physical than the latter.

In Chapters 12 and 13 we introduce potential flow and boundary-layer flows, along with associated concepts. In Chapter 15 the effects of compressibility are introduced. We concentrate on only the most dramatic effects and describe the whole series of flow phenomena that occur when a body accelerates through transonic speeds. In Chapter 16 we have introduced the basic structure of turbulence. This chapter is different from the rest in that it tends to explain and describe only the physical processes and does not attempt to build any problem-solving capability.

We have collected the major engineering applications into three chapters: 8, 10 and 14. This format was chosen after very careful and extensive thought. We found that each of these applications involved more than one concept and hence strained the logic of topic-oriented chapters, if incorporated there. In addition, the present format permits a bit more digression into aspects of problems which are not directly obtained from the usual level of mathematics used in the other chapters, and allows the instructor to select applications of his own interest in his teaching.

In an elementary book like this, we naturally do not claim any originality except in the method of presentation and organization. We think that certain novel (for an elementary textbook) features, like the recasting of the Euler acceleration formula into a physically more meaningful form, a simpler derivation of Reynolds transport theorem, a new approach to modelling, similitude and dimensional analysis, a new introduction to boundary layers and the basis of approximations, a simpler derivation of the Bernoulli equation, etc., will benefit the student. We have borrowed heavily from our teaching experience to anticipate the difficulties, misunderstandings, misapplications and over-generalizations that a student is prone to, and have cautioned against them.

A large number of exercise problems have been included to help the student develop his grasp further. Many problems are new and few are repetitive or simple numericals. Many extend the scope of the treatment presented through carefully guiding the student along. While compiling the problems, we have drawn from the literature of diverse engineering disciplines and have thus provided a flavour of a multidisciplinary approach. *A Teacher's Manual is available on request from the publishers.*

It is possible to cover most of the material presented in this text in a 3-hour-per week, one-semester course, with tutorial support. It is also possible to use this text in a shorter course by omitting some of the material in Chapters 8, 10, 11 and 14, and possibly the whole of Chapters 15 and 16.

Finally, it is our great pleasure to express our most sincere gratitude to Prof. J. Srinivasan for the more than considerable amount of effort and time he spent sitting with us and carefully going through many earlier drafts of this text. We doubt if this text could have evolved into this format without the benefit of his uncompromising stand for exactitude. Prof. T. A. Wilson was kind enough to go through the typescript and offer several suggestions for improvement. We also thank Professors K. S. Gandhi, R. K. Malik, M. M. Oberoi and R. Singh for their critical review of various parts of the manuscript and to a large number of co-teachers and students at IIT Kanpur who have used this text and have given us their valuable comments. We also acknowledge the financial support received from the Curriculum Development Centre of the Quality Improvement Programme, IIT Kanpur for preparation of the manuscript, to Mr. S. J. Gupta and Mr. U. S. Misra for their excellent typing, Mr. B. L. Arora and Mr. S. S. Kushwaha for the drawings and Mr. S. S. Chauhan for help in preparation of the final typescript. And, most pleasantly, we must thank our wives for providing us tea and snacks and joining us at the right moment in order to cut short our often lengthy arguments and for keeping the children from tearing our manuscript to shreds.

Vijay Gupta
Santosh K. Gupta
IIT, Kanpur

CONTENTS

1

INTRODUCTION TO FLUID FLOWS

1.1 INTRODUCTION

Fluid mechanics is concerned with the study of the motions of fluids (i.e., liquids and gases) and with the forces associated with these. The subject is of great interest for two reasons. First, an understanding of fluid mechanics helps us to explain a variety of fascinating phenomena around us. Second, an understanding of this subject is essential to solve many problems encountered by an engineer.

We live in a thin layer of air that blankets the surface of the earth. The local and global movements of the air determine our weather. The origins of tornadoes, hurricanes and the monsoon can be understood only through the use of the laws of fluid mechanics. The availability of water has been associated historically with the development and flourishing of many civilizations. To utilize the available water resources optimally, we need to predict the flow rates of water in the rivers during different seasons. The prediction of floods in rivers is equally important. In recent years, we have realized that large scale discharge of effluents into the atmosphere, sea, lakes and rivers has led to serious problems of pollution. In order to control pollution, one has to know the rates of mixing and dispersion of pollutants in these natural 'sinks' nature has provided us with. The rates of mixing are affected to a large extent by the local flow patterns and, therefore, an understanding of air and water movements in the atmosphere, rivers, etc., is required before these rates of dispersion can be calculated.

We are concerned with fluids in a more intimate sense as well. The flow of blood through our arteries and veins and the flow of air through our respiratory passages into the lungs, are examples of fluid motion which one needs to understand in order to deal with circulatory and pulmonary disorders. Artificial heart-lung machines (Fig. 1.1) have been made possible only after a thorough understanding of the fluid mechanics of blood flow in the heart, and the exchange of oxygen and carbon dioxide between the blood flowing on one side and inhaled air on the other, in the lungs. Similarly, artificial kidneys are now available, which duplicate the flow of blood

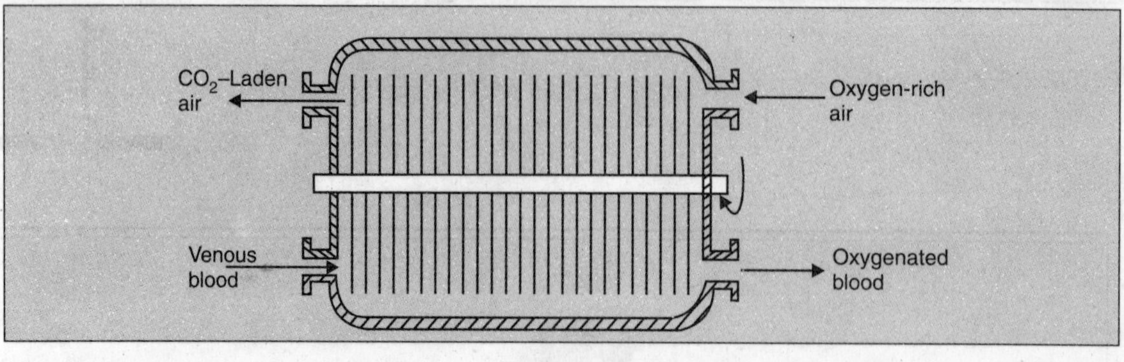

Fig. 1.1. A rotating disc artificial lung. The rotating discs pick up a thin layer of blood which is oxygenated as it comes in contact with oxygen-rich air.

through the kidneys with continuous removal of the waste products from the blood as it passes through the machine. Certain voice disorders can be treated, now, with a thorough understanding of how the exhaled air interacts with the vocal cords and makes them vibrate. The relatively new science named *tribology* is focussing attention, among other things, on the lubrication of various joints in the body.

Most of the engineering applications of fluid mechanics are related to two aspects of fluid motion: one, the forces which cause or result from these motions, and the other, the effect of these fluid motions on the rates of transfer of heat and of mass (e.g., dispersion of pollutants) through the fluid body. The forces in fluids are put to such diverse uses as a sail boat, a wind mill, a hydraulic transmission, and in controlling the motion of an aircraft or a spacecraft, and even in the curving of the trajectory of a tennis ball.

Fluid motion is used to modify heat and mass transfer rates in heat exchangers, cooling towers, boilers, chimneys, artificial kidneys and heart- and- lung machines, and in the manufacture of semiconductor devices, protection of spacecrafts from intense heating during re-entry in the earth's atmosphere, etc.

A knowledge of fluid mechanics is essential in such diverse branches of engineering as aeronautics, astronautics, automotive engineering, biomedical engineering, structural engineering, mining and metallurgical engineering, naval architecture and nuclear engineering.

1.2 FLUIDS

Fluids, as a class of matter, are distinguished from solids on the basis of their response to an applied shear force. If a solid bar is subjected to a torque, it twists (Fig. 1.2). The restoring elastic stresses in a solid (below the yield limit) are proportional to the strains, and therefore, a solid, when subjected to a torque, distorts through an angle θ (equilibrium distortion) such that internal stresses are developed which just balance the applied torque. The magnitude of the angle θ depends on the applied torque as well as on the elastic properties of the solid.

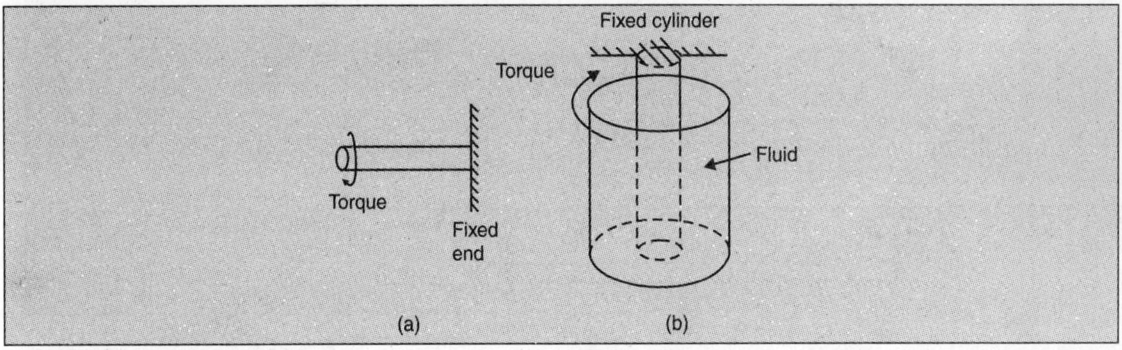

Fig. 1.2. Difference between a solid and a liquid. The solid bar in (a) will acquire an equilibrium deformation, while the fluid in (b) will continue to deform under the action of a torque.

If, however, the torque is applied to a fluid, the behaviour is entirely different. The fluid does not acquire an equilibrium distortion but continues to deform as long as the torque acts. This behaviour is used to define a fluid. Thus, a fluid is a substance which cannot be in equilibrium under the action of any shear force, howsoever small.

Although a fluid does not resist a shear force by acquiring an equilibrium deformation, that is, the outer cylinder in Fig. 1.2 does not have an *equilibrium position* under the action of a torque, it, however, has an *equilibrium velocity*. This equilibrium value increases with the applied torque. This suggests that a fluid does resist a shear force,* not by acquiring an equilibrium deformation but by acquiring an equilibrium *rate* of deformation. Thus, a fluid deforms continuously under the action of a shear force, but at a finite rate determined by the applied shear force and the fluid properties.

1.3 VISCOSITY

The property which characterizes the resistance that a fluid offers to applied shear forces is termed *viscosity*. This resistance does not depend upon the deformation itself (as is the case with solids) but on the rate of deformation. Consider a fluid confined between two parallel plates, with the lower plate stationary and the upper plate moving with a velocity V_0 (Fig. 1.3). The upper plate sets the fluid in motion with a velocity V_x, which is a function of y, the vertical distance measured from the lower plate. Extensive experiments have shown that, for all real fluids possessing any viscosity, however small, the fluid particles in immediate contact with any solid surface, move with the velocity of the surface itself. That is, there is no relative motion between the fluid near the surface and solid surface itself. This condition is termed the *no-slip condition*, and holds good for all fluids except super-cooled helium. We shall require here that $V_x = 0$ at $y = 0$ and $V_x = V_0$ at the moving plate and thus, V_x changes with y. It can be seen that the rate of deformation of the fluid, in such a simple geometry, is (Fig. 1.3)

$$\text{Rate of deformation} \quad = \frac{\partial}{\partial t} \text{ (shear strain)}$$

* Otherwise the cylinder will continue to accelerate.

$$= \frac{\partial}{\partial t}\left[\frac{1}{\delta y}\left\{\left(V_x + \frac{dV_x}{dy}\delta y\right)\delta t - V_x\ \delta t\right\}\right]$$

$$= \frac{dV_x}{dy} \qquad \text{...(1.1)}$$

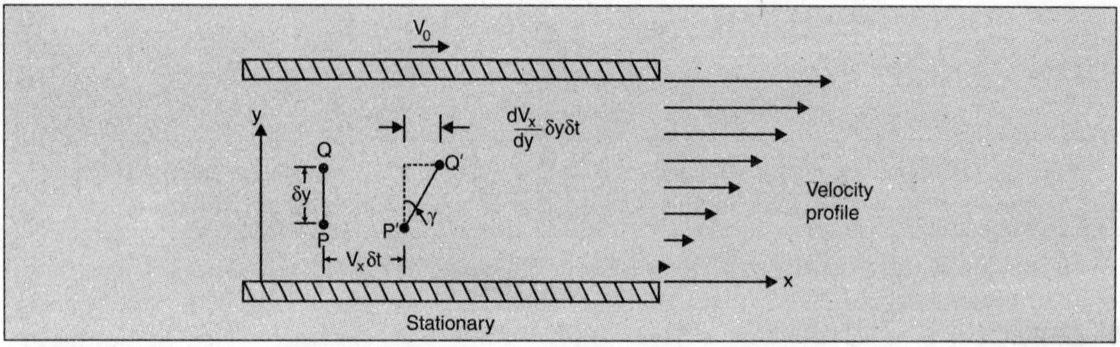

Fig. 1.3. Flow between two parallel plates.
The line PQ moves to P'Q' in time δt resulting in a shear strain γ.

Newton's law of viscosity states that the stresses which oppose the shearing of a fluid are proportional to the rate of shear strain, i.e., the shear stress, τ, is given by

$$\tau = \mu\frac{dV_x}{dy} \qquad \text{...(1.2)}$$

for such simple flows. The coefficient μ is termed the *viscosity** (or the dynamic viscosity) and plays an important role in the study of forces in fluid flows.

The viscosity of some fluids like air, water and glycerin is almost constant over a wide range of rates of deformation. This implies that the shear stress varies linearly with the rate of strain. Such fluids are termed as *Newtonian fluids* and in this book we shall confine our attention to such fluids only. The fluids in which the shear stress does not vary linearly with the rate of strain are termed as non-Newtonian. Blood, grease and sugar solutions are some common non-Newtonian fluids. There are also some substances which cannot be classified as either fluids or solids, but show intermediate behaviour. These are called viscoelastic fluids. Both non-Newtonian and viscoelastic fluids fall outside the scope of this text. Figure 1.4 gives a partial classification of substances based on their rheological (i.e., shear stress *vs.* rate of strain) behaviour. An ideal fluid with zero viscosity plays an important part in the study of fluids. Such a fluid offers no resistance at any rate of strain, and therefore, the upper plate in Fig. 1.3 will move with an ever increasing velocity even with the slightest of forces, if the gap between the two plates is filled with an ideal fluid.

* From Eq. 1.2, it can be seen that the units of viscosity are Pa s (which is the same as kg/ms). A commonly used unit is centipoise (cp) which is equal to 10^{-3} Pa s.

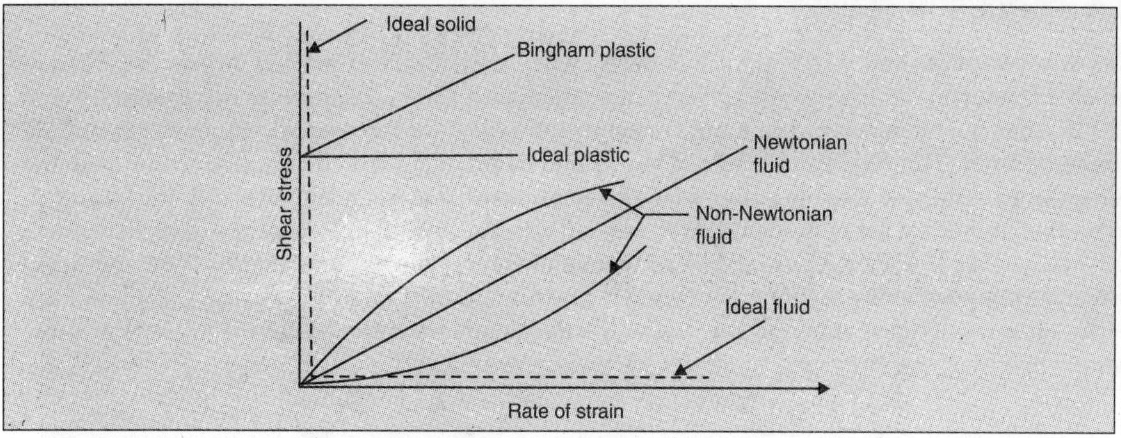

Fig. 1.4. Rheological classification of matter.

1.4 EFFECT OF VISCOSITY

Consider a very large plate initially at rest in a large expanse of a stationary fluid. At time $t = 0$, the plate starts to move with a constant velocity V_0. The layer of fluid in the immediate vicinity of the plate moves with it, so that there is no relative motion between the solid and the fluid in immediate contact with it (by the no-slip condition).

As soon as the fluid in immediate contact with the plate starts moving with a finite velocity, the action of viscosity comes into play. The viscous forces tend to drag other layers of fluid along as well. Figure 1.5 shows the velocity variations normal to the plate at various times. It is noticed that at any given time, the velocity decreases rapidly from its value V_0 as we move away from the plate and soon becomes negligible. The distance over which its value reduces to a fixed fraction of V_0 (usually 1%) is termed as the *penetration depth* and signifies the distance through which the effect of the impulsive plate motion has penetrated into the fluid.

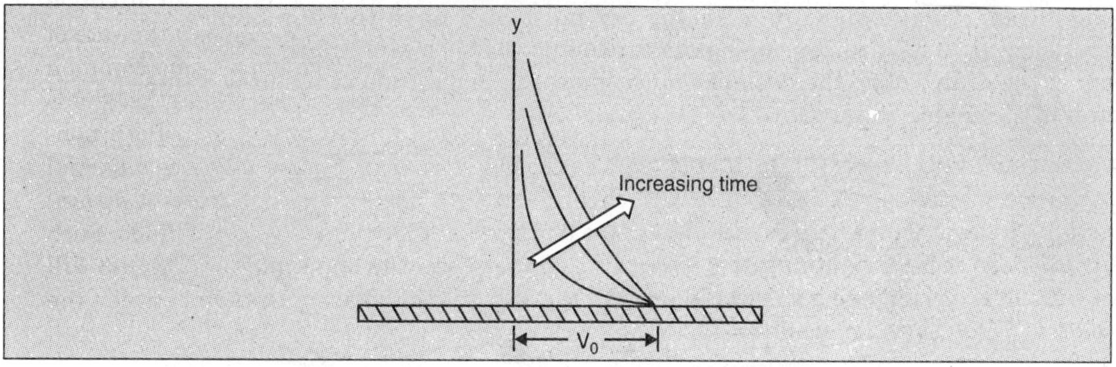

Fig. 1.5. Velocity profiles over a flat plate which is set in motion impulsively.

The penetration depth increases with time. Note that this penetration (of the motion of the plate) is solely due to the action of viscosity, and if the viscosity were zero, this diffusion of fluid velocity (or momentum) into the interior would not have taken place. It can be shown that m can be taken as a measure of the rate at which fluid momentum diffuses.

1.5 FORCES IN FLUIDS

For over two thousand years, man has been aware that fluids in motion or even at rest are capable of exerting forces on solid objects in contact with them. Archimides discovered around 250 BC that a solid immersed in a fluid experiences a buoyant force equal to the weight of fluid displaced by it. The Roman builders of the famous aqueducts which supplied water to Rome across large distances were familiar with the relationship between the rate of flow of water in a channel and the slope of its bed. But it was only in the seventeenth century that the French mathematician B. Pascal clarified the nature of pressure–the force (measured per unit area) which stationary fluids exert on a surface. He postulated that the pressure at a point in a fluid is the same in all directions and this led to the development of the hydraulic press (Fig. 1.6).

Fig. 1.6. Hydraulic ram or press.

It is evident from our common experience of walking against a strong breeze that fluid streams moving past a solid body exert a force on it in the direction of fluid flow. Similarly, a body moving through a stationary fluid experiences a force opposite to the direction of motion. The existence of this force, termed as *drag,* has been known to man for a long time. He had to overcome the drag of water when he propelled his boat or ship. He also found by experience that the shape of the hull, the portion of the boat in contact with the water, controls the drag to a large extent, and that a cusped hull gave the lowest drag (Fig. 1.7). An engineer is often called upon to calculate drag forces and to control them. A ship designer wants a hull leading to the lowest drag. An aeroplane or a racing car must also have the minimum possible drag corresponding to its size, for one has to expend power for maintaining motion of the vehicle against the drag.* Also, the designer must know the magnitude of the drag to prescribe the power of the engine required.

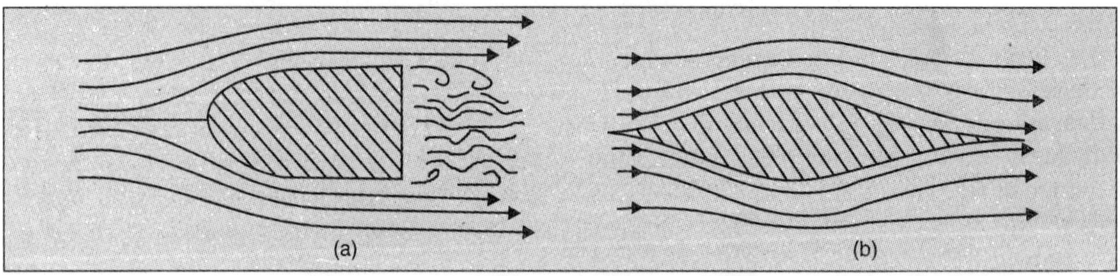

Fig. 1.7. The cusped hull in (b) gives a lower drag.
Lines represent the pattern of water flow as observed from the boat.

* The drag becomes more and more important at high speeds as will be discussed in Chapter 13.

There are several applications where an engineer wants to *maximise* the drag. A parachute exploits the large drag its canopy experiences. Its designer must prescribe a canopy diameter large enough to provide sufficient drag to overcome most of the weight, but not so large that the downward velocity becomes frustratingly low. One part of the wings of aeroplane is raised on landing so as to be perpendicular to the direction of motion and thus, substantially increase the drag force and reduce the landing run (Fig. 1.8). Ships use a similar drag-increasing device for applying brakes.

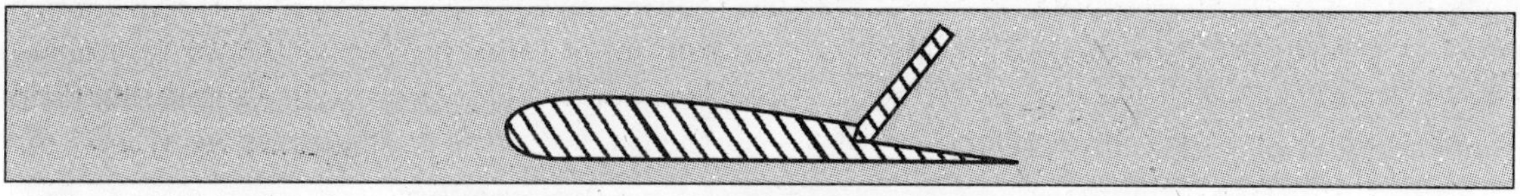

Fig. 1.8. Air brakes in an aircraft. The figure represents the cross-section of the wing. A pivoted flap is raised while landing. It increases the drag and shortens the landing run.

Some equipment used for separating light from heavy solids (or solids of different densities) in chemical and metallurgical industries rely on differential drag forces. In one common separator called gravity settling chamber which is used for pollution control, dust laden gases enter a vessel as shown in Fig. 1.9. The smaller particles experience a smaller transverse drag than the larger particles, and thus, are decelerated less in the horizontal direction. Because of this difference in drags, the larger particles settle to the bottom closer to the point of entry A than the smaller particles which collect at B. The separation of wheat from chaff in a stream of air, as the two are dropped slowly from above the ground, works on the same principle. The chaff, which presents a larger surface area for drag forces, is blown farther than the wheat.

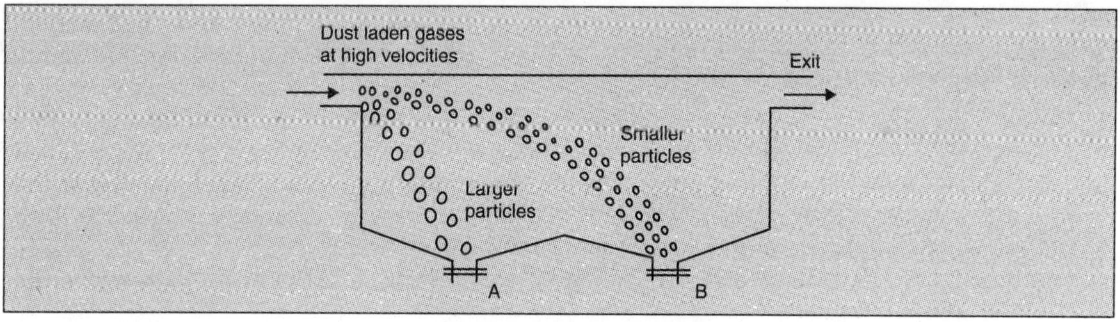

Fig. 1.9. Gravity settling chamber.

The drag that a body experiences while moving through a fluid is opposed to the direction of its motion.* A body can also experience a force perpendicular to its direction of motion. This is demonstrated by the fact that aeroplanes (which are heavier than air and do not have sufficient buoyancy) can fly. The force that balances the weight of the aircraft is perpendicular to the direction of flight (and hence to the direction of relative wind) and is called lift (Fig. 1.10). A cricket ball curves (swings) in flight because of an aerodynamic force acting normal to the direction of flight.

---

* If both the body and fluid move, the drag acts against the *relative* velocity direction.

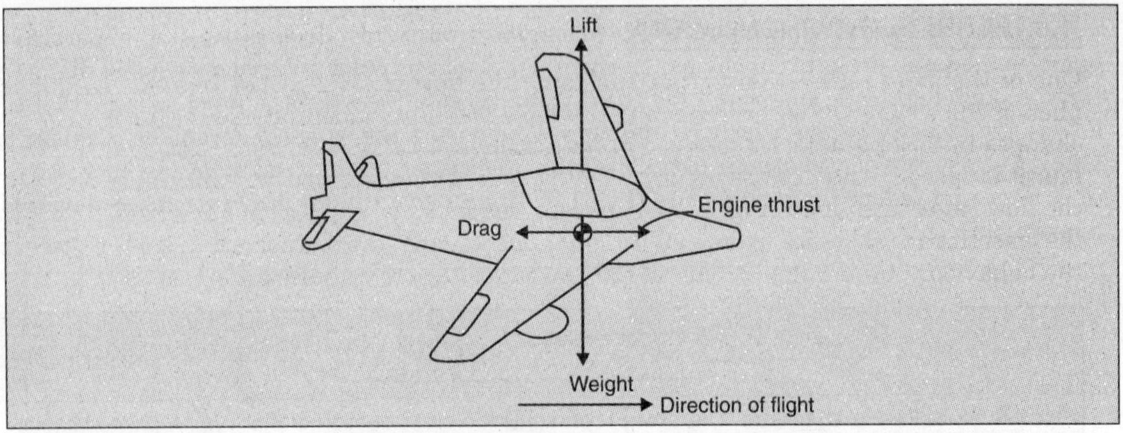

Fig. 1.10. Forces on an aircraft in flight. Aerodynamic lift balances the weight and the engine thrust overcomes the drag due to the forward motion in air.

One of the more intriguing features of the forces due to fluid flow is the possibility of periodic forces *even* when all the imposed conditions are time independent. Examples of the transverse periodic force acting on bodies when air flows past them include the excitation of the vocal cords in sustained vibrations as air from the lungs is exhaled through gaps in between the cords (see Fig. 1.11). These vibrations produce sound (which, when modulated, results in speech). There are several other examples where such fluctuating forces due to steady fluid motion come into play. Overhead telephone wires 'sing' when wind blows steadily past them. In 1940, the suspension bridge at Tacoma Narrows, Washington, USA, started oscillating wildly during a storm and ultimately collapsed. It was obviously a case when the wind, even though largely steady, applied periodic forces on the bridge at a frequency near its natural frequency, causing resonance. To prevent such catastrophies, a bridge designer or a designer of tall or long structures (like skyscrapers, chimneys, etc.) must calculate the frequency and intensity of periodic forces acting on it due to the wind and make the structure strong enough to withstand these oscillating forces.

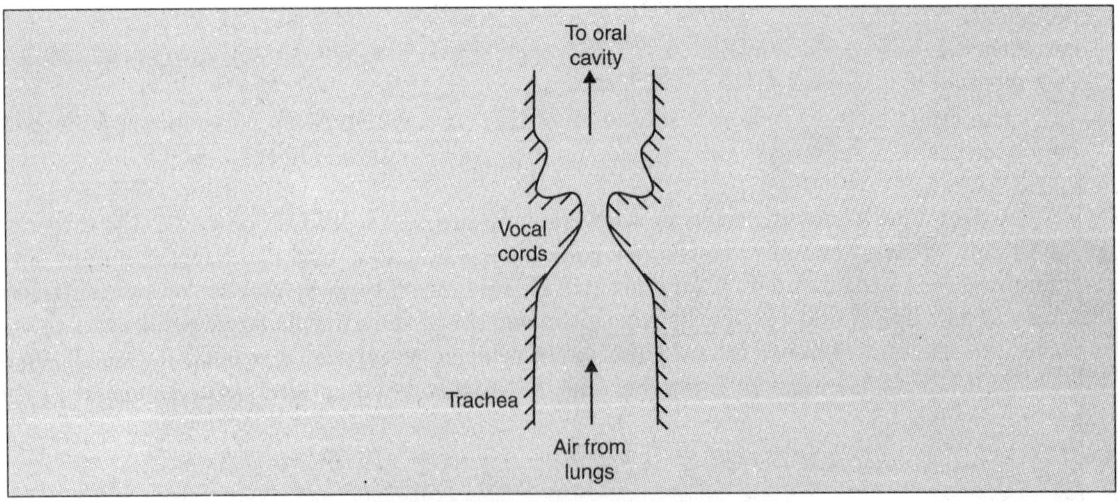

Fig. 1.11. Vocal cords (sectional view).

1.6 FLUID-FLOW PHENOMENA

One of the more serious hurdles in analysing the flow of fluids is the bewildering range of phenomena which may occur in seemingly simple flow situations; and how, at times, very small changes in flow parameters produce drastic changes in flow behaviour. Thus, when a water faucet is opened, water comes out initially in a smooth 'transparent' stream and remains so as the flow rate is increased (Fig. 1.12). But then, suddenly, when a critical flow rate is exceeded, the smooth stream breaks up into an irregular one. Clearly, the mathematical models describing the behaviour of the water stream in the two cases have to be quite different.

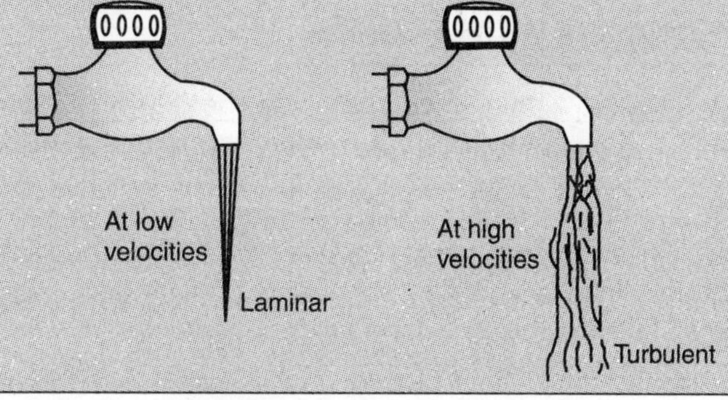

Fig. 1.12. Two types of flow.

Consider next, the drag experienced by a circular cylinder in a uniform air stream. When the drag force is measured at various velocities, varying from very low to very high values and the non-dimensional drag (defined as drag/$\{\frac{1}{2}$ density × velocity^2 × frontal area$\}$ and called *drag coefficient*) is plotted against the flow velocity, we obtain a plot as shown in Fig. 1.13. The striking thing about this non-dimensional drag *versus* velocity curve is its complexity. It will be seen shortly that as the velocity increases, a series of fluid-flow phenomena unfolds itself. Each change in the nature of the curve in Fig. 1.13 corresponds to a major change in the flow behaviour. For this reason, it has not been possible to construct a single model that will predict correctly the drag coefficient over the entire range of velocities. The knowledge of fluid behaviour in a particular range will help in modelling the flow for that range of velocities.

The study of fluid flow is full of such surprises. The following sections are intended to introduce the wide variety of flow phenomena that are encountered in even simple flow situations

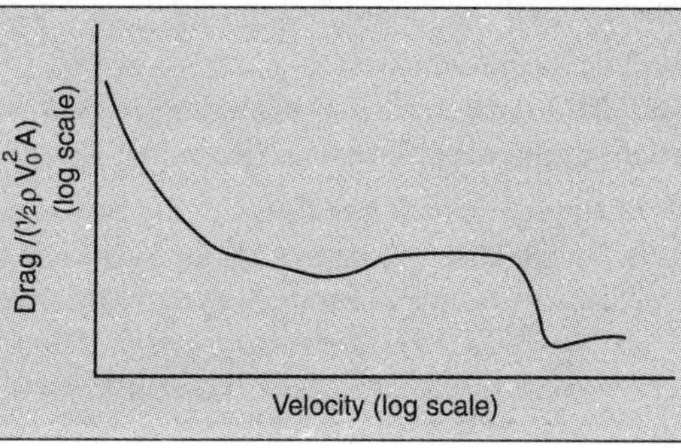

Fig. 1.13. Variation of non-dimensional drag of a cylinder *vs.* free-stream velocity.

which form the subject matter of this text. This introduction will hopefully help in gaining a better insight for modelling of flow problems, and may help develop a perspective of the subject which is constantly threatened by mathematical complexities.

The phenomena described herein are restricted to those observed in flows that are largely incompressible, i.e., in which the density changes are insignificant. This covers almost all flows involving liquids and low velocity flows of gases. Some fluid phenomena peculiar to compressible flows will be discussed later in Chapter 15. Also, it should be noted that only a few of the important phenomena have been discussed here. Others are beyond the scope of this text.

1.7 FLOW PAST A CIRCULAR CYLINDER

One important class of fluid flow phenomena concerns the relative motion between a fluid and the object submerged in it. The flow of air past a telephone wire or a bridge or the motion of an aircraft in a stationary atmosphere are some examples. The last flow situation is unaltered if we fix our reference system with the aircraft so that it appears stationary with air blowing past with the same relative velocity. To study the essential flow phenomena in such situations we take an idealized geometry. This is a long circular cylinder held with its axis normal to a steady stream.

To identify the various fluid flow phenomena associated with this geometry we use a technique termed *flow visualization.* In this direct method of observation of the fluid behaviour an attempt is made to identify some fluid particles and visually follow their motion. One of the ways is to introduce a coloured fluid (e.g. smoke in the case of air, dye in the case of water) at some selected points upstream of the cylinder and then record the motion of these particles, either visually or photographically.

As the speed of the flow is varied from very low to very high values, a series of changes in the type of flow pattern is observed. These changes coincide with the changes in the nature of the drag *vs.* velocity curve (Fig. 1.13). Some of these patterns shall be described. But before doing this, it should be pointed out that the velocities at which transitions from one type of flow pattern to another occur depend upon various flow parameters such as the diameter D of the cylinder and fluid properties like density ρ and viscosity μ. If, however, we non-dimensionalize V_0, the velocity far up-stream, by dividing it by $\mu/\rho D$ (having the dimensions of velocity), it is found that the transitions occur at fixed values of $\rho V_0 D/\mu$ the non-dimensional velocity.* Thus, a given transition in the flow of water over a cylinder will occur at about 1/13 of the velocity at which it occurs in flow of air over the same cylinder because μ/ρ of air is about 13 times that of water. Therefore, in the discussion of the flow patterns we will use $\rho V_0 D/\mu$ as the parameter instead of the velocity V_0. This parameter is termed as *Reynolds number* (and denoted as Re) after the 19th century British physicist Osborne Reynolds, who first discovered its significance.

At very low values of Reynolds number ($\mathrm{Re} \ll 1$) the lines indicating the paths of the fluid particles are shown in Fig. 1.14. The important thing to note in this pattern is its symmetry. The velocity of a fluid decreases along OA as it approaches the cylinder. Point A, where the fluid particles come to rest, is called the *stagnation point.* The velocity increases from A to B (or A to B') attaining the maximum at B (or B'). The fluid decelerates to C which is another

* It will be seen in Chapter 9 that we can use the concept of similitude to arrive at this conclusion.

stagnation point, and then accelerates to O', such that far downstream the velocity has the same value as that far upstream. The effect of the presence of the cylinder is felt over large distances, that is, the local velocity at points many cylinder-diameters away is significantly different from the free-stream value V_0. The flow pattern is completely reversible, that is, by looking at Fig. 1.14, one cannot tell whether the flow is from left or right.

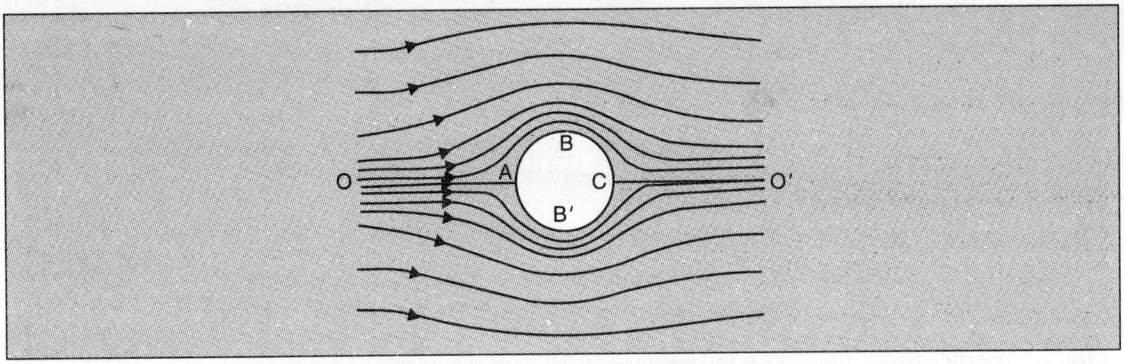

Fig. 1.14. Flow across a circular cylinder at Re $\ll$ 1.

As the flow velocity and consequently Re increases, a number of new phenomena appear. One of them is the development of fore-and-aft asymmetry. Two small regions are formed just behind the cylinder in which the fluid does not flow downstream as it does elsewhere, but whirls around in closed paths. These regions of recirculating flows are termed *attached eddies* and are similar to whirlpools in rivers (Fig. 1.15). One consequence of the existence of these eddies is that the flow no longer 'follows' the contour of the body and is said to be *separated*. In Chapter 13 it shall be seen that such separated flows result in a great increase in the drag experienced by the body. It will also be seen that this tendency to separate from the solid boundary is present whenever the flow is decelerating.

Another major effect that the increasing Re has on the flow picture is the modification of the region of disturbance to the flow stream. The disturbed region (where flow velocity is significantly different from the free stream value V_0) upstream of the body and on its sides contracts, while it elongates in the downstream direction. This tendency continues indefinitely as Re increases to very large values. The region of disturbance downstream of the body is termed its *wake*.

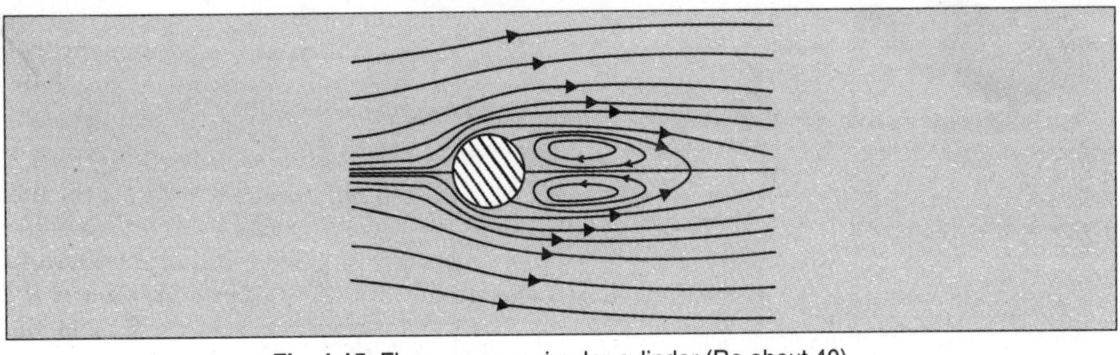

Fig. 1.15. Flow across a circular cylinder (Re about 40).

As Re is further increased, the size of the two attached eddies increases. The large attached eddies in which the fluid recirculates make the flow unstable, and consequently when the Reynolds number exceeds a certain value (of about 40) the eddies are 'shed' from the surface and are swept downstream (Fig. 1.16). Once an eddy is shed and swept away, it is seen that a new one forms in its place and is shed and swept away in due course. The two eddies on either side (top and bottom) of the cylinder are shed alternately, giving rise to an unsteady,

Fig. 1.16. Vortices shed alternately from a circular cylinder form the Karman vortex street (Re about 100).

albeit periodic, wake flow. The periodicity in the flow arises spontaneously even when all the parameters determining the flow are apparently steady. This periodic shedding of eddies (also termed vortices) gives rise to periodic side (transverse) forces on the cylinder. These vortices lead to the singing of telephone wires as described in Sec. 1.5. The wake of such a flow, with a series of staggered vortices shed from either side of the cylinder and swept downstream by the flow, is termed as the Karman vortex street after Th. von Karman who first studied these systematically.

Figure 1.17 shows the variation of velocity with time, at a point some distance away in the wake of the cylinder, for a series of Reynolds numbers. It is noticed that for low values of Re, the velocity at a fixed point varies almost sinusoidally, but, as Re increases, the periodicity

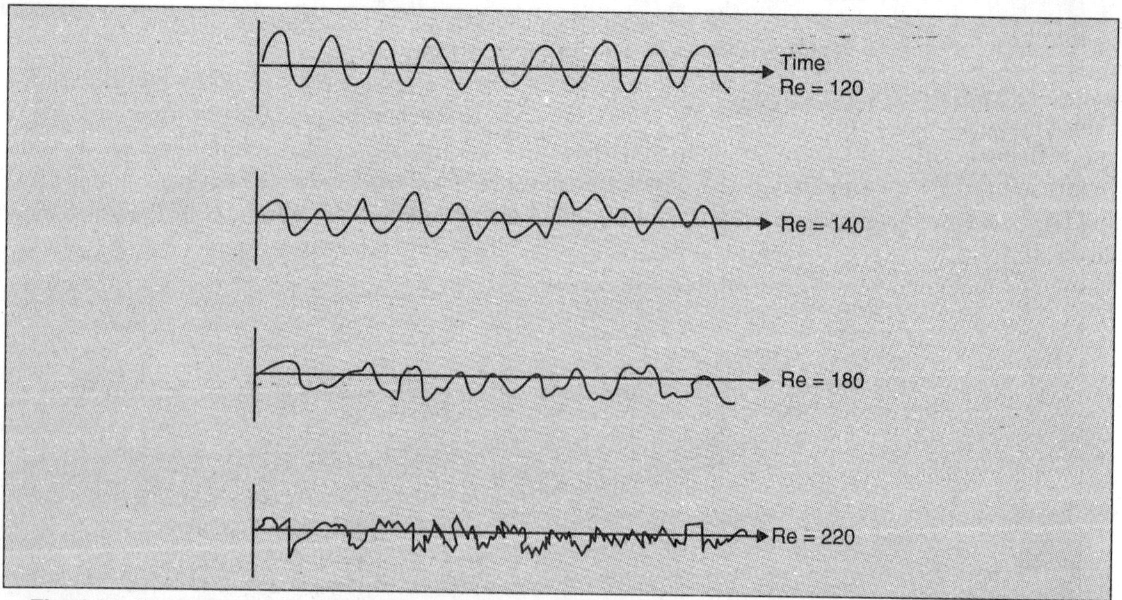

Fig. 1.17. Fluctuations in velocity with time at a fixed point in the wake of a cylinder. The change from periodic variations to random variations denotes that the wake becomes turbulent at Re about 200.

is lost, indicating that the model of flow with regularly shed vortices on either side of the cylinder no longer holds. At Re of about 200, the flow in the wake acquires a random character with highly irregular and rapid fluctuations of velocities, both with time and location, even though the parameters defining the flow are held steady. Such a condition is termed as *turbulence* and the flow is termed as *turbulent*. This flow behaviour contrasts with the earlier behaviour where the flow, termed as *laminar,* varies smoothly. Turbulent flow occurs in all kinds of flow situations including flow past bodies, flow through conduits, flow in atmosphere, etc. In fact, turbulence is the more prevalent mode of flow.

Rapid fluctuations in turbulent flows result in much better mixing of the fluid and this results in higher rates of heat transfer and of dispersal of pollutants, etc. This is, therefore, the preferred mode of flow in heat exchangers and in mass exchangers wherein high rates of transfer are desired. In an artificial lung machine, where blood must exchange carbon dioxide with oxygen in the air, turbulence is created in the air flow to obtain the maximum rates of transfer for the given surface area. Similarly, in the tempering of massive steel parts, turbulence is induced in the cooling water in order to have rapid rates of heat transfer. In fact, our common morning ritual of stirring tea to dissolve sugar rapidly is a good example of turbulence being created to increase mass transfer rates.

Turbulence is not always a desired state of flow from an engineer's stand-point. The increased mixing, which gives high heat and mass transfer rates, also results in larger dissipation of kinetic energy. One effect of this is increased fluid friction on surfaces moving through fluids. Thus, in the motion of aircrafts or ships, turbulence in the flow is best avoided. This, however, is not true of all bodies and we shall soon see that under certain conditions, turbulence may actually reduce the total drag experienced by a body, by reducing the width of the wake behind it. The golf ball is dimpled for this very reason. Since the dimples induce turbulence, the drag force on a common golf ball is less than what it would be in the absence of dimples and, thus, the ball travels through a larger distance in air for the same effort.

Coming back to the flow past a circular cylinder, when the Reynolds number is of the order of 100, it is observed that most of the variations in velocity (in the front portion of the

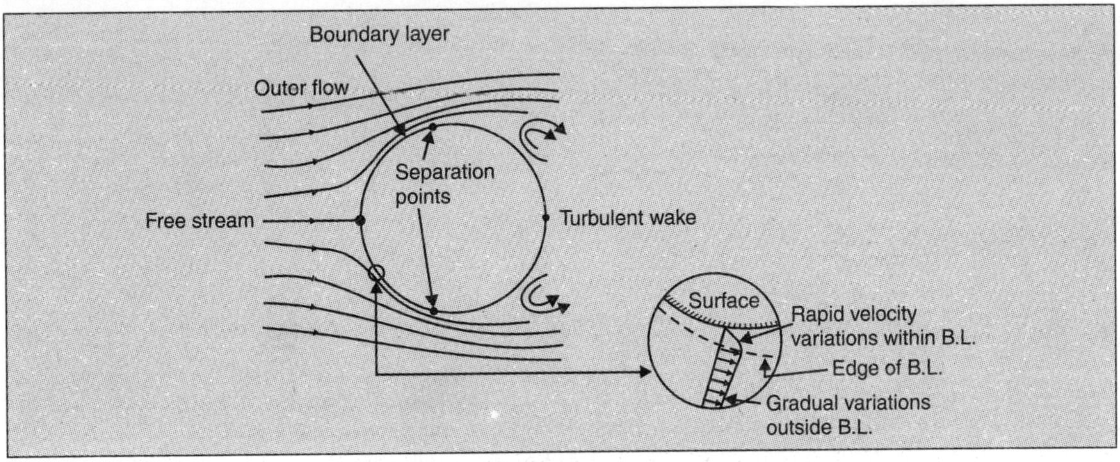

Fig. 1.18. Boundary layer on a cylinder (Re < 10^5).

cylinder) are confined to a thin region near the surface. This thin layer (thin as compared to the cylinder-diameter) is termed the boundary layer. Outside this layer, the velocity variations are relatively moderate (Fig. 1.18). Because of the rapid variations in velocity across the boundary layer, shear stresses are significant in this layer. Outside the boundary layer, the shear stresses are smaller since the velocity variations are gradual. It was shown by Ludwig Prandtl that we can model the flow with the effects of viscosity confined to within the boundary layer. In the region outside this layer the shear stresses can be neglected. Thus, the fluid in the outer region can be treated as ideal, i.e., non-viscous. The flow within the boundary layer is still laminar. At the end of the cylinder (the region of decelerating fluid) this boundary layer separates from the solid surface and gives rise to a wide turbulent wake.

This modelling of flow with the division of the flow field into two regions–one, a thin boundary layer in which viscous stresses are significant, and the other, an 'outer' flow considered as non-viscous, is recognized as a major breakthrough in the understanding of fluid dynamics. The behaviour of this boundary layer helps explain many complex fluid-flow-phenomena, and will be studied in detail in Chapter 13.

When the velocity of the flow is further increased, the flow within the boundary layer also undergoes transition from laminar to turbulent flow. This occurs at around $Re = 2 \times 10^5$. When the boundary layer flow is also turbulent, the flow separation from the solid surface occurs further around the cylinder (Fig. 1.19). This results in a narrower wake. It is seen from Fig. 1.20 that while the pressures on the front portion of the sphere are relatively unaffected, the pressures on the rear are higher, when the boundary layer flow is turbulent. This explains the much lower drag as compared to that when the flow is laminar and the wake is broader. The sudden dip in the drag curve (see Fig. 1.13) is due to this transition.

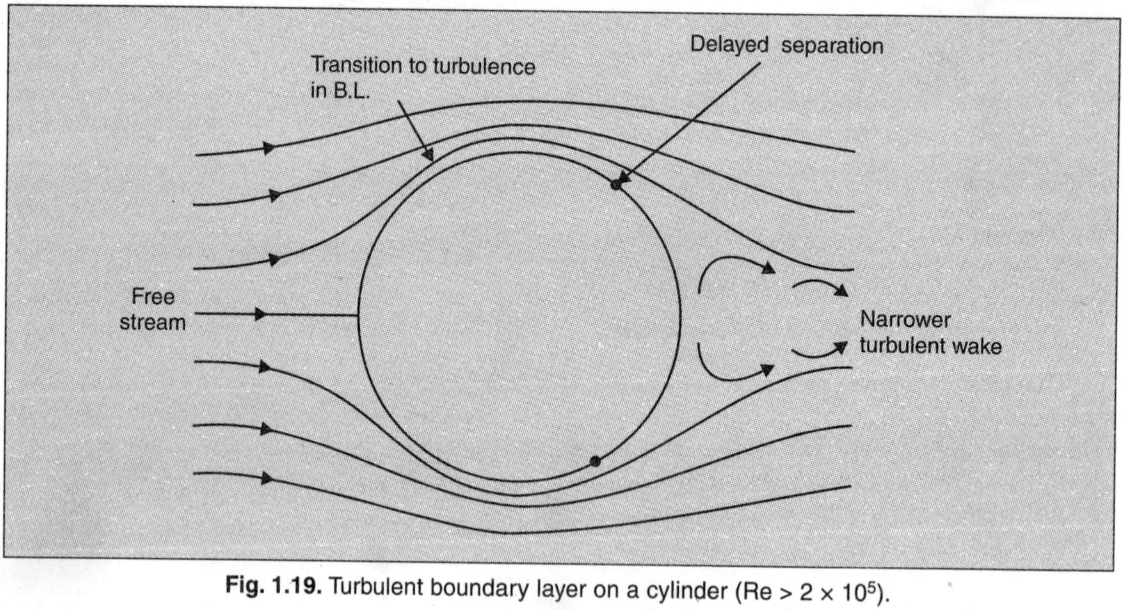

Fig. 1.19. Turbulent boundary layer on a cylinder ($Re > 2 \times 10^5$).

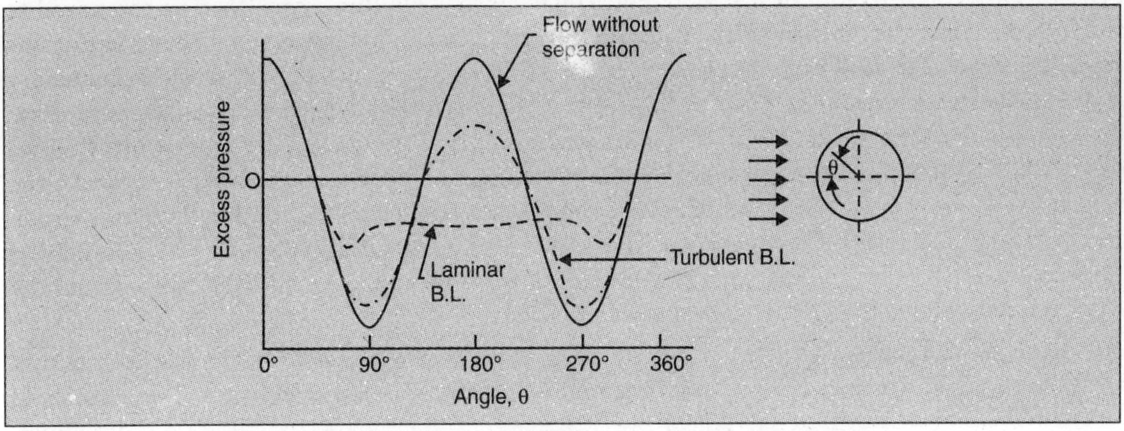

Fig. 1.20. Pressure distributions on a sphere for various conditions.

Turbulence within the boundary layer can also be triggered at much lower Re by such external factors as surface roughness and presence of vibrations. Hence, the drag on a dimpled golf ball is less than that on a smooth one because dimples make the flow turbulent, resulting in a delayed separation and a narrower wake (Fig. 1.21). The swing of a cricket ball is also explained by the same phenomenon. The seam of the ball (inclined to the direction of motion) trips the boundary layer on only one side (Fig. 1.22). Therefore, the flow is turbulent on one side and laminar on the other. This asymmetry of flow results in a lateral force on the ball which curves it in flight.

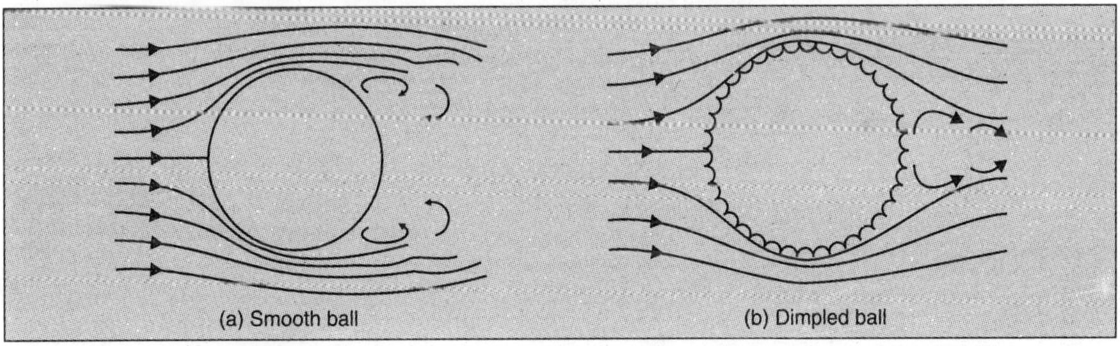

Fig. 1.21. Modification of flow over a golf ball by dimples. This results in a lower drag.

Therefore, to reduce drag in flows at high Reynolds numbers, one attempts to reduce the width of the wake, i.e., to delay the separation of flow from the body surface. An ideal body from this point of view would be the one in which the flow separates, if it does so at all, so close to its rear-end that the resulting wake is very thin. Such a body is termed a *streamlined* body.

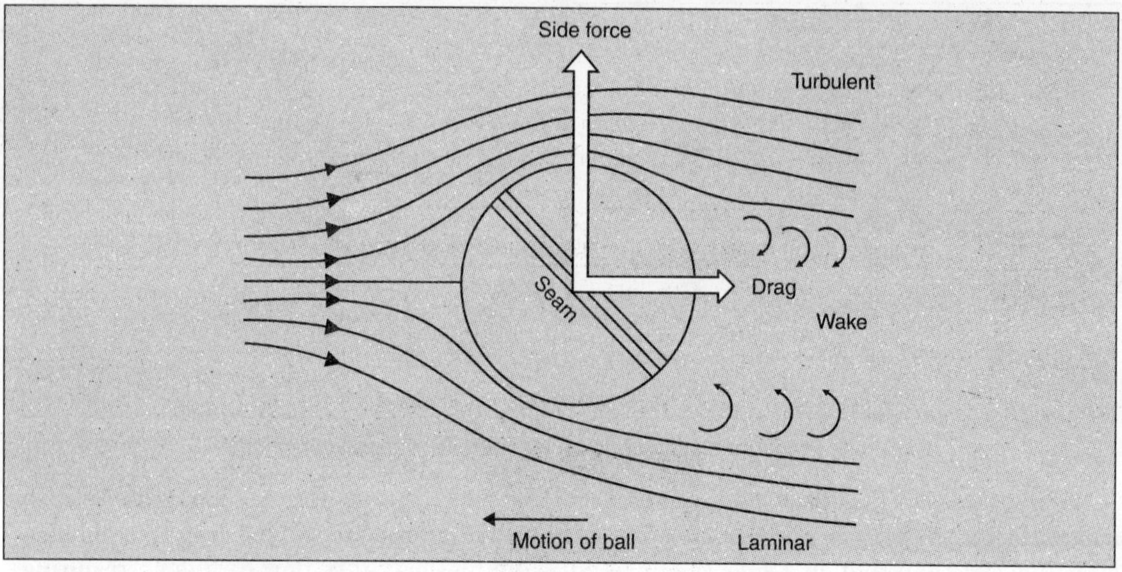

Fig. 1.22. Asymmetric separation on a cricket ball makes it swing in flight.

Experience shows that streamlined bodies have rounded noses and long tapered tails. The profile of a dolphin (Fig. 1.23) is streamlined and so also is the cross-section of an aircraft wing.

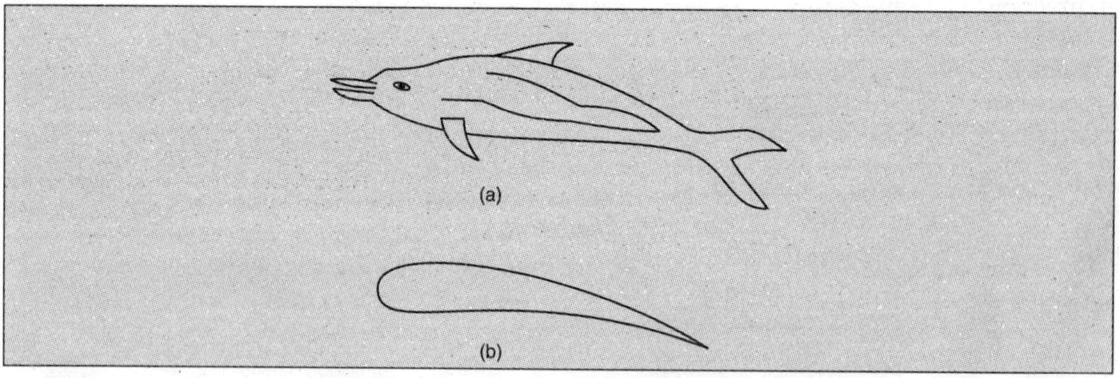

Fig. 1.23. Streamlined profiles. (a) Profile of a dolphin, and (b) profile of NACA 8410 aerofoil.

The peculiar shapes of the cross-sections of aircraft wings (termed aerofoil), are specially designed to give very low drag forces as compared to the lift forces that they generate. Lift is generated because of the asymmetry in the upper and lower surfaces of an aerofoil. The round-nosed, slowly tapering, thin construction of the aerofoil can give lift-to-drag ratios of as high as 40 for well designed glider wings.

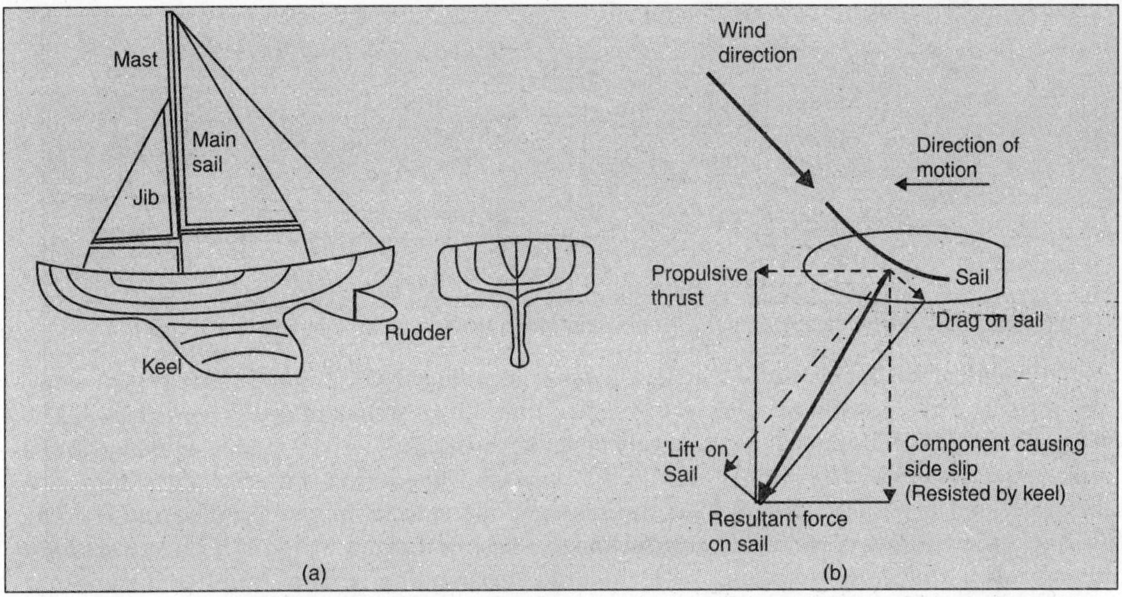

Fig. 1.24. (a) Main elements of a sailboat. Note the large transverse area of the keel. (b) Aerodynamic forces on a sailboat moving at an acute angle to the wind direction.

The high lift-to-drag ratios are also used in sailboats for moving at acute angles to the wind direction. The sails acting as aerofoils generate a drag in the direction of the wind and a force component normal to this direction. This is akin to the lift force in the aeroplane, except that it acts horizontally. The sails are adjusted at such an angle that the resultant of the drag and the normal force has a component towards the front end of the boat, moving it in that direction (Fig. 1.24). The broadside component of this resultant *aerodynamic* force is balanced out by the lateral 'lift' component of the *hydrodynamic* force on the keel. The keel is an aerofoil shaped body which is aligned with the longitudinal axis of the boat. When the boat side-slips, the keel acts as a *hydrofoil* generating a horizontal 'lift'.

Streamlining is used in all situations where high speed performance is desired. Racing automobiles (Fig. 1.25), small missiles, submarines, etc., are all streamlined for this reason.

Fig. 1.25. High performance automobiles are streamlined.

1.8 FLOW THROUGH A PIPE

When a fluid enters a pipe in which a constant axial pressure difference is somehow maintained (one such arrangement shown in Fig. 1.26), the flow is established at a rate at which the viscous drag (due to the velocity gradient across the pipe section) just balances the force due to the axial pressure difference.

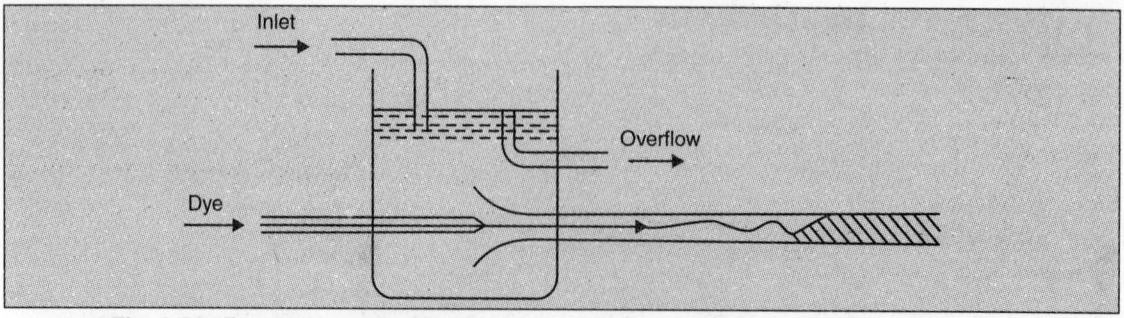

Fig. 1.26. 'Reynolds' experiment. The mixing of the dye indicates that flow is turbulent.

To visualize the behaviour of the flow, a dye is introduced continuously near the entrance of the pipe. The dye injection rate is so controlled that a thin streak of dye is maintained. The behaviour of this dye filament, as it travels down with the fluid, gives visible evidence of the nature of the fluid flow. The state of flow in this case too, depends on the value of the Reynolds number defined as $\rho V_0 D/\mu$, where V_0 is the *average* fluid velocity across the pipe and D is the diameter. At low values of Reynolds number the dye travels down the pipe in a smooth straight line, indicating that it does not mix with the fluid surrounding it. This signifies the laminar state of flow (there is a bit of mixing due to diffusion of dye into the surrounding fluid, but it is an order of magnitude less than what occurs in turbulent flow discussed below). As the flow rate and thus the Reynolds number is increased, the flow behaviour changes. The dye streak becomes sinuous some distance from the entrance and then spreads rapidly over the cross-section indicating that large scale transverse mixing is taking place in the tube (Fig. 1.26). This signifies that turbulent flow has set in.

If we study the variation of velocity across the tube cross-section we see a number of phenomena. In the Reynolds number range which corresponds to laminar flow, the variations in velocities are like those shown in Fig. 1.27. Close to the tube entrance, the action of viscosity at the wall brings the fluid to rest and a thin boundary layer develops there. The velocity varies rapidly from zero at the wall to a constant value at the core. As we move down the pipe the thickness of the boundary layer increases and the extent of this constant velocity core decreases, till at section C the boundary layer spreads all across the tube and we get a parabolic velocity variation. Beyond that section, no changes are observed till very close to the exit. From the, entrance to section C, where the parabolic variation is first established, the flow is said to be *developing* while beyond C the flow is termed as *fully developed.* The length over which it is developing is termed the entrance length and it increases with Reynolds number. Since the velocity pattern does not change in the fully-developed region, it is easier to analyse.

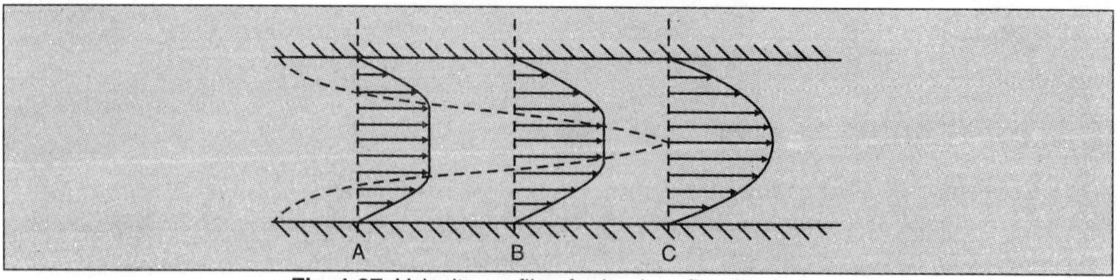

Fig. 1.27. Velocity profiles for laminar flow in pipes.

For higher values of Reynolds number when the flow becomes turbulent, the variation of velocity across the tube section in the fully developed region is quite different from the laminar velocity profile which is parabolic (Fig. 1.28). The velocity increases more sharply at the wall and is almost constant near the centre of the pipe.

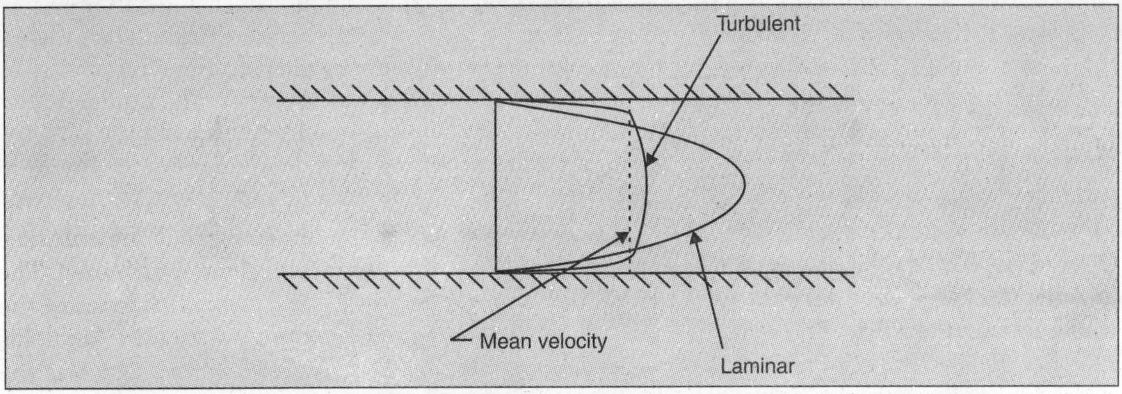

Fig. 1.28. Fully developed turbulent and laminar profiles for flow in a circular pipe.

When an incompressible fluid flows through a pipe whose cross-sectional area is increasing smoothly, i.e., the pipe diverges, the flow behaviour is quite complex and depends on the angle of divergence α. The flow decelerates down the pipe so that the mass flow rate across any section is the same. Because of this deceleration, the boundary layer has a tendency to separate from the wall just as it does on the downstream side of a cylinder. As the divergence angle α increases, the boundary layer at the wall first becomes turbulent and then separates on one side. The flow has a tendency to remain attached to the other side of the wall (Fig. 1.29). If mechanically diverted, the flow may switch from attachment on one side to the other, but it remains asymmetrical. At still larger values of the divergence angle α, a symmetrical separation is achieved and the fluid issues as a jet with regions of recirculating flow surrounding it. In a smoothly converging pipe, the flow accelerates and hence there is no tendency to separate.

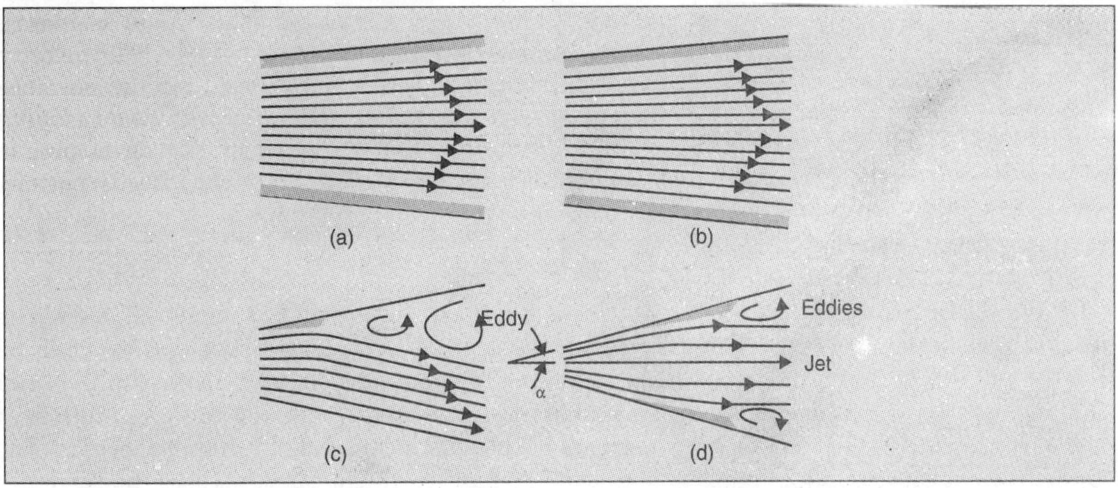

Fig. 1.29. Flow in a diverging section. Divergence angle α increases from (a) to (d). (a) Laminar boundary layer, (b) turbulent boundary layer, (c) separation on one side, and (d) symmetrical separation.

Consider next some flows with abrupt changes in cross-section. Figure 1.30 shows a flow through a sudden expansion. The fluid decelerates as it flows from A to B and, therefore, the boundary layer which develops in the narrower pipe section separates at the corner of the expansion. This results in an annular region at the expansion wherein the fluid does not flow downstream, but recirculates as a turbulent eddy. Such turbulent eddies cause a loss in pressure. This results in the loss of mechanical energy carried by the flow and, therefore, should be avoided. Flow in a sudden contraction is a bit more complex. Fluid flowing close to the wall at A (Fig. 1.31) has to curve in and this also results in a recirculating eddy at the corner. The fluid particles entering the contraction do so at a small inclination to the pipe axis. Further down the pipe the particle paths must be straight and parallel to the pipe axis. Due to the inertia of the fluid particles their paths cannot straighten out immediately on entrance into the narrower pipe and this results in another recirculating annular eddy at B as shown. Thus, in this region the full area of the narrower pipe is not available for the flow. The section at which the area available to the flow is a minimum is termed the *vena contracta*.

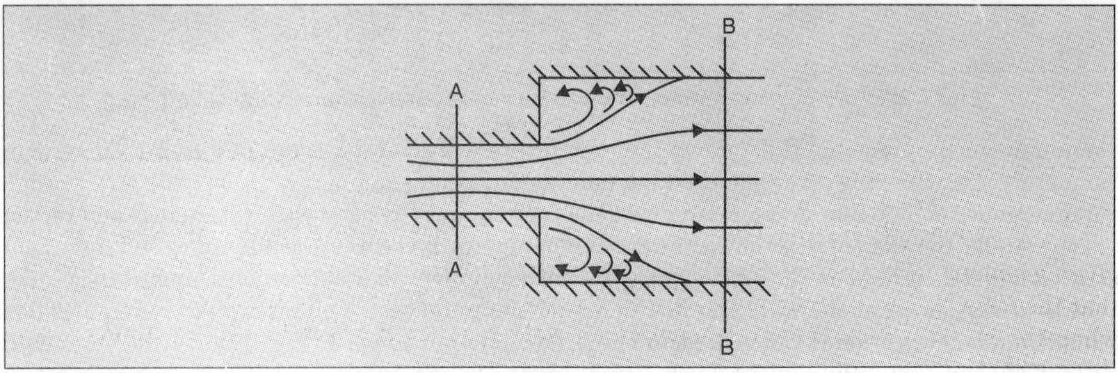

Fig. 1.30. Flow in a sudden expansion.

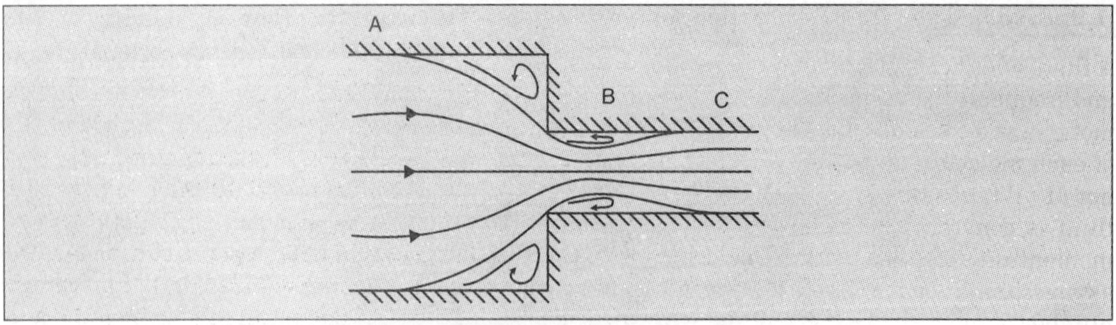

Fig. 1.31. Flow in a sudden contraction.

When water issues out of a reservoir through a sharp-edged orifice, a similar phenomenon occurs. Due to the separation of the boundary layer at the orifice, water issues as a jet. As the fluid particles cannot negotiate a sharp bend, they emerge along curved streamlines and the diameter of the jet decreases for some distance. The steady value of the jet diameter is achieved only at B, some distance downstream of the orifice (Fig. 1.32). A rounded orifice (Fig. 1.33) does not require a fluid particle to execute a sharp turn and, thus, reduces the tendency of the jet to contract. The jet in Fig. 1.33 comes out with essentially no contraction.

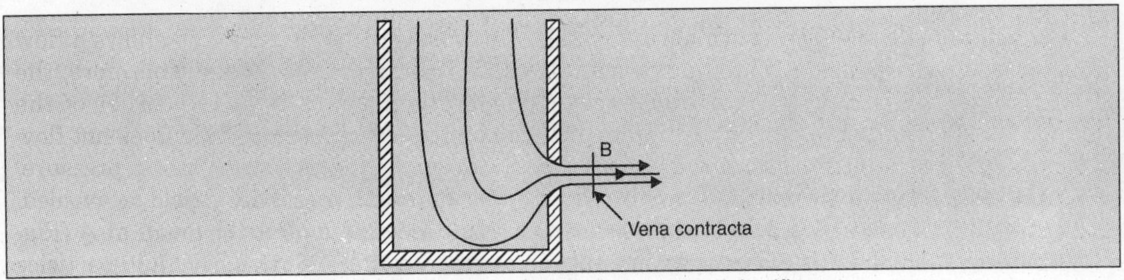

Fig. 1.32. Flow through a sharp-cornered orifice.

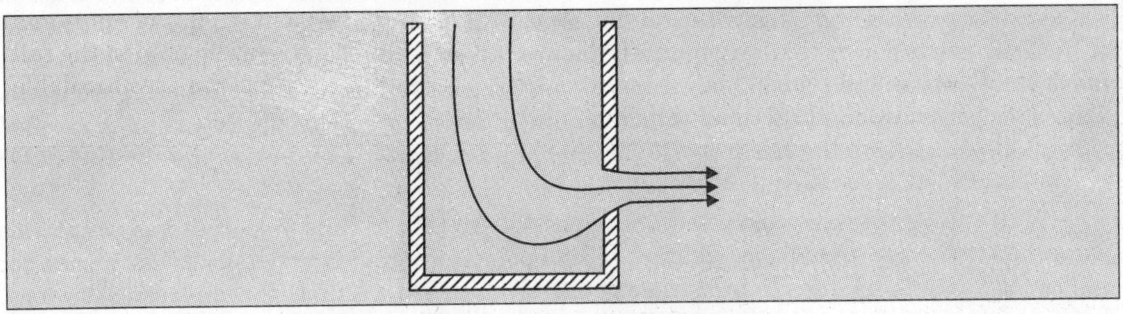

Fig. 1.33. Flow through a rounded orifice.

A comparison of the flow pictures in Figs. 1.30 and 1.31 shows that the flow pattern in a given geometry may change drastically, when the direction of flow is reversed, that is to say that the *fluid flow is not reversible.* This holds good for all flows except for those at very low Re when the inertia effects are quite negligible. Note that the flow pattern about a cylinder at Re << 1 has fore-and-aft symmetry (see Fig. 1.14).

1.9 CONCEPT OF CONTINUUM

A fluid, whether a gas or a liquid, is composed of a large number of molecules in constant motion and frequently colliding with each other. The forces within the fluid are a result of these complex motions and collisions. One approach to study the fluid motions is by accounting for the behaviour of each molecule in the flow. But, apart from the fact that such an approach is unwieldy, it is not at all necessary in most cases. Take, for example, the pressure in a fluid. Any surface in a fluid is constantly bombarded by a large number of molecules and the aggregate rate of momentum exchange of these molecules with the surface results in a force per unit area termed as *pressure.* Even if we take the surface of area 10^{-12} m^2 (i.e., a square of side 1 μm), the average number of molecules which collide with it are of the order of 10^7 per second in a gas at normal temperature and pressure. This is such a large number that the individuality of molecular interaction is of no consequence and what we observe is an aggregate effect. If we insert a pressure probe in the field, the smallest of probes will probably be large enough to interact with millions of individual molecules and, therefore, for all practical purposes, the distribution of matter may be treated as a continuous one. The pressure measured by such a probe is quite insensitive to small changes in the area of the probe's sensor. If, however, the area is reduced considerably, it is conceivable that we may reach the limit when only a few molecules will hit the probe and the averaged molecular interchange is quite meaningless in that limit. In the other limit, as the probe becomes very large, it no longer measures the *local* pressure but a kind of global average.

Thus, for the earth's atmosphere, where the pressure changes with altitude, a very large pressure probe may not measure the vertical variations of pressure faithfully. Note that in such cases there exist two distinct length scales; one, the microscopic, which is of the order of the mean-free-path of molecules, and the other, the macroscopic, which is the length over which the pressure changes appreciably.

The continuum model is possible only when these two length scales are very different. This is because in such a case it is possible to select a measuring length (equal to the linear dimensions of the probe) which is large enough, so that only an aggregate effect of the molecular interaction is reflected in the measurements, but, which (the measuring length) is small enough so that the macroscopic variations are not smothered out. The smaller length scale is of the order of the mean-free-path of the molecules and the larger scale is of the order of the linear dimensions of the body about which or through which the flow takes place. Thus, if the mean-free-path is much smaller than the typical body dimensions, the continuum model would be appropriate. This holds good for almost all flows of engineering interest, except for the flow of some rarefied gases. One example of the latter is the flow about a spacecraft as it enters the rarefied outer reaches of the earth's atmosphere.

With the continuum model, then, it is possible to talk of fluid velocities and properties associated with a continuous distribution of fluid points, where it is understood that we are not interested in the differences in properties of points which are within the mean-free-path range. In this text we shall restrict our attention to those fluid flows where the continuum model is fully adequate.

PROBLEMS

1.1 Different grades of lubricating oils are prescribed for use in cars during different seasons—a 'thinner' oil for winters and a 'heavier' oil for summers. What do these terms signify in the language of fluid mechanics? Can you think of a reason for this difference?

1.2 The velocity of water between two flat stationary parallel plates shown is given by

$$V_x = V_m\left(1-\frac{y^2}{b^2}\right).$$

Draw the shear stress profile on the upper half of the channel. Compute the x-force (per unit area) required to hold the upper plate stationary if $V_m = 2$ m/s and $b = 2$ cm. What is the direction of this force? μ of water = 1 cp.

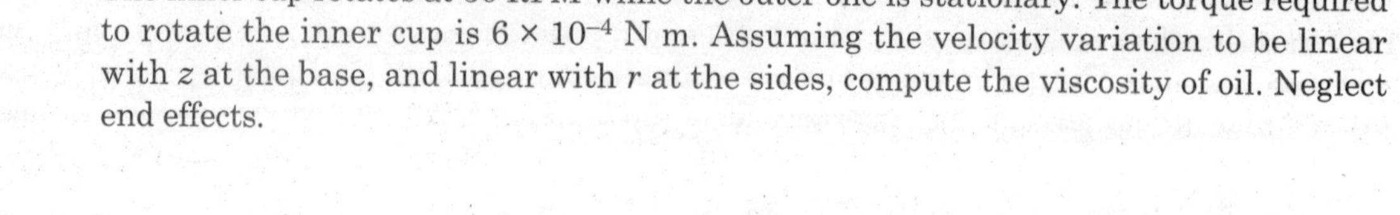

1.3 Oil fills the 1.5 mm gap between two concentric cups having flat bottoms as shown. The inner cup rotates at 50 RPM while the outer one is stationary. The torque required to rotate the inner cup is 6×10^{-4} N m. Assuming the velocity variation to be linear with z at the base, and linear with r at the sides, compute the viscosity of oil. Neglect end effects.

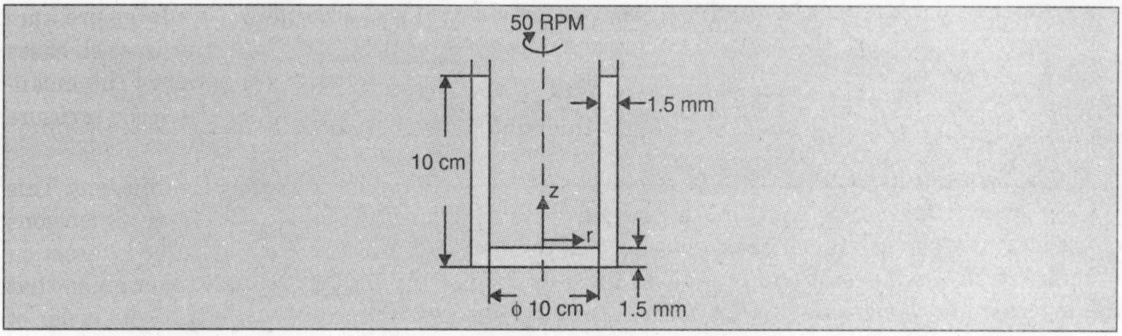

1.4 Consider a fluid between a cone and a plate as shown. The cone is rotated at an angular velocity ω. If the θ-velocity is assumed to vary linearly with *z*, obtain the following expression for the torque required to hold the plate stationary:

$$\text{Torque} = \frac{2\,\pi\mu\omega R^3}{3\tan\alpha}.$$

Assume the only non-zero velocity component to be V_θ.

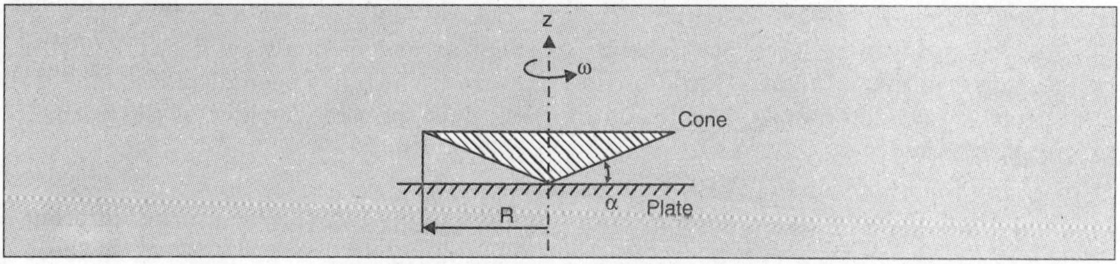

Why does the fluid not flow out of the gap? Take α very small.

1.5 A vertical glass tube, open at both ends, contains a small plug of a suspension of clay in water. It is observed that the suspension does not flow down the tube. However, when a tube having a larger radius is used, the suspension flows. What can you infer from this regarding the stress *vs.* strain–rate behaviour of the suspension? (*Hint*: see Fig. 1.4).

1.6 An engineer specifies the torque rating of the electrical motor required for a mayonnaise blender using the value of viscosity obtained from a single experiment. He finds that at the 'low' setting the blender behaviour is quite as predicted but at the 'high' setting the blender speed is much higher than that predicted. What conclusions do you draw about the stress *vs.* rate-of-strain curve for mayonnaise?

1.7 A liquid is filled between two large stationary, parallel, flat plates. At $t = 0$, the lower plate starts moving at a constant velocity. V_0 to the right. Draw schematically the velocity profiles at various times. What happens at large times? Compare this to when there is no upper plate. What would be the energy expanded till $t \to \infty$ for the latter?

1.8 From the fact that a log of wood floats down a river at the same velocity as the water, deduce that fluids exert a drag force on objects moving through them.

1.9 In the maiden transatlantic voyage of the helium balloon *Double Eagle II* from Presque Isle, Maine to Miserey, France, in 1978, the balloonists carried about 2000 kg of ballast. Of this, 1490 kg was jettisoned. What role did the ballast play?

1.10 The density of clouds is slightly higher than that of the surrounding air. Why, then, do they not come down?

1.11 Calculate the lift-to-drag ratio of the kite shown weighing 100 g, if the tension in the cord (at 30° to the ground) is 5 N and the wind blows horizontally (see *Science Today*, Jan. 1982).

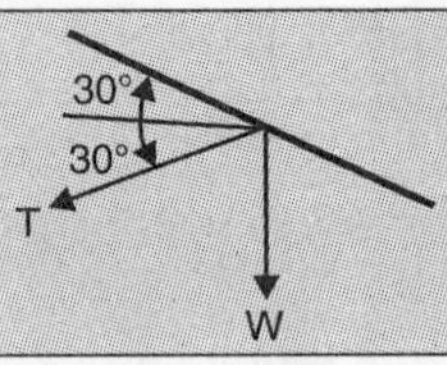

1.12 When a ceiling fan rotates in air, it experiences a force. The reaction of this force acts on air and moves it downwards. Should this force be treated as a lift or a drag according to the definitions of Sec. 1.7?

1.13 Gulliver told the Lilliputian emperor to use roughened stones instead of smooth ones as projectiles to sink enemy ships. Can you explain why?

1.14 An arrow has a heavy nose and a few light feathers attached at the rear end. The heavy nose brings the centre of gravity forward and the feathers give a large surface on which aerodynamic forces act. Show that such an arrangement always keeps the flying arrow pointing straight. (*Hint*: Consider an arrow which has been displaced slightly from the straight course). The fins on a torpedo have the same function as the feathers on an arrow.

1.15 Using the arguments of Prob. 1.14 discuss how

(a) a weather-cock always indicates the wind–direction correctly.

(b) a badminton shuttle-cock always comes down 'head' first.

(c) tail planes of an aircraft keep its longitudinal axis aligned with the (relative) wind direction.

1.16 *Half-cup anemometer:* A half cup anemometer is used to measure wind velocities. It consists of four cups mounted circumferentially as shown. When a wind blows past the anemometer, it rotates about its axis—always clockwise (looking from above), irrespective of the wind direction. The rate of rotation gives a measure of the wind velocity. Show by considering the flow patterns for different cup orientations (use only two extreme cases) that the rotation will indeed be clockwise.

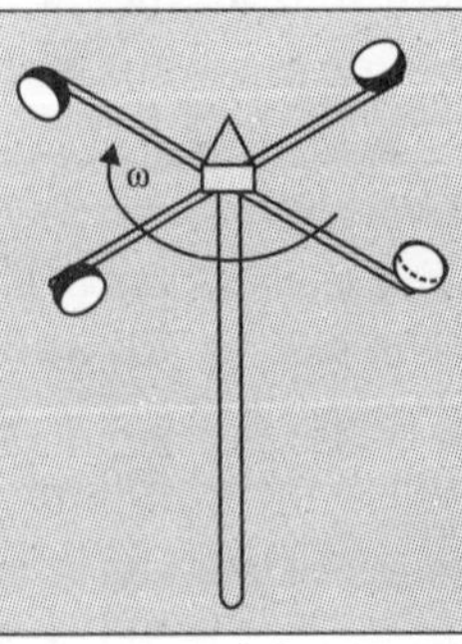

1.17 *Frisbee:* A frisbee flies because of the lift generated when the air flows past it as shown (spin is for stability alone). Show how a frisbee can be thrown to follow a curved path.

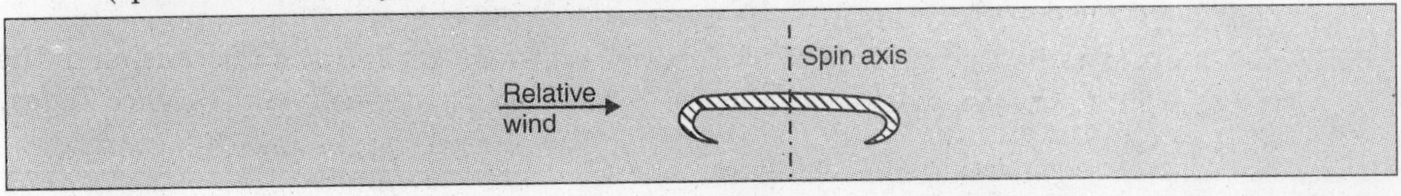

1.18 It is a common observation that 'swing' bowlers in cricket like to maintain the shine on the ball for as long as possible resorting at times to such unethical practices as application of 'Vaseline'. Why? Also, the bowlers seem to concentrate their efforts on only one half of the ball. Which half?

Noting that the transition from laminar to turbulent flow on a sphere occurs at a (somewhat) fixed velocity, show how an early or a late swing can be effected by controlling the speed of delivery (see *Science Today,* Sep. 1976, p. 46).

1.19 Study the motion of the hands and feet during swimming and thus see how the hydrodynamic forces help keep the body afloat.

1.20 Explain the swishing sound produced by a cane when it is swung in air.

1.21 A cold liquid flowing inside tubes is usually heated by high-temperature steam flowing on the outside. A series of baffles are used to make the steam flow essentially crosswise as shown. There is a very small clearance between the holes in the baffles (through which the tubes pass) and the tubes. Sometimes the equipment starts to 'chatter'. Can you explain the fluid-flow phenomenon responsible and a possible remedy? Can you also suggest why the spacing between baffles is generally quite small?

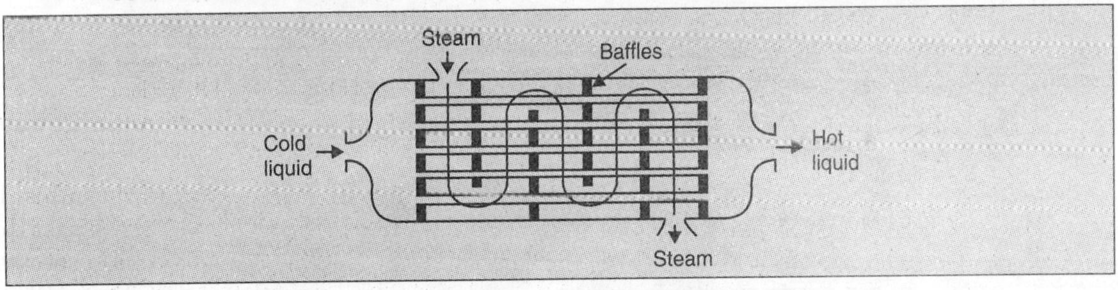

1.22 What flow patterns do you expect (based on information of Sec. 1.7 and 1.8) in the geometries shown on p. 26?

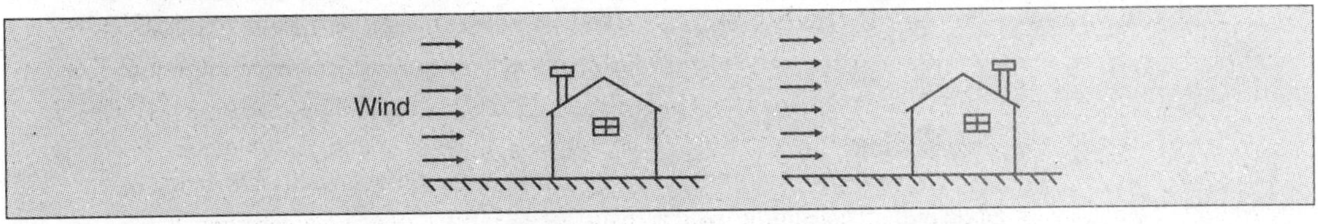

1.23 An engineer suggests that the chimney should be placed on the leeward side instead of the windward side of a house, as shown so that the smoke does not get blown in. Do you agree with this?

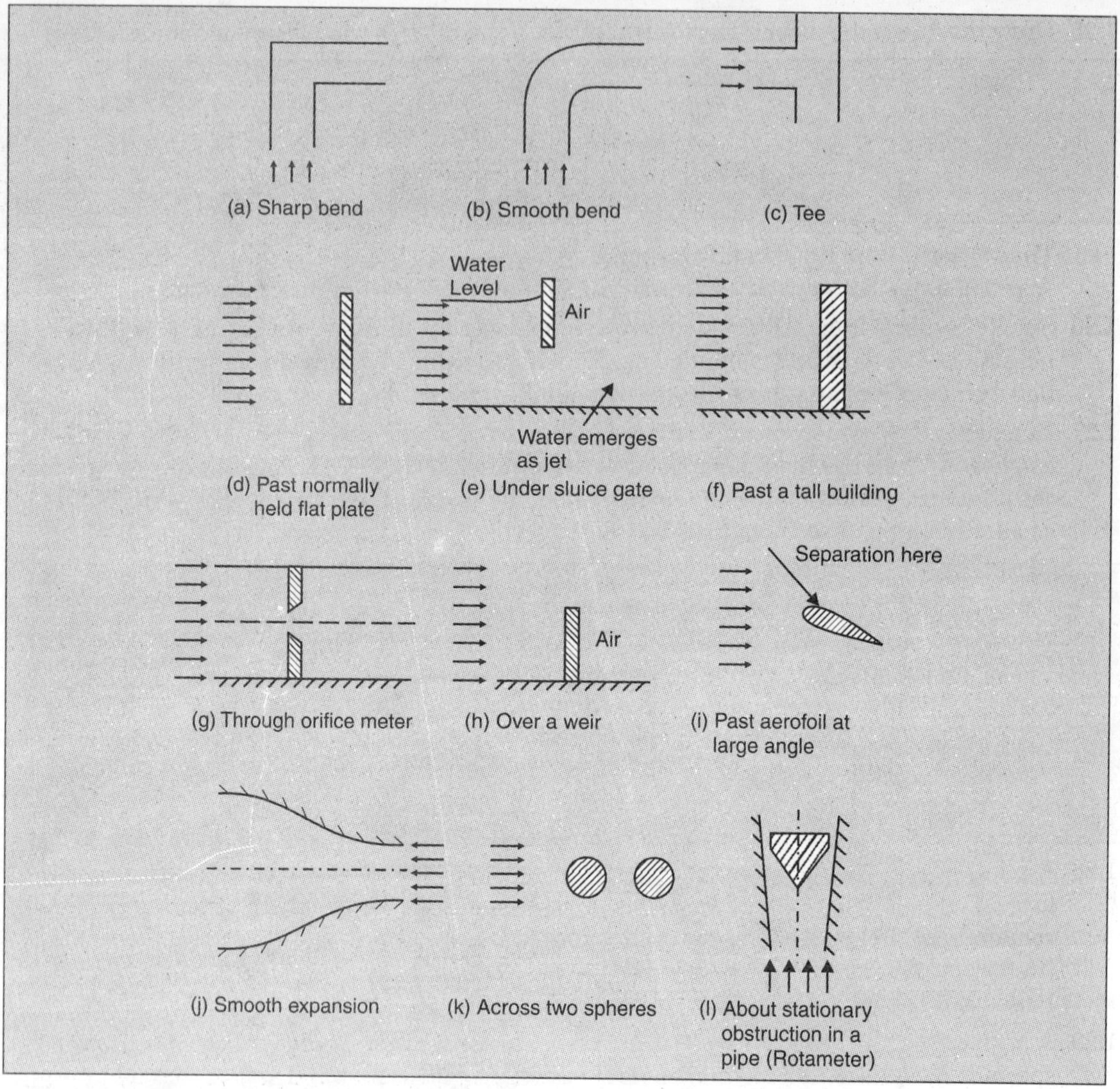

(a) Sharp bend (b) Smooth bend (c) Tee

(d) Past normally held flat plate (e) Under sluice gate (f) Past a tall building

(g) Through orifice meter (h) Over a weir (i) Past aerofoil at large angle

(j) Smooth expansion (k) Across two spheres (l) About stationary obstruction in a pipe (Rotameter)

1.24 A cottage is located far downstream on the leeward slope of a hillock as shown. Draw the wind-flow pattern and find out whether the smoke will enter the house or not? How will your answer change if the house is located at some point A, nearer the top?

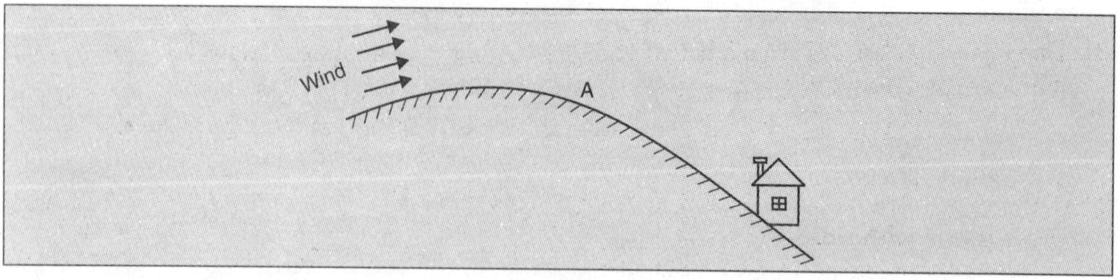

1.25 Draw the flow pattern for a tractor-trailer when the relative wind directions are as shown.

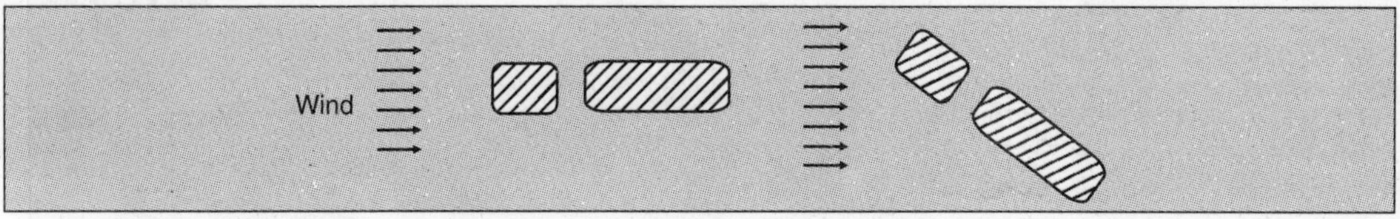

Thus observe how significantly larger drag forces act in the second case. *(Science Today,* Aug. 1980, p. 48). Vestibuled trains will similarly have lower drag coefficients.

1.26 Smoke issuing from a lighted cigarette-end rises up as a smooth streak to a certain height, and then starts meandering. To which regimes of flow do these two types of flow belong? Give another example of a similar transition.

1.27 In a penicillin fermentor the contents inside are heated by means of steam in a jacket as shown. This, however, does not give adequate heating rates. An engineer suggests that the same amount of steam be made to flow through a narrower jacket with baffles so as to give the flow a tortuous path. Would his suggestion work? Why?

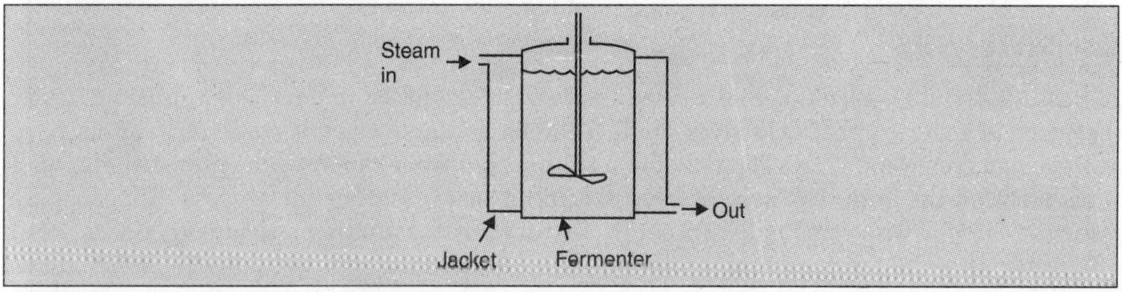

1.28 A plant species has a better chance of survival if the area of dispersal of its seeds is large. For these reasons, seeds of some plants have such 'aerodynamic' design that they remain afloat for quite some time after they come out of pods. Find out how they do it. (Refer: Papanek, V., *Design for Real World—Human Ecology and Social Change,* Palladine, 1974).

1.29 A *boomerang* is a specially carved stick used by Australian aborigines for hunting. Its special aerodynamic shape makes it fly out-wards to a target and return to the hunter if it misses. Find out how it does so. (Refer: *Scientific American,* May 1968, p. 124; Aug. 1979, p. 162).

1.30 A *hovercraft* is a vehicle that travels on a cushion of air maintained between its base and the terrain. Find out how it does so (see Sec. 8.6).

1.31 One can tell a raw egg from a boiled one by spinning it on a table. A boiled egg spins for a longer time. Explain.

2

FORCES IN STATIONARY FLUIDS

2.1 PRESSURE

In Sec. 1.2, a fluid was defined as a substance which cannot be in (static) equilibrium under the action of a shear force, howsoever small. Therefore, a fluid which is stationary cannot have any shear force acting on it. The only force that such a fluid can sustain acts normally on a surface within the fluid. The normal force per unit area (and acting inwards) is termed as the *fluid pressure p*. The famous seventeenth century French mathematician, B. Pascal first established that the pressure at any point within a stationary fluid is the same in all directions. To prove this, consider an infinitesimal triangular prism (having length L normal to the plane of the paper) of fluid within a large expanse of fluid at rest (Fig. 2.1). The pressure at any surface within the fluid is normal to it and varies, in general, from point to point. But if δx and δz are

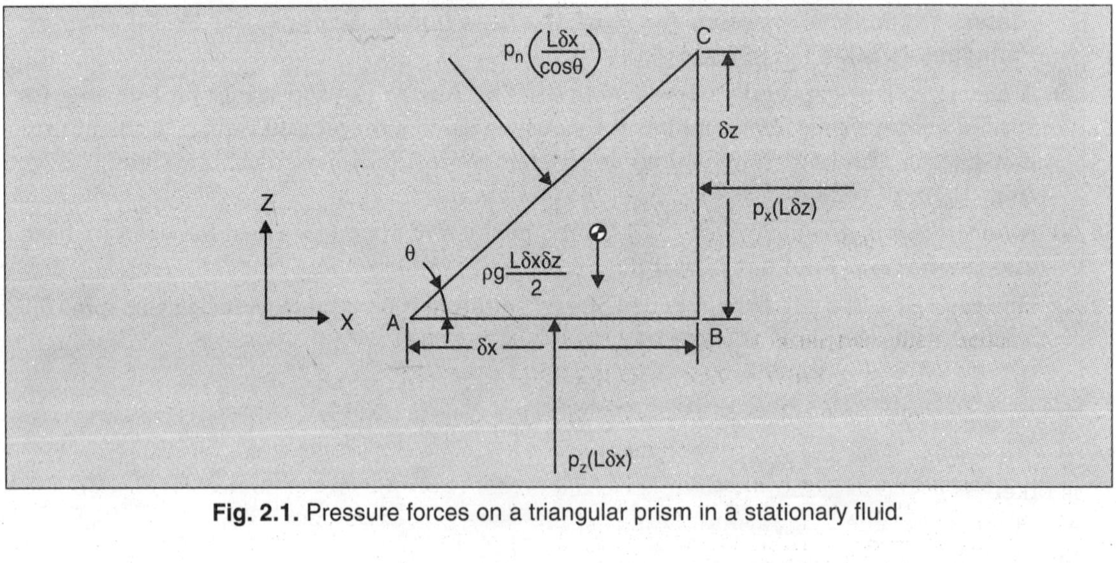

Fig. 2.1. Pressure forces on a triangular prism in a stationary fluid.

infinitesimal, the pressure can be taken as constant along each of the three surfaces shown. Let p_x be the pressure on the face *BC* with its normal in the *x*-direction, p_z on the face *AB* with its normal in the $-z$-direction and p_n on the diagonal plane *AC*. The forces due to these pressures are indicated in the figure. Since the mass of the fluid within the prism is $\rho\ (L\ \delta x\ \delta z/2)$, the gravity force in the *z*-direction is $-\ \rho g(L\ \delta x\ \delta z/2)$. A force balance, then, gives

$$-p_x\ (L\ \delta z) + p_n \left(\frac{L\ \delta x}{\cos\theta}\right) \sin\theta = 0 \qquad ...(2.1)$$

and

$$p_z\ (L\ \delta x) - p_n \left(\frac{L\ \delta x}{\cos\theta}\right) \cos\theta - \rho g(L\ \delta x\ \delta z/2) = 0 \qquad ...(2.2)$$

On noting that $\delta z = \delta x \tan\theta$, and taking the limit as δx and δz tend to zero (i.e., when the triangle *ABC* shrinks to a point), Eqs. (2.1) and (2.2) give

$$p_x = p_z = p_n \equiv \text{(say)}\ p \qquad ...(2.3)$$

for any value of θ. Thus, the pressure at any point in a stationary fluid is the same in all directions. This result has been obtained again in Sec. 6.2 where it is shown that it is valid even when the fluid is moving, provided the shear forces are zero everywhere. Thus, the pressure at a point is the same in all directions in a fluid which is stationary, or moving as a rigid body or has zero viscosity.

CHAPTER 2

2.2 PRESSURE FORCE ON A FLUID ELEMENT

Consider a small element of a stationary fluid (Fig. 2.2), having sides δx, δy and δz. Let it be subjected to a pressure which varies with *x*, *y* and *z*. If the pressure acting on the face *ADHE*

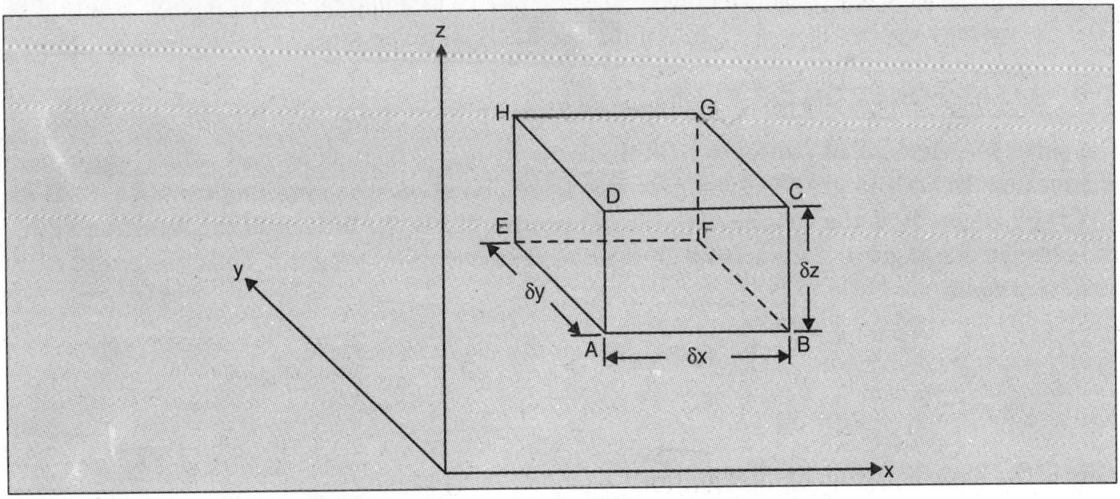

Fig. 2.2. An element of stationary fluid.

is taken as *p*, then for small values of δx, the pressure on the opposite face *BCGF* can be taken as $p + (\partial p/\partial x)\delta x$. The *x*-component of the *net* pressure force is the resultant of these two pressure forces. Thus,

$$dF_{p,x} = p \times \delta y\, \delta z - \left(p + \frac{\partial p}{\partial x}\delta x\right)\delta y\, \delta z = -\frac{\partial p}{\partial x}(\delta x\, \delta y\, \delta z)$$

The pressures acting on the other faces, similarly, give the y- and z-components of the net pressure force acting on the element and

$$d\mathbf{F}_p = -\left(\frac{\partial p}{\partial x}\hat{\mathbf{i}} + \frac{\partial p}{\partial y}\hat{\mathbf{j}} + \frac{\partial p}{\partial z}\hat{\mathbf{k}}\right)(\delta x\, \delta y\, \delta z) \qquad \text{...(2.4)}$$

If $\mathbf{f}_p$ is introduced as the net pressure force *per unit volume*, then

$$\mathbf{f}_p = -\left(\frac{\partial p}{\partial x}\hat{\mathbf{i}} + \frac{\partial p}{\partial y}\hat{\mathbf{j}} + \frac{\partial p}{\partial z}\hat{\mathbf{k}}\right)$$

Using the gradient operator, it can be written as

$$\mathbf{f}_p = -\nabla p \qquad \text{...(2.5)}$$

The net pressure force arises because of the spatial variations of pressure. A uniform pressure acting all around on an element does not result in any net force. This conclusion is significant and is reflected in the way pressures are specified. Engineers interested in the forces that fluids exert on bodies generally specify pressures as deviations from the atmospheric pressure which acts all around a body and hence does not contribute to the net pressure force. This specification of pressure is distinguished from the absolute level by labelling it as the *gauge pressure.* Thus, the gauge pressure at a point is the absolute pressure at that point *minus* the actual atmospheric p_{atm}. Negative values of gauge pressures are often termed as *vacuum pressures.* If the local atmospheric pressure is 101 kPa, then at a point where the absolute pressure is 120 kPa, the pressure may be specified as 19 kPa gauge, and at a point where it is 80 kPa, we may specify it as – 21 kPa gauge or as 21 kPa vacuum.

2.3 BASIC EQUATION OF FLUID STATICS

Consider the element of stationary fluid shown in Fig. 2.2. It is at rest under the action of pressure forces and gravity. The pressure force acting on it is obtained from Eq. (2.4) as – $(\nabla p)\, \delta V$ where δV is the volume of the fluid element. If ρ is the density of the fluid and $\mathbf{g}$, the acceleration due to gravity, the gravity force acting on the element is $\rho \mathbf{g}\, dV$. Thus, the equilibrium condition gives

$$-\nabla p + \rho \mathbf{g} = 0$$

or

$$\nabla p = \rho \mathbf{g} \qquad \text{...(2.6)}$$

This is the basic equation of fluid statics.

If $\mathbf{g}$ is taken as acting in the negative z-direction, i.e., $\mathbf{g} = -g\hat{\mathbf{k}}$, then the three components of Eq. (2.6) are

$$\frac{\partial p}{\partial x} = \frac{\partial p}{\partial y} = 0 \qquad ...(2.7)$$

and $$\frac{\partial p}{\partial z} = -\rho g \qquad ...(2.8)$$

That is, the pressure in a stationary fluid varies only in the vertical direction and is constant in any horizontal plane.

2.4 HYDROSTATIC PRESSURE DISTRIBUTION

If the variations in the density of a fluid can be neglected (as in a liquid), Eq. (2.8) is readily integrated to give

$$p = p_0 - \rho g z \qquad ...(2.9)$$

where p_0 is the pressure at $z = 0$. Thus, in stationary liquids pressure increases linearly with depth (negative of z). This linear pressure distribution is termed as the *hydrostatic pressure distribution*. As shall be seen later in Examples 6.1 and 6.3, this also holds true in moving fluids which do not have any acceleration in the vertical direction. For a column in which fluids of different densities are arranged one on top of the other (Fig. 2.3), the pressure at any point within the column can be found by repeated application of Eq. (2.9) to different layers of constant density. This is illustrated in Fig. 2.3. Thus, pressure at $z = z_4$ is

$$p_4 = p_1 - \rho_{12}\, g\, (z_2 - z_1) - \rho_{23}\, g\, (z_3 - z_2) - \rho_{34}\, g\, (z_4 - z_3)$$
$$= p_1 + g\, (\rho_{12}\, h_{12} + \rho_{23}\, h_{23} + \rho_{34}\, h_{34})$$

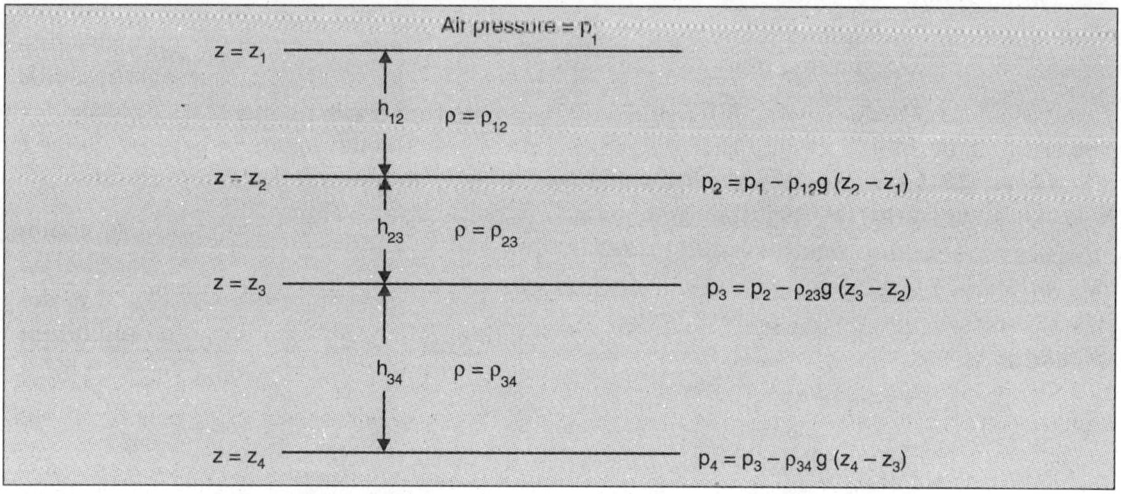

Fig. 2.3. Pressure at various positions in stationary multilayered fluids.

This equation finds application in *manometers* which are instruments used for measuring pressure differences. The following examples illustrate this application.

Example 2.1. The simple *U*-tube manometer shown in Fig. 2.4 is used to measure the steady gauge pressures in a fluid. Find $p_{A,g}$, the gauge pressure at *A* in terms of the fluid densities and the dimensions shown.

CHAPTER 2

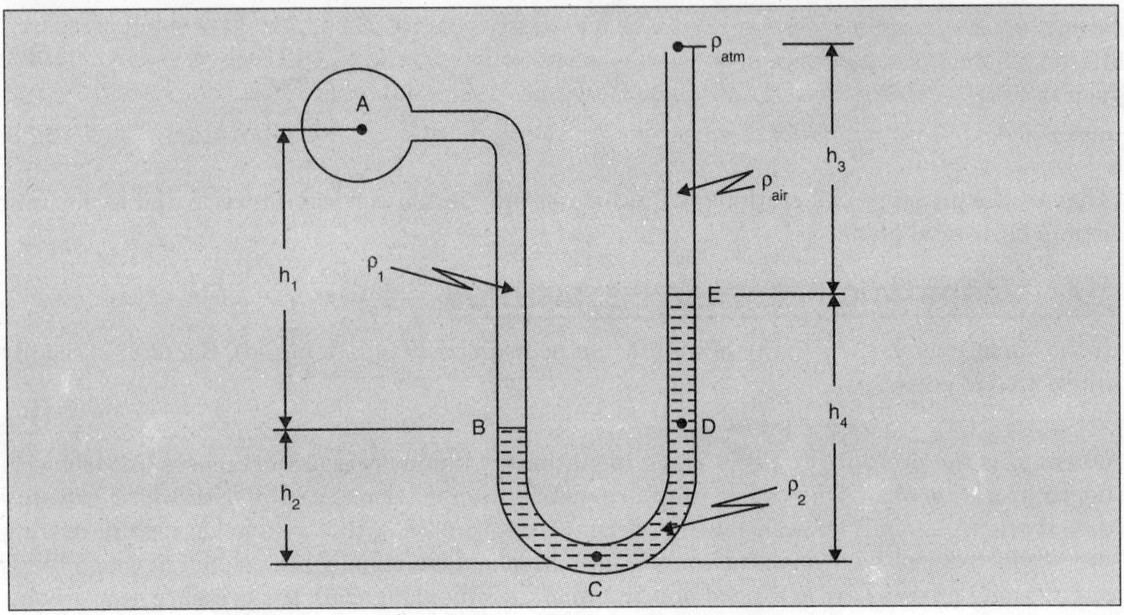

Fig. 2.4. A simple *U*-tube manometer to measure $p_{A,g}$.

Perhaps the most straightforward way of dealing with such problems is to calculate the pressures at each fluid interface starting with a point of known pressure. Thus

$$p_E = p_{atm} + \rho_{air}\, gh_3$$

or

$$p_{E,g} = \rho_{air}\, gh_3$$

From point *E* move on to point *B*. To calculate $p_{B,g}$, note that pressure increases linearly down the right leg of the *U*-tube and then decreases linearly up the left leg up to point *B*. *Since the fluid from E to B has the same density,* the pressure variation from *D* to the bottom-most point *C* is cancelled out by the variation from point *C* to point *B*. Therefore,

$$p_{B,g} = p_{E,g} + \rho_2\, g\, (h_4 - h_2)$$

or

$$p_{B,g} = \rho_{air}\, gh_3 + \rho_2\, g\, (h_4 - h_2)$$

Similarly,

$$p_{A,g} = p_{B,g} - \rho_1\, gh_1$$

and so,

$$p_{A,g} = g[\rho_{air}\, h_3 + \rho_2\, (h_4 - h_2) - \rho_1 h_1]$$

If h_1, h_2, h_3 and h_4 are comparable, and ρ_1, $\rho_2 \gg \rho_{air}$, then $p_{A,g}$ can be approximated by

$$p_{A,g} = g[\rho_2\, (h_4 - h_2) - \rho_1 h_1]$$

and $p_{A,g}$ can be obtained by measuring h_1 and $(h_4 - h_2)$. If the fluid at *A* is air, the above equation further simplifies to

$$p_{A,g} = \rho_2 g\, (h_4 - h_2) \equiv \rho_2\, g\, (\Delta h)$$

where Δh is termed as the *manometer reading*.

Example 2.2. A micromanometer (Fig. 2.5) is used for accurate measurement of small pressure differences. Find the difference between the gas pressures p_A and p_B for the system shown. It is given that the difference h in the levels of the liquid in the two reservoirs is zero when $p_A = p_B$.

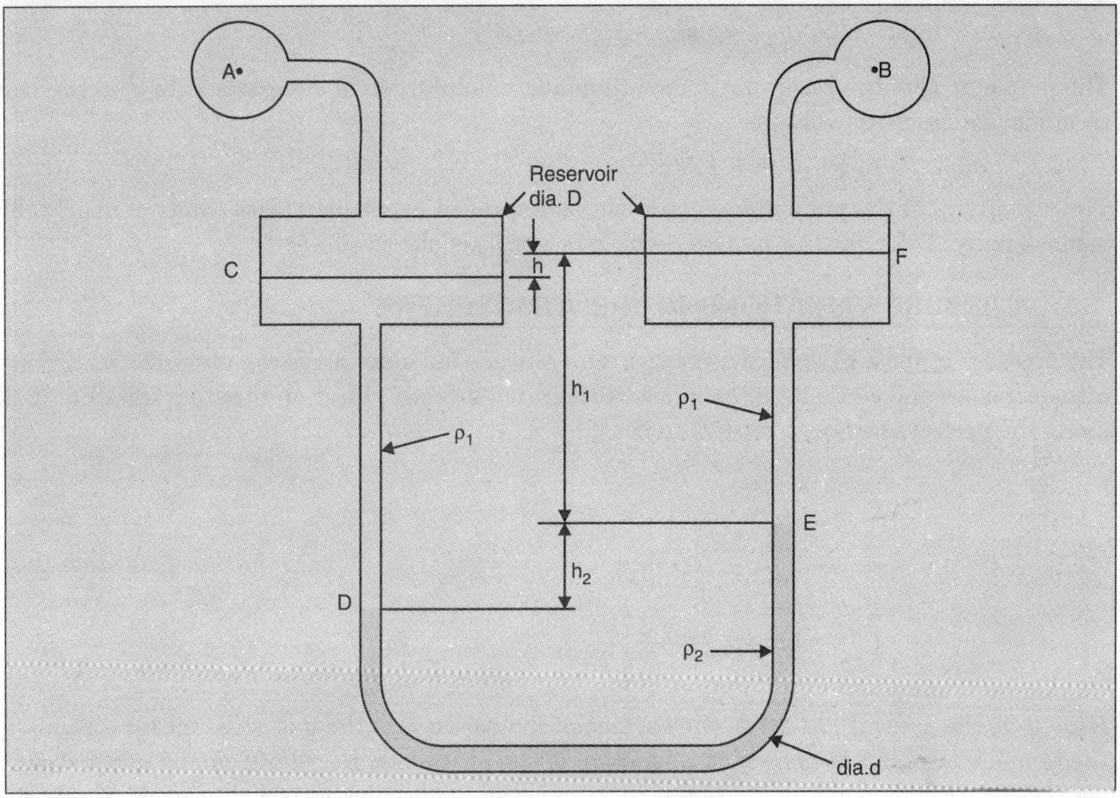

Fig. 2.5. A micromanometer.

Since the flasks A and B contain gases whose density is much smaller than those of the two liquids used, ignore the pressure variations in the two gas columns. Thus,

$$p_B = p_F \text{ and } p_A = p_C$$

Repeated applications of Eq. (2.9) give

$$p_E = p_B + \rho_1 g h_1$$

$$p_D = p_E + \rho_2 g h_2$$

and

$$p_C = p_A = p_D - \rho_1 g (h_1 + h_2 - h)$$

Therefore

$$p_A = p_B + \rho_1 g h_1 + \rho_2 g h_2 - \rho_1 g (h_1 + h_2 - h)$$

or

$$p_A - p_B = \rho_2 g h_2 - \rho_1 g (h_2 - h)$$

Further, the length h_2 is related to h. When $p_A = p_B$ both h and h_2 are zero, otherwise the liquid column will not be in equilibrium. As the two liquids are incompressible, it can easily be seen that the volume $(\pi D^2/4)\, h$ equals the volume $(\pi d^2/4)\, h_2$ or

$$h = h_2 d^2/D^2$$

and then

$$p_A - p_B = (\rho_2 - \rho_1)\, gh_2 + \rho_1 g h_2\, d^2/D^2$$

The diameter D of the reservoir is usually made much larger in comparison to d so the last term may be dropped to obtain

$$p_A - p_B = (\rho_2 - \rho_1)\, gh_2$$

The sensitivity of the micromanometer can be improved by choosing two fluids of nearly the same density. Thus, for the same $p_A - p_B$ a larger h_2 is observed.

2.5 PRESSURE VARIATIONS IN THE ATMOSPHERE

The pressure in the atmosphere decreases with altitude but since air is compressible, its density also decreases and so Eq. (2.8) cannot be integrated directly. First eliminate ρ from Eq. (2.8) using the perfect gas law, $p = \rho RT$, to get

$$\frac{\partial p}{\partial z} = \frac{-gp}{RT}$$

or

$$\int_{p_0}^{p} \frac{dp}{p} = \ln(p/p_0) = -\frac{g}{R}\int_0^z \frac{dz}{T} \qquad \text{...(2.10)}$$

Here p_0 is the pressure at $z = 0$, the surface of the earth, R is the gas constant for air and T, the temperature. If T is known as a function of the altitude z, Eq. (2.10) can be integrated to get $p(z)$. One common approximation is to assume the atmosphere to be isothermal, i.e., $T = T_0$, and then

$$p(z) = p_0 \exp\left(\frac{-gz}{RT_0}\right) \qquad \text{...(2.11)}$$

The pressure in an isothermal atmosphere decreases exponentially. This is a fairly close approximation to the earth's atmosphere. If the variation of temperature with altitude is also taken into account, a more accurate result will be obtained. Figure 2.6 shows typical variations of temperature and pressure with respect to altitude.

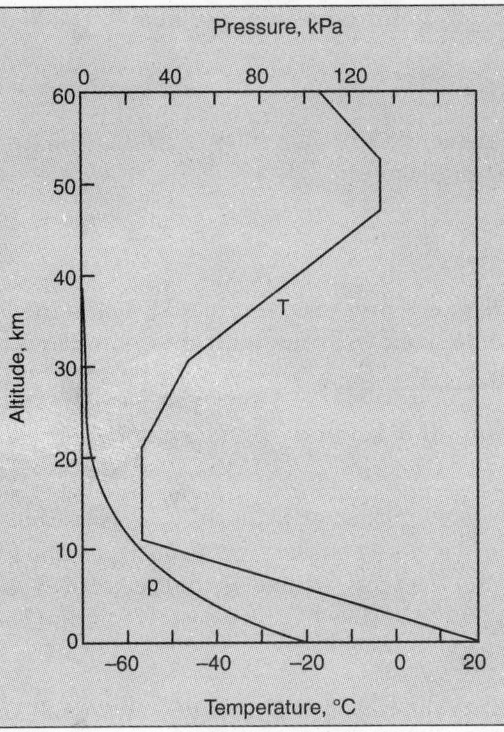

Fig. 2.6. Typical variations of temperature and pressure with altitude (U.S. Standard atmosphere).

2.6 HYDROSTATIC FORCES ON SUBMERGED SURFACES

Consider a two-dimensional (2-*D*, i.e., extending uniformly in the third dimension, *y*) curved plate submerged in a stationary liquid (of density ρ) with its free surface at $z = z_0$ (Fig. 2.7). Consider the forces on the upper surface of the plate. The pressure *p* is normal to the surface everywhere. If we take an element $d\mathbf{A}$ the pressure force acting on it is $-p\, d\mathbf{A}$ (i.e., inwards). The value of *p* can be obtained in terms of *z* from Eq. (2.9) as

$$p = p_{atm} - \rho g\,(z - z_0)$$

or

$$p = p_{atm} + \rho g\,(z_0 - z)$$

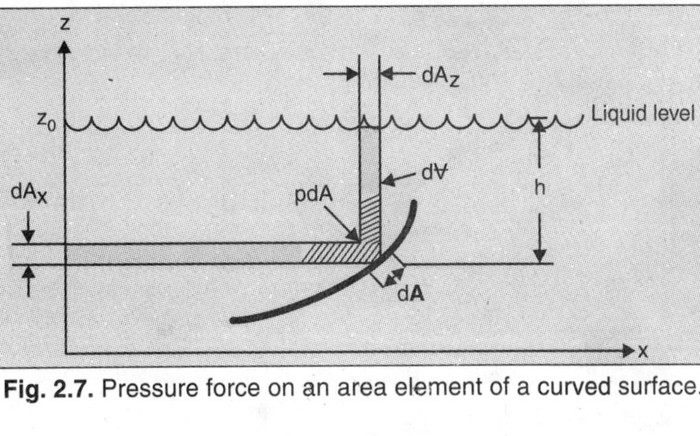

Fig. 2.7. Pressure force on an area element of a curved surface.

In hydrostatics, it is common to introduce the variable h, the depth from the free surface, to write

$$p_g = \rho g h$$

The force $d\mathbf{F}$ on the area element $d\mathbf{A}$ is then $d\mathbf{F} = -\rho g h\, d\mathbf{A}$. The total force on the whole surface is obtained by integration as

$$\mathbf{F} = -\rho g \iint_{\text{Area}} h\, d\mathbf{A} \qquad \text{...(2.12)}$$

This is the force due to the gauge pressure, i.e., in addition to the force due to the atmosphere. It is convenient, at times, to work with the horizontal and vertical components of this force. The vertical component is

$$F_z = -\rho g \iint_{\text{Area}} h\, \hat{\mathbf{k}} \cdot d\mathbf{A} = -\rho g \iint_{\text{Area}} h dA_z \qquad \text{...(2.13)}$$

where dA_z is the area projected in the vertical direction. Note that $h\, dA_z$ is the volume $d\forall$ of the liquid prism that stands vertically on the area element dA and extends upto the free surface. The integral then represents the total volume $\forall$ of the vertical prism bounded on one side by the plate surface area and on the other by the free surface, and then

$$F_z = -\rho g \forall \qquad \text{...(2.14)}$$

The negative sign represents the fact that F_z acts downwards. If the plate is thin, the vertical force on the bottom surface is the same, except that it acts upwards.

The x-component of the force is

$$F_x = -\rho g \iint_{\text{Area}} h\, \hat{\mathbf{i}} \cdot d\mathbf{A} = +\rho g \iint_{\text{Area}} h dA_x \qquad \text{...(2.15)}$$

where dA_x is the area projected in the x-direction. The construction of Fig. 2.8 shows that if the curved surface is replaced by its projection on the y-z plane, the force obtained is the same as in Eq. (2.15). Thus, the *horizontal* forces on curved surfaces are conveniently calculated by replacing them by their projected areas.

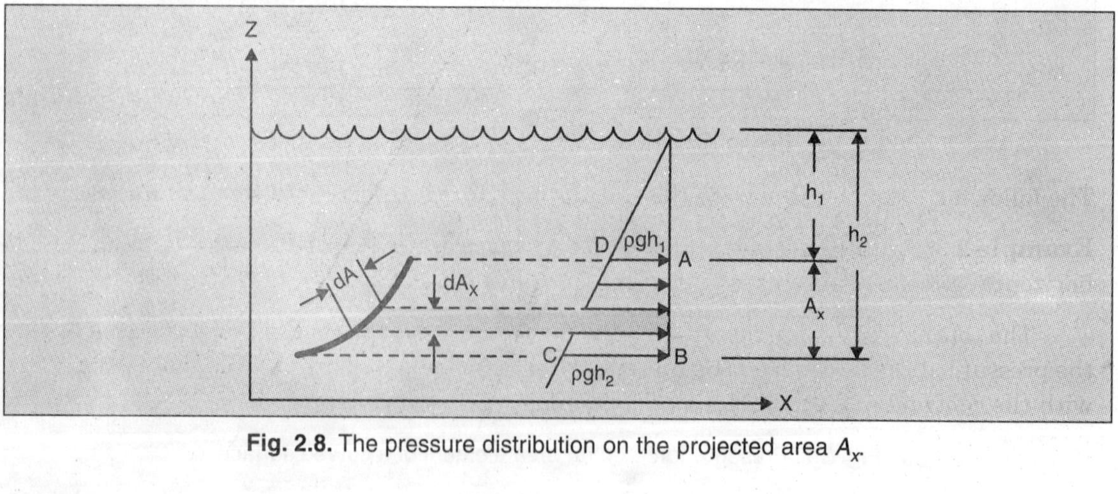

Fig. 2.8. The pressure distribution on the projected area A_x.

An interesting result can be obtained from Eq. (2.15) by multiplying and dividing it by the total projected area A_x. Then

$$F_x = \rho g \frac{\iint_{A_x} h\, dA_x}{A_x} \cdot A_x = \rho g h_c \times A_x \qquad ...(2.16)$$

where $h_c = \left(\iint_{A_x} h\, dA_x\right) / A_x$ is the depth of the centroid of the area A_x. Since $\rho g h_c$ is the gauge pressure at the centroid, Eq. (2.16) gives the general result that the *horizontal component of the pressure force on a surface equals the product of the horizontal projected area and the pressure at the centroid of that area.* This result is also valid for surfaces where the width b is not constant.

Another way to look at the integral in Eq. (2.15) is to consider the pressure distribution on the projected area as shown in Fig. 2.8. The total horizontal force of the surface is the integral of this pressure distribution and is seen as the *volume of the pressure prism* ABCD. For the case of the plate of constant width b, it is easy to see that this volume is equal to the pressure at the centroid *times* the projected area of the plate. The construction also makes it clear that the *point of application of the resultant horizontal force cannot be at the centroid of the projected area, but below it* (it should, in fact, be at the centroid of the pressure prism. See Prob. 2.9).

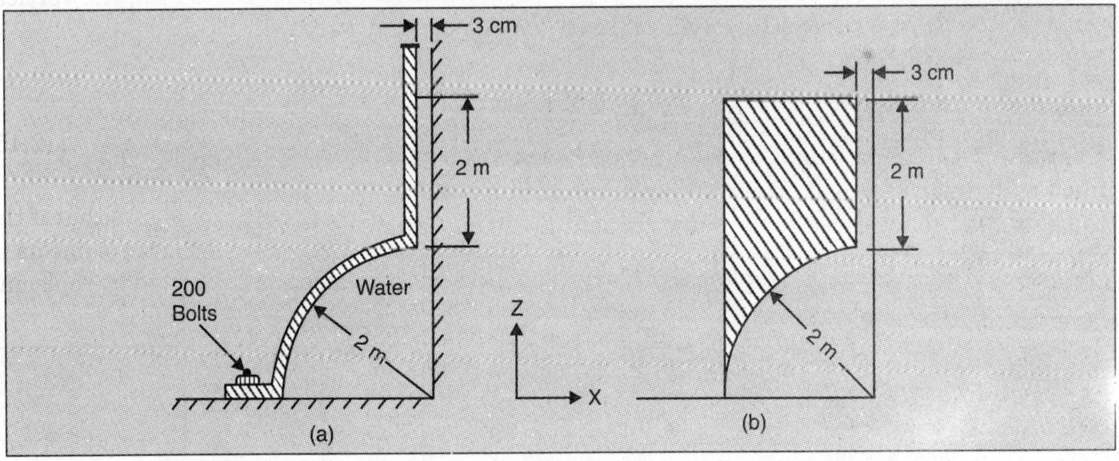

Fig. 2.9. (a) Channel for Example 2.3, and (b) the construction to compute $\forall$.

The following examples illustrate the calculation of pressure forces in some simple cases.

Example 2.3. A quarter-cylindrical channel is shown in Fig. 2.9(a). Its width is 40 m. Find the horizontal and vertical components of the force in each of the bolts due to the water pressure.

The total horizontal pressure force on the channel is equal to the projected area A_x *times* the pressure at the centroid (of the projected area). The projected area in this case is 4 m × 40 m, with the centroid at a depth of 2 m. Therefore,

$$F_x = \rho g h_c \cdot A_x = 10^3\left(\frac{\text{kg}}{\text{m}^3}\right)\times 9.81\left(\frac{\text{m}}{\text{s}^2}\right)\times 2(\text{m})\times 160(\text{m}^2)$$

$$= 3.140 \times 10^6 \text{ N}$$

This acts to the left.

The total vertical force is equal to $\rho g \forall$, where $\forall$ is the volume of the prism that stands on the surface and extends to the free surface. This hatched volume in Fig. 2.9(b) is seen to be the volume of the 4 (m) × 40 (m) × 2 (m) box *minus* the volume of the unhatched portion. Thus

$$\forall = 4\times 40\times 2(\text{m}^3) - \left[\frac{\pi}{4}(2)^2(\text{m})^2 + 0.03(\text{m})\times 2(\text{m})\right]\times 40(\text{m})$$

$$= 191.9 \text{ m}^3$$

The vertical component of the pressure force then becomes

$$F_z = +\rho g \forall = 10^3\left(\frac{\text{kg}}{\text{m}^3}\right)\times 9.81\left(\frac{\text{m}}{\text{s}^2}\right)\times 191.9(\text{m}^3)$$

$$= 1.883 \times 10^6 \text{ N}$$

which acts upwards.

The total force carried by the 200 equidistant bolts is

$$\mathbf{F} = \left(-3.14\times 10^6\,\hat{\mathbf{i}} + 1.883\times 10^6\,\hat{\mathbf{k}}\right)\text{N}$$

and so, the force on each bolt is $\left(-15.7\,\hat{\mathbf{i}} + 9.41\,\hat{\mathbf{k}}\right)$kN

Example 2.4. Figure 2.10(a) shows a pressure vessel with a hemispherical dome. The vessel is filled with an oil of density 0.8×10^3 kg/m^3 at such a pressure that a manometer connected at the bottom of the vessel has mercury rising upto 20 cm. Find the net force on the dome.

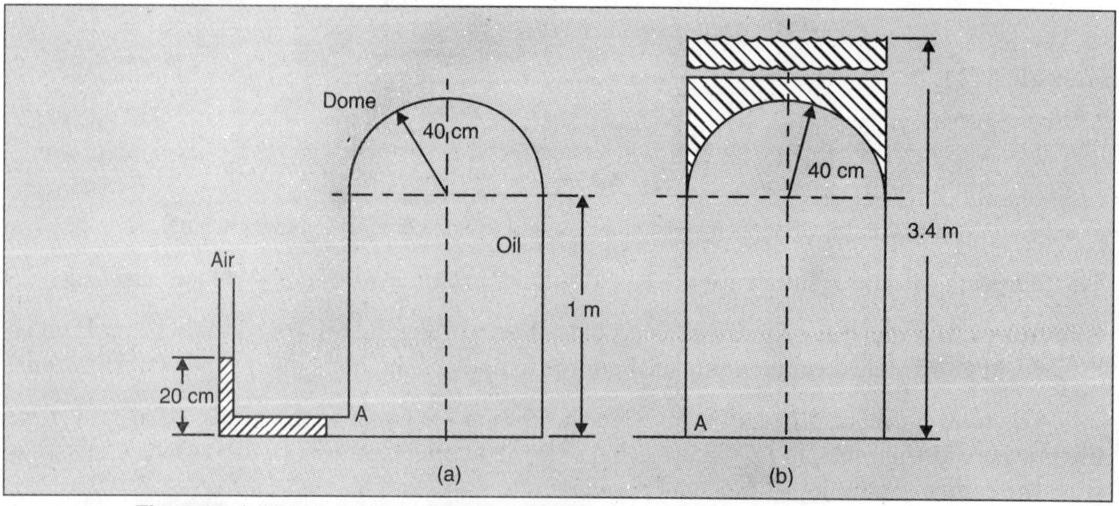

Fig. 2.10. (a) Pressure vessel for Example 2.4. and (b) the oil prism to compute $\forall$.

Since the pressure force acts normally, the symmetry of the dome precludes the possibility of any net horizontal force acting on it. To find the vertical force, we may use the result obtained in Eq. (2.14) that the net vertical force is equal to the weight of the fluid in the prism bounded by the surface and the free surface of the fluid. In this example, there is no free surface of oil. But we know the (gauge) pressure at the bottom of the vessel (point *A*). Using the procedure illustrated in Example 2.1, we obtain

$$p_{A,g} = 13.6\times10^3\left(\frac{\text{kg}}{\text{m}^3}\right)\times9.81\left(\frac{\text{m}}{\text{s}^2}\right)\times0.20(\text{m}) = 26.7\,\text{kPa}$$

where 13.6×10^3 kg/m^3 is the density of mercury.

From the pressure at point *A*, the pressure at every point on the dome can be calculated and the net force found by integrating the vertical component of the pressure force all around it. However, Eq. (2.14) can also be used by placing the hypothetical free surface of oil at a location which gives a (gauge) pressure of 26.7 kPa at point *A*. This free surface will then exert a hydrostatic pressure distribution which matches with the actual pressure distribution. Thus,

$$\rho g h_f = 26.7 \times 10^3 \text{ Pa}$$

or

$$h_f = 26.7\times10^3\,(\text{Pa})\times\frac{1}{0.8\times10^3\left(\frac{\text{kg}}{\text{m}^3}\right)\times9.81\left(\frac{\text{m}}{\text{s}^2}\right)} = 3.4\text{ m}$$

The hatched portion in Fig. 2.10(b) shows the prism that extends from the dome to this equivalent free surface. Thus,

$$V = 2.4(\text{m})\times\pi\times(0.4)^2\left(\text{m}^2\right)-\frac{1}{2}\times\frac{4\pi}{3}\times(0.4)^3\left(\text{m}^3\right) = 0.87\text{ m}^3$$

and the net vertical pressure force is

$$F_z = 0.8\times10^3\left(\frac{\text{kg}}{\text{m}^3}\right)\times9.81\times\left(\frac{\text{m}}{\text{s}^2}\right)\times0.87\left(\text{m}^3\right) = 6.83\text{ kN}$$

Example 2.5. Calculate the horizontal hydrostatic force on the vertical gate (width 0.5 m) shown in Fig. 2.11 (a). The density of the oil is 0.8×10^3 kg/m^3.

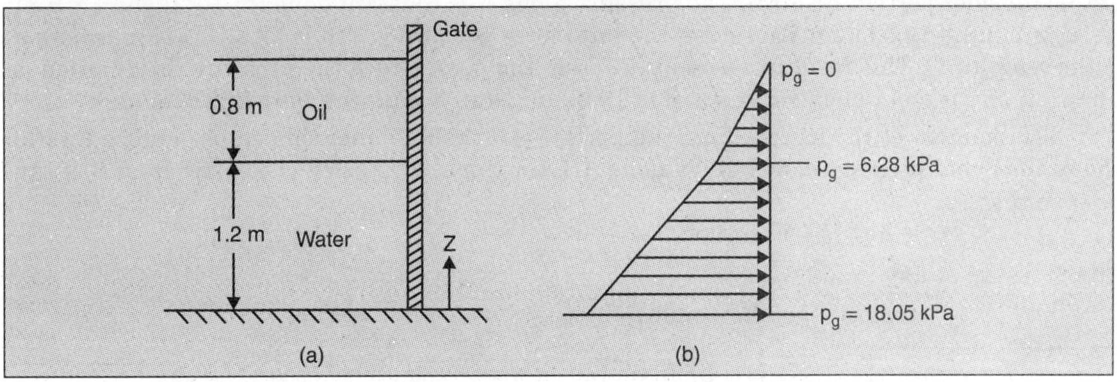

Fig. 2.11. (a) Gate for Example 2.5, and (b) the pressure distribution on it.

In the pressure prism shown (Fig. 2.11b), the pressure increases linearly with depth in the oil according to $\rho_{oil} gh$ till it reaches a value 0.8×10^3 (kg/m^3) × 9.81 (m/s^2) × 0.8 (m) = 6.28 kPa at the interface. The pressure at the centroid of this area in contact with oil is one-half of this, i.e., 3.14 kPa. Thus the pressure force due to the oil is 3.14 (kPa) × 0.8 (m) × 0.5 (m) = 1.256 kN.

The pressure in water increases linearly from 6.28 kPa at the interface till it reaches 6.28 (kPa) + 10^3 (kg/m^3) × 9.81 (m/s^2) × 1.2 (m) or 18.05 kPa. The pressure at the centroid of the water portion is (6.28 + 18.05)/2 kPa, or 12.165 kPa. This results in a hydrostatic force of 12.165 (kPa) × 1.2 (m) × 0.5 (m) or 7.3 kN. The same result is obtained if the oil layer is replaced by a hypothetical water layer with its free surface located such that the hydrostatic pressure due to it at z = 1.2 m is 6.28 kPa. The total* hydrostatic force on the gate is then the sum of the two, equal to 8.56 kN.

Note that in these calculations only p_g has been used. This is because of the fact that the net pressure force on a body depends on the gradient of pressure and, therefore, p_{atm}, the constant part of the pressure distribution does not contribute to the *net* force on it.

Example 2.6. The automatic gate *ABC* shown in Fig. 2.12(a) is pivoted at *B* and opens by itself when the level *H* exceeds 2 m. What is the length *BC* ?

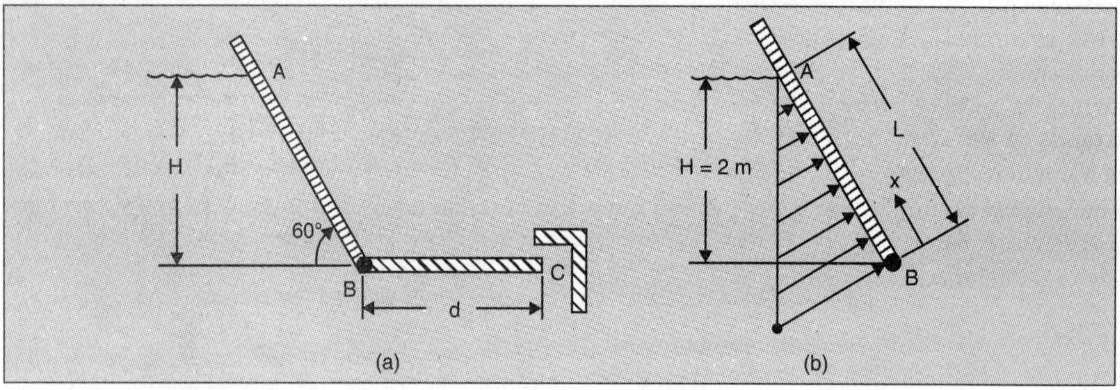

Fig. 2.12. (a) Automatic gate for Example 2.6, and (b) pressure distribution on it.

When the gate is just on the verge of opening, the anticlockwise moment due to the pressure force on the portion *BC* just balances the clockwise moment due to the force on the portion *AB*. Let us consider portion *BC* first. The pressure along it is constant equal to 10^3 (kg/m^3) × 9.81 (m/s^2) × 2 (m) = 19.62 kPa. Therefore, the total force per unit width is 19.62 d kN/m, where d is the length *BC*. This resultant acts at $d/2$ from the pivot (since the pressure distribution is uniform) and the net moment about *B* is 19.62 d(kN/m) × $d/2$(m) = 9.81 d^2 kN m/m.

The moment of the forces along the portion *AB* should equal this value. Figure 2.12(b) shows the linear pressure distribution along *AB*. If x is measured along the plate from *B*, it can be seen that

$$h = H - x \sin 60°$$

The pressure at any location x is

$$p(x) = \rho g (H - x \sin 60°)$$

* The artifice of replacing the oil layer by an equivalent water layer cannot be used for direct calculation of *total* force.

and the moment per unit width of the plate about B is

$$M = \int_0^L x\, p(x)\,dx = \int_0^L x\, \rho g (H - x \sin 60°)\,dx$$

$$= \frac{1}{6}\rho g L^3 \sin 60°$$

For $\rho = 10^3$ kg/m^3 and $L = 2$ m/sin 60°, we obtain

$$M = \frac{1}{6} \times 10^3 \left(\frac{\text{kg}}{\text{m}^3}\right) \times 9.81 \left(\frac{\text{m}}{\text{s}^2}\right) \times \left(\frac{2}{0.866}\right)^3 (\text{m}^3) \times (0.866)$$

$$= 17.44\ \frac{\text{kN m}}{\text{m}}$$

Equating this to 9.81 d^2 kN m/m we obtain d as 1.33 m.

2.7 BUOYANCY

Consider a three-dimensional body of arbitrary shape immersed in a fluid (of density ρ) (Fig. 2.13). Imagine the body surface to be divided into an upper and a lower surface such that the body occupies the volume between the two. Then by Eq. (2.14), the downward force on the upper surface is $\rho g_1 \forall_1$ where $\forall_1$ is the volume of the prism that stands on the upper surface and extends to the free surface of the liquid. Similarly, the upward force on the lower surface is $\rho g \forall_2$, where $\forall_2$ is the volume of the prism standing on the lower surface and extending upto

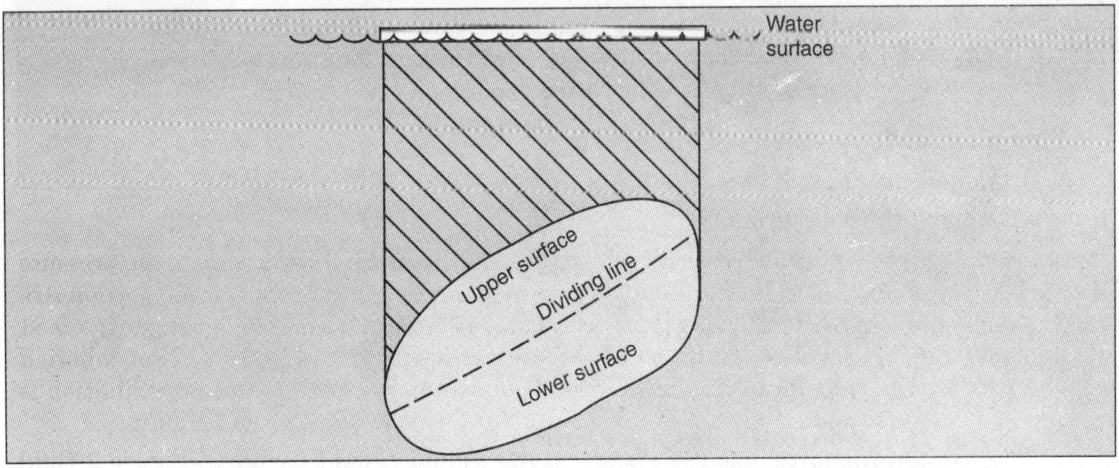

Fig. 2.13. Two liquid prisms of volumes $\forall_1$ and $\forall_2$ (hatched and tinted respectively) for a solid body immersed in a liquid.

the free surface. Since the net upward force F_B on the body, termed as the buoyancy, equals the algebraic sum of these two forces

$$F_B = \rho g (\forall_2 - \forall_1)$$

But $\forall_2 - \forall_1 = \forall_s$, the volume of the body, and so,

$$F_B = \rho g \forall_s \qquad ...(2.17)$$

Eq. (2.17) is the mathematical form of the law of buoyancy due to Archimedes (*ca.*, 3rd century B.C.) which states that a body immersed in a fluid experiences a buoyancy force equal to the weight of the fluid displaced.

Note that the buoyancy force does not depend on the depth of submergence *provided the density is assumed constant.*

Example 2.7. Find the horizontal and vertical components of the hydrostatic forces on the two cylinders (length 3 m, dia. 2 m) shown in Figs. 2.14(a) and (b).

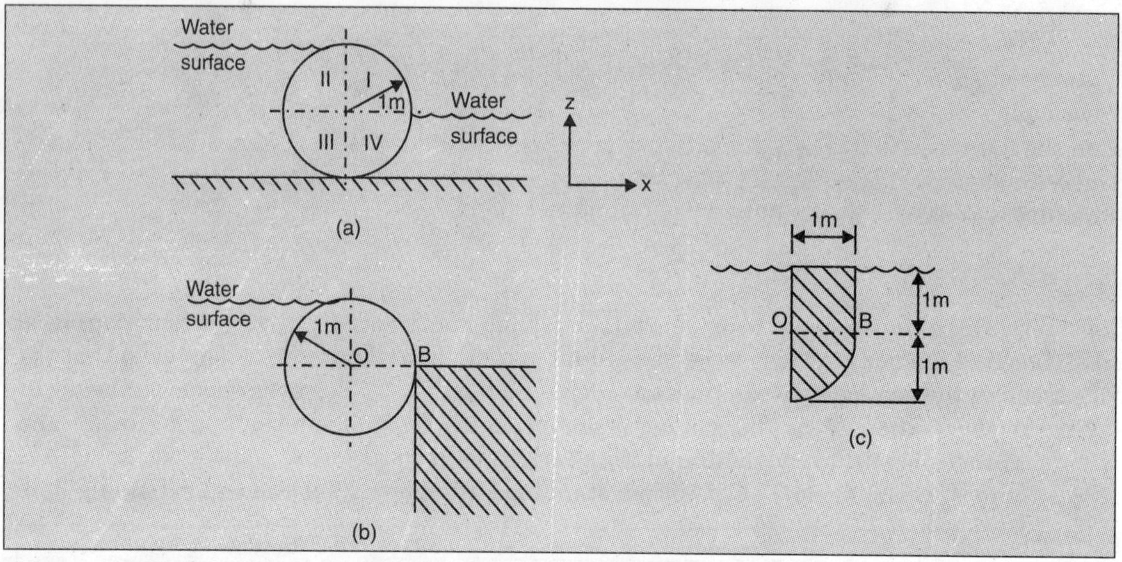

Fig. 2.14. (a), (b) Two cylinders partially immersed in water for Example 2.7, and (c) the prism to obtain $\forall$ for the fourth quadrant of Fig. 2.14 (b).

First consider the geometry of Fig. 2.14 (a).

On the first quadrant: $F_x = F_z = 0$

On the second and third quadrants:

$$F_z = pg\,\forall_s \text{ where } \forall_s \text{ is the volume of the 'body'}$$

$$= 10^3\left(\frac{\text{kg}}{\text{m}^3}\right)\times 9.81\left(\frac{\text{m}}{\text{s}^2}\right)\times\left[\frac{1}{2}\times\frac{\pi}{4}\times(2)^2\times 3\right]\left(\text{m}^3\right)$$

$$= 46.23 \text{ kN upwards}$$

$$F_x = \text{projected area} \times \text{pressure at the centroid}$$

$$= \left[2(\text{m})\times 3(\text{m})\right]\times\left[10^3\left(\frac{\text{kg}}{\text{m}^3}\right)\times 9.81\left(\frac{\text{m}}{\text{s}^2}\right)\times 1(\text{m})\right]$$

$$= 56.86 \text{ kN to the right}$$

On the fourth quadrant

$$F_z = \rho g\,\forall_s = 10^3\left(\frac{\text{kg}}{\text{m}^3}\right)\times 9.81\left(\frac{\text{m}}{\text{s}^2}\right)\times\left[\frac{1}{4}\times\frac{\pi}{4}\times(2)^2\times 3\right]\left(\text{m}^3\right)$$

$= 23.12$ kN upwards

$$F_x = [1(\text{m}) \times 3(\text{m})] \times \left[10^3 \left(\frac{\text{kg}}{\text{m}^3}\right) \times 9.81 \left(\frac{\text{m}}{\text{s}^2}\right) \times 0.5(\text{m})\right]$$

$= 14.71$ kN to the left

Thus, the net force is

$F_z = 69.35$ kN upwards, $F_x = 44.15$ kN to the right.

Next consider the geometry of Fig. 2.14(b). The forces on the first, second and third quadrants are the same as above, but that on the fourth is different. This is because the free surface no longer occurs at *B*. A perfect seal is assumed at point *B*. Since the pressure distribution on the fourth quadrant is identical to that on the third quadrant, the equivalent free surface may be taken at the same level as on the left-half. The volume of the 'prism' on the fourth quadrant is then (Fig. 2.14 c).

$$\forall = [1 \times 1 \times 3](\text{m}^3) + \left[\frac{1}{4} \times \frac{\pi}{4} \times (2)^2 \times 3\right](\text{m}^3)$$

$= 5.36$ m

and the vertical force on this quadrant is

$$F_z = 10^3 \left(\frac{\text{kg}}{\text{m}^3}\right) \times 9.81 \left(\frac{\text{m}}{\text{s}^2}\right) \times 5.36(\text{m}^3) = 52.58 \text{ kN upwards}$$

For the horizontal force, the projected area is still 1 (m) × 3 (m) but the depth of the centroid is 1.5 m (instead of 0.5 m) and so

$$F_x = (1 \times 3)(\text{m}^2) \times \left[10^3 \left(\frac{\text{kg}}{\text{m}^3}\right) \times 9.81 \left(\frac{\text{m}}{\text{s}^2}\right) \times 1.5(\text{m})\right]$$

$= 44.15$ kN to the left

The net vertical and horizontal forces for Fig. 2.14(b) are, therefore,

$F_z = 98.81$ kN upwards, $F_x = 14.17$ kN to the right.

2.8 STABILITY OF FLOATING BODIES

A body is said to be (statically) stable when a slight perturbation from its equilibrium position brings into action forces that tend to restore its equilibrium. A floating body is normally in equilibrium under the action of two forces: the weight mg and the buoyancy F_B. The weight acts vertically downwards from the centre of gravity G of the body (Fig. 2.15) while the buoyancy acts upwards from the *centre of buoyancy B,* which coincides with the centre of mass of the displaced fluid. Equilibrium requires that these two forces have the same line of action and be equal. Hence, points G and B must lie on the same vertical line, as in Fig. 2.15 (a).

CHAPTER 2

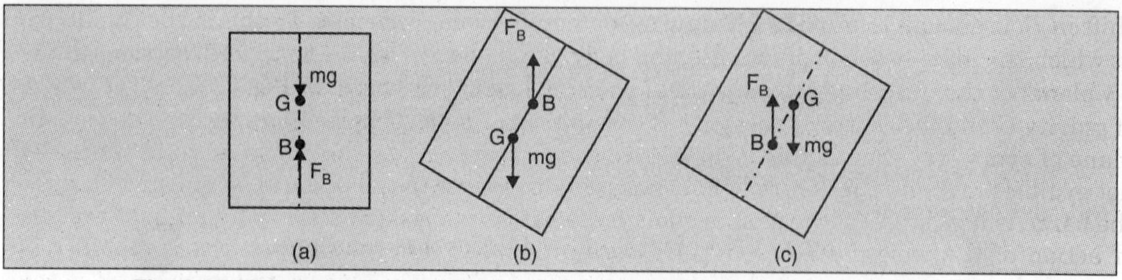

Fig. 2.15. Vertical forces acting on submerged bodies: (a) at equilibrium, $mg = F_B$, G and B on same vertical line, (b) G below B represents stable equilibrium while, (c) G above B represents unstable equilibrium.

When a *fully submerged* body is perturbed slightly from its position of equilibrium, the locations of G and B (relative to the body) cannot change.* It is clear from Figs. 2.15(b) and (c) that the weight mg and the buoyancy F_B constitute a couple which tends to restore the equilibrium if and only if the centre of gravity G lies below the centre of buoyancy B. If G lies above B, the resulting couple tends to topple the body over and is said to be unstable. The special case of neutral equilibrium occurs when G coincides with B, as for example, for fully submerged homogeneous bodies.

If, however, a body is partially submerged, and is floating at the interface of a liquid, such a strict requirement on the locations of G and B is not necessary: In such cases, under certain conditions, there may be stability even when G is above B. This happens because in such a case, the centre of buoyancy B (relative to the body) shifts as the body tilts, and the

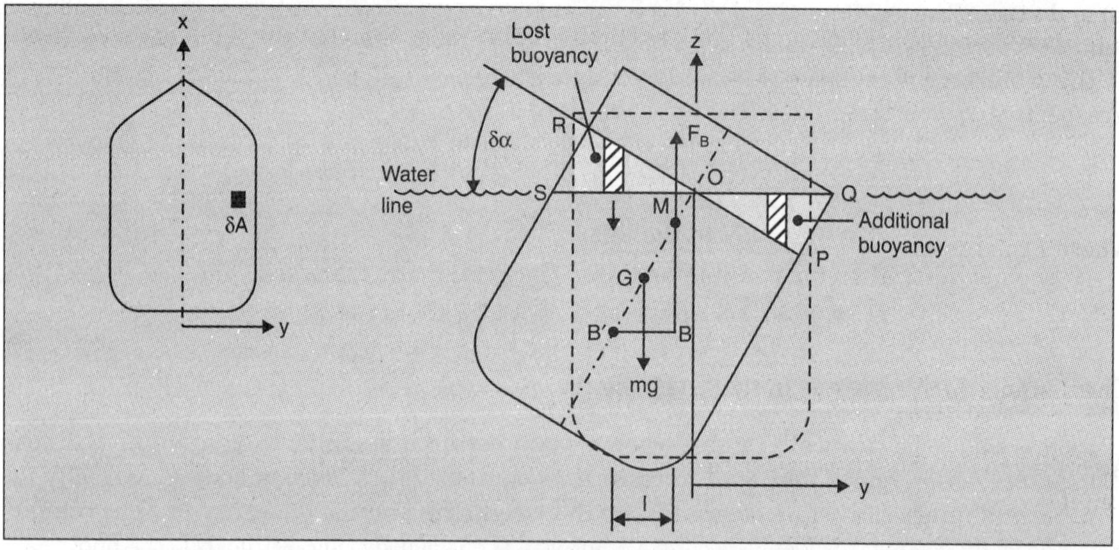

Fig. 2.16. Water 'wedges' representing additional and lost buoyancies when a barge is tilted.

* The locations of points G and B will differ only if the body is non-homogeneous so that its centre of mass does not coincide with its geometrical centre.

shift in B is enough to provide a 'restoring' or 'equilibrating' moment. To obtain the conditions at which this happens, consider a floating body, e.g., a barge, whose *horizontal* section (in the x-y plane) at the waterline is symmetrical about the x-axis, as shown in Fig. 2.16(a). The centre of gravity G and the centre of buoyancy B' when the barge floats upright are both in the vertical plane of symmetry. But as the barge tilts to the right as shown, the displaced fluid volume is not symmetrical and the centre of buoyancy B is no longer in the plane of symmetry. Clearly, the barge will be stable if the shift is enough so that point B is now clear to the right of the line of action of the weight mg through G. Conventionally, this condition of static stability is expressed as the requirement that the *metacentre* M be above G, where the metacentre is the point of intersection of the line of action of the buoyancy with the vertical plane of symmetry.

The shift of the centre of buoyancy from B' to B occurs because of the changed displacement of fluid. The line RP on the barge section represents the position of the waterline when it is upright. Therefore, on tilting, the 'wedge' ORS of the body comes out of the waterline and the 'wedge' OPQ goes under as shown in Fig. 2.16(b). The shift l in the centre of buoyancy may be visualized as that caused by an additional weight of the water wedge OPQ acting upwards, and the weight of the water wedge ORS acting downwards—this representing a loss of buoyancy.

Thus, the new centre of buoyancy is the point where the net moment because of the force F_B' ($= mg$) acting at the original centre of buoyancy B' and the couple acting because of the 'wedges' OPQ and ORS is zero. If the centre of buoyancy is displaced through a distance l to point B, then taking moments about B, we get

$$F_B\, l = \text{net couple of the additional and lost buoyancies}$$

The couple due to the two wedges can be taken at any convenient point. We take it about O. The weight of the liquid displaced by the slice of area δA shown shaded in Figs. 2.16(a) and (b) and at a distance y from O is $\rho_w\,(y\delta\alpha)\,\delta A\, g$ and the moment contribution of this force about point O is $y\,(\rho_w y\, g\delta\alpha\delta A)$. Here, δA is an area element in the (horizontal) section *in the x-y plane,* at the waterline. Note that this is always counter-clockwise, i.e., a positive contribution, irrespective of whether δA is to the left or right of O. Thus,

$$F_B l = \rho_w g\,\delta\alpha \iint\limits_{\text{Area}} y^2 dA = \rho_w g\,\delta\alpha I_{xx}$$

where I_{xx} is the second moment of the *waterline area* about the *axis of tilt.* Therefore,

$$l = \frac{\rho_w g\,\delta\alpha I_{xx}}{F_B}$$

The distance MB' is then $l/\sin(\delta\alpha)$ or $l/\delta\alpha$ for small $\delta\alpha$,

$$MB = \rho_w g\, I_{xx}/F_B$$

According to the above discussion, the body is in stable equilibrium if $MG = MB' - GB'$ is positive, i.e., if

$$\frac{\rho_w g\, I_{xx}}{F_B} - G\,B' > 0$$

It is clear from this formulation that the lesser the value of I_{xx}, the more unstable is the body. Therefore, one needs to investigate the stability about the longest axis of the body only.

Example 2.8. A ship weighs 2×10^7 kg and its waterline cross-section is as shown in Fig. 2. 17. The centre of buoyancy is 2 m below the surface. Find the maximum permissible height of the C.G. for static stability.

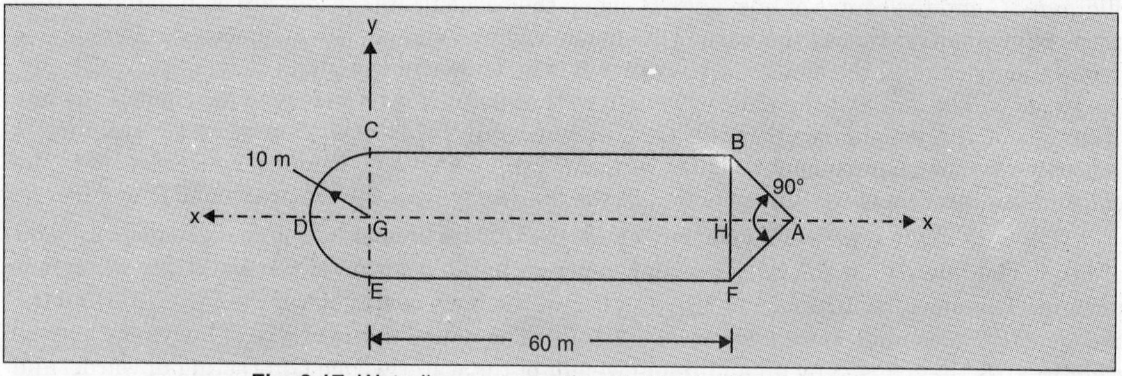

Fig. 2.17. Waterline cross-section of ship, for Example 2.8.

The critical mode for instability would be rolling about the longitudinal axis *x-x* about which the second moment of inertia is the lowest. I_{xx} is obtained as below*:

$$I_{xx} = \iint_{\text{Area}} y^2 dA$$

$$= I_{xx,\, CDEG} + I_{xx,\, CBFE} + I_{xx,\, BAFH}$$

$$= \frac{1}{2} \times \frac{\pi \times 10^4\,(\text{m}^4)}{4} + \frac{60 \times (20)^3\,(\text{m}^4)}{12} + \frac{10 \times (20)^3\,(\text{m}^4)}{48}$$

$$= 4.56 \times 10^4 \text{ m}^4$$

Therefore, the height of the metacentre above the centre of buoyancy is

$$\frac{\rho_w g\, I_{xx}}{F_B} = \frac{10^3\,(\text{kg/m}^3) \times 9.81\,(\text{m/s}^2) \times 4.56 \times 10^4\,(\text{m}^4)}{2 \times 10^7\,(\text{kg}) \times 9.81\,(\text{m/s}^2)}$$

$$= 2.28 \text{ m}$$

Since the metacentre must lie above the centre of gravity, the maximum permissible height of the C.G. is 0.28 m above the waterline.

2.9 SURFACE TENSION

From Eqs. (2.7) and (2.8) governing the variations of pressure in a stationary fluid, it can be shown that the free surface of a liquid must be horizontal. But there are some situations when this fact does not hold. If a thin glass tube open at both ends is pushed in water (Fig. 2.18), water rises in the tube. However, when it is pushed in mercury, the liquid level is depressed

---

* I_{xx} for a rectangle of size b in the x direction and l in the y direction is $bl^3/12$. That for a circle of radius r is $\pi r^4/4$ while that for a triangle having base b in the y direction and *height* l in the x direction is $lb^3/48$.

inside the tube. Looking carefully at the liquid surfaces within the thin tubes in Fig. 2.18, it is noticed that these levels are not horizontal but curved, concave upwards in the case of water and convex upwards in the case of mercury. Similarly, water can be poured upto a level slightly above the edges of a glass beaker whose rim is slightly greased (Fig. 2.19).

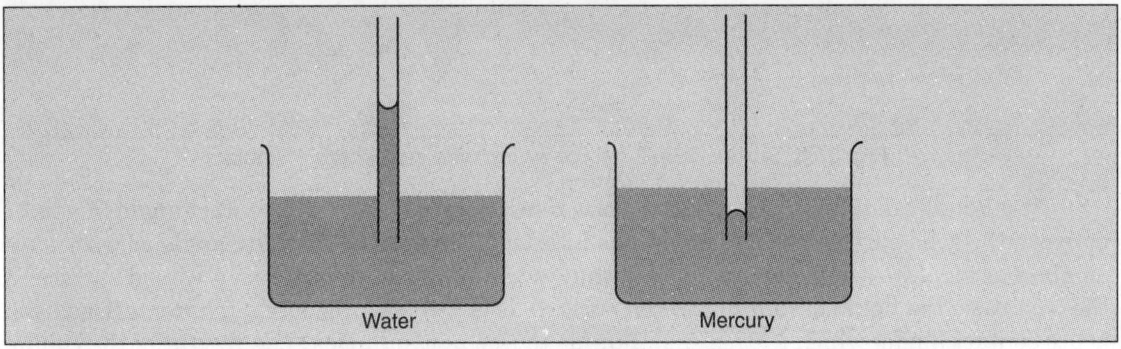

Fig. 2.18. Rise and depression of liquid level when a capillary is immersed in water and mercury.

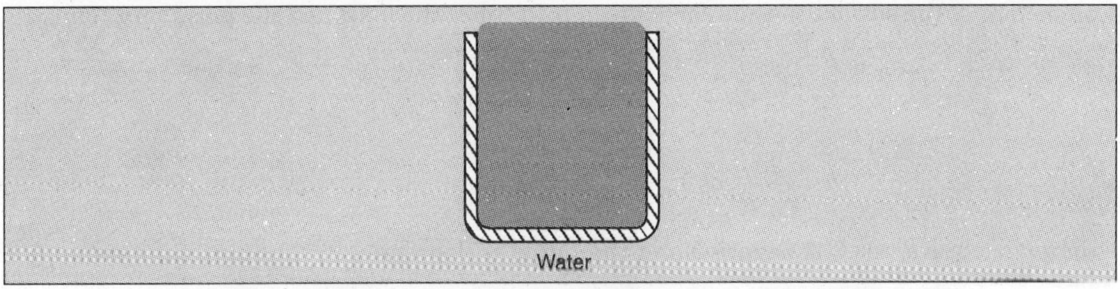

Fig. 2.19. Level of water above the edges in a greased beaker.

The apparent contradictions of the requirement that a free-surface be horizontal is explained by the introduction of a force termed as *surface tension*. This results from the forces of cohesion between like molecules. A molecule well inside a fluid is attracted by like molecules equally from all sides so that the net force on it is zero. But for a molecule on the surface, there is a net force pulling the molecule into the fluid. This gives rise to a force acting on the interface. If a cut of length dl is made on the interfacial surface, a pair of opposite forces, each equal to $\sigma\, dl$ results. These act normal to the cut and are parallel to the surface. The coefficient σ is termed as surface tension and has the dimensions of force per unit length.

One consequence of this surface tension is that it creates a jump in the normal pressure across a curved free-surface. Consider a liquid drop of radius R as shown in Fig. 2.20. If we cut this drop and consider one half of the sphere, the surface tension σ will act per unit length on the circumference of length $2\pi R$. This must be balanced by a pressure difference Δp between the inside and the outside, which gives a net force equal to $(\Delta p)\pi R^2$. Therefore, the pressure difference due to surface tension is

$$\Delta p = \frac{\sigma 2\pi R}{\pi R^2} = \frac{2\sigma}{R} \qquad ...(2.18)$$

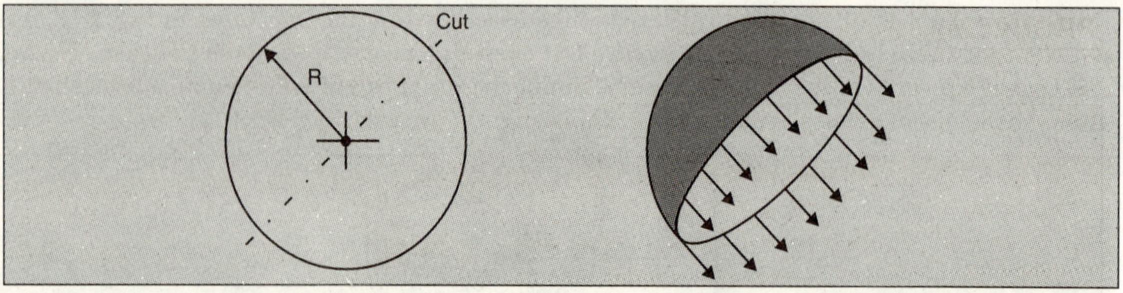

Fig. 2.20. Surface tension force acting on a cut-sphere of liquid.

At a liquid-gas-solid interface the surface tension forces result in a contact angle (Fig. 2.21) between the solid surface and the liquid-gas interface. This can be an acute angle as with water on glass or an obtuse angle as happens with mercury on glass or water on a waxed surface. In the first case, the liquid is said to *wet* the surface. It is this wetting action (characterised by an acute contact angle) which causes such liquids to rise in capillaries. The weight of the column of water which stands above the level in the reservoir (Fig. 2.22) is balanced by the vertical component of the surface tension force given by σ *times* the length of the periphery. Thus

$$\sigma \times 2\pi R \cos\theta = \rho g \pi R^2 h$$

or

$$h = \frac{2\sigma}{\rho g R} \cos\theta \qquad \text{...(2.19)}$$

when θ is obtuse, $\cos\theta$ is negative and the liquid is depressed in the tube, as is mercury in glass.

Fig. 2.21. Contact angle θ for non-wetting and wetting liquids.

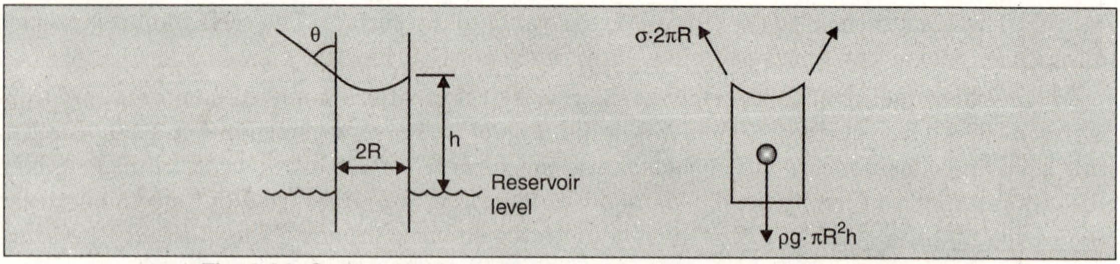

Fig. 2.22. Surface tension balancing the weight of the liquid column.

PROBLEMS

2.1 Obtain $p_A - p_B$ in terms of the quantities shown. Also obtain $p_C - p_A$.

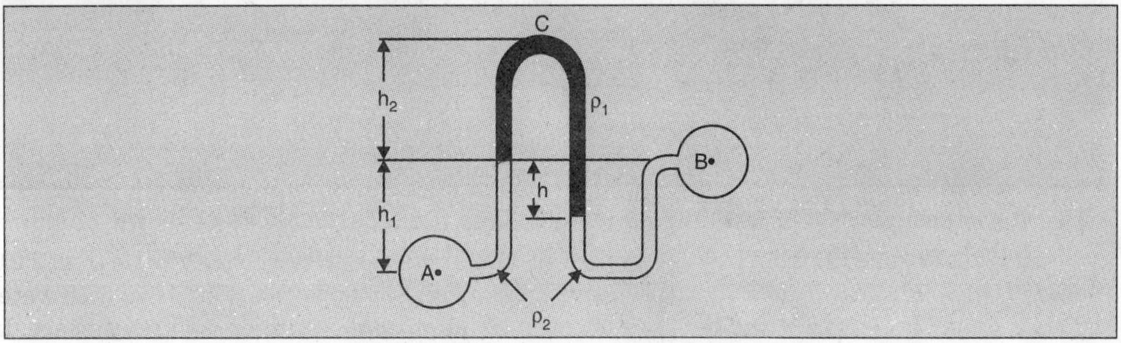

2.2 Find the difference h in the levels of mercury in the manometer tubes shown. Explain your result.

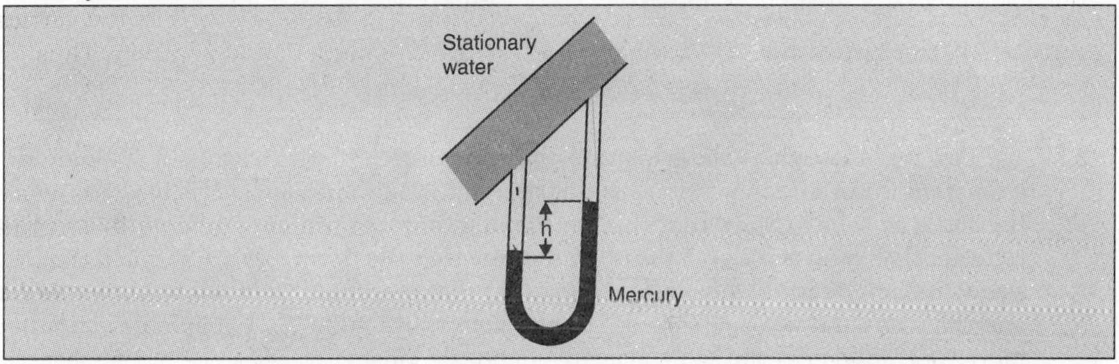

2.3 In the hydraulic jack shown the load calculations are made by setting $F/A_1 = W/A_2$. Does this mean $p_1 = p_2$? Explain.

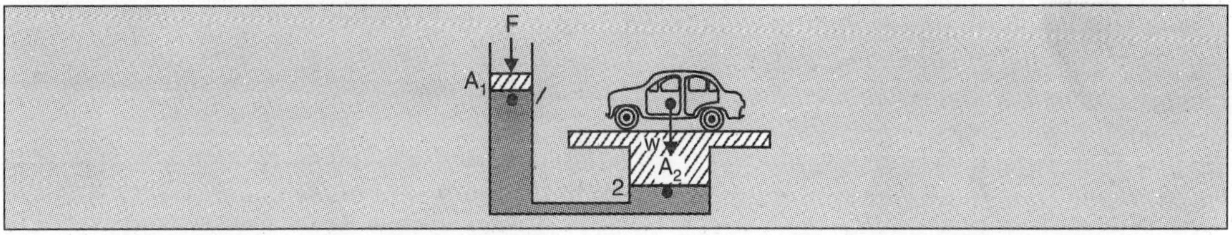

2.4 The towing of Antarctica icebergs to meet the increasing demand of fresh water is generating considerable interest. An inventor proposes that icebergs can be made to propel themselves by shaping them into wedges as shown. He argues that since the pressure at a depth is constant, the larger area AB experiences a larger force than area CB and so there is a net propulsive force acting on the iceberg. Does his proposal 'carry any water'?

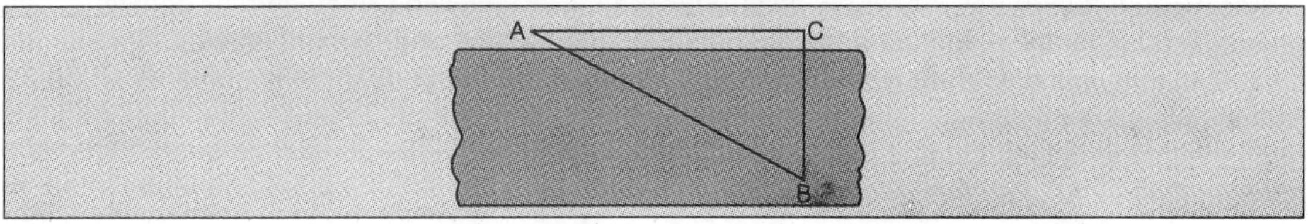

2.5 Obtain the gauge pressure p_1 in terms of the two densities ρ_1 and ρ_2 and the level differences h_1 and h_2 for the arrangement shown. Is the pressure at point A atmospheric?

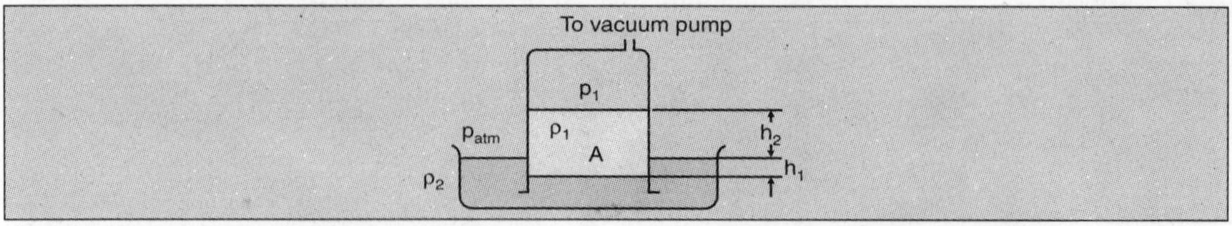

2.6 A common method of levelling large fields is to use a garden hose as shown. If an air bubble gets entrapped in the hose, will the operation of the device be affected?

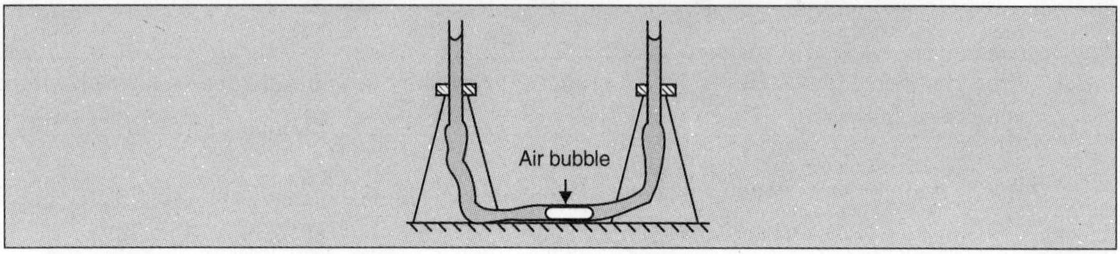

2.7 The flow rate through a tube changes with the pressure difference across it. To maintain a constant drain rate in a glucose-saline drip, a hypodermic needle B is inserted inside the bottle as shown. Show that this results in a constant pressure difference (and so a constant flow rate) across the tube CD. Assume that the flow rates are so small that the equations for hydrostatics apply. Show also that the pressure at point A is subatmospheric, and that the fluid will not ooze out of any hole above point E. A simple device called *Mariotte's bottle* uses the same principle to puzzle kids.

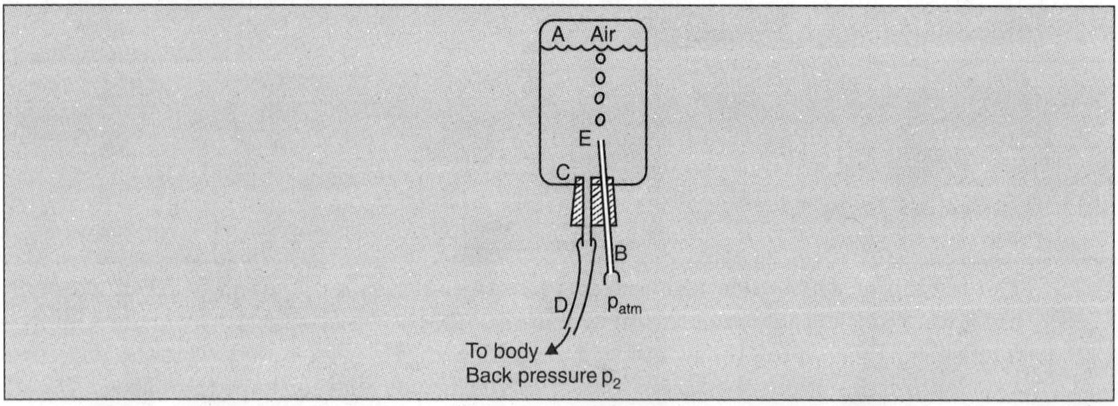

2.8 A diving bell is idealized as a cylinder of length L closed at one end, as shown. At sea level it is full of air. As it sinks the air is compressed and the water rises into the bell to a height h. Obtain a relation between h and the depth H if the compression of air is assumed isothermal. Simplify for $L \ll H$..

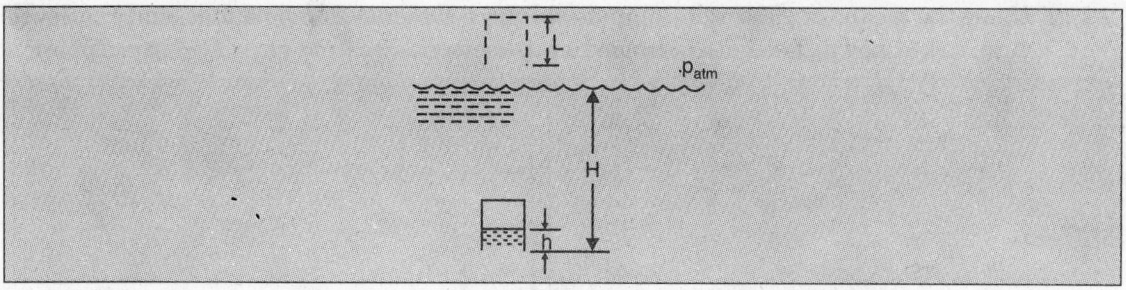

2.9 *Centre of pressure*: Prove that the resultant pressure force acting on the flat rectangular plate *AB* of width *W*, as shown, acts at the centroid *C* of the pressure prism.

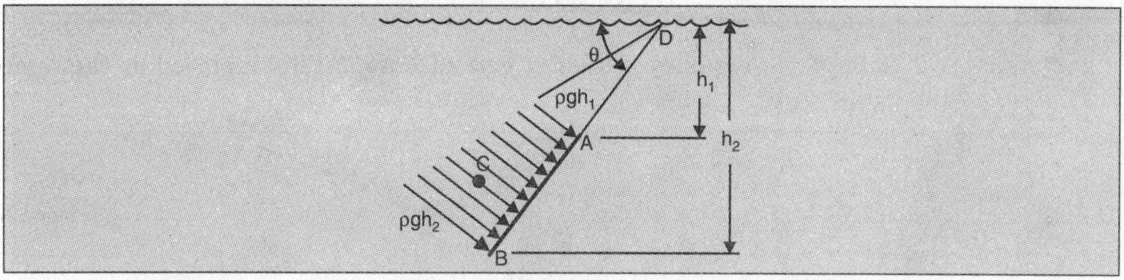

2.10 Find the pressure at the bottom of the tank shown. Also, find the force on the gate per unit width.

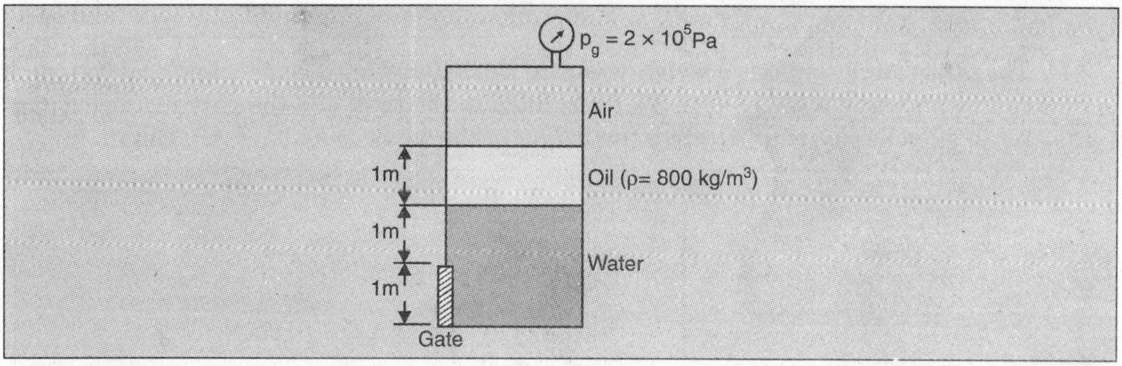

2.11 Compute the force required to support the gate shown at the bottom of the tank. Take the width of the gate to be 1 m.

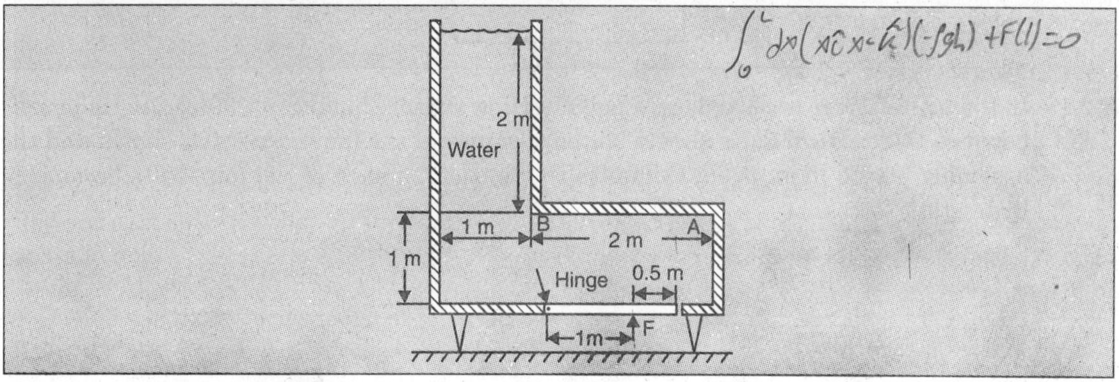

2.12 Using the results of Prob. 2.9, compute the force F required to hold stationary the gate (1 m wide) shown. The gate is hinged at its centre.

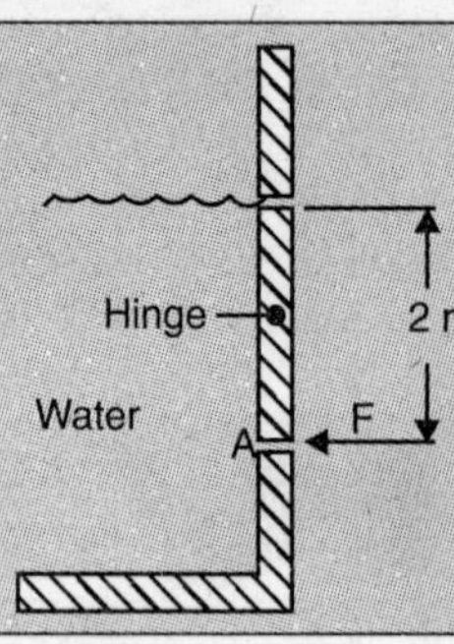

2.13 What will be the force required if the flat gate of Prob. 2.12 is replaced by the semi-cylindrical gate shown?

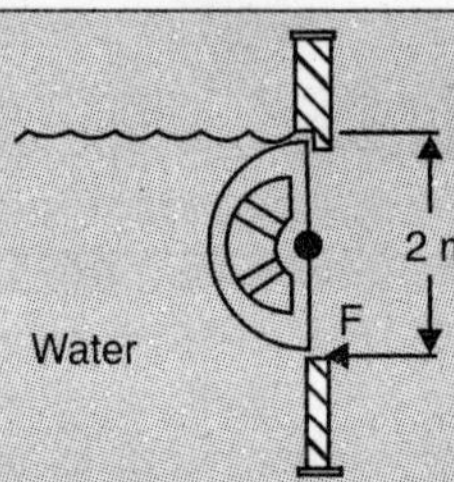

2.14 The tank shown contains a weightless gate *EABCD* having a semi-cylindrical bottom. It can rotate about a hinge at *A*. Water is filled in the tank and the gate just opens (i.e., rotates clockwise about *A*) when the height of the water is *h* above *AB*. Obtain *h*.

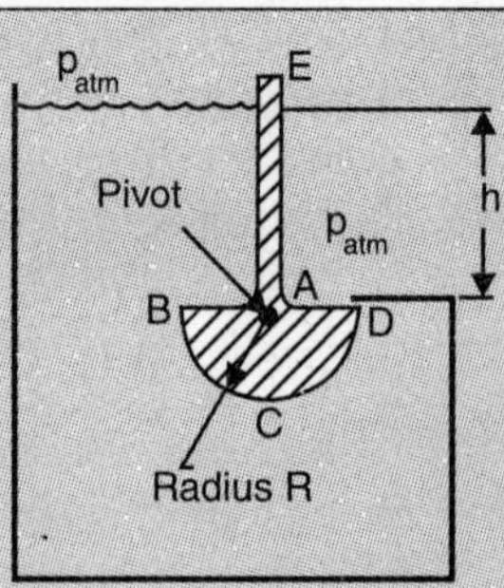

2.15 In the manufacture of plexiglass, a polymerizing mixture having $\rho = 900$ kg/m^3 is poured between two vertical glass sheets bolted together at the four corners as shown and the assembly placed in an oven. Compute the tension on each of the four bolts because of hydrostatic forces.

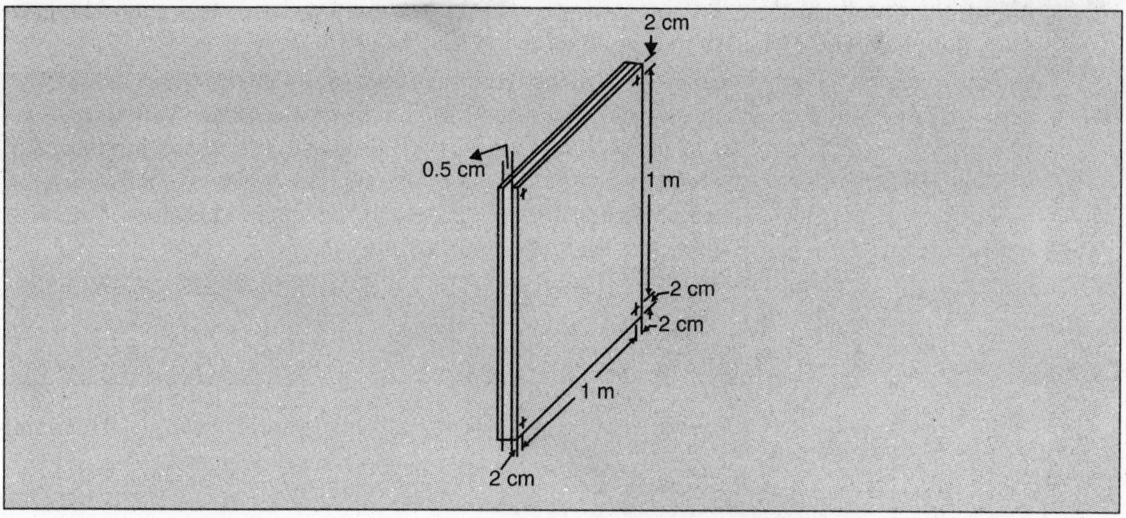

2.16 An inexpensive design of a fuel gauge for a scooter measures the pressure at the bottom of the tank as shown.

(a) Obtain the relationship between the level h of gasoline in the tank and the pressure reading p_{gauge}.

(b) If 1 cm layer of water is present at the bottom of the tank as shown, find the quantity of gasoline in the tank when the gauge indicates it to be full (total capacity = 8 litres).

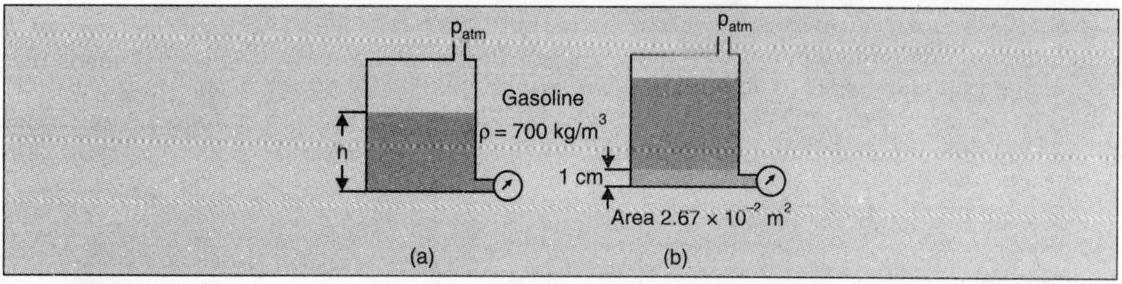

2.17 The pressure gauge of Prob. 2.16 is proposed to be replaced by a mercury manometer as shown. Obtain the relation between h and h_1, the level of mercury above its zero reading.

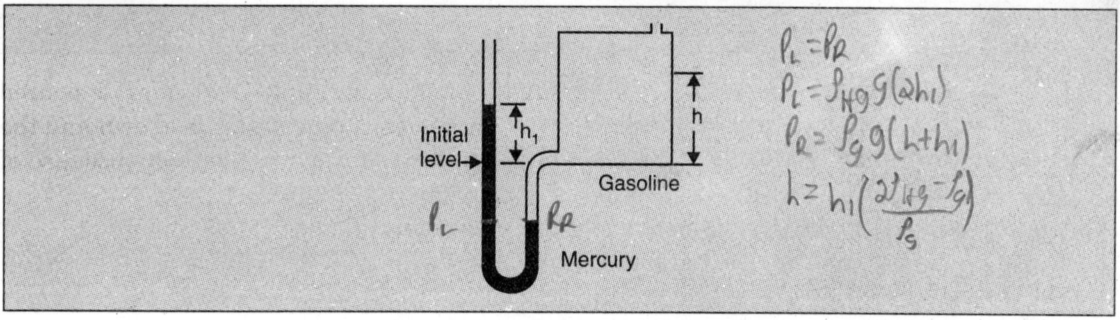

2.18 A hydrostatic 'uplift' due to the seepage of water acts on the base of a gravity dam shown. A common approximation is to assume this uplift to vary linearly from p_0 (where p_0 is the hydrostatic gauge pressure at the bottom of the reservoir) at the inner edge O to zero at the outer edge A. The normal reaction from the ground is also assumed to vary linearly along the base. As the water fills up in the reservoir, the water pressure as well as the seepage 'uplift' tend to topple the dam about point A. The limiting case of acceptable design occurs when the normal reaction at point O just vanishes. Find the maximum height h to which water may be stored safely.

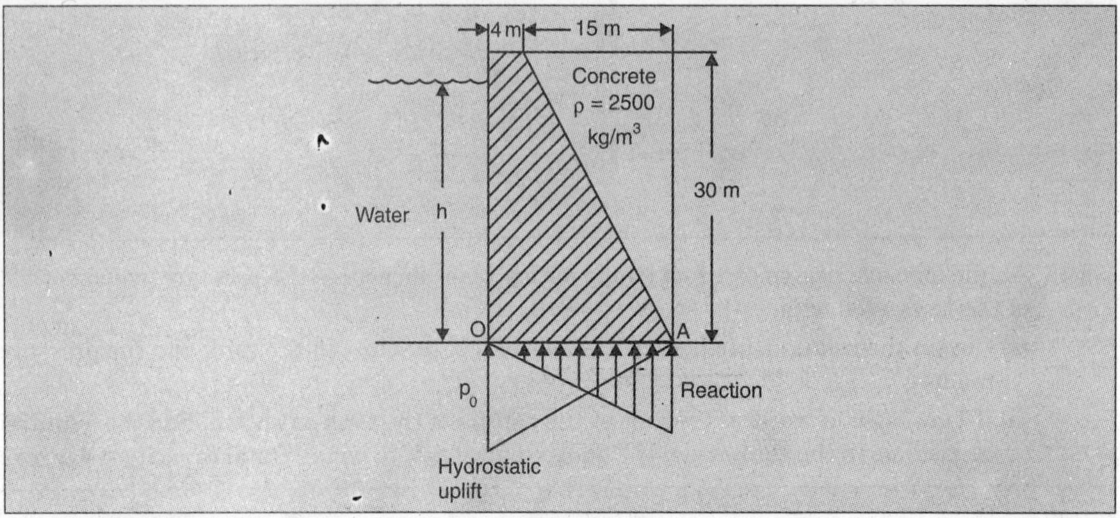

2.19 A gate of size 2 m × 1 m can slide without friction as shown. It is held in place by a thin cable which pulls it to the left due to buoyancy on a balloon of diameter D and of negligible weight. The gate just opens when the level of water is as shown. Obtain D. Where should the cable be attached so that the gate slides smoothly (i.e., without rotation) inside the channel?

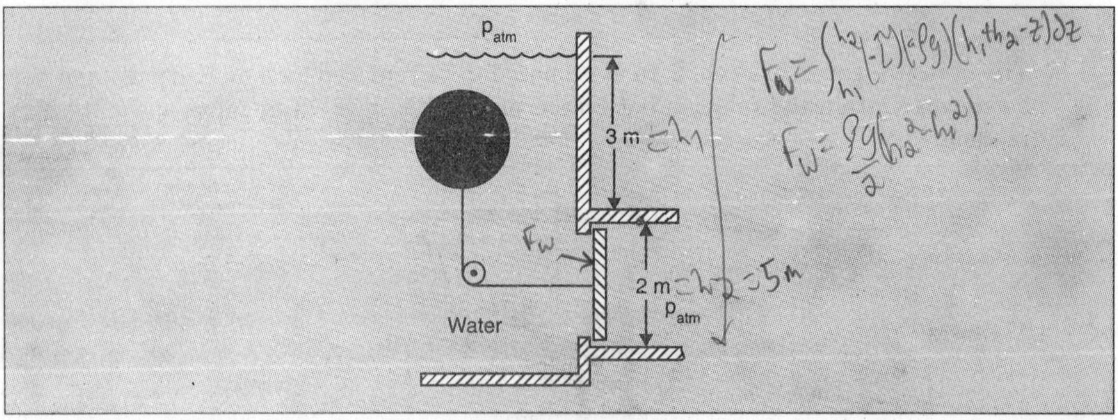

2.20 A 1 m × 1 m × 1 m steel cube is to be floated in mercury in the container shown with a clearance of 5 mm on each side. Find the weight of mercury required. Note that the amount of mercury required to float the cube is a very small fraction of its weight.

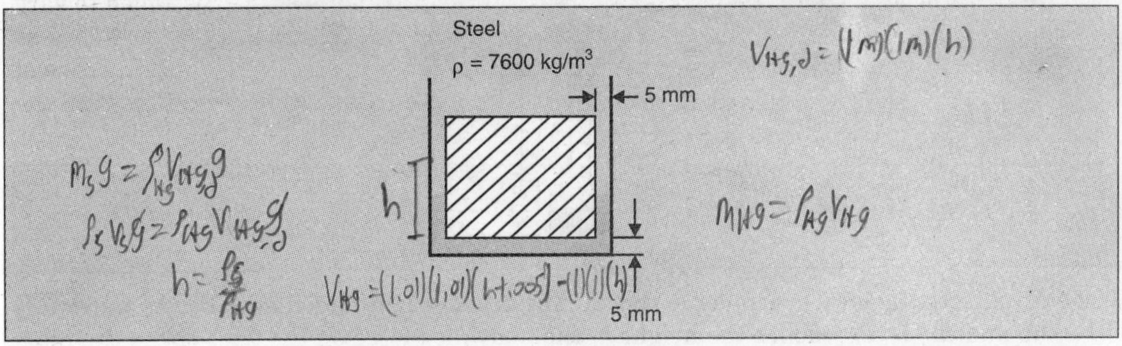

2.21 A scooter tube inflated to the dimensions shown is used as a life-preserver in a lake. Find the maximum weight that can be supported by it when it is immersed fully in water. A child sitting on this tube floats with 80% of his body immersed. If the effective density of the human body is 1.073×10^3 kg/m³ estimate the we ight of the child.

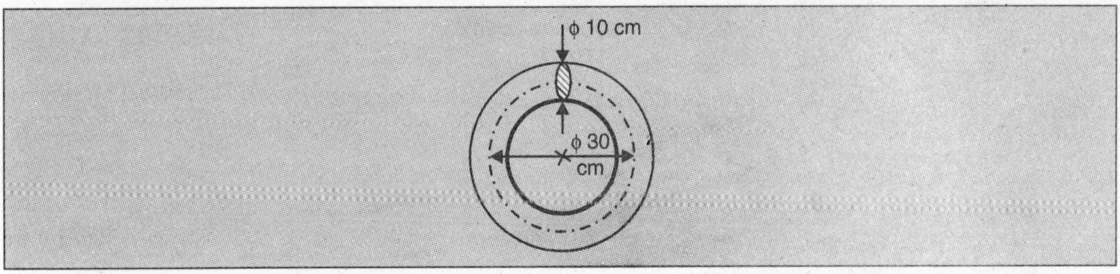

2.22 In an estuary where the fresh water of a river meets the sea it is observed that fresh water maintains a distinct layer above the saline water for quite some length into the sea, after which it loses its identity. Thus, close to the shore we can model the density variation as shown. A sphere of 40 cm diameter weighing 33.8 kg floats near the interface of the fresh and saline waters. Find the location of its centre with respect to this interface.

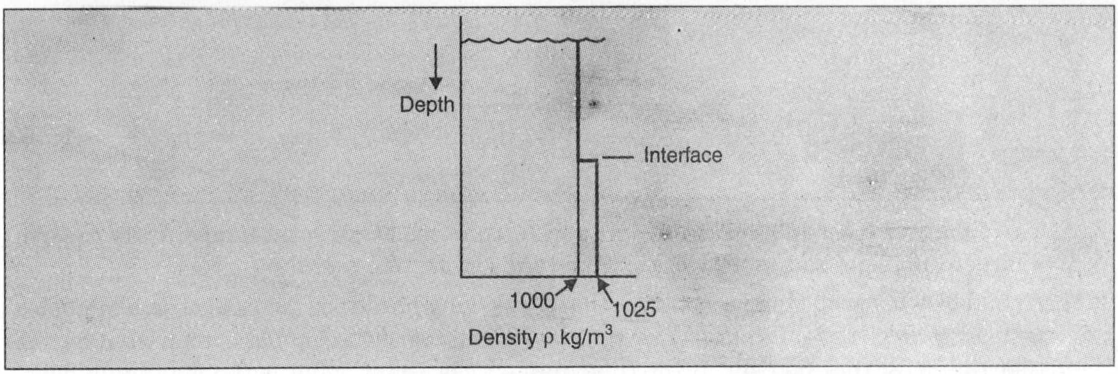

2.23 A lactometer has dimensions as shown. Its weight is W. Obtain a relationship between the reading h and the density of the milk ρ. Also obtain an expression indicating the sensitivity of the instrument (i.e., Δh as a function of $\Delta\rho$).

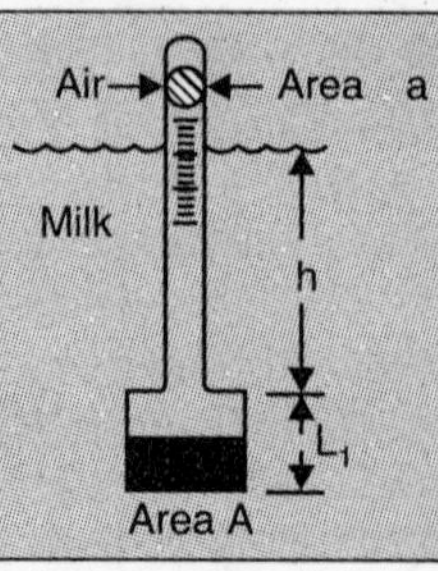

2.24 An interesting level controller is shown. The drain plug is closed initially. As water fills up and the level reaches the height h, the buoyancy force on the float opens the plug. Find h.

As soon as the plug opens, it is observed that the plug-float assembly jumps upwards and attains a floating position. Explain why? Determine the level in the reservoir when the plug closes again. Can the plug diameter be larger than the float diameter? Explain.

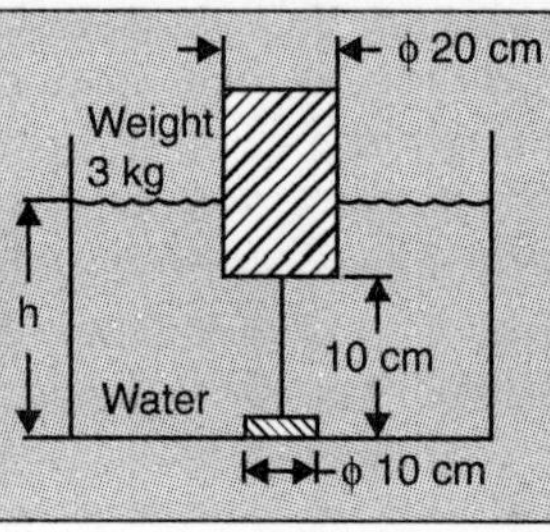

2.25 Obtain an expression for the rise h of the fluid between the two vertical parallel plates held at a distance d apart in terms of σ and the contact angle θ.

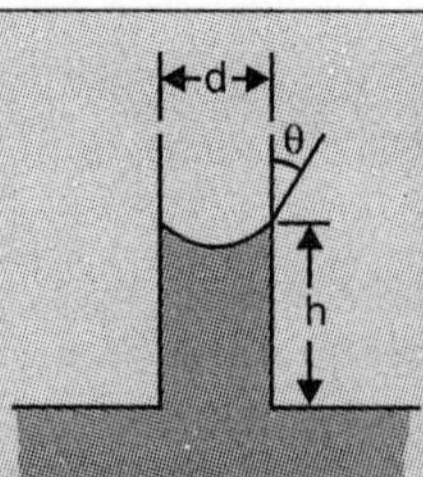

2.26 Two spherical soap bubbles of radii R_1 and R_2 coalesce at constant temperature to form a larger bubble of radius R. Obtain R in terms of R_1, R_2, σ and p_{atm}.

2.27 A child blowing soap bubbles observes that on blowing harder he gets lots of small bubbles instead of large ones. Explain. The smallest diameter of the bubble is restricted by the tube diameter that he uses. If the tube diameter is 1 mm, what is the maximum over-pressure that he should exert to get good bubbles? Take $\sigma = 2 \times 10^{-2}$ N/m. What happens if he blows much harder than this?

2.28 To estimate the clotting time of blood, a sample is drawn into a fine glass capillary of diameter 1 mm by touching one end of it to a freshly punctured fingertip. If the surface tension of the blood is approximately 5×10^{-2} N/m, its contact angle with glass 0° and its density 1056 kg/m^3, estimate the volume of the blood sample so drawn.

2.29 Imagine an astronaut in a pressurised vehicle moving in gravity-free space. He wants to drink orange-juice from a cup and finds he has a problem on his hands. The ground-command suggests using a straw to suck it up. He wonders whether the lack of gravity would not foul it up. Can you clear his doubt?

2.30 Water flows at rate $\overset{\circ}{Q}$ through a pipe of diameter a into a vessel of diameter c placed on a weigh-bridge. The scale reading S is used to estimate $\overset{\circ}{Q}$. Assume a is very small compared to c and that the velocity of water in the pipe is so small that we can neglect all dynamic effects.

Obtain a relationship between dS/dt and $\overset{\circ}{Q}$ when

(a) the surface of the water is below the pipe as shown.

(b) the pipe is submerged in water. Assume the pipe to be supported at A and make suitable approximations.

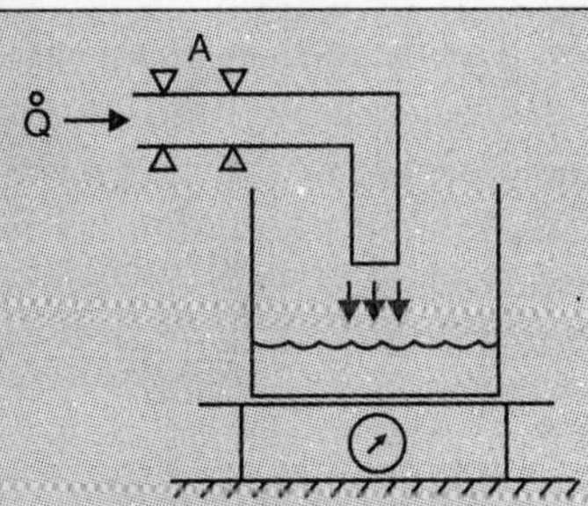

2.31 The mild steel plate forming the quarter-cylindrical channel of Example 2.3 (Fig. 2.9a) rests unbolted on the floor. Its mass is 900 kg per metre run. Find the height of water when the plate just lifts up from the floor.

3

DESCRIPTION AND ANALYSIS OF FLUID MOTION

3.1 DESCRIPTION OF PROPERTIES IN A MOVING FLUID

The continuum hypothesis enables us to assume that at any time *t*, there is a fluid 'particle' corresponding to every point in space occupied by the fluid. If a property, say temperature *T* of the fluid were to be specified, it could be done in two distinct ways. In one of these, the property is specified as a function of the position in space (denoted by vector **x**) and of time *t*. This is termed as the *Eulerian* or *field* description and, in effect, specifies the property of *that* fluid particle which happens to be at the location **x** at the given time *t*. At time $t + dt$, the temperature $T(\mathbf{x}, t + dt)$ is the temperature, not of the same particle, but of a different particle—the one that

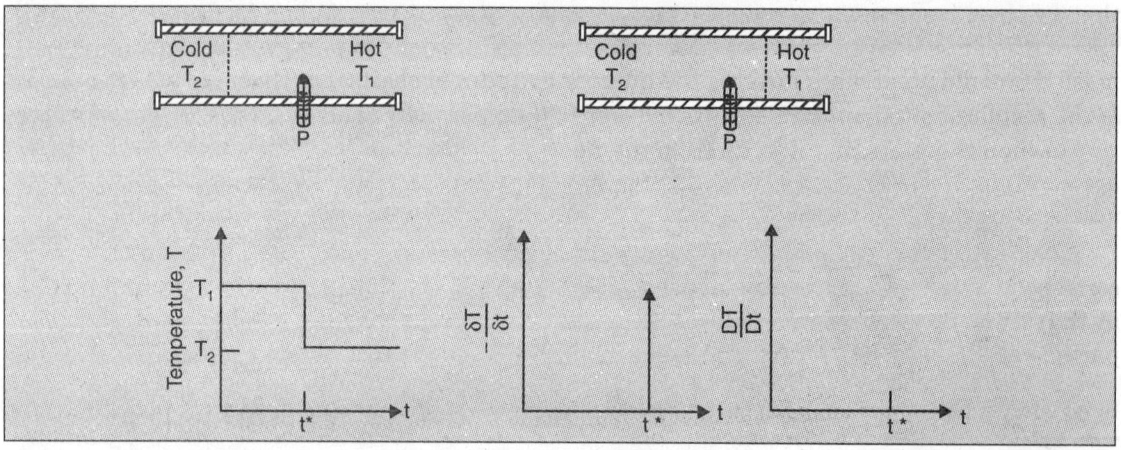

Fig. 3.1. Measurement of temperature at position *P* in a pipe with a hot liquid at temperature T_1 being pushed by a cold liquid at temperature T_2.

happens to be at location **x** at $t + dt$. Obviously, the field approach does not describe the temperature-time history of any particular particle. Thus, if in a tube, a hot liquid is being pushed by a cold liquid (Fig. 3.1) then, at a given location *P* the temperature measured at one instant of time will be the higher T_1. When the insulated surface of separation of the two liquids has passed beyond P (i.e., at $t > t^*$), the temperature recorded will be the lower T_2. Thus,

$$T(\mathbf{x}, t_1) = T_1; \qquad \text{for } t_1 < t^*$$

$$T(\mathbf{x}, t_2) = T_2; \qquad \text{for } t_2 > t^*$$

This, of course, does not mean that the liquid, or any fluid particle, has cooled. It is only that, at two different times, the temperature measured is of two different fluid particles.

The other approach of specifying a property in a moving fluid consists of identifying the fluid particles with some labels, following them around, and specifying their properties as a function of time. Usually, the particles are labelled by the space point they occupied at some initial time t_0. Thus, $T(\mathbf{x}_0, t)$ refers to the temperature at time t of a particle which was at location $\mathbf{x}_0$ at time t_0. This approach of identifying material points and following them along is termed as the *Lagrangian* or the *particle* or the *material* description.

In the study of fluids, the Eulerian (or field) approach is preferred because it is difficult to follow a fluid particle (which cannot be differentiated visually from its neighbours), as also the fact that measurements are easier made at fixed locations. In contrast to this, the Lagrangian approach is usually preferred in the description of moving solids, as for example, in describing the motion of a projectile.

Thus, if the fluid is in motion, a point fixed in space is occupied by different material particles at different instants of time. The time rate of change, which is measured by a probe stationed at a fixed location, therefore, does not give the rate of change of property of any material particle. To measure this rate the probe will have to, so to say, move with the fluid.* The rates of change measured by probes at fixed locations are referred to as *local* rates of change as opposed to the rates of change experienced by a material particle. The latter are termed as the *material* or the *substantive* rates of change.

The local rate of change of a property η is denoted by $\partial\eta(x, t)/\partial t$ where it is understood that $\mathbf{x}$ is held constant. The material rate of change of a property η shall be denoted by $D\eta/Dt$, where it is understood that this is the rate of change of η for a *particular* material particle. If η is the velocity $\mathbf{V}$, then, $D\mathbf{V}/Dt$ is the rate of change of velocity for a *fluid particle* and, thus, is the acceleration that the fluid particle experiences. On the other hand, $\partial\mathbf{V}/\partial t$ is just a local rate of change of velocity recorded by a stationary probe and is in no sense a measure of particle acceleration. Following the above mentioned nomenclature, $D\mathbf{V}/Dt$ is the particle or material acceleration and $\partial\mathbf{V}/\partial t$ is the local rate of change of velocity, or the local acceleration.

Note that the values of the velocity (or any property) recorded by a stationary probe correspond to the velocity (or any property) of the fluid particle that happens to be at that location at that time, but the rates of change recorded by the probe may not be associated with that particle at all.

3.2 RELATION BETWEEN THE LOCAL AND THE MATERIAL RATES OF CHANGE

Consider a tube of uniform cross-section (Fig. 3.2). The fluid is flowing to the right with a uniform velocity V_0. The positions of fluid particles at time $t = t_0$ are used as their material labels. Thus, $x_0 = b$ identifies that particle which was at $x = b$ at $t = t_0$.

---

* Probes mounted on a neutral-buoyancy balloon, floating freely with air measure such rates of change for atmospheric particles.

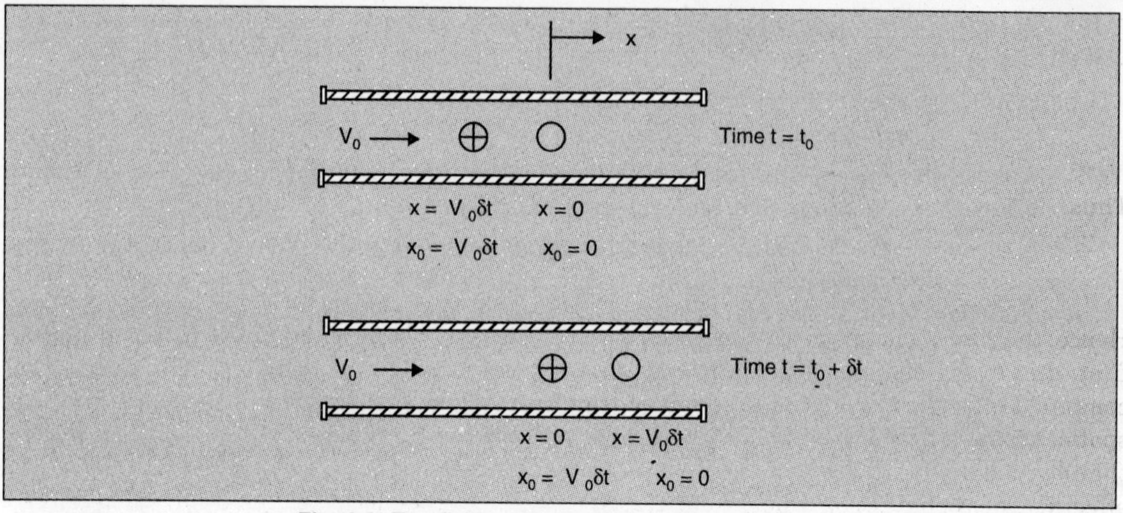

Fig. 3.2. Two fluid particles at two different times.

At a later time $t_0 + \delta t$, all fluid particles have moved a distance $V_0\, \delta t$. Thus, the particle labelled $x_0 = -V_0\, \delta t$, i.e., the particle which was at $x = -V_0\, \delta t$ at $t = t_0$, occupies the location $x = 0$ at $t = t_0 + \delta t$. If a probe measuring property η was stationed at $x = 0$ at time $t = t_0$ it will measure η associated with the particle $x_0 = 0$. But at time $t = t_0 + \delta t$ it will measure the value for the particle $x_0 = -V_0 \delta t$ (Fig. 3.2). Therefore, the local rate of change as measured by the probe stationed at $x = 0$ is

$$\frac{\partial \eta}{\partial t} = \frac{\eta(x=0, t=t_0+\delta t) - \eta(x=0, t=t_0)}{\delta t}$$

$$= \frac{\eta(x_0=-V_0\delta t, t=t_0+\delta t) - \eta(x_0=0, t=t_0)}{\delta t}$$

This can be rewritten as

$$\frac{\partial \eta}{\partial t} = \left\{\frac{\eta(x_0=-V_0\delta t, t=t_0+\delta t) - \eta(x_0=-V_0\delta t, t=t_0)}{\delta t}\right\}$$

$$+ \left\{\frac{\eta(x_0=-V_0\delta t, t=t_0) - \eta(x_0=0, t=t_0)}{\delta t}\right\}$$

The first term in brackets on the right-hand side is the rate of change of temperature of a material particle labelled $x_0 = -V_0\, \delta t$ and is, therefore, $D\eta/Dt$. The second term in brackets, which involves conditions of two different particles at time t_0, can be simplified using Fig. 3.2 as follows:

$$\frac{\eta(x_0=-V_0\delta t, t=t_0) - \eta(x_0=0, t=t_0)}{\delta t}$$

$$= \frac{\eta(x = -V_0\delta t, t = t_0) - \eta(x = 0, t = t_0)}{\delta t}$$

$$= \frac{\dfrac{\partial \eta}{\partial x}(-V_0\delta t)}{\delta t} = -V_0 \frac{\partial \eta}{\partial x}$$

Thus,

$$\frac{\partial \eta}{\partial t} = \frac{D\eta}{Dt} - V_0 \frac{\partial \eta}{\partial x} \quad \text{...(3.1)}$$

Hence, the rate of change of η experienced by a stationary probe or observer is due to two effects: first, due to the change of the property of each particle with time, and secondly due to the combined effect of the spatial gradient of that property and the motion of the fluid. When a spatial gradient exists, the fluid motion brings a particle with a different value of η to the probe, thereby, modifying the rate of change observed. This latter effect is termed as the convective effect (cf. heat convection where heat transfer is due to the motion of the fluid).* Therefore, $V_0 \times \partial\eta/\partial x$ is referred to as the convective rate of change of η.

Even though Eq. (3.1) has been obtained under the assumption of uniform velocity V_0, note that in the limit of $\delta t \to 0$ it is only the local velocity $V(x)$ which enters into the analysis and so

$$\frac{\partial \eta}{\partial t} = \frac{D\eta}{Dt} - V \frac{\partial \eta}{\partial x} \quad \text{...(3.2)}$$

Eq. (3.2) can be generalized for a three-dimensional (3-D) space as

$$\frac{\partial}{\partial t} = \frac{D}{Dt} - (\mathbf{V} \cdot \nabla) \quad \text{...(3.3)}$$

where ∇ is the gradient operator $\left(= \hat{\mathbf{i}}\frac{\partial}{\partial x} + \hat{\mathbf{j}}\frac{\partial}{\partial y} + \hat{\mathbf{k}}\frac{\partial}{\partial z} \text{ for rectangular coordinates}\right)$ and $(\mathbf{V}\cdot\nabla)$, the scalar dot product $\left(V_x \frac{\partial}{\partial x} + V_y \frac{\partial}{\partial y} + V_z \frac{\partial}{\partial z}\right)$. Equation (3.3) is usually written in an alternate manner as

$$\frac{D}{Dt} = \frac{\partial}{\partial t} + (\mathbf{V} \cdot \nabla) \quad \text{...(3.4)}$$

which, though operationally convenient, gives lesser physical insight. The corresponding expanded forms of Eq. (3.4) in cylindrical and spherical coordinate systems are given in Section B.5 (in Appendix B).

When the property of concern is the velocity of the fluid particle, then $D\mathbf{V}/Dt$ gives the acceleration of a fluid particle and the resultant equation

$$\frac{D\mathbf{V}}{Dt} = \frac{\partial \mathbf{V}}{\partial t} + (\mathbf{V} \cdot \nabla)\mathbf{V} \quad \text{...(3.5)}$$

---

* The more recent practice is to use the terms *advection* for the contribution due to fluid motion. The term convection is nowadays reserved for the combined effect of conduction and advection in heat transfer. We, however, will continue to use convection in this narrower sense.

CHAPTER 3

is known as *Euler's acceleration formula.* Note that Eq. (3.5) is really three equations in one, one corresponding to each of the three spatial directions.

Example 3.1. In winter, the Gangetic plains experience cold winds blowing from the Himalayas which bring about dramatic decreases in the local temperatures. At a given time, the temperature at Dehradun is 2°C and at Kanpur, 300 km directly downstream, it is 17°C. The average wind velocity is 40 km/hr and it is observed that the air heats up at the rate of 1°C/hr as it flows past the hotter plains. Find the rate of fall of temperature at Kanpur assuming the temperature gradient between the two cities to be constant.

Here the material rate of change of temperature DT/Dt is given as 1°C/hr. To find the local rate of change, use

$$\frac{\partial T}{\partial t} = \frac{DT}{Dt} - (\mathbf{V}\cdot\nabla)T$$

In this simple case, $(\mathbf{V}\cdot\nabla)T = V\dfrac{\partial T}{\partial x}$

$$= 40\times10^3\left(\frac{\text{m}}{\text{hr}}\right)\frac{(17-2)(°\text{C})}{300\times10^3\,(\text{m})} = 2°\text{C/hr}$$

Thus, the local rate of change at Kanpur is

$$\frac{\partial T}{\partial t} = 1(°\text{C/hr}) - 2(°\text{C/hr}) = -1(°\text{C/hr})$$

or, the temperature at Kanpur falls at the rate of 1°C/hr due to the cold winds. The particle at Kanpur at $t_0 + \delta t$ was at a distance $V_0\delta t$ upstream at time t_0, and therefore, cooler by $V_0\partial t$ $(\delta T/\partial x)$. During its journey, it heats up (material description) by an amount $(DT/Dt) \times \delta t$. The net effect is the local rate obtained above.

Example 3.2. Consider a velocity field given as

$$\mathbf{V} = 2xy\hat{\mathbf{i}} + \left(x^2 + z^2\right)t^2\hat{\mathbf{j}} + y^2t^2\hat{\mathbf{k}}$$

with x, y, z in metres, t in seconds and $\mathbf{V}$ in m/s. Find the acceleration of a particle located at (0, –1, –1) at time $t = 1$.

Euler's acceleration formula Eq. (3.5) gives

$$\frac{D\mathbf{V}}{Dt} = \frac{\partial \mathbf{V}}{\partial t} + (\mathbf{V}\cdot\nabla)\mathbf{V}$$

For the given data

$$\frac{\partial \mathbf{V}}{\partial t} = 2t\left(x^2 + z^2\right)\hat{\mathbf{j}} + 2ty^2\hat{\mathbf{k}} = 2\times1\times(0+1)\hat{\mathbf{j}} + (2\times1\times1)\hat{\mathbf{k}}$$

$$= 2\hat{\mathbf{j}} + 2\hat{\mathbf{k}}$$

$$(\mathbf{V}\cdot\nabla)\mathbf{V} = \left(V_x\frac{\partial}{\partial x} + V_y\frac{\partial}{\partial y} + V_z\frac{\partial}{\partial z}\right)\times\left(V_x\hat{\mathbf{i}} + V_y\hat{\mathbf{j}} + V_z\hat{\mathbf{k}}\right)$$

$$= \left[2x\,y\frac{\partial}{\partial x} + t^2\left(x^2+z^2\right)\frac{\partial}{\partial y} + y^2t^2\frac{\partial}{\partial z}\right]\times\left[2xy\hat{\mathbf{i}} + t^2\times\left(x^2+z^2\right)\hat{\mathbf{j}} + y^2t^2\hat{\mathbf{k}}\right]$$

$$= 2xy\left[2y\,\hat{\mathbf{i}} + 2xt^2\hat{\mathbf{j}}\right] + t^2\left(x^2+z^2\right)\left[2x\,\hat{\mathbf{i}} + 2yt^2\hat{\mathbf{k}}\right] + y^2t^2\times 2t^2z\hat{\mathbf{j}}$$

$$= \hat{\mathbf{i}}\left[4xy^2 + 2x^3t^2 + 2xt^2z^2\right] + \hat{\mathbf{j}}\left[4x^2y\,t^2 + 2y^2t^4\,z\right] + \hat{\mathbf{k}}\left[2yt^4\left(x^2+z^2\right)\right]$$

$$= 0\hat{\mathbf{i}} - 2\hat{\mathbf{j}} - 2\hat{\mathbf{k}}$$

Therefore, the total acceleration is

$$\frac{D\mathbf{V}}{Dt} = \left(2\mathbf{j} + 2\hat{\mathbf{k}}\right) + \left(-2\hat{\mathbf{j}} - 2\hat{\mathbf{k}}\right) = 0$$

Thus, the particle located at (0, –1, –1) at $t = 1$ does not experience any acceleration. But, a probe located at (0, –1, –1) would record a changing velocity, with the y and z components of velocity increasing at a rate of 2 m/s^2. This *local* acceleration is only an observed effect due to the spatial gradient of velocity. In other words, particles of differing velocities are being convected past the point (0, –1, –1) around time $t = 1$ without undergoing any acceleration.

3.3 STEADY AND UNSTEADY VELOCITY FIELDS

As seen in Sec. 1.9, the continuum assumption allows one to treat a fluid mass as a continuous distribution of matter and to imagine that each point of the physical space corresponds to a fluid particle—an aggregation of millions of molecules. By the velocity of such a fluid particle, then, is meant a velocity which is averaged over these molecules which constitute a fluid particle. Such a velocity, then, results in a bulk transfer of fluid from one place to another and is quite different from the random motion of molecules. This velocity, which shall be termed as *fluid velocity,* will, in general, vary from fluid particle to particle, and for a given particle, from time to time. In the Eulerian description, the *velocity field* $\mathbf{V}(\mathbf{x}, t)$ is specified as a function of the location in space $\mathbf{x}$ and time t, and this measures the velocity of the particle which is at location $\mathbf{x}$ at time t.

If the velocity field $\mathbf{V}$ is not a function of time, i.e., if the velocity at any location does not vary with time, the velocity field is termed as *steady,* otherwise it is termed as unsteady. Thus, for steady flow of a fluid in a nozzle (Fig. 3.3), the velocity at a point A remains constant with time as also that at a point B. But the field being steady or the velocities being constant at fixed locations does not imply that fluid particles are not accelerating. In Fig. 3.3, the velocity at B is larger than that at A and, therefore, a particle which moves from point A to B experiences acceleration as it flows. Thus, a steady field only implies that the local acceleration $\partial\mathbf{V}/\partial t$ is zero, and the particle acceleration $D\mathbf{V}/Dt$ is equal to the convective acceleration $(\mathbf{V}\cdot\nabla)\mathbf{V}$.

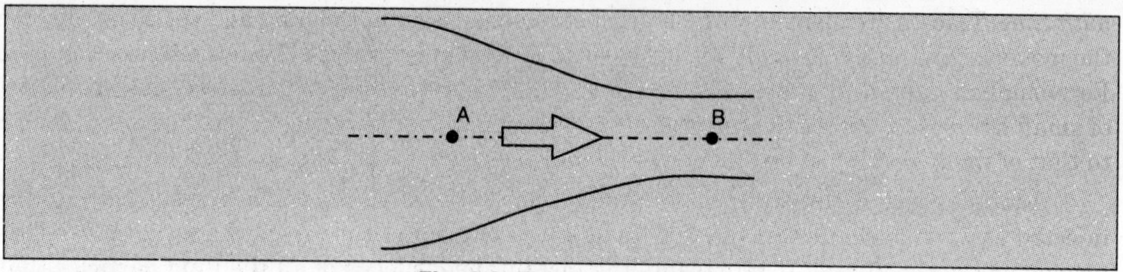

Fig. 3.3. Flow in a nozzle.

A flow, in which the velocity field as well as *all other property fields* remain unchanged with time, is termed a *steady flow,* otherwise, it is referred to as an unsteady flow.

Quite often, a flow which is unsteady in one coordinate system becomes steady when the frame of reference is changed. Consider a boat moving to the left with a constant velocity V_0 (Fig. 3.4). Also, consider a location A (x_0, y_0) fixed with respect to a coordinate system x-y attached to the lake-shore (e.g., as observed by a person on the shore). When the boat is far away from A, the velocity at A will be zero (or very small). As the boat approaches A and passes by, the velocity at A increases and then falls back to zero again. Clearly, the field with respect to the x-y coordinate system is unsteady. If, however, we attach the coordinate system x'-y' with the boat (i.e., as observed by a person sitting on the moving boat), at a fixed point B (x'_0, y'_0) the velocity is the same at all times and the flow field is steady with respect to the $x'-y'$ coordinate system (Note that point B will be moving to the left at a velocity V_0 as observed from the shore). In this coordinate system the points far upstream and down stream of the boat have constant velocities V_0.

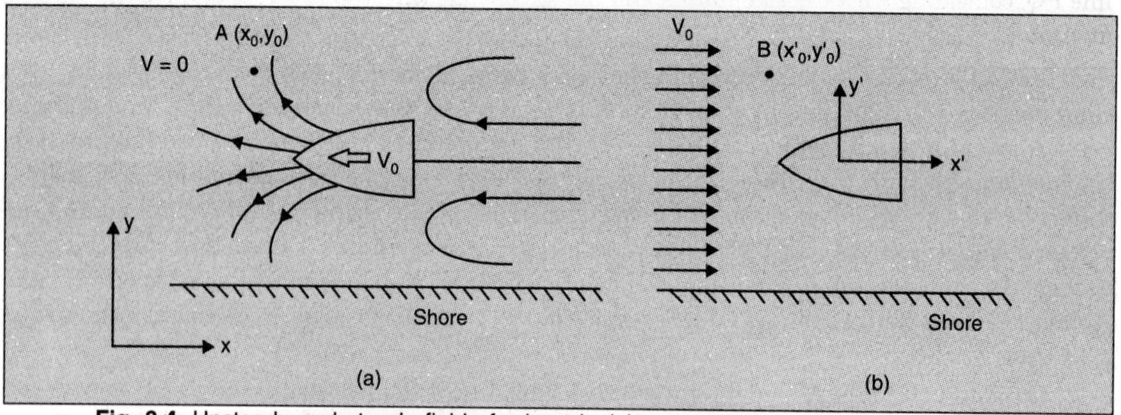

Fig. 3.4. Unsteady and steady fields for boat in (a) a coordinate system fixed to shore, and (b) a coordinate system fixed to boat.

3.4 GRAPHICAL DESCRIPTION OF FLUID MOTION

A number of methods are used for visualizing the complex motions of fluid particles and describing them graphically. Perhaps, the most straight-forward of these employs the concept of *pathline.* It is a line in the flow field describing the trajectory of a given fluid particle. These lines may be obtained experimentally by putting some tracers in the fluid and following their progress *with time.* Tracers are elements which are distinguishable from the fluid and yet do not disturb the motion significantly. A small puff of smoke in air or a pulse of dye in liquids are two examples. For visualizing the flow of water, a simple and elegant technique consists of releasing a burst of small hydrogen bubbles by immersing an electrode in it. The concept of pathline is identical to that of the trajectory of a solid body.

Another concept employed quite often is that of the *streakline.* If a tracer is *continually* injected at a fixed location in the flow field and a photograph is taken, the trace obtained is termed as the streakline. It represents the instantaneous locus of all the fluid particles that have passed the point of injection at some *earlier* time. The visible smoke-plume from a chimney represents such a locus because all the smoke particles have passed through the chimney mouth.

The most commonly used graphical concept is that of the *streamline.* It is a line in the flow field drawn in such a manner that the tangent at every point of it is in the direction of the local velocity vector. This concept is akin to the concept of the lines of force in a magnetic field (including the fact that these lines are closer together where the velocity or magnetic intensity is higher) and can be visualized in a similar manner. If we mix small, slender and neutrally buoyant needles with the fluid, each one of these will align itself with the direction of the local fluid velocity and the overall picture will give the pattern of streamlines.

To clarify the three concepts further, consider the flow of air from west to east with a uniform velocity V_0. It can be seen that the streamlines, as defined above, will be straight lines from left to right (Fig. 3.5). If P (x_0, y_0) represents the mouth of a chimney, all the fluid particles that pass (x_0, y_0) will move in a straight line and the streakline will coincide with the streamline through P. The path of the particle which was at (x_0, y_0) at time $t = 0$, will also be the straight line PQ, coinciding with the streamline and the streakline, where Q is the position of the particle at time t.

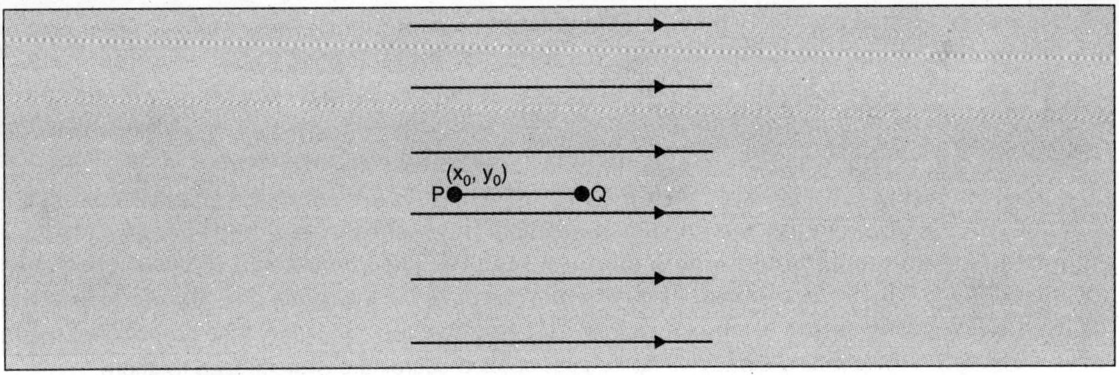

Fig. 3.5. Streamlines, pathlines and streaklines in air flowing from west to east.

Now, let the wind direction change suddenly at time $t = t_0$, when it starts blowing in the south-easterly direction. All fluid particles start moving in that direction thereafter. The streamlines for any time $t > t_0$ will then be straight lines in the north-west to south-east direction (Fig. 3.6). It may be noted that a streamline represents an 'instantaneous' picture of the flow field and does not depend upon the history of the fluid flow. In contrast, the particle path and the streakline

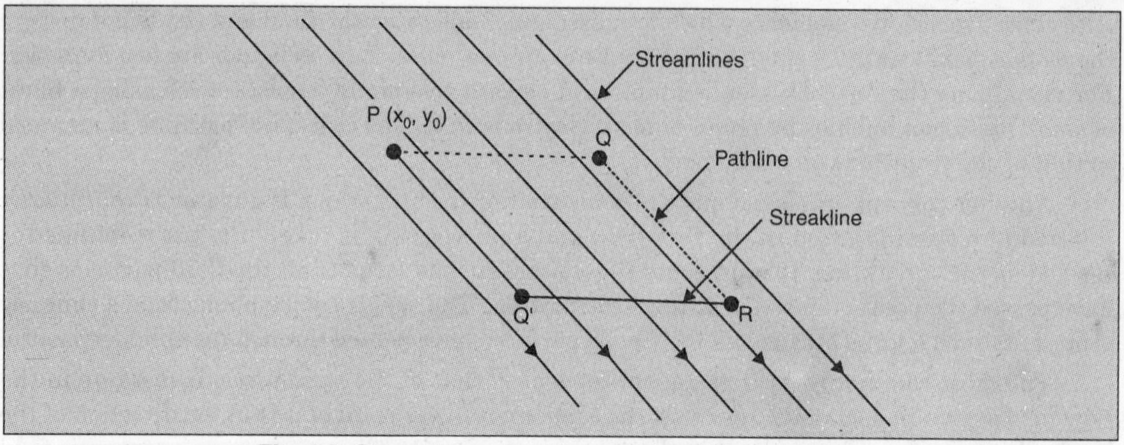

Fig. 3.6. Streamlines, streaklines and pathlines for $t > t_0$.

reflect a cumulative picture. The path of the particle which was at $P(x_0, y_0)$ at $t = 0$ will be PQR, where Q was the position reached at $t = t_0$ and $QR = V_0(t - t_0)$. It can be seen easily that the streakline through P will be represented by $PQ'R$, because the particles which passed through P at all $t > t_0$ will occupy positions on PQ' and all the particles which had passed P earlier than $t = t_0$ will lie along $Q'R$ (having moved partly along PQ and then parallel to PQ').

In steady flows streamlines, streaklines and particle-paths are all identical. Since a particle moves instantaneously along a streamline, it will continue moving along it if the streamline does not change (because the flow is steady) and so the particle-path coincides with the streamline. Similarly, all particles starting from a point at different times will take the same paths and thus streaklines and pathlines will also coincide. If the flow is unsteady, however, the pathlines, streaklines and streamlines are all different, as illustrated above.

It may be pointed out that while the streamline is an Eulerian (or field) concept, the pathline is a Lagrangian (or particle) concept. Since the Eulerian description is preferred in the study of fluid flows, streamlines are the natural choice for graphical description. On the other hand, streaklines are easier to obtain experimentally. The fact that for steady flows streamlines and streaklines are identical is exploited for visualizing streamlines in such flows.

As seen earlier, an unsteady flow may be made steady by a change of reference. For a cylinder moving at constant velocity in a stationary fluid, one can fix the reference with the ambient to get a streamline pattern as shown in Fig. 3.7. The arrows at A, B and C represent the directions in which the fluid particles are moving at these locations. For the same flow the paths of a few particles are as shown in Fig. 3.8. If, however, the reference is fixed with the moving cylinder, streamlines, pathlines and streaklines, all coincide (since the flow is now steady), giving the pattern shown in Fig. 3.9.

3.5 ANALYSIS IN FLUID MECHANICS

Now that we have learnt to *describe* the flow of fluids, we turn our attention to its *analysis*. The reader will recall that the mechanics of solids was studied through the use of Newton's laws of motion alongwith constitutive relations such as Hooke's law. Newton's second law applies to a body of constant mass and postulates that (in an inertial or non-accelerating frame of

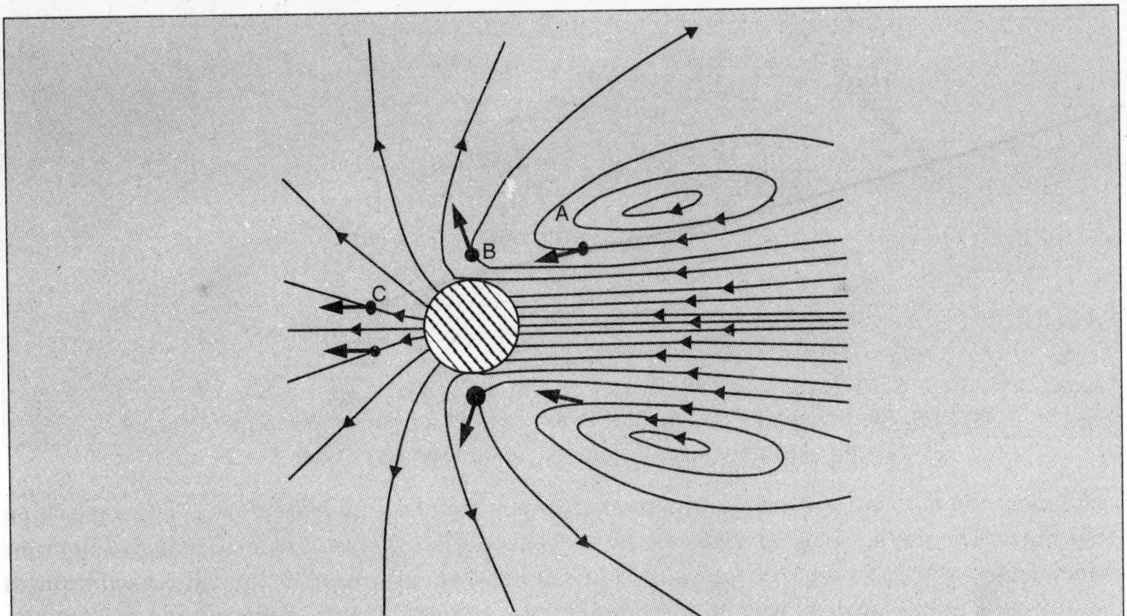

Fig. 3.7. Streamline pattern as observed in a stationary frame of reference for a cylinder moving at constant velocity in a fluid at rest.

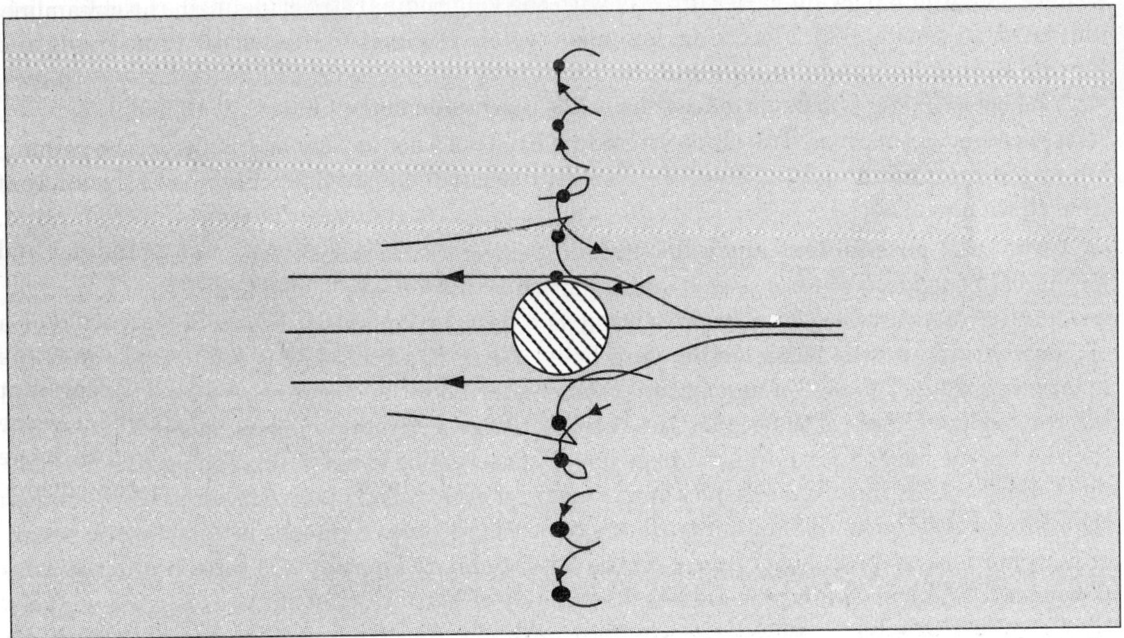

Fig. 3.8. Particle paths with respect to a stationary observer for a cylinder moving at constant velocity in stagnant fluid. Points show the positions of the fluid elements at the time when the cylinder is just below or above, as shown.

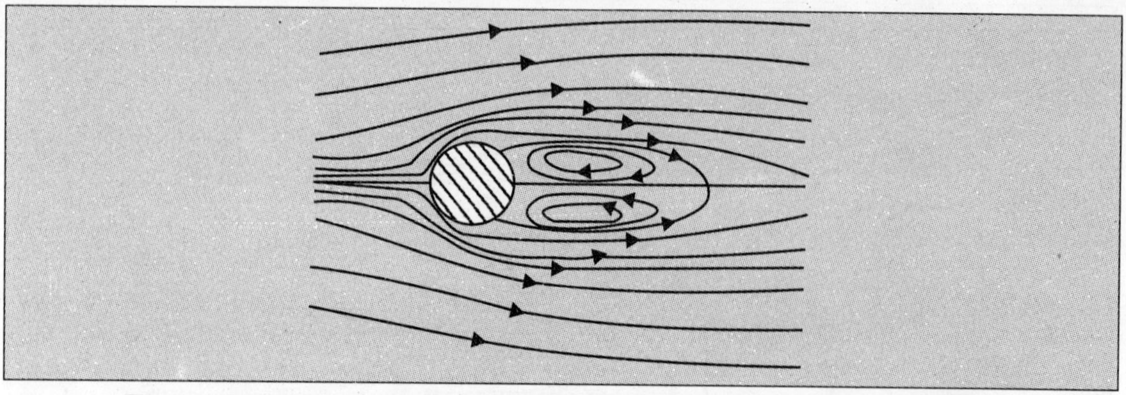

Fig. 3.9. Streamlines for steady flow in the case of a cylinder moving at constant velocity in a stationary fluid. Reference frame fixed to cylinder (Re $\simeq$ 40).

reference) the net rate of change of its momentum is equal to the net external force acting on that mass. The method of analysis uses the concept of a *free body,* an identified body of matter which is acted upon by surface forces (i.e., normal and shear forces) at the bounding surfaces and by body forces (such as weight, electromagnetic attraction, etc.). For example, for a mass-spring system the free-body diagrams of the mass and spring when a force F is applied are as shown in Fig. 3.10. The constitutive equations relate the surface forces with the state of deformation produced within the body. For example, Hooke's law states that for unidirectional loading, longitudinal strain varies directly with the longitudinal stress (defined as the force per unit cross-sectional area). For the spring-mass system of Fig. 3.10 the constitutive equation is Force = Spring constant × compression.

As noted in Sec. 1.2, fluids differ from solids essentially in the kind of relationships between stresses and deformations. The shear stress in a fluid does not depend on the deformation itself, but on the *rate* of deformation. Thus, the constitutive equations for fluids are markedly different from those for solids.*

The basic physical laws applicable in fluid mechanics are exactly the same as those in the mechanics of solids. Newton's second law of motion is, indeed, applicable to any class of matter, irrespective of its physical state, representing as it does, an equality between the rate of change of momentum and the forces acting on a body. When Newton's law is applied to fluids the appropriate rate of change of momentum is the material or the substantive rate of change and not the local one. This is so because the rate of change envisaged in Newton's law refers to a specific body of matter. Similarly, the principle of conservation of mass also applies to both solids and fluids, though, in solid mechanics one is rarely, if ever, called upon to use it explicitly unless the body is breaking up or different parts are coalescing to form a composite body during motion or deformation. In the study of fluids, on the other hand, the principle of conservation of mass is used as an essential analytic aid because a body of matter is not that readily identifiable.

* As a matter of fact, in Sec. 1.2 the distinction between a solid and a fluid was just on the basis of the constitutive relations.

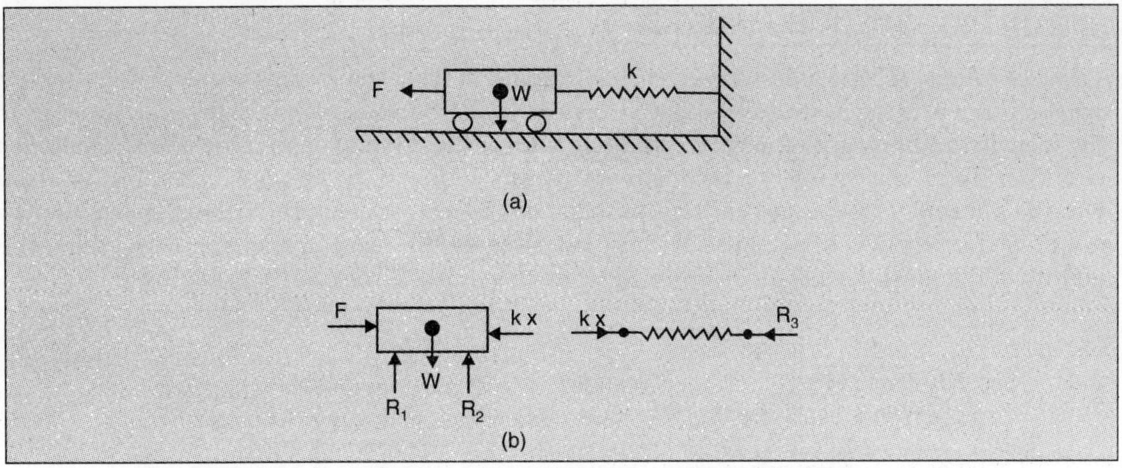

Fig. 3.10. Free body diagram of a spring-mass system on frictionless support. x is the compression of the spring.

3.6 CONTROL MASS ANALYSIS

As indicated above it is conceivable, at least in principle, that even with fluids we may select a body of matter, treat it as a free body identifying all the external forces acting on it, write Newton's second law together with all the constitutive relationships and solve the problem, as is done in the mechanics of solids. This is termed as the *control mass* or *system* approach. A control mass or system may change in size, shape or location within the reference frame, but it must always consist of the same material. For example, to study the pulsation of a gas-filled balloon one may consider the entire mass of the gas in the balloon as a control mass. The volume of this control mass changes with time, as also the shape and location of the bounding surface (the similarity of control mass and a closed system in thermodynamics may be noted).

In fluid mechanics the control mass analysis can be used only in some specialized situations, for example, in a fluid at rest or in a body of fluid contained in a vessel. For a moving fluid, in general, such an approach does not take one far, for this approach involves following the material particles around and studying the rates of change of properties associated with them. This is unsatisfactory on at least two counts, the first being that it is difficult, both analytically and experimentally, to follow the material particles around in most flow situations (i.e., using the Lagrangian approach). The second shortcoming is a little more serious. As stated in Chapter 1, one is usually interested in obtaining the forces that a fluid exerts on *fixed* bounding surfaces. If a fluid particle were to be followed around as it moves, the analysis will give the history of force experienced by this particle. Only that part of this information which pertains to the instant when the particle is next to the surface of interest is relevant. Hence it is a wasteful exercise. For such problems, then, one is much better off considering a fixed volume adjacent to the surface of interest and studying the rates of changes of properties associated with the matter contained therein. This is termed as the *control volume* approach.

3.7 CONTROL VOLUME ANALYSIS

A control volume *(CV)* is a designated region in space of fixed shape and volume.* Its bounding surface, termed as the *control surface (CS),* is fixed in the frame of reference being considered.** The identity and amount of matter contained within the *CV* may, and most often do, change with time. Thus, the casing of a jet engine can be a *CV* (Fig. 3.11) for a reference system fixed with the aircraft. The engine continually sucks in air, mixes fuel with it, there is combustion and then the products of combustion are exhausted at the rear. In this approach, then, the identity of the matter contained within the control volume is constantly changing.

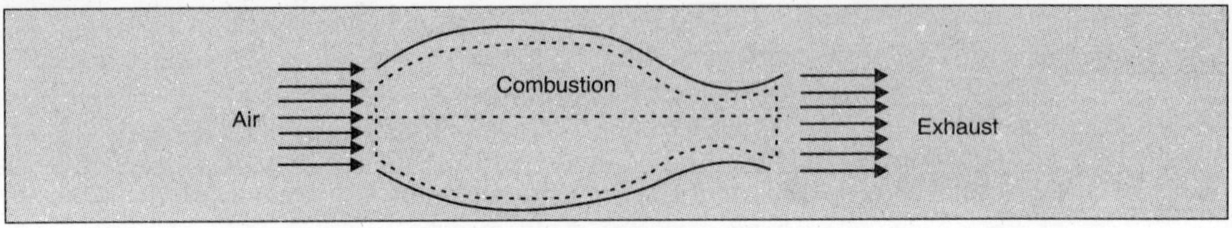

Fig. 3.11. The jet engine. *CV* shown by dotted lines. Flow shown relative to it. *CV* moves at the velocity of the aircraft.

The use of control volumes does not permit a direct application of physical laws as known from mechanics. Since the second law of Newton relates the force to the rate of change of momentum of *a* body of matter, i.e., a control mass, the rate of change of momentum contained in a *CV* is not directly and simply related to the forces acting on it. An example will make this clear. In the case of the jet engine, when the flow is steady, i.e., when the conditions of flow are not changing with time, the momentum of the matter within the *CV* must be constant. Does it imply, then, that there is no force acting on the *CV*? Obviously this is not the case as the engine is indeed developing a thrust. The crux of the matter lies in the fact that Newton's law, as any other physical law, is applicable only to a control mass formulation, i.e., the rates of change envisaged in the physical laws are the material rates of change with time and not the local rates. The rates of change of properties associated with *CV's* are, in fact, local rates of change (refer to Sec. 3.1). Before we can apply the physical laws we require a relationship between the material rates of change and the local† rates of change for a *CV*. In Sec. 3.2 we developed a similar relationship as applicable to a *point* in space. Here we want a relation for a *finite region.*

Another example will help clarify the situation further. Consider a tank which is supplied with a fluid at a rate $\mathring{m}_1$ kg/s and from which the fluid is drained at a rate $\mathring{m}_2$ kg/s. For the *CV* shown by dotted lines in Fig. 3.12, which includes the interior of the tank, the rate of accumulation of mass is the local rate of change, $\partial M/\partial t$, where M is the total mass inside the *CV*. The mass of *a particle* does not change and, therefore, the material derivative of the mass of the system DM/Dt is zero. The same, however, cannot be said for $\partial M/\partial t$, the (local) rate of change for the control volume. In this simple case, $\partial M/\partial t = \mathring{m}_1 - \mathring{m}_2$, i.e., the rate of accumulation of mass is equal to the net rate of *influx* of mass into the *CV*.

---

* In certain cases even the shape and size of the control volume is allowed to change. But we will restrict ourselves in this book to simpler problems in which the size and shape of the *CV* are not permitted to vary at all.

** The frame of reference itself may move, however.

† Local here does not refer to the value at a point but to the integrated value over the entire control volume.

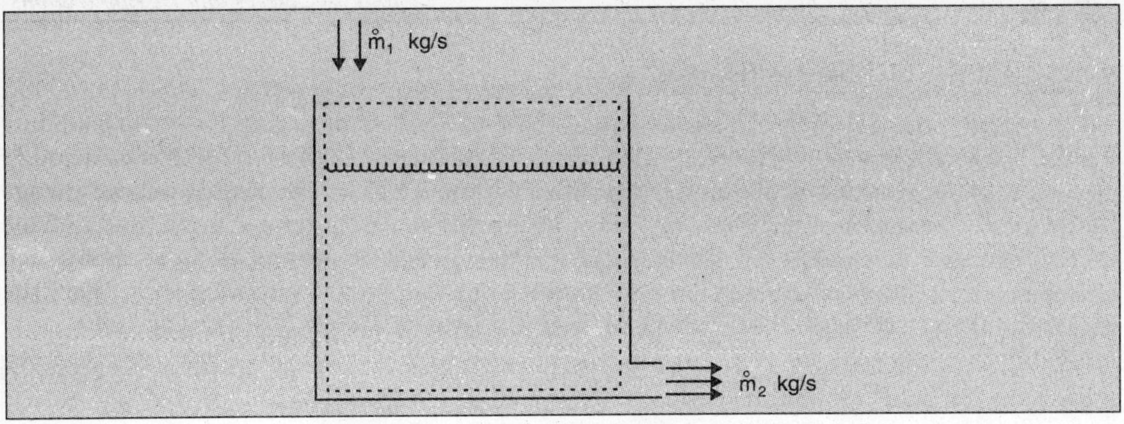

Fig. 3.12. Tank with fluid flowing in and out.

3.8 REYNOLDS TRANSPORT THEOREM

Reynolds transport theorem relates the rate of change of a property associated with a *CV* (or the rate of accumulation) to the material rate of change of that property (or the rate of growth). Consider the chamber in Fig. 3.13 having two ports *AB* and *CD* of areas δA_1 and δA_2 respectively. Consider a *CV* coinciding with its interior so that the *CS* is as shown by broken lines. The fluid crosses the *CS* at *AB* and *CD* only.

Suppose we are interested in studying the rate of change of a property whose specific value per unit mass is η. The value of η in the field description is specified by η (**x**, *t*). If we are interested in the mass contained in the *CV* then $\eta = 1$, while for momentum the specific value is **V** (**x**, *t*), and for kinetic energy the value of η is $\frac{1}{2} V^2$ (**x**, *t*).

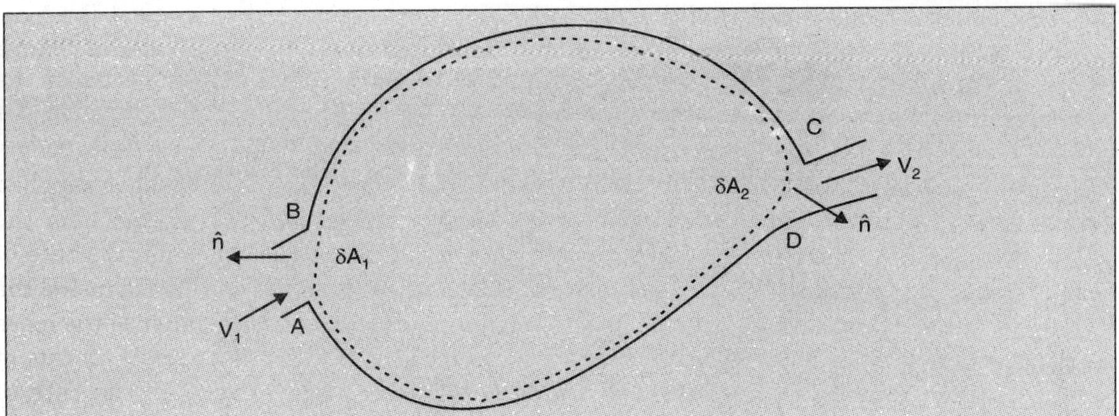

Fig. 3.13. A *CV* with one inlet and one exit port.

If N represents the total value (over CV) of the property whose specific value is η, then

$$N(t)=\iiint_{CV}\eta\rho d\forall$$

where $d\forall$ represents a small volume element and the integration is taken over the entire CV.

If the ports were closed, the matter contained within the CV will be the same at all times. Therefore, the rate of accumulation of N, i.e., $\partial N/\partial t$, will be solely due to the changes in the property of each fluid particle, and $\partial N/\partial t$ will be identical to DN/Dt, the rate of change associated with the material. If, however, the ports were open, allowing fluid to flow in and out, the rate of accumulation of N in the CV (i.e., $\partial N/\partial t$) will not only be made up of the rate of change associated with the material (DN/Dt), but also of the rate at which the property N enters or leaves *with* the fluid through the ports.

In other words,

Accumulation = Growth + Influx − Efflux

$$\begin{bmatrix}\text{Rate of increase}\\ \text{of } N \text{ in } CV\end{bmatrix}\begin{bmatrix}\text{Rate of increase}\\ \text{of } N \text{ of matter}\end{bmatrix}\begin{bmatrix}\text{Contribution}\\ \text{due to inflow}\end{bmatrix}\begin{bmatrix}\text{Contribution}\\ \text{due to outflow}\end{bmatrix}$$

The term on the left hand side is $\partial N/\partial t$, the first term on the right is DN/Dt, the second is η *times* the mass entering across AB, which is $[\eta(-\rho\mathbf{V}\cdot\delta\mathbf{A})]_1$, and the third is $[\eta(+\rho\mathbf{V}\cdot\delta\mathbf{A})]_2$. Thus,

$$\frac{\partial N}{\partial t}=\frac{DN}{Dt}-[\eta(\rho\mathbf{V}\cdot\delta\mathbf{A})]_1-[\eta(\rho\mathbf{V}\cdot\delta\mathbf{A})]_2$$

where subscripts 1 and 2 refer to the values at ports AB and CD respectively.*

If the mass were crossing the CS at more locations, appropriate terms would be added to the above equation, and for n ports the equation becomes

$$\frac{\partial N}{\partial t}=\frac{DN}{Dt}-\sum_{i=1}^{n}[\eta(\rho\mathbf{V}\cdot\delta\mathbf{A})]_i$$

The last summation can be replaced by an integral over the entire CS so that the equation reads**

$$\frac{\partial}{\partial t}\iiint_{CV}\eta\rho\, d\forall=\frac{DN}{Dt}-\oiint_{CS}\eta(\rho\mathbf{V}\cdot d\mathbf{A}) \quad \text{...(3.6)}$$

---

* Note that $\delta\mathbf{A}$ represents $(\delta A)\,\hat{\mathbf{n}}$ where $\hat{\mathbf{n}}$ is the unit normal vector pointing outwards from the system or the CV. This is why a minus sign has been used with the influx term.

** For a moving CV (where the frame of reference is moving because it is attached to the CV), $\mathbf{V}$ is the relative velocity of the fluid with respect to the frame of reference.

This is the Reynolds transport theorem* and it states that the local rate of change of a physical quantity associated with a *CV* (i.e., the rate of accumulation) is equal to the (material) rate of change of that quantity associated with the control mass *whose boundaries coincide with the control volume at that time* (i.e., the rate of growth) *minus* the net quantity leaving the control volume in association with the mass crossing the control surface (i.e., *net* efflux):

(rate of accumulation) = (rate of growth) – (*net* efflux)

For the example in Sec. 3.7 the conservation of mass requires that *DM/Dt* be zero and the rate of change of mass contained in the *CV*, $\partial M/\partial t$ is then given by

$$\frac{\partial M}{\partial t} = -\oiint_{CS} 1 \times (\rho \mathbf{V} \cdot d\mathbf{A}) = \mathring{m}_1 - \mathring{m}_2$$

the same as seen earlier.

- The Reynolds transport Eq. (3.6) is written in the alternate form

$$\frac{DN}{Dt} = \frac{\partial}{\partial t}\iiint_{CV} \eta\rho\, d\forall + \oiint_{CS} \eta(\rho \mathbf{V} \cdot d\mathbf{A}) \qquad ...(3.7)$$

This enables us to evaluate the material rate of change from the *observable* local rate of change, and the laws of physics can be applied to the material rate of change thus obtained. Eq. (3.7) forms the basis of the control volume analysis and it shall be used extensively.

CHAPTER 3

3.9 INTEGRAL AND DIFFERENTIAL ANALYSIS

The laws of physics are as applicable to individual particles as to a collection of particles. Similarly, the laws obtained for large *CV's* are also applicable when a *CV* shrinks to a point in the flow field. The method of analysis, where large *CV's* are used to obtain aggregate forces or transfer rates, is termed as the *integral analysis.* On the other hand, when the analysis is applied to individual points in the flow field, the resulting equations are differential equations and the method is termed as the *differential analysis.* Both these approaches are used in fluid mechanics, depending upon the requirements of the problem. If we require a detailed description of the flow field, as we would if we were designing the wing section of a supersonic aeroplane, we would resort to differential analysis. If only the overall forces are required, as for example, while calculating the power requirements of a compressor, the simpler integral analysis suffices. Unfortunately, in several cases, it is too complicated to obtain final solutions using the differential analysis. Integral analysis, thus, can be used for a wider range of problems than can the differential analysis, even though it may not give as detailed a solution as the latter.

---

* Note that in Eq. 3.6 while we have replaced *N* in the local rate of change by its integral representation, we have not done so for the material rate. This is because to obtain the latter we must follow the matter as it moves. After an infinitesimal time δt, this matter (constituting the system) occupies a slightly different region in space. Therefore, for obtaining the material rate of change we take the difference between the values of *N* for these two different regions at t and $t + \delta t$ respectively. This is equivalent to saying that for obtaining *DN/Dt* we take the value of *N* for the *system which coincides with the CV at t.* On the other hand, to get the local rate we have to obtain the difference in the values of *N* at two different times over the *same* space (*CV*) and therefore we may replace *N* on the LHS by its integral representation.

PROBLEMS

3.1 When a beaker of water is rotated at a constant angular velocity ω, it moves as a solid body after a while. Write the expression for the velocity vector **V** in the r, θ coordinates. Also obtain the acceleration vector.

3.2 A small thermocouple probe is floating with the fluid whose temperature and velocity fields are given by

$$T = x^2 + 2yz + 3t$$

$$\mathbf{V} = 3x\ \hat{\mathbf{i}} + (2y + 6t^2y)\ \hat{\mathbf{j}} + 5\ \hat{\mathbf{k}}$$

Find the rate of change of temperature recorded by the probe when it is at the position

$$5\ \hat{\mathbf{i}} + 2\ \hat{\mathbf{j}} + \hat{\mathbf{k}} \text{ at time } t = 1.$$

3.3 A gas flows in a pipe with a velocity of 10 cm/s and has an axial temperature gradient of 10°C/m in the flow direction. On account of absorption of thermal radiation the temperature of each gas 'particle' is increasing at a rate of 1°C/s. Find the rate of change of the gas temperature as recorded by a stationary probe. Is the temperature field steady?

3.4 Will there be any fluid acceleration in a *steady* flow of a liquid through the nozzle shown? If yes, obtain an expression for it in terms of the volumetric flow rate $\overset{\circ}{Q}$ if the area of cross-section is given by $A(x) = A_0(1 + e^{-x})$. Assume 1-*D* flow. Note that the value of $\overset{\circ}{Q}$ is independent of x.

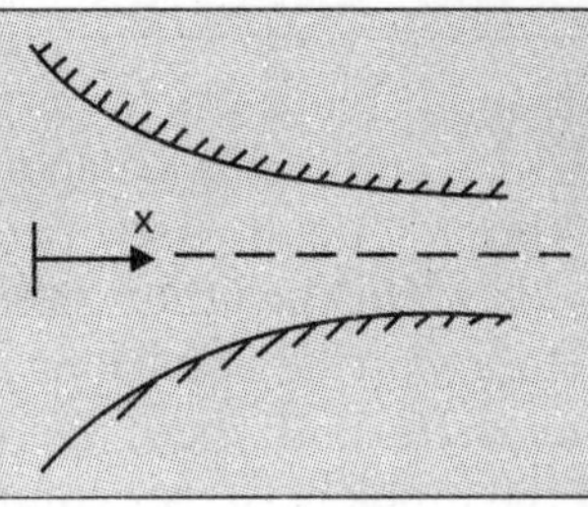

3.5 For the velocity field $\mathbf{V} = 2y\ \hat{\mathbf{i}} + x\ \hat{\mathbf{j}} + 0\ \hat{\mathbf{k}}$, obtain the (material) acceleration vector at $x = 3$, $y = 1$. At this position, obtain the components of acceleration parallel and normal to **V**.

3.6 Air is sucked in through the wet packing of a desert cooler as shown and is cooled by 15°C. The cooled air then flows across a hall and picks up heat from the roof at a rate of 0.1°C/s. The air speed in the room is 0.5 m/s. At some time after switching on, it is seen that the temperature gradient may be *assumed* as constant equal to 1°C/m. Obtain DT/Dt and $\partial T/\partial t$ at point A, 20 m from the cooler.

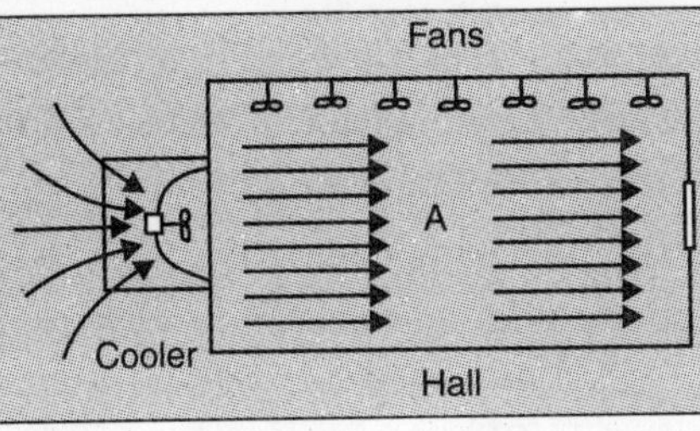

3.7 The price of a car is known to change as it is driven out from the factory because of two reasons: (i) it reaches a better market, and, (ii) it becomes older. The price is given by:

$$\text{Price (Rs)} = 50{,}000 + 1.0\, x - 100\, t$$

where x is the distance from the factory in km and t the time (in days) it takes to sell the car. Use the Euler relationship to find the rate of increase of the value of a car per day if it is driven out at a speed of 400 km per day from the factory.

3.8 Extend the derivation of Sec. 3.2 to show that the rate of change of η as measured by a probe moving with velocity V_p (different from the fluid velocity V) in the x-direction is

$$\frac{\partial \eta}{\partial t} + V_p \frac{\partial \eta}{\partial x}$$

3.9 A sensitive electronic instrument on board a balloon must not experience a rate of change of temperature larger than ± 0.01 K/s for proper functioning. The atmospheric temperature is given by

$$T = (288.2 - 6.5 \times 10^{-3}\, z)\,(2 - e^{-0.01t})\text{K}$$

where z is the height in m above the ground and t is the time in hr after sunrise. Find the maximum permissible rate of ascent when the balloon is at the ground at $t = 1$ hr.

[*Hint*: Use the results of Prob. 3.8.]

3.10 In a 2-D channel water flows with a velocity given by

$$V_x = V_0\left[1 - \left(\frac{y}{b}\right)^2\right]$$

A thin wire stretches across the channel as shown.

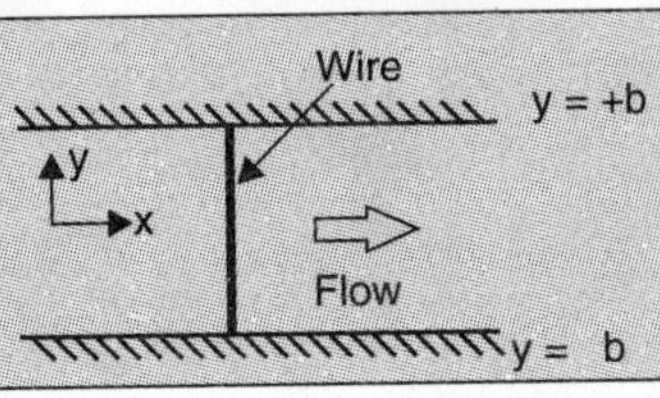

At time $t = 0$ a short pulse of electrical current passes through the wire releasing a line of very small hydrogen bubbles across the channel. The location of this line at any later time t is another visual indicator of the flow pattern and is called the *timeline*. Plot the timelines for $t = 1/V_0$ and $2/V_0$. What will the streamlines look like for this flow?

3.11 Hurricanes are indicated in satellite photographs by spiralling cloud formations. What do these represent: streaklines, pathlines, streamlines or timelines (Prob. 3.10)?

CHAPTER 3

3.12 The velocity at a fixed point in a flow field is given by $2\hat{\mathbf{i}} + 6t\,\hat{\mathbf{j}}$. Fluid particles passing through that point continue to move with the velocity they had at that point.

(a) Draw the pathline of the particle passing through the point at $t = 0$.

(b) Draw the pathline of the particle passing through the point at $t = 1$.

(c) If a dye is injected continuously at this point starting at $t = 0$, what would the dye-pattern look like at $t = 3$? What does this dye-pattern represent?

3.13 Consider a small town as a *CV*. Write down an expression for its population P in terms of the birth and death rates B and D (per thousand of population per year) respectively, and the immigration and emigration rates I and E (per year). Identify the accumulation, growth and efflux terms with reference to Eq. (3.6).

3.14 A chemical reaction $A \rightarrow B$ occurs in a well-mixed reactor of constant volume $\forall$ as shown. A is converted to B at a rate R kg/m^3 s. Apply Reynolds transport theorem to the *CV* to obtain the mass fraction m_A (= kg of A/kg of mixture) of A at the exit of the reactor if the mass fraction at the inlet is $m_{A,0}$. Assume the density of the reaction medium in the reactor to be constant at ρ. Simplify your results for steady state. $\overset{\circ}{m}$ is the mass flow rate (kg/s) of the mixture.

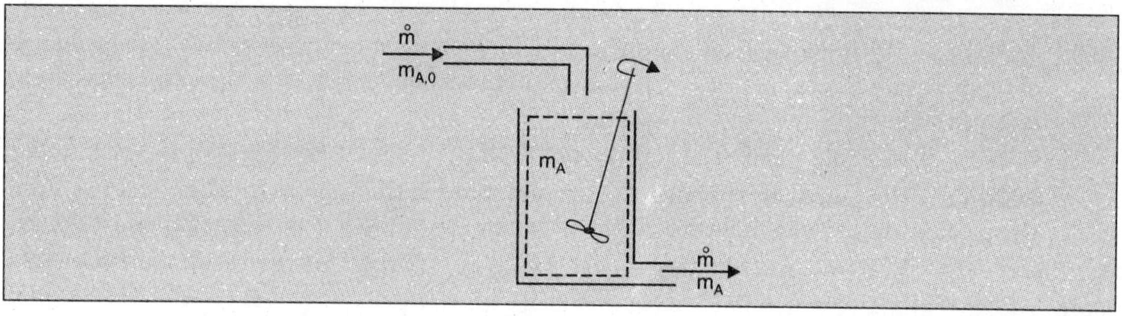

3.15 An exothermic chemical reaction takes place in the liquid-phase-reactor shown. If N is the thermal energy [= mass × C_p × $(T - T_{ref})$] interpret the various terms in Reynolds transport equation. Assume steady state, i.e., T_2 is independent of time. The contents of the reactor are well-mixed.

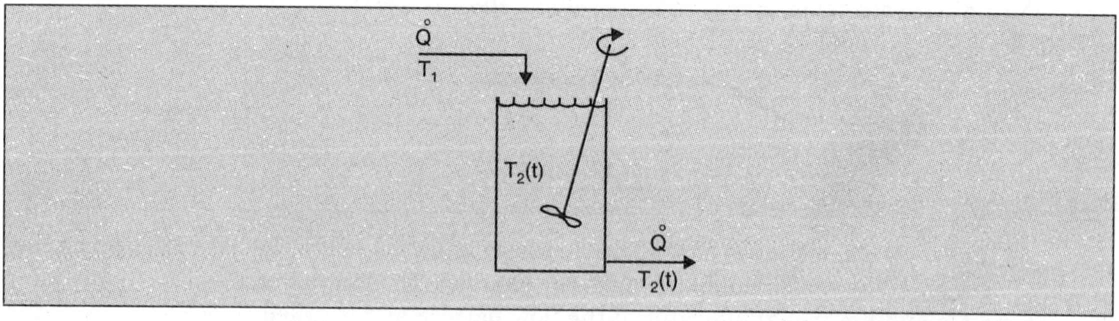

3.16 The reactor of Prob. 3.15 is changed so that there is no outlet as shown. How will your interpretation of the various terms in Reynolds transport equation change?

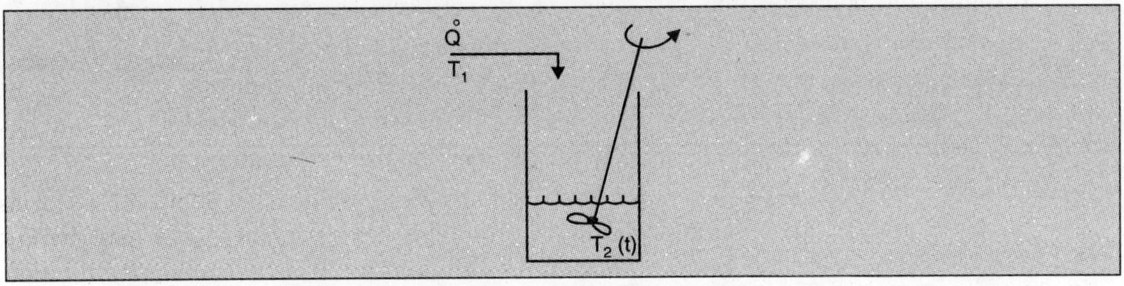

3.17 A jet engine sucks in the atmospheric air for combustion. The products of combustion and the excess hot air are exhausted at the rear. Identify the quantities represented by each term in the Reynolds transport theorem applied to (a) mass, and (b) thermal energy.

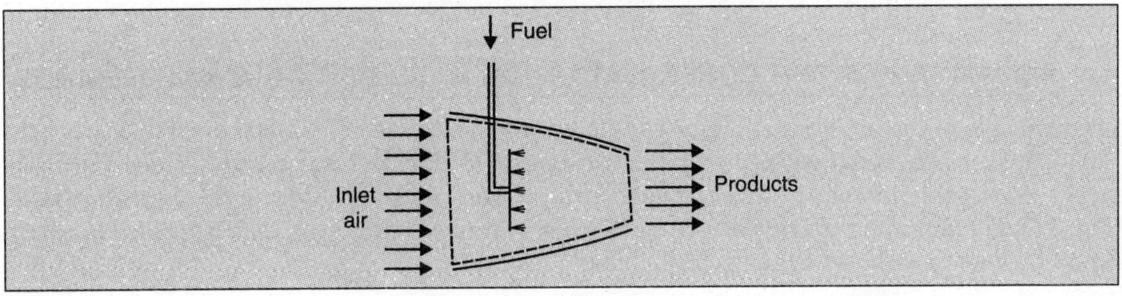

4

CONSERVATION OF MASS

4.1 EQUATION FOR THE CONSERVATION OF MASS FOR CONTROL VOLUMES

The principle of conservation of mass stipulates that the mass of a *system* remains constant. But the mass contained within a control volume *(CV)* may not be constant since the fluid moves across the bounding surface, *the control surface (CS).* Consider a *CV* in a flow field (Fig. 4.1). For obtaining the rate of change of mass contained within the *CV,* Reynolds transport Eq. (3.6) is used with $\eta = 1$ (i.e., specific mass), to obtain

$$\frac{\partial}{\partial t}\iiint_{CV} \rho \, d\forall = \frac{DM}{Dt} - \oint\!\!\oint_{CS} \rho \, \mathbf{V} \cdot d\mathbf{A} \qquad \text{...(4.1)}$$

where M is the mass contained within the CV at time t. The first term on the right hand side is, as before, the rate of change of mass of the system which coincides with the control volume at the given time t. This is the rate of growth (generation) of mass which must be zero*, and, therefore,

$$\frac{\partial}{\partial t}\iiint_{CV} \rho \, d\forall = -\oint\!\!\oint_{CS} \rho \, \mathbf{V} \cdot d\mathbf{A} \qquad \text{...(4.2)}$$

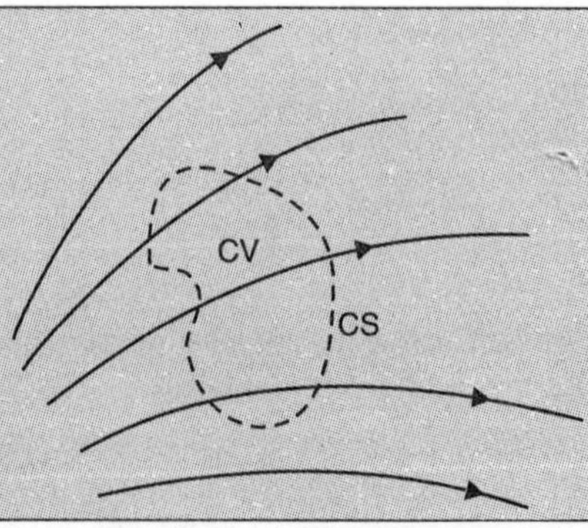

Fig. 4.1. A control volume in a flow field.

* Except when a nuclear reaction converts mass into energy and *vice versa*, through Energy = Δ (mass) × (velocity of light)^2.

This equation holds for any CV and states that the rate of accumulation of mass within a control volume equals the *net influx* across the control surface.

The following examples illustrate the applications of this equation.

Example 4.1. Consider a water storage tank (Fig. 4.2) fed by a pipe of dia. 10 cm and drained by a pipe of dia. 6 cm. The water enters at a uniform velocity of 3 m/s and leaves at a velocity of 2 m/s. Calculate the rate at which the water level in the tank rises, if the tank diamter is 5 m.

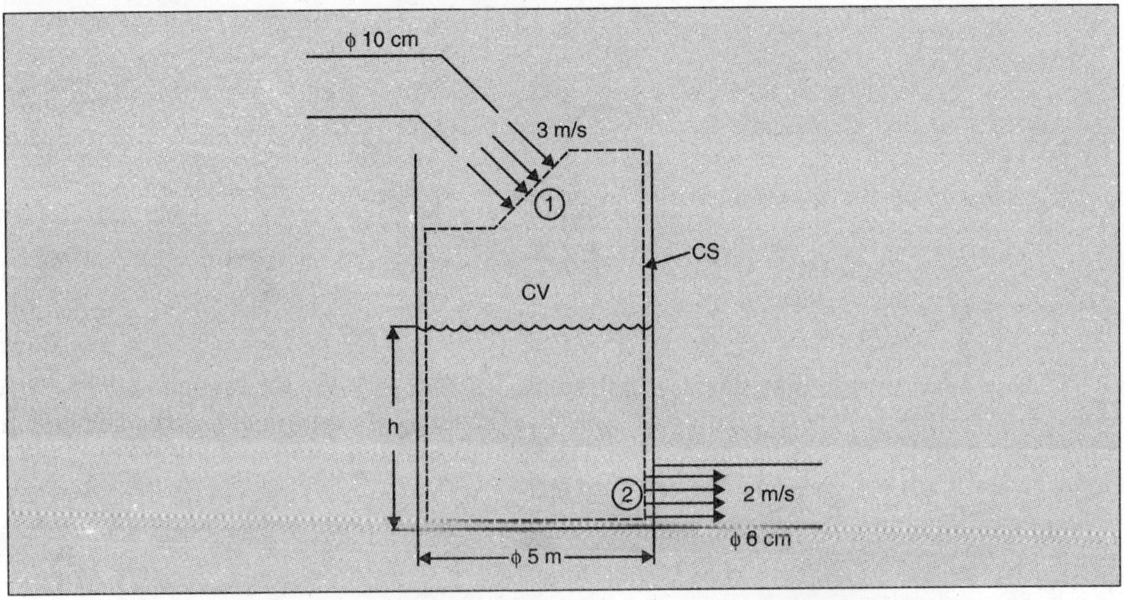

Fig. 4.2. Water storage tank and associated control volume for Example 4.1.

The choice of CV plays an important part in the solution of such problems. The CV should be so chosen that most of the CS lies adjacent to the rigid walls across which there is no flow, so that the contribution of these regions to the surface integral in Eq. (4.2) is zero. Also, the control surface should preferably be taken normal to the flow velocity (as at 1 in Fig. 4.2) to save on computations. The choice of CV for this problem is shown in Fig. 4.2 by broken lines. Water crosses the CS at the inlet port 1 [area = π $(0.1)^2/4$ m^2; velocity = 3 m/s] and at the outlet port 2 [area = π $(0.06)^2/4$ m^2; velocity = 2 m/s] only. If the height of the water surface at time t is h, then the rate of accumulation of water is

$$\frac{\partial}{\partial t}\iiint_{CV} \rho_{\text{water}}\, d\forall = \rho_{\text{water}}\, A_{\text{tank}} \frac{dh}{dt}$$

$$= 10^3 \left(\frac{\text{kg}}{\text{m}^3}\right)\frac{\pi}{4}5^2\left(\text{m}^2\right)\frac{dh}{dt}$$

$$= \pi\, 6250 \frac{(\text{kg})}{\text{m}}\frac{dh}{dt}$$

CHAPTER 4

The net efflux of water is

$$\oint\!\!\oint_{CS} \rho \mathbf{V} \cdot d\mathbf{A} = -[\rho_{\text{water}} VA]_1 + [\rho_{\text{water}} VA]_2$$

$$= -10^3 \left(\frac{\text{kg}}{\text{m}^3}\right) 3 \left(\frac{\text{m}}{\text{s}}\right) \frac{\pi}{4} (0.1)^2 \left(\text{m}^2\right)$$

$$+10^3 \left(\frac{\text{kg}}{\text{m}^3}\right) 2 \left(\frac{\text{m}}{\text{s}}\right) \frac{\pi}{4} (0.06)^2 \left(\text{m}^2\right)$$

$$= -5.7\pi (\text{kg/s})$$

By Eq. (4.2)

$$\pi \times 6250 \left(\frac{\text{kg}}{\text{m}}\right) \frac{dh}{dt} = 5.7 \left(\frac{\text{kg}}{\text{s}}\right) \times \pi$$

or $$\frac{dh}{dt} = 9.12 \times 10^{-4} \text{ (m/s)}$$

The reader can see that the alternate choices of *CVs* in Figs. 4.3 and 4.4 also give the same final results. It may be mentioned here that the mass of air has not been considered at all in this analysis. This is not an assumption and the mass of water and air are conserved independently (as water cannot convert into air).

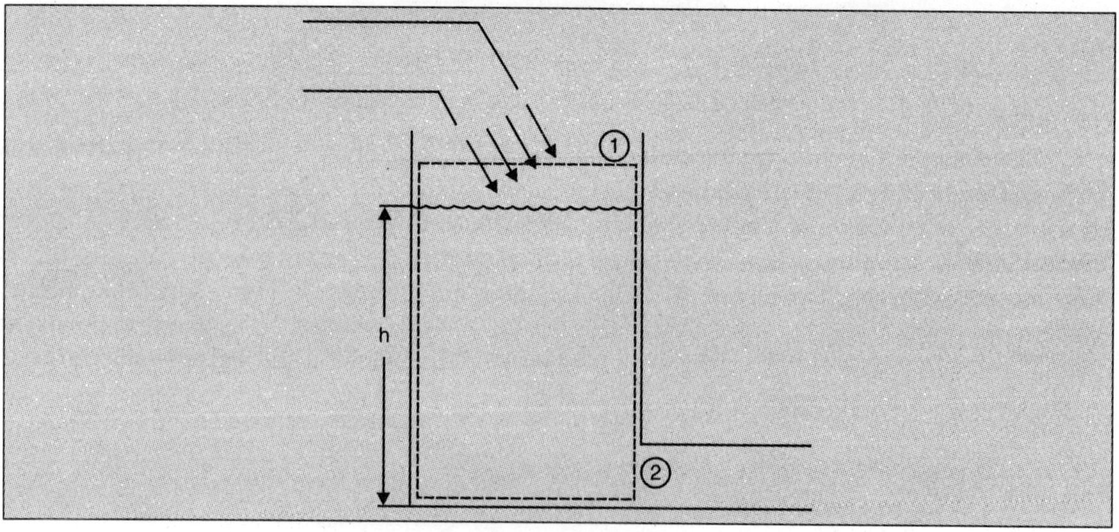

Fig. 4.3. An alternate *CV* for Example 4.1.

Example 4.2. In long distance mail trains, water in tank *T* (Fig. 4.5) can be replenished by lowering down a scoop *S* to dip into a long stretch of water in between the rails. Compute the rate of collection of water in the tank if the train is moving at 108 kmph and the scoop dips 4 cm into the water. The width of the scoop is 0.8 m.

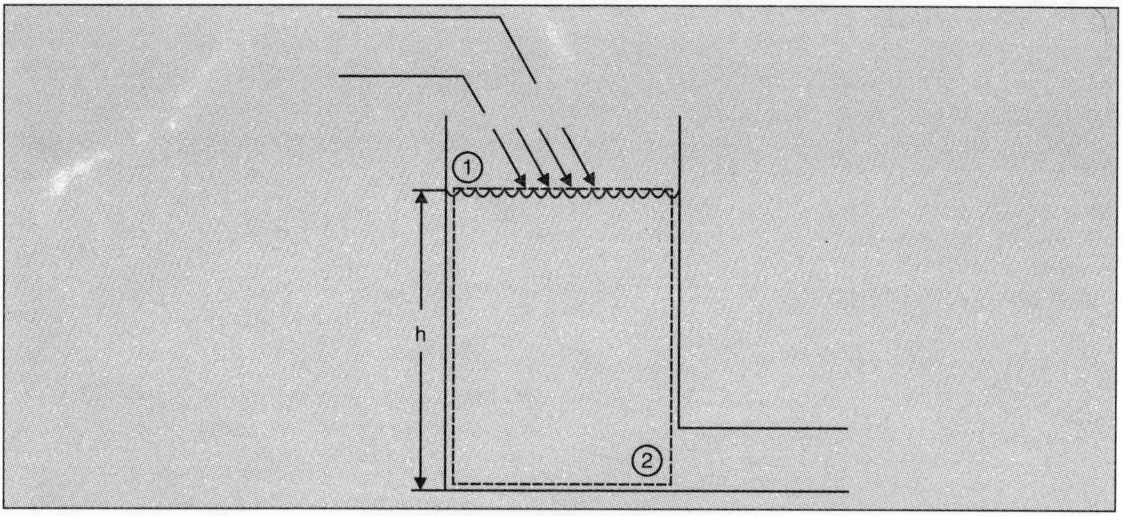

Fig. 4.4. Another *CV* for Example 4.1.

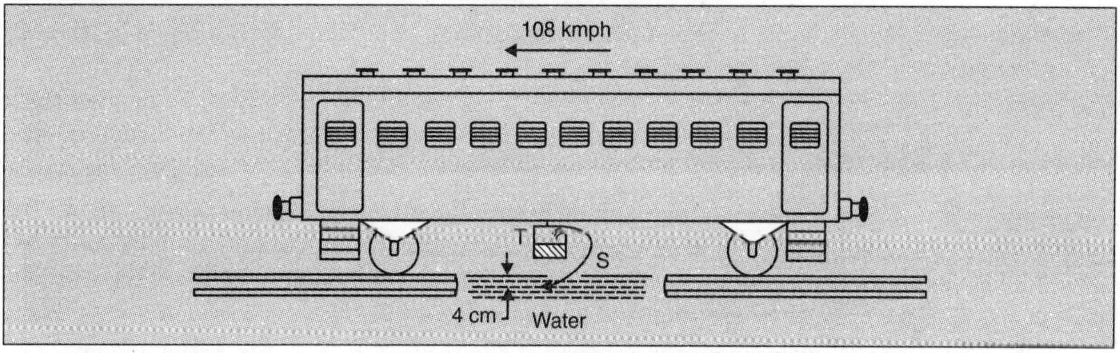

Fig. 4.5. Train scooping up water.

In this problem it would be useful to consider a frame of reference fixed to, and so moving with the train. It may be noted that in Reynolds transport Eq. (3.6), **V** is the relative velocity of the fluid with respect to the control volume. Thus, water enters the *CV* (Fig. 4.6) at a velocity of 108 kmph. Application of Eq. (4.2) to the *CV* shown gives the rate of accumulation of water in the tank as $-\oint\!\!\oint_{CS} \rho_{\text{water}}\ (\mathbf{V}\cdot d\mathbf{A})$. The only contribution to this surface integral comes from the section where water enters the *CV*. There is no flow across any other section of the *CS*. Thus,

$$\oint\!\!\oint_{CS} \rho\mathbf{V}\cdot d\mathbf{A} = -\rho_{\text{water}} \times \left[108\left(\frac{\text{km}}{\text{hr}}\right)\times\frac{1\,\text{hr}}{3600\ \text{s}}\times\frac{10^3\,\text{m}}{1\ \text{km}}\right]$$

$$\times\ [0.04\ (\text{m}) \times 0.8\ (\text{m})]$$

$$= -10^3\left(\frac{\text{kg}}{\text{m}^3}\right)\times 30\left(\frac{\text{m}}{\text{s}}\right)\times 0.032\left(\text{m}^2\right)$$

$$= -\ 960\ \text{kg/s}$$

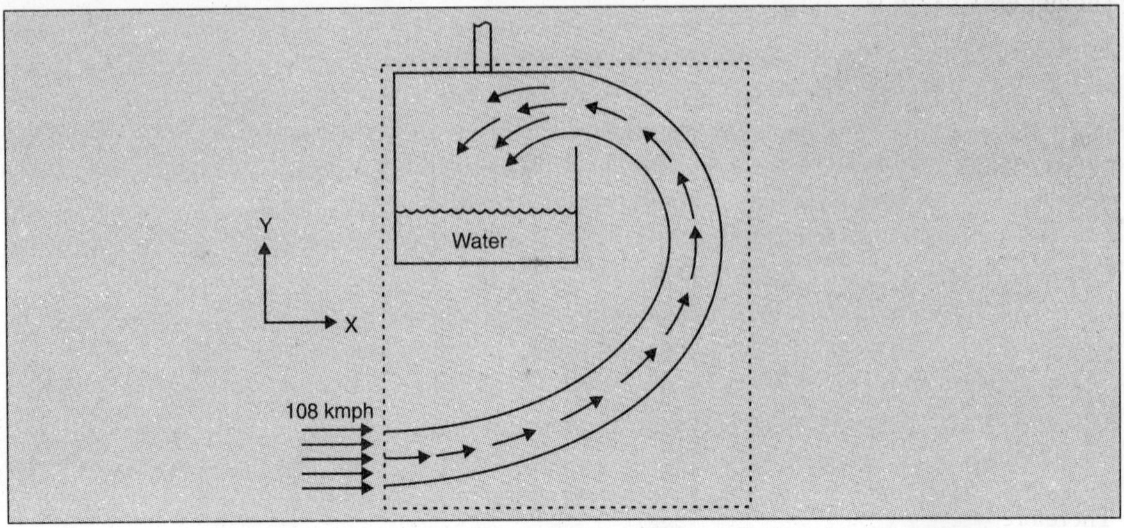

Fig. 4.6. *CV* for Example 4.2. Reference system fixed to the train.

The minus sign appears because **V** and $d\mathbf{A}$ are in opposite directions. Using Eq. (4.2), the rate of accumulation of water can be obtained as

$$\overset{\circ}{m} = -\oint\!\!\oint_{CS} \rho \mathbf{V} \cdot d\mathbf{A} = 960 \text{ kg/s}$$

Example 4.3. Consider a spherical tank of air originally at atmospheric pressure (1.013 × 10^5 N/m^2). Air is being supplied at a uniform velocity of 2 m/s through a tube of diameter 2 cm. The inlet pressure is maintained at 3 × 10^5 N/m^2 and the temperature of the tank and inlet pipe is 300 K. Determine the time required for the pressure in the tank to reach 3 × 10^5 N/m^2.

Choose the control volume as the interior of the tank (Fig. 4.7). Since the density of air within the *CV* at any time t may be assumed to be independent of position, the rate of accumulation of mass is

$$\frac{\partial}{\partial t}\iiint_{CV} \rho d \forall = \frac{d\rho}{dt} \forall_{CV} = \frac{4\pi}{3}(0.5)^3 \left(\text{m}^3\right)\frac{d\rho}{dt} = 0.523\left(\text{m}^3\right)\frac{d\rho}{dt}$$

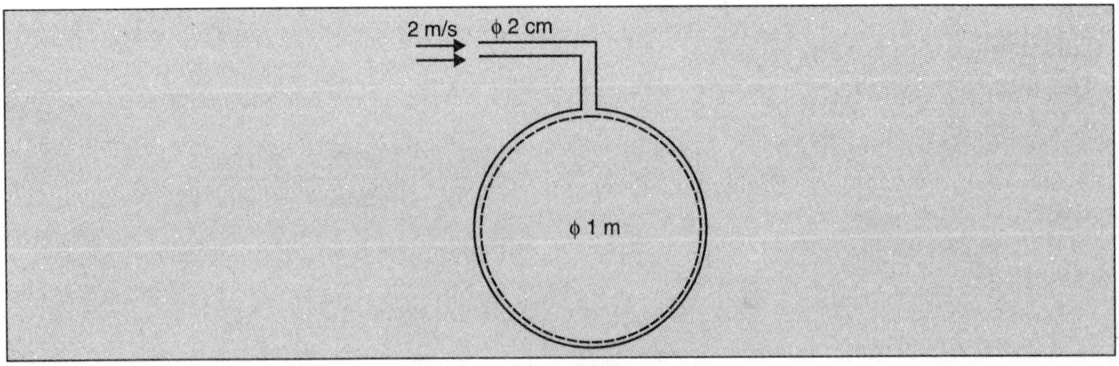

Fig. 4.7. *CV* for Example 4.3.

The net efflux rate is

$$\oint\!\!\oint_{CS} \rho \mathbf{V} \cdot d\mathbf{A} = -(\rho VA)_{\text{inlet}}$$

The density of inlet air is obtained from the ideal gas law, $p = \rho RT$, where R is the gas constant for air. Thus,

$$R = \frac{\text{Universal gas constant}, R^*}{\text{Molecular weight of air}}$$

$$= \frac{8.314 \times 10^3}{29.0} \frac{\text{N m}}{\text{kg K}}$$

$$= 286.96 \frac{\text{N m}}{\text{kg K}}$$

Therefore,

$$\rho_{\text{inlet}} = \frac{p}{RT} = 3 \times 10^5 \left(\frac{\text{N}}{\text{m}^2}\right) \times \frac{1}{286.69} \left(\frac{\text{kg K}}{\text{N m}}\right) \times \frac{1}{300} \times \left(\frac{1}{\text{K}}\right)$$

$$= 3.49 \text{ kg/m}^3$$

and

$$(\rho VA)_{\text{inlet}} = 3.49 \left(\frac{\text{kg}}{\text{m}^3}\right) 2 \left(\frac{\text{m}}{\text{s}}\right) \frac{\pi}{4} (0.02)^2 (\text{m})^2$$

$$= 2.19 \times 10^{-3} \text{ kg/s}$$

From Eq. (4.2),

$$0.523 \text{ (m}^3\text{)} \left(\frac{d\rho}{dt}\right) = 2.19 \times 10^{-3} \text{ (kg/s)}$$

or

$$\left(\frac{d\rho}{dt}\right) = \frac{2.19 \times 10^{-3}}{0.523} \left(\frac{\text{kg}}{\text{m}^3\text{s}}\right) = 4.19 \times 10^{-3} \left(\frac{\text{kg}}{\text{m}^3\text{s}}\right)$$

This implies that density changes at a constant rate.

The total density change corresponds to a change in pressure from 1.013×10^5 (N/m²) to 3×10^5 (N/m²). Or,

$$\rho_2 - \rho_1 = \frac{p_2 - p_1}{RT} = (3 - 1.013) \times 10^5 \left(\frac{\text{N}}{\text{m}^2}\right)$$

$$\times \frac{1}{286.69} \left(\frac{\text{kg K}}{\text{N m}}\right) \times \frac{1}{300} \left(\frac{1}{\text{K}}\right)$$

$$= 2.31 \text{ (kg/m}^3\text{)}$$

Therefore,

$$\text{Time required} = \frac{\rho_2 - \rho_1}{d\rho/dt} = 2.31\left(\frac{\text{kg}}{\text{m}^3}\right)\times\frac{1}{4.19\times10^{-3}}\left(\frac{\text{m}^3\text{s}}{\text{kg}}\right)$$

$$= 551.3 \text{ s} = 9.2 \text{ min}$$

4.2 SPECIAL FORMS OF THE MASS CONSERVATION EQUATION

If the flow field is steady such that its properties do not change with time, the rate of accumulation of mass is zero and Eq. (4.2) becomes

$$\oiint_{CS} \rho\mathbf{V}\cdot d\mathbf{A} = 0 \qquad ...(4.3)$$

This equation is further simplified if the control surface is so chosen that at ports of entry and exit the velocity vector is normal to the area:

$$\oiint_{CS} \rho\, V\, dA_n = 0 \qquad ...(4.4)$$

where A_n represents the area normal to the velocity. If we knew the variation of ρV across the inlets and outlets* we could carry out the integration.

Consider the flow of a fluid through a conduit (Fig. 4.8). If we plot the distribution of fluid velocity across a cross-section of this conduit, the resulting 3-dimensional pattern is termed as the *velocity profile*. In general, this velocity profile may change from one cross-section to another.

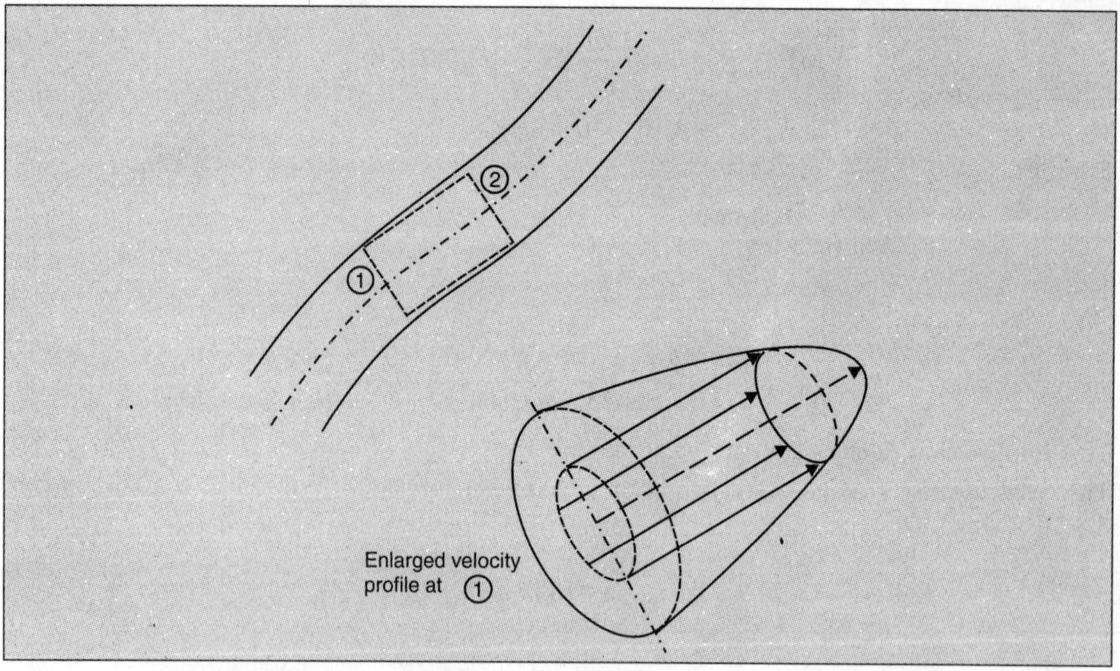

Fig. 4.8. Portion of a conduit and distribution of fluid velocity at section 1.

---

* In the earlier examples, we assumed the velocity to have constant values across a port.

There are, however, situations where the velocity profiles are far simpler. For example, in flow between two infinite parallel plates (Fig. 4.9) the flow pattern does not change in the z-direction. The velocity profile *at a section* then, is a 2-dimensional pattern* (the two dimensions being V_x and y) and the flow is termed as 2-*D*. For flow through a circular pipe, the angular symmetry of the flow may be exploited to give a 2-*D* velocity profile (V_z and r) as shown in Fig. 4.10. When the flow velocity across a section is constant, i.e., the velocity profile is flat, the flow is termed

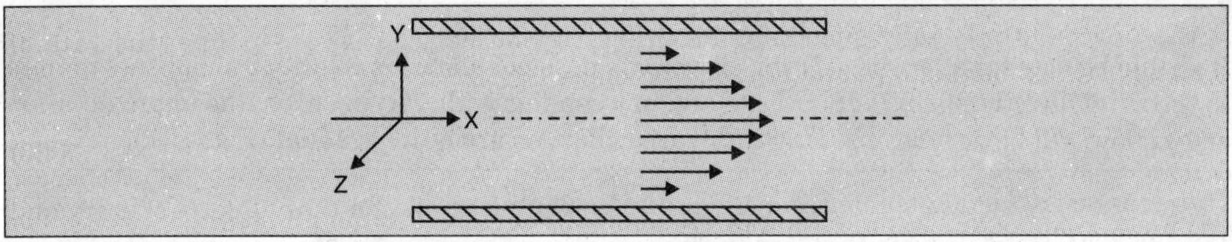

Fig. 4.9. 2-*D* flow between two flat, parallel plates

as one-dimensional (1-*D*) [see Figs. 4.11 (a) and (b)]. In such a flow the velocity may or may not change in the flow direction.

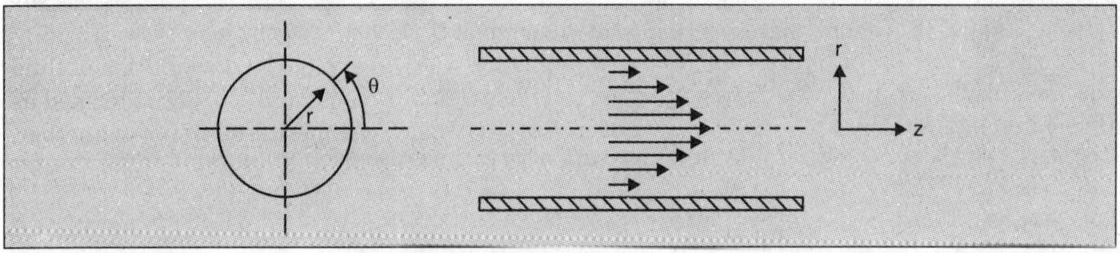

Fig. 4.10. 2-Dimensional flow in a pipe.

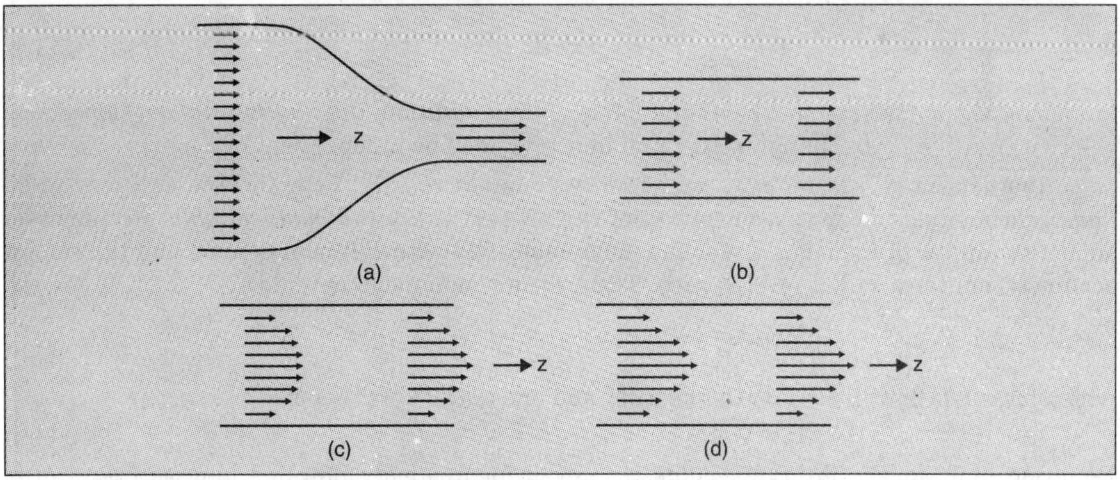

Fig. 4.11. Dimensionality and full development of velocity profiles: (a) 1-*D* non-fully-developed, $V_z = V_z\ (z)$; (b) 1-*D* fully-developed, (c) 2-*D* non-fully-developed $V_z = V_z\ (r, z)$; and (d) 2-*D* fully-developed, $V_z = V_z\ (r)$.

CHAPTER 4

* Some books in fluid mechanics define one-, two- and three-dimensional flows differently.

If the velocity profile (one-, two- or three-dimensional) does not change along the length of the conduit, the flow is termed *fully developed.** Such a condition obtains in flow through a long, narrow pipe far downstream of the entrance. Figure 4.11(d) illustrates this situation.

The 1-*D* approximation makes possible great simplifications in flow analysis and, therefore, is used quite often. If sections 1 and 2 in Fig. 4.8 are normal to the respective velocity vectors and 1-*D* flow is assumed, Eq. (4.4) reduces to

$$(\rho VA)_1 = (\rho VA)_2 \qquad \text{...(4.5)}$$

It should be noted that a true 1-*D* velocity profile does not allow for velocity components normal to the axial direction (why?) and, therefore, in a conduit with varying area, the approximation of 1-*D* flow will be permissible only if the area changes are quite gradual (Fig. 4.12).

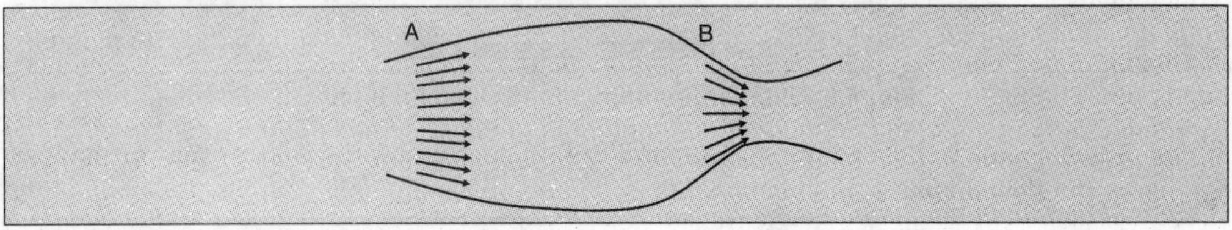

Fig. 4.12. 1-*D* flow may be *assumed* at *A* but not at *B*. Neither is strictly a 1-*D* flow.

Equation (4.5) can also be applied in 2-*D* flow situations with *V* replaced by the average velocity defined by

$$V_{av} = \text{(volume flow rate across the cross-section)/(area of flow)}$$

$$= \frac{1}{A}\iint_A V dA \qquad \text{...(4.6)}$$

which, for a circular pipe of radius *R* is

$$V_{av} = \frac{1}{\pi R^2}\int_0^R V_z(r, z)\, 2\pi r\, dr \qquad \text{...(4.7)}$$

This V_{av}, in general, will be a function of *z*. If, in addition, the flow is fully developed, the velocity profile will not change with *z* and hence V_{av} will be independent of *z*.

Another simplification in the continuity equation results if the fluid is assumed to be *homogeneous* (i.e., no spatial variation of the properties) *and incompressible.* In this case, since the volume of the *CV* is fixed, the mass contained within it is also fixed and the rate of accumulation term in Eq. (4.2) is zero. Thus, for a constant density flow,

$$\oiint_{CS} \mathbf{V} \cdot d\mathbf{A} = 0 \qquad \text{...(4.8)}$$

which for a 1-*D* flow in a *CV* with one inlet and one outlet port becomes

$$V_1A_1 = V_2A_2 \qquad \text{...(4.9)}$$

Equation (4.9), unlike Eq. (4.5), holds true even for unsteady flows as long as the fluid is homogeneous and incompressible.

Equation (4.9) is also applicable to a *mixture* of incompressible fluids, i.e., even when the condition of homogeneity does not exist. This can be seen by noting that for such systems, the

* The origin of the term will become clear in chapter 6.

total fluid volume remains constant. Using $\eta = 1/\rho$ (i.e., specific volume) in the Reynolds transport Eq. (3.6), we obtain

$$\frac{\partial}{\partial t}\iiint_{CV} d\forall = \frac{D\forall}{Dt} - \oiint_{CS} \mathbf{V}\cdot d\mathbf{A} \qquad ...(4.10)$$

In this equation both the accumulation and the growth terms are zero and, thus, for incompressible flow (even non-homogeneous and unsteady),

$$\oiint_{CS} \mathbf{V}\cdot d\mathbf{A} = 0 \qquad ...(4.11)$$

which gives Eq. (4.9) for a one-dimensional flow through a two-port control volume.

Example 4.4. Figure 4.13 shows a hydraulic cylinder used as a shock absorber. When a force is applied to the piston and it moves, the incompressible fluid flows through the metering orifice. A relatively large velocity through the metering orifice creates large viscous forces which cushion the piston movement. If the expected maximum piston velocity is 2 m/s, what is the maximum velocity of fluid at A?

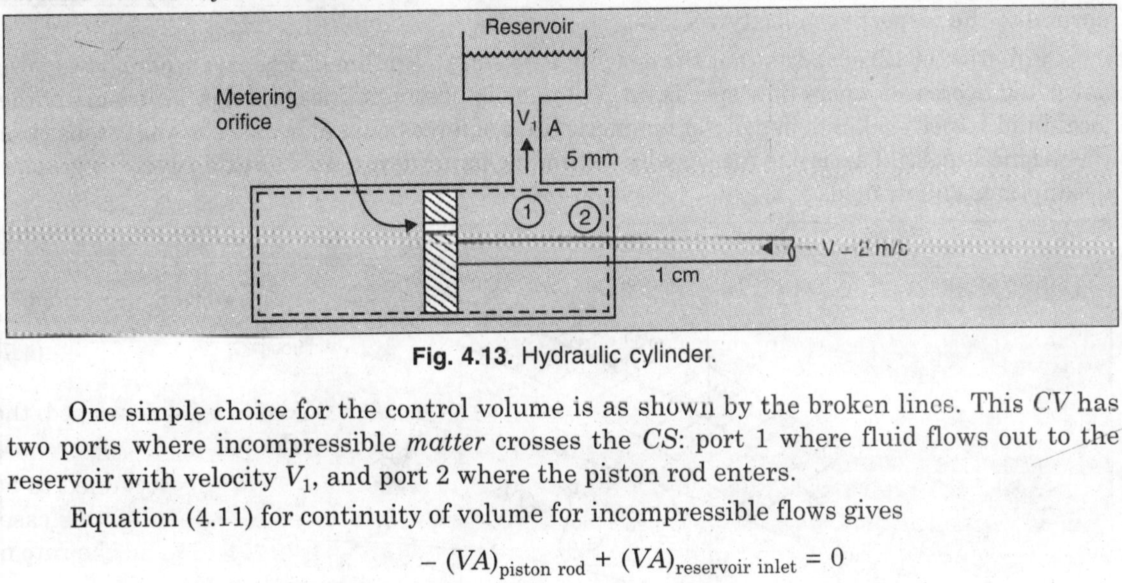

Fig. 4.13. Hydraulic cylinder.

One simple choice for the control volume is as shown by the broken lines. This CV has two ports where incompressible *matter* crosses the CS: port 1 where fluid flows out to the reservoir with velocity V_1, and port 2 where the piston rod enters.

Equation (4.11) for continuity of volume for incompressible flows gives

$$-(VA)_{\text{piston rod}} + (VA)_{\text{reservoir inlet}} = 0$$

or

$$-2\left(\frac{\text{m}}{\text{s}}\right)\frac{\pi\times(0.01)^2}{4}(\text{m})^2 + V_1\frac{\pi\times(0.005)^2}{4}(\text{m}^2) = 0$$

which gives $V_1 = 8$ m/s as the maximum expected velocity in the tube connection A to the reservoir.

Note that the conditions inside the CV are unsteady and the LHS of Eq. (4.2) is not zero. Application of the volume continuity Eq. (4.11) circumvents this problem.

However, Eq. (4.2) could still be used profitably by realizing that this equation is also a statement of conservation of *mass of a species* in a heterogeneous system with ρ and $\mathbf{V}$ replaced by the appropriate values corresponding to that species. The LHS of Eq. (4.2) then gives the rate of accumulation of the fluid in the *CV* which, in this case, equals –(density of fluid) × $(VA)_{\text{piston rod}}$. Only port *A* now contributes to the RHS, and therefore, the proper use of Eq. (4.2) gives the same result as above.

4.3 STREAM FUNCTION

Consider a 2-*D* flow of an incompressible fluid. At some time *t*, one can draw a set of streamlines which, as defined in Sec. 3.4, are tangent everywhere to the local velocity vector. Since the flow is 2-*D* the streamlines will be identical in every plane parallel to the *x*-*y* plane (Fig. 4.14). If we draw a surface normal to the *x*-*y* plane and passing through a streamline, the velocity vector will be tangent to it at every point and no fluid will cross this surface. Such a surface is termed as a *stream surface.* Since no fluid can move across a stream surface, the space bounded by *any* two stream surfaces can be visualised as a 2-*D* conduit for fluid flow.* If we take two stream surfaces passing through streamlines S_1 and S_2 in the *x*-*y* plane, and define a *CV* as shown by broken lines in Fig. 4.14, we may write the equation of volume continuity (Eq. 4.11) as

$$(VA)_1 = (VA)_2 \qquad \text{...(4.12)}$$

where V_1 and V_2 are averaged velocities at cross-sections 1 and 2 and A_1 and A_2 are selected normal to the respective velocity vectors.

Equation (4.12) suggests that the spacing between streamlines increases where flow slows down and decreases where flow speeds up. The density of streamlines, thus, is a measure of the local fluid velocity (akin to density of magnetic lines of force being a measure of the intensity of the magnetic field). This is one reason why streamline patterns are used so extensively as graphic descriptions of flow fields.

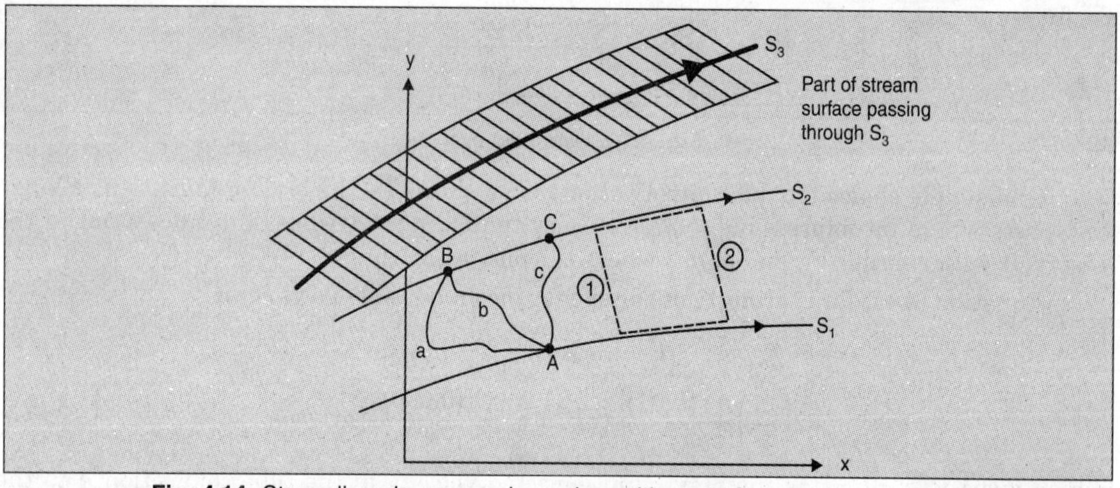

Fig. 4.14. Streamlines in an *x-y* plane alongwith part of one stream surface.

Next, take a point *A* on the streamline S_1 and a point *B* on S_2. It can readily be seen that the volume flow rate (per unit depth) across any line joining *A* and *B* must be the same. Further

* In a general flow the streamline which pass through a *closed curve* form a tubular stream surface. As for a 2-*D* flow, there is no flow across these stream surfaces which, therefore, can be visualised as conduits. These are termed as *stream tubes* and have the obvious property that flow across any section of these is constant.

by taking any other point C on S_2, it can be shown that the flow across any line from A to B must be the same as across any line from A to C, since no fluid crosses the streamline segment BC. This suggests the possibility of defining a function which takes constant values for a given streamline such that the difference of the values of this function for two streamlines gives the flow rate (per unit depth) between them. This is termed as the *stream function* ψ (x, y, t). A streamline is thus represented by

$$\psi\ (x, y, t) = \text{constant}$$

and the flow rate (per unit depth) between two streamlines S_1 and S_2 is $\psi\ (S_2) - \psi\ (S_1)$.

The stream function ψ is simply related to the velocity **V**. Consider two streamlines (Fig. 4.15), S_1 with stream function ψ and another, S_2, slightly away from it with stream function $\psi + \delta\psi$. This implies that the volume flow rate per unit depth across any line connecting the two streamlines is $\delta\psi$.

First, consider a line AB, parallel to the y-axis, between S_1, and S_2. Let the coordinate of B be $y + \delta y$. If the average x-component of velocity across this infinitesimal line-segment is V_x, the flow rate per unit depth is $V_x\ \delta y$, which should be equal to $\delta\psi$. Thus

$$\psi(B) - \psi(A) = \frac{\partial \psi}{\partial y}\delta y = V_x \delta y$$

or
$$\frac{\partial \psi}{\partial y} = V_x \qquad \text{...(4.13)}$$

Similarly, by taking a line-segment AC, parallel to the x-axis, where the x-coordinate of C is $x - \delta x$, we obtain

$$\psi(C) - \psi(A) = \frac{\partial \psi}{\partial x}(-\delta x) = V_y \delta x \quad \text{or} \quad \frac{\partial \psi}{\partial y} = -V_y \qquad \text{...(4.14)}$$

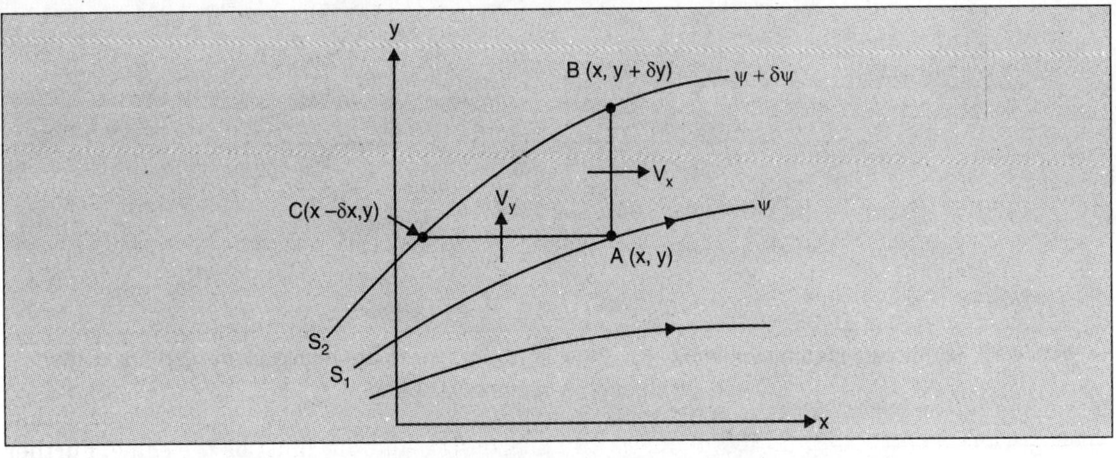

Fig. 4.15. Two streamlines having stream functions ψ and $\psi + \delta\psi$.

If the velocity field **V** (x, y, t) is specified, one may obtain the stream function ψ (x, y, t) by integration of Eqs. (4.13) and (4.14). It should be noted that ψ may be determined in this manner only upto an undetermined constant.

The above derivation implicitly uses the fact that if the flow between S_1 and S_2 is shown from left to right $\psi(S_2) > \psi(S_1)$. This is summarized by the following sign convention: *if an observer looks from one streamline to another, the* ψ *of the second is larger than that of the base streamline if the flow crosses from the observer's left to his right, otherwise it is smaller.*

The stream function ψ is defined for 2-*D* flows above. For 3-*D* flows, a vector function ψ, is used.

Example 4.5. The stream function of a flow is $\psi = 2xy$. Sketch the streamlines and determine the velocity at (0, 0), (2, 0) and (5, 0).

The streamline pattern is shown in Fig. 4.16a. Note that the arrows have been marked according to the sign convention stated above.

Parts of this streamline pattern represent the flow through various geometries generated by replacing stream surfaces by solid walls. This is possible because there is no flow across a stream surface and thus a stream surface is akin to an impervious boundary.* Thus, $\psi = 2xy$

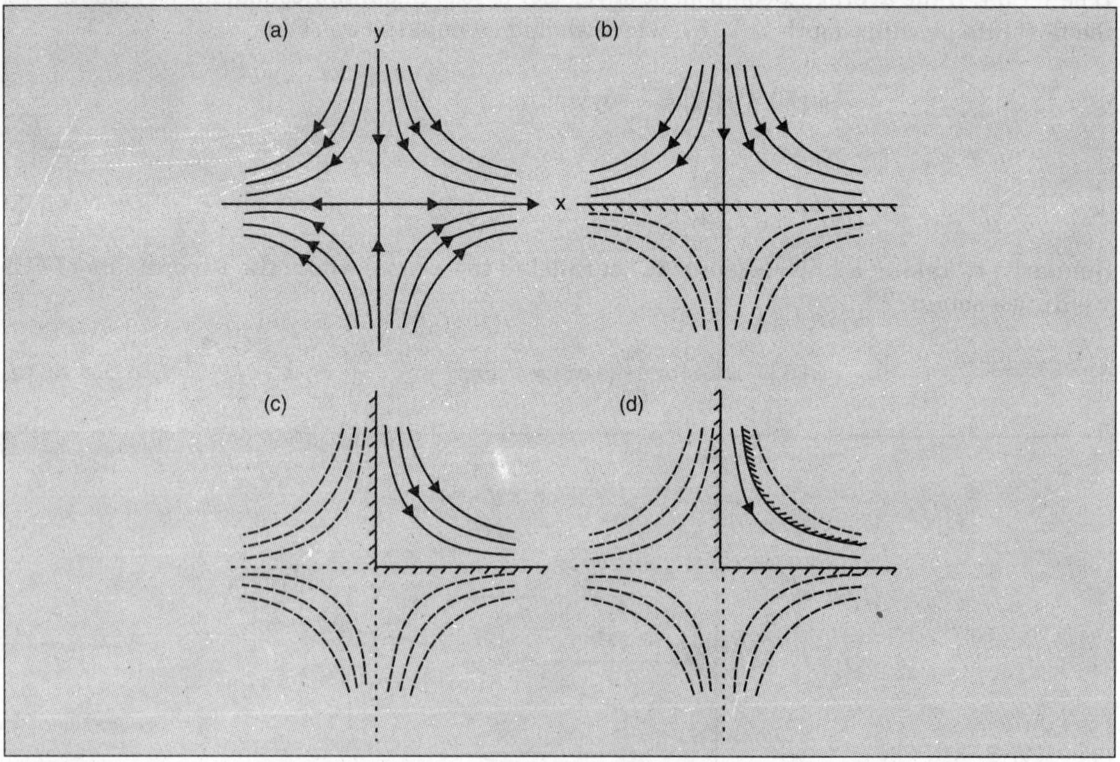

Fig. 4.16. (a) Streamlines for Example 4.5. Flow is (b) normal to an infinite plane; (c) in a corner; and (d) through a rectangular bend.

* Any real fluid must satisfy one more condition, namely, the no slip condition at the wall. This formulation, therefore, applies strictly only to the case when the no slip condition can be ignored, i.e., in the case of ideal fluids alone. It will be seen in Chapter 13 that in real fluids the region of flow affected by the viscous action is restricted to a very thin boundary layer. Hence, this formulation approximates the real flow over *most* of the flow field.

may represent the flow normal to an infinite plane [Fig. 4.16 (b)], in a corner [Fig. 4.16 (c)] or through a rectangular bend [Fig. 4.16 (d)]. Eqs. (4.13) and (4.14) give the velocity vector as

$$\mathbf{V}(x, y) = \frac{\partial \psi}{\partial y}\hat{\mathbf{i}} - \frac{\partial \psi}{\partial x}\hat{\mathbf{j}}$$
$$= 2x\hat{\mathbf{i}} - 2y\hat{\mathbf{j}}$$

Therefore, $\mathbf{V}(0, 0) = 0$. The origin is thus a *stagnation* point (defined as a point where flow comes to a stop). Similarly, $\mathbf{V}(2, 0) = 4\,\hat{\mathbf{i}}$ and, $\mathbf{V}(5, 0) = 10\,\hat{\mathbf{i}}$.

The flow accelerates as we move away from the origin. This is also apparent from the fact that the spacing between streamlines decreases as x increases.

4.4 DIFFERENTIAL FORM OF THE CONTINUITY EQUATION

In Sec. 4.1 and 4.2, the integral form of the continuity equation was obtained for use with finite-sized control volumes. Consider here an infinitesimal *CV* in 2-*D* fluid flow (Fig. 4.17) to obtain

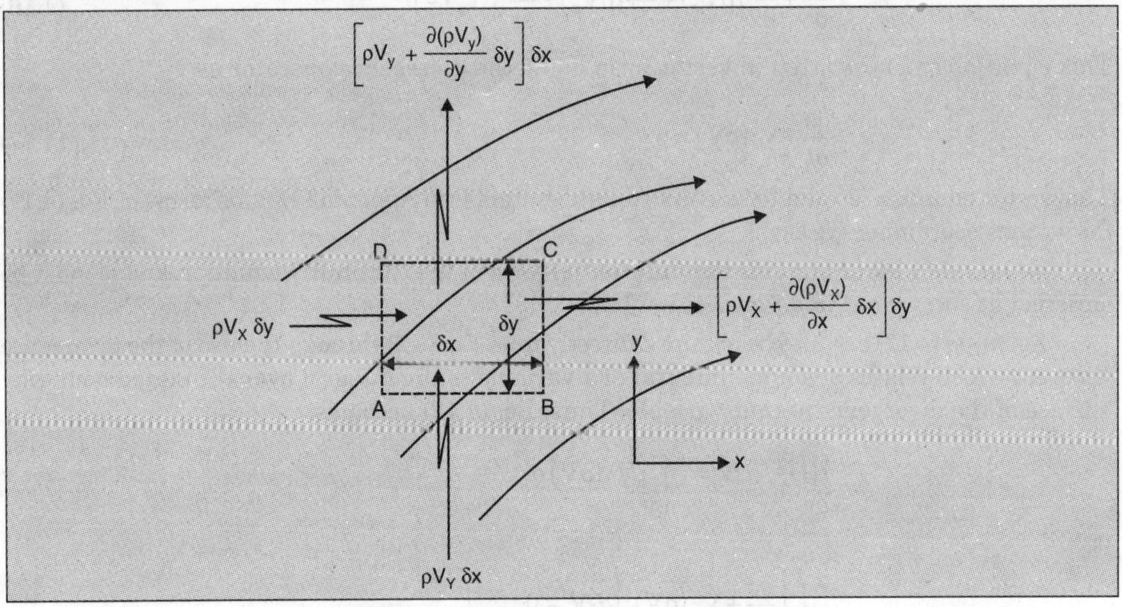

Fig. 4.17. A differential *CV* in 2-*D* flow.

the differential form of the continuity equation. The mass efflux across face *AD* (of unit depth) is $-\rho V_x\,\delta y \times 1$, the negative sign signifying that mass crosses into the *CV*. The efflux across face *CB*, a distance δx away, is obtained by a Taylor expansion as $\left[\rho V_x + \frac{\partial}{\partial x}(\rho V_x)\delta x\right] \times \delta y \times 1$

Similarly, one can write the mass flow rates across faces *AB* and *CD*. The net efflux across the control surface is, then,

$$\left[\frac{\partial}{\partial x}(\rho V_x)+\frac{\partial}{\partial y}(\rho V_y)\right]\delta x\ \delta y$$

The mass contained within the CV is $\rho\ (\delta x\ \delta y \times 1)$ and, therefore, the rate of accumulation is $\frac{\partial \rho}{\partial t}\delta x\ \delta y$. From Eq. (4.2), we get

$$\frac{\partial \rho}{\partial t}\delta x\ \delta y+\left[\frac{\partial}{\partial x}(\rho V_x)+\frac{\partial}{\partial y}(\rho V_y)\right]\delta x\ \delta y=0$$

or

$$\frac{\partial \rho}{\partial t}+\frac{\partial}{\partial x}(\rho V_x)+\frac{\partial}{\partial y}(\rho V_y)=0 \qquad ...(4.15)$$

On generalizing it to three-dimensions, we get

$$\frac{\partial \rho}{\partial t}+\frac{\partial}{\partial x}(\rho V_x)+\frac{\partial}{\partial y}(\rho V_y)+\frac{\partial}{\partial z}(\rho V_z)=0 \qquad ...(4.16)$$

This equation can be written in vector form using the divergence operator as

$$\frac{\partial \rho}{\partial t}+\nabla\cdot(\rho \mathbf{V})=0 \qquad ...(4.17)$$

This vector equation is valid for any coordinate system [see Appendix B-8 for forms of Eq. (4.17) for various coordinate systems].

Equation (4.16) or (4.17) is the differential form of the continuity equation and it must be satisfied *at every point* in a flow field.

Equation (4.17) can also be obtained directly from Eq. (4.2) through the use of the divergence theorem which relates a volume integral of a variable to the integral over the closed bounding surface of the divergence of that variable. Thus, Eq. (4.12) becomes

$$\iiint_{CV}\frac{\partial \rho}{\partial t}\,d\forall=-\iiint_{CV}\nabla\cdot(\rho \mathbf{V})\,d\forall$$

or

$$\iiint_{CV}\left[\frac{\partial \rho}{\partial t}+\nabla\cdot(\rho \mathbf{V})\right]d\forall=0$$

This holds for any arbitrary control volume and, therefore, requires that

$$\frac{\partial \rho}{\partial t}+\nabla\cdot(\rho \mathbf{V})=0$$

For steady flows, the special form of the differential continuity equation is obtained as

$$\nabla\cdot(\rho \mathbf{V})=0 \qquad ...(4.18)$$

Equation (4.17) may be written in an alternate form as,

$$\frac{\partial \rho}{\partial t}+\mathbf{V}\cdot\nabla\rho+\rho\nabla\cdot\mathbf{V}=0$$

Realizing that the first two terms represent the substantial derivative [Eq. (3.4)], one can write

$$\frac{D\rho}{Dt} + \rho\nabla \cdot \mathbf{V} = 0 \qquad ...(4.19)$$

For incompressible fluids, the density ρ of a particle does not change and, therefore, $\frac{D\rho}{Dt}$ in Eq. (4.19) is zero and we obtain

$$\nabla \cdot \mathbf{V} = 0 \qquad ...(4.20)$$

which, for a 2-D flow in cartesian coordinates, gives

$$\frac{\partial V_x}{\partial x} + \frac{\partial V_y}{\partial y} = 0 \qquad ...(4.21)$$

Note that the stream function ψ introduced in Sec. 4.3 with properties given by Eqs. (4.13) and (4.14), automatically satisfies the continuity Eq. (4.21). In fluid flow analysis it is convenient, at times, to work with the stream function instead of the velocity components V_x and V_y since it reduces the number of dependent variables by one.

Example 4.6. Fluid flows past a flat plate shown in Fig. 4.18. Due to the viscous action the fluid in the immediate vicinity of the plate slows down. The thin region within which the velocity is appreciably different from the free-stream value V_0, is termed as boundary layer (see Sec. 1.7). Outside this layer, then, the velocity may be assumed to be unaffected by the presence of the plate. One approximation for the velocity profile within this layer is

$$V_x = V_0 \left[\frac{3}{2}\left(\frac{y}{\delta}\right) - \frac{1}{2}\left(\frac{y}{\delta}\right)^3 \right] \text{for } y \le \delta$$

and $\quad V_x = V_0$ for $y \ge \delta$

where V_0 is a constant and δ, the boundary-layer thickness, is a function of x given as

$$\delta = 5\sqrt{\frac{\mu x}{\rho V_0}}$$

Here ρ is the density and μ the viscosity of the fluid. Obtain how V_y varies across the boundary layer.

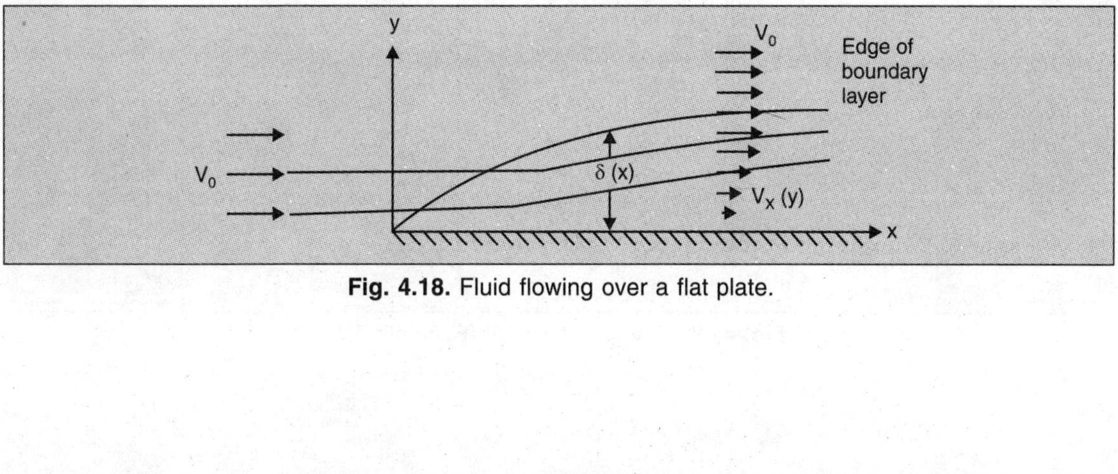

Fig. 4.18. Fluid flowing over a flat plate.

CHAPTER 4

For 2-D incompressible flows, the continuity Eq. (4.21) gives

$$\frac{\partial V_x}{\partial x} + \frac{\partial V_y}{\partial y} = 0$$

or

$$\frac{\partial V_y}{\partial y} = -\frac{\partial V_x}{\partial x}$$

$$= V_0 \left[\frac{3}{2}\frac{y}{\delta^2} - \frac{3}{2}\frac{y^3}{\delta^4} \right] \times \frac{5}{2}\sqrt{\frac{\mu}{\rho V_0 x}}$$

or

$$\frac{\partial V_y}{\partial y} = \frac{3\,V_0}{4x}\left[\left(\frac{y}{\delta}\right) - \left(\frac{y}{\delta}\right)^3\right]$$

Integrating with respect to y gives

$$V_y = \frac{3\,V_0}{8}\left(\frac{y}{x}\right)\left[\left(\frac{y}{\delta}\right) - \frac{1}{2}\left(\frac{y}{\delta}\right)^3\right] + C(x)$$

where $C(x)$ is the constant of integration to be determined by the boundary condition. At $y = 0$, V_y is 0 since the plate is impervious, and so $C(x) = 0$. Hence

$$V_y = \frac{3}{8}V_0\left(\frac{y}{x}\right)\left[\left(\frac{y}{\delta}\right) - \frac{1}{2}\left(\frac{y}{\delta}\right)^3\right]$$

Example 4.7. Consider a channel of constant width b, one end of which is closed off (Fig. 4.19). An incompressible fluid falls uniformly into the channel at the rate of $\mathring{q}$ m³/s per metre *length* of the channel. The profile of the free surface of the fluid in the channel is defined by the depth h, which changes with x. One can make the 1-D approximation if the variation of h with x is slow. Develop a continuity equation involving h for the flow in the channel assuming

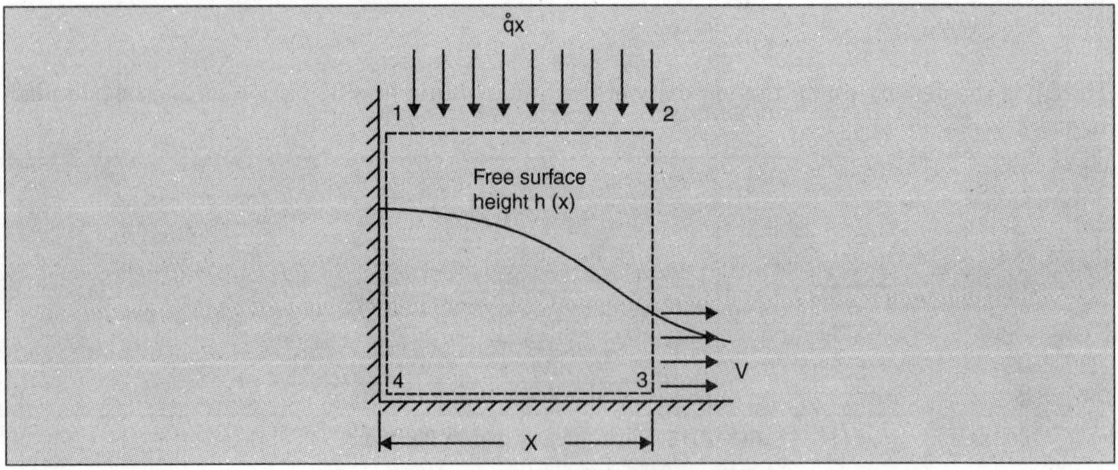

Fig. 4.19. Flow in a channel (Example 4.7).

that $\mathring{q}$ does not vary with time and that the flow is steady. Next, develop a continuity equation for the case when $\mathring{q}$ varies with time such that the depth h (and velocity V) is a function of both x and t.

Select a *CV* of length x as shown. In steady flow there is no accumulation of fluid within the *CV*, and Eq. (4.2) gives the net efflux to be zero. For 1-*D* flow the efflux of fluid across segment 2-3 of the *CS* is ρVhb, where V and h are the velocity and the depth at x respectively. The influx across 1-2 is $\rho\mathring{q}\,x$, and thus,

$$\rho Vhb - \rho\mathring{q}\,x = 0$$

or

$$Vh = \mathring{q}\,x/b \tag{a}$$

This is the form of the continuity equation which governs the flow in the channel.

When $\mathring{q}$ is varying with time and the flow is unsteady, it is more convenient to use a *CV* of infinitesimal width δx at x (Fig. 4.20). Since δx is small, the mass of fluid contained within the *CV* can be approximated as $\rho bh\,\delta x$, and therefore, the rate of accumulation is $\rho b\,(\partial h/\partial t)\,\delta x$.

The effluxes across various segments of the *CS* are:

across 1-2 : $-\rho\mathring{q}\;\delta x$

across 1-4 : $-\rho V hb$

across 2-3 : $+\rho Vhb + [\rho b\,\partial(Vh)/\partial x]\,\delta x$

Equation (4.2) then gives

$$\frac{\partial h}{\partial t} + \frac{\partial (Vh)}{\partial x} = \frac{\mathring{q}}{b} \tag{b}$$

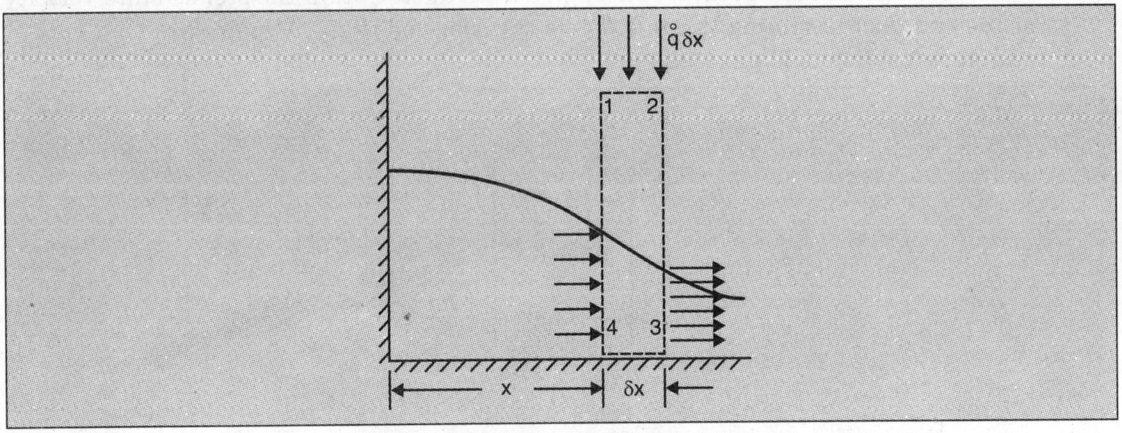

Fig. 4.20. Differential *CV* for Example 4.7.

It is easy to see that when $\mathring{q}$ does not vary with time and the flow is steady, the first term in Eq. (b) is zero, and upon integration Eq. (a) is obtained.

To obtain $h\,(x, t)$ or $V\,(x, t)$ one more relation (as obtained in Example 5.5) is required.

CHAPTER 4

PROBLEMS

4.1 The velocity V of water measured at various vertical positions over a 2-*D* obstacle shown is given below:
Compute the flow rate per unit width.

y (m)	$V\left(\frac{m}{s}\right)$	θ (degree)
0.5	9	4
1	7.95	6.5
1.5	7.0	9
2	6.4	12
2.5	5.9	16
3	5.4	18
3.5	5	20

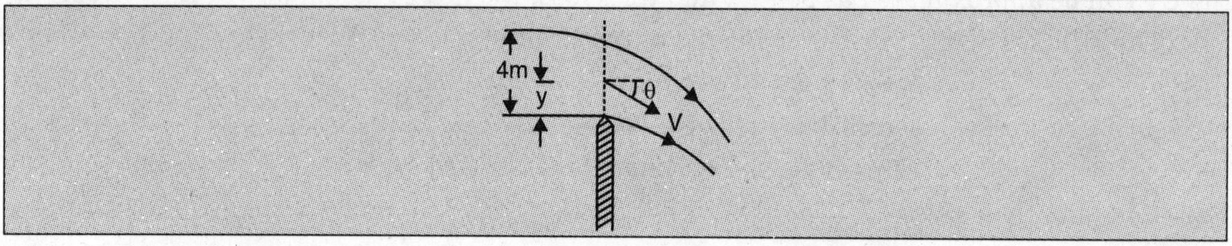

4.2 A reactor converts *A* to A_2 by the reaction $2A \rightarrow A_2$. Pure (gaseous) *A* at 100 kPa and 400 K enters at a velocity of 20 m/s through a duct of dia. 4 cm. The gases leave at the same pressure and temperature through a 2 cm duct. Obtain the steady state velocity at the exit duct assuming that all *A* gets transformed to A_2. Also assume that both *A* and A_2 follow the ideal gas law.

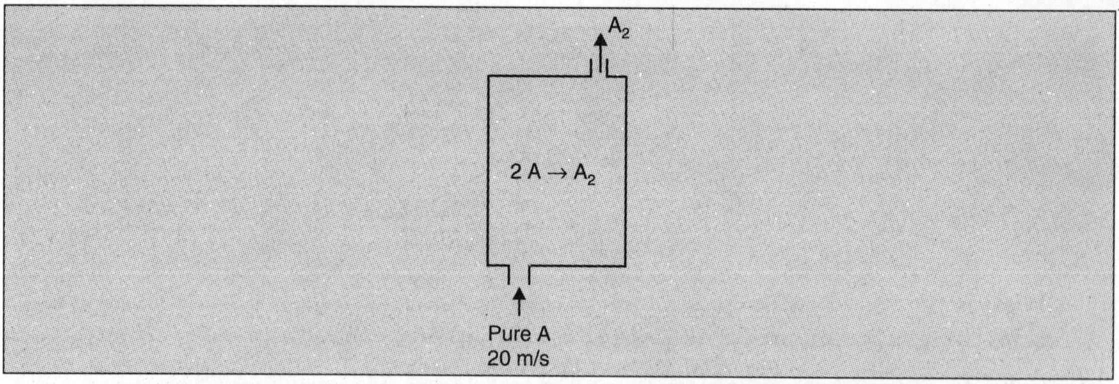

4.3 A highway has three lanes for traffic in either direction. Cars (of length approx. 5 m) are piled up at an obstruction where one lane is closed due to repairs. The maximum safe speed is 90 kmph. If the spacing between cars is maintained at 1 m for every 4 kmph speed, obtain the maximum speed upstream of the obstruction.

4.4 A student applies the continuity equation to the *CV* shown and simplifies the accumulation term $\frac{\partial}{\partial t}\iiint_{CV} \rho d\forall$ as $\iiint_{CV} \frac{\partial \rho}{\partial t} d\forall$. He argues that since water is incompressible, $\frac{\partial \rho}{\partial t} = 0$ and that there is no accumulation. Where is the fallacy in his argument?

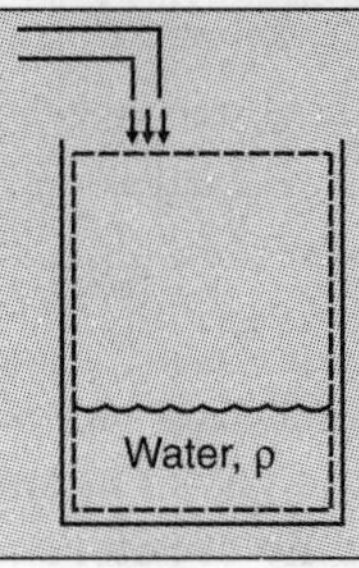

4.5 A piston moves vertically downwards with a constant velocity of 0.1 m/s. What is the velocity *V* of the fluid? Take the stationary *CV* shown.

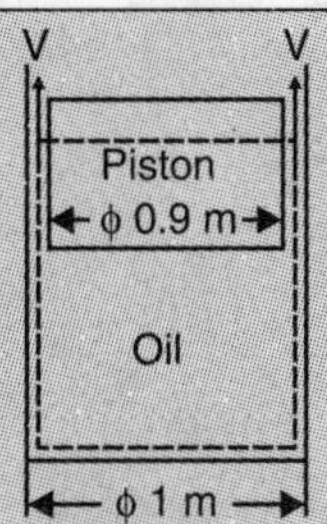

4.6 A circular plate of 1 m diameter is pushed towards another stationary circular plate below it at a velocity V_p which may vary with time.

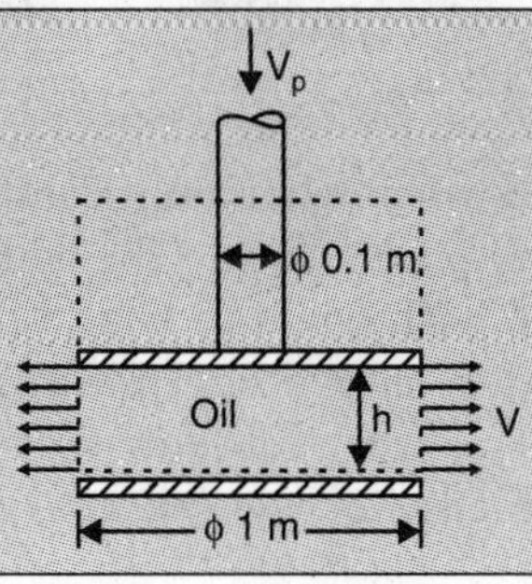

What is the velocity *V* of the fluid at the edge of the plates when the gap between them is *h*? Use the stationary *CV* indicated. (*Hint*: Use species-specific continuity equation).

4.7 A stirred tank has two inlets *A* and *B* and an outlet port *O* as shown. Two incompressible liquids enter the tank and are mixed instantaneously. A piston of area A_p pushes the mixture at a velocity V_p. Obtain the expressions for the velocity V_0 and density ρ at the exit.

CHAPTER 4

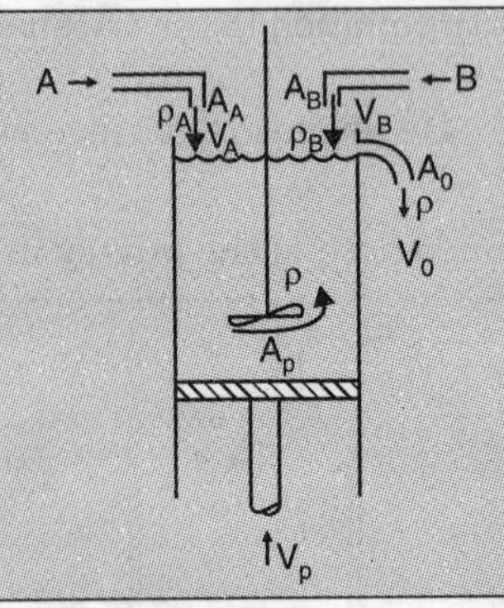

4.8 In a water pump, water enters the rotating vanes at the centre at a rate of 2×10^{-3} m^3/s. It then flows outwards and leaves the vanes at their outer edges in a direction tangent to the vanes (as observed from a frame of reference moving with the vanes) as shown. If the width of the vanes is 1 cm and they rotate at 200 rad/s, compute the exit velocity as seen from a stationary frame of reference.

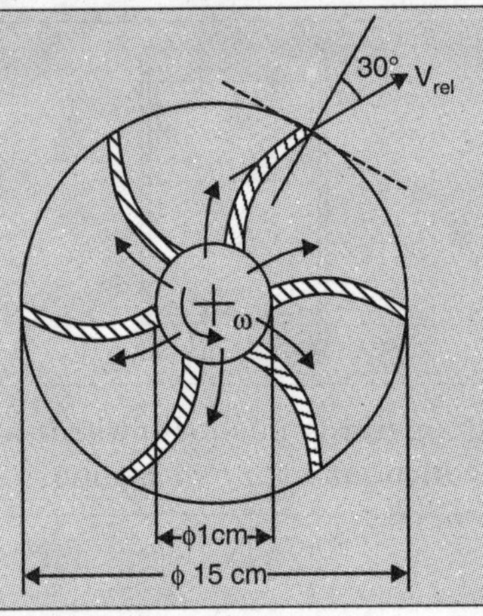

4.9 Consider the boundary layer formed when an incompressible fluid flows over a flat plate [see Example 4.6]. The velocity profile within the boundary layer at any section x is given by $V_x = V_0 f(\eta)$ where $\eta = y/\delta$ with δ as the boundary layer thickness at that location. Outside the boundary layer the x-component of velocity is V_0 everywhere. (a) Using $ABCD$ as the CV, show that mass crosses into the boundary layer at CD. (b) To find V_y near the edge of the boundary layer, use $ABCE$ as the CV. For $f(\eta) = \frac{3}{2}\eta - \frac{1}{2}\eta^3$, determine V_y at $\eta = 1$.

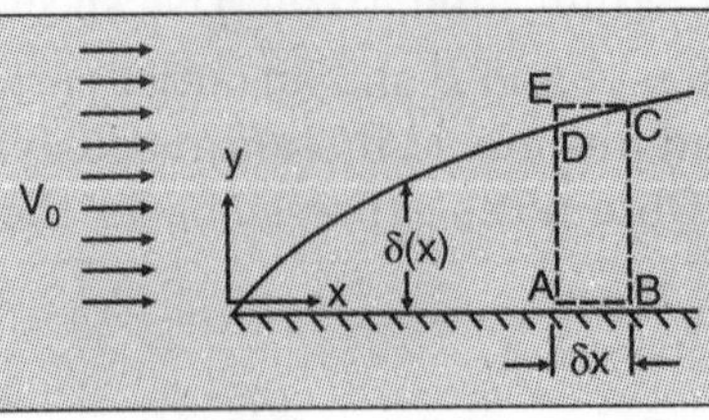

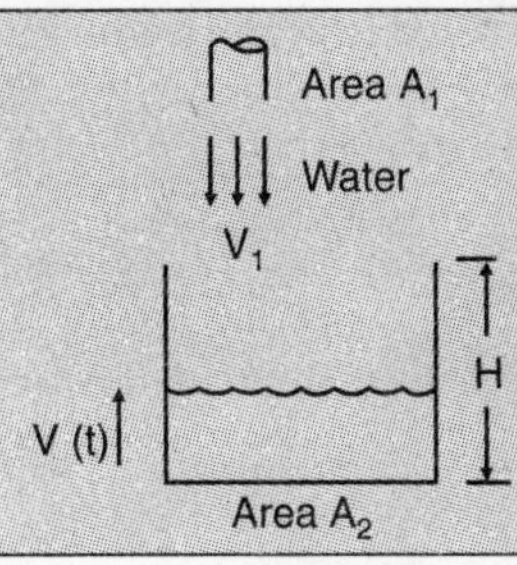

4.10 A bucket is placed on an open elevator which starts moving upwards at time $t = 0$ with a constant acceleration a. A stationary hose discharges water into the bucket at a constant rate as shown. Find the time required to fill the bucket if it is empty at $t = 0$.

4.11 *Jet entrainment:* When a jet of air issues from an orifice, it drags the surrounding air along with it due to viscous action as shown. Thus the mass of air moving downstream increases as we move farther from the orifice. This phenomena is called *entrainment.* Experimental studies show that the air motion is confined within a conical region, with the centreline velocity V_c varying inversely with the axial location z. The non-dimensional velocity profile $V_z(r, z)/V_c(z)$ is found to be a function of a single composite non-dimensional variable $\eta = r/R$, where R is the radius of the jet at z. Since the jet is conical, $R = bz$ with b constant depending on the cone angle. By applying the mass balance to a CV of thickness dz, show that the rate of entrainment per unit axial distance dQ/dz is independent of z.

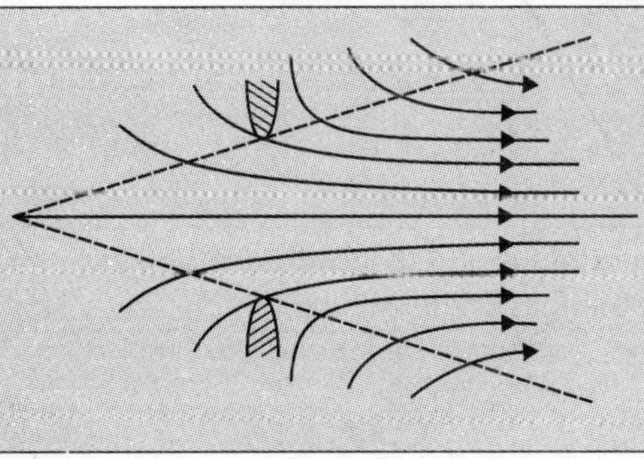

4.12 It is commonly observed that it is easier to put out a candle by blowing than by sucking in the air present around it. This is because of the fact that while blowing the flow separates (see Sec. 1.7) at the lips forming a conical jet, while during suction a "sink" is created with air moving in from all directions. The velocity in a jet varies as $1/z$ with z measured from the apex of the jet cone as described in Prob. 4.11. If the air velocity at the lips in a short puff may be assumed to be 40 m/s with a circular opening of dia. 8 mm, and if it is estimated that we need a velocity of at least 2 m/s for putting out a candle, estimate the maximum distances from which we can *blow out* or *suck out* a flame. Assume the cone angle of the exhaled air jet to be 45°.

4.13 Indicate the arrows on the streamline patterns shown.

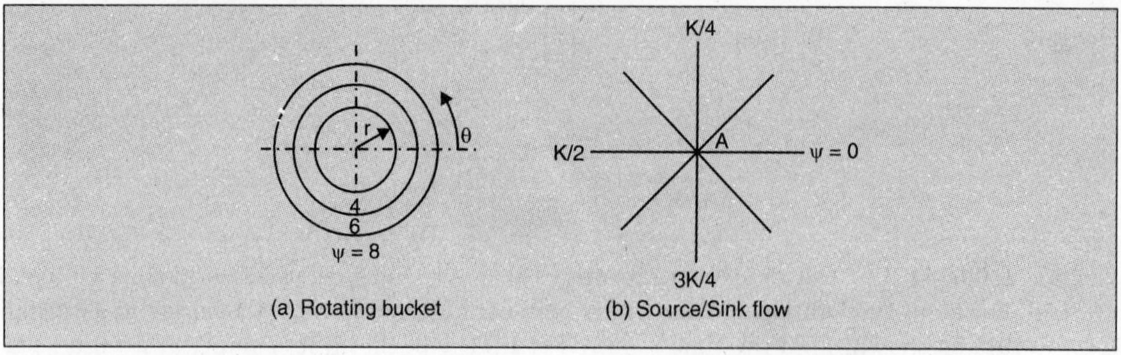

4.14 Show that the streamlines must cross the line marking the edge of the boundary layer in Fig. 4.18. Can these be parallel to the plate inside the boundary layer?

4.15 Classify the flows shown as 1-*D*, 2-*D* or 3-*D*, and fully developed or developing at the sections marked.

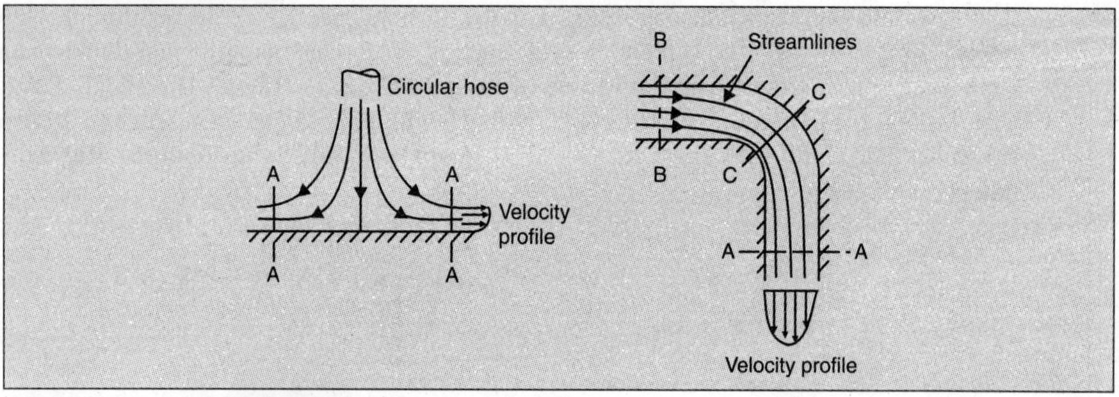

4.16 Develop the differential form of the continuity equation for cylindrical polar coordinates by taking an infinitesimal *CV* as shown.

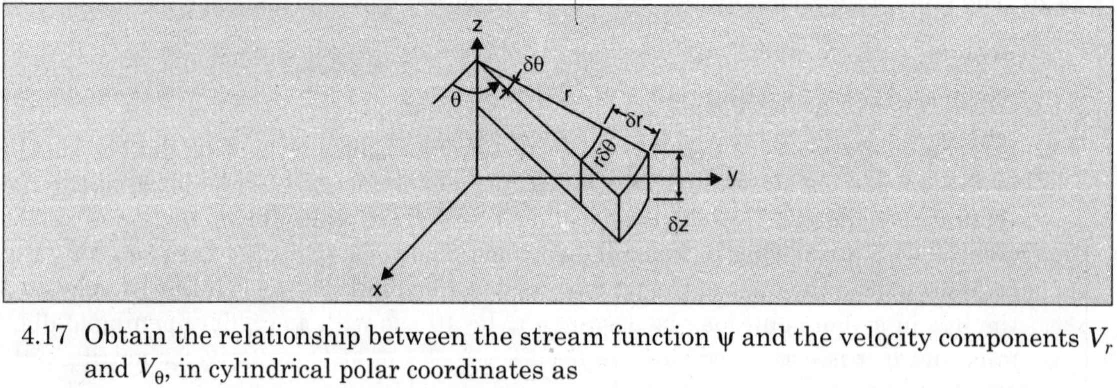

4.17 Obtain the relationship between the stream function ψ and the velocity components V_r and V_θ, in cylindrical polar coordinates as

$$V_r = \frac{1}{r}\frac{\partial \psi}{\partial \theta} \text{ and } V_\theta = -\frac{\partial \psi}{\partial r}$$

4.18 Sketch the streamline patterns for

(a) $\psi = \dfrac{q}{2\pi}\tan^{-1}\dfrac{y}{x}$

(b) $\psi = \dfrac{q}{2\pi}\tan^{-1}\dfrac{y}{x+a}$

(c) $\psi = \dfrac{q}{2\pi}\left[\tan^{-1}\dfrac{y}{x-a} - \tan^{-1}\dfrac{y}{x+a}\right]$

[*Hint*: Patterns (b) and (c) can be obtained by suitably modifying (a)].

4.19 For flow of a fluid about a stationary cylinder of radius R shown, the function ψ is given by

$$\psi = V_0\left(r - \frac{R^2}{r}\right)\sin\theta + \text{constant}$$

Sketch the streamlines qualitatively. Also obtain V_r and V_θ (see Prob. 4.17) and check their values at $r = R$. See if V_r and V_θ satisfy the differential form of the continuity equation.

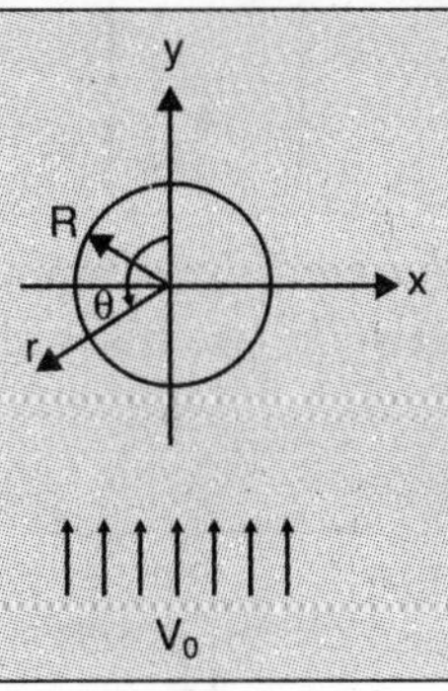

4.20 The stream function for flow of water in a pump is given by $\psi = \dfrac{\mathring{Q}\theta - \Gamma \ln r}{2\pi}, R_1 \le r \le R_2$ where R_1, R_2, $\mathring{Q}$ and Γ are positive constants. Draw streamlines in the r, θ plane. Also compute V_r and V_θ (see Prob. 4.17). Interpret $\mathring{Q}$ physically. Observe how streamlines change as Γ increases. Thus, give a physical interpretation of Γ.

4.21 A stream function is given by $\psi = x^2 + 2xy - 4t^2y$. What is the flow rate per unit width across the path ABC shown at $t = 2$? Indicate the flow direction.

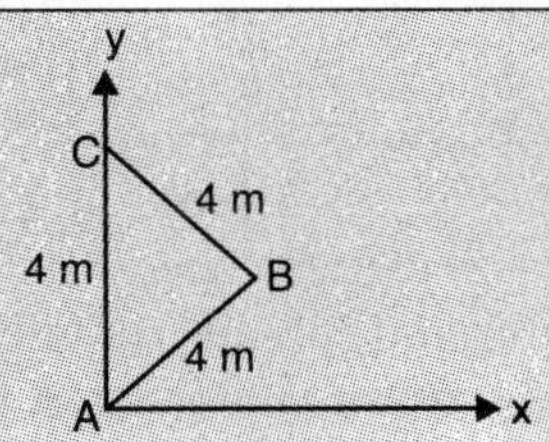

CHAPTER 4

4.22 An unsteady velocity field is given by

$$\mathbf{V} = -2x\,\hat{\mathbf{i}} + (2y + 3t)\,\hat{\mathbf{j}}$$

Obtain ψ and plot the streamline pattern at $t = 3$. Give a physical interpretation of the flow field.

4.23 Find V_y and ψ when

$$\text{(a)}\ V_x = e^{-x}\cosh y + 1 \qquad \text{(b)}\ V_x = V_0\left(\frac{2y}{ax} - \frac{y^2}{a^2x^2}\right)$$

Use $V_y = 0$ at $y = 0$.

4.24 Consider a fully developed 2-*D* flow of an incompressible fluid between two impervious flat plates as shown. Use the continuity equation to show that $V_x = 0$ everywhere.

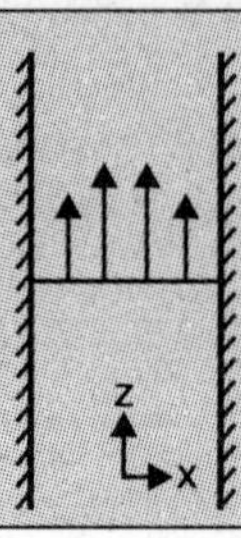

4.25 Apply the differential form of the continuity equation to obtain the variation of V_r with r for large r for radial flow in-between two discs shown.

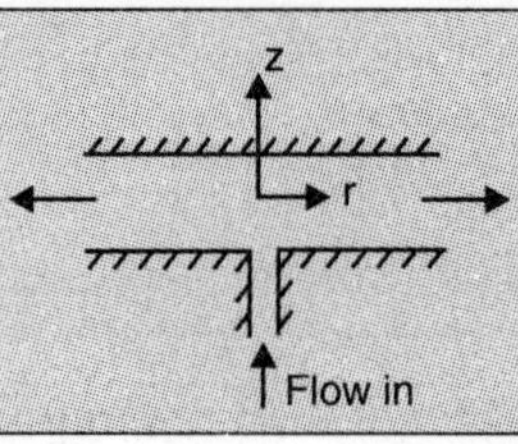

4.26 The sluice gates of a dam are opened in an emergency so that water drains out. If the flow at point x is assumed to be almost 1-*D* as shown, obtain a partial differential equation relating $\delta\ (x, t)$ to $V_x\ (x, t)$.

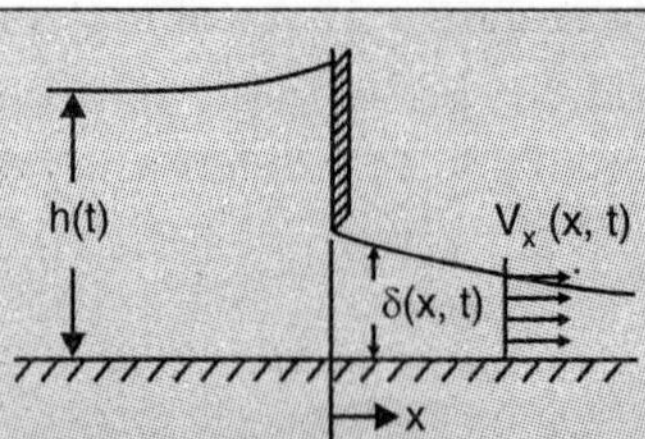

4.27 Oil spills from a pipe onto water and spreads radially as shown. Using 1-*D* approximation, simplify the species-specific continuity equation for oil to relate δ and V, both being functions of time and the radial location r.

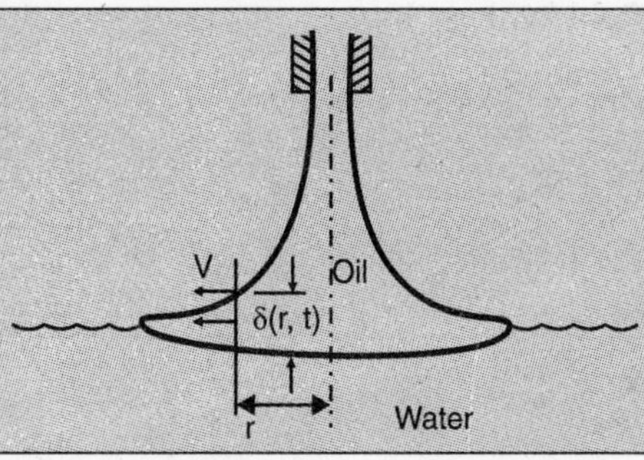

4.28 During the downward stroke of the upper plate of a Blacksmith's bellows shown, the air inside, assumed incompressible, moves towards the delivery port. If the flow is assumed 1-*D*, obtain the differential equation for the variation of the air velocity *V* with *x*.

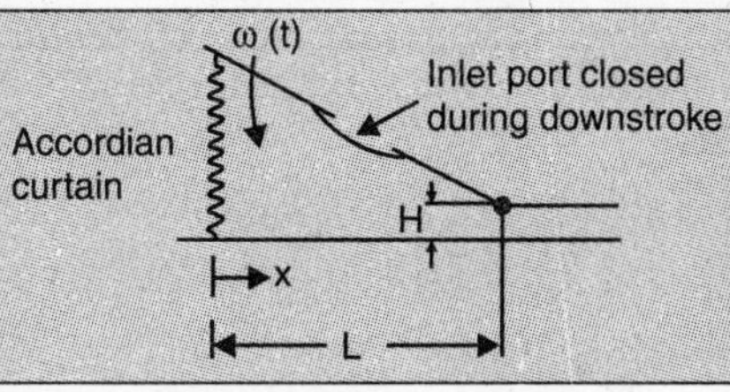

4.29 The semi-digested food is driven forward in the small intestines by a wave like motion (called *peristalsis)* as shown. A simple description of this motion is given by

$$h\ (x,\ t) = A\cos\ (\omega t - kx)$$

As the food passes through, a part of it is absorbed through the intestinal walls and goes to the blood stream and the lymphatic vessels. This removal rate may be assumed as $\mathring{q}\ m^3/m$ length of intestine. Assuming a 1-*D* unsteady velocity profile, obtain a differential equation for $V_x\ (x, t)$.

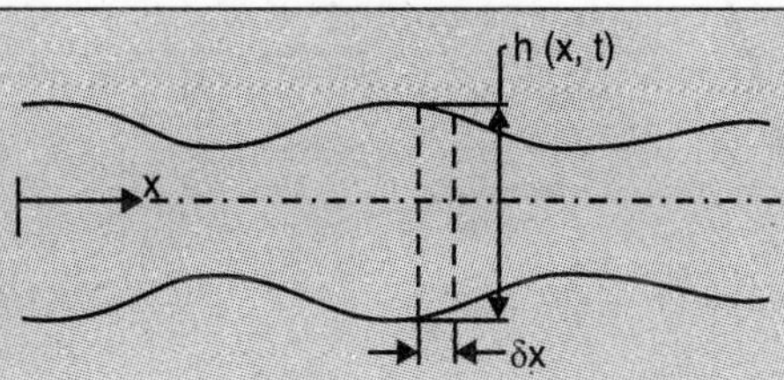

4.30 Deduce the following relationship between **V** and the stream function ψ for a 2-*D* steady compressible flow

$$V_x = \frac{\rho_0}{\rho}\frac{\partial \psi}{\partial y}$$

$$V_y = -\frac{\rho_0}{\rho}\frac{\partial \psi}{\partial x}$$

where ρ_0 is the density of the fluid at some reference condition. Show that this satisfies the differential form of the continuity equation.

5

MOMENTUM THEOREMS

5.1 EXTERNAL FORCES

Newton's second law states that the time rate of change of momentum of a body of matter (measured in a non-accelerating, i.e., an *inertial* frame of reference) is proportional to the net external force applied on the body. The two types of external forces are *surface* and *body forces.* Surface forces are due to the interaction between the body and the matter in immediate contact with it and act on its bounding surfaces. Their intensity is expressed in terms of *stress* and is defined as the force per unit surface area. The stress vector in a stationary fluid acts normal to the bounding surface (Chapter 2). In a fluid in motion this *may not* be true and there may be both normal and tangential components of stress (Fig. 5.1). The pressure in a fluid is an example of *normal stress.* The tangential component is termed *shear stress.*

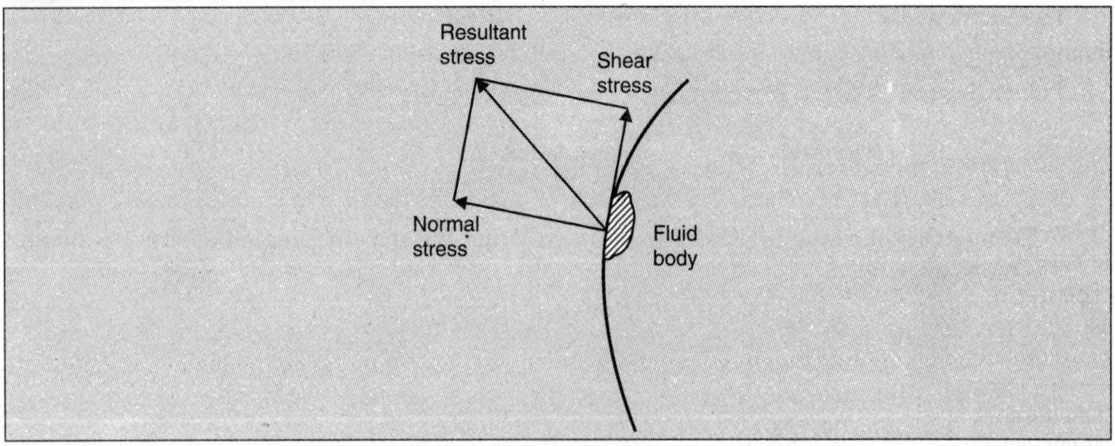

Fig. 5.1. Resolution of a surface force acting on an area-element into shear and normal components.

Body forces, on the other hand, do not act on the surface but act throughout the bulk of the body. These forces, such as due to gravity or electromagnetism, act from a distance and their intensity is expressed as force *per unit mass*. The only body force which will concern us in this book is that due to gravity, the intensity of which is **g**, the acceleration due to gravity.

5.2 MOMENTUM THEOREM

As was stated in Sec. 3.1 control volume (*CV*) formulations are preferred in fluid mechanics. Newton's second law cannot be applied directly to a *CV* since the matter contained within it changes with time. We can, however, use Reynolds transport theorem (Eq. 3.7) to relate the rate of change of momentum in a *CV* to the rate of change of momentum of the fluid body (i.e., **DP**/*Dt*) which occupies the *CV* at that time. Thus,

$$\frac{\partial}{\partial t}\iiint_{CV} \mathbf{V}\,\rho\, d\forall + \oiint_{CS} \mathbf{V}(\rho\mathbf{V}\cdot d\mathbf{A}) = \frac{D\mathbf{P}}{Dt} \qquad \text{...(5.1)}$$

By Newton's second law the *net* external force **F** acting on the *CV* is equal to **DP**/*Dt*. This gives

$$\mathbf{F} = \frac{\partial}{\partial t}\iiint_{CV} \mathbf{V}\,\rho\, d\forall + \oiint_{CS} \mathbf{V}(\rho\mathbf{V}\cdot d\mathbf{A}) \qquad \text{...(5.2)}$$

Equation (5.2) is known as the momentum theorem and states that *the net external force acting on a CV is equal to the rate of change of momentum contained in the CV (i.e., the rate of accumulation) plus the net efflux of momentum across the CS.** The rate of accumulation is zero for steady flows.

Note that, since Newton's second law of motion applies to inertial systems, the momentum theorem 5.2 is valid *only for inertial systems.* (This restriction is not applicable to other equations such as the continuity Eq. (4.2), see Problem 4.10). This means that the *CV* (which is fixed with the reference system) should be so chosen that it is either stationary or moving in a *straight line* at a constant speed. This rules out the application of Eq. (5.2) to a rocket taking off (which is accelerating) or to a spinning water sprinkler (which is not moving in a straight line). Equations for such cases are available but are beyond the scope of this text.

The momentum Eq. (5.2) applies to flows that are steady or unsteady, homogeneous or inhomogeneous, compressible or incompressible, and one-, two-, or three-dimensional.

The scalar components of Eq. (5.2) in the cartesian system are given by

$$F_x = \frac{\partial}{\partial t}\iiint_{CV} V_x \rho\, d\forall + \oiint_{CS} V_x(\rho\mathbf{V}\cdot d\mathbf{A}) \qquad \text{...(5.3)}$$

with similar equations written for the *y*- and *z*-directions. Here, the first term on the RHS is the rate of change of the *x*-momentum in the *CV* (i.e., the rate of accumulation) and the second is the net rate of *x*-momentum efflux.

---

* This form is valid only for the MLT systems of units such as **SI** units (see Appendix A). For FMLT systems, such as British or MKS, the two terms on the RHS of Eq. (5.2) are modified by the factor *k* which is $(1/32.2)\dfrac{\text{lb}_f}{\text{lb-ft/s}^2}$, or $(1/9.81)\ \dfrac{\text{kg}_f}{\text{kg-m/s}^2}$ respectively in these two systems.

The following examples illustrate the use of the momentum theorem to some simple situations. It should be noted that the choice of the control volume, though arbitrary, has to be made judiciously so as to simplify the calculations.

Example 5.1. A jet of water of diameter 1 cm impinges on a stationary vane which deflects it through 60° as shown in Fig. 5.2(a). If the jet velocity remains constant at 10 m/s throughout the flow over the vane, find the force exerted by the jet on the vane. Assume that the pressure does not vary across a straight free jet, i.e., a jet which is open to the atmosphere on all sides. (It will be shown in Sec. 7.6 that this is a reasonable assumption.)

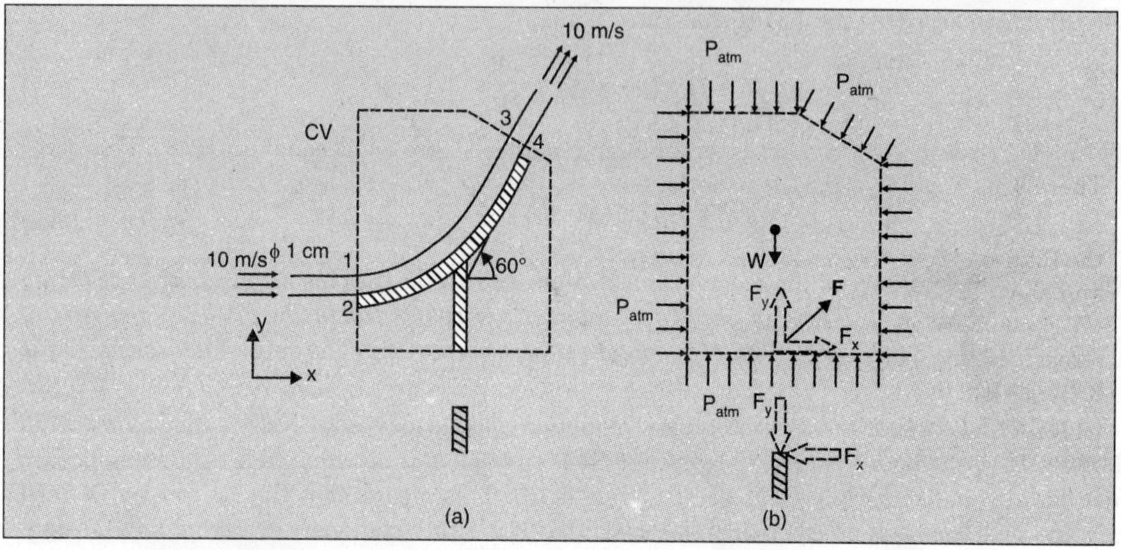

Fig. 5.2. Flow of a water jet over a vane (Example 5.1). (a) The *CV* chosen, and (b) external forces acting on the *CV*.

First consider the *CV* shown in Fig. 5.2 (a) with the control surface so chosen that it cuts the jet of water normally at 1-2 and 3-4. The flow through this *CV* is steady. The external forces acting on the *CV* are the body force *W* due to gravity on the mass contained in the *CV*, and the surface forces on the *CS*. The surface forces include the atmospheric pressure force all around and the force **F** where the *CS* cuts through the support. Let F_x and F_y be the components of this force acting on the *CV*. Since the force on the vane must be balanced by the force at the support, **F** is equal and opposite to the force due to the water jet.

The force due to gravity is ignored here as the prime interest is in the dynamic force due to the jet. Also, since a uniform pressure acts all around a closed surface, its *net* contribution is zero. The momentum equations in the *x*- and *y*-directions are

$$F_x = \frac{\partial}{\partial t}\iiint_{CV} V_x \rho \, d\forall + \oiint_{CS} V_x (\rho \mathbf{V} \cdot d\mathbf{A})$$

$$F_y = \frac{\partial}{\partial t}\iiint_{CV} V_y \rho \, d\forall + \oiint_{CS} V_y (\rho \mathbf{V} \cdot d\mathbf{A})$$

Since the flow is steady the first term on the RHS of both these equations is zero. Mass crosses the *CS* at segments 1-2 and 3-4 only. At 1-2, $\rho\mathbf{V} \cdot d\mathbf{A}$ is negative since **V** and the outward normal are in opposite directions, and at 3-4, $\rho\mathbf{V} \cdot d\mathbf{A}$ is positive. The *value* at either section is equal to the mass flow rate = 10^3 (kg/m^3) × 10 (m/s) × π/4 × 0.01^2 (m)^2 = 0.786 (kg/s). Thus,

$$F_x = 10\ (\text{m/s}) \times (-\ 0.786)\ (\text{kg/s}) + 10\ (\text{m/s}) \times \cos 60° \times (+\ 0.786)\ (\text{kg/s})$$
$$= -\ 3.93\ \text{N}$$

and

$$F_y = 0\ (\text{m/s}) \times (-\ 0.786)\ (\text{kg/s}) + 10\ (\text{m/s}) \times \sin 60° \times (+\ 0.786)\ (\text{kg/s})$$
$$= 6.81\ \text{N}$$

or

$\mathbf{F} = (-\ 3.93\hat{\mathbf{i}} + 6.81\hat{\mathbf{j}})$ N is the force *on the CV* (reaction at the support).

Therefore, $(+\ 3.93\hat{\mathbf{i}} - 6.81\hat{\mathbf{j}})$ N is the force *on the vane* due to the action of the jet.

In order to illustrate how the choice of the *CV* affects the computations (without altering the final results), we consider an alternative *CV* shown in Fig. 5.3. On surface 2-4 the pressure and shear stress distributions are unknown. Figure 5.4 shows that the external forces on the *CV* can be viewed as a sum of two distributions, p_{atm} all around and p_{gauge} (= p–p_{atm}) and shear stress τ on the surface 2-4 alone. The first distribution gives zero net force on the *CV* (see Sec. 2.2). Let **R** represent the resultant of the p_{gauge} and τ distributions. Application of the momentum equation 5.2 to this *CV* gives **R** equal to the net momentum efflux from this *CV*. This is the same as that computed for the *CV* of Fig. 5.2a. Thus,

$$R_x = -\ 3.93\ \text{N}$$
$$R_y = +\ 6.81\ \text{N}$$

Figure 5.5 shows that **R** should equal the net force acting on the vane support in Fig. 5.2.

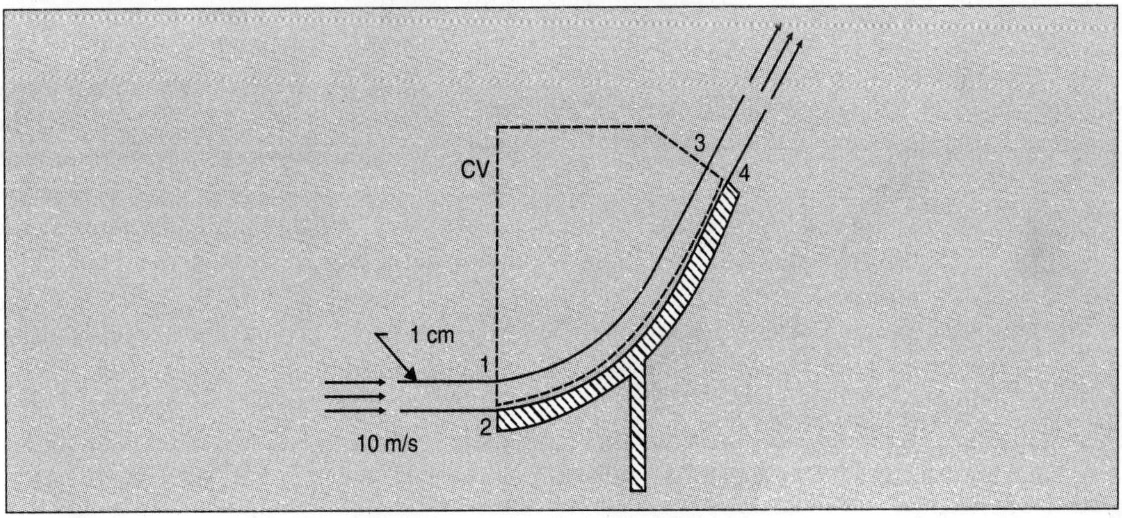

Fig. 5.3. An alternate *CV* for Example 5.1.

CHAPTER 5

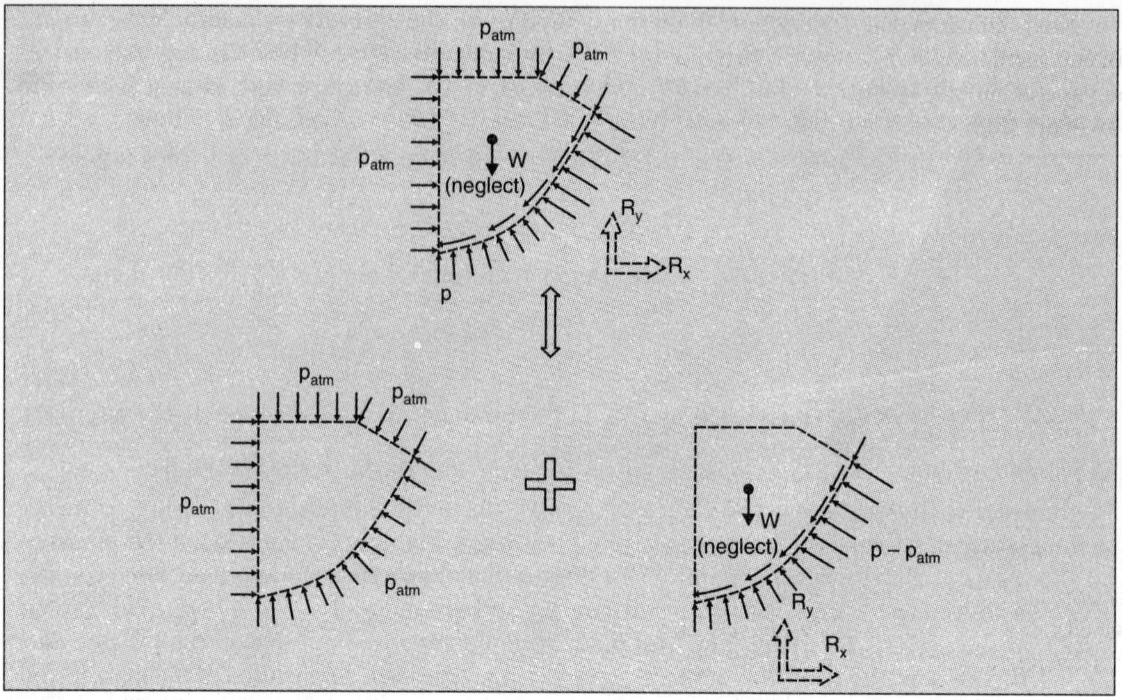

Fig. 5.4. Forces on the *CV* shown in Fig. 5.3 and its resolution into two components.

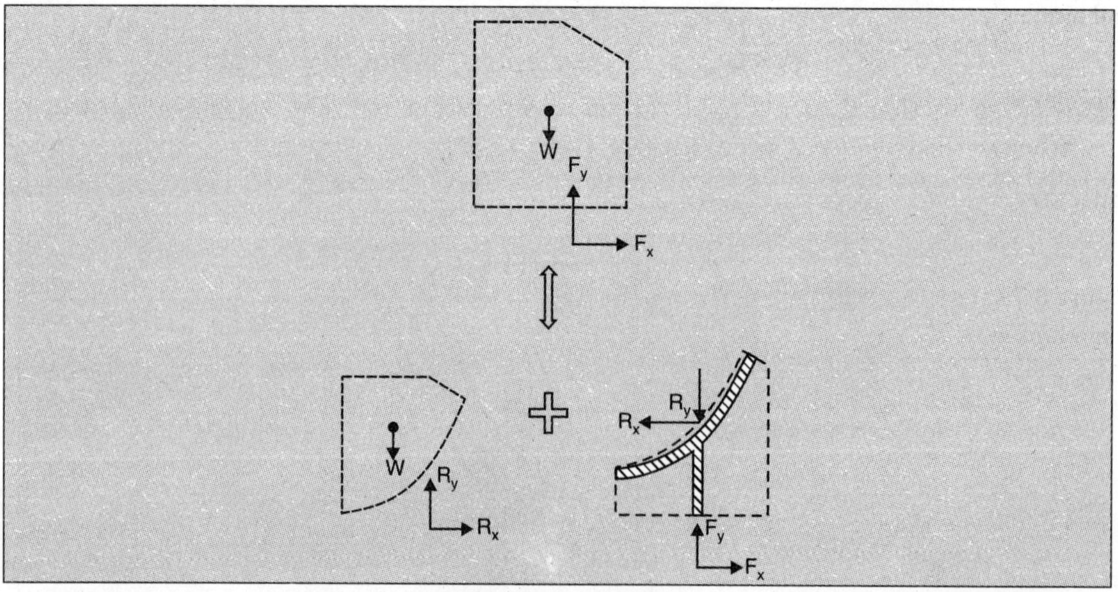

Fig. 5.5. Forces on the *CV* of Fig. 5.2 as related to forces on the *CV* of Fig. 5.3. p_{atm} all around.

Thus, the choice of the control volume does not change the result but may lead to considerable complexities. Also since a constant pressure acting all around a closed surface gives zero resultant force, in most cases, it is convenient to subtract the atmospheric pressure from all pressure forces and work in terms of gauge pressures.

Next, consider the case when the vane is moving to the right at a constant velocity V_0. Fix the frame of reference with the moving vane and choose a CV as shown in Fig. 5.6. Since the vane is moving at a constant velocity, the reference frame is inertial and Eq. (5.2) still applies. However, the velocities are now measured in this moving frame of reference. The velocity at

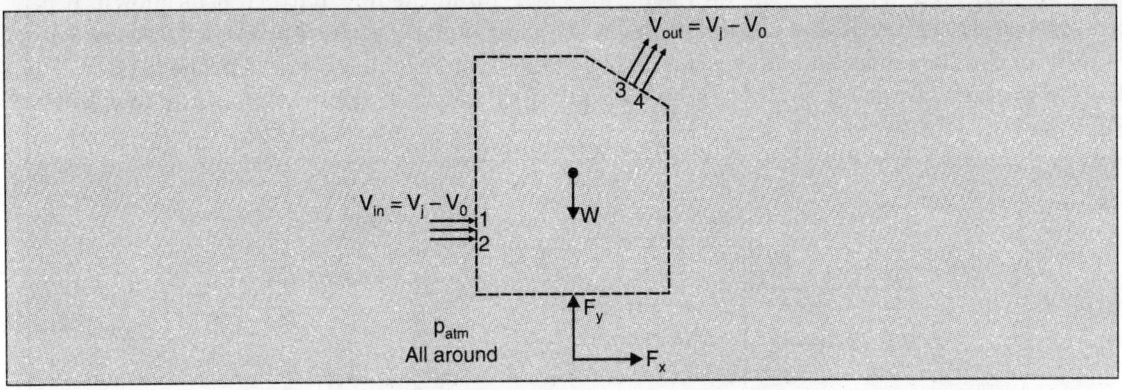

Fig. 5.6. Forces and relative velocities on a *CV* enclosing the vane moving to right at V_0.

section 1-2 is $(V_j - V_0)$, where V_j is the jet velocity. When the fluid flows over the vane its relative velocity does not change (see Chapter 7), and thus, the (relative) velocity at 3-4 is $(V_j - V_0)$ inclined at an angle of 60° to the horizontal. Since the flow is steady within this CV, Eq. (5.3) gives

$$F_x = -(V_j - V_0)\,\rho\,(V_j - V_0)\,A_j + (V_j - V_0)\,(\cos 60°)\,\rho\,(V_j - V_0)\,A_j$$
$$= \rho\,(V_j - V_0)^2\,A_j\,(\cos 60° - 1)$$

and similarly,

$$F_y = \rho\,(V_j - V_0)^2\,A_j\,\sin 60°$$

These are the components of the force on the CV. The force on the vane is $(-F_x\hat{\mathbf{i}} - F_y\hat{\mathbf{j}})$.

When the vane moves to the right at velocity V_0 it does work at the rate of $(-F_x)\,V_0$ and, thus, the power $\mathring{W}$ developed by the water jet acting on the moving vane is

$$\mathring{W} = \rho\left(V_j - V_0\right)^2 A_j\,(1 - \cos 60°)\,V_0$$

Figure 5.7 shows the variation of $\mathring{W}$ with V_0. The value of $\mathring{W}$ is zero at $V_0 = 0$ and $V_0 = V_j$, with a maximum at $V_0 = V_j/3$.

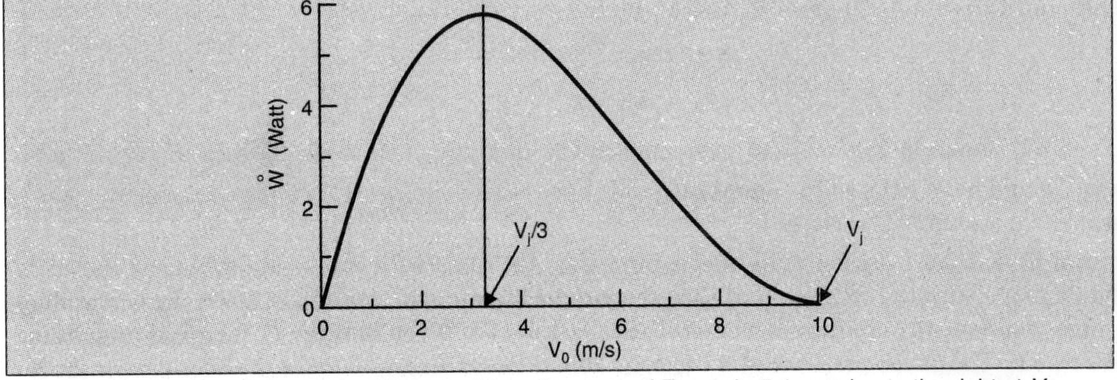

Fig. 5.7. Power developed by the water jet on the vane of Example 5.1, moving to the right at V_0.

The principle of a moving vane is used in a Pelton-wheel turbine (see Sec. 8.4) to extract power in hydroelectric installations.

Example 5.2. Figure 5.8(a) shows a sudden expansion in a pipeline. Water flows from left to right. As indicated in Sec. 1.8 the flow separates at 1-1 and the flow pattern is as shown. It has been observed experimentally that the pressure is almost uniform across any section and decreases rapidly in the flow direction till section 2-2, beyond which the decrease is much slower. This drop in pressure from 1-1 to 2-2 can be estimated by a simple application of the momentum Eq. (5.2).

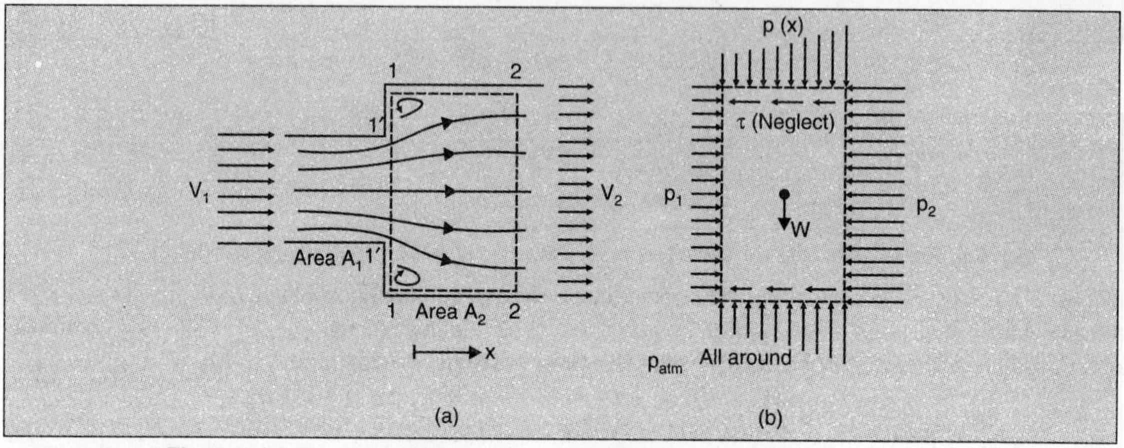

Fig. 5.8. (a) Flow in an expansion, and (b) External forces acting on the *CV*.

For this purpose the region 1-2-2-1 should be taken as the *CV*. The external forces acting on this *CV* are shown in Fig. 5.8(b). The shear stresses are relatively small in this region and are, therefore, neglected in the following analysis. Water flows across segments 1'-1' and 2-2 of the *CS*. Since the flow is steady, Eq. (5.3) in the *x*-direction gives

$$p_1 A_2 - p_2 A_2 = -\mathring{m}\, V_1 + \mathring{m}\, V_2$$

or

$$p_1 - p_2 = \frac{\mathring{m}\,(V_2 - V_1)}{A_2}$$

The continuity Eq. (4.2) gives $V_2 A_2 = V_1 A_1$, so

$$p_1 - p_2 = \frac{\mathring{m}\, V_1}{A_2}\left(\frac{A_1}{A_2} - 1\right)$$

If the *CV* shown in Fig. 5.9 had been chosen the unknown reactions F_1' and F_2' at the pipe walls would have entered the equations and thwarted a solution. This shows once again that a proper choice of *CV* is essential.

Example 5.3. An *L*-shaped tube of constant area *A* is filled with water and is held as shown in Fig. 5.10(a). With the stopper at 4, the pressure increases hydrostatically down the vertical leg and is constant through the horizontal leg. At time $t = 0$ the stopper is removed and water drains out. The gauge pressure at 4 becomes zero. Find an expression for the acceleration of the fluid in terms of the level *h* in the vertical leg.

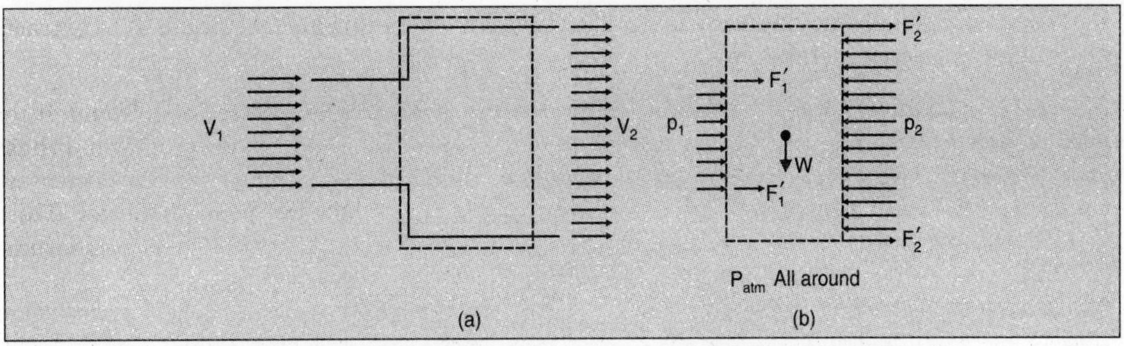

Fig. 5.9. (a) An alternate *CV* for Example 5.2, and (b) *x*-forces acting on it.

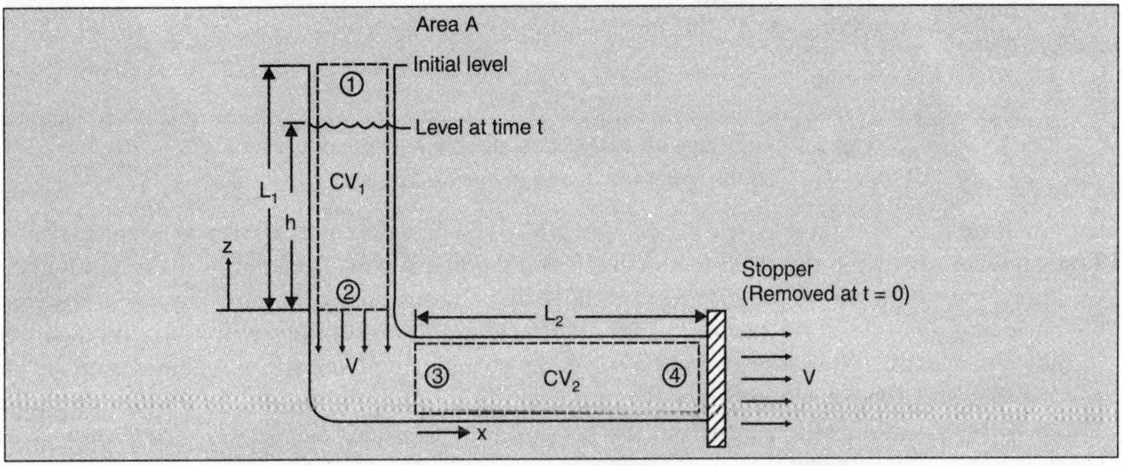

Fig. 5.10. (a) CV_s for the *L*-shaped tube of Example 5.3.

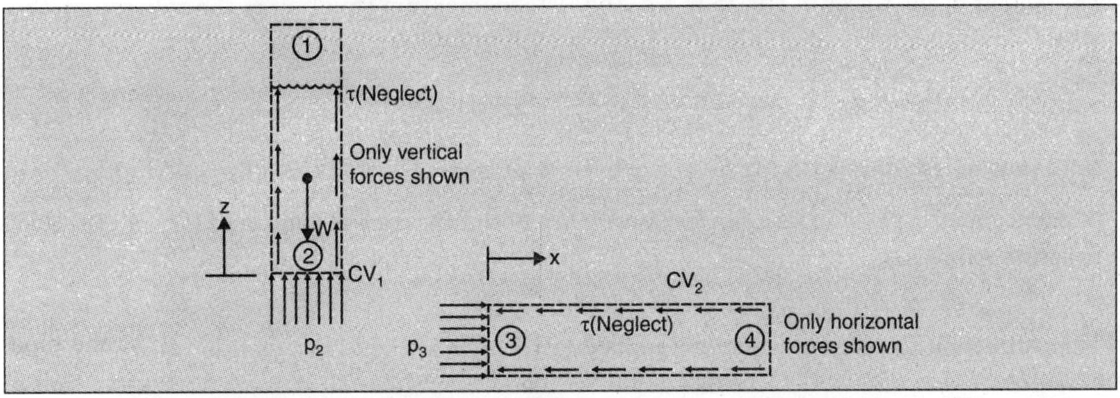

Fig. 5.10. (b) Vertical external forces on CV_1 and horizontal external forces on CV_2.

If we neglect shear stresses the external forces acting on the fluid are the gravity and the (unknown) gauge pressure distribution at the tube walls. If a *CV* encompassing the entire fluid is chosen, this pressure distribution complicates the problem. However, throughout the vertical

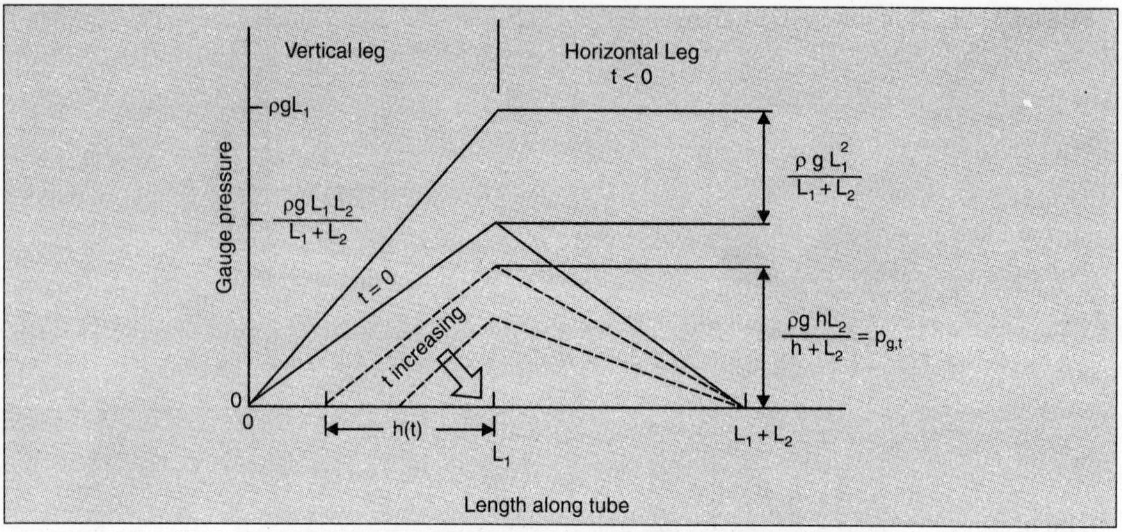

Fig. 5.10. (c) Pressure distribution along the length of the tube as the water drains out of the vertical tube.

leg, the fluid acceleration is in the z-direction while the pressure at the walls acts horizontally. Thus, if a CV enclosing the fluid in the vertical leg alone is considered only the z-momentum equation is used, in which the unknown pressure distribution does not occur. Similarly, a second CV enclosing the fluid in the horizontal leg avoids the use of the unknown pressure distribution in that region. Thus, the problem can be solved by considering two control volumes, one for the vertical leg and the other for the horizontal one. The *relevant* external forces are shown in Fig. 5.10(b). The following approximations are made: (i) shear stresses are negligible, and (ii) p_2 is approximately equal to p_3. The flow is unsteady. At time t, the water level in CV_1 is at h and the velocity everywhere is V. The z-momentum contained in CV_1 is thus, $(-V) \times (\rho hA)$. The weight of the water in CV_1 is $W = \rho\, Ahg$. The momentum equation in the z-direction, thus, gives,

$$Ap_2 - \rho Ahg = \frac{\partial}{\partial t}(-V\rho hA) - V(\rho VA)$$

Note that mass crosses CS_1 at section 2 but not at section 1, and the only contribution to the momentum efflux is $V(\rho VA)$. Since V and h are both functions of time, and $V = -\mathring{h}$, the above equation reduces to

$$p_2 = \rho gh + \rho h \ddot{h} \tag{a}$$

The x-direction momentum equation applied to CV_2 gives

$$Ap_3 = \frac{\partial}{\partial t}(V\rho L_2 A)$$

since the momentum influx and efflux are same. Thus,

$$p_3 = -\rho L_2 \ddot{h} \tag{b}$$

Since $p_3 \simeq p_2$, Eqs. (a) and (b) give

$$\rho gh + ph\ddot{h} = -\rho L_2 \ddot{h}$$

or
$$\ddot{h} = \frac{-gh}{(h+L_2)}$$

This can be integrated numerically to give the time taken by the water to drain out of the vertical leg.

The pressure at the junction is then obtained from either Eq. (a) or (b) as

$$\rho gh - \frac{\rho gh^2}{h+L_2}$$

or
$$\frac{\rho g L_2 h}{h+L_2}$$

The pressure variation along each leg is linear (why?).

Figure 5.10(c) shows the pressure distribution along the tube at $t < 0$ as well as at $t = 0$ when the stopper is removed. At this instant, the pressure difference $\rho g L_1^2/(L_1+L_2)$ accelerates the fluid in the vertical leg and $\rho g\, L_1 L_2/(L_1 + L_2)$ accelerates it in the horizontal leg. At larger times, the pressure distributions are shown by broken lines in Fig. 5.10(c).

Example 5.4. Example 4.2 described a fresh-water scoop of long-distance trains. Find the drag force on the carriage because of the scooping action.

Here the *CV* attached to the train in Fig. 5.11(a) should be considered. The external *horizontal* forces on the *CV* are shown in Fig. 5.11 (b). F_x represents the net force on the *CV* at the attachment to the carriage. Note that the flow is unsteady since more and more water is accumulating in the *CV*, but the rate of accumulation of momentum is still zero. This is because

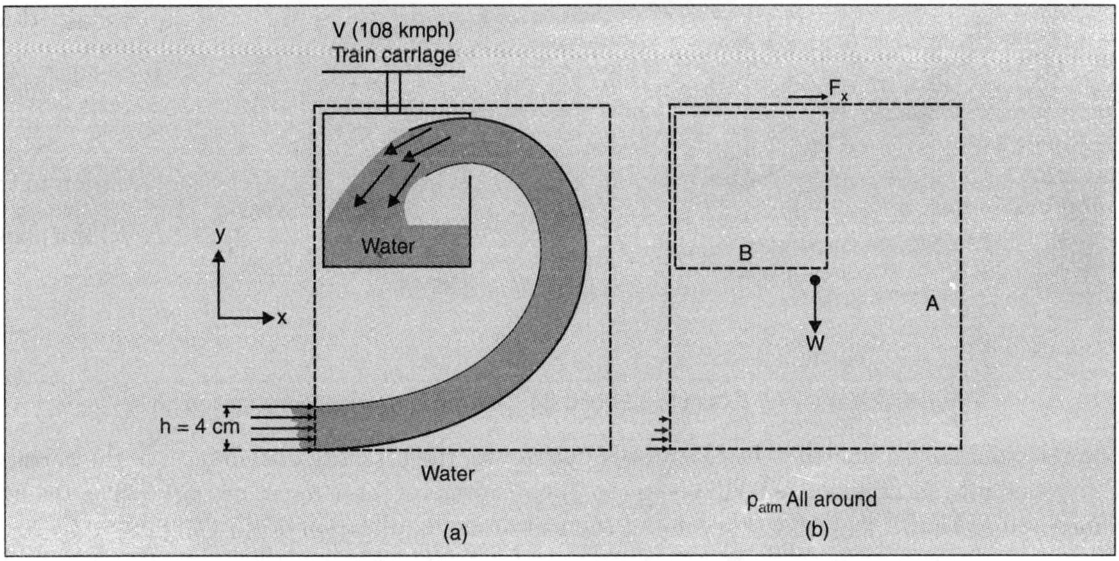

Fig. 5.11. (a) *CV* for Example 5.4, and (b) external horizontal forces acting on *CV*.

over portion A of the CV where the velocity of water (in the frame of reference fixed with the CV) is significant, the flow is steady, while over region B, where the mass of water is increasing with time, the water is at rest. Eq. (5.2) then gives

$$F_x + \frac{\rho g h}{2} \cdot A = \left[V(-\rho V A)\right]_{\text{inlet}} = -\rho V^2 A$$

$$F_x = -\frac{1}{2} \times 10^3 \left(\frac{\text{kg}}{\text{m}^3}\right) \times 9.81 \left(\frac{\text{m}}{\text{s}^2}\right) \times (0.04)\ (\text{m}) \times (0.04 \times 0.8)\ (\text{m}^2)$$

$$-10^3 \left(\frac{\text{kg}}{\text{m}^3}\right) \times (30)^2 \left(\frac{\text{m}}{\text{s}}\right)^2 (0.04 \times 0.8)\ (\text{m}^2)$$

$$= -6.28\ \text{N} - 2.88 \times 10^4\ \text{N}$$

$$= -2.88 \times 10^4\ \text{N}$$

The drag *on the train* is thus 28.8 kN in the positive x-direction.

Example 5.5. Consider the channel described in Example 4.7. The continuity equation was obtained as

$$\frac{\partial h}{\partial t} + \frac{\partial (hV)}{\partial x} = \frac{\mathring{q}}{b} \tag{a}$$

As mentioned earlier, the momentum equation gives another relation between V and h. Obtain this relationship and solve for h as a function of x when the flow is steady, i.e., $\mathring{q}$ = constant.

Take a CV of thickness δx at x as shown in Fig. 5.12(a). The *horizontal* external forces are shown in Fig. 5.12(b). The shear stresses acting on the CV are neglected. If the flow velocity

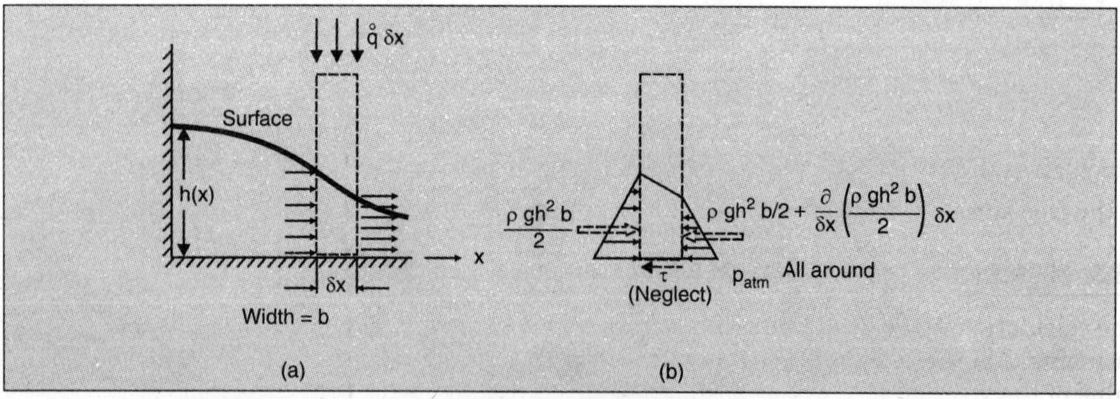

Fig. 5.12. (a) *CV* for Example 5.5, and (b) External horizontal forces acting on it.

in the channel is assumed to be essentially horizontal, the pressure distribution in the vertical direction may be assumed to be hydrostatic. The resultant of these forces are $\rho g h^2 b/2$ on the left face and $\rho g h^2 b/2 + \partial(\rho g h^2 b/2)/\partial x \times \delta x$ on the right face. Application of Eq. (5.3) gives

$$\frac{\rho g h^2 b}{2} - \left[\frac{\rho g h^2 b}{2} + \frac{\partial}{\partial x}\left(\frac{\rho g h^2 b}{2}\right)\delta x\right] = \frac{\partial}{\partial t}[V\rho h b\,\delta x] - V(\rho h V b)$$

$$+\left[\rho h b V^2 + \frac{\partial}{\partial x}\left(\rho h b V^2\right)\delta x\right] \quad \text{(b)}$$

since there is no horizontal momentum flux associated with the fluid entering the top surface. On simplification Eq. (b) gives

$$\frac{\partial (Vh)}{\partial t} + \frac{\partial \left(hV^2\right)}{\partial x} = -gh\frac{\partial h}{\partial x} \quad \text{(c)}$$

Eqs. (a) and (c) are two equations for the two unknowns, V and h.

When $\mathring{q}$ is a constant and the flow is steady, Eq. (a) integrates to (see Example 4.7)

$$Vh = \mathring{q}\, x/b \quad \text{(d)}$$

and Eqs. (c) and (d) give

$$\frac{\partial \left(hV^2\right)}{\partial x} = -g\frac{\partial}{\partial x}\left(h^2/2\right)$$

or

$$hV^2 = -\frac{gh^2}{2} + C_1 \quad \text{(e)}$$

At $x = 0$, $V = 0$ and $h = h_0$ and, therefore, C_1 is $gh_0^2/2$. Eqs. (d) and (e) lead to

$$h^3 - h_0^2\ h + \left(\frac{2\,\mathring{q}^2}{gb^2}\right)x^2 = 0$$

which gives the free surface profile. It can be confirmed that initially (till $h = h_0/\sqrt{3}$) the slope of the free surface is negative, i.e., the level decreases with x!

5.3 MOMENTUM CORRECTION FACTOR

The calculation of the momentum flux across a tube section in Example 5.2 was simplified by assuming that the velocity is constant across the section, i.e., the flow is one-dimensional. The value of the velocity used in the calculations is the average velocity at the given section. The correct momentum flux is *larger* than the value so obtained. When more accurate results are required a correction for this difference is usually done through the use of a *momentum correction factor*. Obtained below is an expression for the momentum correction factor for incompressible flows where the velocity is normal everywhere to the cross-sectional area. This condition holds when the flow is uni-directional.

The actual momentum flux across a cross-section in such a flow is given by $\iint_A V(\rho V\, dA)$, while the one based on the average velocity is $\rho V_{av}^2 A$, where $V_{av} = \frac{1}{A}\iint_A V\, dA$. The momentum correction factor β is defined as the ratio of the actual momentum flux to the one obtained with the 1-*D* approximation. Thus,

$$\beta = \frac{\iint_A \rho V^2\, dA}{\rho V_{av}^2 A} = \frac{1}{A}\iint_A \left(\frac{V}{V_{av}}\right)^2 dA \qquad \text{...(5.4)}$$

since ρ is constant.

The use of this expression to compute β for fully-developed flow through a circular pipe of radius R is illustrated below. The velocity profile in this case is (see Fig. 1.28)

$$V = V_m\left(1 - \frac{r^2}{R^2}\right), \text{ for laminar flows}$$

and

$$V = V_m\left(1 - \frac{r}{R}\right)^{1/7} \text{ for turbulent flows}$$

where V_m is the maximum velocity (at the centre line).

For laminar flows

$$V_{av} = \frac{1}{A}\iint_A V\, dA = \frac{1}{\pi R^2}\int_0^R V_m\left(1 - \frac{r^2}{R^2}\right) 2\pi r\, dr = V_m/2$$

and

$$\beta = \frac{1}{A}\iint_A \left(\frac{V}{V_{av}}\right)^2 dA = \frac{1}{\pi R^2}\int_0^R 4\left(1 - \frac{r^2}{R^2}\right)^2 2\pi r\, dr = 1.33 \qquad \text{...(5.5)}$$

For turbulent flows

$$V_{av} = \frac{1}{A}\iint_A V\, dA = \frac{1}{\pi R^2}\int_0^R V_m\left(1 - \frac{r}{R}\right)^{1/7} 2\pi r\, dr = 49\, V_m/60$$

and

$$\beta = \frac{1}{A}\iint_A \left(\frac{V}{V_m}\right)^2 dA = \frac{1}{\pi R^2}\int_0^R \left(\frac{60}{49}\right)^2 \left(1 - \frac{r}{R}\right)^{2/7} 2\pi r\, dr = 1.020 \qquad \text{...(5.6)}$$

Thus, one-dimensionality assumption is far more acceptable when the flow through a circular tube is turbulent than when it is laminar. This is because the velocity profiles in turbulent flows are much flatter (Fig. 1.28) and, thus, are closer to the 1-*D* approximation.

Example 5.6. In Example 5.2 the pressure drop across a sudden expansion assuming 1-*D* flow was calculated. If the flow in the narrower pipe is assumed to be turbulent and fully developed and that in the wider pipe, laminar and fully developed, obtain the correct expression.

Application of momentum equation to the *CV* in Fig. 5.8 (b) gives

$$p_1A_2 - p_2A_2 = -\beta_1\left(\mathring{m}\ V_{1,\text{av}}\right) + \beta_2\left(\mathring{m}\ V_{2,\text{ av}}\right)$$

where $V_{1,\text{av}}$, and $V_{2,\text{av}}$ are the average velocities at sections 1 and 2 respectively, and β_1 and β_2 are the corresponding momentum correction factors. Here $\beta_1 = 1.33$ and $\beta_2 = 1.02$ by Eqs. (5.5) and (5.6). Thus,

$$p_1 - p_2 = \frac{1.33\ \mathring{m}\ V_{1,\text{av}}}{A_2}\left[\frac{0.77A_1}{A_2} - 1\right]$$

This is larger in magnitude than that obtained earlier.

5.4 MOMENT-OF-MOMENTUM EQUATION

In many applications such as rotary pumps and turbines, we are interested in torques rather than forces. In these cases it is convenient, at times to use the concept of moment of momentum (or angular momentum). One starts with the application of Newton's law to a single particle of fluid in a flow field to write

$$\delta\mathbf{F} = \frac{D}{Dt}(\delta m\ \mathbf{V})$$

where $\delta\mathbf{F}$ is the external force acting on it, $\mathbf{V}$ is the velocity in an inertial frame of reference and *D*/*Dt* is the material derivative. The torque about a fixed point *O* (Fig. 5.13) is, then,

$$\delta\mathbf{T} = \mathbf{r}\times\delta\mathbf{F} = \mathbf{r}\times\frac{D}{Dt}(\delta m\ \mathbf{V}) \qquad ...(5.7)$$

where **r** is the position vector of the particle with origin at *O*. Application of the chain rule* gives

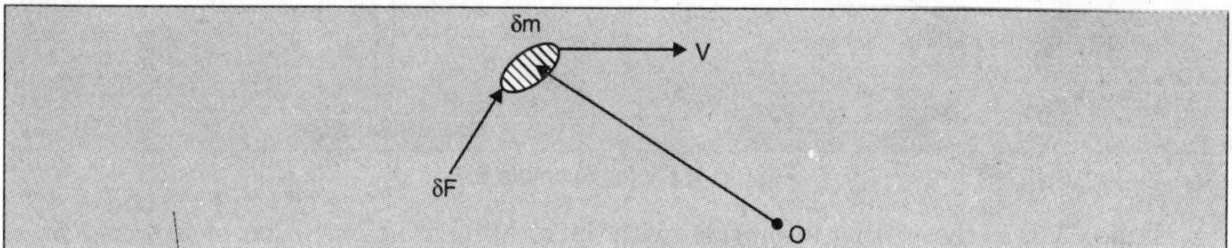

Fig. 5.13. Force δF acting on a fluid particle.

$$\frac{D}{Dt}(\mathbf{r}\times\delta m\ \mathbf{V}) = \mathbf{r}\times\frac{D}{Dt}(\delta m\ \mathbf{V}) + \frac{D\mathbf{r}}{Dt}\times(\delta m\ \mathbf{V})$$

Since $D\mathbf{r}/Dt$ is the velocity **V**, the second term on the right hand side is zero. Therefore, Eq. (5.7) becomes

$$\delta\mathbf{T} = \frac{D}{Dt}(\mathbf{r}\times\delta m\ \mathbf{V}) \qquad ...(5.8)$$

* It can be shown easily that the chain rule of differentiation is applicable to material derivatives as well.

CHAPTER 5

where $\mathbf{r} \times \mathbf{V}\, \delta m$ is the *moment of momentum* $\delta\mathbf{M}$ of the fluid particle.

On integrating Eq. (5.8) over the fluid body one obtains

$$\mathbf{T} = \frac{D\mathbf{M}}{Dt} \quad \text{...(5.9)}$$

where **M** is the moment of momentum of the whole body. The specific value η of the moment of momentum is **r** × **V**.

For the control-volume formulation, Reynolds transport theorem (with η = **r** × **V**) is used to obtain *D***M**/*Dt* in terms of the rate of accumulation and the net efflux. This gives

$$\mathbf{T} = \frac{\partial}{\partial t}\iiint_{CV} (\mathbf{r}\times\mathbf{V})\rho\, d\forall + \oint\!\!\oint_{CS} (\mathbf{r}\times\mathbf{V})(\rho\mathbf{V}\cdot d\mathbf{A}) \quad \text{...(5.10)}$$

Thus, the net external torque acting on a *CV* is equal to the rate of change of the moment of momentum contained within it (i.e., the rate of accumulation) *plus* the net efflux of angular momentum across the *CS*.

Given below is an application of this equation. Another application appears in Sec. 8.3.

Example 5.7. The lawn spinkler (Fig. 5.14) has two jets of water (diameter 5 mm) issuing at 3 m/s at 60° to the tangent. The arms of the sprinkler rotate because of the jet reaction. Find the *steady state* angular velocity ω of rotation if the pivot is assumed frictionless. Assume water enters the spinkler axially through a central pipe.

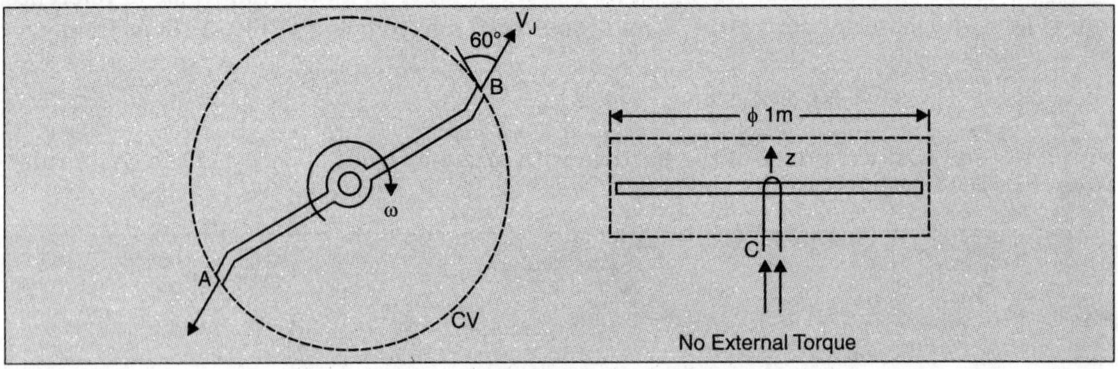

Fig. 5.14. *CV* for Example 5.7.

If the *CV* is so chosen that it coincides with the rotating arms, the frame of reference fixed with the *CV* will be non-inertial and we will not be able to apply the momentum Eq. (5.2) to obtain the forces. The problem is considerably simplified, however, if we take a stationary *CV* enclosing the entire region swept by the arms (Fig. 5.14). As the arms rotate, the water issues in different directions at different times and so the *(linear)* momentum flux changes with time. But the flux of the moment-of-momentum has a constant direction *(along the axis of rotation)* and magnitude. Thus, it is more convenient to work with the *z*-component of the moment-of-momentum Eq. (5.10), rather than with the momentum Eq. (5.2).

Although the flow within the *CV* is unsteady, the total moment of momentum within the *CV* is constant with time. This is because **r** × **V** for any segment of the sprinkler arms is independent of the angular position of the arms. Thus, the rate of accumulation of the moment of momentum is zero and, since the external torque in the z-direction is also zero (due to the pivot being frictionless), we have

$$T_z = 0 = \oint\!\!\oint_{CS} (\mathbf{r} \times \mathbf{V})_z (\rho \mathbf{V} \cdot d\mathbf{A})$$

Note that the velocity here is measured in the frame of reference fixed with the stationary CV.

Water crosses the CS at the two jet orifices A and B and at the centre C. At C, there is no contribution to the moment of momentum flux since $\mathbf{V}$ is normal to $\mathbf{r}$. Thus, the efflux of the moment of momentum at the two jet orifices is zero.

The jet velocity of 3 m/s is with respect to the jet orifice and, since the jet orifice itself is moving, the two velocities should be added vectorially to obtain the velocity with respect to the stationary CS. This is done using the velocity triangle (Fig. 5.15). The tangential component V_{tan} of the absolute velocity, $\mathbf{V}_{abs}$ is

$$V_{tan} = V_j \cos 60° - \omega R$$
$$= (3 \cos 60° - 0.5\,\omega) \text{ m/s}$$

and the radial component V_r is

$$V_r = V_j \sin 60° = 3 \sin 60° \text{ m/s}$$

Only the tangential component of $\mathbf{V}_{abs}$ contributes to the moment of momentum in the z-direction. Thus

$$2 \times \left[\frac{1}{2}(\text{m}) \times (3 \cos 60° - 0.5\,\omega)(\text{m/s}) \times \rho V_j A_j\right] = 0$$

where A_j is the area of the jet. (The reader can verify that $\mathbf{V}_{abs} \cdot \mathbf{A}$ is indeed equal to $V_j A_j$). This gives the steady state angular velocity ω as 3 rad/s.

The torque T_z on the sprinkler arm is zero because the arm has acquired a rotational velocity such that the water coming out of the jets does not have any tangential velocity as seen by a stationary observer. Such an observer, therefore, sees water issuing radially from the sprinkler.

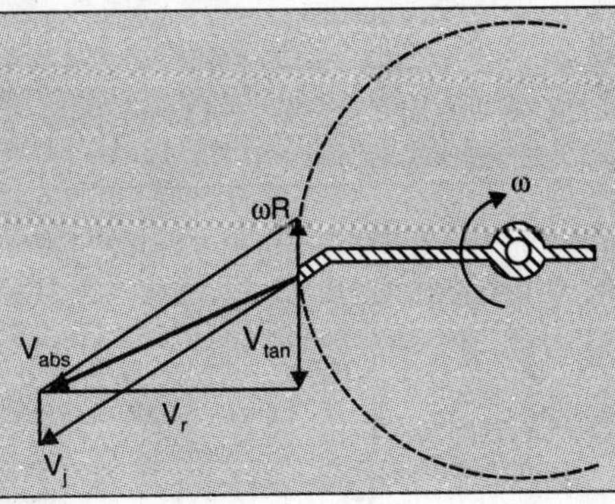

Fig. 5.15. Velocity triangle to obtain the velocity of water with respect to the stationary CV.

PROBLEMS

5.1 Water flows through constant area pipes of the forms shown. Mark the directions of the resultant forces, if any, on the pipes if friction is neglected.

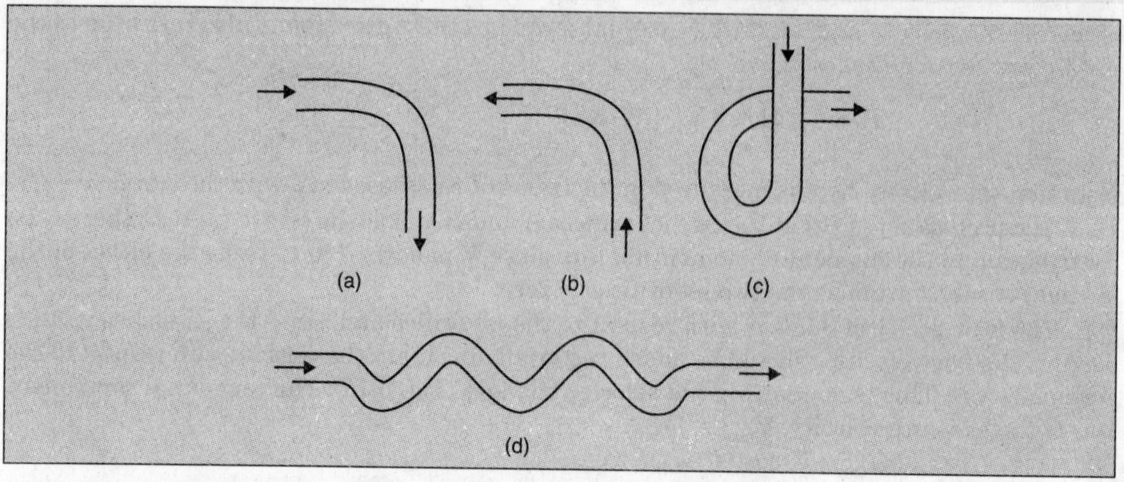

5.2 If the pressure across a straight free jet of water is atmospheric everywhere, why does a person feel a force when he places his hand against such a jet.

5.3 It is argued that a person on a fireboat feels a higher reaction when he directs the water jet against a solid surface on the coast than when he discharges it in the air. Is this correct?

5.4 Obtain the horizontal force acting at the flange *AA* of the nozzle assembly shown if the pressure at point 1 is 10^5 Pa *gauge* and the water issues as a free jet into the atmosphere. Take $\mathring{Q} = 2\ m^3/s$.

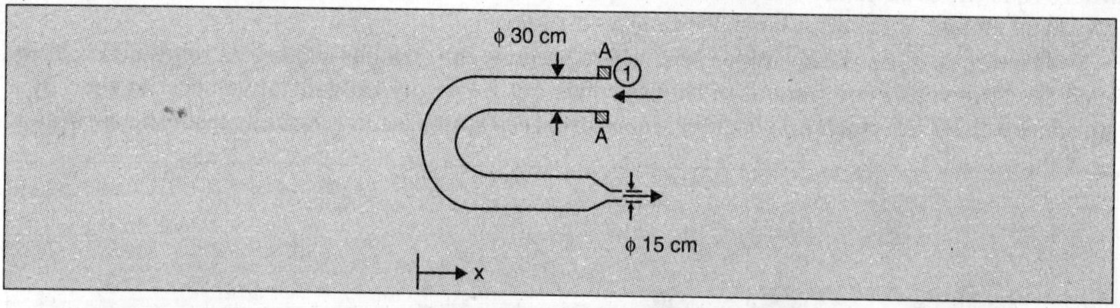

5.5 Is the spring under tension or compression in the system shown?

5.6 A newly graduated engineer dreams up of an ingeneous way of propelling fresh water tankers in the *Thar* desert. A jet of water issuing from a specially designed tanker as shown is deflected back into it by a vane. The thrust by the jet propels the railroad with no loss of precious water. Is his idea feasible?

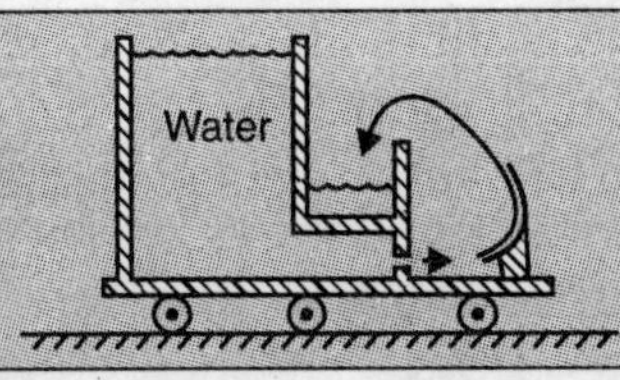

5.7 A person holds a hose through which a liquid of density ρ flows. If the fluid issues as a free jet into the atmosphere, obtain the vertical component of the force experienced by the person. Does this act upwards or downwards?

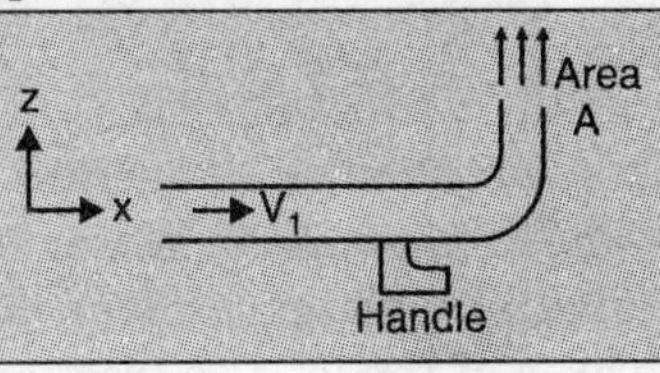

5.8 Water flows through the reducing elbow shown at the rate of 1 m^3/s. The gauge pressure at 1 is 0.1 MPa and that at 2 is 0.09 MPa. What is the resultant force on the elbow? Neglect the weight of water.

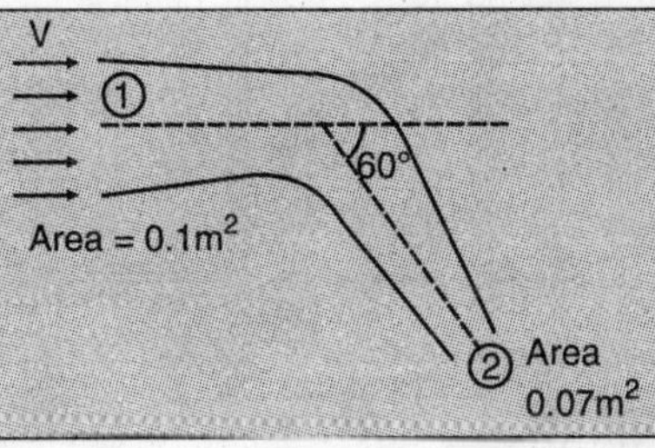

5.9 A jet of water issues out of a fireman's nozzle at 6 m/s. If the gauge pressure at section 1 is 1.8×10^4 Pa, estimate the force in each of the eight bolts connecting the nozzle to the pipe. Note that each bolt is in tension, i.e., the nozzle has a tendency to pull on the hose. Does this mean that the fireman holding the hose is tugged forward? Explain.

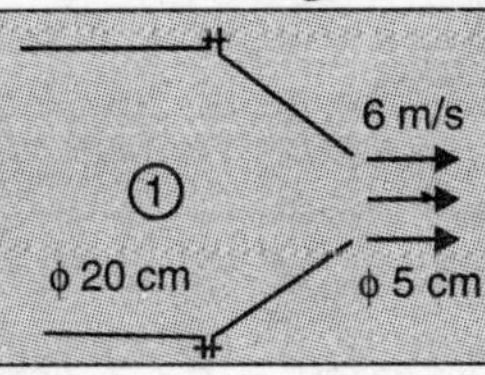

5.10 Assuming that $V_1 = V_2 = V_3$ and that the force exerted by the water on the stationary plate shown acts *normally*, obtain the volume flow rates $\mathring{Q}_2$ and $\mathring{Q}_3$ in terms of the incoming flow rate $\mathring{Q}_1$. Also obtain the normal force F_n.

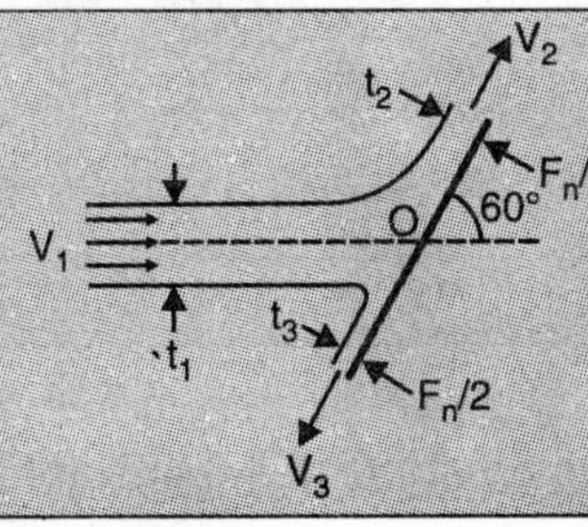

CHAPTER 5

5.11 The sluice gate on a dam is raised to allow the flow of water as shown. Estimate the force acting on the gate per unit width. Assume 1-*D* flow downstream of the gate and the pressure distributions to be hydrostatic far upstream and downstream.

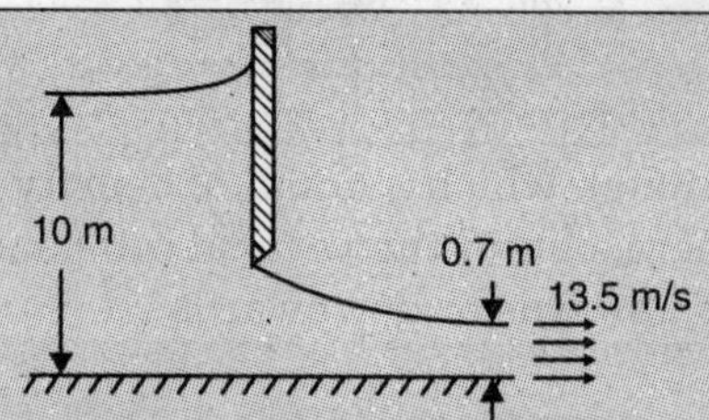

5.12 *Borda's mouthpiece:* A tank has a circular re-entrant outlet near its bottom as shown. Such an outlet is called Borda's mouthpiece. Water issues from this as a jet having a uniform velocity V of 3.13 m/s. The water jet does not fill the tube completely and is surrounded by air. Find the area of the jet as a fraction of the mouthpiece area. Use the fact that with such a mouthpiece, the velocity along all the walls is negligible.

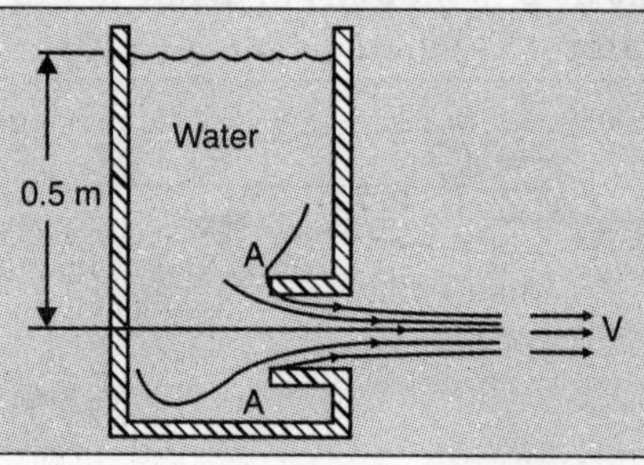

5.13 Consider a tank with a simple sharp-edged orifice. In this case, the velocity at the wall near the orifice is no longer negligible. Show that the 'contraction' of the jet from this orifice is smaller than of that from the Borda's mouthpiece (Prob. 5.12).

5.14 *Hydraulic damper:* A hydraulic damper consists of a piston (area A_p) moving in a slightly larger cylinder (area A_c). Find the force F resisted by the piston when it moves at a constant velocity V_p. Take the gauge pressure at the bottom of the cylinder to be $F/A_p - \rho V_p^2/2$ (Prob. 7.52). Neglect viscous effects. Note that the flow is unsteady.

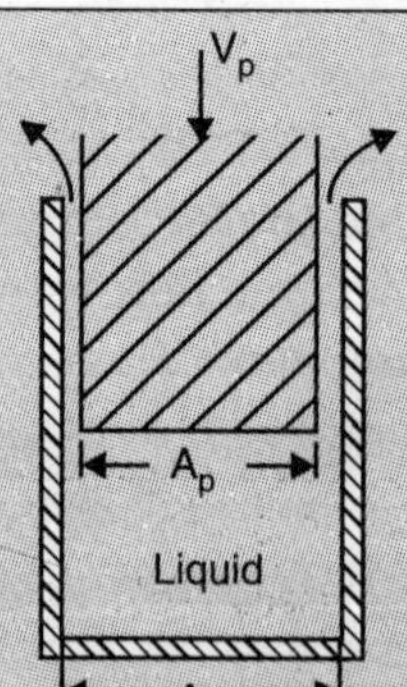

5.15 (a) Obtain the forces F_1 and F_2 required to prevent motion of the tank and the vane showr The water velocity in the jet is constant at 4.5 m/s. (b) Find F_1 and F_2 if the vane is he stationary but the tank moves to the left at a constant speed of 2 m/s. The velocity of water *with respect to the tank* is 4.5 m/s.

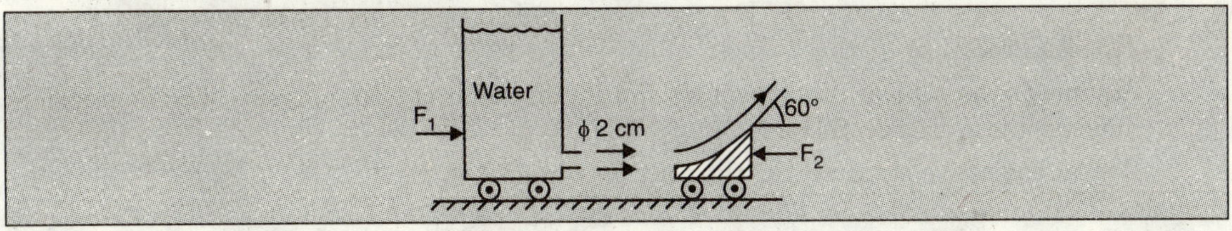

5.16 Consider the tank of Prob. 5.15. The jet of water issues at a constant relative velocity V_j. Show by applying the momentum theorem that the tank will move to the left at a constant velocity *only* if a finite horizontal force is applied to it (otherwise it accelerates). Contrast this behaviour with that of a rotating sprinkler. (Example 5.7).

5.17 An incompressible fluid is supplied to a large tank from where it flows out through a long pipe of length 12 m and diameter 5 cm. At time $t = 0$, a valve is closed slowly so that for a short interval of time thereafter, the velocity is given by $(10 - 5\sqrt{t})$ m/s, where t is in seconds. Find the horizontal force F_x (as a function of time) required to hold the tank in place.

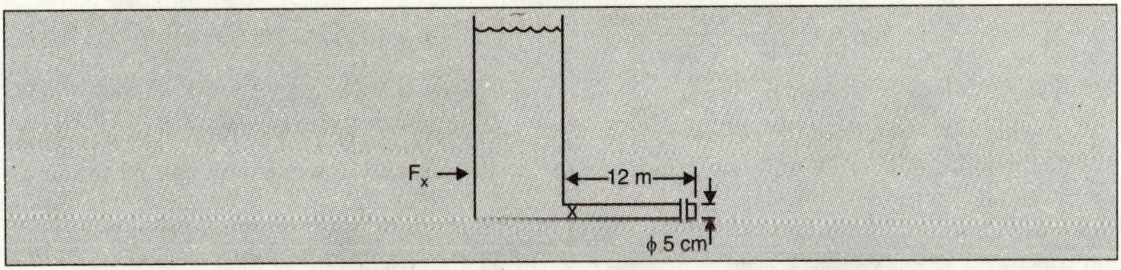

5.18 A partitioned tank on frictionless wheels as shown contains a gas at high pressure in one chamber and a gas at much lower pressure in the other. The plug separating the two is removed at $t = 0$ when the tank is at rest. The tank accelerates to the left for a short time and thereafter moves at a constant velocity for some time. If we assume that the opening between the two chambers is so small that the pressures in them are essentially constant for a short duration, show that the above behaviour is consistent with the momentum equation.

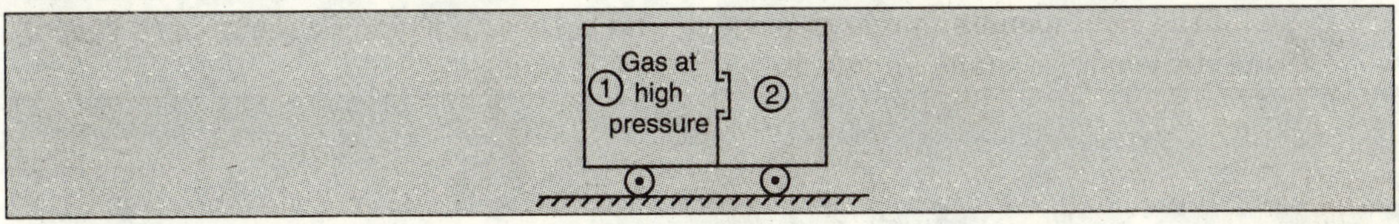

5.19 *Boundary layer:* Fluid flowing at a uniform velocity V_0 at constant pressure encounters a flat plate as shown and a boundary layer results (see Sec. 1.7). The x-component of the fluid velocity at any section BC is given by $V_x = V_0 f$, where f is a function of y/δ and is less than *one* between points C and B. At point B and above, $f = 1$. Since δ is small, the pressure inside the boundary layer may be assumed to be constant everywhere. (This will be shown in Sec. 13.2). Using the control volume DBC show that the drag force on

CHAPTER 5

the plate per unit width is $-\int_0^\delta \rho(V_x - V_0)V_x dy$. Note that mass crosses the *CS* across *BD* (see Prob. 4.9).

Assume *f* to be a linear function of *y*/δ and obtain the drag force in terms of an appropriate dimensionless drag coefficient.

Repeat the above steps taking the control volume *ABCD*.

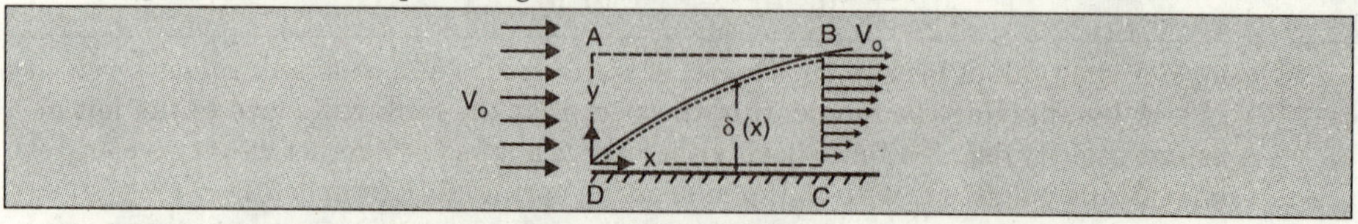

5.20 *Wake-survey method:* An experimental method of measuring the force exerted on a solid body consists of placing it in a uniform stream of fluid and studying the velocity pattern downstream. Using the *CV* shown, obtain the following dimensionless equation for the drag force on a cylinder of diameter *D:*

$$C_D = \frac{x\text{-force on cylinder}}{\frac{1}{2}\rho V_0^2\,(WD)} = 4\int_0^K V_x^*\left(1 - V_x^*\right)dy^*$$

where W = width of the cylinder, $V_x^* = V_x/V_0$ and $y^* = y/D$. (Hint: The fluid bleeds through the sides of the *CV* and the pressures far upstream and downstream can be taken as equal.

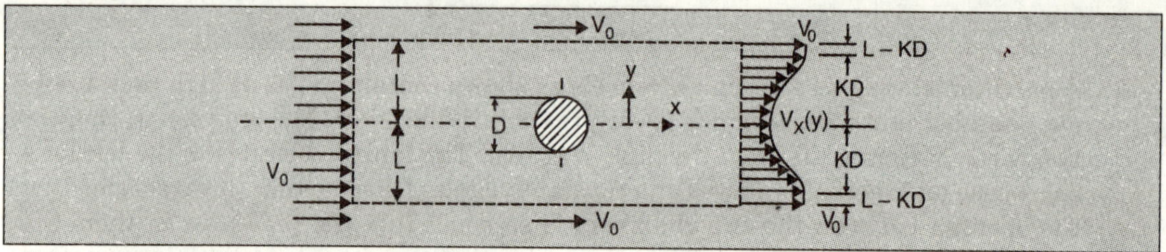

5.21 A series of identical turning-vanes are used in the wind-tunnel bends to keep the flow smooth and 1-*D*. If the velocities at points just upstream and downstream are as shown, and if the corresponding pressures are p_1 and p_2, obtain the force required to keep a vane stationary. Assume an infinite array of 2-*D* vanes.

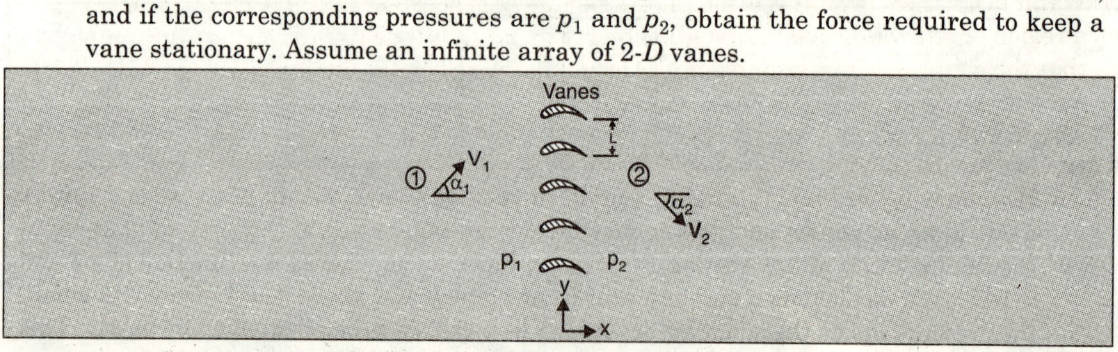

5.22 A jet of fluid at velocity V_1 is directed towards the vane shown (which is moving at velocity V_0) such that the vane sees the fluid entering tangentially to it. The fluid leaves with the same *relative* velocity at the exit of the vane, again tangentially (as observed by the moving vane). Obtain

(a) V_1 in terms of V_0, β_1 and α_1,

(b) the magnitude and direction of the exit velocity (you may not be able to write this explicitly), and

(c) the x-force exerted by the fluid on the vane.

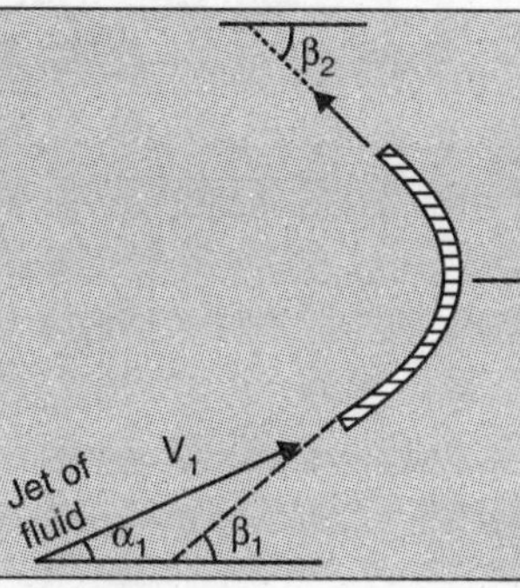

5.23 A child playing with a water pistol directs the jet of water on the circular base of an inverted glass vase floating with its open end down in a bucket as shown. The vase weighs 200 gm and has a cross-sectional area of 50 cm^2. Assume that the velocity V_2 equals V_1. Obtain the vertical force on the vase due to the jet of water. Neglect the weight of water in the CV.

Obtain V_1 required to just submerge the vase as shown. Assume the vase walls to be thin. What is the height to which water rises inside the vase?

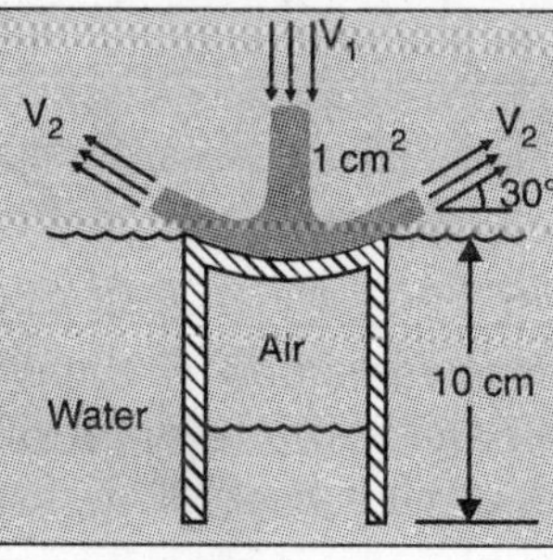

5.24 Water discharges over a weir into a channel having the same width. It is observed that a region of still water backs up to height a at the back of the weir as shown. Assuming that the water discharges horizontally over the weir and that the pressure variation at AB and CD are hydrostatic, obtain a in terms of V, h and h_0. Neglect frictional stresses. Also take the pressure across PQ as atmospheric (see Sec. 7.6).

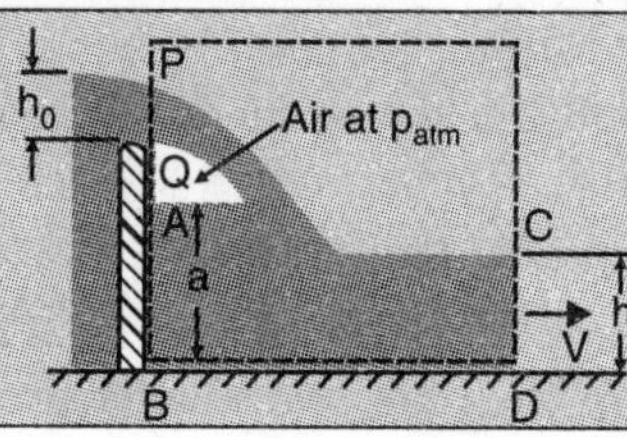

CHAPTER 5

5.25 *Poiseuille flow:* Consider a cylindrical *CV* in a circular pipe carrying a fully-developed laminar flow as shown. A uniform pressure p_1 acts on face *AB* and similarly a pressure p_2 acts on face *CD*. A shear stress τ acts on the curved surface. Apply the momentum equation to obtain $-\dfrac{p_1 - p_2}{2}\dfrac{r}{L} = \tau$

Using the relationship $\tau = \mu dV_z/dr$ between the shear stress and the velocity gradient, and assuming that $(p_1 - p_2)/L$ = constant, obtain the velocity profile. (Hint: Convective term is zero. Why?)

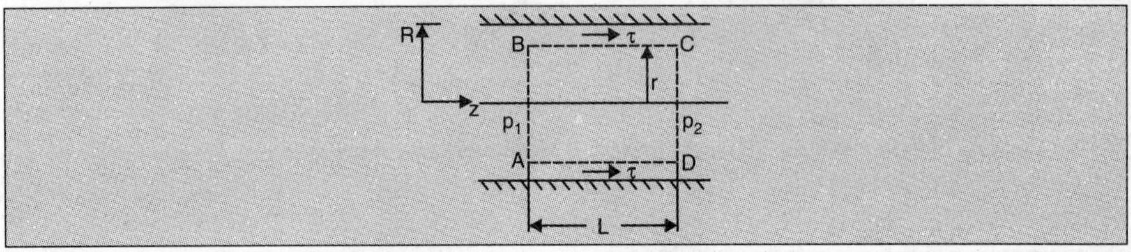

5.26 If the cylindrical *CV* in Prob. 5.25 is replaced by a cylindrical shell as shown, obtain the relationship between p and τ.

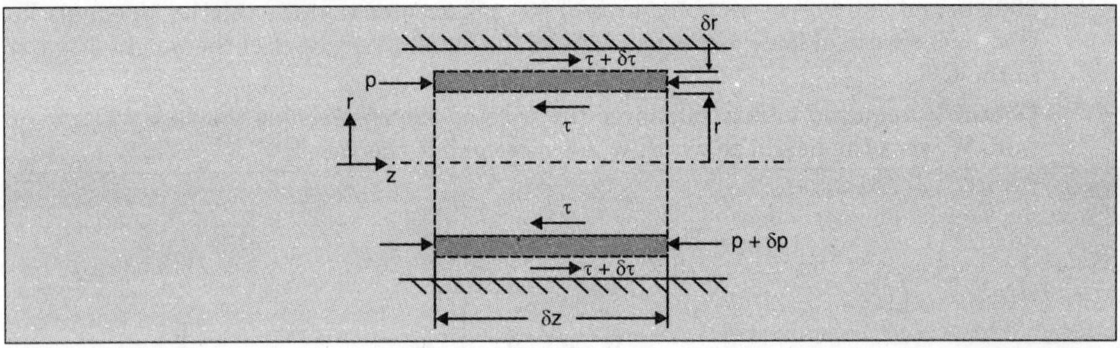

5.27 *Hydraulic jump:* A high speed channel flow at 1 may 'jump' to a low speed condition at 2. The pressure variations at 1 and 2 may be approximated as hydrostatic. Obtain h_2 in terms of h_1, V_1 and g. Neglect wall friction. Discuss the significance of the three mathematical solutions.

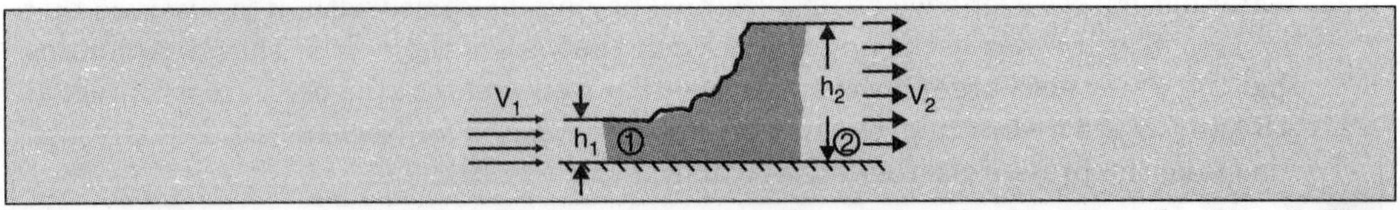

5.28 *Jet contraction:* Consider the liquid jet coming out of a circular pipe of radius *R*. The velocity distribution at section 1-1 is assumed to be parabolic

$$V_z = 2\bar{V}\left(1 - \frac{r^2}{R^2}\right)$$

After the liquid emerges from the pipe, its velocity profile changes and the jet diameter decreases as shown, till at section 2-2 the velocity is uniform across the jet. Assuming that the pressure is approximately atmospheric over the *CV* shown, use the continuity and momentum equations to show that $R_j = \sqrt{3}\, R/2$.

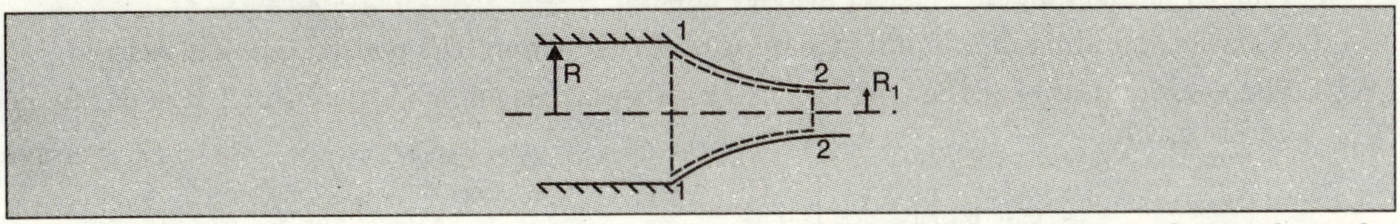

5.29 *Ejector pump:* A high-speed water jet issuing from a pipe of area A_j drags along the surrounding water such that the device shown can be used as a pump. If the velocity profiles are assumed as 1-*D* at sections 1 and 2, relate V_2 to V_j and V_1. If shear stresses at the pipe wall are neglected, and the pressure is assumed uniform across the entire section 1, use the momentum equation to show that

$$p_2 - p_1 = \rho\left(\frac{A_j}{A_p}\right)\left(1-\frac{A_j}{A_p}\right)(V_1 - V_j)^2$$

Note that p_1 is lower than p_2 confirming that the device is indeed a pump.

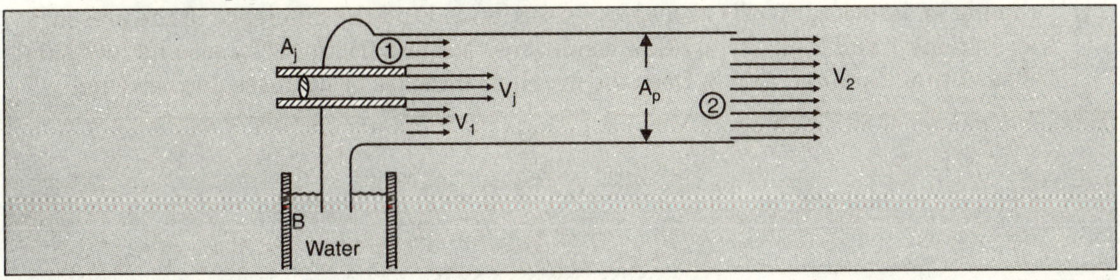

5.30 In Example 5.3, we have drawn the pressure profile at time *t* along the vertical and horizontal legs as linear. Show that this must be so.

5.31 Consider the flow of water below the sluice gate as discussed in Prob. 4.26. Again assuming 1-*D* flow, obtain another relationship between $\delta\,(x, t)$ and $V_x(t)$. Assume hydrostatic pressure variation downstream of the gate and neglect frictional losses.
For steady flow, show that δ cannot vary continuously with *x*. The depth of water δ admits only two values, one δ_0 and the other related to δ_0 by the hydraulic jump relation of Prob. 5.27.

5.32 *Coanda effect:* When a jet of water just touches a curved surface it attaches itself to the

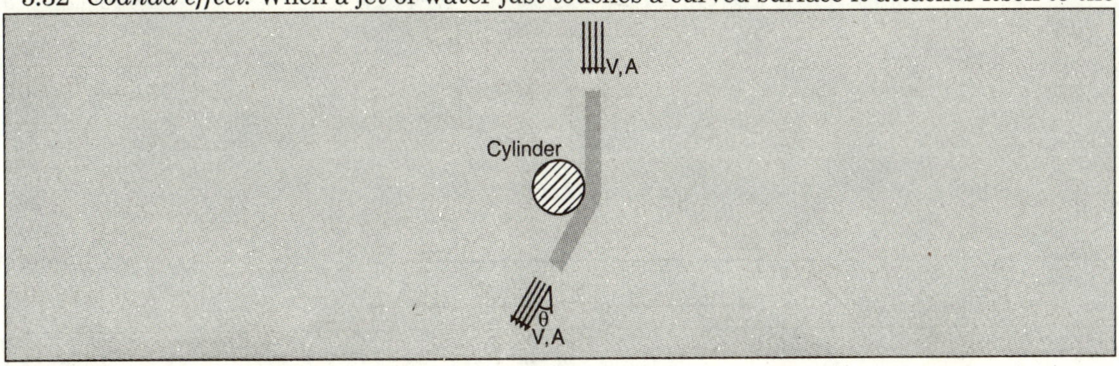

CHAPTER 5

surface and is bent through an angle θ as shown. This is called the Coanda effect. Obtain the magnitude and direction of the force acting on the cylinder assuming that the jet velocity is unchanged. Neglect gravity effects.

A similar bending of a liquid stream is observed when it is poured out of a vessel.

5.33 The velocity profile at the entrance of a pipe is flat, as shown. At section 2, it is parabolic and is given by

$$V = V_m\left(1 - \frac{r^2}{R^2}\right)$$

Obtain the drag force F acting on the fluid in terms of the pressures p_1 and p_2 and ρ, V_0 and R using the momentum correction factor. Verify the results by direct integration.

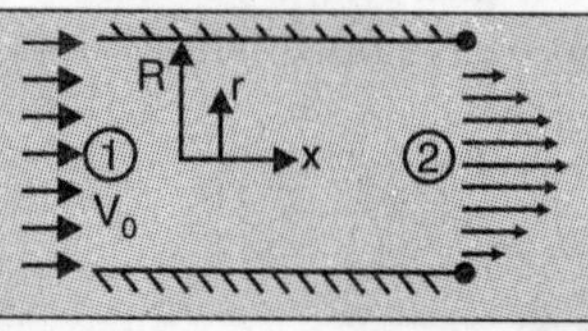

5.34 *Manifolds:* Water flows through a pipe with a hole at the side. One third of the water coming in, issues *vertically* as a spray at section 3. Find the difference in the pressures at sections 1 and 2 for steady flow conditions. Neglect frictional forces and any axial momentum lost at section 3. Does the pressure increase or decrease downstream?

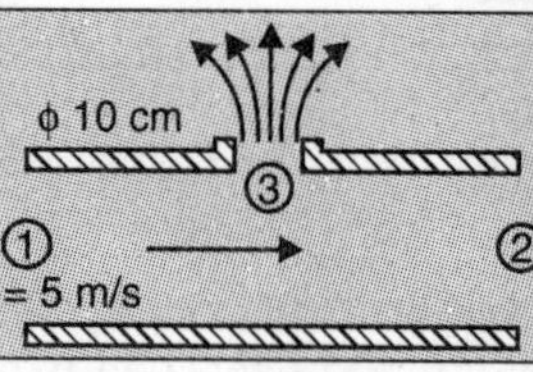

5.35 Compute the torque required to prevent the sprinkler of Example 5.7 from rotating.

5.36 Show that the contribution to $\oiint_{CS} \rho \mathbf{V}_{abs} \cdot d\mathbf{A}$ of any jet in Example 5.7 is $\rho \mathbf{V}_j A_j$.

5.37 A pump takes in water axially near the centre and delivers it at a higher pressure from an exit port at 2.5 m/s. Two bolts on each side, as shown, fasten it securely to the base. Compute the tensile and compressive loadings in the bolts due to the unbalanced torque alone.

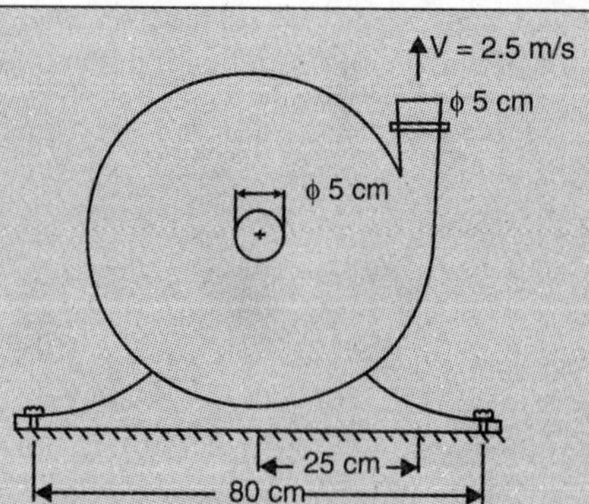

5.38 A cooling system for a central air-conditioning plant uses a 5 cm dia. pipe. Water enters at A and issues vertically at the six 2 cm dia. nozzles as shown. Assuming that the water velocity at each of the nozzles is *approximately* 6 m/s, compute the bending moment at the flange at A due to the flow of water alone.

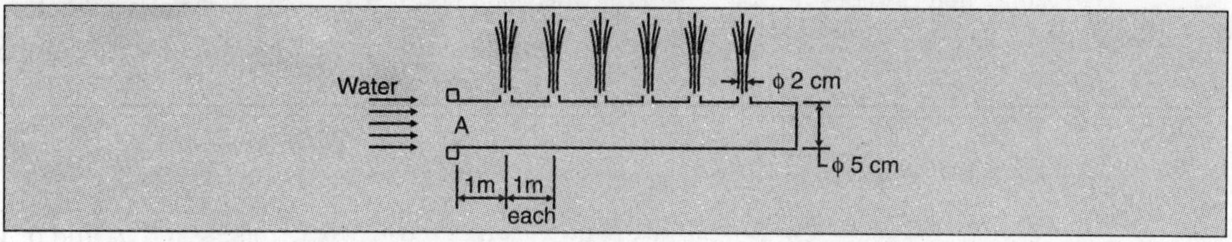

5.39 A toy cracker, "chakri", consists of a thin tube containing a combustible powder. The tube is wound spirally about O as shown. If the combustible material burns at the rate of $\mathring{m}$ and if the density of the combustion products is ρ find the torque required to prevent rotation when the cracker is ignited at A. Also estimate the initial rate of rotation of the wheel. Assume $\mathring{m}$ small so that unsteady, non-inertial effects are negligible.

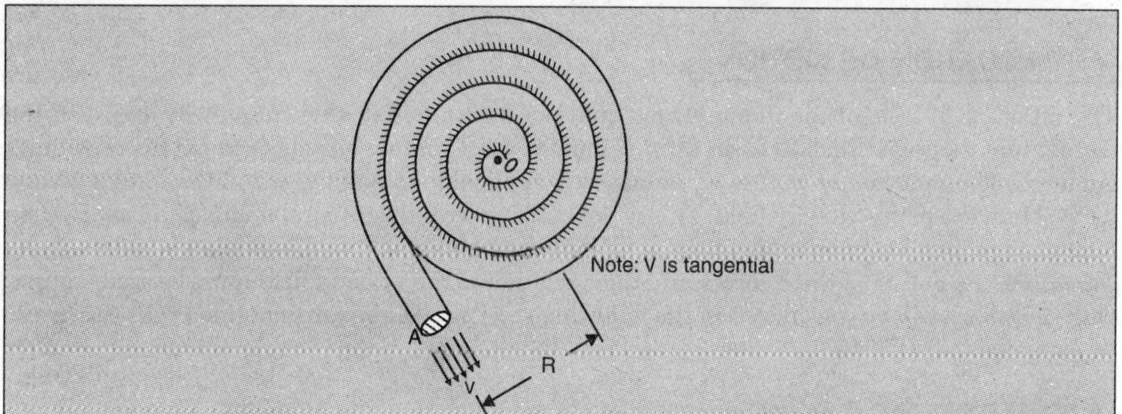

5.40 Find the point of action of the normal force F_n due to the water jet in Prob. 5.10. This is the point at which an external force (equal to F_n) must be applied to prevent rotation of the plate.

6

EQUATION OF MOTION

6.1 EQUATION OF MOTION

The differential form of the momentum equation is obtained in this chapter by applying the momentum theorem (Eq. 5.2) to an infinitesimal control volume as was done for the continuity equation. The *equation of motion* so obtained is applicable to every point in the fluid enabling us to obtain the entire velocity field. This is not possible with the integral approach. The equation of motion requires a knowledge of the forces on the surface of a small element within a fluid. Therefore, the nature of these forces and their relation to the velocity field must be studied first. Only 2-*D* flows will be considered in the following sections. The results obtained thus can easily be extended to 3-*D* flows.

6.2 STRESS AT A POINT

In Sec. 5.1 it was seen that the surface force at a point in a fluid is expressed in terms of a stress defined as the force per unit area. In a stationary fluid the pressure (which is a compressive

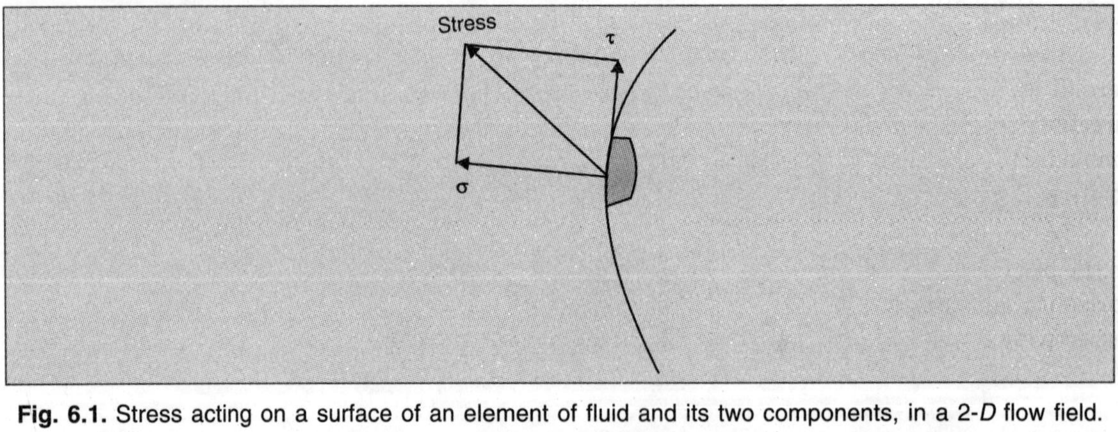

Fig. 6.1. Stress acting on a surface of an element of fluid and its two components, in a 2-*D* flow field.

stress) acts normal to a surface (Chapter 2). In the presence of fluid motion, the direction of stress on a surface does not, in general, coincide with the normal to the surface. We resolve the stress vector into a normal component σ and a tangential component τ (Fig. 6.1). The magnitude and direction of the stress vector may change from point to point on a surface within a fluid. Also, if two surfaces of different orientations passing through a point are taken together, the stress vectors on them will, in general, be different. Thus, the stress vector within a fluid is a function of both the position $\mathbf{x}$ as well as the orientation $\hat{\mathbf{n}}$ of the surface. A similar situation was encountered in the study of mechanics of solids and, as seen there, the stress vector on any surface $\hat{\mathbf{n}}$ through a point $\mathbf{x}$ can be obtained from the stresses on three mutually perpendicular planes through $\mathbf{x}$. The relationship among stresses on different planes passing through a point in a 2-*D* flow is obtained below.

Consider a small triangular element *OAB* in a 2-*D* flow (Fig. 6.2). The surface *OA* (of unit depth) has its normal in the $+y$ direction. The stress on this face has two components, the normal and the shear. A *double-index notation* is used to identify the stress components. The first index denotes the direction of the normal to the plane, and the second denotes the direction of the stress component itself. Thus, σ_{yy} is the normal stress on the face *OA* and τ_{yx} is the shear stress on it (in 3-*D* flow, there will be another shear stress component τ_{yz} acting on face *OA*). Similarly, σ_{xx} and τ_{xy} denote the normal and shear stress components respectively on the face *OB*, whose normal is in the $+x$ direction.

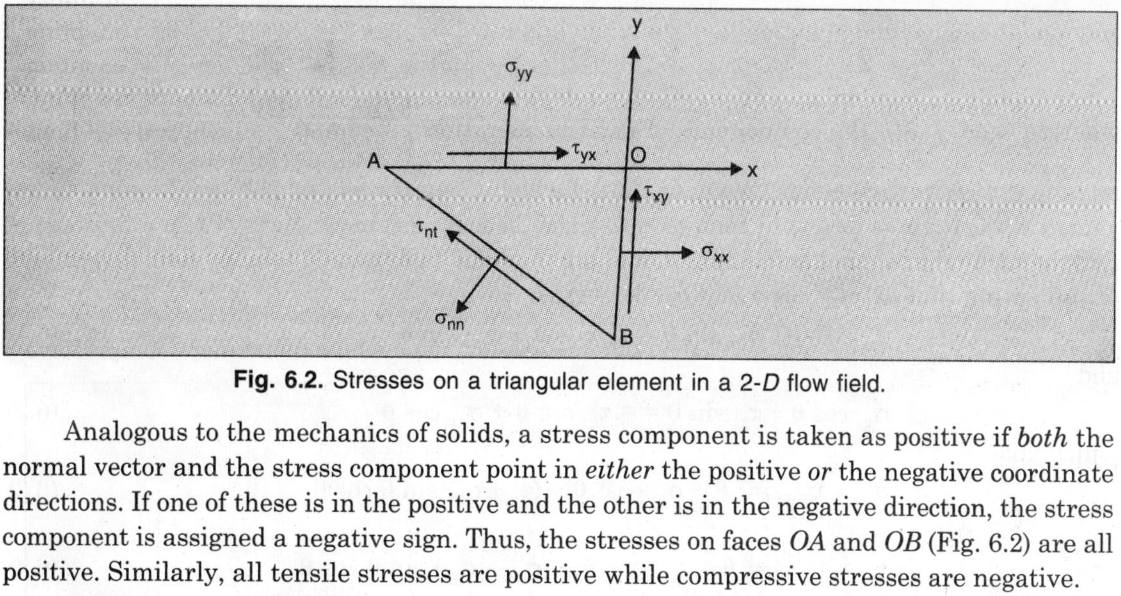

Fig. 6.2. Stresses on a triangular element in a 2-*D* flow field.

Analogous to the mechanics of solids, a stress component is taken as positive if *both* the normal vector and the stress component point in *either* the positive *or* the negative coordinate directions. If one of these is in the positive and the other is in the negative direction, the stress component is assigned a negative sign. Thus, the stresses on faces *OA* and *OB* (Fig. 6.2) are all positive. Similarly, all tensile stresses are positive while compressive stresses are negative.

To obtain the stresses σ_{nn} and τ_{nt} on face *AB*, a force balance is needed. The forces acting on the prismatic fluid element are shown in Fig. 6.3. These include the body forces as well. The horizontal and vertical components of the body force per unit mass are denoted by f_x and f_y respectively.

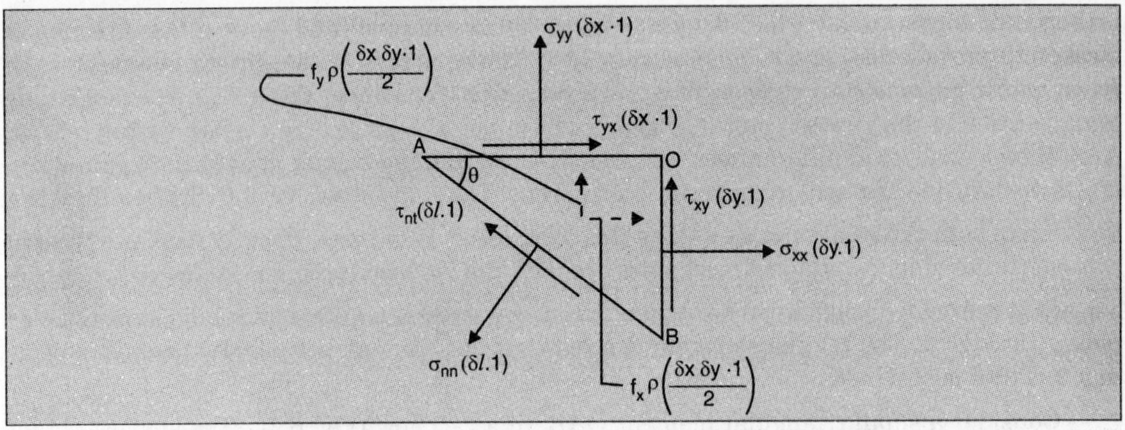

Fig. 6.3. Forces on the element shown in Fig. 6.2. *AB* = *l*.

Application of Newton's second law gives

$$\tau_{yx}\delta x + \sigma_{xx}\delta y - \tau_{nt}\delta l\cos\theta - \sigma_{nn}\delta l\sin\theta$$
$$+ \rho f_x \frac{\delta x\delta y \cdot 1}{2} = \rho\left(\frac{\delta x\delta y \cdot 1}{2}\right)a_x$$

$$\sigma_{yy}\delta x + \tau_{xy}\delta y + \tau_{nt}\ \delta l\sin\theta - \sigma_{nn}\delta l\cos\theta$$
$$+ \rho f_y \frac{\delta x\delta y \cdot 1}{2} = \rho\frac{\delta x\delta y \cdot 1}{2}\ a_y$$

where a_x and a_y are the components of fluid acceleration $\left(\frac{DV_x}{Dt}\text{ and }\frac{DV_y}{Dt}\text{ respectively}\right)$, and include the convective terms. We note that the body force terms and the acceleration terms vanish in the limit as δx and δy tend to zero since these consist of products of two infinitesimal lengths while the others have only one. Taking the limit as δx and δy tend to zero, dividing by δl and noting that $\delta x/\delta l = \cos\theta$ and $\delta y/\delta l = \sin\theta$, we get

$$\tau_{yx} \cos\theta + \sigma_{xx} \sin\theta = \tau_{nt} \cos\theta + \sigma_{nn} \sin\theta \qquad ...(6.1)$$

and

$$\sigma_{yy} \cos\theta + \tau_{xy} \sin\theta = -\tau_{nt} \sin\theta + \sigma_{nn} \cos\theta \qquad ...(6.2)$$

which give

$$\sigma_{nn} = \sigma_{xx} \sin^2\theta + \sigma_{yy} \cos^2\theta + (\tau_{xy}+\tau_{yx}) \sin\theta \cos\theta \qquad ...(6.3)$$

and

$$\tau_{nt} = \tau_{yx} \cos^2\theta - \tau_{xy} \sin^2\theta + (\sigma_{xx} - \sigma_{yy}) \sin\theta \cos\theta \qquad ...(6.4)$$

Thus, given the four stress components σ_{xx}, σ_{yy}, τ_{xy} and τ_{yx} on *two orthogonal planes*, the stress components on any other plane at that point* can be found. Therefore, the *state of stress* at any point in a 2-*D* flow field is determined fully by prescribing four stress components on two

---

* Though plane *AB* does not pass through the point *O* when δx and δy are finite, it does so in the limit as they approach zero.

orthogonal planes. By extension, nine stress components are required in a 3-*D* flow field (three each on three mutually orthogonal planes) to describe the state of stress. These nine components (each a function of position **x**) are conventionally written as

$$\begin{pmatrix} \sigma_{xx} & \tau_{xy} & \tau_{xz} \\ \tau_{yx} & \sigma_{yy} & \tau_{yz} \\ \tau_{zx} & \tau_{zy} & \sigma_{zz} \end{pmatrix}$$

A row represents the stress components on a particular plane and a column represents the stress components in a given direction on the three orthogonal planes (Fig. 6.4).

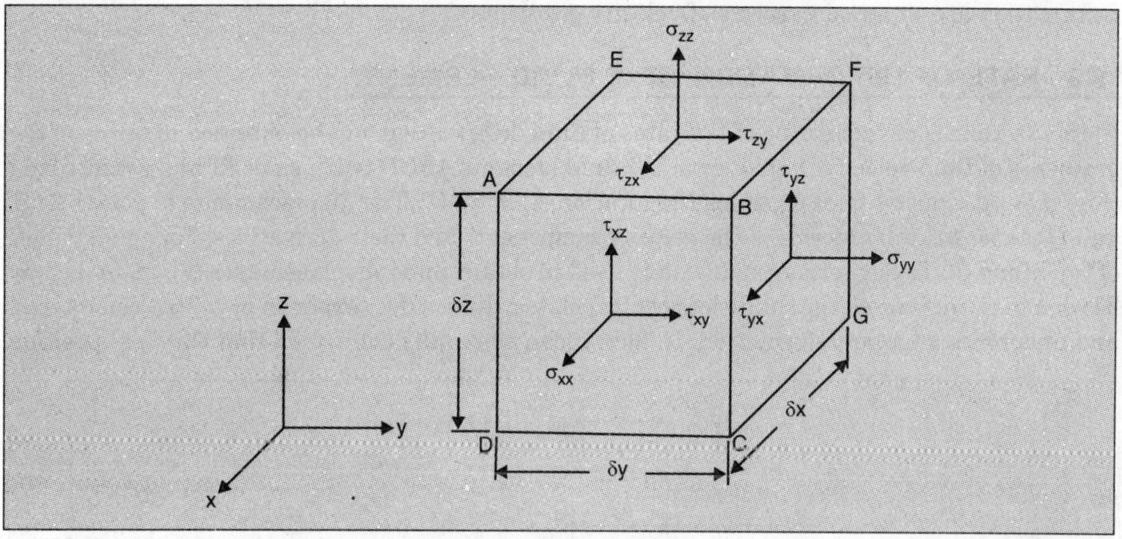

Fig. 6.4. Nine components of stress in a 3-*D* flow field. Stress components shown on faces, *ABCD*, *BFGC* and *EFBA*.

A fluid cannot sustain a shear stress if it is stationary, or is moving as a rigid body so that there is no deformation of the fluid. A fluid with zero viscosity also will not sustain a shear stress. In these cases, then, all τ's are zero and Eq. (6.4) gives $\sigma_{xx} = \sigma_{yy}$. Further, from Eq. (6.3),

$$\sigma_{nn} = \sigma_{xx} = \sigma_{yy} \; (= \sigma, \text{ say}) \qquad \text{...(6.5)}$$

Thus, the normal stress at a point for a flow in which shear stresses are absent must be the same in all directions.

Another property of stress components can be established by considering the rotational equilibrium of the fluid element *OAB* of Fig. 6.3. The moments of all the forces shown about the mid-point of *AB* are equated to the product of the moment of inertia and the angular acceleration of the element. Of the surface forces, only $\tau_{xy} \cdot \delta y$ and $\tau_{yx} \cdot \delta x$ contribute, since the moment arms of all the others are zero. In the limit as δx and δy tend to zero, the contributions of the moment of inertia and the body-force terms vanish, since these contain higher order infinitesimals than do surface force terms. Thus, the angular momentum equation gives

$$\tau_{yx} \cdot (\delta x \cdot 1) \cdot \frac{\delta y}{2} - \tau_{xy} \cdot (\delta y \cdot 1) \cdot \frac{\delta x}{2} = 0$$

or

$$\tau_{yx} = \tau_{xy} \qquad \text{...(6.6)}$$

A general 3-*D* state of stress gives, in addition,

$$\tau_{yz} = \tau_{zy} \text{ and } \tau_{xz} = \tau_{zx} \qquad \text{...(6.7)}$$

Thus, shear stresses occur as pairs of equal magnitudes, one rotating an element clockwise and the other, counterclockwise.

The stresses in a fluid are related not to the net deformation that the fluid elements undergo, but to the *rates* of deformation (Chap. 1). In Sec. 6.3 expressions for the rates of deformation are obtained in terms of velocity gradients.

6.3 RATE OF DEFORMATION OF A FLUID ELEMENT

In this section, some measures of the rates of fluid deformation will be obtained in terms of the gradients of fluid velocity. Consider a 2-*D* fluid element *ABCD* (with sides δx and δy) at time t (Fig. 6.5). At time $t + \delta t$, it occupies the location *A′ B′ C′ D′*. The displacements of points *A, B,* and *D* can be written in terms of the velocity components and their derivatives. For small values of δx, δy and δt, Taylor series expansion is used to obtain these displacements (Fig. 6.5). It may be seen that, in general, the fluid element translates, dilates (i.e., expands or contracts), rotates, and undergoes angular deformation. In fact, it can be readily confirmed that the final position

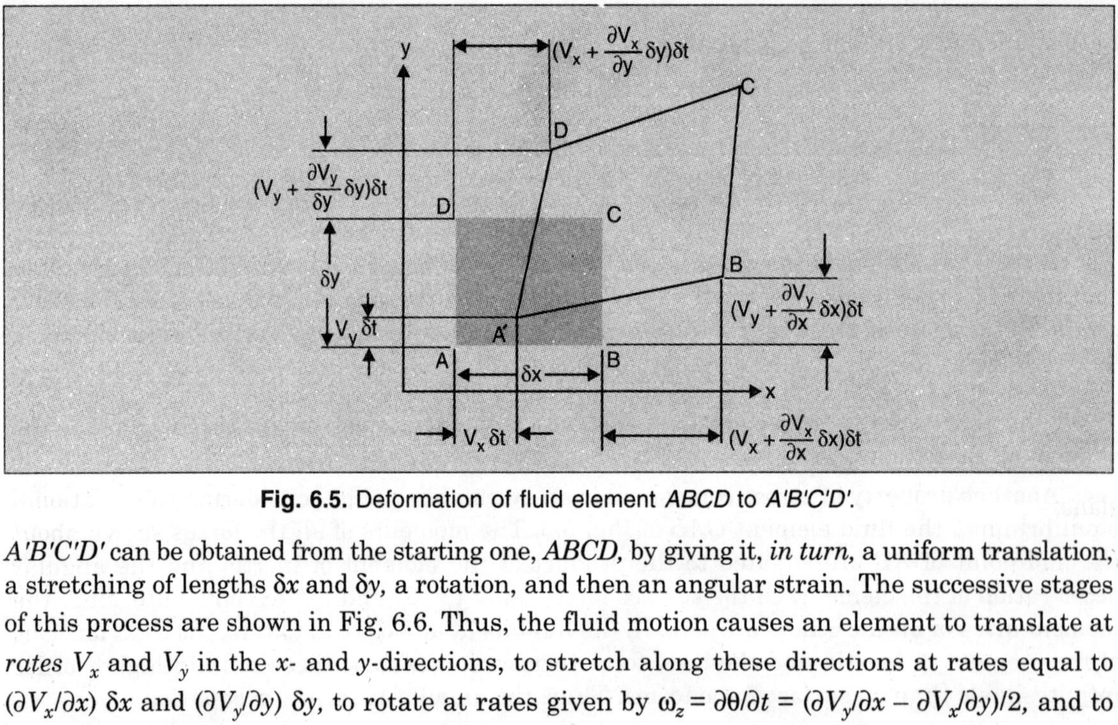

Fig. 6.5. Deformation of fluid element *ABCD* to *A′B′C′D′*.

A′B′C′D′ can be obtained from the starting one, *ABCD,* by giving it, *in turn,* a uniform translation, a stretching of lengths δx and δy, a rotation, and then an angular strain. The successive stages of this process are shown in Fig. 6.6. Thus, the fluid motion causes an element to translate at *rates* V_x and V_y in the *x*- and *y*-directions, to stretch along these directions at rates equal to $(\partial V_x/\partial x)\ \delta x$ and $(\partial V_y/\partial y)\ \delta y$, to rotate at rates given by $\omega_z = \partial\theta/\partial t = (\partial V_y/\partial x - \partial V_x/\partial y)/2$, and to

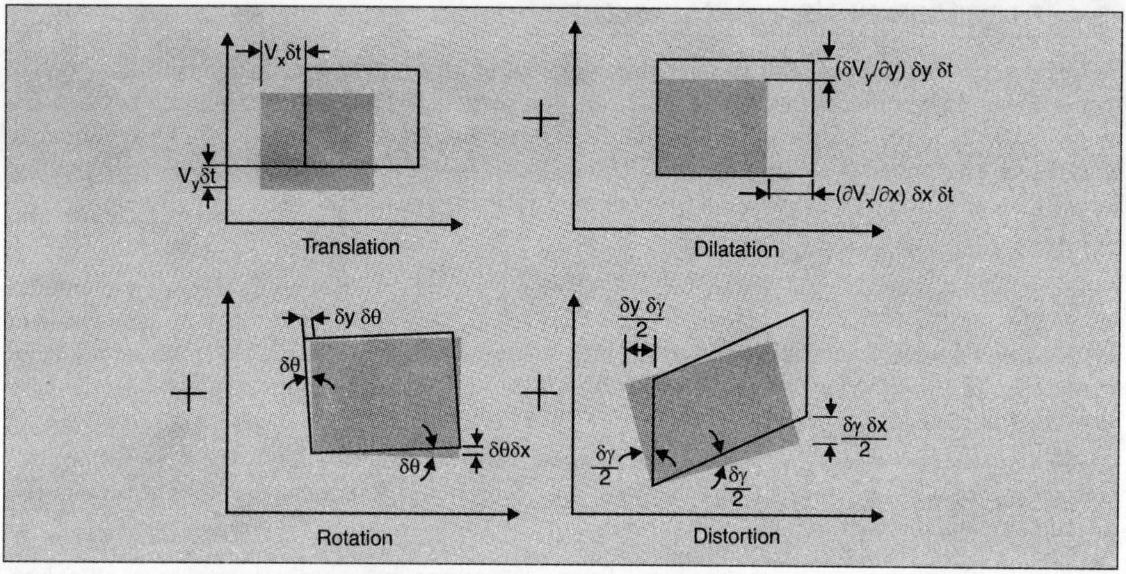

Fig. 6.6. Resolution of the deformation of the fluid element of Fig. 6.5.

undergo a shear deformation at a rate given by $\mathring{\gamma}_{xy} = \partial\gamma/\partial t = (\partial V_y/\partial x + \partial V_x/\partial y)$. Therefore, the following *rates of strain* are obtained in terms of the velocity gradients for a 2-*D* flow field*:

Rate of linear strain in the *x*-direction, $\mathring{\epsilon}_x = \dfrac{\text{change in length}}{\delta x}$

$$= \frac{\partial V_x}{\partial x} \quad \text{...(6.8)}$$

Rate of linear strain in the *y*-direction, $\mathring{\epsilon}_y = \dfrac{\partial V_y}{\partial y}$ ...(6.9)

Rate of rotation about the *z*-axis, $\omega_z = \dfrac{1}{2}\left(\dfrac{\partial V_y}{\partial x} - \dfrac{\partial V_x}{\partial y}\right)$ ...(6.10)

and, rate of shear strain in the *x*-*y* plane, $\mathring{\gamma}_{xy} = \left(\dfrac{\partial V_y}{\partial x} + \dfrac{\partial V_x}{\partial y}\right)$ ...(6.11)

* For the more general 3-*D* motion, additional components of the rates of strain may be written by exploiting symmetry.

6.4 STRESSES IN NEWTONIAN FLUIDS

So far, the concept of stress at a point in a flow field and expressions for the deformation rates in terms of velocity gradients have been given. In this section, the stresses are related to the rates of strain for a *Newtonian fluid.* Newtonian fluids are those in which the stresses vary linearly with the rates of strain (see Sec. 1.3). Thus the shear stresses are proportional to the corresponding rates of shear strain, the constant of proportionality being termed as viscosity μ. Thus,

$$\tau_{xy} = \tau_{yx} = \mu \mathring{\gamma}_{xy} = \mu\left(\frac{\partial V_y}{\partial x} + \frac{\partial V_x}{\partial y}\right) \quad \text{...(6.12)}$$

with similar results written for the other shear stresses in a 3-*D* flow. Eq. (1.2) is a special form of Eq. (6.12) valid for flows where the streamlines are straight and parallel, as in the case described in Sec. 1.3.

The normal stress σ_{xx} is related not only to the rate of strain $\mathring{\varepsilon}_x$ in the *x*-direction, but also to the strains $\mathring{\varepsilon}_y$ and $\mathring{\varepsilon}_z$ in the other two directions. For most cases of interest, the relationship can be written as

$$\sigma_{xx} = -p + 2\mu\frac{\partial V_x}{\partial x} - \frac{2}{3}\mu\left[\frac{\partial V_x}{\partial x} + \frac{\partial V_y}{\partial y} + \frac{\partial V_z}{\partial z}\right] \quad \text{...(6.13)}$$

where *p* is the pressure. This result is due to the famous British mathematician G.G. Stokes (1845). Similar equations can be written down for σ_{yy} and σ_{zz}. Relations for rates of strains and stress components in cylindrical and spherical coordinates are given in Sec. B·6 and B·7 (Appendix B).

It may be observed that for non-viscous fluids or in flows with no velocity gradients, the normal stress in all directions is the same (equal to $-p$). The same result was obtained earlier as Eq. (6.5). For incompressible fluids, the term in brackets in Eq. (6.13) drops out to give

$$\sigma_{xx} = -p + 2\mu\frac{\partial V_x}{\partial x} \quad \text{...(6.14)}$$

In this text we confine our attention only to incompressible Newtonian fluids where stresses are given by expressions of the types (6.12) and (6.14). We proceed to obtain the equation of motion for such a fluid.

6.5 EQUATION OF MOTION FOR INCOMPRESSIBLE FLUIDS

Consider an infinitesimal control volume *ABCD* (of unit depth) in a 2-*D* flow field with its centre located at *(x, y)* (Fig. 6.7). If the state of stress at *(x, y)* in this 2-*D* flow field is represented by σ_{xx}, σ_{yy}, τ_{xy} and τ_{yx}, then the surface forces on the four faces of the *CV* can be written in terms of these stresses and their derivatives using Taylor series. A few of these are shown in the figure. The net surface forces acting on this element in the *x*- and *y*-directions are easily seen to be

$$\delta F_{\text{surface},\,x} = \left(\frac{\partial \sigma_{xx}}{\partial x} + \frac{\partial \tau_{yx}}{\partial y}\right)\cdot\delta x\cdot\delta y\cdot 1 \quad \text{...(6.15)}$$

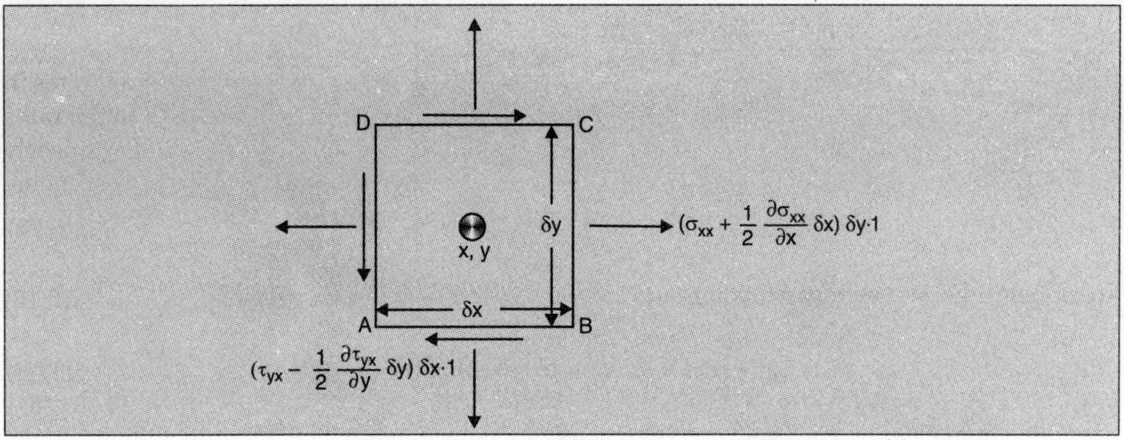

Fig. 6.7. Forces acting on the four faces of an infinitesimal *CV* of unit depth in a 2-*D* flow field.

and

$$\delta F_{\text{surface},y} = \left(\frac{\partial \sigma_{yy}}{\partial y} + \frac{\partial \tau_{xy}}{\partial x}\right) \cdot \delta x \cdot \delta y \cdot 1 \qquad ...(6.16)$$

If f_x and f_y are the components of the body force per unit mass, then

$$\delta F_{\text{body},x} = \rho f_x \cdot (\delta x \cdot \delta y \cdot 1) \qquad ...(6.17)$$

and

$$\delta F_{\text{body},y} = \rho f_y \cdot (\delta x \cdot \delta y \cdot 1) \qquad ...(6.18)$$

By Newton's law of motion

$$\rho(\delta x \cdot \delta y \cdot 1)\frac{DV_x}{Dt} = \rho f_x \delta x \cdot \delta y \cdot 1 + \left(\frac{\partial \sigma_{xx}}{\partial x} + \frac{\partial \tau_{yx}}{\partial y}\right)\delta x \cdot \delta y \cdot 1$$

so that

$$\rho \frac{DV_x}{Dt} = \rho f_x + \frac{\partial \sigma_{xx}}{\partial x} + \frac{\partial \tau_{yx}}{\partial y} \qquad ...(6.19)$$

and, similarly,

$$\rho \frac{DV_y}{Dt} = \rho f_y + \frac{\partial \sigma_{yy}}{\partial y} + \frac{\partial \tau_{xy}}{\partial x} \qquad ...(6.20)$$

DV_x/Dt and DV_y/Dt may be expressed in terms of the field derivatives using the Euler acceleration formula Eq. (3.5). Eq. (6.19) then gives

$$\rho\left[\frac{\partial V_x}{\partial t} + V_x \frac{\partial V_x}{\partial x} + V_y \frac{\partial V_x}{\partial y}\right] = \rho f_x + \frac{\partial \sigma_{xx}}{\partial x} + \frac{\partial \tau_{yx}}{\partial y} \qquad ...(6.21)$$

with a similar equation obtained from Eq. (6.20). We can now use the expressions for τ_{yx} (Eq. 6.12) and σ_{xx} (Eq. 6.14) to obtain

$$\rho\left[\frac{\partial V_x}{\partial t}+V_x\frac{\partial V_x}{\partial x}+V_y\frac{\partial V_x}{\partial y}\right]=\rho f_x-\frac{\partial p}{\partial x}+\mu\left[\frac{\partial^2 V_x}{\partial x^2}+\frac{\partial^2 V_x}{\partial y^2}\right] \quad ...(6.22)$$

and

$$\rho\left[\frac{\partial V_y}{\partial t}+V_x\frac{\partial V_y}{\partial x}+V_y\frac{\partial V_y}{\partial y}\right]=\rho f_y-\frac{\partial p}{\partial y}+\mu\left[\frac{\partial^2 V_y}{\partial x^2}+\frac{\partial^2 V_y}{\partial y^2}\right] \quad ...(6.23)$$

For a general 3-*D* flow these components can be written in the vector form

$$\rho\left[\frac{\partial \mathbf{V}}{\partial t}+(\mathbf{V}.\nabla)\mathbf{V}\right]=\rho\mathbf{f}-\nabla p+\mu\nabla^2\mathbf{V} \quad ...(6.24)$$

known as *Navier-Stokes (NS) equation.* Note that this contains three equations, for V_x, V_y and V_z. The LHS of Eq. (6.24) represents the fluid acceleration (local and convective) and is referred to as the inertial term. The terms on the right represent the body force, the pressure and the viscous forces respectively. The component equations for various coordinate systems are given in Appendix B-9.

The three components of the equation of motion (6.24) and the continuity Eq. (4.17) for incompressible fluids form a system of four equations for the four unknown variables, namely, the three components of velocity and pressure. This is a pretty formidable set of non-linear second-order coupled partial-differential equations and have been solved analytically only for a few simple geometries. In most of these, there are some simplifying features that reduce the complexities of these equations. For instance, in Examples 6.1 to 6.3 which follow, the inertial terms on the left are identically zero (because of full development) rendering the equations linear. However, there are many situations in which one or more terms, though not exactly zero, are so small compared to the other terms that they can be neglected. One common simplification arises from the fact that in many flows of engineering interest the viscous terms in the major part of the flow region are much smaller than the inertial terms (Chapter 11) and, thus, can be neglected, giving

$$\rho\left[\frac{\partial \mathbf{V}}{\partial t}+(\mathbf{V}\,.\,\nabla)\mathbf{V}\right]=\rho\mathbf{f}-\nabla p$$

This is known as the *Euler equation* which is applicable to non-viscous flows and plays an important role in the study of fluid motion (see Chapter 12).

6.6 BOUNDARY CONDITIONS IN VISCOUS FLOWS

Consider next the type of boundary conditions that apply to these systems. At the solid boundaries of the flow field, it is obvious that there should be no *relative* velocity normal to the boundary, for otherwise the fluid will penetrate the boundary. Thus, $\mathbf{V}\cdot\hat{\mathbf{n}}=0$ at stationary, impervious solid surfaces. The situation regarding the relative tangential velocity, however, is not that obvious, and has been a subject of intense debate. Extensive experiments have established that for *real* fluids under the continuum approximation (except for Helium at very low temperatures), the relative tangential velocity at a solid boundary is also zero. This is called the *no-slip condition.*

Thus, the velocity of the fluid at a solid, impervious boundary is the same as the velocity of the bounding surface itself. Another way of expressing this is to say that for real fluids, there is no jump (or, step change) in the velocity at a solid boundary, i.e. the velocity field is continuous with that of the solid.

At an interface between two fluids (including a free surface between a liquid and a gas), the shear stresses too do not experience a jump in their values. Thus, both the velocity and the shear stress fields are continuous at such an interface. The pressure is also continuous across a fluid-fluid interface, except for cases where the surface-tension effects are appreciable (see Sec. 2.8). In this text, we restrict ourselves to cases where surface tension effects are negligible and, therefore, in the following chapters we assume that the pressure, shear stress and the velocity are continuous at a fluid-fluid interface.

At liquid-gas interfaces, the above-mentioned boundary conditions on the stresses can usually be simplified because of the fact that the density ρ and the viscosity μ of gases are much smaller than the corresponding values for liquids. Thus, the surface of an ocean with waves (Fig. 6.8) can be approximated as a constant pressure surface (with pressure equal to atmospheric) because the variations in the pressure on the air-side are negligible*. This is because the ρ of air is much smaller than that of water. The picture is slightly more complicated as far as shear stresses are concerned. Consider the situations where the velocities are comparable on the gas and liquid sides (Fig. 6.9). The fact that the shear stresses on the two sides at the interface are equal, and

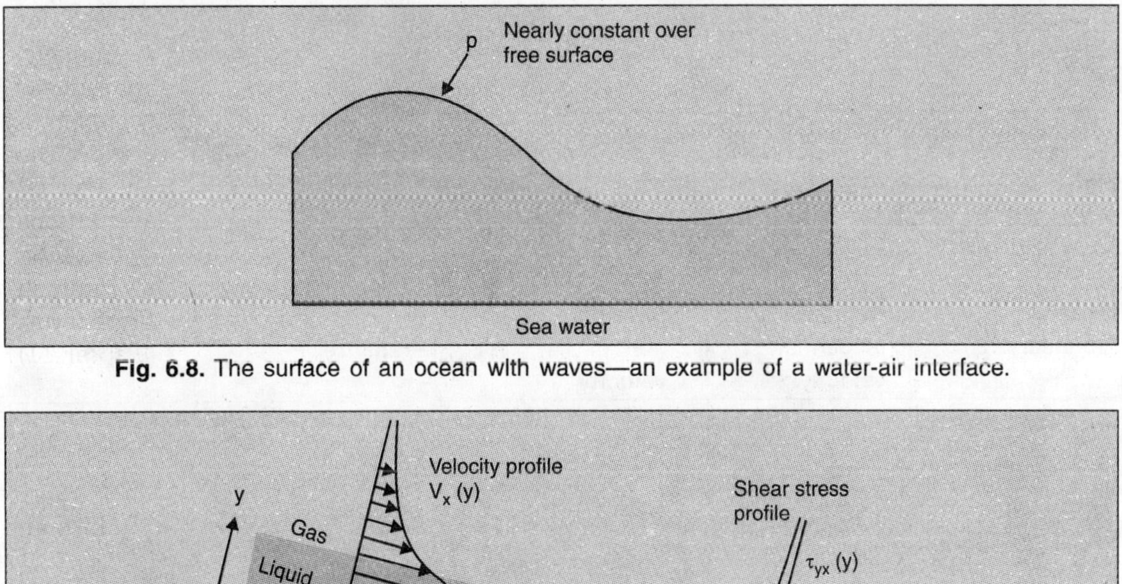

Fig. 6.8. The surface of an ocean with waves—an example of a water-air interface.

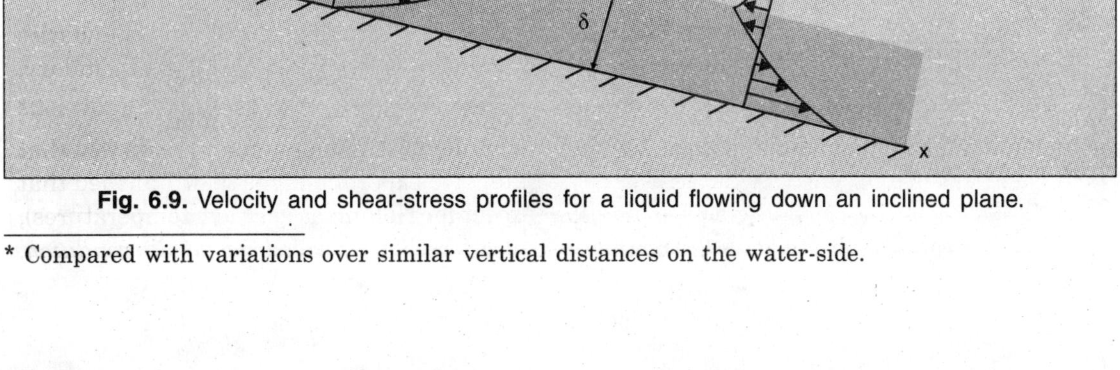

Fig. 6.9. Velocity and shear-stress profiles for a liquid flowing down an inclined plane.

* Compared with variations over similar vertical distances on the water-side.

that the viscosity of gases is much smaller than that of liquids requires that the shear stress at the interface be smaller than elsewhere within the liquid. On the liquid-side, therefore, the velocity gradient near the interface will be much smaller than elsewhere. If we were interested in studying the liquid-side motion, we could neglect the velocity gradient at the interface and set $\partial V_x/\partial y = 0$ at $y = \delta$ on the liquid-side. If, on the other hand, we were interested in studying the motion of the gas, we would compare the shear stress at the interface with that elsewhere within the gas, and this would not be negligible at all. In fact, the gradient near the interface (on the gas-side) dominates.

This simplification does not apply to cases where the velocities on the gas-side are much larger than in the liquid, as, for example, in wind-driven ocean waves (Fig. 6.10), where the shear stress at the interface dominates on either side.

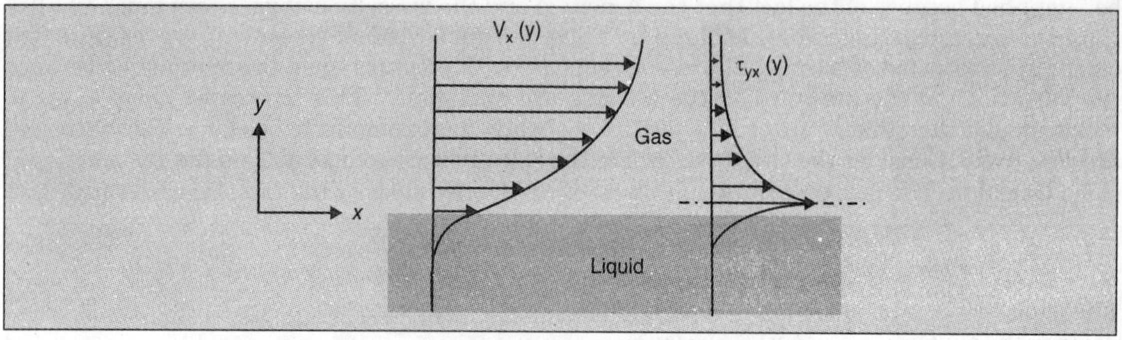

Fig. 6.10. Velocity and shear-stress profiles for wind-driven ocean waves.

These results are summarized in Table 6.1.

Table 6.1. Summary of boundary conditions

Type of boundary	Boundary conditions on: Relative tangential velocity	Relative normal velocity	Pressure	Shear stress
Impervious Solid-Fluid	Zero (no slip)	Zero	–	–
Fluid-Fluid	Zero	Zero	Same on either side (negligible surface tension case)	Same on either side $\left(\mu \frac{\partial V}{\partial n}\right)_1 = \left(\mu \frac{\partial V}{\partial n}\right)_2$
Approximate for Liquid-Gas (for liquid-side eqns. only)	Zero	Zero	$p_{interface} = p_{gas} \simeq$ constant	$\frac{\partial V}{\partial n} = 0$ if velocities in gas are not large

Example 6.1. *Poiseuille-Couette Flow in a 2-D Channel*

Consider the steady laminar flow of a viscous, incompressible fluid through a channel (Fig. 6.11) made up of two flat plates, infinite in the x- and z-directions with the lower plate (at $y = -b$) stationary and the upper plate (at $y = b$) moving at a velocity V_0. Obtain the velocity profile.

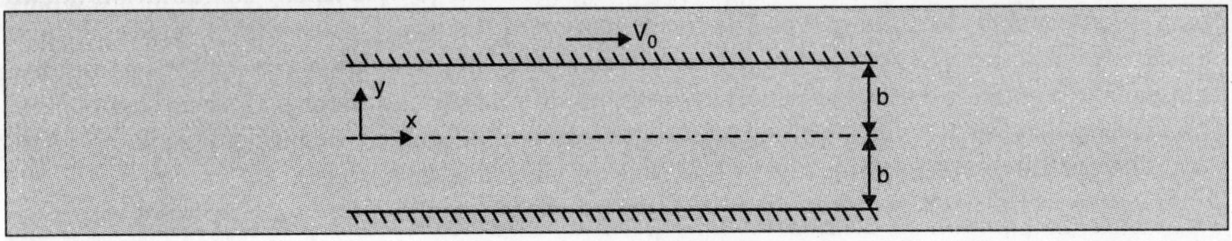

Fig. 6.11. Flow in a 2-*D* channel.

The symmetry of the arrangement suggests that $V_z = 0$ and $\partial/\partial z$ of any variable is zero. Since the plates are infinitely long in the x-direction, it is reasonable to assume full development of flow, and consequently, $\partial V_x/\partial x = 0$. Thus, V_x is a function of y alone.

Use of the continuity Eq. (4.16) gives $\partial V_y/\partial y = 0$.

or

$$V_y = \text{constant}$$

and since V_y is zero at either plate, it must be zero everywhere. V_x is, therefore, the only component of velocity present.

Since gravity is the only body force, $f_x = 0$ and $f_y = -g$, the body force per unit mass. The components of the NS Eqs. (6.22) and (6.23) then reduce to

$$0 = -\frac{\partial p}{\partial x} + \mu \frac{d^2 V_x}{dy^2} \tag{a}$$

$$0 = -\rho g - \frac{\partial p}{\partial y} \tag{b}$$

Equation (b) gives

$$p = -\rho g y + F(x)$$

and therefore, $\partial p/\partial x$ is a function of x alone. But from Eq. (a), $\partial p/\partial x = \mu d^2 V_x/dy^2$, and since V_x is a function of y alone, $\partial p/\partial x$ cannot be a function of x and must be a constant, say, p', Then

$$\frac{d^2 V_x}{dy^2} = \frac{p'}{\mu} \tag{c}$$

Equation (c) is to be solved with the no-slip conditions at the lower and upper boundaries, i.e., with

$$V_x = 0 \text{ at } y = -b$$

and

$$V_x = V_0 \text{ at } y = +b \tag{d}$$

This gives

$$V_x = -\frac{p'b^2}{2\mu}\left(1-\frac{y^2}{b^2}\right)+\frac{V_0}{2}\left(1+\frac{y}{b}\right) \tag{e}$$

Note that Eq. (c) is linear and, thus, the resulting equation (e) is a linear combination of two results—one, where the fluid moves due to the motion of the upper plate with a velocity V_0 and no pressure gradient *(Couette problem),* and two, where the two plates are stationary and the fluid motion is entirely due to the constant longitudinal pressure gradient p' *(Poiseuille problem).* The velocity profiles for a few values of p', both positive as well as negative are given in Fig. 6.12. The profile corresponding to $p' = 0$ is that for the Couette problem.

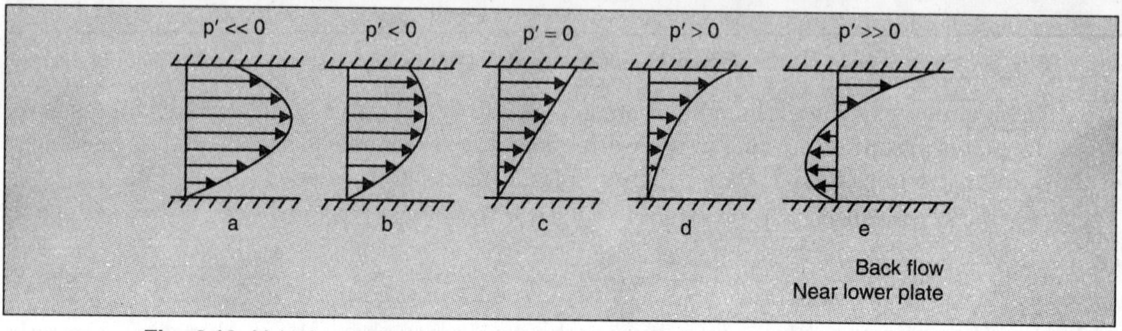

Fig. 6.12. Velocity profiles for the Poiseuille-Couette flow in a 2-*D* channel for various values of p' ($\equiv \partial p/\partial x$).

The shear-stress at any location y is

$$\tau_{yx} = \mu\frac{dV_x}{dy} = p'y + \mu\frac{V_0}{2b} \tag{f}$$

Note that the shear stress varies linearly for the Poiseuille problem (represented by the first term in Eq. (*f*)) while for the Couette problem, it is constant across any section (the second term).

The above results are valid only for laminar flows. In turbulent flows the velocities are functions of time (see Sec. 1.7), and the unsteady terms in both the continuity and NS equations cannot be dropped. Also, V_z and V_y are no longer zero.

Example 6.2. *Thrust Bearing.* A hydrodynamic thrust bearing prevents a rotating shaft from coming into direct contact with the base by a build-up of pressure due to the dynamics of flow of a viscous fluid. Figure 6.13(a) shows schematically the top view of a self-adjusting *Michell pad thrust bearing.* A lubricating fluid is used between the pads and the horizontal base. When the shaft rotates, the pads move relative to the base, forcing the fluid to move. Each pad of the bearing adjusts itself such that the gap between its under surface and the base forms a narrow converging passage for the fluid (Fig. 6.13(b)). Seen in a reference frame fixed with the pad, a thin oil film is dragged through this converging passage by the base plate moving to the right at a velocity ΩR. Pressure builds up under the pad and this supports the axial load F (per pad).

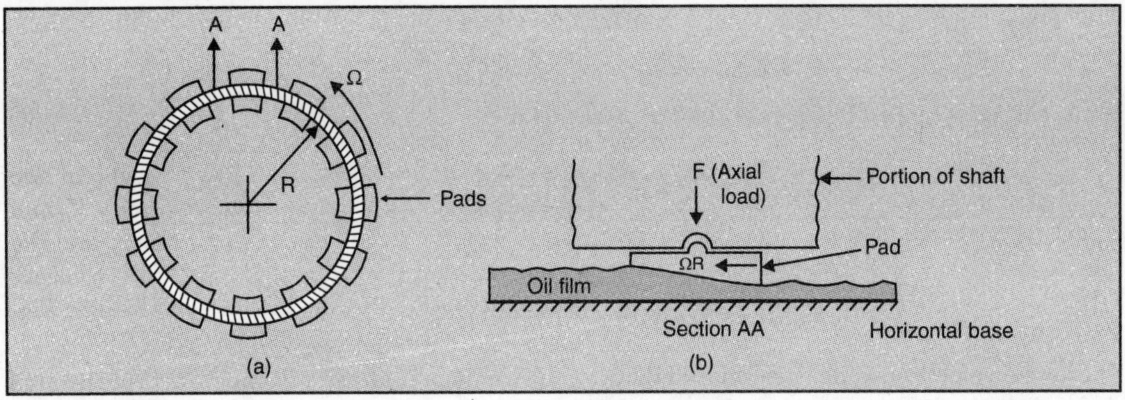

Fig. 6.13. A Michell pad thrust bearing. (a) top view, and (b) fluid flow situation for a single pad.

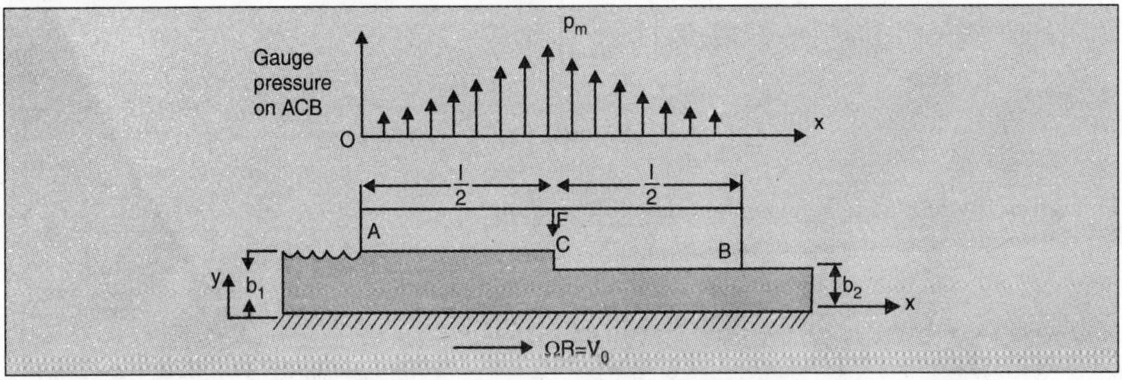

Fig. 6.14. A stepped channel model for the converging section of Fig. 6.13(b).

One highly simplified model of this flow replaces the converging channel by a stepped channel as shown in Fig. 6.14. If the gaps b_1 and b_2 are much smaller than the length l, regions AC and CB may be approximated as 'infinitely long' parallel channels with the lower plate moving at a velocity $V_0 = \Omega R$. Example 6.1 showed that for fully developed flow in such a channel the pressure gradient dp/dx is constant. The velocity profile may be obtained from Eq. (c) of that example

$$\frac{d^2V_x}{dy^2} = \frac{p'}{\mu}$$

by integration as

$$V_x = \frac{p'}{2\mu}y^2 + C_1y + C_2$$

where C_1 and C_2 are constants of integration and p' corresponds to the pressure gradient in the *respective channel.* The boundary conditions are

For channel AC: $V_x = V_0$ at $y = 0$

$V_x = 0$ at $y = b_1$

For channel CB: $V_x = V_0$ at $y = 0$

$V_x = 0$ at $y = b_2$

Thus, the velocity profiles across the two sections are

$$AC: V_x = \frac{p_1'}{2\mu}(y^2 - b_1 y) + V_0\left(1 - \frac{y}{b_1}\right)$$

and

$$CB: V_x = \frac{p_2'}{2\mu}(y^2 - b_2 y) + V_0\left(1 - \frac{y}{b_2}\right)$$

These two velocity profiles are related by the requirement that the volume flow rate through channel AC is the same as that through channel CB. Thus

$$\int_0^{b_1}\left[\frac{p_1'}{2\mu}(y^2 - b_1 y) + V_0\left(1 - \frac{y}{b_1}\right)\right] dy \times 1$$

$$= \int_0^{b_2}\left[\frac{p_2'}{2\mu}(y^2 - b_2 y) + V_0\left(1 - \frac{y}{b_2}\right)\right] dy \times 1$$

This gives the following relation between p_1' and p_2'.

$$-\frac{p'_1 b_1^3}{12\mu} + \frac{V_0 b_1}{2} = -\frac{p'_2 b_2^3}{12\mu} + \frac{V_0 b_2}{2} \tag{a}$$

As shown in Example 6.1 the pressure gradient remains constant along the length of the channel formed by two parallel plates. The pressures at the two ends A and B are equal to the atmospheric pressure and hence, the maximum (gauge) pressure p_m must occur at C. The pressure profile for the stepped channel is thus as shown in Fig. 6.14. Clearly, $p_1' = \frac{p_m}{l/2}$ and $p_2' = -\frac{p_m}{l/2}$. Using these in Eq. (a) above we obtain a simple equation for p_m

$$-\frac{p_m b_1^3}{6\mu l} + \frac{V_0 b_1}{2} = \frac{p_m b_2^3}{6\mu l} + \frac{V_0 b_2}{2}$$

or

$$p_m = \frac{3\mu l V_0 (b_1 - b_2)}{(b_1^3 + b_2^3)}$$

The axial load F supported by one pad can be obtained from the area under the pressure diagram and is given by

$$F = \frac{1}{2} p_m l w$$

where w is the width of the pad. If b_1 is $2b_2$, we get

$$F = \frac{1}{6}\frac{\mu V_0 w l^2}{b_2^2} = 0.1667\frac{\mu V_0 w l^2}{b_2^2}$$

It may be mentioned that the fully-developed flow approximation will not hold near points A, B and C. But if $l \gg b_1$, the lengths over which this approximation cannot be used will be small compared to the lengths over which this analysis applies and, thus, the errors will be small.

The above result has been obtained using a very crude model in which the converging channel of the Michell bearing has been replaced by a one-step channel. The model can be improved by taking a series of infinitesimal steps and the pressure profile obtained is as shown in Fig. 6.15. The load carried by the bearing (with $b_1 = 2b_2 \equiv 2b$) is

$$F = 0.1587\, \mu V_0 w \frac{l^2}{b^2}$$

Figure 6.15 also shows the velocity profiles at various sections across the gap. Note that for $x < 2l/3$, the velocity profiles are typical for positive pressure gradients, and for $x > 2l/3$, they are typical for negative pressure gradients. At $x = 2l/3$ the profile is linear corresponding to zero pressure gradient (Couette flow). Also note that the larger the value of μ, the more is the load that the bearing can support. However, a large μ results in increasing drag on the pad and thus requires more torque. An optimal value of μ is, therefore, indicated.

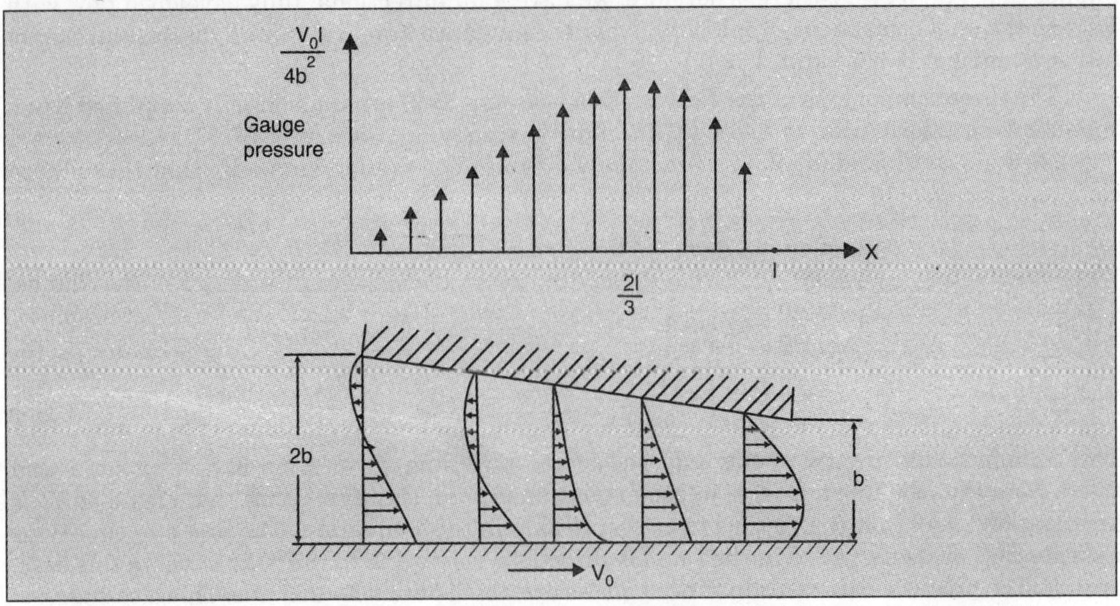

Fig. 6.15. Pressure and velocity profiles for the converging channel shown in Fig. 6.13(b).

Example 6.3. *Steady Laminar Flow through a Horizontal Pipe.* Consider now the steady laminar flow of an incompressible fluid of density ρ and viscosity μ through a long horizontal circular pipe of radius R (Fig. 6.16). Let us restrict our attention to flows in which there is no swirl, i.e., the tangential component of velocity V_θ is absent. The cylindrical symmetry of the geometry suggests that the velocity profile should be axially symmetric, i.e., $\partial V/\partial \theta = 0$. In a very long pipe it is reasonable to assume that far away from the entrance and exit, the velocity profile does not change from section to section, i.e., the flow *is fully developed* (see Sec. 1.8).

Fig. 6.16. Flow through a horizontal circular pipe.

The flow is governed by the continuity and the Navier-Stokes equation. In cylindrical polar coordinates, the continuity equation (App. B.8) for steady, incompressible flows is

$$\frac{1}{r}\frac{\partial}{\partial r}(rV_r)+\frac{1}{r}\frac{\partial V_\theta}{\partial \theta}+\frac{\partial}{\partial z}V_z=0$$

Since the velocity profile is fully developed, $\partial V_z/\partial z$ is zero and since $V_\theta = 0$, this gives

$$\frac{1}{r}\frac{\partial (rV_r)}{\partial r}=0$$

Thus, rV_r does not vary with r. At $r = R$, V_r has to be zero because of the impervious wall. This implies that $V_r = 0$ everywhere. Therefore, we have a uni-directional fully developed flow with $V_z(r)$ as the only component of velocity. (Near the entrance and exit, where the assumption of full development is not valid, $V_r \neq 0$.)

The three components of the NS equation (see App. B-9) are considerably simplified when the following information is used: (a) the flow is steady, i.e., $\partial/\partial t = 0$; (b) $V_r = 0$; (c) $V_\theta = 0$ because of no swirl, and (d) V_z is a function of r alone. The reader can verify that these three components reduce to

$$0=-\frac{\partial p}{\partial r}-\rho g\sin\theta$$

$$0=-\frac{1}{r}\frac{\partial p}{\partial \theta}-\rho g\cos\theta$$

$$0=-\frac{\partial p}{\partial z}+\frac{\mu}{r}\frac{\partial}{\partial r}\left(r\frac{\partial V_z}{\partial r}\right)$$

This assumes that gravity is the only body force and thus, $f_r = -g \sin \theta$, $f_\theta = -g \cos \theta$ and $f_z = 0$. Note that the inertial term in each equation is zero, signifying that there is no *particle* acceleration*, and that the viscous term occurs only in the z-equation. The first two equations indicate that pressure varies in the r and θ directions solely due to the body force, in this case, gravity. The pressure variations in the r–θ plane are, thus, purely hydrostatic. This also suggests the introduction of a new variable $\mathcal{P} = p + \rho g r \sin \theta$, the *non-gravitational pressure.*** It can be seen easily that $\partial\mathcal{P}/\partial r$ and $\partial\mathcal{P}/\partial\theta$ are both zero. This implies that $\mathcal{P}$ remains constant across any section of the pipe and is a function of z alone. The pressure gradient in the third equation is then replaced by $d\mathcal{P}/dz$. Thus

$$0=-\frac{d\mathcal{P}}{dz}+\frac{\mu}{r}\frac{\partial}{\partial r}\left(r\frac{\partial V_z}{\partial r}\right)$$

---

* This is true of all fully developed flows.

** The significance of the nomenclature becomes clear when we recast the definition as $p = \mathcal{P} - \rho g r \sin \theta$. The variations of p are due both to the dynamics of the flow as well as to the effcts of gravity. The term $\rho g r \sin \theta$ represents the latter, while the variations of $\mathcal{P}$ are solely due to the flow. Hence the name.

Since $\mathcal{P}$ is a function of z alone and V_z a function of r alone, this equation can be satisfied only if

$$\frac{\mu}{r}\frac{d}{dr}\left(r\frac{dV_z}{dr}\right)=\frac{d\mathcal{P}}{dz}=\text{constant}=\mathcal{P}'$$

that is, the pressure gradient is constant along the pipe axis. It may be noted that $\mathcal{P}$ is the centre-line pressure. Integration of this equation yields

$$V_z=\frac{r^2\mathcal{P}'}{4\mu}+C_1\ln r+C_2$$

where the constants of integration C_1 and C_2 are evaluated from the boundary conditions. There is only one boundary condition in this problem, viz., the no-slip condition: $V_z\,(r = R) = 0$. The other condition is the requirement of a finite velocity at $r = 0$. These conditions lead to

$$V_z=-\frac{\mathcal{P}'}{4\mu}\left(R^2-r^2\right)$$

which is parabolic in shape, with the maximum velocity at the axis.

It can be confirmed that

$$V_{av}=\frac{\int_0^R 2\pi rV_z dr}{\pi R^2}=-\frac{\mathcal{P}'R^2}{8\mu}=\frac{V_{\max}}{2}$$

and that the volume flow rate is

$$\mathring{Q}=\pi R^2V_{av}=\frac{\pi R^4\left(-\mathcal{P}'\right)}{8\mu} \qquad \text{(a)}$$

Equation (a) is known as the *Hagen-Poiseuille law.*

It should be noted that the scope of this 'law' is quite limited because of the two conditions under which it is derived. These are the conditions of laminar flow and of full development. The velocity profile is not parabolic at the pipe entrance but develops into one over a length termed as the *entrance length* (see Secs. 1.8 and 10.1). As the velocity of the flow increases, this entrance length increases and it can become a significant fraction of the total length of the pipe. Also, at larger velocities the flow tends to become turbulent and is no longer steady. Equation (a) is, thus, valid only over a small range of velocities.

6.7 EQUATION OF MOTION FOR STEADY NON-VISCOUS FLOWS IN NATURAL COORDINATES

At times it is more convenient to work in a coordinate system where one distance s is measured along the streamlines, and the other n is measured normal to them. These are termed as the *natural coordinates* and form, in general, a curvilinear network (Fig. 6.17). The advantage of working in this system lies in the fact that the velocity vector has a component only along the streamline. The acceleration, however, has components both along s and n because of the curvature of the streamlines.

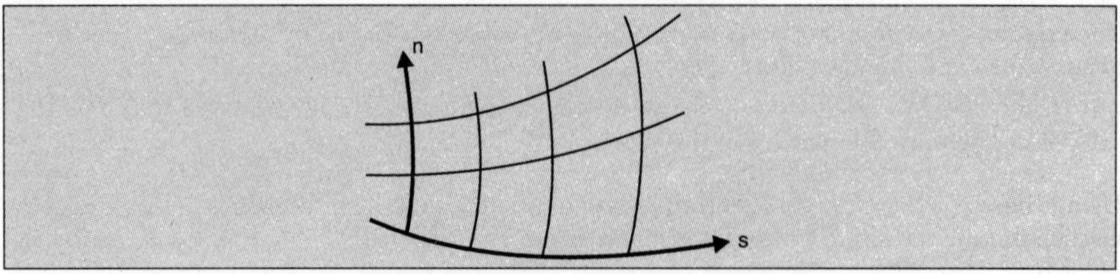

Fig. 6.17. The natural coordinate system.

The component of acceleration a_s along the streamline for steady flows is obtained from Euler acceleration formula Eq. (3.5) as

$$a_s = V_s \frac{\partial V_s}{\partial s} + V_n \frac{\partial V_s}{\partial n}$$

and since $V_s = V$, the fluid speed, and $V_n = 0$,

$$a_s = V \frac{\partial V}{\partial s} = \frac{\partial\left(V^2/2\right)}{\partial s} \qquad \text{...(6.25)}$$

Obtaining the normal component of the acceleration is slightly more complicated. A *particle* moving through a distance δs undergoes a change in velocity δV_n in the normal direction because

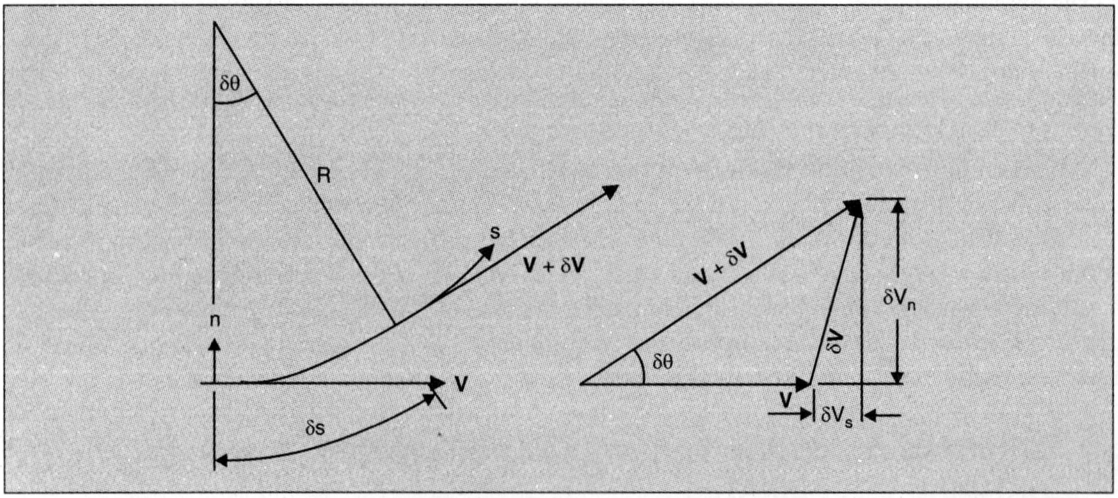

Fig. 6.18. Components of **δV** in natural coordinates.

of the curvature of the streamline (Fig. 6.18). Thus, the normal acceleration due to a particle moving through a distance δs (which is the convective contribution) is

$$a_n = \lim_{\delta t \to 0} \frac{\delta V_n}{\delta t} = \lim_{\delta t \to 0} \frac{\delta V_n}{\delta s} \cdot \frac{\delta s}{\delta t}$$

The velocity triangle in Fig. 6.18 shows that $\delta V_n = V\,\delta\theta$ and $\delta s = R\,\delta\theta$, where R is the radius of curvature at the given point*. Thus,

$$a_n = \lim_{\delta t \to 0} \frac{V}{R} \frac{\delta s}{\delta t} = \frac{V^2}{R} \qquad \text{...(6.26)}$$

Using these components of acceleration the equation of motion for steady 2-*D* flow of a non-viscous fluid in natural coordinates can be written as

$$\rho \frac{\partial \left(V^2/2\right)}{\partial s} = \rho f_s - \frac{\partial p}{\partial s} \qquad \text{...(6.27)}$$

and

$$\frac{\rho V^2}{R} = \rho f_n - \frac{\partial p}{\partial n} \qquad \text{...(6.28)}$$

where f_s and f_n are the components of the body force per unit mass in the s- and n-directions respectively.

Equations (6.27) and (6.28), though obtained for non-viscous flows, have wider applicability because viscous terms are small compared to inertial terms in many engineering flows (Sec. 6.4).

Equation (6.28) shows that the curvature of a streamline is related to the normal component f_n of the body force and to the pressure gradient $\partial p/\partial n$ in the normal direction. In the absence of f_n, the pressure gradient provides the centripetal force which curves the streamlines. Thus, in a flow in a curved horizontal channel, the pressure increases towards the outside of the bend (Fig. 6.19). In flows where the streamlines are straight, there can be no pressure variation in the normal direction in the absence of f_n (this fact has already been made use of in Example 5.1).

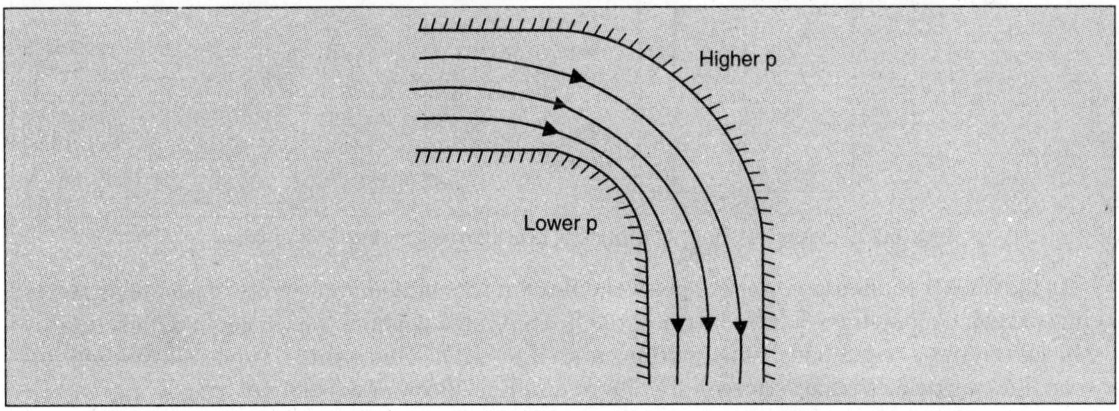

Fig. 6.19. Direction of increasing pressure in flow through a curved horizontal channel.

* R is taken positive for streamlines which are convex when viewed in the direction of increasing n, as in Fig. 6.17.

When a jet issues out of a sharp-edged orifice, the streamlines are curved (Fig. 6.20). The jet must be curved because a straight jet will require the outer streamlines to have a sharp bend (radius of curvature $R = 0$) at the orifice lip and this would call for an infinitely large pressure gradient. At the periphery the pressure is atmospheric and Eq. (6.28) shows that the pressure increases towards the centre of the jet. At the *vena contracta* where the diameter of the jet is the minimum, the streamlines are necessarily parallel, causing no normal variation in the pressure. At such a section the pressure is atmospheric everywhere.

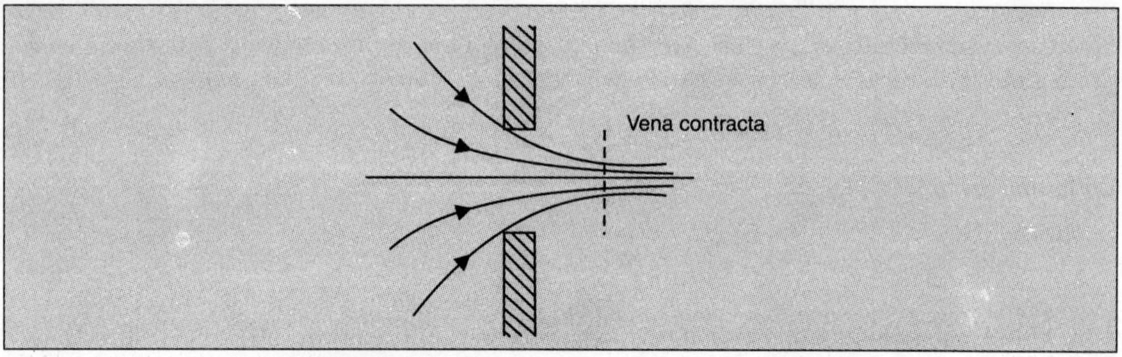

Fig. 6.20. Streamlines for flow through a sharp-edged orifice.

Example 6.4. *Free Surface in a Rotating Liquid.* A cylindrical bucket partially filled with water rotates about its own axis at an angular speed Ω. As it rotates the free surface of the water assumes a shape (Fig. 6.21). Find the equation characterizing the free surface.

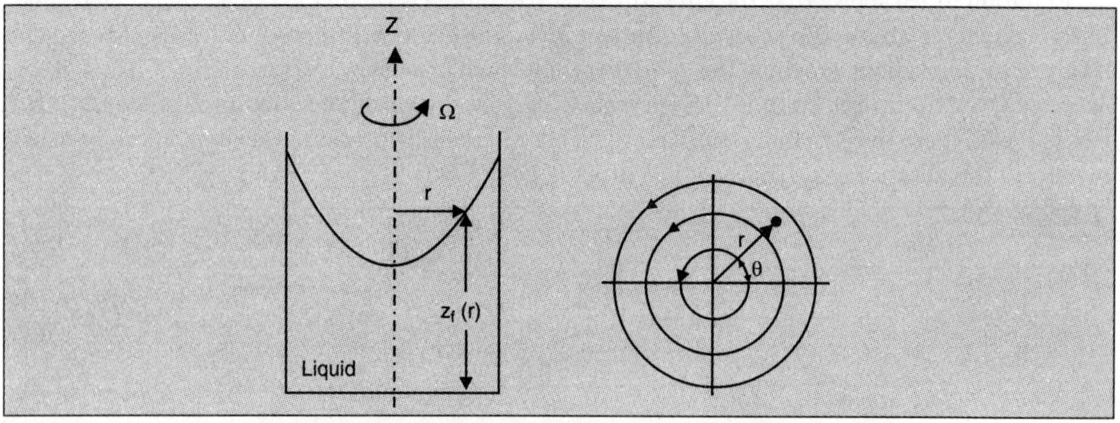

Fig. 6.21. Liquid rotating in a bucket and corresponding streamlines.

As the bucket rotates at a steady speed, the fluid in the immediate vicinity of the walls starts moving, dragging along the adjacent layers of the fluid. At steady state, the whole fluid body rotates as a block (giving a 'rigid body' type motion) and $V_\theta(r) = \Omega r$. Since there is no relative motion between fluid particles, viscous stresses are absent and Eqs. (6.27) and (6.28) are exactly applicable.

The radius of curvature of the streamline at point (r, θ) is $-r$, and Eq. (6.28) gives

$$\frac{\rho V^2}{r} = \frac{\partial p}{\partial r}$$

This can be integrated to give

$$p = \frac{\rho\Omega^2 r^2}{2} + f_1(z) \tag{a}$$

where f_1 is an unknown function of z. The streamlines do not curve in the z direction and, therefore,

$$0 = -\rho g - \frac{\partial p}{\partial z}$$

or

$$p = -\rho g z + f_2(r) \tag{b}$$

Equations (a) and (b) lead to

$$p = -\rho g z + \frac{\rho\Omega^2 r^2}{2} = \text{constant} \tag{c}$$

The shape of the free surface of water can be obtained by noting that the pressure along it must be a constant (equal to the atmospheric pressure). If $z_f(r)$ represents the level of the free surface at radius r, Eq. (c) implies

$$-\rho g z_f + \frac{\rho\Omega^2 r^2}{2} = \text{constant}$$

or

$$z_f = \text{constant} + \frac{\Omega^2 r^2}{2g}$$

Thus the free surface of the liquid in a rotating bucket is a paraboloid of revolution.

PROBLEMS

6.1 Does flow of mercury follow the no-slip condition?

6.2 A velocity field is given as

$$V_x = 3 + 6y - 5z$$
$$V_y = 2 - 6x + 7z$$
$$V_z = 8 + 5x - 7y$$

Show that the normal and shear components of deformation of a fluid element are zero everywhere and, thus, the field represents a rigid-body type motion.

6.3 Consider the following flow fields in cylindrical coordinates:

(a) $V_r = 0$, $V_\theta = K/r$; $V_z = 0$. Such a flow occurs in cyclones outside a small 'core' near the axis.

(b) $V_r = 0$, $V_\theta = \omega r$, $V_z = 0$. Such a flow occurs in a liquid in a bucket rotating about its axis.

Show that a fluid element in the flow field (a) does not rotate but shears as it flows while that in field (b) undergoes a rigid body like motion.

6.4 An axi-symmetric jet issuing from an orifice in a plane wall spreads out into a cone entraining surrounding air due to viscous action (see Prob. 4.11). The axial component of velocity within the cone is given by

$$V_z = \frac{C_1}{\mu z} \frac{1}{\left(1 + \frac{1}{4}\xi^2\right)^2}$$

with $$\xi = \frac{C_2}{\mu} \cdot \frac{r}{z}$$

where C_1 and C_2 are constants depending on the efflux from the orifice. The radial component of velocity is much smaller than the axial component. Obtain expressions for the rate of shear strain $\mathring{\gamma}$ and rate of rotation ω_z as a function of r and z, and show that the shear stress varies as $1/z^2$ for points lying on straight lines emanating from the orifice. Also show that the maximum shear stress for a given section (i.e., for constant z) lies on a line with slope given by $\dfrac{2\mu}{\sqrt{5}\, C_2}$.

6.5 *Stokes law:* The velocity and pressure fields around a solid sphere of radius R located at point O as shown, are given for *low speeds* of flow by

$$V_r = V_\infty \left[1 - \frac{3}{2}\frac{R}{r} + \frac{1}{2}\left(\frac{R}{r}\right)^3\right] \cos\theta$$

$$V_\theta = -V_\infty \left[1 - \frac{3}{4}\frac{R}{r} - \frac{1}{4}\left(\frac{R}{r}\right)^3\right] \sin\theta$$

$$V_\phi = 0$$

$$p = p_\infty - \rho g(z + z_\infty) - \frac{3}{2}\frac{\mu V_\infty}{R}\left(\frac{R}{r}\right)^2 \cos\theta$$

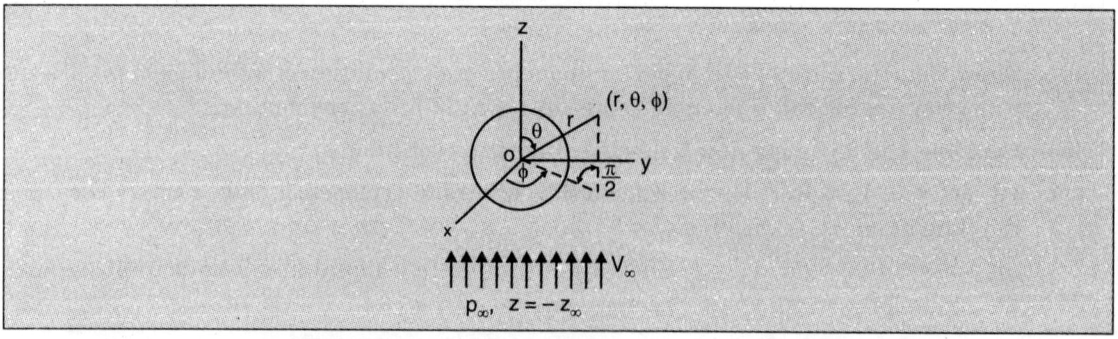

Obtain the components of the pressure and shear stress on an element of area on the surface of the sphere. Integrate over the entire surface to obtain the total pressure- and shear-forces acting on the sphere in the z-direction as $(\frac{4}{3}\pi R^3 \rho g + 2\pi\mu R V_\infty)$ and $(4\pi\mu R V_\infty)$ respectively. Thus, show that, in addition to the buoyancy force, a net z-direction drag force acts on the sphere and is given by

$$F_{\text{drag}} = 6\pi\mu R V_\infty$$

This is known as Stoke's law and is valid only for speeds of flow which give $\text{Re} = \rho V_\infty D/\mu$ less than *one*.

6.6 Separation of boundary layers can be avoided by removing the slow moving boundary-layer fluid through the porous walls by suction as shown. If the suction is at a uniform rate of $\mathring{Q}$ for a plate of area A, write the appropriate boundary conditions on velocity at $y = 0$.

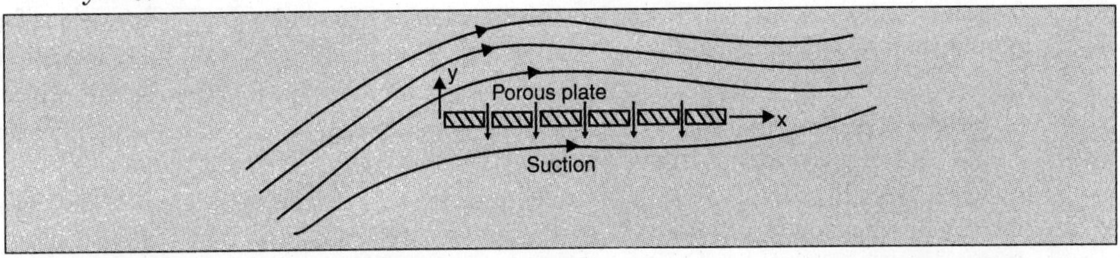

6.7 For flows in the situations shown, indicate whether the stresses marked are positive or negative.

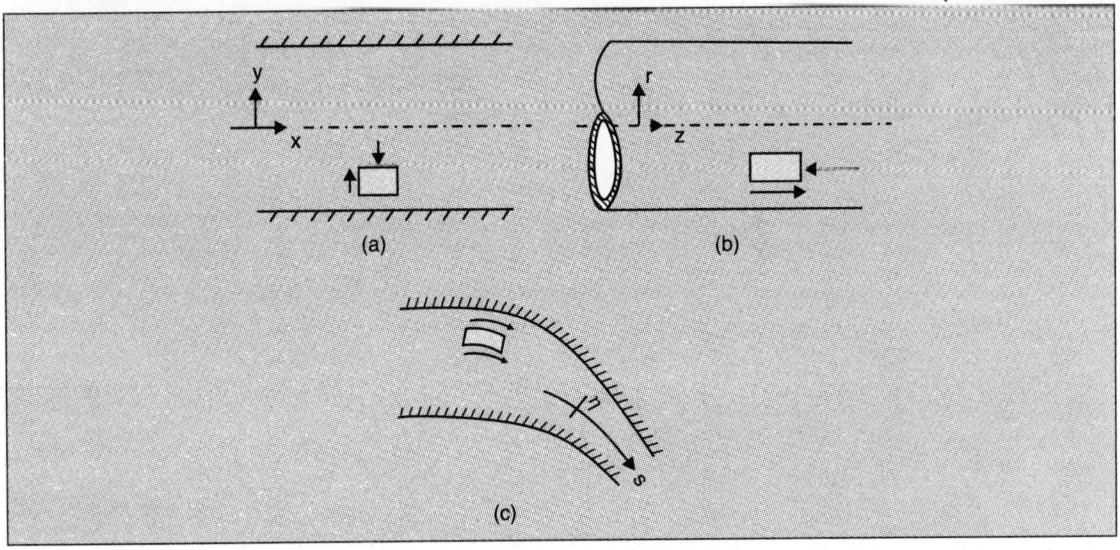

6.8 For each of the flows shown, indicate the direction of shear stress on the surface AA of the fluid element. Indicate also whether the stress is positive or negative.

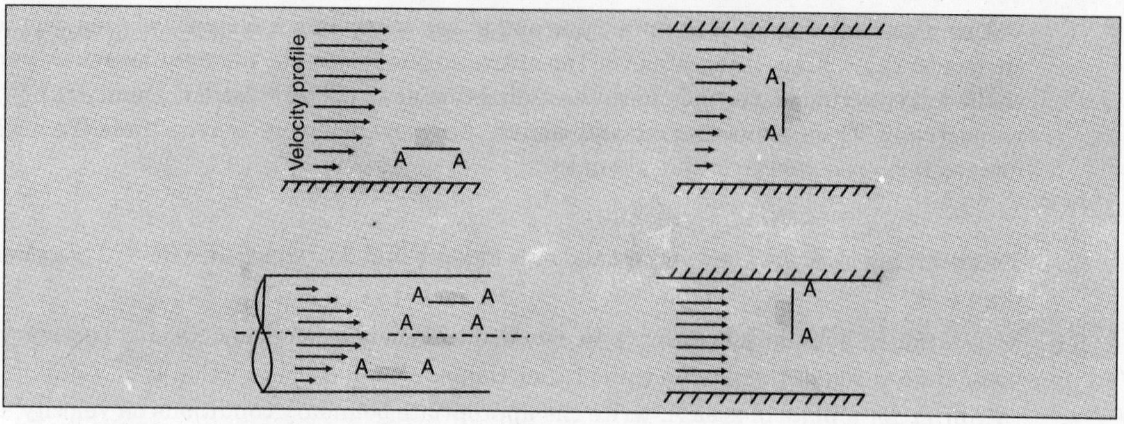

6.9 Consider a 2-*D* flow between two parallel plates caused by moving one plate with velocity V_0 relative to the other as shown. In case I a student calculates the force on the fixed plate using the equation

$$F = (\text{area of plate}) \times \mu \left.\frac{dV}{dy}\right|_w$$

and concludes that as $(dV/dy)_w$ is positive, the force acts to the right. He now uses this formula to calculate the force on the fixed plate in case II and concludes that as $(dV/dy)_w$ is negative, the force on the upper plate is to the left. Is he correct? If not, point out the error in his analysis.

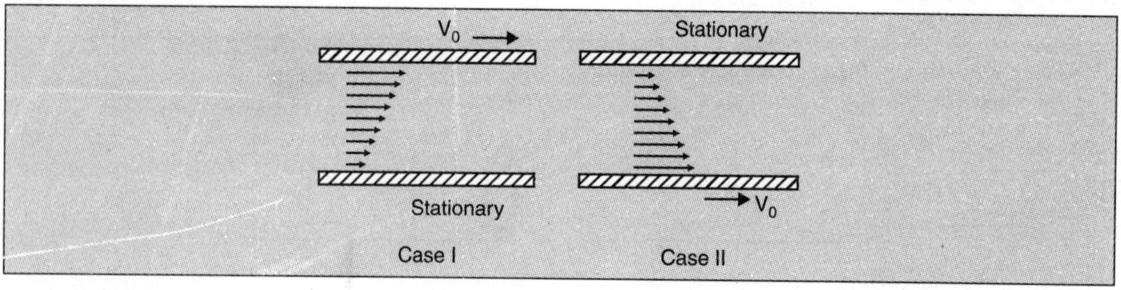

6.10 In Example 4.6, the velocity profile inside the thin boundary layer over a flat plate was given as

$$V_x = V_0\left[\frac{3}{2}\left(\frac{y}{\delta}\right) - \frac{1}{2}\left(\frac{y}{\delta}\right)^3\right], \; 0 \leqslant y \leqslant \delta$$

with

$$\delta = 5\sqrt{\frac{\mu x}{\rho V_0}}$$

Obtain an expression for ω_z as a function of x and y and show thereby that a fluid element within the boundary layer has a rotational component of motion. Assume that V_y is negligible compared to V_x.

Also obtain an expression for τ_{yx} acting on the plate as a function of the position x.

Thus estimate the dimensionless drag force, $\dfrac{\text{Drag}}{\left(\frac{1}{2}\rho V_0^2\right)\cdot(LW)}$, on the plate of length L and width W.

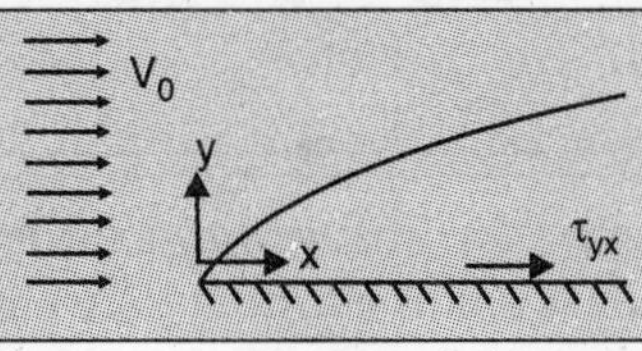

6.11 Obtain the pressure and shear stress distributions for fully developed laminar flow of a fluid between two infinite, parallel and *stationary* flat plates. Why are the results independent of ρ? Also obtain the rotational component of motion ω_z.

6.12 The stress components at a point in a 2-*D* flow are given as

$$\sigma_{xx} = 10 \text{ Pa} \qquad \tau_{xy} = 3 \text{ Pa}$$
$$\sigma_{yy} = -2 \text{ Pa}$$

Obtain the stress vector on an area element characterized by the normal vector $5\,\hat{\mathbf{i}} + 2\,\hat{\mathbf{j}}$.

6.13 *Rigid-body like motion:* When a tank containing a liquid moves at a constant acceleration **a**, each element of the liquid moves with the tank with the same velocity and acceleration. Show that Navier-Stokes equation for such a case reduces to

$$\rho\mathbf{f} - \nabla p = \rho\mathbf{a}$$

with the viscous terms dropping out. The same equation is obtained for general fluid motion when viscosity μ is zero, and is known as the *Euler equation.* Solve this equation for the tank shown to obtain the shape of the free surface (shown by the broken line). Note that the pressure is constant on the free surface.

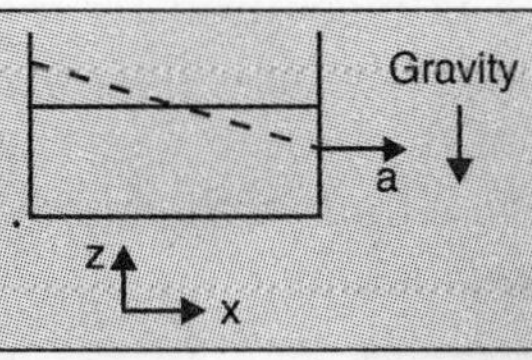

6.14 In designing the fuel tank of a rocket, care must be taken to make the bottom of the tank strong enough to bear the higher pressures during acceleration. If a tank carries liquid fuel (ρ = 800 kg/m^3) 1.5 m deep and the maximum upward acceleration of the rocket is 40 m/s^2, find the stress for which the tank bottom must be designed.

6.15 Compute the horizontal acceleration a_x of the tank of Prob. 2.11 such that the pressure at point *A* becomes atmospheric. What happens as the acceleration increases beyond this value? Also compute the pressure at *A* and the acceleration a_x when the water level is just above point *B*. Note that this is the minimum pressure reached at *A*.

6.16 Compute the acceleration *a* of the tank shown so that the liquid just spills out.

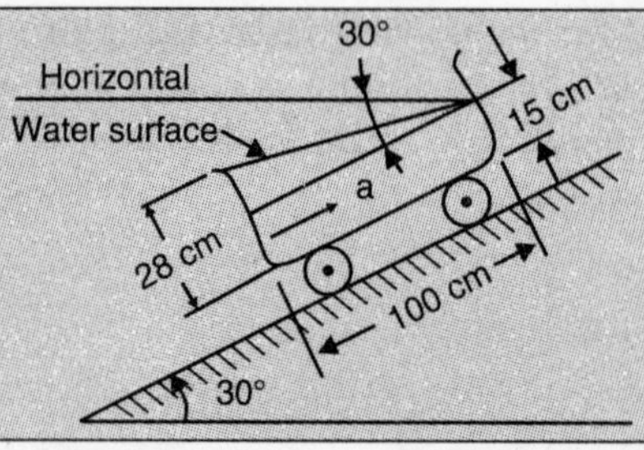

6.17 *Persian wheel:* A simple device for lifting water for irrigation is shown. Semi-cylindrical buckets are fixed to an endless chain running between two wheels. Estimate the slope of the water surface in a bucket at the position marked *A*.

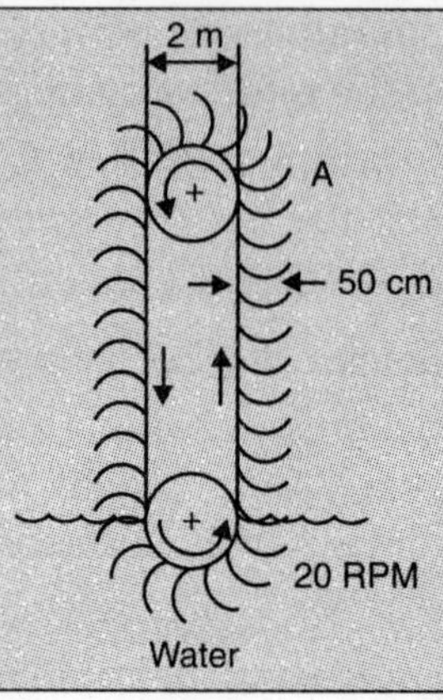

6.18 *Couette viscometer:* In the Couette viscometer shown, the fluid, whose viscosity is to be measured, is filled between two concentric cylinders, the inner rotating at a uniform angular velocity Ω while the outer one is restrained by a torsional spring. The torque acting on the outer cylinder is measured by means of the deflection of the spring. Assuming the flow to be steady, laminar, purely circular, and independent of z, obtain an expression for μ in terms of the torque T on the outer cylinder, angular speed Ω and the geometric parameters. Show that, if the gap between the two cylinders is small, the velocity profile is approximately linear.

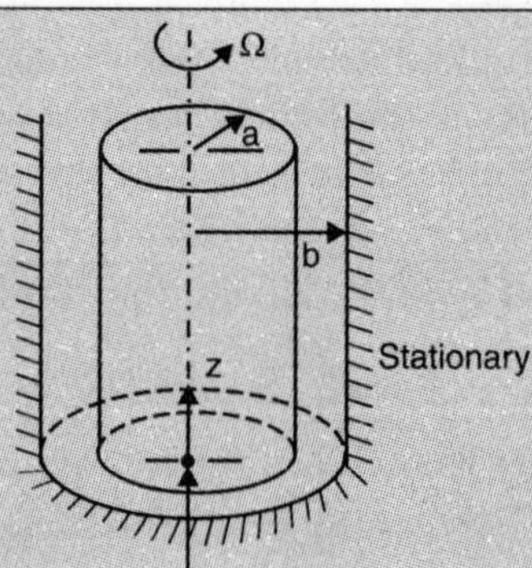

6.19 A common method used for measuring viscosity of a fluid *in situ* consists of measuring the torque required for rotating a cylinder at constant angular speed in the fluid (assumed to be of infinite expanse). Find the relationship between the torque and the viscosity in terms of the cylinder parameters. Assume laminar flow.

6.20 A Couette viscometer described in Prob. 6.18 has the following dimensions

$a = 3$ cm, $b - a = 2.5$ mm, length $L = 10$ cm

It gives the following readings for a polymer solution

Speed of rotation (rpm)	*Torque* (N m)
0.08	0.17
0.8	1.2
8	1.8
80	6.0

Using the approximation of a thin annulus, compute the viscosity at the different values of the rate of shear strain, $\mathring{\gamma}$. Is the fluid Newtonian?

6.21 A liquid of density ρ and viscosity μ flows laminarly down a wide flat inclined plate as shown. After an initial developing flow region, the depth of the liquid becomes constant at h. Show that the pressure within the fluid in the fully developed region is a function of y alone, and is given by the hydrostatic pressure distribution with g replaced by $g \cos \theta$. Note that the pressure variation in the vertical direction is *not* hydrostatic. Why? Obtain the velocity profile and the flow rate per unit width of the plate.

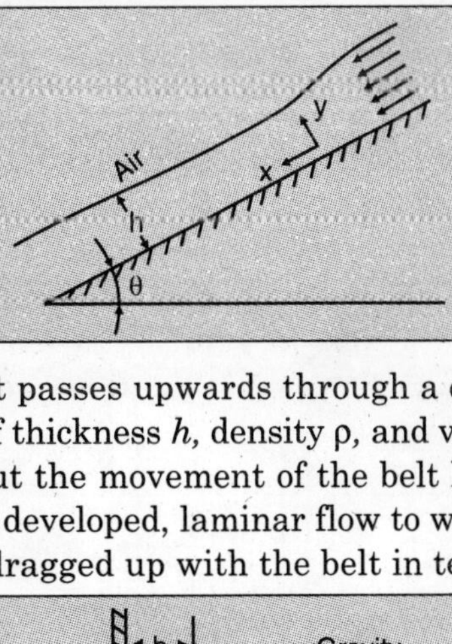

6.22 A continuous moving belt passes upwards through a chemical bath at velocity V_0 and picks up a film of liquid of thickness h, density ρ, and viscosity μ. Gravity tends to make the liquid drain down, but the movement of the belt keeps the fluid from running off completely. Assume fully developed, laminar flow to write an expression for the rate at which the fluid is being dragged up with the belt in terms of ρ, μ, h and V_0.

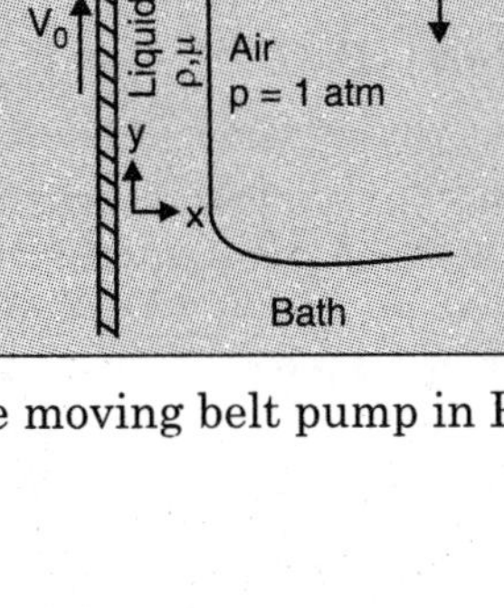

6.23 How is h determined for the moving belt pump in Prob. 6.22?

6.24 Water and oil flow down a vertical plane. The flow is steady, laminar and fully developed. Simplify the Navier-Stokes equation separately for water and oil and write the relevant boundary conditions. Obtain the two velocity profiles. Sketch these qualitatively, taking care near the oil-water interface.

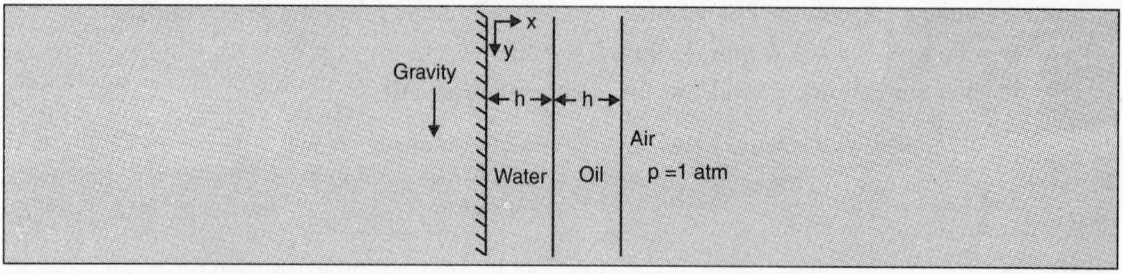

6.25 Consider the steady, laminar, incompressible flow between two large parallel plates as shown. The upper plate moves with velocity V_0 to the right and the lower plate is stationary. The pressure gradient in the flow direction is zero. The lower half of the region between the plates is filled with fluid *A*, and the upper half is filled with fluid *B*. Assume fully developed laminar flow to obtain the velocity profile.

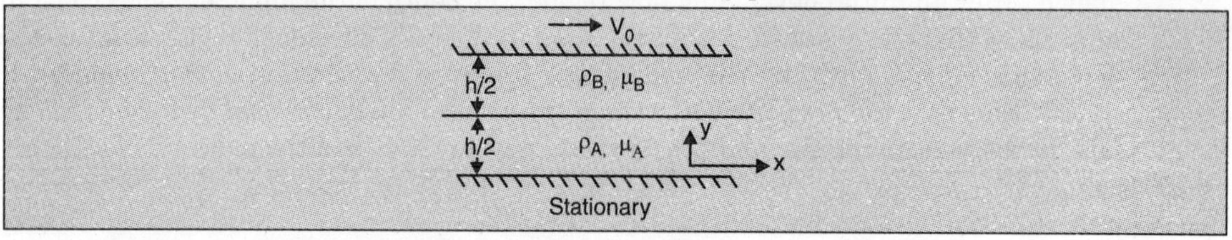

6.26 A wetted wall column is used to measure mass transfer coefficients. A liquid of density ρ and viscosity μ flows down inside of a tube of radius *R* shown. After an initial region, the flow becomes fully developed and the thickness of the fluid layer is constant at *h*. Simplify the Navier-Stokes and continuity equations to obtain $V_z(r)$ for laminar conditions.

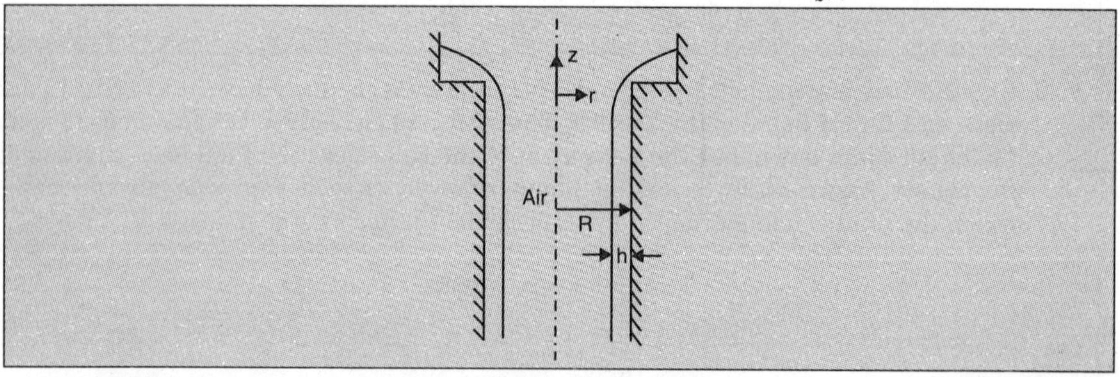

6.27 An incompressible fluid flows radially through a long and thin inner porous cylinder of radius R_1 across to the outer, concentric porous cylinder of radius R_2. The non-gravitational pressure at R_1 is $\mathcal{P}_1$ ($= p_1 + \rho g z$) and the radial velocity at the inner cylinder is V_1. For steady, laminar flow between the two cylinders, simplify the continuity equation and the three components of the Navier-Stokes equation and then solve them to obtain $\mathcal{P}(r)$ in terms of $\mathcal{P}_1$ and V_1.

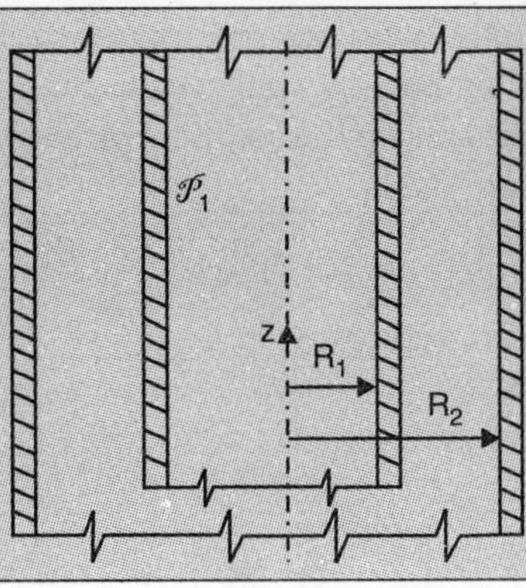

6.28 *Coating of wires*: A wire to be coated moves at a velocity V_0 through a long cylindrical die filled with an incompressible fluid. Obtain the velocity profile $V_z(r)$ inside the die, and the power required to pull the wire. Neglect end effects. Assume $\partial p/\partial z$ to be zero, and flow to be laminar.

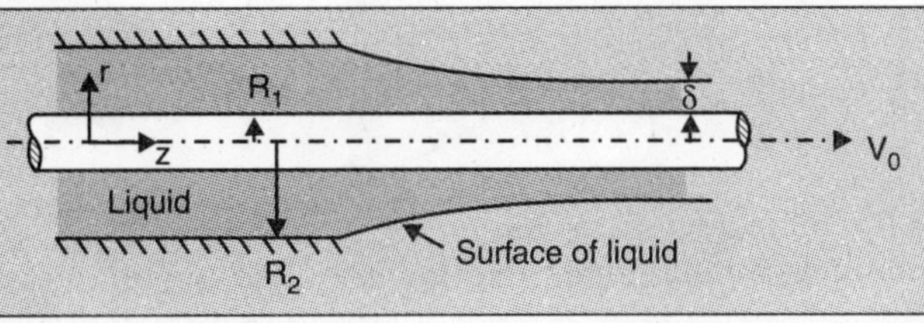

6.29 In Prob. 6.28, the velocity profile outside the cylindrical tank changes till at some distance far downstream it becomes uniform. Show that this is consistent with the Navier-Stokes equation. Also obtain an expression for the thickness δ of the coating far downstream.

6.30 A viscous incompressible liquid flows down a long, cylindrical rod of radius R as a thin, fully-developed laminar film. Simplify the continuity and Navier-Stokes equations and solve for $V_z(r)$.

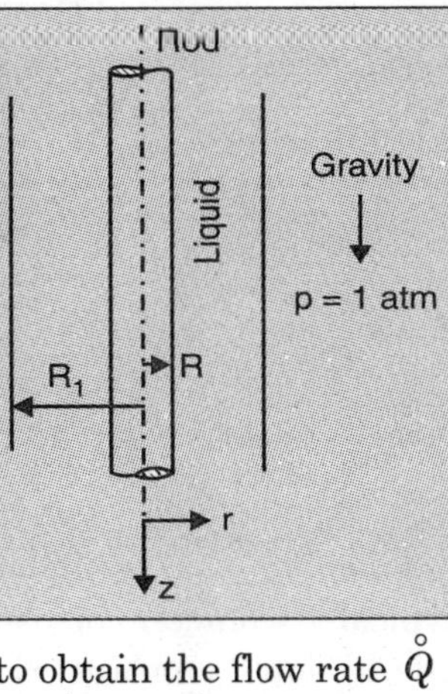

6.31 In Example 6.1, use Eq. (e) to obtain the flow rate $\mathring{Q}$ across any section of width w. Also obtain an expression for the drag force on the lower plate per unit length in the x-direction.

6.32 The lower plate of a lubricated thrust bearing moves to the right at velocity V_0. The stop at the right prevents fluid flow beyond that point. Find the weight W supported by the fluid (of viscosity μ and of density ρ). Assume l/b large so that the end effects can be neglected. Use results of Prob. 6.31.

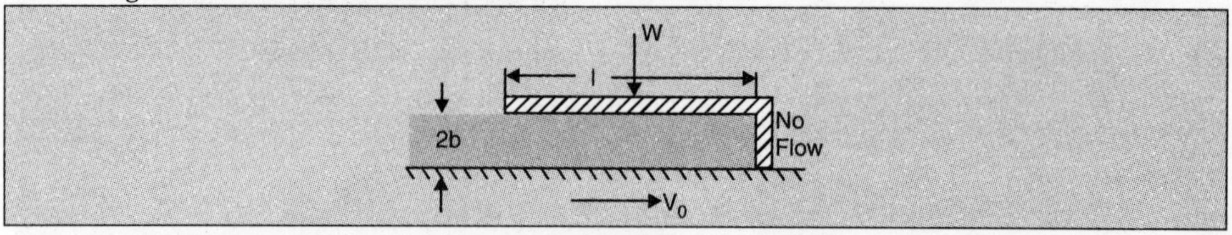

6.33 A belt of width w moves at velocity V_0 as shown. A liquid fills the gap between the belt and an upper plate and is dragged down to tank B. After some time, tank B gets filled up and steady conditions prevail. What force must be applied on the upper plate to hold it in place? Do not neglect the effect of gravity.

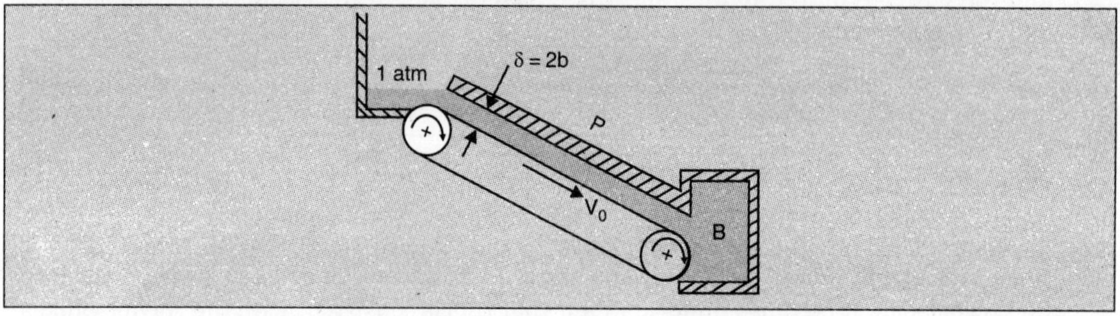

6.34 A weigh-bridge is designed based on the system in Prob. 6.33. The object to be weighed is placed on the *horizontal* platform P and the belt is started at velocity V_0. The film thickness δ is directly calibrated to give the load. Is the scale linear?

6.35 A viscosity pump shown takes in liquid at A and delivers it at a higher pressure at B. If the thin gap δ between the rotating 'piston' and the outer casing is constant, find the pressure rise and the power required by the pump as a function of ω, ρ, μ, width w, δ, R and the flow rate $\mathring{Q}$. Assume laminar flow.

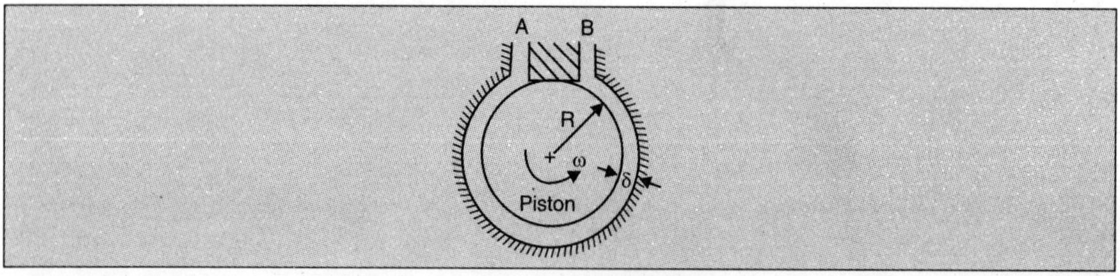

6.36 *Capillary-flow viscometer:* A typical viscometer consists of a capillary of 2 mm diameter through which a liquid is made to flow by means of a piston. A liquid of density 800 kg/m^3 flows out at a rate of 15 ml/min (computed from the dimensions of the piston and its velocity). Compute the viscosity of the liquid. Assume tank diameter to be very large and the pressure inside it to be uniform at 0.5×10^7 Pa *gauge* (as measured by a 'load-cell' on the piston).

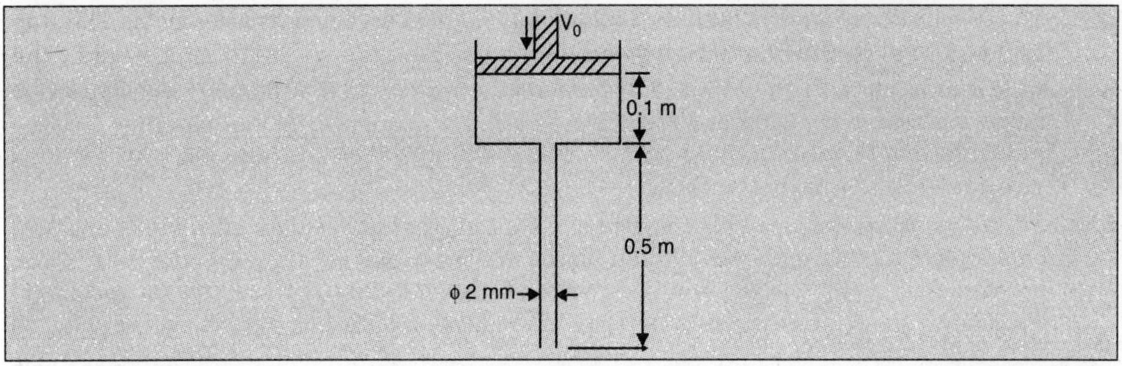

6.37 *Penetration depth:* Consider a large flat plate initially at rest, with an incompressible fluid above it. At $t = 0$, the plate starts moving to the right at a constant velocity V_0. Simplify Navier-Stokes equation for $V_x\ (y, t)$. Use the fact that the pressure as $y \to \infty$ is independent of x. Using the transformation $\eta = \dfrac{y}{2\sqrt{\nu t}}$, show that

$$\frac{V_x}{V_0} = 1 - \frac{2}{\sqrt{\pi}} \int_0^{\eta} e^{-\xi^2} d\xi \equiv \operatorname{erfc}(\eta)$$

Though at any time the velocity everywhere is non-zero (see Sec. 1.4) a common approximation is to regard the region where the velocity is less than one per cent of V_0 as 'undisturbed'. As t increases the velocity profile stretches outwards from the plate. Thus, the effect of plate motion *penetrates* further with time. Show that the penetration depth (defined as above) is $3.6\sqrt{\nu t}$ given that erfc (η) = 0.01 at η = 1.8.

6.38 *Cone and plate viscometer:* It consists of a cone rotating at a constant angular velocity ω over a stationary plate. Liquid fills the gap between the cone and plate and the torque required to keep the plate stationary is measured through the deflection of a torsional spring. Simplify the Navier-Stokes equation assuming that flow is laminar and tangential (i.e., only $V_\phi \neq 0$), and that the intertial and gravity terms are negligible. Assuming V_ϕ of the form: $V_\phi = r\,f(\theta)$, show that $f(\theta)$ must satisfy

$$f'' + f'\cot\theta + f\left(2 - \frac{1}{\sin^2\theta}\right) = 0$$

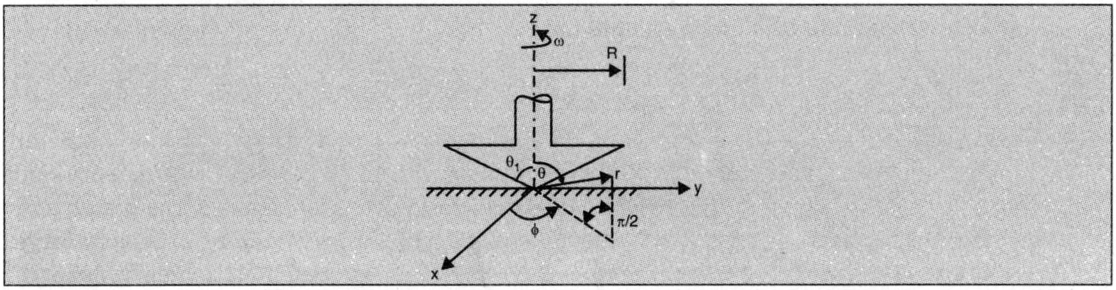

Since this is an ordinary differential equation it is much easier to programme on a digital computer than the full Navier-Stokes equation.

6.39 If the cone angle $\pi/2-\theta_1$ in Prob. 6.38 is small, the velocity V_ϕ may be fairly well approximated as linear in the vertical direction (see Prob. 1.4). Compute the torque acting on the lower plate in terms of μ, R, ω, and θ_1. If $R = 5$ cm, $\theta_1 = 89°$, torque $= 5 \times 10^{-3}$ N m, rotational speed 2 rpm, determine μ.

6.40 *Parallel plate viscometer:* This consists of a stationary circular plate over which another similar plate rotates as shown. Fluid is placed in the gap and the torque on the lower plate is measured. Simplify the Navier-Stokes equation and show that $V_\theta = \omega rz/(2h_0)$ is a solution. Then obtain an expression for μ in terms of the measured torque T on the lower plate.

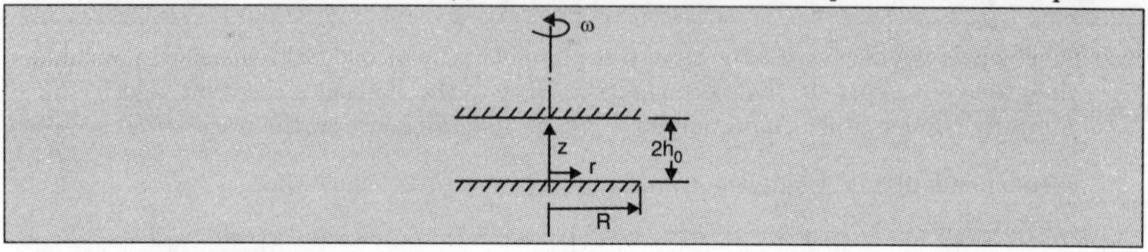

6.41 A thrust bearing shown is lubricated by pumping oil at high pressure p_0. Find the vertical load that the bearing can support and the flow rate of oil required. Neglect inertial and body force terms. Note that under laminar conditions both V_r and V_θ in the thin gap will be non-zero (such a bearing is used in the famous Mt. Palomar telescope). Also show that $V_\theta = \dfrac{\omega r}{2}\left(1+\dfrac{z}{h}\right)$ satisfies the θ component of the Navier-Stokes equation.

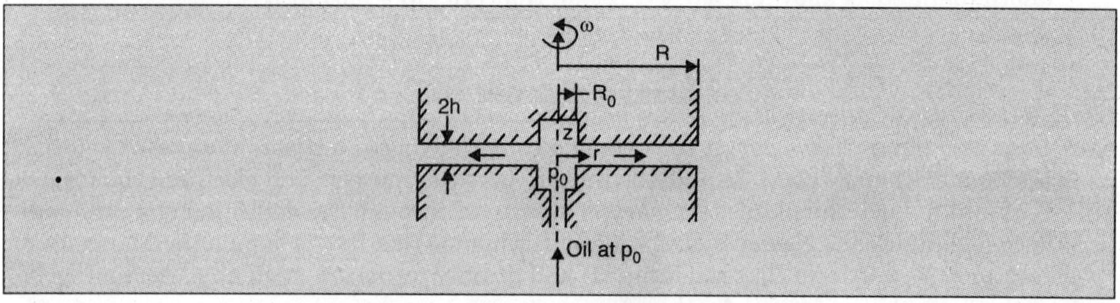

6.42 Consider the flow of a fluid in the conical tube shown. The cone angle is small, i.e., $(D_2 - D_1)/L \ll 1$. Both V_r and V_z are non-zero, but V_r is small enough so that we can assume a quasi 1-D situation to obtain dp/dz in terms of the flow rate $\mathring{Q}$ and the diameter D at any z. Integrate this expression to obtain

$$\frac{p_1 - p_2}{L} = \frac{128\,\mu \mathring{Q}}{3\pi(D_2 - D_1)}\left(\frac{1}{D_1^3} - \frac{1}{D_2^3}\right)$$

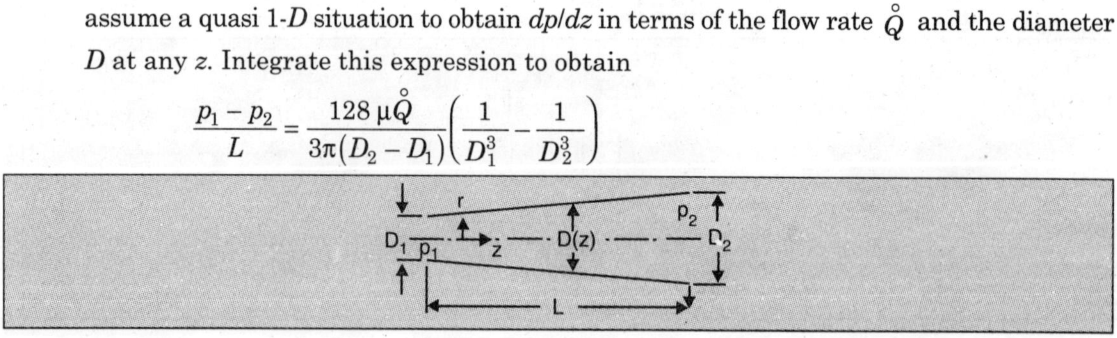

6.43 *Centrifugal buoyancy:* In several pollution control devices (like *cyclones)* particle-laden fluid enters a chamber in which it is made to swirl. The swirl produces a radial pressure gradient which results in a net radial pressure force acting on the solid particles. Consider a small 'cube' of volume in a liquid of density ρ rotating with constant angular velocity Ω. Show that the net radial pressure force on the particle is

$$F = -\rho\Omega^2 R \forall$$

where R is the location of the particle from the axis of rotation. Note that this is akin to the buoyancy force and is termed as centrifugal buoyancy. Thus, a particle denser than the liquid moves outwards while a lighter one moves towards the axis.

Estimate the *net* radial force on a sand particle of volume 10^{-9} m^3 situated at 5 cm from the axis in water rotating at 5000 rpm. Density of sand may be taken as 2500 kg/m^3. Compare this with the weight of the particle.

6.44 A cylindrical bucket with a shaft at the centre carries a solid sphere of mass m and density ρ_s on a string of length L as shown. When the assembly rotates at ω, the sphere acquires a steady location. Compute θ and the tension in the string.

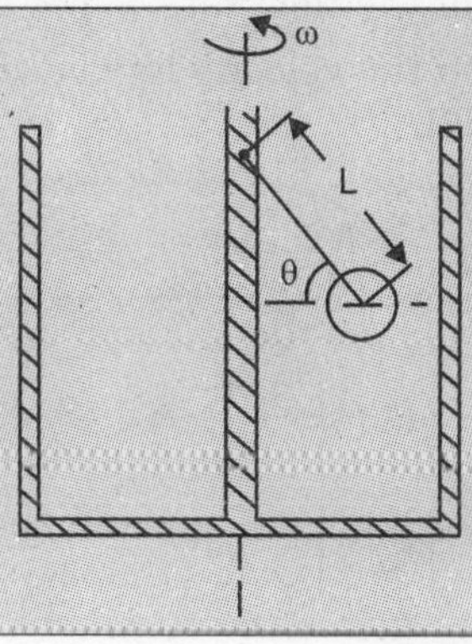

When the tank is filled up with a liquid of density ρ_f so that the sphere is submerged and the entire assembly is again rotated at ω, what is the new value of θ and the tension?

6.45 In Prob. 5.32 the bending of a jet of liquid touching the sides of a cylinder was discussed. Show using the equation of motion in natural coordinates that the direction of the force acting on the cylinder is indeed correct.

6.46 *Inertial impaction:* The capture of airborne particles such as dust, fog droplets and pollen on solid surfaces is an important phenomenon. It is governed essentially by inertial and pressure forces. Show that

(a) while particles denser than air tend to be deposited on the body, lighter than air particles cannot be caught by the wind-facing side of the body.

(b) for bodies of a given shape, smaller ones are more efficient in catching airborne particles, and therefore, while snow piles up upon thin poles and wires, it hardly accumulates on large wind-facing walls.

(c) bodies with sharp corners gather more particles.

(d) the presence of viscosity tends to reduce the efficiency of catch.

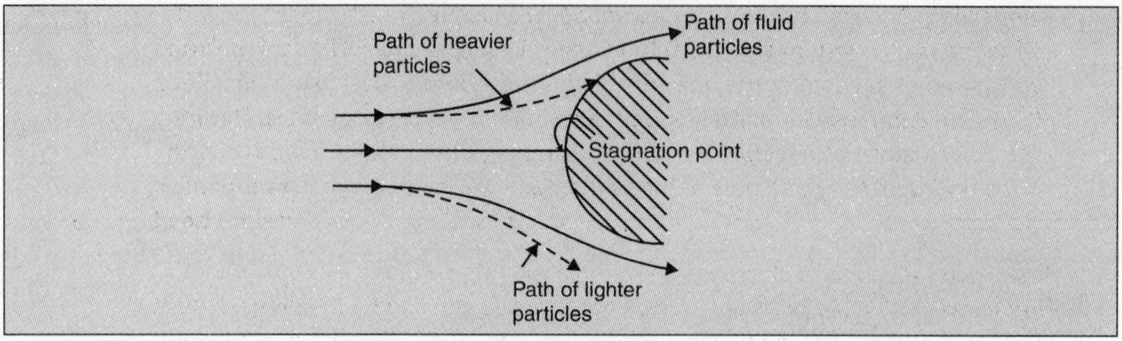

6.47 A hydrogen-filled balloon is less likely to collide with a building than an air-filled one. Why? Consider only the horizontal motion.

6.48 A rectangular channel has a 90° smooth bend. Two pressure taps are located as shown and measure the pressure difference $p_0 - p_1$. As a very crude approximation take the streamlines to be circular and concentric and velocity independent of the location r to obtain $p_0 - p_1$ as a function of the flow rate $\mathring{q}$ per unit width. In actual flows both the assumptions are invalid, with the velocity higher at the inner edge compared to that at the outer edge (see Prob. 7.56).

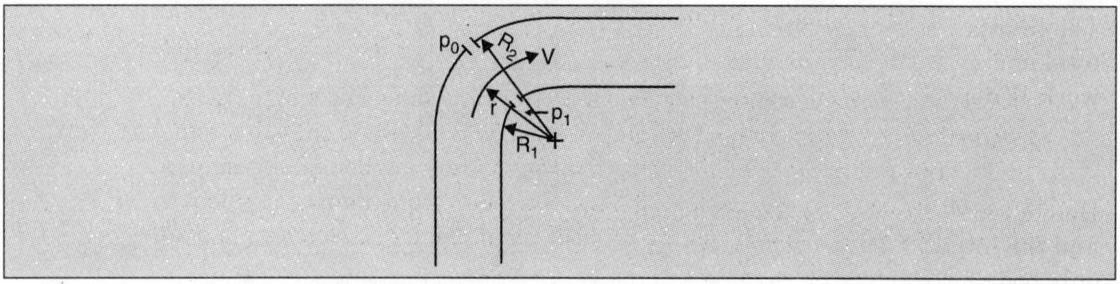

6.49 In measuring the clotting time of blood (see Prob. 2.28), a plug of blood is taken in the capillary tube and is allowed to flow down the capillary (because of its own weight). The time t taken by the plug to traverse the full length of the tube is taken as a measure of the blood viscosity. Show that t is directly proportional to the blood viscosity for such a flow. Assume that during flow the surface tension forces are unimportant and that the velocity V_z at any radial position is given by

$$V_z(r) = 2\bar{V}\left(1 - \frac{r^2}{R^2}\right).$$

Discuss the change in the shape of the plug with time. (Hint: the no slip condition must be satisfied).

6.50 Derive Eq. 6.6 using the procedure described.

7

ENERGY EQUATIONS

7.1 FIRST LAW OF THERMODYNAMICS

The first law of thermodynamics for a system of fluid particles states that the increase in the total energy E of the system is equal to the energy added to the system as heat Q *minus* the work W done *by* it on the surroundings. In terms of the rates of change,*

$$\frac{DE}{Dt} = \mathring{Q} - \mathring{W} \qquad ...(7.1)$$

Here E represents the total energy of the system which includes the kinetic and potential energies and the internal energy associated with the random motion of the molecules. If gravity is the only body force acting on the system, the specific energy (per unit mass) is given by

$$e = \frac{V^2}{2} + gz + u \qquad ...(7.2)$$

where $V^2/2$ is the specific kinetic energy, gz represents the specific potential energy with z measured from some arbitrary datum, and u represents the internal energy per unit mass. In Eq. (7.1), $\mathring{Q}$ is the rate of heat added to the system *by conduction* across the boundaries, and $\mathring{W}$ is the rate at which the system does work on the surroundings through the *surface forces.* The contribution of body forces is already accounted for by the inclusion of the potential energy term in E and so is excluded from $\mathring{W}$ to avoid duplication.

For a control volume fixed in an inertial frame of reference we can use Reynolds transport theorem Eq. (7.3) to evaluate DE/DT in terms of the local and convective rates of change. Thus,

$$\frac{DE}{Dt} = \frac{\partial}{\partial t}\iiint_{CV} e\rho \, d\mathcal{V} + \oiint_{CS} e(\rho \mathbf{V} \cdot d\mathbf{A}) = \mathring{Q} - \mathring{W} \qquad ...(7.3)$$

---

* $\mathring{Q}$ has also been used for the volumetric flow rate. The context will make it clear as to which of the two is intended.

On moving the surface integral term to the right, Eq. (7.3) states that the rate of change of energy 'contained' in a *CV* (i.e., the rate of accumulation) is equal to the rate of heat addition $\overset{\circ}{Q}$ across the *CS* *minus* the rate of work done by the *CV* (through surface forces) *minus* the energy convected across the *CS* in association with the mass flowing across it.

7.2 WORK DONE BY SURFACE FORCES

As stated in Sec. 7.1, the work term in Eq. (7.3) includes the contribution of all the surface forces acting on the *CS*. A force does work only if the element on which it acts undergoes a displacement along the line of action of the force. Therefore, not all surface forces do work. Figure 7.1 shows various types of *CS* elements from the point of view of work done:

(a) a portion A_i of the *CS* may lie *adjacent* to an impervious and stationary solid surface. Even though there are both normal and shear stresses acting on A_i, no work is done since there is no corresponding motion.

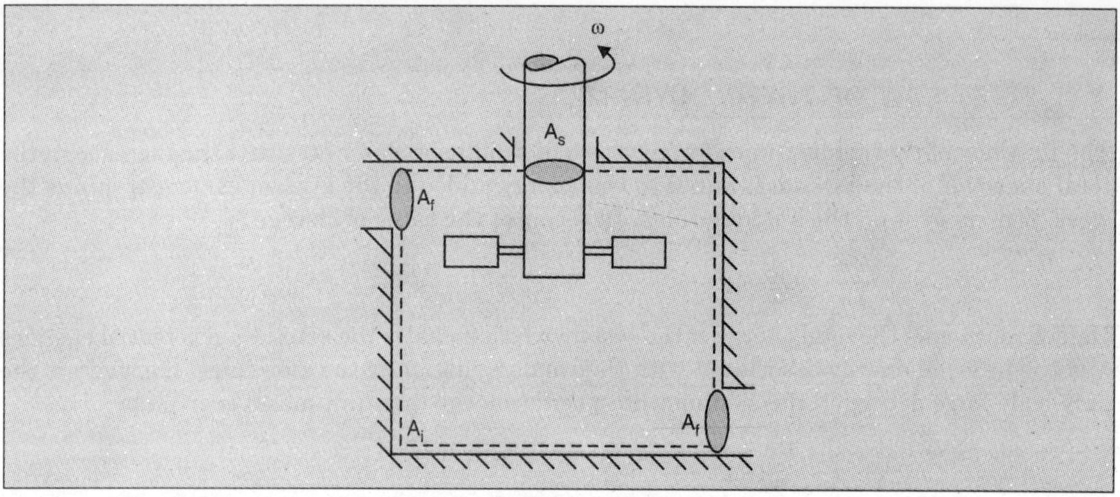

Fig. 7.1. Various types of *CS* elements from the point of view of work done.

(b) A portion A_s of the *CS* may *cut across a* solid body. In general, there will be stresses within the solid body. Usually, two types of such cuts are of engineering interest: one, in which the solid is at rest (and does not contribute to the work done) and the other, when the solid rotates, as does a shaft. On a rotating shaft, shear stresses act on section A_s (Fig. 7.2) and do work because of the associated motion. This work is termed *shaft work.* We will denote the rate of doing shaft work *on* the surroundings as $\overset{\circ}{W}_s$. Thus, if the *CS* encloses a turbine, $\overset{\circ}{W}_s$ is positive and equal to the turbine power output, and if the *CS* encloses a pump or a blower, $\overset{\circ}{W}_s$ is negative, equal to the power input.

(c) A portion A_f of the *CS* across which the fluid flows. At such a surface, both shear as well as normal stresses do work, since, in general, both normal and tangential components of velocities are present at these locations. The total rate of work done *by* the system at A_f is seen to be

$$-\iint_{A_f} \mathbf{V}.\left(\sigma\hat{\mathbf{n}} + \tau\hat{\mathbf{t}}\right) dA$$

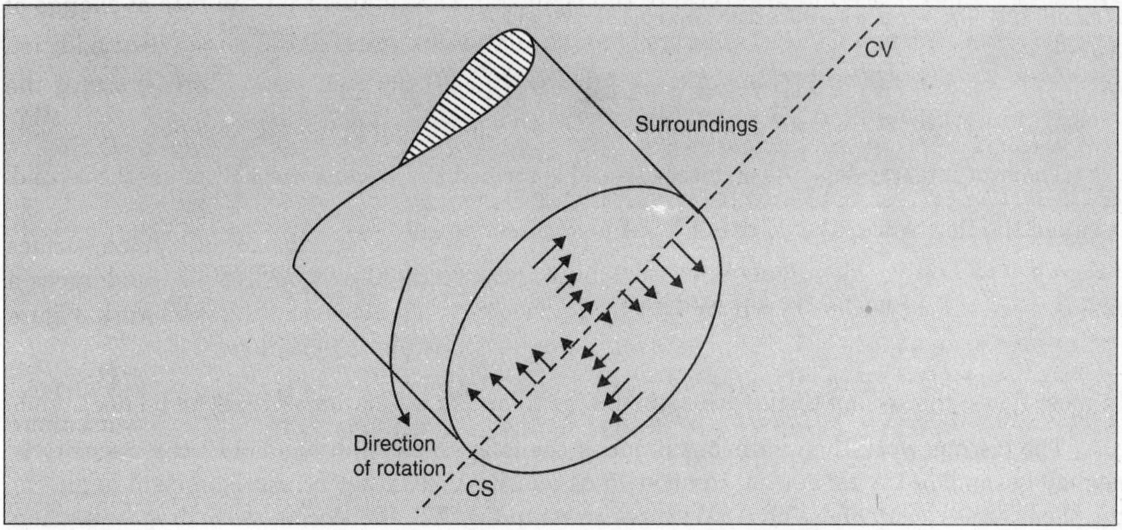

Fig. 7.2. Shear stress distribution on a rotating shaft. Stresses *on* the *CV* will have the opposite sense.

where $\left(\sigma\hat{\mathbf{n}}+\tau\hat{\mathbf{t}}\right)$ is the stress vector acting *on* the *CS*. With reference to Fig. 7.3,

$$\mathring{W}_{A_f} = -\iint_{A_f} \tau\ V_t \cos\alpha dA - \iint_{A_f} \sigma V_n dA$$

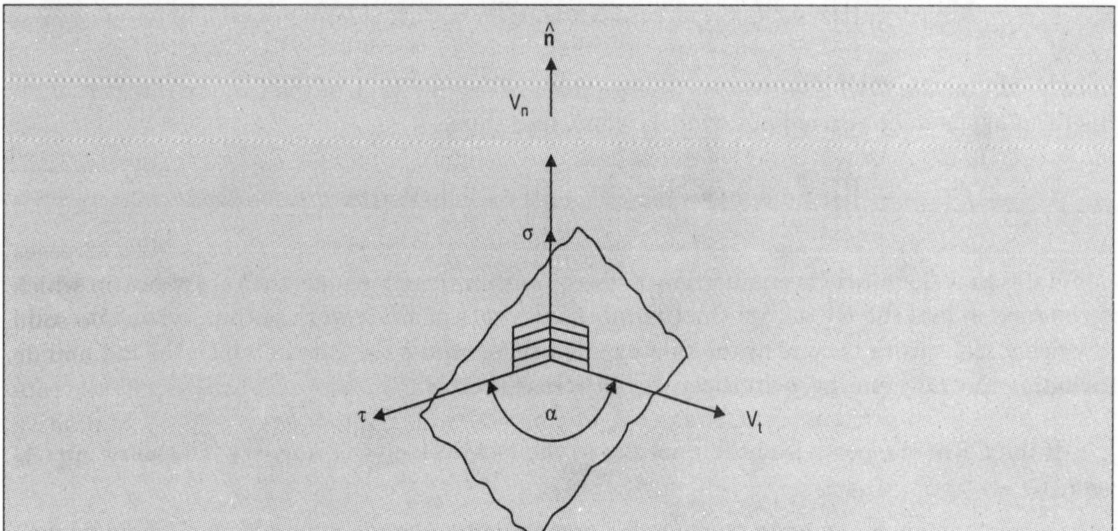

Fig. 7.3. Stress and velocity components *on* an area element $d\mathbf{A}$ of the *CS*. Stress components on the surroundings have opposite signs.

Writing $\sigma = -p + 2\mu\, \partial V_n/\partial n$ after Eq. (6.14), we get

$$\overset{\circ}{W}_{A_f} = -\iint_{A_f}\left(\tau V_t \cos\alpha + 2\mu \frac{\partial V_n}{\partial n} V_n\right) dA + \iint_{A_f} p V_n dA$$

The first integral depends on viscosity and is termed the viscous work W_τ while the second is called the flow work, $\overset{\circ}{W}_f$. Writing $V_n dA$ as $\mathbf{V} \cdot d\mathbf{A}$, we get

$$\overset{\circ}{W}_{A_f} = \overset{\circ}{W}_\tau + \iint_{A_f} p\mathbf{V} \,.\, d\mathbf{A}$$

In most flows, the second term within the integral for $\overset{\circ}{W}_\tau$ is very small and can be neglected.

The viscous work then vanishes if one of the following conditions holds: the viscosity is negligibly small or the tangential component of velocity V_t is either zero or is at right angles to the shear stress everywhere on A_f. With reference to Fig. 7.3, the viscous work drops out when either $\mathbf{V}$ is along $\hat{\mathbf{n}}$ (i.e., $V_t = 0$), or the angle α between τ and V_t is $\pi/2$. We can usually arrange for V_t, and therefore $\overset{\circ}{W}_\tau$, to be zero by choosing the *CS* normal to the velocity vector at the ports.

7.3 THE ENERGY EQUATION

Equation (7.3) can now be rewritten in terms of the various types of work as

$$\frac{\partial}{\partial t}\iiint_{CV} e\, \rho d\forall + \oint\!\!\oint_{CS} e\rho\mathbf{V} \cdot d\mathbf{A} = \overset{\circ}{Q} - \overset{\circ}{W}_s - \overset{\circ}{W}_\tau - \iint_{A_f} p\mathbf{V} \cdot d\mathbf{A} \qquad \text{...(7.4)}$$

Since $\mathbf{V} \cdot d\mathbf{A}$ on the LHS is zero everywhere except at segments A_f of the *CS*, the integration in this term needs to be carried out over A_f alone, and thus,

$$\frac{\partial}{\partial t}\iiint_{CV} e\rho\, d\forall = \overset{\circ}{Q} - \overset{\circ}{W}_s - \overset{\circ}{W}_\tau - \iint_{A_f}\left(e + \frac{p}{\rho}\right)\rho\mathbf{V} \cdot d\mathbf{A} \qquad \text{...(7.5)}$$

This states that the rate of accumulation of energy within the *CV* equals the rate of heat transfer by *conduction* into the *CV* across the *CS*, *minus* the rate of shaft work output, *minus* the rate at which viscous work is done at the inlets and outlets, *minus* the rate at which the net energy (including the 'flow energy' p/ρ) is convected across the *CS*.

If the *CS* at the ports is chosen normal to the local velocity vector, $\overset{\circ}{W}_\tau = 0$ then Eq. (7.5) becomes

$$\frac{\partial}{\partial t}\iiint_{CV} e\rho\, d\forall = \overset{\circ}{Q} - \overset{\circ}{W}_s - \iint_{A_{f,o}}\left(e + \frac{p}{\rho}\right)\rho V\, dA + \iint_{A_{f,i}}\left(e + \frac{p}{\rho}\right)\rho\, V dA \qquad \text{...(7.6)}$$

where $A_{f,i}$ represents the area of the inlet ports and $A_{f,o}$ represents the area of the outlet ports. Equation (7.6) is valid for viscous fluids as well. The only assumption made in the derivation of this equation is that the CS segments $A_{f,i}$ and $A_{f,o}$ are so chosen that the velocity of the fluid is everywhere normal to them and, therefore, $\overset{\circ}{W}_\tau = 0$. Of course, if the viscous stresses are negligible Eq. (7.6) holds without this restriction on the choice of the CS.

7.4 SPECIAL CASES

One of the most important special cases of Eq. (7.6) occurs when the flow is steady, so that the time rate of change of energy contained within the CV is zero. Equation (7.6) then reduces to

$$\iint_{A_{f,o}} \left(e + \frac{p}{\rho}\right) \rho V \, dA = \iint_{A_{f,i}} \left(e + \frac{p}{\rho}\right) \rho V \, dA + \overset{\circ}{Q} - \overset{\circ}{W}_s \qquad \text{...(7.7)}$$

Thus, the convective efflux of energy (including the 'flow energy') is equal to the *sum* of the convective influx of energy and the energy conducted in as heat, *minus* the shaft work done by the system.

A general class of fluid mechanical problems (viz., those involving fluid machinery like pumps, turbines, etc.) consists of cases where the fluid field is not exactly steady but is changing *cyclically*. For example, with the paddle stirrer of Fig. 7.1 the flow field changes with time but it is reasonable to suppose that the variations are cyclic with the time period of rotation of the stirrer. If we average the energy equation over one time period in this case, the averaged quantities should not change with time. Thus, Eq. (7.7) should hold with the understanding that all quantities are now averaged over a rotational cycle.

Further simplification of Eq. (7.7) results if one-dimensionality is assumed at the inlet and exit ports, i.e., the flow variables do not change across the entry and exit areas. Equation (7.7) then takes the following form for one inlet and one outlet port:

$$\left[\left(e + \frac{p}{\rho}\right) \rho V A\right]_o = \left[\left(e + \frac{p}{\rho}\right) \rho V A\right]_i + \overset{\circ}{Q} - \overset{\circ}{W}_s \qquad \text{...(7.8)}$$

On realizing that $[\rho VA]_o$, = $[\rho VA]_i = \overset{\circ}{m}$, the steady mass flow rate through the control volume, Eq. (7.8) becomes

$$\left(e + \frac{p}{\rho}\right)_o = \left(e + \frac{p}{\rho}\right)_i + q - w_s \qquad \text{...(7.9)}$$

where $q(\equiv \overset{\circ}{Q} / \overset{\circ}{m})$ and w_s $(\equiv \overset{\circ}{W}_s / \overset{\circ}{m})$ are the heat input (by conduction) and shaft work output *per unit mass throughput* respectively. Subscripts o and i imply that the quantities are to be evaluated at the outlet and inlet conditions respectively. On using Eq. (7.2) for the specific energy e, we get

$$\left[u + \frac{V^2}{2} + gz + \frac{p}{\rho}\right]_o = \left[u + \frac{V^2}{2} + gz + \frac{p}{\rho}\right]_i + q - w_s \qquad \text{...(7.10)}$$

CHAPTER 7

This equation states that the 'energy' of the fluid at the outlet is equal to the 'energy' of the fluid at the inlet modified by the energy added as heat and the energy extracted as shaft work. The 'energy' of the fluid consists of both the thermal energy (through u) as well as the mechanical energy (through $V^2/2$, gz and p/ρ). For an *incompressible* fluid, u is purely a thermodynamic quantity, and p is purely a mechanical one. Thus, for an incompressible fluid, we can separate the mechanical energy terms from the rest, and write

$$\left[\frac{V^2}{2}+gz+\frac{p}{\rho}\right]_o=\left[\frac{V^2}{2}+gz+\frac{p}{\rho}\right]_i-w_s-[(u_o-u_i)-q] \quad ...(7.11)$$

In a flow situation where viscous stresses are operative, some mechanical energy is being continually converted into thermal energy. The decrease in the mechanical energy as the fluid flows from the inlet to the outlet is always greater than the shaft work extracted from the fluid. This 'loss' is seen to be $[(u_o - u_i) - q]$ from Eq. (7.11) and is, therefore, always positive. The increase in internal energy, $u_o - u_i$ (and hence the temperature) of an incompressible fluid can take place in two ways: through the heat conduction q and through the irreversible viscous action. Thus, $[(u_o - u_i) - q]$ represents the conversion of mechanical energy into thermal energy due to viscous action. Equation (7.11) can then be rewritten as

$$\left[\frac{V^2}{2}+gz+\frac{p}{\rho}\right]_o=\left[\frac{V^2}{2}+gz+\frac{p}{\rho}\right]_i-w_s-w_l \quad ...(7.12)$$

which is termed the *mechanical energy equation* for incompressible fluids. Here, $w_l = (u_o - u_i) - q$ represents the loss of mechanical energy (per unit mass throughput) by viscous action*. In a non-viscous fluid, the loss w_l is zero, $q = u_o - u_i$, and, therefore, any heat added to the fluid raises its internal energy and temperature.

Note that Eq. (7.12) is valid only for an incompressible fluid, flowing through a two-port *CV*. In compressible flows, the pressure p has both mechanical and thermodynamic roles to play and, therefore, we cannot separate u and p/ρ as thermal and mechanical energies. Thus, we have to use the energy equation in the form given by Eq. (7.10) for such cases.

In hydraulic engineering practice, it is conventional to divide each term of Eq. (7.12) by g, the acceleration due to gravity:

$$\left[\frac{V^2}{2g}+z+\frac{p}{\rho g}\right]_o=\left[\frac{V^2}{2g}+z+\frac{p}{\rho g}\right]_i-h_s-h_l \quad ...(7.13)$$

---

* It may be noted that w_l represents a *conversion* of energy from one form to another (mechanical to thermal) and so does not appear in the *total energy equation* (7.10). However, it occurs as a loss in the mechanical energy equation and as a term representing 'generation' in the thermal energy equation (see, for example, V. Gupta, "*Elements of Heat and Mass Transfer*", New Age International, Publishers, New Delhi, 1995, Eq. (5.23). Also the viscous action, being internal to the *CV*, does not appear in the energy equation for a finite *CV* (Fig. 7.1). But if we write the energy equation for an infinitesimal *CV*, the velocity component V_t, at the surface is not, in general, zero and therefore, the viscous action has to be accounted for.

where $h_s = w_s/g$ and $h_l = w_l/g$. Each of the terms in this equation has dimensions of length and is termed as a *head.* Thus, z is called the *elevation head,* $p/\rho g$, the (static) *pressure head,* $V^2/2g$, the *velocity head,** and h_l is the *head loss.* The term h_s is the head corresponding to the shaft work. The sum of the terms in brackets is termed as the *total head* h_t, and Eq. (7.13) may be rewritten as

$$h_{t,o} = h_{t,i} - h_s - h_l \qquad \text{....(7.14)}$$

This states that the total head on the downstream side is the upstream total head *minus* the head corresponding to the shaft work *minus* the head loss due to viscous action. This energy equation finds widespread use in engineering practice.

Another term commonly used in hydraulic engineering practice is the *piezometric head* which is the sum of the pressure and elevation heads, $h_{pz} = \dfrac{p}{\rho g} + z$. It derives its name from the pressure measuring tubes called piezometers. If a tube is attached to a pipe in which a fluid is flowing (see Fig. 7.4), the fluid rises in the tube to a level given by $p/\rho g$, and thus the level of fluid in this tube above the elevation datum is $z + p/\rho g$, or, the piezometric head.

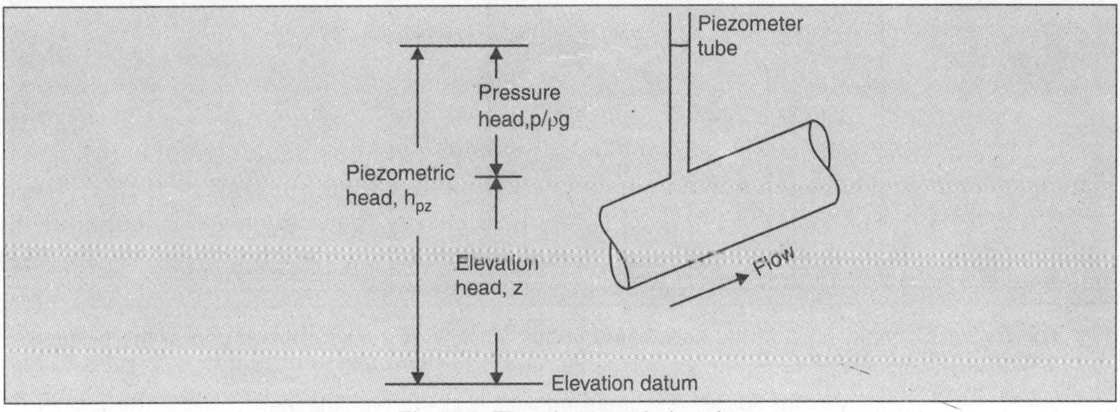

Fig. 7.4. The piezometric head.

Example 7.1. *Energy Balance for a Pump.* Consider a common household centrifugal pump used for lifting water. It draws up water from a line and delivers it at a point 2 m higher at an enhanced pressure (Fig. 7.5). It is given that $p_2 - p_1 = 0.1$ MPa and the volume flow rate is 0.01 m^3/s. If the power consumption is 1.5 kW, find the viscous losses. Neglect the purely mechanical losses in the transmission, bearings, etc.

We choose the *CV* coinciding with the inside boundary of the pump casing and intersecting the inlet and outlet sections normally, and use Eq. (7.13) (in a time-averaged sense). From the given data, we have

$$V_1 = \frac{0.01\left(m^3/s\right)}{\frac{\pi}{4}(0.1)^2\,(m)^2} = 1.27 \text{ m/s} \; ; \; V_1^2/(2g) = 0.0822 \text{ m}$$

---

* This form of the energy equation is valid for a 1-*D* flow across the inlet and outlet ports only. When the flow is not 1-*D*, we may still use this equation by replacing V by V_{av} and use a kinetic energy correction factor as in Prob. 7.13.

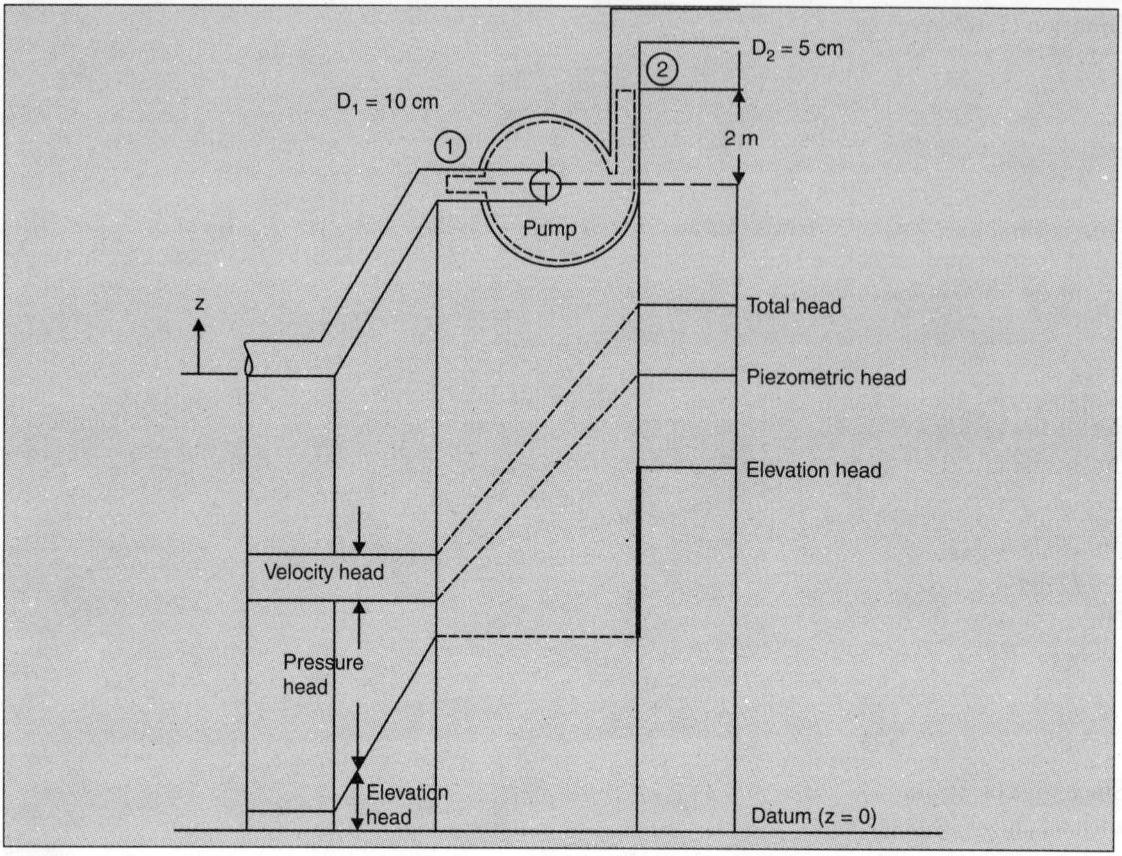

Fig. 7.5. *CV* for Example 7.1 and the associated heads for flow of a fluid through part of the pipeline. Note that the pump increases the total head of the fluid flowing. Viscous losses assumed negligible in the pipelines.

$$V_2 = \frac{0.01\left(\text{m}^3/\text{s}\right)}{\frac{\pi}{4}(0.05)^2\,(\text{m})^2} = 5.09 \text{ m/s} \; ; \; V_2^2/(2g) = 1.32 \text{ m}$$

$$h_s = \frac{w_s}{g} = \frac{\overset{\circ}{W}_s}{\overset{\circ}{m}\,g} = -1500\,(\text{J/s})\frac{1}{0.01\left(\text{m}^3/\text{s}\right)\times 10^3\left(\text{kg/m}^3\right)}\times\frac{1}{(9.81)\left(\text{m/s}^2\right)}$$

$$= -\,15.29 \text{ m}$$

This is negative since work is done on the *CV*.

$$\frac{p_2 - p_1}{\rho g} = \frac{0.1\times 10^6\,(\text{Pa})}{10^3\left(\text{kg/m}^3\right)\times 9.81\left(\text{m/s}^2\right)} = 10.19 \text{ m}$$

$$z_2 - z_1 = 2 \text{ m}$$

Equation (7.13) gives

$$h_l = \frac{V_1^2}{2g} - \frac{V_2^2}{2g} + \frac{p_1 - p_2}{\rho g} + (z_1 - z_2) - h_s$$
$$= 1.86\,\text{m}$$

This is the head loss by viscous action. It represents a power loss, $\mathring{W}_l = h_l g \mathring{m}$, or

$$\mathring{W}_l = 1.86(\text{m}) \times 9.81\left(\frac{\text{m}}{\text{s}^2}\right) \times 10^3 \left(\text{kg/m}^3\right) \times 0.01\left(\text{m}^3\text{/s}\right)$$
$$= 182.3\ \text{W}$$

Recall that the loss term $w_l (= \mathring{W}_l / \mathring{m})$ is given by $(u_2 - u_1) - q$, where $(u_2 - u_1)$ is the change in the internal energy of the incompressible fluid and q is the heat transfer per unit mass throughput. If the pump casing is insulated and it is assumed that the heat conduction at the two ports 1 and 2 are negligible (compared to the heat convected across the *CS* there), then w_l is the change in the internal energy of the fluid which is reflected as a rise in its temperature. Thus,

$$u_2 - u_1 = \frac{\mathring{W}_l}{\mathring{m}} = \frac{182.3(\text{W})}{10(\text{kg/s})} = 18.2\ \text{J/kg}$$

Since the specific heat of water is 4.18×10^3 J/kg K, this increase in internal energy is associated with a temperature rise of

$$18.2\ (\text{J/kg}) \frac{1}{4.18 \times 10^3\ (\text{J/kg K})} \quad \text{or a negligible } 0.0043\ \text{K}.$$

Example 7.2. *Maximum Power from a Hydroelectric Plant.* Water from behind a dam passes through a tube (called the *penstock)* and runs a turbine (Fig. 7.6). If the difference in elevation between the reservoir level and the discharge point is *H,* find the maximum power the turbine can generate for a given penstock diameter.

Figure 7.6 illustrates the *CV* we use in this problem. 'The water flows into the *CV* at 1 and out at 2. Since the area of cross-section is much larger at 1 than at 2, we may neglect the velocity at 1. Equation (7.12) applied to this *CV* gives

$$\left(\frac{V_2^2}{2} + 0 + \frac{p_{atm}}{\rho}\right) = \left(0 + gH + \frac{p_{atm}}{\rho}\right) - w_s - w_l$$

As the pressure is atmospheric at 1 as well as 2 where water issues out as a free jet, we have

$$w_s = gH - \frac{V_2^2}{2} - w_l$$

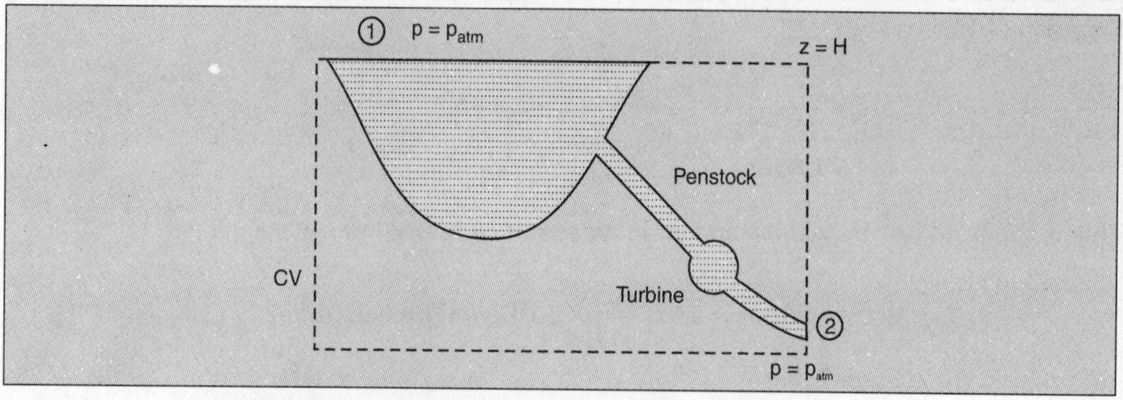

Fig. 7.6. Power generation from water flowing through a penstock.

If w_l is assumed negligible* compared to gH,

$$w_s = gH - \frac{V_2^2}{2}$$

where V_2 is the discharge velocity of the jet. This equation expresses the fact that out of the potential energy gH available per unit mass of water, w_s is extracted in the turbine and $V_2^2/2$ goes down as the kinetic energy of the water at the discharge. We could increase w_s by reducing V_2 but since the power output of the turbine, $\mathring{W}_s$ is w_s *times* the mass throughput, i.e., $w_s\,(\rho VA)_2$, a reduction in V_2 may reduce the rate at which power is extracted from the water by the turbine. Thus, the power,

$$\mathring{W}_s = w_s\,(\rho VA)_2 = \rho A_2\left(gHV_2 - \frac{V_2^3}{2}\right)$$

The velocity V_2^* at which this power is maximum is obtained by setting $d\mathring{W}_s/dV_2 = 0$, or

$$gH - \frac{3V_2^{*2}}{2} = 0$$

or
$$V_2^* = \sqrt{\frac{2}{3}gH}$$

and the maximum power output then is

$$\mathring{W}_{s,\max} = \rho A_2\left(\frac{2}{3}gH\right)^{3/2}$$

Figure 7.7 shows a plot of $\mathring{W}_s$ as a function of V_2. The above result is based on the assumption of no losses, whether in the flowing fluid or in the mechanical operation of the turbine.

---

* We shall study the validity of such approximations in Chapter 10.

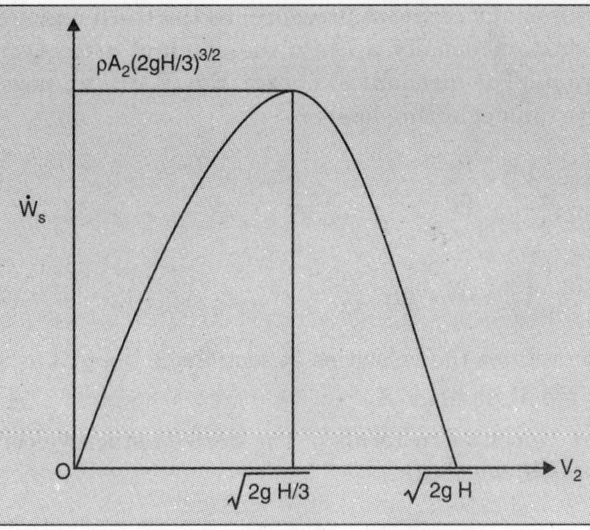

Fig. 7.7. Power developed by the turbine in Example 7.2, as a function of V_2.

Example 7.3. *Natural Convection in a Chimney.* As air heats up in a fire-place it expands so that the gas flowing inside the chimney stack is lighter than the air outside (Fig. 7.8). The

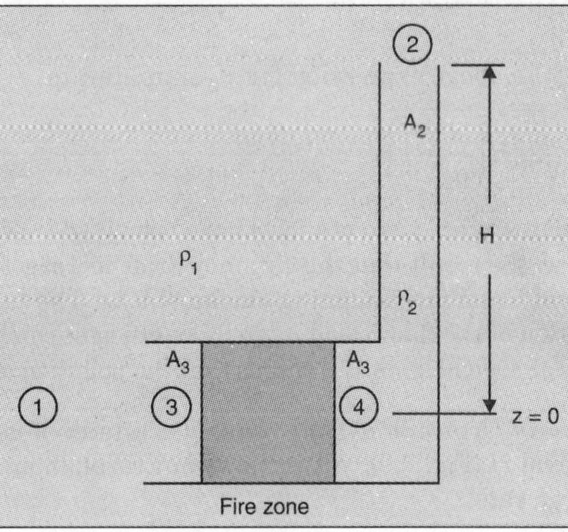

Fig. 7.8. Natural convection in a chimney.

pressures at the lower and upper ends of the stack are determined by the relatively denser outside air, so that $p_2 - p_1 = -\rho_1 gH$, where H is the height of the chimney stack and ρ_1 is the density of the cooler air outside. The lighter gas inside the chimney stack, therefore, experiences buoyancy and moves upwards. Since the density of the fluid is not constant everywhere, the mechanical energy Eq. 7.12 cannot be applied between sections 1 and 2. The problem is simplified, however, if we divide the flow into three regions. The first region extends from 1 to 3 where the colder air flows at constant density ρ_1 from the relatively stagnant region into the fire zone of cross-sectional area A_3. The second region is the fire zone itself where heat is added to the air such that its temperature rises at essentially constant pressure. In the third region from section 4 to 2 the lighter gas flows (at constant density ρ_2) into the stack of area A_2 finally exiting into the atmosphere at pressure p_2. The mechanical energy Eq. (7.12) can now be applied to the first and third regions to give (on neglecting losses):

$$\frac{p_1}{\rho_1} = \frac{1}{2}V_3^2 + \frac{p_3}{\rho_1}$$

and

$$\frac{1}{2}V_4^2 + \frac{p_4}{\rho_2} = \frac{1}{2}V_2^2 + \frac{p_2}{\rho_2} + gH$$

The continuity equation relates the velocities at sections 2, 3 and 4 as

$$\rho_1 V_3 A_3 = \rho_2 V_4 A_3 = \rho_2 V_2 A_2$$

Expressing all the velocities in terms of the desired stack velocity V_2 and noting that $p_3 = p_4$ and $p_2 - p_1 = -\rho_1 gH$, we obtain

$$V_2 = \sqrt{\frac{2\left(\frac{\rho_1}{\rho_2} - 1\right)gH}{\left[1 - \left(\frac{A_2}{A_3}\right)^2\left(1 - \frac{\rho_2}{\rho_1}\right)\right]}}$$

For $A_2 \ll A_3$ and $\frac{\rho_2}{\rho_1} \approx 1$, the above expression for V_2 simplifies to

$$V_2 = \sqrt{2gH\left(\frac{\rho_1}{\rho_2} - 1\right)}$$

This equation is approximate in that viscous effects have been neglected and the flow is assumed one-dimensional. However, the result that the chimney draft increases as the square root of its height, is used as a rule of thumb in designing smoke stacks. The use of such stacks results not only in better ventilation of the flame leading to more efficient combustion, but also permits the discharge of flue gases at such heights and speeds that the pollution at ground level is reduced.

Example 7.4. *Flow from an Orifice in a Tank.* Consider a tank of large cross-sectional area filled with water up to level H (Fig. 7.9). Water flows out through an orifice of area A_0 at the bottom. Find the discharge rate.

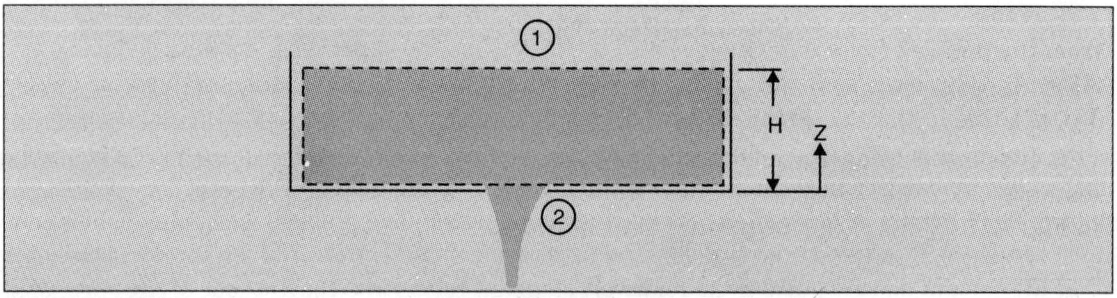

Fig. 7.9. *CV* for Example 7.4.

Consider the *CV* in Fig. 7.9. The flow may be assumed to be 1-*D* at sections 1 and 2. On neglecting viscous losses, Eq. (7.12) gives

$$\frac{V_1^2}{2} + \frac{p_{\text{atm}}}{\rho} + gH = \frac{V_2^2}{2} + \frac{p_{\text{atm}}}{\rho}$$

where it is recognized that the pressure at 2 is atmospheric since water issues as a free jet. Because the area at 1 is much larger than that at 2, the velocity V_1 will be negligible compared to V_2 and so

$$V_2 = \sqrt{2gH}$$

The discharge rate is, thus

$$\mathring{Q} = A_0\sqrt{2gH}$$

This equation* has been obtained on the assumptions that the losses are negligible and that the flow is 1-*D*. In actual fact, there will be some losses and the flow at 2 will not be strictly 1-*D*. The actual velocity profile at 2 will depend on the nature of the orifice. A sharp-edged orifice will result in much larger 2-*D* effects than will a smooth one. In engineering practice these non-ideal effects are usually taken care of by introducing an experimentally obtained correction factor termed the *discharge coefficient* C_d. Thus,

$$\mathring{Q} = C_d \mathring{Q}_{\text{ideal}} = C_d A_0 \sqrt{2gH} \qquad \text{(a)}$$

The value of the coefficient C_d depends upon the orifice geometry, and Fig. 7.10 gives typical values of this coefficient for some orifices.

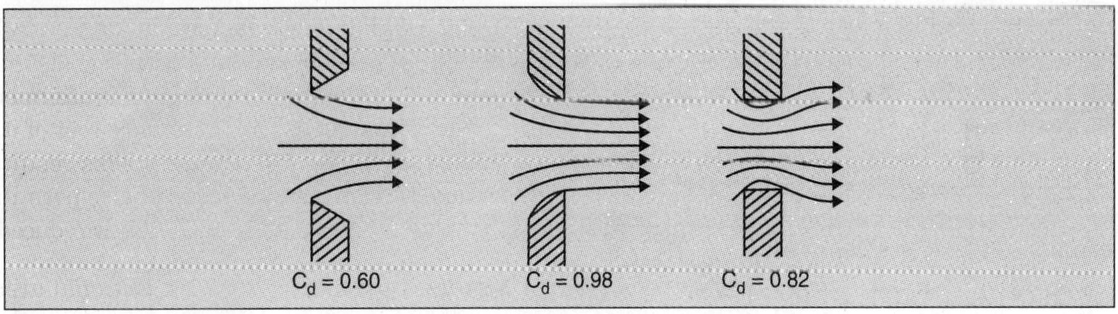

Fig. 7.10. C_d for some typical orifices.

Another source of error in Eq. (a) is the fact that as the water issues out of the orifice from the tank the level in it changes and the flow is, strictly speaking, not steady. However, if $A_1 \gg A_0$, this effect (as measured by the ratio of the rate of change in the level and the velocity V_2), is so small that the error made is insignificant.

It is interesting to study the velocity and cross-section of the water jet at location 3, a distance *h* downstream of the orifice. We can apply the mechanical energy Eq. 7.12 to the *CV* in Fig. 7.11. Using the same assumptions as before,

---

* This formula is named after E. Torricelli (1608–1647).

$$V_3 = \sqrt{2g(H+h)} \tag{b}$$

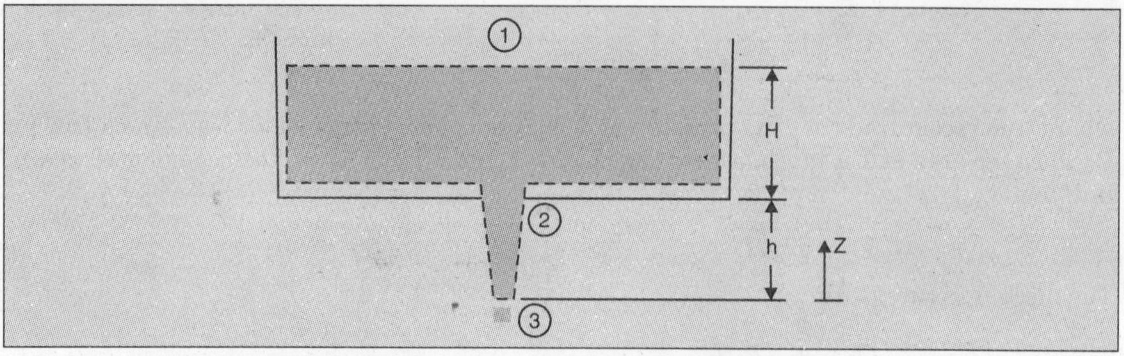

Fig. 7.11. *CV* to obtain the area of the water jet as a function of position.

Thus, the velocity of the water keeps on increasing as h increases, *as if the water was falling freely.* The continuity equation dictates that at steady state the discharge at 3 must be the same as that at 2. Thus,

$$V_3A_3 = V_2A_2$$

or

$$A_3 = A_0\sqrt{\frac{H}{H+h}}$$

Consider, on the other hand, a tank to which an outlet tube of constant area A_0 is connected at the bottom (Fig. 7.12). If viscous losses are neglected (though neglecting it in this case is far less realistic than earlier), the mechanical energy equation reveals that the velocity at 3 is still given by Eq. (b) but the discharge in this case would be larger since the area of flow at 3 is larger.* Hence

$$\mathring{Q} = A_0\sqrt{2g(H+h)}$$

compared to the previous value $A_0\sqrt{2gH}$

An interesting result is obtained if the mechanical energy Eq. 7.12 is applied to the control volume *CV*2 in Fig. 7.12, between sections 4 and 3. Since the velocities at these points must be the same,

$$p_4 = p_3 - \rho g\,(h - x) = p_{atm} - \rho g\,(h - x) \tag{c}$$

The variation of pressure p_4 with x is shown in Fig. 7.12. It is observed that the flow takes place from a region of lower pressure to one of higher pressure! There is nothing contradictory in this result. As is clear from Eq. (7.14), it is not the pressure head but the total head which is responsible for flow and should be greater at the upstream end than at the downstream end.

* Note that the mechanical energy equation has been applied in the case without the outlet tube at a general point 3. In the presence of an outlet tube, point 3 must be at the exit of the pipe. Anywhere inside the pipe the pressure is not known.

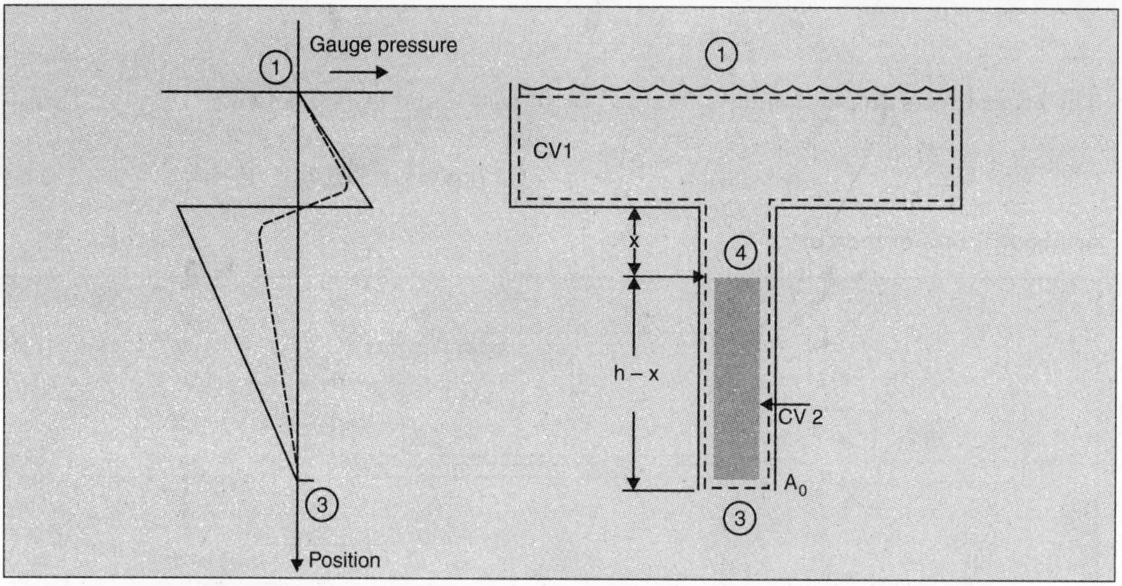

Fig. 7.12. Pressure variation in flow through an outlet tube connected to a tank. Solid lines correspond to Eq. (c). Broken lines are in the presence of viscous losses.

Example 7.5. *A Ducted Fan.* Consider a fan in a duct (Fig. 7.13). It sucks in air from region 1 and expels it as a jet at 3. The flow pattern is as shown. At the entrance, the air is sucked in from all directions while at the exit, it issues as a jet because of separation of the boundary layer (see Sec. 1.8). Because of the much larger flow area at section 1 (some distance upstream of the duct inlet) the velocity there is very small. If the area of the duct is 0.1 m^2 and the power consumed is 0.5 kW, determine the velocity of the air in the duct, taking the density of air as 1.5 kg/m^3, and neglecting losses.

If we assume that changes in the pressure of the air as it flows through are much smaller than the absolute level of the pressures, the changes in density can be neglected (it will be shown in Sec. 15.1 that this condition holds when the velocity of flow is much smaller than the speed of sound in the fluid). We can apply Eq. (7.12) to the *CV* between sections 1 and 3. Here, V_3 equals the velocity V of air in the duct, $p_1 = p_3 = p_{atm}$, $V_1 \simeq 0$, $w_l = 0$, $z_1 = z_3$, and

$$w_s = \mathring{W}_s/\mathring{m} = -500\ (\text{W})/\left[0.1\left(\text{m}^2\right)\times V\left(\frac{\text{m}}{\text{s}}\right)\times 1.5\left(\text{kg/m}^3\right)\right] = (3.33 \times 10^3/V)\ \text{J/kg}$$

Then Eq. (7.12) gives

$$\frac{V^2}{2} = \frac{3.33\times 10^3}{V}$$

or

$$V = 18.77\ \text{m/s}$$

This velocity will result in an appreciable suction being set up inside the duct. To calculate this suction, we apply the mechanical energy equation to the *CV* between sections 1 and 2 to get

$$\frac{p_{atm}}{\rho} = \frac{V^2}{2} + \frac{p_2}{\rho}$$

The other terms drop out since $z_1 = z_2$, $V_1 \simeq 0$, $w_l = 0$ and $w_s = 0$ between 1 and 2. Thus,

$$p_2 - p_{atm} = -\frac{\rho V^2}{2} = -1.5\ (\text{kg/m}^3) \times (18.77)^2\ (\text{m}^2/\text{s}^2)/2 = 264.2\ \text{N/m}^2$$

or about 2 mm of mercury.

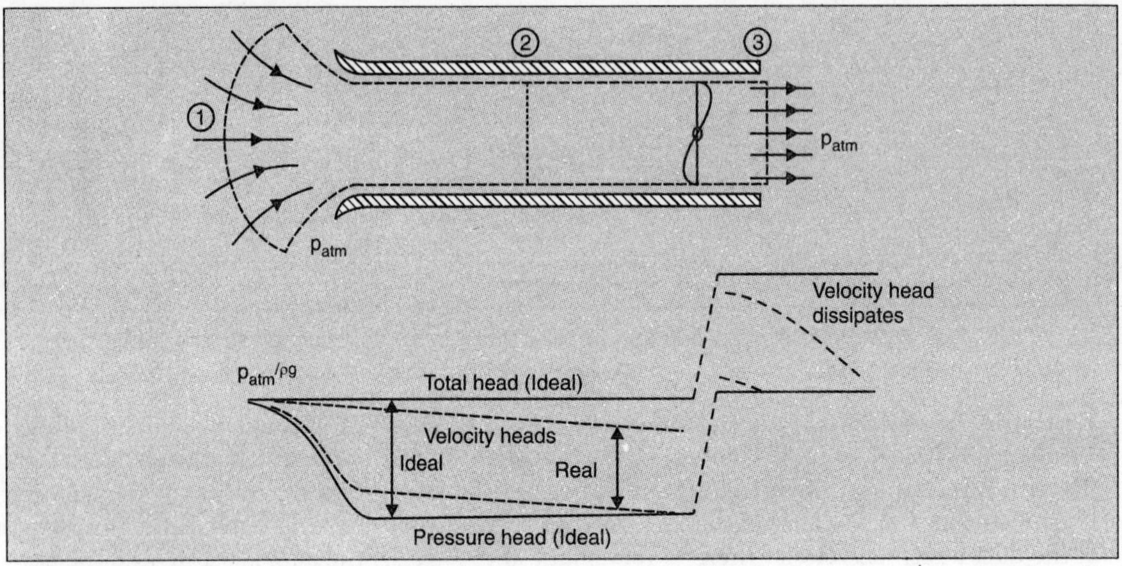

Fig. 7.13. A ducted fan with the various heads shown as a function of position. Solid lines show the ideal flow case $(w_l = 0)$ and broken lines, the real flow. Losses are shown exaggerated.

The total and the pressure heads for the ideal conditions (i.e., neglecting w_l) are shown in Fig. 7.13 as a function of position along the duct. Note that the total and pressure head lines are parallel all along the duct, signifying the constant velocity head. For such an ideal flow, the kinetic energy at the exit equals the fan power input. For the same power input, the velocity head and the suction for the real flow are smaller than those for the ideal flow. Beyond section 3, the pressure in the real flow remains constant at p_{atm} but the velocity head dissipates due to viscous action.

Example 7.6. *Flow in a Slowly Converging Channel.* Consider the flow of an incompressible fluid in a slowly converging channel. Flow in a converging channel is not strictly 1-*D*, but if the convergence is slow enough, we may model it to be so (see Fig. 4. 12). Application of the continuity equation to the *CV* in Fig. 7.14 gives

$$V_2 = V_1 \frac{A_1}{A_2}$$

where A_1 and A_2 are the areas and V_1 and V_2, the velocities at sections 1 and 2 respectively. Application of the mechanical energy Eq. 7.12 gives

$$\frac{p_2 - p_1}{\rho g} = \frac{V_1^2 - V_2^2}{2g} = \frac{V_1^2}{2g}\left[1 - \left(\frac{A_1}{A_2}\right)^2\right] \quad \text{(a)}$$

assuming that the viscous losses are negligible. Thus, the pressure decreases as A_2 decreases.

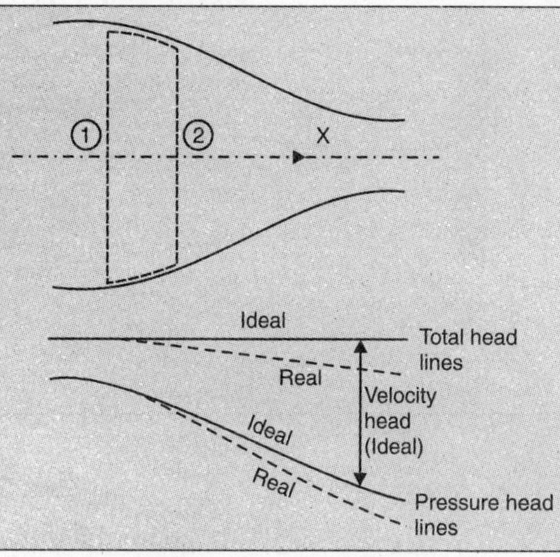

Fig. 7.14. Flow in a converging section along with total- and pressure-head curves. Same flow rate assumed for real and ideal cases.

The variation of the heads with position for both the ideal and real flows are shown in Fig. 7.14. The loss of total head equals the loss in the pressure head since the flow rate is being held constant.

It would appear that Eq. (a) should also apply when A_2 is larger than A_1, i.e., when the channel is diverging. But the decrease in the velocity results in an increasing pressure downstream, causing the flow to separate (Fig. 7.15) at the boundaries (see Sec. 1.8). The actual flow area is then less than the channel area and the pressure downstream is much lower than that predicted by Eq. (a).

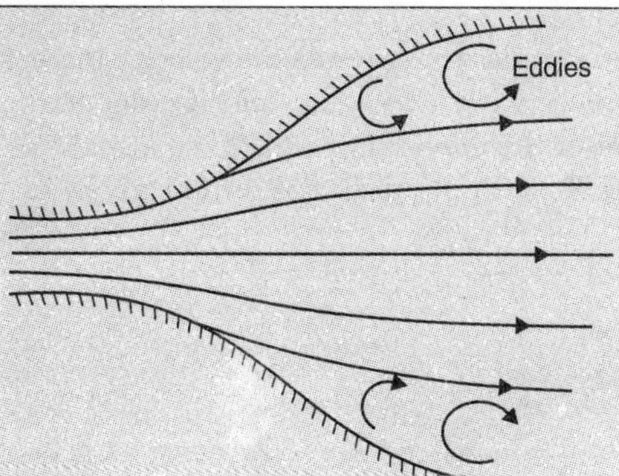

Fig. 7.15. Separation in a diverging channel.

The separation tendency can be suppressed by making the divergence very slow. This is used in the design of wind tunnels (Fig. 7.16) which are employed for the study of the flow over *models* of aeroplanes, etc. In these tunnels, since the kinetic energy of the fluid at the exit represents the power required by the fan an attempt is made to keep the velocity of the fluid at the exit as low as possible by avoiding separation.

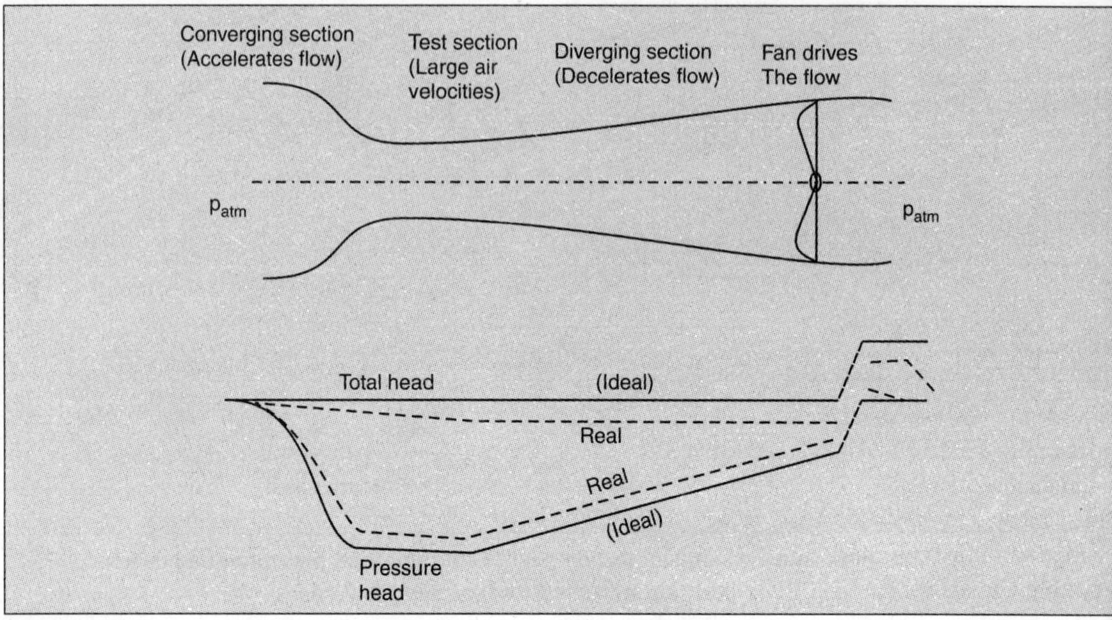

Fig. 7.16. Model of a wind tunnel with the various heads. Velocity heads and losses are exaggerated. Same fan power assumed for real and ideal flows.

If the converging section was inclined (Fig. 7.17), the energy equation (neglecting losses) gives

$$\left(\frac{p_2}{\rho g}+z_2\right)-\left(\frac{p_1}{\rho g}+z_1\right)=h_{pz,2}-h_{pz,1}=\frac{V_1^2}{2g}\left[1-\left(\frac{A_1}{A_2}\right)^2\right]$$

meaning that the piezometric head decreases as the velocity increases. In this case, the pressure head may increase or decrease with increasing velocity, depending upon the inclination of the channel, i.e., the variation of the elevation head. If we attach piezometric tubes to a channel carrying a liquid, the liquid will rise up to the height given by the piezometric head curve in Fig. 7.17.

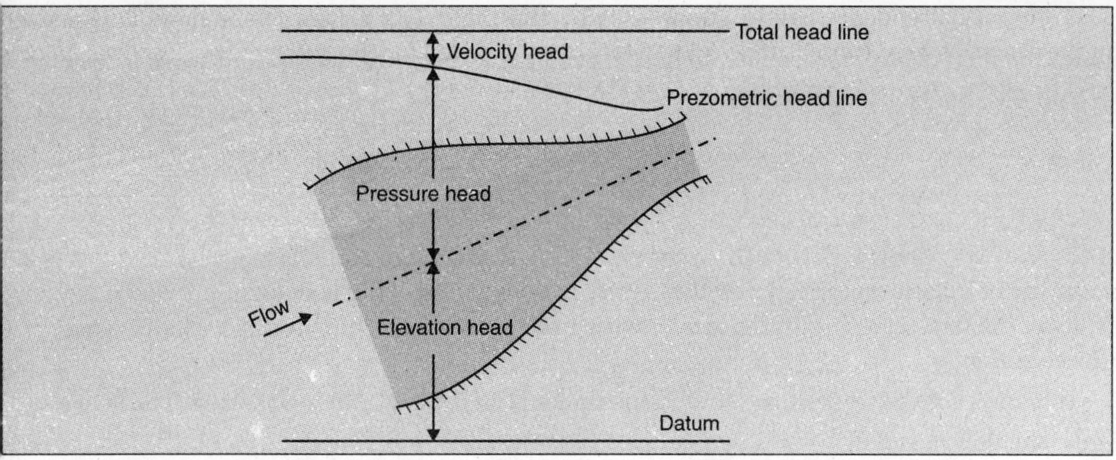

Fig. 7.17. Flow in an inclined converging section.

7.5 ENERGY EQUATION FOR A STREAMTUBE—BERNOULLI EQUATION

Equations (7.8)–(7.13) have been obtained for the cases when the flow across the inlet and exit ports may be assumed one-dimensional. There are many situations where this may not be a valid assumption. In those cases, we may divide the flow region into a number of thin streamtubes (Fig. 7.18). Since the surface of a streamtube is made up of streamlines, a streamtube can be regarded as a *CV* with one inlet and one outlet port and if the cross-sectional area is small enough, the 1-*D* approximation can be made for this thin *CV*. In this *CV*, the *CS* is surrounded by the fluid and therefore, the velocity of the fluid at the *CS* is not zero but is finite. Thus, if there are shear stresses present, there will be some work done by the *CV* on the surroundings. This contribution equals

$$\overset{\circ}{W}_{\tau} = - \iint_{\substack{\text{streamtube}\\ \text{surface}}} \tau V \, dA$$

where τ and V are necessarily collinear.

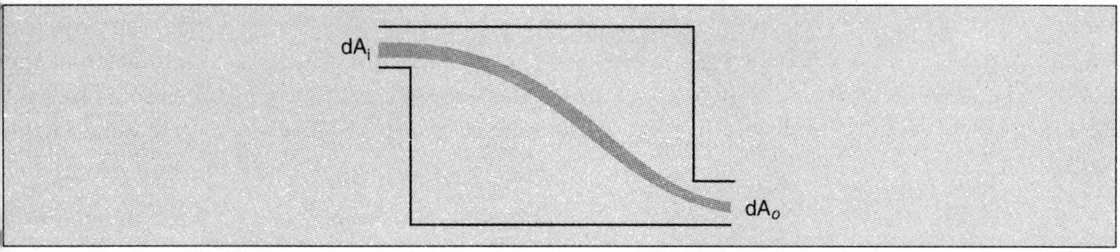

Fig. 7.18. Flow region subdivided into thin streamtubes when the flow is not 1-*D* at inlet and outlet ports.

For the special case when the fluid is non-viscous, $\overset{\circ}{W}_\tau$ will be zero. If we further assume that (a) the fluid is incompressible, (b) heat transfer is absent, and (c) no shaft work is done by this thin *CV*, Eq. (7. 10) reduces to

$$\left[\frac{V^2}{2}+gz+\frac{p}{\rho}\right]_i=\left[\frac{V^2}{2}+gz+\frac{p}{\rho}\right]_o \quad \text{...(7.15)}$$

This is the well known *Bernoulli equation* which states that the total mechanical energy (consisting of kinetic, potential and flow energy) is constant along a streamline. Thus, as the velocity of the flow increases in the streamwise direction, the pressure decreases (at constant z) and vice versa.

It is instructive to recall the conditions under which Eq. (7.15) is obtained. These are:

(a) the flow is steady

(b) the flow is incompressible

(c) no shaft work is done or heat conducted

(d) points i and o are on the same streamline (i.e., thin streamtube approximation)

(e) the flow is non-viscous.

It will be shown in Chapter 11 that for many flows of engineering interest, the viscous forces are negligible compared to other forces. Such flows may be considered non-viscous and the Bernoulli equation in the form of Eq. (7.15) may be applied.

The Bernoulli equation can be obtained alternatively by direct integration of the equation of motion of an *inviscid* fluid. We start with Eq. (6.27) in the streamline direction, for steady flow

$$\rho\frac{\partial\left(V^2/2\right)}{\partial s}=\rho f_s-\frac{\partial p}{\partial s}$$

This is integrated between two points i and o (Fig. 7.19) lying on a streamline.

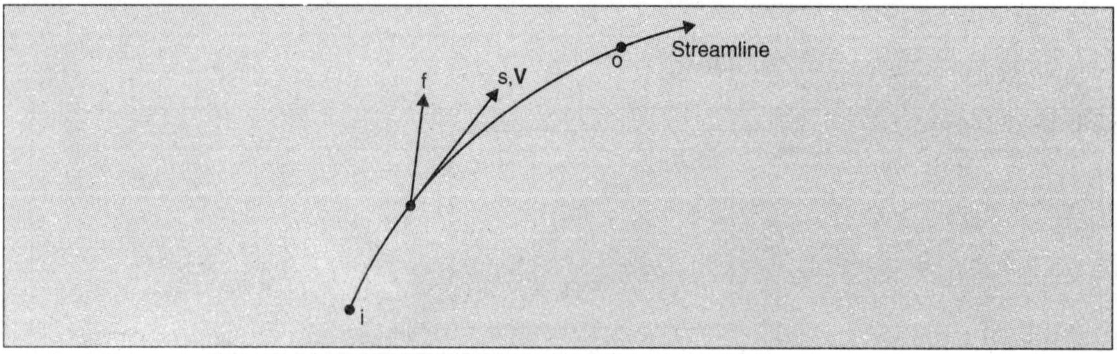

Fig. 7.19. Two points *i* and *o* on a streamline between which the inviscid-flow equation is to be integrated.

$$\rho\int_i^o \frac{\partial\left(V^2/2\right)}{\partial s}ds = \rho\int_i^o f_s\, ds - \int_i^o \frac{\partial p}{\partial s}ds$$

Note that $f_s\, ds = \mathbf{f}\cdot \mathbf{ds}$. When $\mathbf{f}$ is the force due to gravity, i.e., $\mathbf{f} = -g\hat{\mathbf{k}}$,

$$\mathbf{f}\cdot ds = \left(-g\hat{\mathbf{k}}\right).\left(dx\ \hat{\mathbf{i}} + dy\ \hat{\mathbf{j}} + dz\ \hat{\mathbf{k}}\right) = -g\, dz$$

and the above integration gives

$$\rho\frac{V_o^2}{2} - \rho\frac{V_i^2}{2} = -\rho g\left(z_o - z_i\right) - \left(p_o - p_i\right)$$

or

$$\left(\frac{V^2}{2} + gz + \frac{p}{\rho}\right)_i = \left(\frac{V^2}{2} + gz + \frac{p}{\rho}\right)_o$$

Example 7.7. *Flow about a cylinder.* When an incompressible, non-viscous fluid flows past a long circular cylinder, the velocity field is given by (see Sec. 12.7)

$$V_r = V_0\left(1 - \frac{a^2}{r^2}\right)\cos\theta$$

$$V_\theta = -V_0\left(1 + \frac{a^2}{r^2}\right)\sin\theta$$

where V_0 is the velocity of the fluid far away from the cylinder (Fig. 7.20) and a is its radius. Obtain the pressure distribution on this cylinder and the net force acting on it.

Consider a streamline between point 1 far away and a point 2 on the surface of the cylinder. The velocity components at 2 are

$$V_r = 0 \text{ and } V_\theta = -2V_0 \sin\theta$$

The velocity, V_2 is then $\sqrt{V_r^2 + V_\theta^2} = 2V_0 \sin\theta$. If the pressure at 1 is p_0 and if the variation in z is neglected, Eq. (7.15) gives

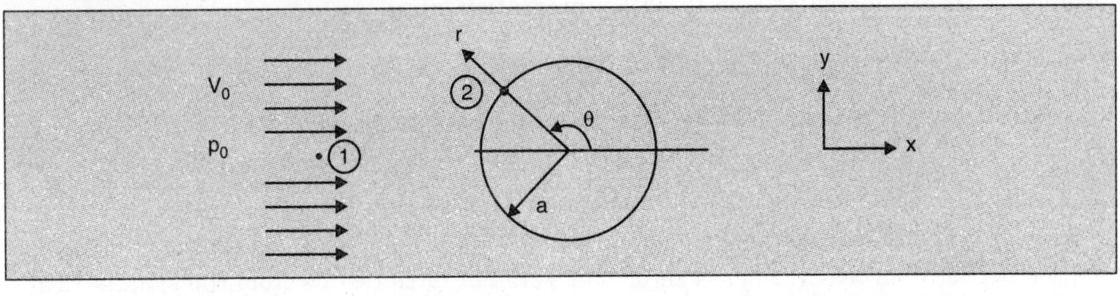

Fig. 7.20. Flow of a uniform stream of fluid around a long, stationary cylinder.

$$\frac{4V_0^2 \sin^2 \theta}{2} + \frac{p_2}{\rho} = \frac{V_0^2}{2} + \frac{p_0}{\rho}$$

or
$$p_2 - p_0 = \frac{\rho V_0^2}{2}\left(1 - 4\sin^2 \theta\right)$$

Figure 7.21 shows the variation of $p_2 - p_0$ at the surface. Note that this gives a maximum positive pressure fore and aft of the cylinder, i.e., at the stagnation points, $\theta = \pi$ and 0. The points of minimum pressure lie at $\theta = \pi/2$ and $3\pi/2$, the locations of maximum fluid speed.

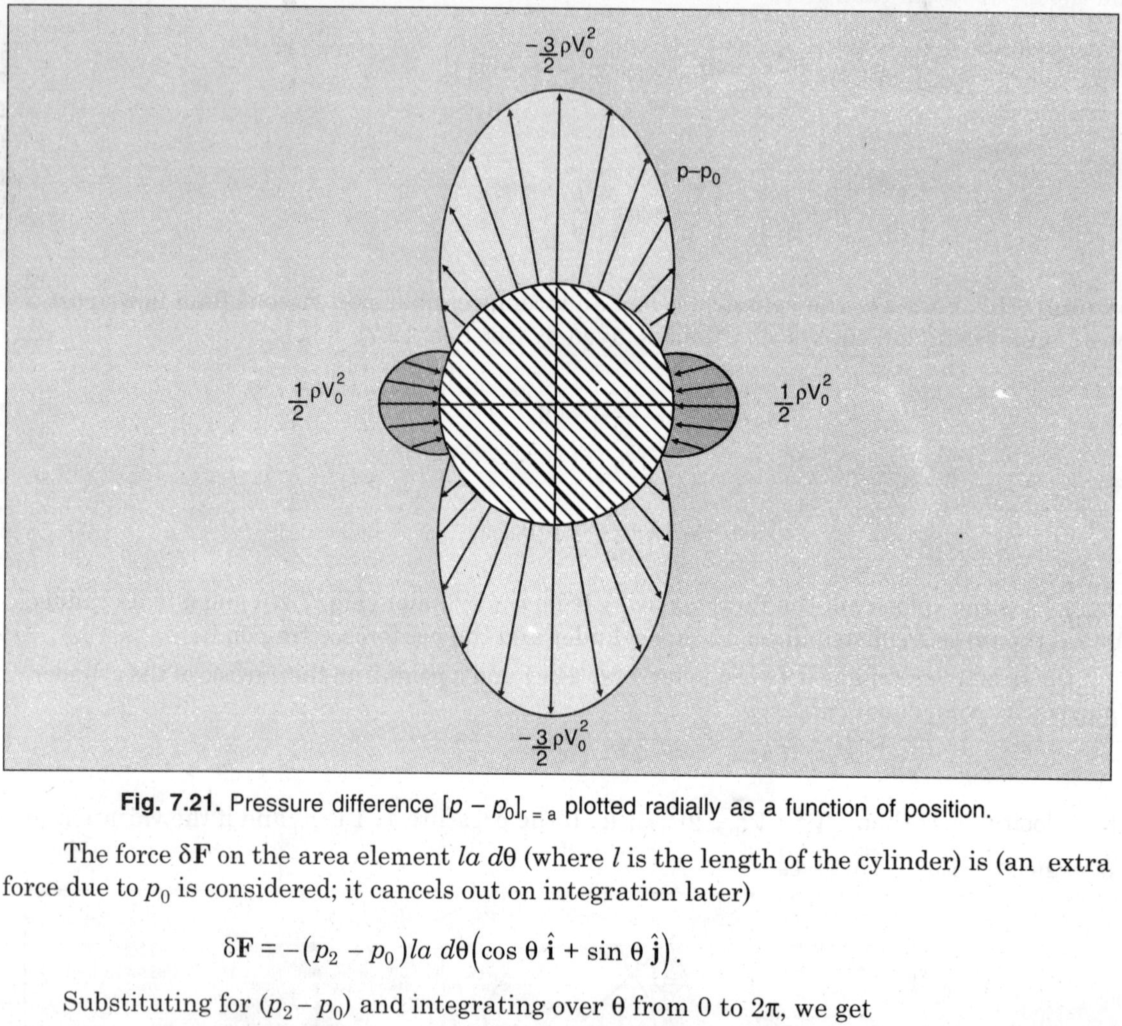

Fig. 7.21. Pressure difference $[p - p_0]_{r=a}$ plotted radially as a function of position.

The force $\delta\mathbf{F}$ on the area element $la\, d\theta$ (where l is the length of the cylinder) is (an extra force due to p_0 is considered; it cancels out on integration later)

$$\delta\mathbf{F} = -(p_2 - p_0)la\, d\theta\left(\cos\theta\,\hat{\mathbf{i}} + \sin\theta\,\hat{\mathbf{j}}\right).$$

Substituting for $(p_2 - p_0)$ and integrating over θ from 0 to 2π, we get

$$\mathbf{F} = -\int_0^{2\pi} \frac{\rho V_0^2}{2}\left(1 - 4\sin^2\theta\right) la\left(\cos\theta\,\hat{\mathbf{i}} + \sin\theta\,\hat{\mathbf{j}}\right)d\theta = 0$$

Thus, the net force on the cylinder is zero! This zero net force results from the symmetrical distributions of velocity and pressure around the cylinder which is typical of non-viscous flows.

Fluid dynamical forces on the body can appear only if this symmetry is destroyed. It will be seen in Chapter 11 that at high speeds (characterized by large values of the Reynolds number, $\rho V_0\, a/\mu$) the forces due to viscosity are much smaller than pressure forces. But even in these cases, the drag on a cylinder is significant. This is due to the fact that the presence of viscosity, howsoever small, results in the formation of a very thin boundary layer on the cylinder, which separates from the surface. The pressure and velocity distributions in the front portion of the cylinder then are close to those predicted above, but the distributions at the rear are quite different (see Fig. 1.20). This fore- and aft- asymmetry in the presence of viscosity results in a net *drag* force on the cylinder. In a cambered (i.e., curved centreline) aerofoil (Fig. 7.22), the presence of viscosity causes a pressure asymmetry in the vertical direction also, giving rise to a *lift* force on the body. (See Sec. 14.1)

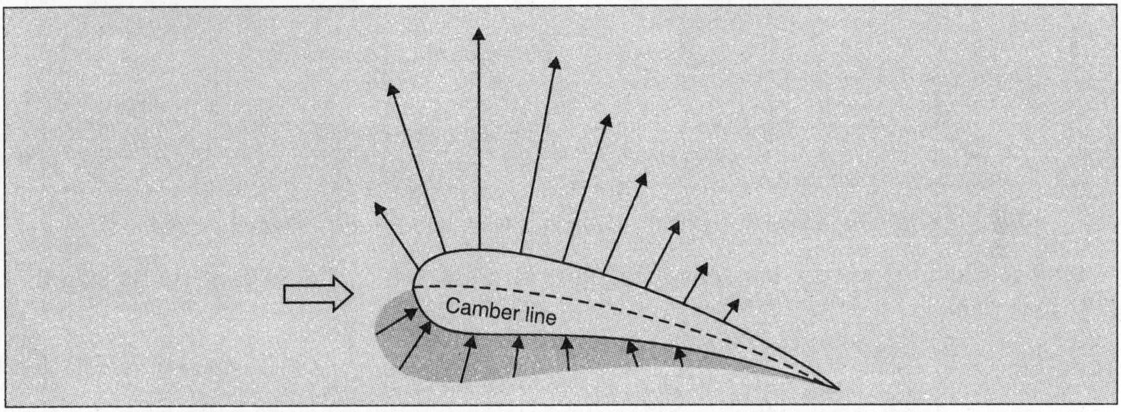

Fig. 7.22. A cambered aerofoil and the pressure distribution around it.

Example 7.8. *Flow over a Weir.* Flows through open channels are sometimes measured by inserting a plate containing a cut-out (called a notched weir) (Fig. 7.23). The volume flow rate is related to the height H of the fluid stream through the cut-out. Obtain this relationship, assuming that the velocities through the weir are much larger than the velocities in the main channel.

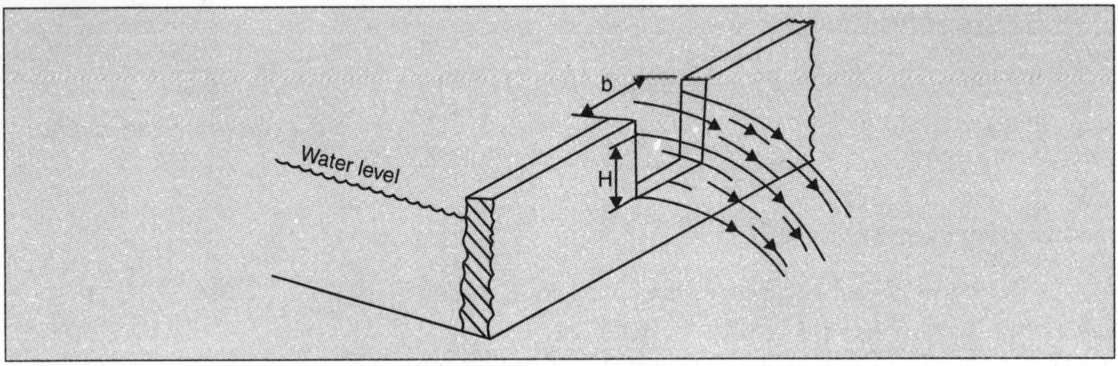

Fig. 7.23. Flow through a rectangular weir.

The actual flow picture looks something like that shown in Fig. 7.24(a) and is quite complicated. It is possible to idealize this flow by the one in Fig. 7.24(b) if the curvature of the streamlines is small and if the drop d of the water surface over the weir from that in the main channel is small compared to H. The velocity at height y above the crest of the weir can be found by applying the Bernoulli equation on a streamline from point 1 to 2. Before doing this it should be noted that the vertical velocity and acceleration components are small in the main channel and therefore, the pressure variation in the vertical direction is essentially hydrostatic.* That is,

$$p_1 = p_{atm} + \rho g h$$

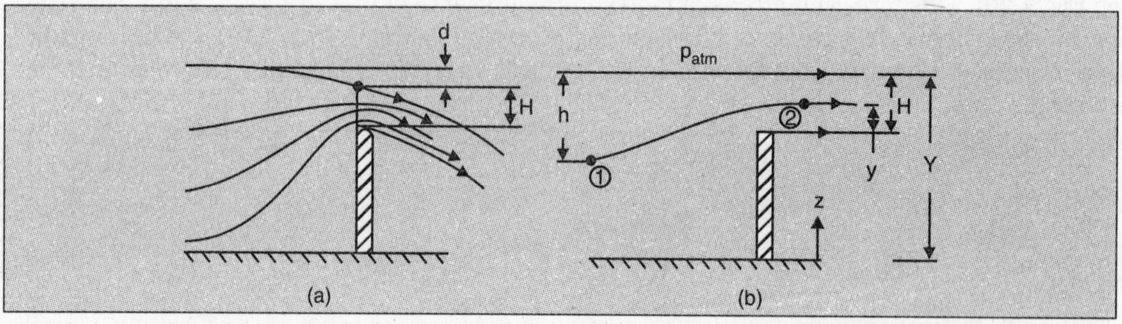

Fig. 7.24. (a) The actual flow pattern through a weir, and (b) the idealized picture.

The total mechanical energy per unit mass flow at point 1 (i.e., the LHS of the Bernoulli Eq. 7.15) is then

$$\frac{V_1^2}{2} + gz_1 + \frac{p_1}{\rho} \simeq 0 + g(Y-h) + \frac{p_{atm} + \rho gh}{\rho} = \frac{p_{atm}}{\rho} + gY$$

Since the water issues as a free jet from the cut-out, the pressure at its surface is atmospheric. As the curvature of the streamlines is negligible, the pressure can be taken as p_{atm} throughout the jet. The RHS of the Bernoulli equation then becomes

$$\frac{V_2^2}{2} + gz_2 + \frac{p_2}{\rho} \simeq \frac{V_2^2}{2} + g(Y - H + y) + \frac{p_{atm}}{\rho}$$

and the Bernoulli equation gives

$$\frac{p_{atm}}{\rho} + gY = \frac{V_2^2}{2} + g(Y - H + y) + \frac{p_{atm}}{\rho}$$

or

$$V_2 = \sqrt{2g(H-y)}$$

This derivation neglects viscous losses.

The discharge $\mathring{Q}$ across the width b of the weir is then

$$\mathring{Q} = \int_0^H V_2(y) b\, dy = \int_0^H b\sqrt{2g(H-y)}\, dy = \frac{2b}{3}\sqrt{2gH^3}$$

* As can be verified from the vertical component of the Navier-Stokes equation.

Using a discharge coefficient C_d to account for deviations from the ideal conditions, the equation becomes

$$\mathring{Q} = \frac{2}{3} C_d b \sqrt{2gH^3}$$

The value of C_d varies with the height H above the crest and lies typically around 0.6.

The various heads at the weir are shown in Fig. 7.25.

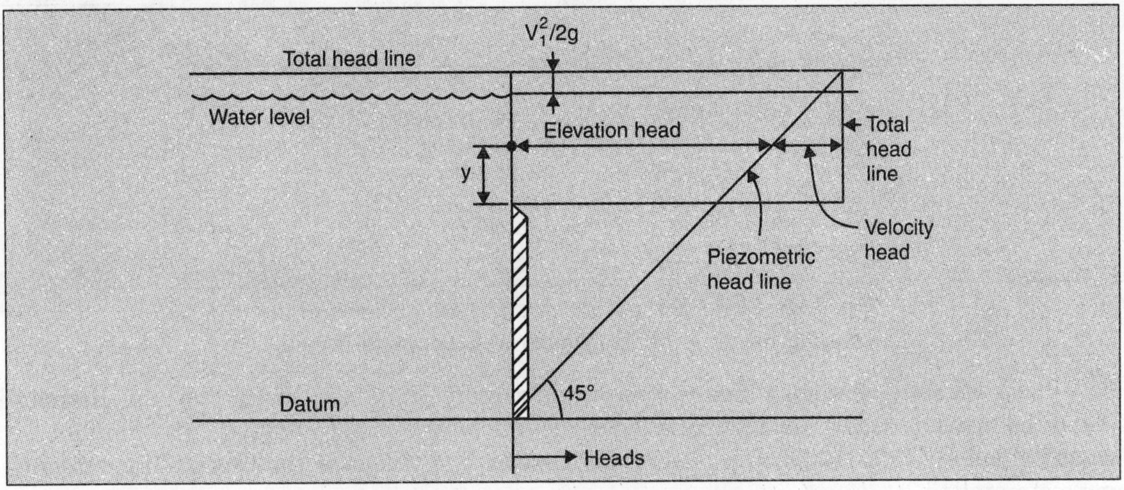

Fig. 7.25. Various heads at the weir.

7.6 PRESSURE VARIATIONS NORMAL TO STREAMLINES

The Bernoulli equation can be used to obtain the pressure variations along a streamline (Sec. 7.5). In flows which are not 1-*D*, the pressure variations normal to the streamlines may be obtained using the normal component of the equation of motion

$$\frac{\rho V^2}{R} = \rho f_n - \frac{\partial p}{\partial n} \qquad \text{...(6.28)}$$

where R is the radius of curvature of the streamline.

For $\mathbf{f} = -g\hat{\mathbf{k}}$,

$$f_n = \mathbf{f} \cdot \hat{\mathbf{n}} = \left(-g\hat{\mathbf{k}}\right) \cdot \left(\frac{\partial x}{\partial n}\hat{\mathbf{i}} + \frac{\partial y}{\partial n}\hat{\mathbf{j}} + \frac{\partial z}{\partial n}\hat{\mathbf{k}}\right) = -g\frac{\partial z}{\partial n}$$

and so

$$\frac{\rho V^2}{R} = -\rho g \frac{\partial\left(z + \frac{p}{\rho g}\right)}{\partial n} \qquad \text{...(7.16)}$$

Here, $z + \dfrac{p}{\rho g}$ is the piezometric head h_{pz}. Equation (7.16) can, thus, be written as

$$\frac{\partial h_{pz}}{\partial n} = -\frac{V^2}{gR} \quad ...(7.17)$$

For the streamline pattern shown (Fig. 7.26), R is positive and $\partial h_{pz}/\partial n$ is negative, i.e. the piezometric head (or the pressure if elevation differences are small) decreases with increasing n.

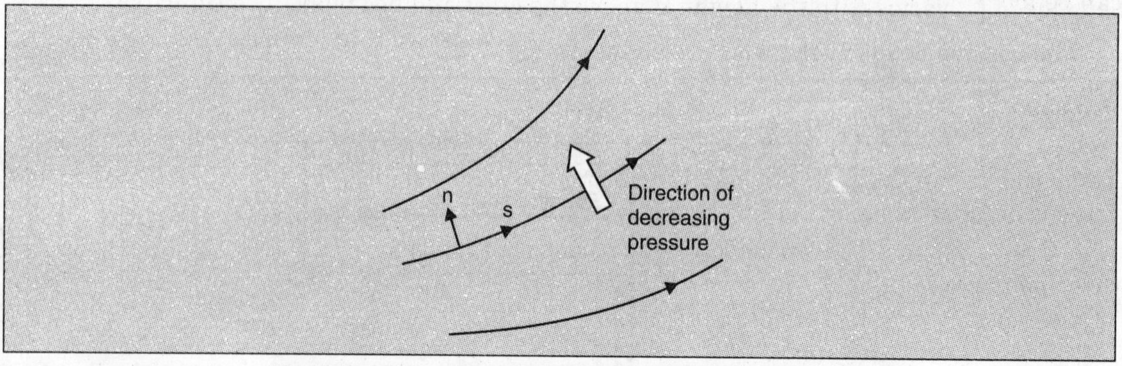

Fig. 7.26. Streamline coordinates and the direction of decreasing pressure (for negligible elevation differences).

A flow in a horizontal channel which is converging rapidly (Fig. 7.27), cannot be modelled as 1-D and the pressure variations along the length cannot be obtained by applying the simplified mechanical energy Eq. 7.12 (with $w_s = w_l = 0$). Instead, one must use the Bernoulli equation along a streamline.

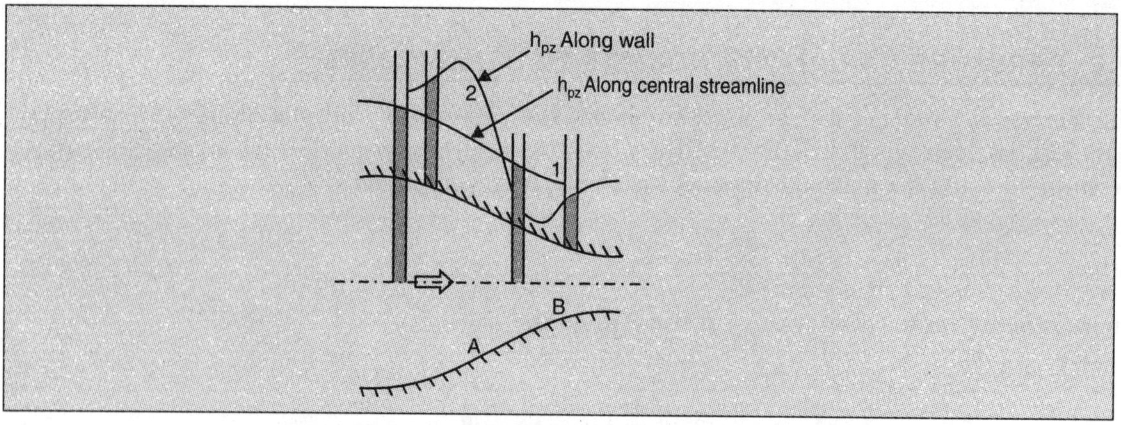

Fig. 7.27. Pressure variations along central streamline and along wall for flow through a converging channel.

The pressure variation along the central streamline would be similar to that in Fig. 7.14 of Example 7.6, with the pressure decreasing monotonically downstream. The pressure distribution along any other streamline can be obtained by modifying this basic distribution to account for the effect of curvature using Eq. (7.17). At the upper wall near section A, the streamlines are concave, R is negative and hence the pressure increases towards the wall. Similarly, at section B the pressure decreases towards the wall. Figure 7.27 shows plots of the pressure variations along the central streamline and at the outer wall. Piezometric tubes tapping on the axis of the channel will stand upto curve 1 while tubes tapping on the wall will stand up to curve 2.

Example 7.9. *Pressure Variations in a Forced Vortex.* The flow within the core of a tornado can be modelled roughly as a *forced vortex* with velocity components

$$V_\theta = \omega r$$

and

$$V_r = 0$$

Thus, the streamlines are circular (Fig. 7.28), and the fluid rotates like a rigid body. Since there is no relative motion of fluid particles, shear stresses are zero everywhere.

One cannot use the Bernoulli equation to relate the pressures at points 1 and 2 because these points are *not* on the same streamline. One can, however, use Eq. (7.17) to get

$$\frac{\partial h_{pz}}{\partial r} = \frac{V^2}{gr}$$

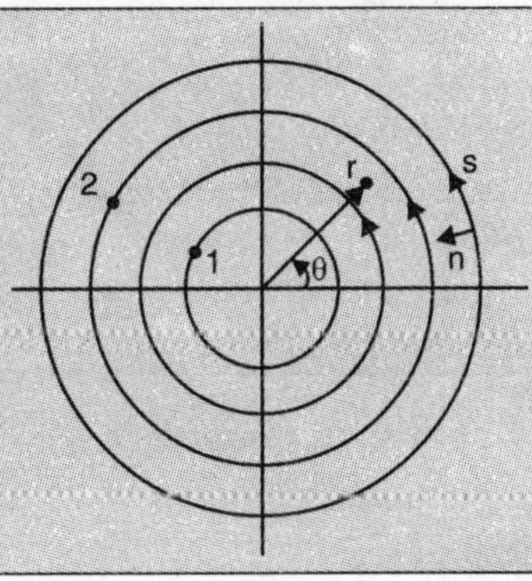

Fig. 7.28. Streamlines inside the core of a tornado.

since $n = -r$ and $R = r$. This gives

$$h_{pz,2} - h_{pz,1} = \int_{r_1}^{r_2} \frac{V^2 dr}{gr} = \int_{r_1}^{r_2} \frac{\omega^2 r}{g} dr$$

or

$$h_{pz,2} - h_{pz,1} = \frac{\omega^2 \left(r_2^2 - r_1^2\right)}{2g}$$

If the two points are at the same horizontal level, their pressures are related as

$$p_2 - p_1 = \frac{\rho\omega^2 \left(r_2^2 - r_1^2\right)}{2}$$

with the pressure increasing outwards. Thus the pressure at 2 is higher than that at 1 even though the velocity at 2 is higher (a result contrary to what a careless application of the Bernoulli equation would lead to). This is the same result as obtained in Example 6.4.

CHAPTER 7

PROBLEMS

7.1 Give an example where flow occurs from a low pressure region to a higher pressure one. Explain the apparent contradiction.

7.2 The famous Roman aqueducts carried water to the city of Rome from distant rivers. These were laid on tall stilts and sloped down throughout their lengths because the engineers argued that if the pipes were laid on the ground and followed the natural terrain, the water would have to flow upwards at places. Find the flaw in the argument.

7.3 A shaft of diameter d rotates at an angular velocity ω inside a lubricated bearing as shown. The bearing diameter is D and the shaft may be assumed to rotate concentrically inside it. If the bearing is insulated, at what rate would the oil heat up if its specific heat is C_p? Assume that the clearance is very small.

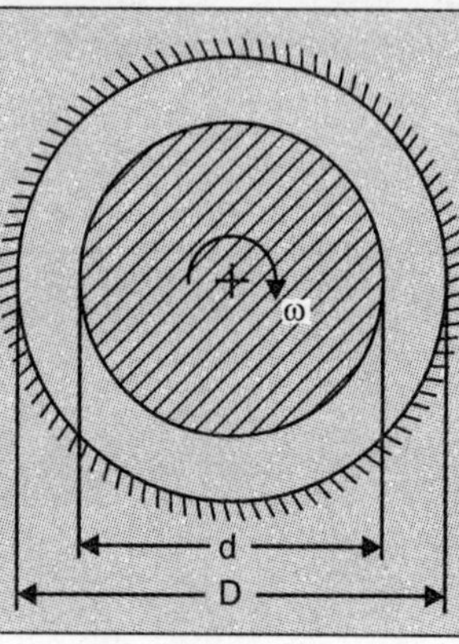

7.4 Justify the following assumptions in Chapter 5:

(a) $V_1 = V_2 = V_3$ in Prob. 5.10.

(b) $V = 3.13$ m/s in Prob. 5.12.

(c) $V_2 = V_1$ in Prob. 5.23.

7.5 Use the mechanical energy equation and the results of Example 6.3 to show that the head loss for laminar, fully-developed flow in a straight circular pipe is given by

$$h_l = \frac{64}{\text{Re}} \cdot \frac{L}{D} \cdot \frac{V_{av}^2}{2g}$$

where Re is the Reynolds number $\rho V_{av} D/\mu$.

7.6 The human heart behaves like a pump. The pressures at various positions are as shown. If the cardiac output is 5 lit/min of blood, estimate the power of the heart (no pun intended). Thus verify that the heart does enough work to lift 30 tons from sea level to the top of Mt. Everest in a lifetime. If during exercise, $\mathring{Q}$ increases by a factor of four and the Δp by 1.5, estimate the power required.

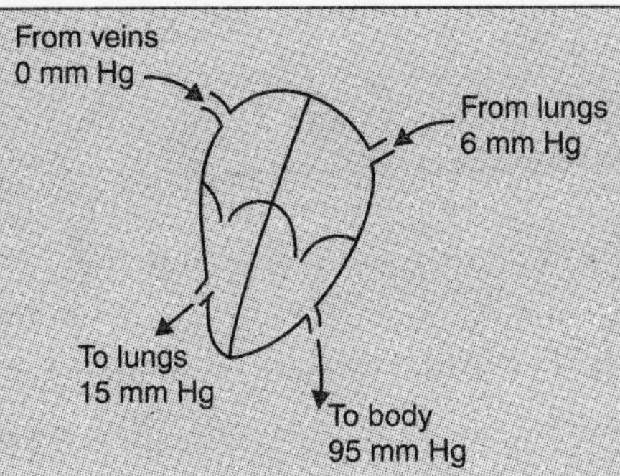

7.7 Deduce the direction of flow of water in the conduit shown if $V_2 = 1$ m/s, $p_1 = 100$ kPa *gauge*, $p_2 = 125$ kPa *gauge* and the cross-sectional areas are $A_1 = 5 \times 10^{-4}$ m^2 and $A_2 = 2 \times 10^{-3}$ m^2.

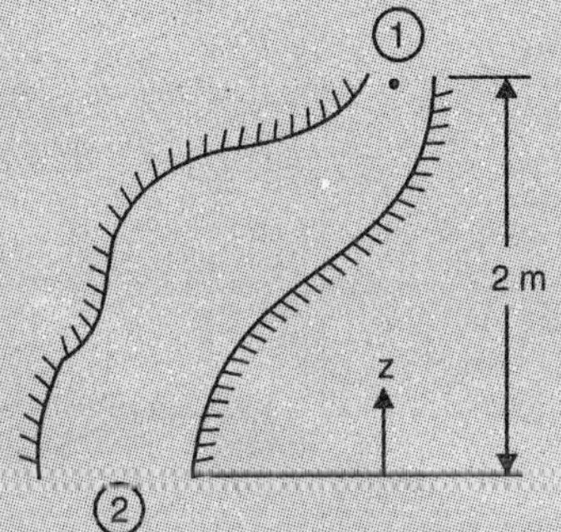

7.8 Is the device shown a pump or a turbine? The pipe diameter is constant throughout.

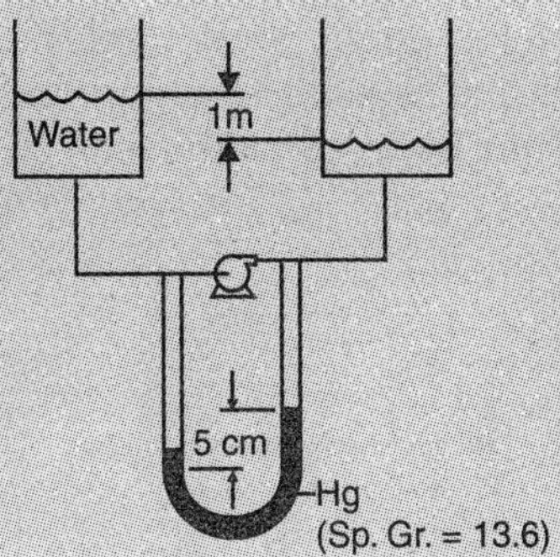

7.9 Compute the power required to drive the pump shown if the pressure at point 1 is 0.5×10^5 Pa and the atmospheric pressure is 10^5 Pa. The areas of both the inlet and outlet pipes are 10^{-3} m^2 and the velocity of the water is 5 m/s.

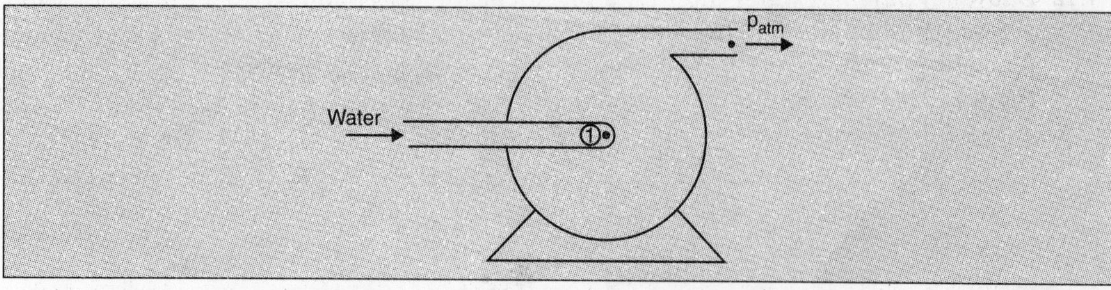

7.10 Consider a household pump driven by a 25 W motor. It is connected by a 12.7 mm plastic tube to a tank where the water surface is 3 m below the pump inlet as shown. The discharge is open to atmosphere. If the pipe is filled with water before the pump starts, determine the pressure at point 1 and the discharge velocity. As a first good approximation, neglect frictional losses and assume that all the energy put into the pump goes to the fluid. Note that the pressure p_1 is subatmospheric and that there exists a likelihood of it being lower than the vapour pressure of water, leading to vaporization. (This is called *cavitation* and is avoided).

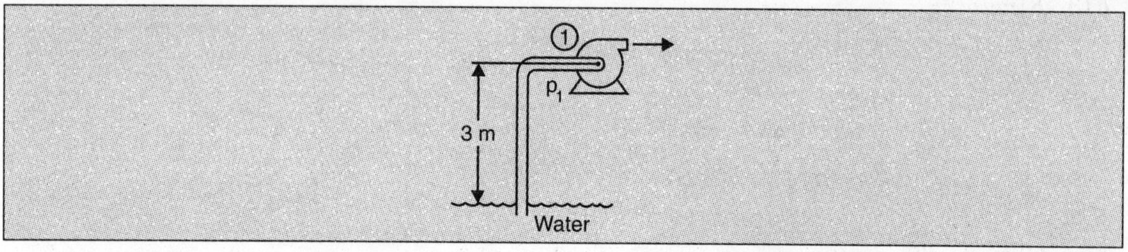

7.11 The tank shown contains water up to height z_i and air at p_{atm} for a length L. It is sealed at time $t = 0$ and a constant power $\mathring{W}_s$ is supplied to the pump. The pump sucks out the water from the tank at a mass flow rate $\mathring{m}(t)$ and discharges it to the atmosphere. As the water level drops the pressure above it decreases opposing the action of the pump. After some time the flow stops and the water level is stationary at z_f. If the head generated by the pump under these conditions is h_s^*, obtain an expression for z_f assuming that air expands isothermally. What happens to the energy supplied to the pump under these conditions?

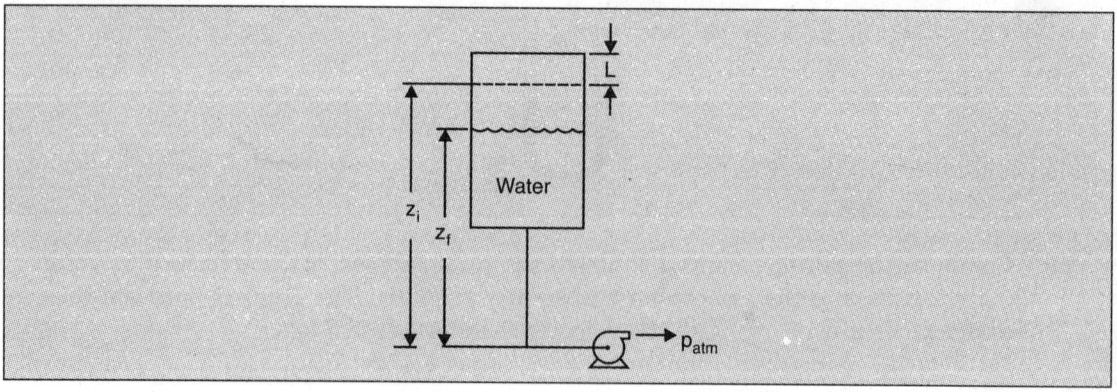

7.12 *Buoyancy pump:* A water filter for a toy aquarium uses compressed air to pump water across the filter screen. The compressor delivers a 2 mm jet of air at 1 m/s into a 5 mm diameter pipe as shown. If the flow rate of water through this device is 100 ml/min find the pressure p_1 inside the tube at the air-tube exit.

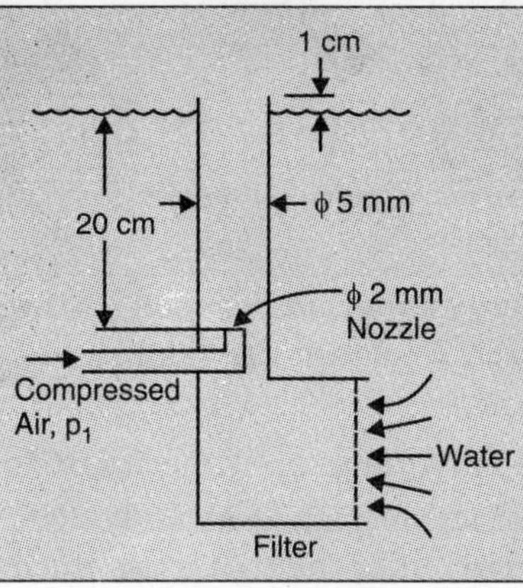

7.13 *Kinetic energy correction factor:* If the velocity in a tube is not uniform across the cross-section, the average velocity V_{av} can still be used in the energy equation provided a kinetic energy correction factor γ (defined analogous to β of Sec. 5.3) is used:

$$\gamma = \frac{\text{net kinetic energy flowing across a section}}{\frac{1}{2}\rho V_{av}^2 \left(\pi R^2 V_{av}\right)}$$

Obtain γ for laminar flow through a circular pipe.

7.14 For a liquid jet emerging out of a pipe in which it was flowing laminarly (Prob. 5.28), obtain R_j using the continuity and Bernoulli equations. Neglect viscous losses. Use the kinetic energy correction factor and the average velocity. The use of momentum equation in Prob. 5.28 gives a better prediction than that obtained here. Why?

7.15 Water flows steadily in a 10 cm diameter pipe at 3 m/s. The pipe has a sudden enlargement where eddies are formed and energy is dissipated. Compute the pressure increase between sections 1 and 2 using the results of Example 5.2. Then apply the mechanical energy equation to obtain the head loss h_l. What is the rise in the temperature of water? Specific heat of water is 4.2 kJ/kg K.

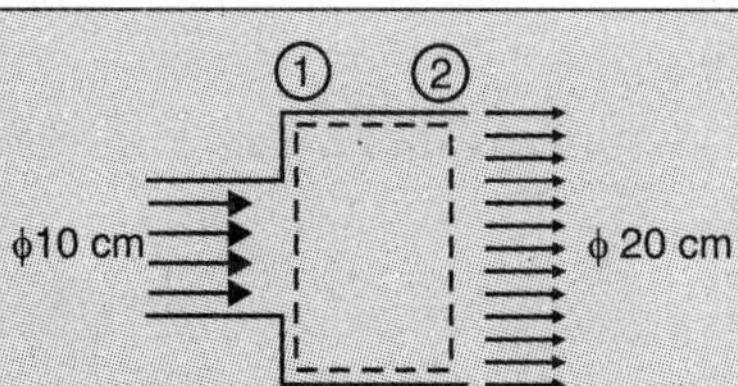

7.16 In Prob. 5.11, the level of water is shown to be slightly higher at the gate than far upstream. Explain. Also verify that at the downstream point where the height of water is 0.7 m, the velocity is 13.5 m/s. Neglect frictional losses.

7.17 Explain why the pressure distributions at the gate in Prob. 5.11 and at the weir in Example 7.8 are as shown

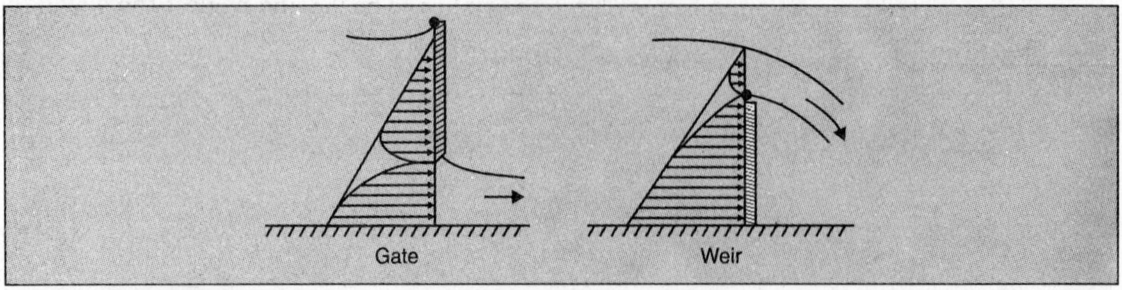

7.18 Apply the mechanical energy equation between sections 1 and 2 on the surface of the *hydraulic jump* of Prob. 5.27 to obtain the head loss as

$$h_l = (h_2 - h_1)^3 / 2h_1h_2$$

and argue that the liquid must *jump up* rather than down at a discontinuity.

7.19 Water drains out of a container with a hole at the bottom. A student argues that a heavier fluid like mercury will drain out of the container much faster. Do you agree?

7.20 Obtain the velocity of water in the nozzle for the tank containing oil and water as shown. Also compute the coefficient of static friction to prevent the tank from slipping. Neglect the weight of the tank. How would your results change if the tank had sloping sides at an angle of 45°, but having the same area of the base as before? (Note that the mechanical energy equation as derived in Chapter 7 is applicable for single fluids only.)

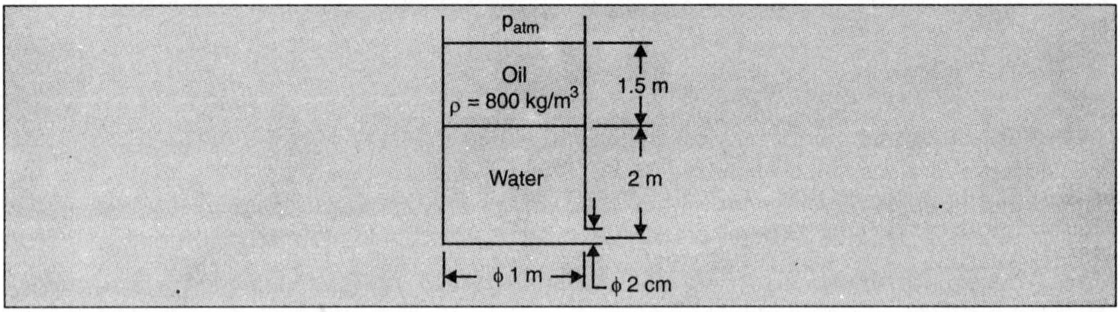

7.21 The ancient water clock *Clepsydra* shown schematically had such a shape that the water level descended at a constant rate at all times. Obtain the shape of the axi-symmetrical jar, i.e., specify R as a function of y, if the water level falls by 4 cm every hour. The drain hole diameter is 2 mm and can be assumed to be very small compared to R.

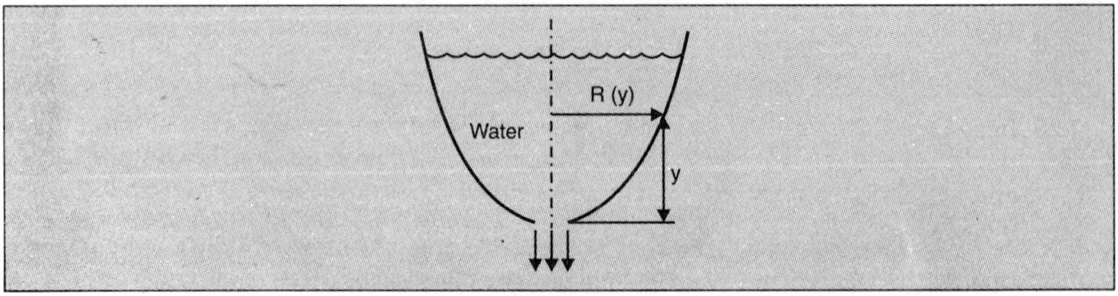

7.22 A 1100 kg car stands on a hydraulic jack as shown. In order to bring the car down, oil is drained out of a 5 cm diameter orifice. Using the quasi-steady state approximation, estimate the time taken. Assume C_d for the orifice to be 0.8 and neglect the weight of the piston.

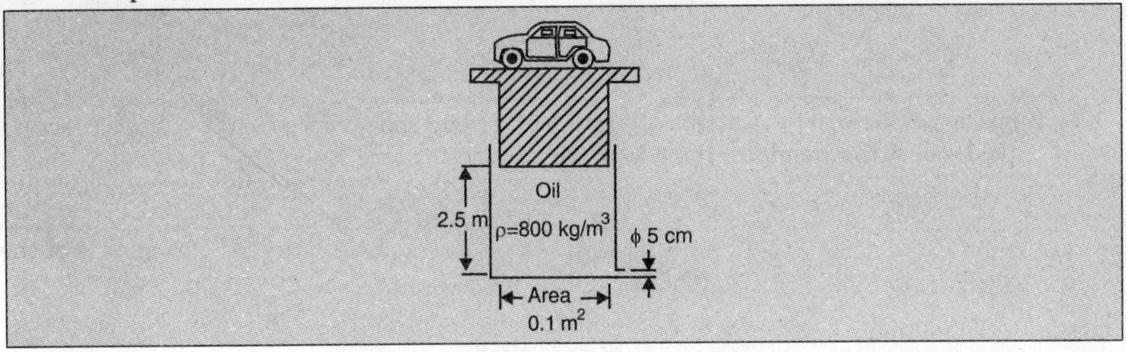

7.23 Consider two tanks with outlet tubes of equal diameter but unequal lengths as shown. The heights h_a and h_b of liquid in the tanks are adjusted such that the flow rates are the same. Show that this set-up can be used to obtain the viscosity using the equation:

$$\rho g(h_b - h_a) = 32\mu\,(L_b - L_a)V_{av}/D^2$$

Note that this experimental technique gets around the error introduced by the developing flow regions (see Sec. 10.1). How? Use expression in Prob. 7.5 for h_l.

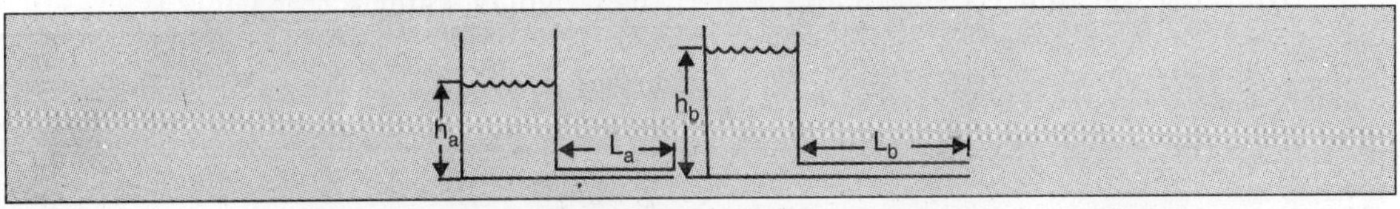

7.24 A jar with a small circular outlet contains oil on top of water, as shown. It is immersed in a large tank of the same oil. Water flows through the outlet. What is the velocity of this flow at time $t = 0$? What would be the position of the oil-water interface in the jar when the flow stops if the density of oil is 800 kg/m³?

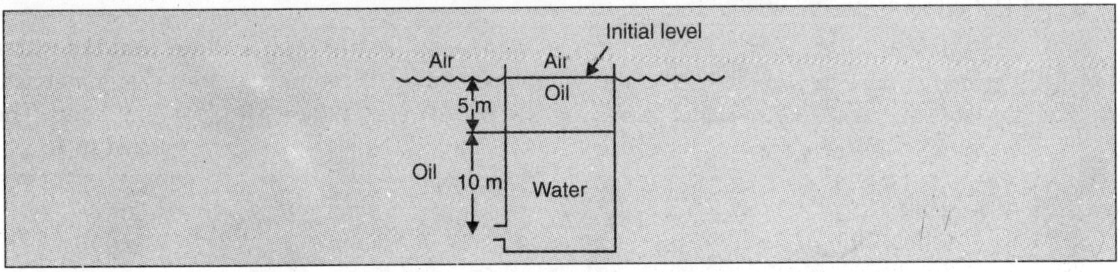

7.25 In a *disc centrifuge* (used for removing fine solid particles from liquids), a bowl containing the liquid is rotated at 6000 rpm. The heavier solids move towards the walls of the bowl and are discharged intermittently through A as shown. If the liquid moves like a solid (i.e., $V_\theta = \omega r$), estimate the velocity of the jet coming out when A is first opened. (Note that the gz term in the Bernoulli equation must be replaced by $\frac{-\omega^2 r^2}{2}$ in this case. Why?)

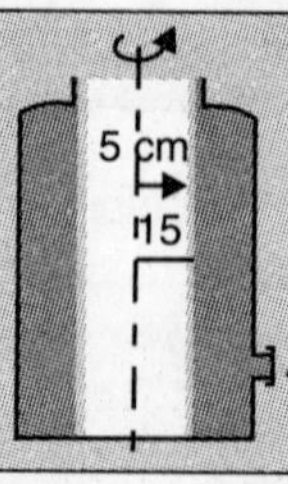

7.26 Find the net vertical force acting on the circular plate shown if the water spreads radially on it. Neglect the weight of the water on the plate.

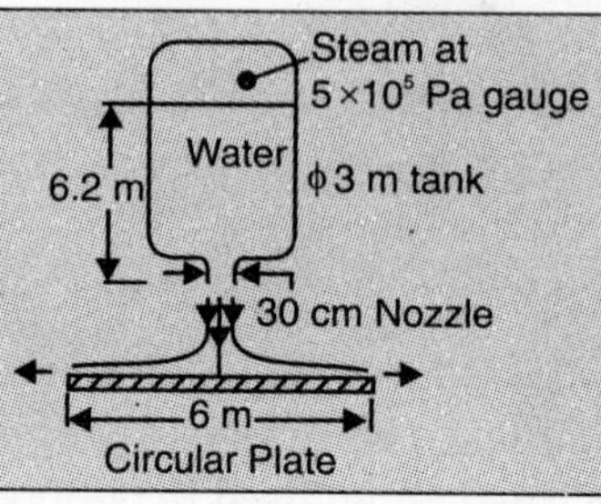

7.27 A circus elephant supports a circular flat plate of mass M horizontally on the top of the water-jet from its trunk. If the velocity of water coming out from the trunk is 5 m/s, the area of the jet at point 1 is 1 cm^2 and the plate is 1 m above point 1, compute the mass of the plate.

(*Hint*: Use the 'thin' *CV* shown and neglect the weight of water within it).

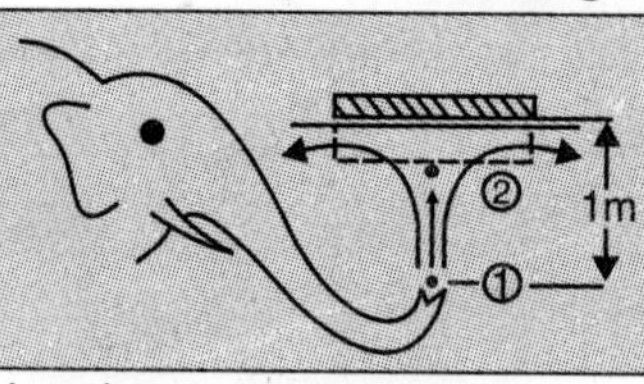

7.28 Consider the jet of water issuing out at the bottom of the large tank shown. The jet impinges on the flat top of a block of wood (diameter 20 cm and height 15 cm). The block is just submerged in water. Find the density of wood. Neglect the change in the velocity of the jet as it moves down.

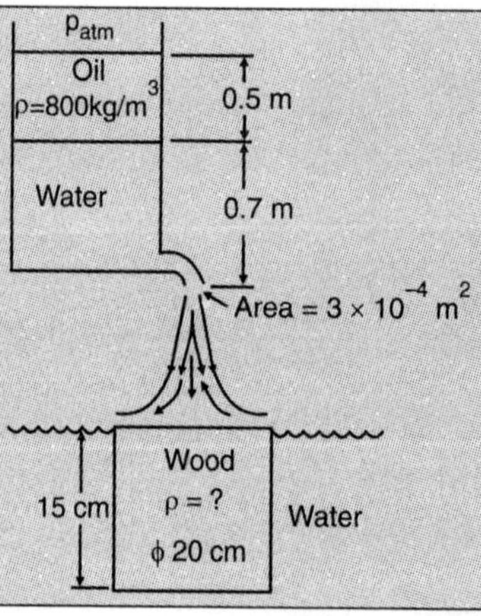

7.29 *Flapper valve:* When a plate is held very close to an orifice, the distance between the two controls the flow rate. This is so because the lesser the distance, the more curved are the streamlines and the higher is the pressure at the orifice, reducing the flow rate. It is found that for very small gaps H, the flow rate $\mathring{Q}$ varies directly as H. Deduce that in the arrangement shown, the force F required to hold the plate stationary is

$$F = \frac{\pi D_0^2}{4} p_0 \left\{ 1 + \left(\frac{4D_p h}{D_0^2} \right)^2 \right\}$$

where D_0 and D_p are the diameters of the orifice and the plate respectively, p_0 is the stagnation pressure of the fluid of density ρ, and h is the thickness of the water at the rim of the disc. Note that force F decreases as the plate moves towards the orifice! Such an arrangement is called a flapper valve. Assume pseudo 1-D flow at the orifice—not a very good assumption for small H.

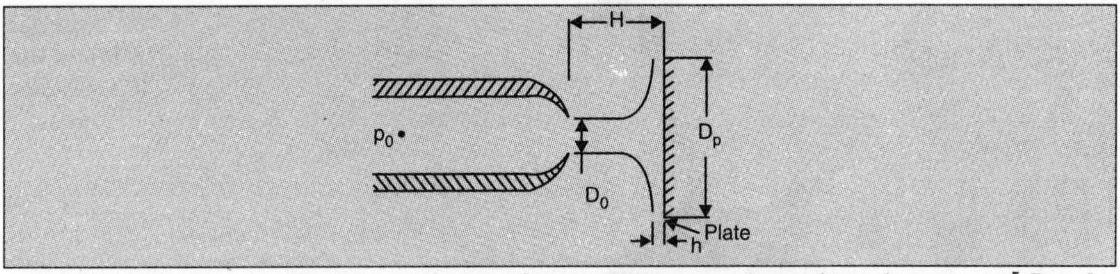

7.30 For the pipe-nozzle system shown, the gauge pressure at point 1 is 1.5×10^5 Pa, the velocity at the nozzle is 6 m/s, and the head loss between points 3 and 2 is estimated as 0.6 m. Compute the x-force on the bolts connecting the nozzle to the pipe. Do not neglect losses in the pipe.

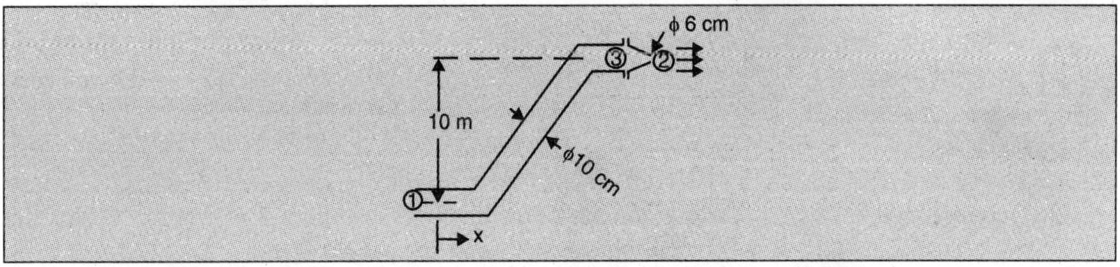

7.31 Two designs for a toy rocket propulsion system are to be studied. In one, pressurized air issues from the nozzle and in the other, air at the same pressure forces water out through an identical nozzle. In which case would there be larger initial acceleration? Assume quasi-steady behaviour and neglect the effect of acceleration of the CV.

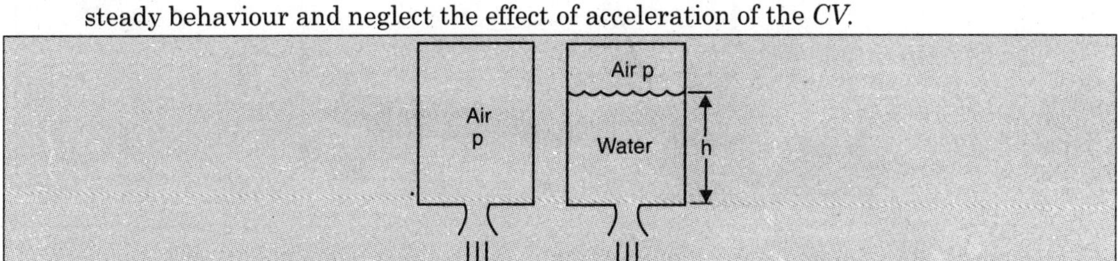

CHAPTER 7

7.32 A boat is propelled by a jet of air issuing from a tank which is fed by a compressor delivering air at a pressure p_0. Show that if the tank is filled with water such that the compressed air pushes out water as a jet, the net thrust is approximately the same but the power consumed by the compressor is drastically reduced.

7.33 Water flows through the pipeline system shown at the rate of 2.5 m^3/min. Obtain the power consumed by the pump neglecting losses.

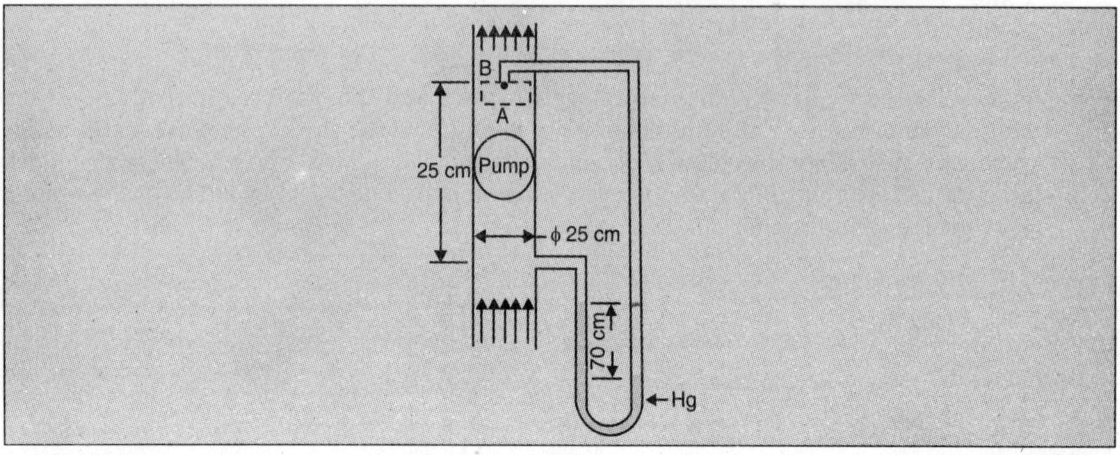

7.34 The propulsion mechanism for a motorboat consists of a pump that takes in water from the lake (inlet pipe area A_1) and forces it out as a jet of area A_j as shown. The boat experiences a drag force F when it moves at a speed V. Obtain an expression for the drag force F in terms of w_s and the other quantities. Also obtain expressions for V_1 and V_j, the velocities relative to the boat.

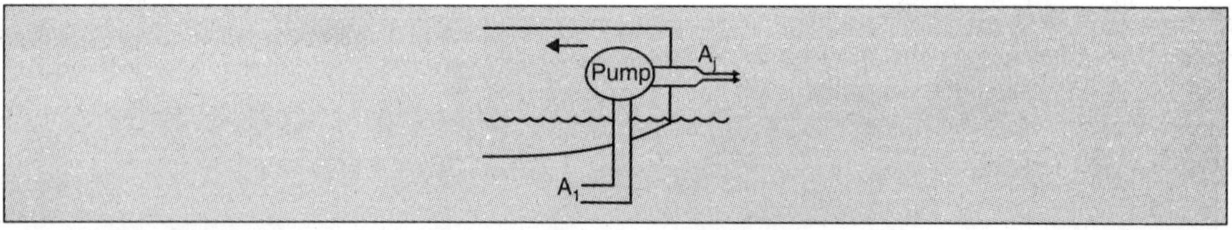

7.35 Two immiscible liquids of same density ρ are flowing through a 2-D nozzle. Neglecting the effect of gravity and friction, obtain h_1/h and $(p_1 - p_2)/\rho V_1^2$. Assume pressure to be constant across any section.

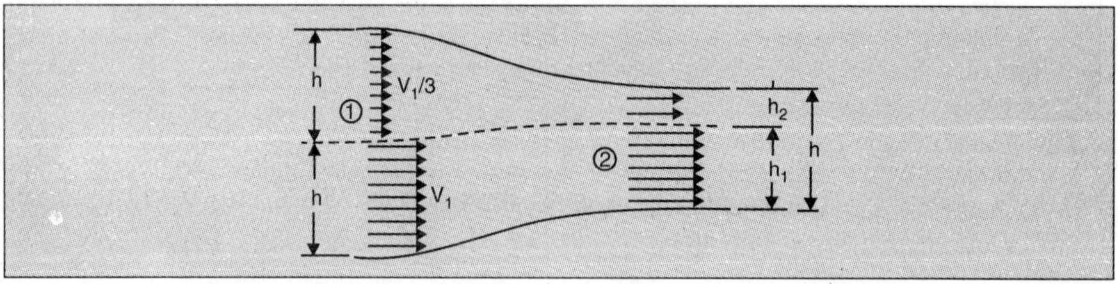

7.36 The expression for $p_2 - p_1$ obtained for the liquid-liquid ejector of Prob. 5.29 contained an unknown velocity V_1. Assuming the head loss to be negligible, use the Bernoulli equation to obtain an implicit equation for V_1. Note that a kinetic energy correction factor may be used at section 1.

7.37 Water flows at 0.05 m³/s through the *Y*-connector shown and splits into two equal streams. The pressure at point 1 is 2×10^5 Pa. Compute p_2 and p_3 and then obtain the *x* and *y* components of the force required to hold the *Y* in place. Neglect the weight of liquid.

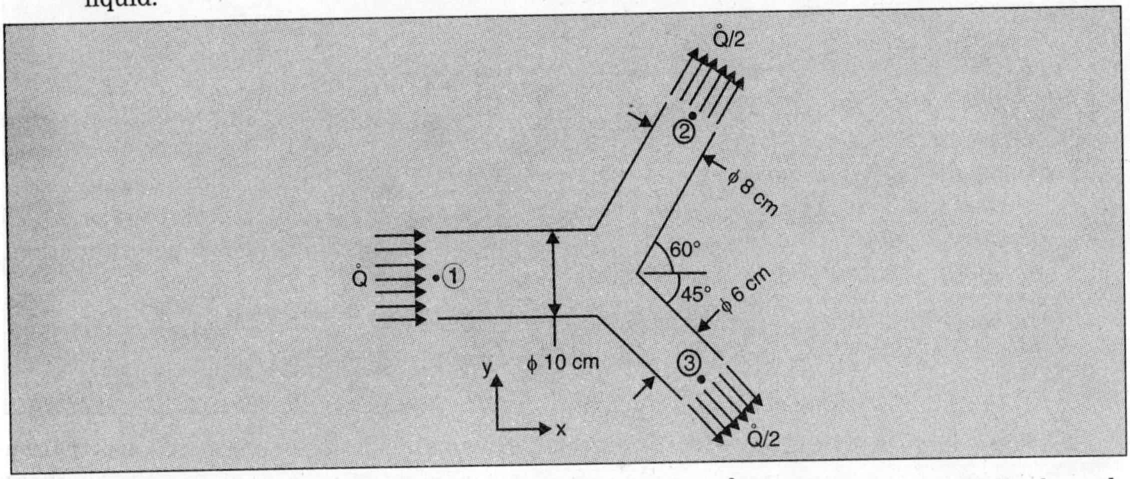

7.38 A fluid flows up the central pipe at a volumetric rate $\mathring{Q}$ and then flows radially through the space between the two circular discs (of radius R_o) as shown. Apply the continuity and Bernoulli equations between points *B* and *C* to obtain the pressure at *B* and show that it is less than atmospheric. This explains why the plates tend to come together. Neglect losses and assume 1-*D* flow. If $V_C \ll V_A$ and *h* is small, compute the pressure at *A*. Is p_A also sub-atmospheric?

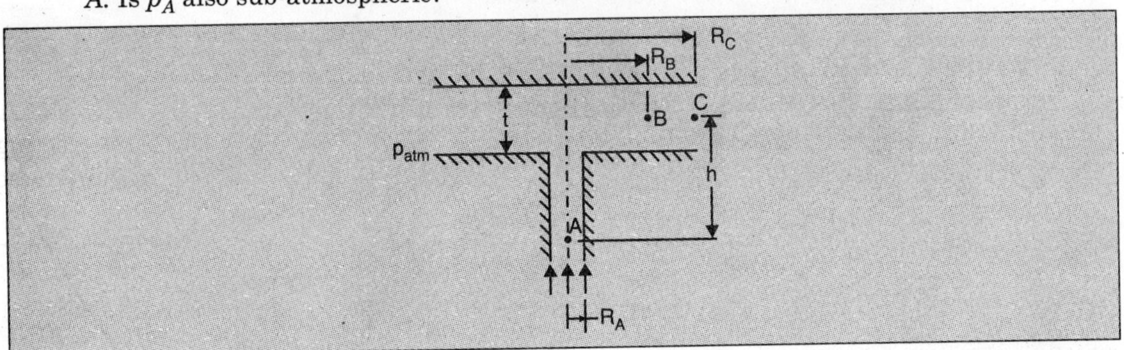

7.39 The fan located in a duct sucks in air from the atmosphere as shown. If the displacement of the fan is 1 m³/s, find the power consumed by it. What is the maximum length *h* through which water will be sucked up a tube by the flowing air? What is the force required to hold the fan in place?

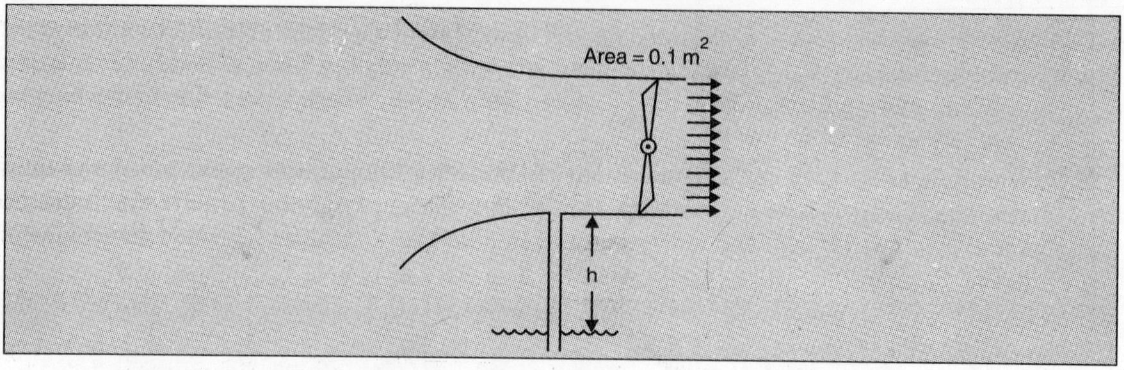

7.40 A 2-*D* body is moving at velocity V_0 in water in a channel of width h as shown. The pressure far upstream is p_0. Under certain conditions it is observed that the pressure in the wake is low enough for water to vaporize and a large bubble of water vapour exists behind the body. The flow far downstream can be assumed to be 1-*D* and the pressure there as p_v, the vapour pressure. Use the continuity, momentum and Bernoulli equations to obtain the drag force per unit depth of the body as

$$F' = (p_0 - p_v)h + \rho V_0^2 h\left\{1 - \sqrt{\frac{2(p_0 - p_v)}{\rho V_0^2} + 1}\right\}$$

where ρ is the density of water. As a first approximation, one can neglect viscous effects.

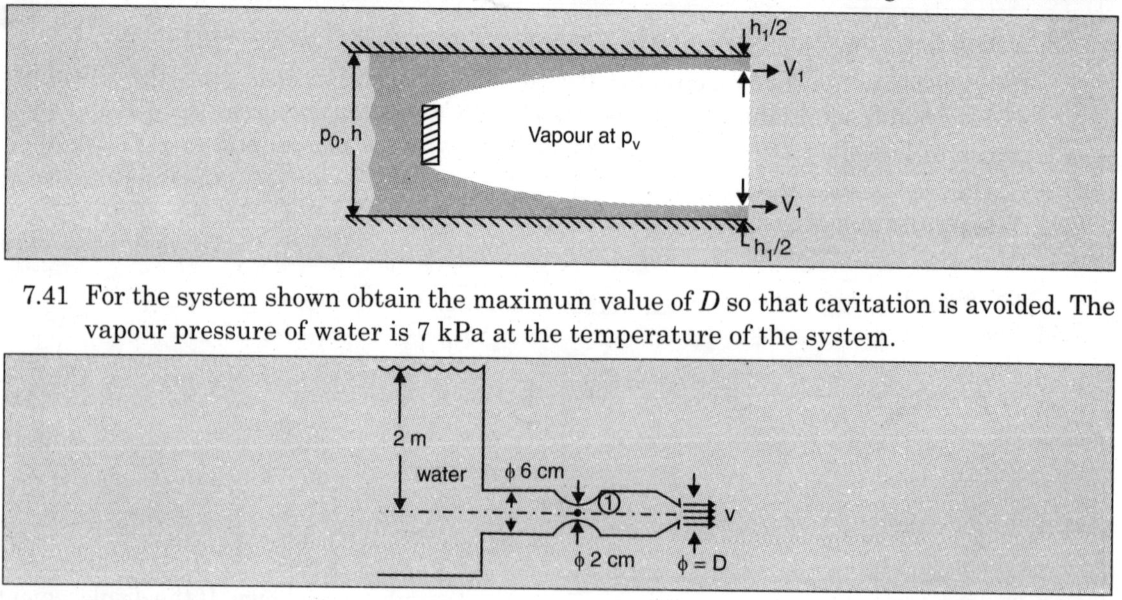

7.41 For the system shown obtain the maximum value of *D* so that cavitation is avoided. The vapour pressure of water is 7 kPa at the temperature of the system.

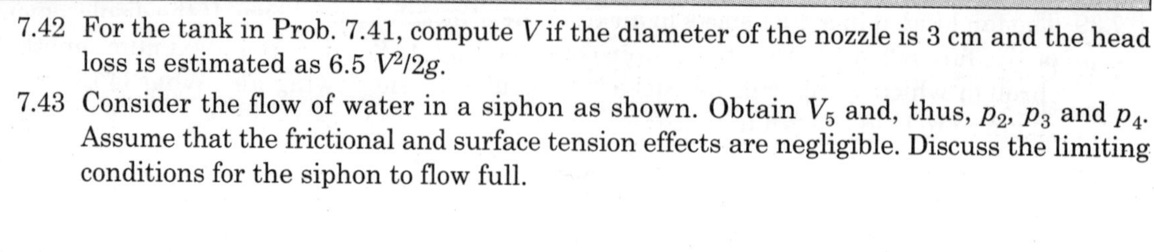

7.42 For the tank in Prob. 7.41, compute *V* if the diameter of the nozzle is 3 cm and the head loss is estimated as $6.5\ V^2/2g$.

7.43 Consider the flow of water in a siphon as shown. Obtain V_5 and, thus, p_2, p_3 and p_4. Assume that the frictional and surface tension effects are negligible. Discuss the limiting conditions for the siphon to flow full.

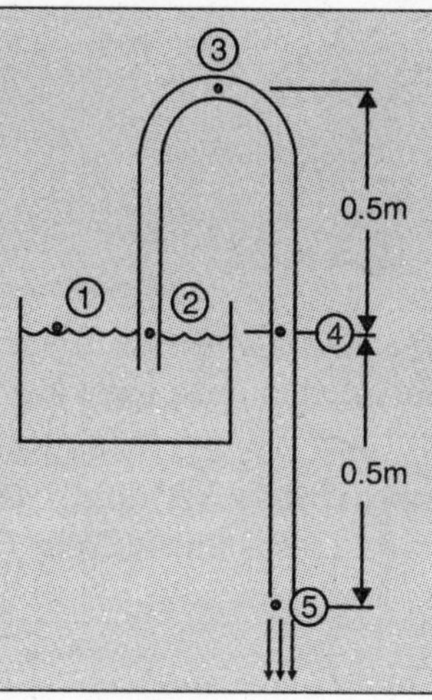

7.44 A pipe of diameter D is immersed to depth h_1 in a stream flowing at velocity V_1, as shown. Obtain the flow rate $\mathring{Q}$ through the tube. Neglect friction.

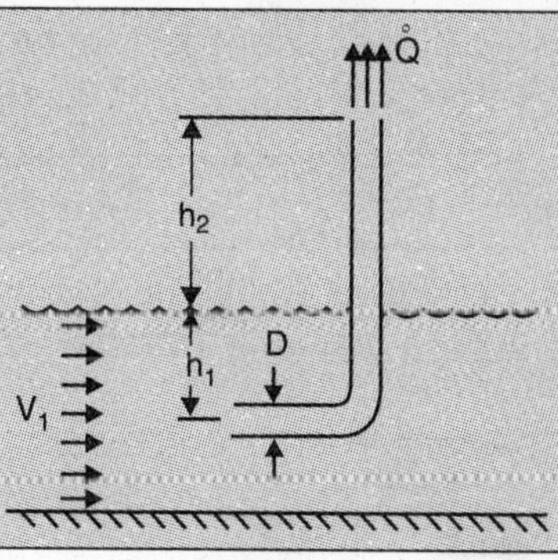

7.45 Water flows at a rate of 60 m^3/s in a 6 m wide channel. The height of the water surface at one station is 3 m while that at another is 2.75 m. If the channel bed is horizontal, estimate the loss due to viscous action.

7.46 *Venturi flume:* A common device used for flow measurement in open channels is a venturi flume as shown. On such a flume with a 10 cm high obstruction, and with the incoming velocity being 3.162 m/s, find the dip in the water level. Use $g = 10\ ms^{-2}$.

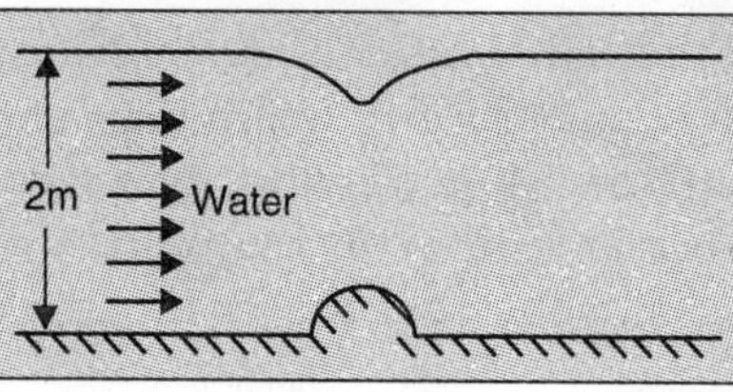

7.47 Calculate the discharge from a triangular notch in a weir in terms of α and h. For a triangular notch with $\alpha = 90°$, the discharge coefficient C_d is 0.58. What is the flow rate when h = 20 cm?

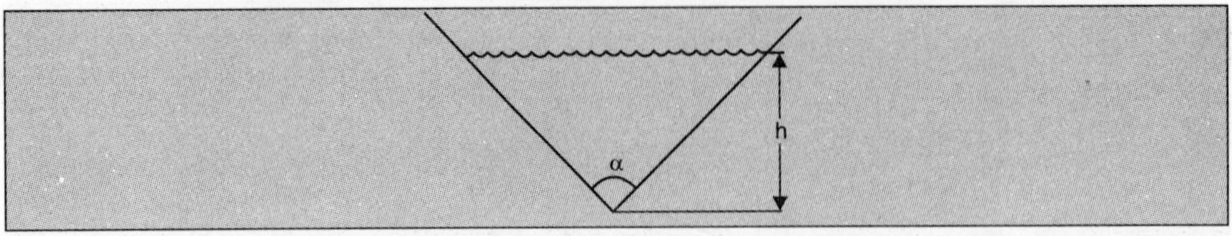

7.48 Water flows uphill as shown. Use the continuity and mechanical energy equations to obtain the three solutions for h. Compute the x-force of water on the ramp, assuming hydrostatic pressure distributions when h is the larger of the two feasible values.

Note that in the hydraulic jump (Probs. 5.27 and 7.18) the use of Bernoulli equation was not indicated—in fact there was a loss of head. Why then, can Bernoulli equation be used here?

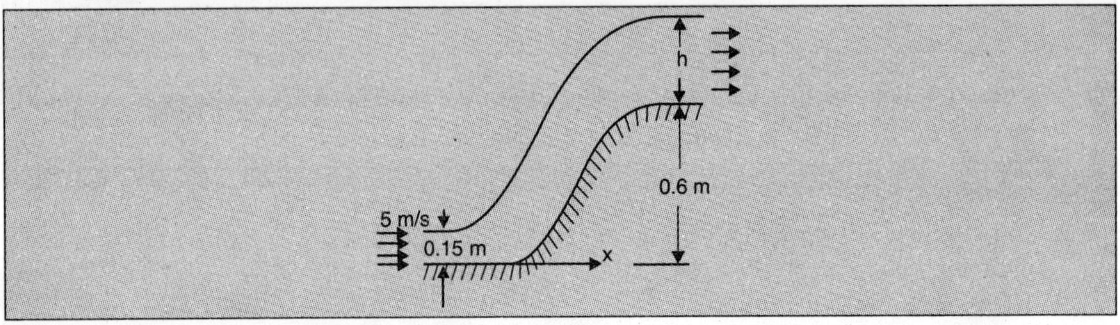

7.49 Water flows down a slope as shown. Apply the continuity and mechanical energy equations to obtain three solutions for h (assume 1-D flow and neglect frictional losses). Discuss the three solutions.

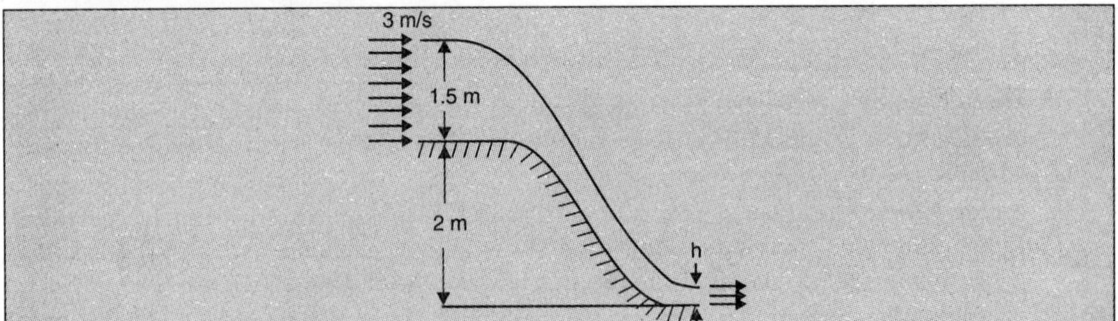

7.50 Consider the flow of water out of a fire hydrant as shown. Apply the continuity and x-momentum equations to the CV marked to show that the x velocity of the water jet remains unchanged. (This is similar to the motion of projectiles.) Now, using the mechanical energy equation, show that the jet hits the ground at a distance l given by

$$l = 2\sqrt{\frac{(p_1 - p_{atm})}{\rho g} y_1}$$

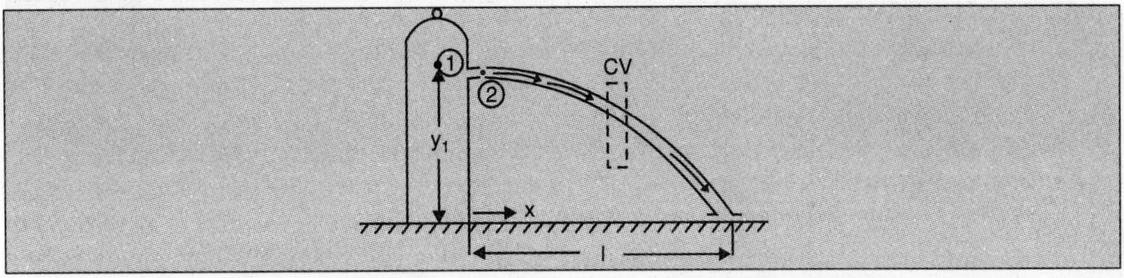

7.51 A blunt-nosed body travels to the left with a speed of *V* in a stationary atmosphere. The ambient pressure is p_0. If the density of air is ρ, use the Bernoulli equation to determine the pressure at point 1 where the fluid moves with the body.

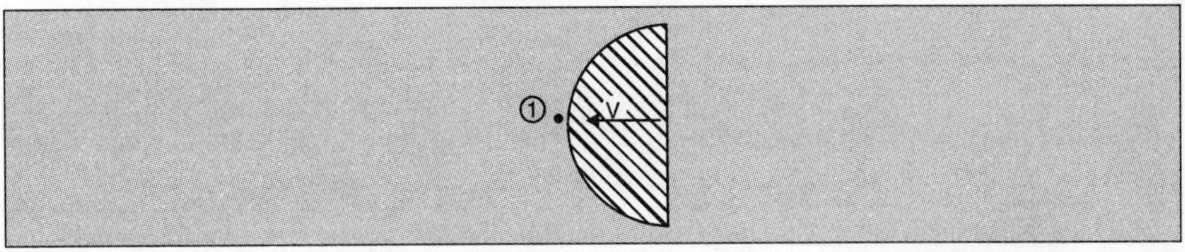

7.52 In the hydraulic damper described in Prob. 5.14, the gauge pressure at the bottom of the cylinder was taken as $F/A_p - \rho V_p^2/2$. Derive this using the mechanical energy equation. [*Hint*: Use a frame of reference in which the flow is steady. The pressure on the piston face is determined by its free-body diagram.]

7.53 In Prob. 7.33 a student applies the momentum equation over the *small CV* shown (assuming the velocity at *B* to be zero) and obtains

$$\frac{p_B - p_A}{\rho} = V_A^2$$

The Bernoulli equation on a streamline, however, leads to

$$\frac{p_B - p_A}{\rho} = \frac{1}{2} V_A^2$$

Which of these is correct?

7.54 Consider the manifold described in Prob. 5.34 with a third of the water flowing out through the side. Apply the Bernoulli equation between points 1 and 2 to obtain

$$p_2 - p_1 = \frac{5}{18} \rho V_1^2 .$$

This equation gives a better representation of experimental data than does the expression derived in Prob. 5.34. Why?

7.55 Show that at the *vena contracta* in a jet (Fig. 1.32), where the streamlines are parallel and straight, the velocity must be uniform across the jet.

7.56 A 90° smooth bend in a rectangular channel is shown in Fig. 6.19. The velocity is uniform at the upstream and downstream sections. Show that the streamlines at the bend are closer together near the inside than at the outside. Neglect gravity and viscous effects.

7.57 Consider the 90° bend of Prob. 7.56 (see Fig. 6.19). Obtain expressions for velocity V and pressure p at the bend as a function of the radial position r. As a first approximation assume that the streamlines are all circular and concentric. This derivation is an improvement over that of Prob. 6.48.

7.58 The pressure profiles for an orifice-plate flow meter (see Prob. 1. 22g) obtained by two sets of piezometric tubes, one reading at the centre-line and the other at the wall, are as shown. Explain the main features of the curves.

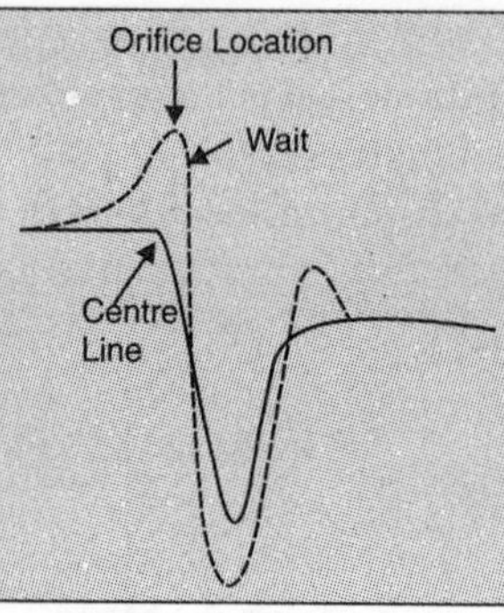

7.59 A popular book (Ya Perelman, *Physics for Entertainment,* Mir Pub., Moscow) describes and explains the following experiment:

"Figure depicts a small pith ball floating in a jet of air. The air strikes the ball and

prevents it from falling, Meanwhile, should the ball pop aside, the outer air—whose pressure is greater since its velocity is smaller—returns it to the jet".

What is the flaw in this argument? The phenomenon described is, however, true. Can you explain? [*Hint*: The answer lies in the bending of streamlines.]

7.60 Explain how the liquid rises up in the atomizer (e.g., a *Flit* pump) shown.

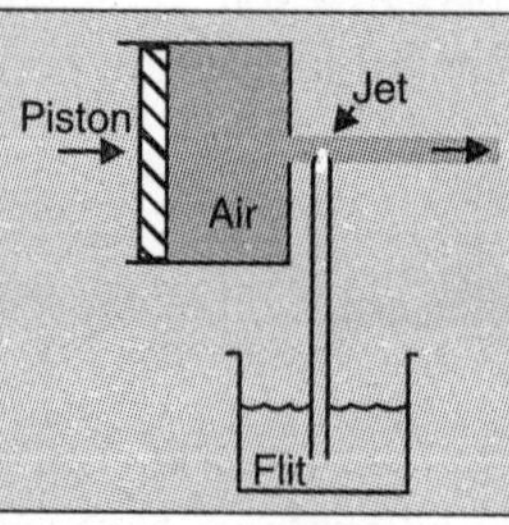

7.61 In 1912, a small cruiser *Hawk* approaching a big ocean liner *Olympic* on a parallel course suddenly veered off its course and rammed into the bigger ship as if attracted to it. Explain this attraction. Then argue that the mean water level in the channel between the two ships must have been lower than elsewhere. The sea was calm on that fateful day.

8

SOME ENGINEERING APPLICATIONS-I

8.1 TURBOJET ENGINE

A jet engine develops a thrust by the reaction of hot gases exhausted at large velocities. Such an engine needs to *breathe* air at the front end. In a turbojet engine (Fig. 8.1a) this is accomplished through the use of a compressor driven by a turbine. The inlet air is taken through

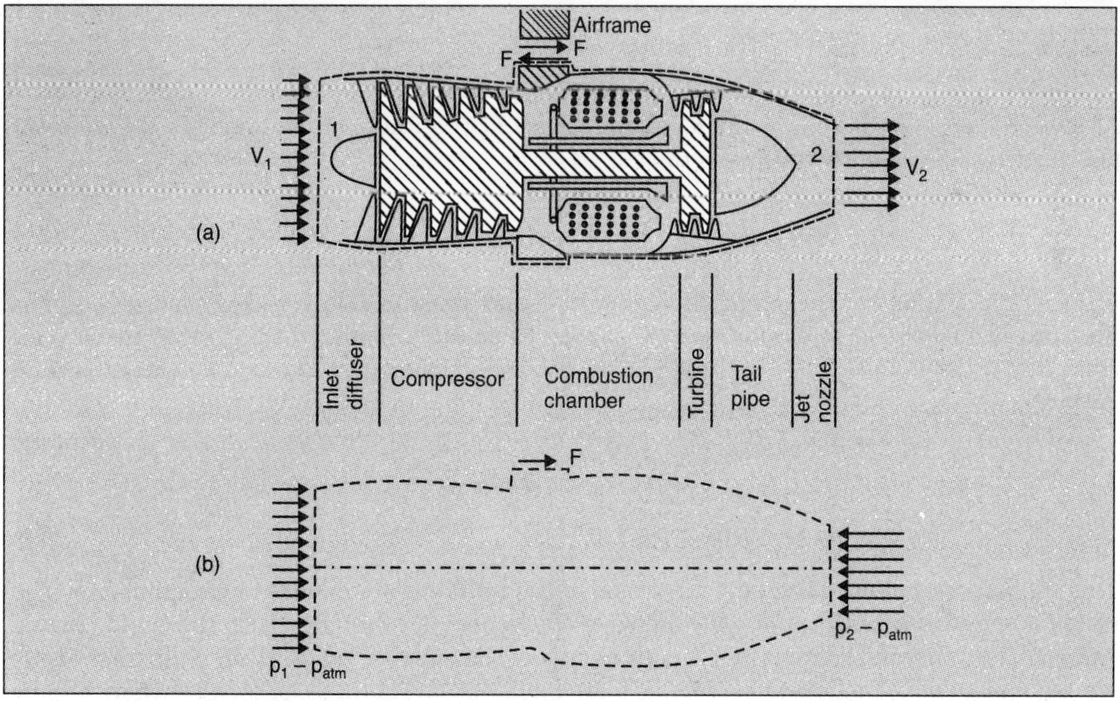

Fig. 8.1. (a) Schematic of a turbojet engine. *CV* shown. Reference frame fixed to *CV;* (b) External *x*-forces acting on *CV*. p_{atm} all around.

a diverging section, called the diffuser, which slows it down, increasing its pressure. An axial-flow compressor thereafter further increases the air pressure. Atomized fuel (essentially kerosene) is then mixed with the compressed air and is burnt in circumferentially arranged combustion chambers. A part of the energy of combustion is extracted through a turbine (hence the name turbojet) which runs the compressor. Finally, the gases expand through the exit nozzle into the atmosphere, the high velocity jet giving the reaction thrust to the engine. It is this thrust that is calculated below.

Consider the *CV* in Fig. 8.1 (b) where F is the force applied by the support on the *CV*. The force applied by the engine on the airframe is F acting in the opposite (i.e., the forward) direction. This force F *on* the aircraft is the thrust. Let the engine shown be moving to the left with a *constant velocity* V_1, the speed of the aircraft. Considering a reference frame fixed with the aircraft, the flow through the *CV* is steady and the reference frame is inertial. The air enters the *CV* with speed V_1 to the right. Let V_2 be the exhaust velocity of the combustion gases. If subscripts 1 and 2 denote conditions at the inlet and outlet respectively, and if fuel is mixed with air in the ratio of 1 : N, the mass balance Eq. 4.2 gives

$$\rho_1 V_1 A_1 + \frac{1}{N}(\rho_1 V_1 A_1) = \rho_2 V_2 A_2 \qquad \text{...(8.1)}$$

The thrust F is obtained by the momentum Eq. 5.2 applied to the *CV*:

$$F + (p_1 - p_{atm})A_1 - (p_2 - p_{atm})A_2 = (\rho_2 V_2 A_2)V_2 - (\rho_1 V_1 A_1)V_1 \qquad \text{...(8.2)}$$

which assumes that the fuel enters the *CV* with negligible x-momentum. Using Eq. (8.1),

$$F = \rho_1 V_1 A_1 \left[\left(1 + \frac{1}{N}\right)V_2 - V_1\right] + (p_2 - p_{atm})A_2 - (p_1 - p_{atm})A_1 \qquad \text{...(8.3)}$$

At subsonic speeds of the aircraft p_1, the pressure at the inlet can be taken as the ambient pressure p_{atm}, and Eq. (8.3) reduces to

$$F = \rho_1 V_1 A_1 \left[\left(1 + \frac{1}{N}\right)V_2 - V_1\right] + (p_2 - p_{atm})A_2 \qquad \text{...(8.4)}$$

In most applications of turbojets, the gases in the exhaust nozzle are expanded so that the pressure at the exhaust is as close to atmospheric as possible. Thus, $p_2 \simeq p_{atm}$. Also, the air-fuel ratio N is usually quite large and $1/N$ may be neglected compared to one. The thrust is then given by

$$F = \rho_1 V_1 A_1 (V_2 - V_1) \qquad \text{...(8.5)}$$

For the aircraft travelling at a speed V_1 the *useful* power developed by the engine is

$$\mathring{W}_{out} = F V_1 = \rho V_1^2 A_1 (V_2 - V_1) \qquad \text{...(8.6)}$$

This useful power represents only a fraction of the total energy released by combustion. The rest of it is lost in various ways, which include convection and radiation from the engine casing and increased thermal content and kinetic energy of the exhaust gases. Even if all other losses are reduced to zero, the losses as kinetic energy must still remain or, otherwise, there will be no thrust. The lost kinetic energy, thus, represents the absolute minimum losses that occur in a turbojet if it is to generate any thrust. In a frame of reference fixed with the ground, the inlet air has zero velocity. After the engine passes by, the velocity of the gases is $(V_2 - V_1)$. Thus, $\frac{1}{2}\overset{\circ}{m}\,(V_2 - V_1)^2$ is the kinetic energy remaining in the air after the aircraft has passed by and so represents this minimum loss of energy. Thus,

$$\overset{\circ}{W}_{l,\,\min} = \frac{1}{2}\overset{\circ}{m}\,(V_2 - V_1)^2 = \frac{1}{2}(\rho_1 V_1 A_1)(V_2 - V_1)^2$$

The *minimum* rate of energy input (by combustion) should then be

$$\overset{\circ}{W}_{\text{in, min}} = \overset{\circ}{W}_{\text{out}} + \overset{\circ}{W}_{l,\min}$$

$$= \rho_1 V_1^2 A_1 (V_2 - V_1) + \frac{1}{2}\rho_1 V_1 A_1 (V_2 - V_1)^2$$

and the *maximum possible* efficiency of a turbojet is

$$\eta_{\max} = \frac{\overset{\circ}{W}_{\text{out}}}{\overset{\circ}{W}_{\text{in, min}}} = \frac{\rho_1 V_1^2 A_1 (V_2 - V_1)}{\rho_1 V_1^2 A_1 (V_2 - V_1) + \frac{1}{2}\rho_1 V_1 A_1 (V_2 - V_1)^2}$$

$$= \frac{2}{1 + \dfrac{V_2}{V_1}} \qquad \text{...(8.7)}$$

Full efficiency can be obtained only when V_2 equals V_1 but this leads to zero thrust according to Eq. (8.5).

The value obtained in Eq. (8.7) is only the maximum possible propulsive efficiency. In practice, the other losses are quite significant. Typical values of the overall energy efficiency are around 10 per cent for well-designed turbojets.

To increase the efficiencies of the basic turbojet engine various modifications have been suggested. They include a bypass engine, a turbofan engine, or some combination thereof. Figure 8.2 shows a bypass engine in which a portion of the air from the first stage of the compressor

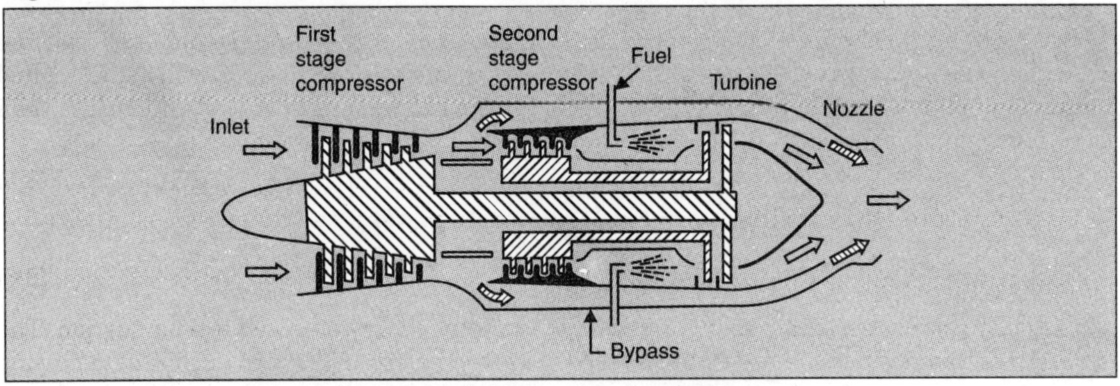

Fig. 8.2. Bypass turbojet engine.

bypasses the combustion chamber and the turbine, and mixes with the rest of the gases in the jet pipe outlet. Figure 8.3 shows a turbofan engine in which the exhaust stream from the main

CHAPTER 8

turbine drives the fan-turbine which, in turn, is attached to a fan handling additional cold air and developing a thrust. By a proper combination of these two principles, much higher efficiencies may be obtained.

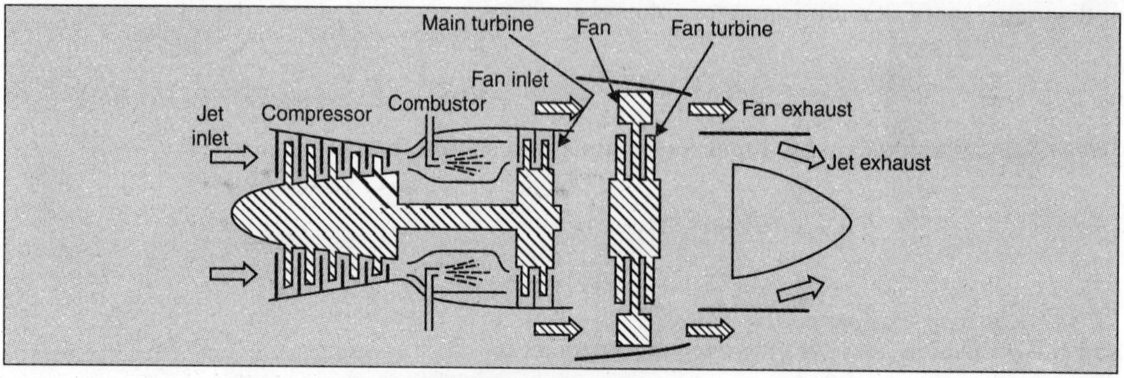

Fig. 8.3. Turbofan engine.

8.2 PROPELLERS AND WINDMILLS

A propeller or a windmill consists of several rotating blades attached to a hub connected to a shaft. In a propeller torque is applied to the shaft, resulting in the motion of air (or water in the case of a marine propeller) past the blades (Fig. 8.4). On the other hand, the natural motion of air past the blades of a windmill develops a torque at the shaft, which can be made to do useful work (Fig. 8.5). In this section a simple analysis is presented which is applicable to both types of devices.

The propeller or the windmill is replaced by what is termed as an *actuator disc* (3-4-4′-3′ in Figs. 8.4 and 8.5) which is a thin cylindrical *CV* enclosing the propeller (or windmill). The reference frame is fixed with the actuator disc which imparts a steady (in an averaged sense) change in momentum to the fluid stream passing across it, without giving it any swirl or radial velocity.

Consider the streamlines which pass through the periphery of the actuator disc. In this simplified analysis, it is assumed that these streamlines define the extent of the fluid that is

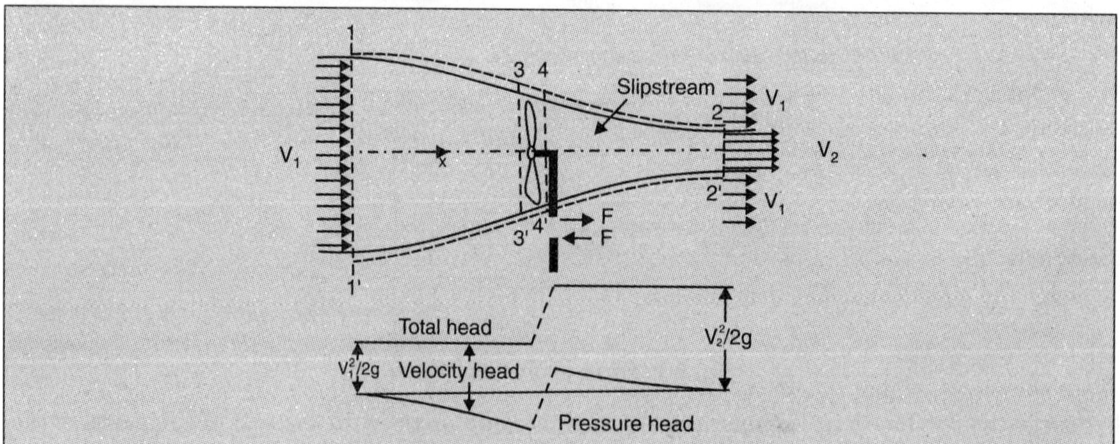

Fig. 8.4. A propeller and the associated heads.

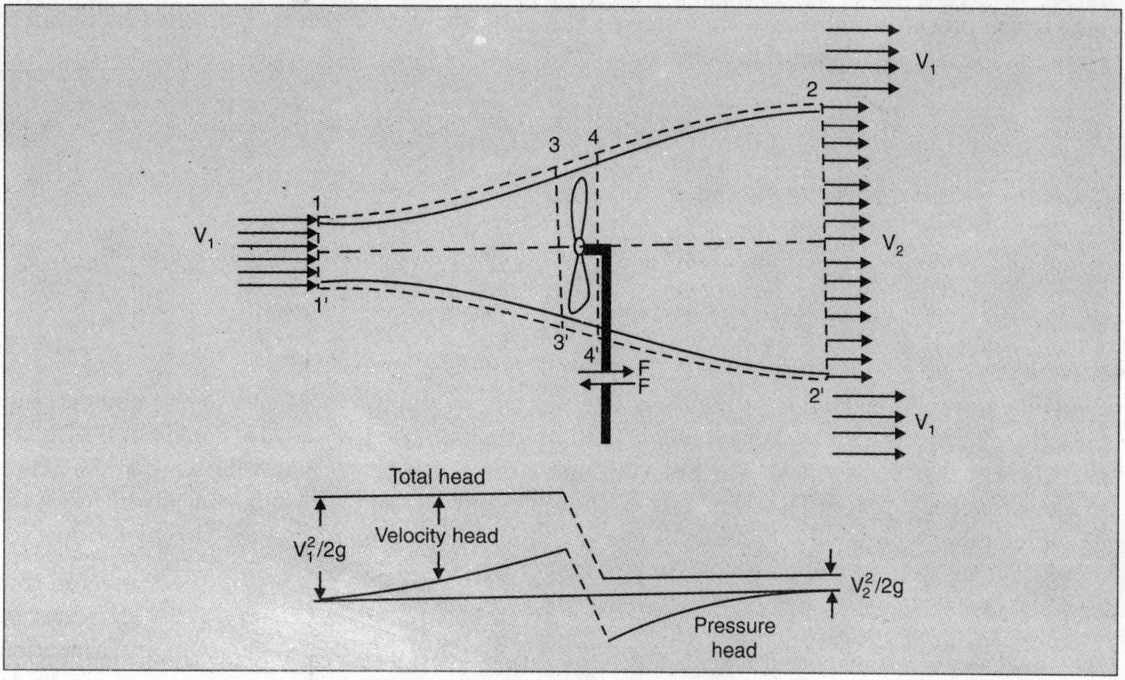

Fig. 8.5. A windmill and the associated heads.

disturbed by the actuator disc, such that the velocity at any point outside these limits is the undisturbed velocity V_1. The flow region between these streamlines is termed as the *slipstream*. Consider the region bounded by these streamlines to be the *CV*. There is no flow across segment 1-2 of the axisymmetric *CS*.

To find the thrust *F* apply the momentum equation to the control volume 1-2-2'-1'. If the flow is assumed to be steady, incompressible, 1-*D* and with negligible viscosity, Eq. (5.2) gives

$$F = \overset{\circ}{m}(V_2 - V_1) \quad ...(8.8)$$

where

$$\overset{\circ}{m} = \rho V_1 A_1 = \rho V_2 A_2 \quad ...(8.9)$$

One can also apply the Bernoulli equation or the mechanical energy Eq. (7.15) between sections 1-1′ and 3-3′ to get

$$\frac{1}{2}\rho V_1^2 + p_1 = \frac{1}{2}\rho V_3^2 + p_3 \quad ...(8.10)$$

Similarly, the Bernoulli equation between sections 4-4′ and 2-2′ gives

$$\frac{1}{2}\rho V_4^2 + p_4 = \frac{1}{2}\rho V_2^2 + p_2 \quad ...(8.11)$$

Note that $p_1 = p_2 = p_{atm}$ and $V_3 = V_4$ (since $A_3 \simeq A_4$). Thus, Eqs. (8.10) and (8.11) give

$$p_4 - p_3 = \frac{1}{2}\rho\left(V_2^2 - V_1^2\right) \quad ...(8.12)$$

This is the pressure difference that acts on the actuator disc which if multiplied by its area, gives the force F on the actuator disc as

$$F = (p_4 - p_3)A_3 = \frac{1}{2}\rho A_3 \left(V_2^2 - V_1^2\right) = \rho A_3 (V_2 - V_1)\frac{(V_2 + V_1)}{2} \quad ...(8.13)$$

Comparing this with Eq. (8.8), and noting that

$$\overset{\circ}{m} = \rho V_3 A_3$$

we obtain

$$V_3 = \frac{V_1 + V_2}{2} \quad ...(8.14)$$

Consider first the propeller in Fig. 8.4. The velocity of the air increases in the downstream direction. The pressure at section 3 is less than atmospheric (this is also consistent with the curvature of the streamlines). The pressure increases across the actuator disc due to the action of the propeller blades. The slip-stream then straightens out and narrows down, with air accelerating to velocity V_2. The useful power output of the propeller is the thrust F *times* the forward velocity V_1 of the propeller (in still air). Thus,

$$\overset{\circ}{W}_{out} = \overset{\circ}{m}(V_2 - V_1)V_1 \quad ...(8.15)$$

In a frame of reference fixed with the ground, air is initially at rest. After the propeller has passed by the velocity of the air there is $(V_2 - V_1)$.

Thus, the kinetic energy of the air increases at a rate of $\frac{1}{2}\overset{\circ}{m}\,(V_2 - V_1)^2$ due to the action of the propeller. This represents a loss of energy from the point of view of the propulsion system. Thus, even if there were no other losses, the minimum power input must be equal to $\overset{\circ}{W}_{out} + \overset{\circ}{W}_{l,\,min}$ with $\overset{\circ}{W}_{l,\,min} = \frac{1}{2}\overset{\circ}{m}\,(V_2 - V_1)^2$. The maximum efficiency of the propeller is, therefore,

$$\eta_{max} = \frac{\overset{\circ}{W}_{out}}{\overset{\circ}{W}_{out} + \overset{\circ}{W}_{l,\,min}} = \frac{\overset{\circ}{m}(V_2 - V_1)V_1}{\overset{\circ}{m}(V_2 - V_1)V_1 + \frac{1}{2}\overset{\circ}{m}(V_2 - V_1)^2}$$

$$= \frac{2}{1 + \frac{V_2}{V_1}} \quad ...(8.16)$$

which is of the same form as Eq. (8.7) for a turbojet engine.

The power input $(-\overset{\circ}{W}_s)$ can be calculated by applying the mechanical energy equation between sections 1 and 2. Equation (7.12) in the absence of viscous losses gives

$$\overset{\circ}{w}_s = \frac{V_1^2 - V_2^2}{2}$$

and so $$\overset{\circ}{W}_{in} = -\overset{\circ}{W}_s = -\overset{\circ}{m} w_s = \frac{1}{2}\overset{\circ}{m}\left(V_2^2 - V_1^2\right)$$

which gives the same result for η_{max} as Eq. (8.16).

In windmills (see Fig. 8.5) the speed of air decreases as it passes through the actuator disc. The slipstream diverges and p_3 is larger than p_4. The maximum work output in this case is the change in the kinetic energy of the fluid stream. Thus,

$$\overset{\circ}{W}_{\text{out, max}} = \frac{1}{2}\overset{\circ}{m}\left(V_1^2 - V_2^2\right) = \frac{1}{2}\rho V_1 A_1 \left(V_1^2 - V_2^2\right) \quad ...(8.17)$$

The efficiency of a windmill may be defined by regarding $\overset{\circ}{W}_{\text{in}}$ as the energy of the free stream which passes through the actuator-disc area *without any energy being taken out,* i.e., with the windmill not rotating. Thus,

$$\overset{\circ}{W}_{\text{in}} = \frac{1}{2}(\rho V_1 A_3) V_1^2$$

The maximum efficiency of the windmill is then

$$\eta_{max} = \frac{\frac{1}{2}\rho V_1 A_1 \left(V_1^2 - V_2^2\right)}{\frac{1}{2}(\rho V_1 A_3) V_1^2} = \frac{A_1}{A_3}\frac{V_1^2 - V_2^2}{V_1^2} \quad ...(8.18)$$

The continuity equation and Eq. (8.14) give

$$\frac{A_3}{A_1} = \frac{V_1}{V_3} = \frac{2V_1}{V_1 + V_2}$$

and so

$$\eta_{max} = \frac{(V_1 + V_2)\left(V_1^2 - V_2^2\right)}{2V_1^3} \quad ...(8.19)$$

The highest value of η_{max} of 59.3 per cent is obtained when $V_2 = V_1/3$. Under thes conditions, the maximum work output, $\overset{\circ}{W}_{\text{out, max}}$ is $8\,\rho A_3 V_1^3/27$, i.e., it varies as the cube of the wind velocity. Efficiencies as high as 40 per cent have been achieved.

8.3 TURBOMACHINERY

A turbomachine consists of an *impeller* or a *runner* attached to a rotating shaft. The impeller or runner interacts with the fluid flowing through the machine changing its mechanical energy. In a turbine, the mechanical energy of the fluid decreases generating a torque which can be made to do useful work, while in a pump, compressor or a blower, the torque applied at the shaft increases the mechanical energy of the fluid. The increased mechanical energy of the fluid

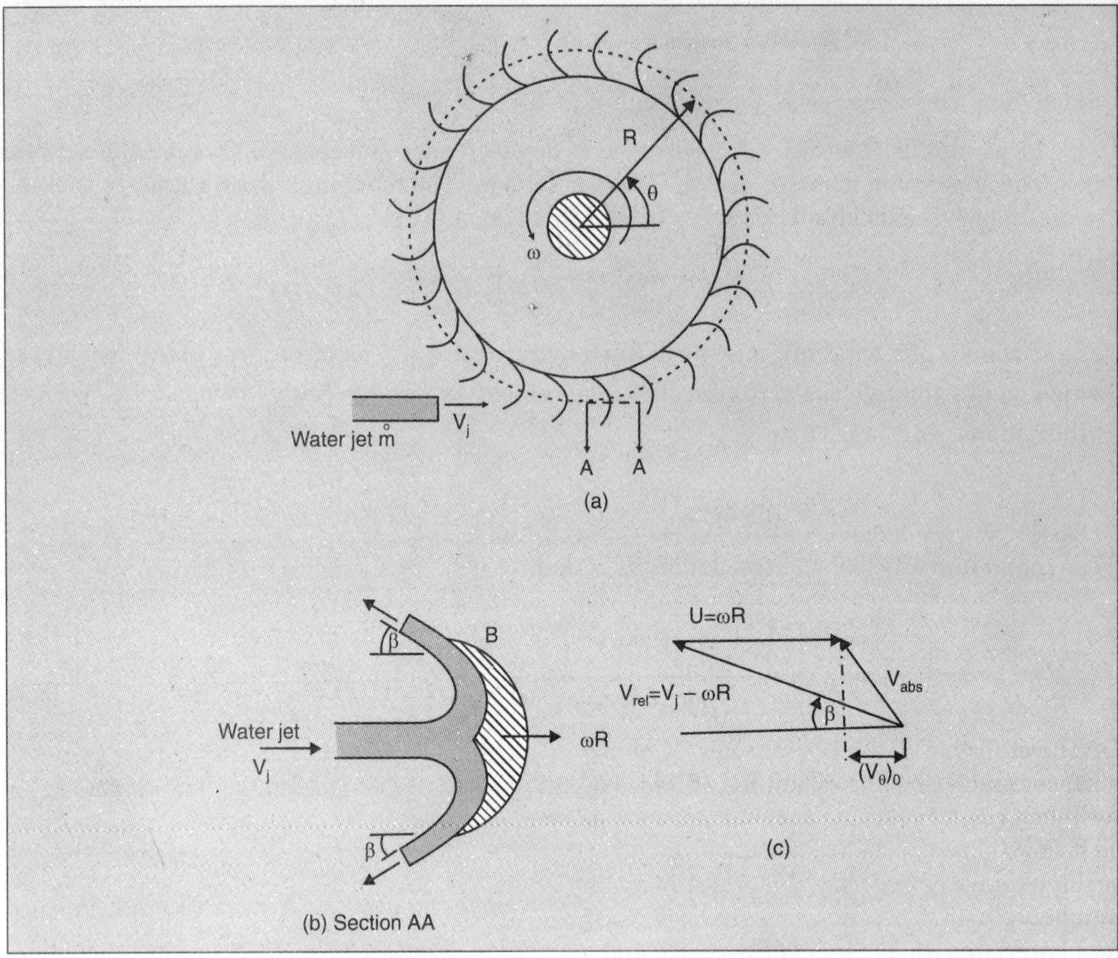

Fig. 8.6. (a) A Pelton-wheel runner, (b) flow of the water jet across a horizontal section *AA* of a bucket, and (c) velocity triangle at exit point *B*.

can manifest itself either as increased fluid pressure or as increased flow velocity. A propeller or a fan is a turbomachine of the second type while a windmill is a turbine which extracts power from the prevailing wind.

Based on its mode of operation, a turbomachine is classified either as an *impulse* or as a *reaction* machine. In an impulse machine, the change in the mechanical energy of the fluid comes entirely from the change in its kinetic energy, the pressure remaining constant throughout. A *Pelton-wheel* turbine (Fig. 8.6) is the most common example of such a machine. It consists of a series of buckets attached to the periphery of a disc (termed as runner). The fluid enters the machine as a jet, impinges on the buckets which split the flow into two streams and deflect it back towards the incoming jet. The flow is open to the atmosphere (that is, the machine is not filled completely with water) and, therefore, the pressure is constant everywhere. The impulsive action at the bucket results in a decreased kinetic energy of the flow which appears as power at the shaft.

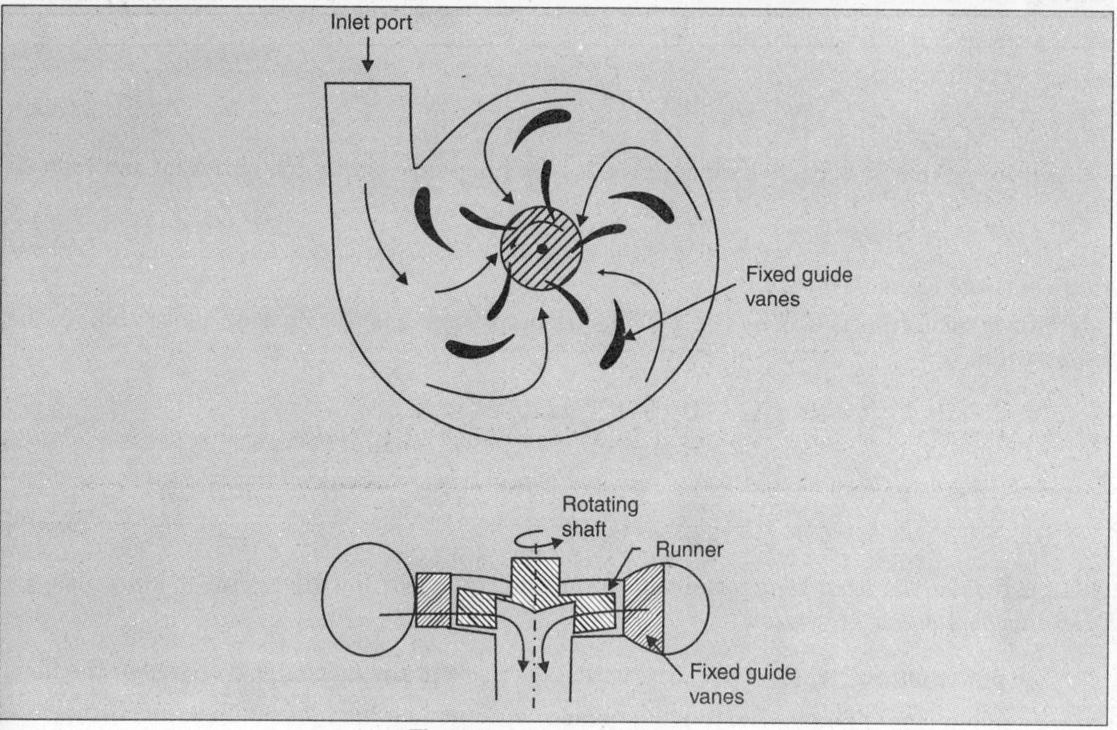

Fig. 8.7. A reaction turbine.

On the other hand, the flow through a *reaction* machine (see Fig. 8.7) is confined within solid surfaces (such that the machine is filled completely) and there is a change in the pressure as well. In a reaction turbine the fluid enters at the periphery of the runner. Its pressure is reduced as it flows through the runner towards the exit port at the centre. The energy extracted by the turbine comes partly from the pressure energy and partly from the kinetic energy of the fluid, and hence both these energies are lower at the outlet than at the inlet. In a centrifugal pump (or blower), the fluid enters at the centre and, as it flows outwards through the rotating impeller, its pressure and velocity increase.

This section gives the general equation for turbomachines. Consider a stationary cylindrical control volume (Fig. 8.8) which encloses the machine. The choice of such a *CV* permits the application of equations obtained for inertial frames of reference. It is convenient to work in

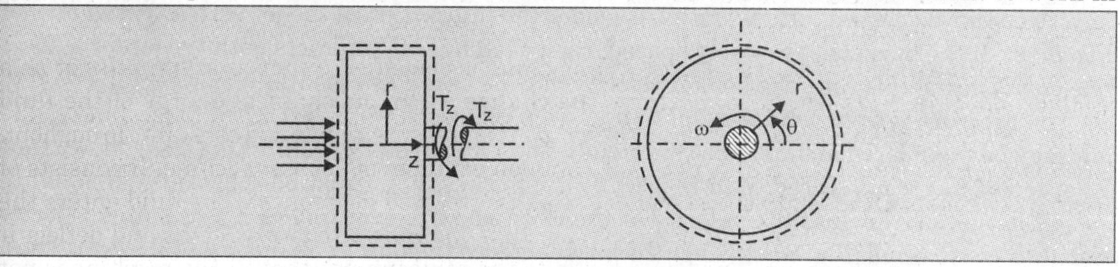

Fig. 8.8. *CV* for a turbomachine.

cylindrical polar coordinates. Since the z-component of the torque is to be calculated, the z-component of the moment of momentum Eq. (5.10) will be used here. Since the flow is steady (when averaged over a rotational cycle),

$$T_z = \oiint_{CS} (\mathbf{r} \times \mathbf{V})(\rho \mathbf{V} \cdot d\mathbf{A}) \qquad ...(8.20)$$

Assuming that $(\mathbf{r} \times \mathbf{V})_z$, which equals $r\,V_\theta$, does not vary across the outlet (*o*) and inlet (*i*) ports*,

$$T_z = \mathring{m}\,[(rV_\theta)_o - (rV_\theta)_i] \qquad ...(8.21)$$

where $\mathring{m}$ is the mass flow rate.

This is the torque acting *on* the *CV*. The output torque has the opposite sense and, so, the power output is

$$\mathring{W}_s = -T_z\omega = \mathring{m}\omega\left[(rV_\theta)_i - (rV_\theta)_0\right] \qquad ...(8.22)$$

or

$$\mathring{W}_s = \mathring{m}\left[(UV_\theta)_i - (UV_\theta)_0\right] \qquad ...(8.23)$$

where $U = \omega r$ is the *local* tangential velocity of the runner or impeller. This is known as the *Euler turbomachinery equation.*

The power output $\mathring{W}_s$ is associated with a decrease in the mechanical energy of the flow. Expressed in terms of heads, Eq. (8.23) becomes

$$h_s = h_i - h_0 = \frac{\mathring{W}_s}{\mathring{m}g} = \frac{(UV_\theta)_i - (UV_\theta)_0}{g} \qquad ...(8.24)$$

In Secs. 8.4 and 8.5, this equation is applied to an impulse turbine and a reaction-type blower respectively.

8.4 PELTON WHEEL TURBINE—AN IMPULSE MACHINE

A Pelton wheel turbine (Fig. 8.6) finds extensive use in high-head hydroelectric installations as in the *Koyna* hydroelectric project. The power developed by a Pelton wheel is obtained from Eq. (8.23) for which we need to calculate UV_θ at the inlet and outlet ports. Since the jet enters in the tangential direction (Fig. 8.6a), $(V_\theta)_i = V_j$ and $(U)_i = \omega R$.

To obtain the velocity at the outlet, consider the interaction of the jet with a typical bucket (Fig. 8.6b). If the buckets are closely spaced, then each bucket interacts with the jet for a short time only. During this time the bucket velocity may be taken as essentially horizontal, equal to ωR. The pressure is everywhere equal to p_{atm}. In a reference frame attached to the bucket the flow may be considered steady. For this steady flow, then, the Bernoulli Eq. (7.15) shows that the speed of water measured with respect to the bucket, V_{rel}, must be constant and equal to $V_j - \omega R$. To obtain the absolute velocity at the exit (which is required in Eq. (8.23)), a velocity triangle is constructed (Fig. 8.6c). Since the bucket guides the flow at an angle β to the horizontal, the relative velocity is in that direction. Clearly

* Most turbomachines have either axial ports or peripheral ports. At the former, $V_\theta = 0$ and at the latter, r and V_θ are constant. Therefore, rV_θ = constant is usually true at all ports.

$$(V_\theta)_0 = U - V_{rel} \cos \beta$$

or

$$(V_\theta)_0 = \omega R - (V_j - \omega R) \cos \beta$$

Equation (8.23) then gives

$$\overset{\circ}{W}_s = \overset{\circ}{m} \left[\omega R V_j - \omega R\left\{\omega R - \left(V_j - \omega R\right)\cos\beta\right\}\right]$$

or

$$\overset{\circ}{W}_s = \overset{\circ}{m}\ \omega R\left(V_j - \omega R\right)\left(1 + \cos\beta\right) \qquad \text{...(8.25)}$$

Since a Pelton wheel is a turbine, $\overset{\circ}{W}_s$ should be positive and V_j should be larger than ωR. It can be shown easily that $\overset{\circ}{W}_s$ is maximum when $\omega R = V_j/2$ and, then

$$\overset{\circ}{W}_{s,\max} = \overset{\circ}{m}\ (\omega R)^2\left(1 + \cos\beta\right) \qquad \text{...(8.26)}$$

The power developed also depends on the turning angle β and is maximum when $\beta = 0$. However, this is not practical because the return jet interferes with other buckets. A typical value of β is 15°.

CHAPTER 8

8.5 A CENTRIFUGAL BLOWER—A REACTION MACHINE

A centrifugal blower is next taken as an example of a reaction machine. The fluid enters at the centre of a rotating impeller which consists of thin curved vanes attached to a disc (Fig. 8.9). The flow enters the impeller radially, but since the impeller is rotating, the velocity with respect to it has a tangential component as well. The vanes are curved such that their slopes at the inlet are in the direction of the *relative* inlet velocity of the fluid so that it can glide onto them smoothly (thereby suppressing separation). As the fluid passes over the vanes, its mechanical energy increases because of the action of the rotating impeller.

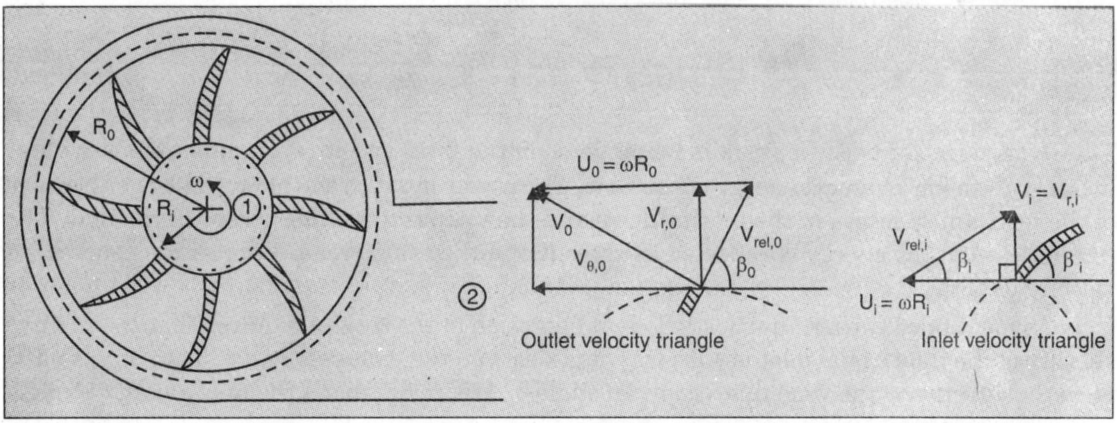

Fig. 8.9. *CV* for a reaction machine along with velocity triangles at inlet and outlet points.

For analysing the flow consider a *CV* which encloses the annular region between the inner and outer peripheries of the impeller. The Euler turbo machinery Eq. (8.23) is

$$\overset{\circ}{W}_s = \overset{\circ}{m}\left[(UV_\theta)_i - (UV_\theta)_0\right] \quad ...(8.23)$$

The velocities $V_{\theta,i}$ and $V_{\theta,0}$ in Eq. (8.23) are the absolute velocities. Since there are no guide vanes at the inlet in this machine, the flow velocity at the inlet is radial and $V_{\theta,i}$ is zero. The radial component $V_{r,i}$ is related to the flow rate $\overset{\circ}{Q}$ by

$$\overset{\circ}{Q} = 2\pi R_i b\, V_{r,i} \quad ...(8.27)$$

where b is the width of the impeller.

Evaluate $V_{\theta,0}$ by constructing the velocity triangle at the outlet tips of the vanes as in Fig. 8.9. The relative velocities at the inlet and outlet are not equal in this case (unlike that in the impulse turbine) because the pressure is not constant in a reaction machine. The flow glides off the vanes at the outlet and, therefore, $V_{rel,0}$ is at an angle β_0, the vane-tip angle at the outlet. In this velocity triangle, U_0 (= ωR_0) and the angle β_0 are known. Thus the radial velocity component $V_{r,0}$ (with respect to the stationary *CV*) can be determined using the continuity equation as

$$V_{r,0} = \frac{\overset{\circ}{Q}}{2\pi R_0 b} \quad ...(8.28)$$

Then, the use of velocity triangle at the outlet gives

$$V_{\theta,0} = U_0 - V_{r,0} \cot \beta_0$$

$$= \omega R_0 - \frac{\overset{\circ}{Q} \cot \beta_0}{2\pi R_0 b} \quad ...(8.29)$$

The Euler turbomachinery equation then gives

$$\overset{\circ}{W}_s = -\rho \overset{\circ}{Q}\,(UV_\theta)_0 = -\rho \overset{\circ}{Q}\,\omega R_0 \left\{\omega R_0 - \frac{\overset{\circ}{Q} \cot \beta_0}{2\pi R_0 b}\right\} \quad ...(8.30)$$

This $\overset{\circ}{W}_s$ is negative because work is being done on the fluid and it represents the theoretical increase in the mechanical energy of the flow. The power input to the blower is larger because of the mechanical losses in the blower drive and the viscous and other losses in the flow. The net increase in the energy is reflected at the outlet either as increased pressure or increased velocity or both.

As mentioned earlier, the most efficient operation of the blower requires that the relative velocity of the fluid at the inlet is such that it glides onto the vanes smoothly. Thus $V_{\text{rel},i}$ should be in the direction of the vane tips, i.e., at an angle β_i. Figure 8.9 shows the inlet velocity triangle and it is seen that

$$V_{r,i} = \omega R_i \tan \beta_i$$

But $V_{r,i} = \mathring{Q}/(2\pi R_i b)$ and so the volume flow rate for most efficient operation of the blower is

$$\mathring{Q} = 2\pi\omega R_i^2\, b \tan\beta_i \qquad \text{...(8.31)}$$

The following numerical example will clarify the calculations involved.

Example 8.1. Consider a blower having the following dimensions: I.D. = 300 mm, O.D. = 400 mm and width = 100 mm. It rotates at 720 revolutions per minute to deliver 12 m^3 per minute ($\mathring{Q}$) of atmospheric air (of $\rho = 1.2$ kg/m^3) through a 100 mm × 120 mm rectangular duct (point 2 in Fig. 8.9) to atmospheric pressure. Find β_i, $\mathring{W}_s$ and β_0.

$$V_{r,i} = \frac{\mathring{Q}}{2\pi\, R_i b} = \frac{12}{60}\left(\frac{m^3}{s}\right) \times \frac{1}{2\pi \times 0.15 \times 0.1\left(m^2\right)} = 2.12 \text{ m/s}$$

The inlet angle required for efficient operation is given by

$$\tan\beta_i = V_{r,i}/(\omega R_i) = \frac{2.12(\text{m/s})}{\frac{720 \times 2\pi}{60}\left(\frac{1}{s}\right) \times 0.15(\text{m})}$$

$$= 0.1874$$

or

$$\beta_i = 10.62°$$

To obtain $\mathring{W}_s$, note that the blower inlet (point 1 in Fig. 8.9) and the duct exhaust (point 2) are both at atmospheric pressure. Therefore, the Bernoulli Eq. (7.15) indicates that the entire blower power goes to increase the kinetic energy of the flow. The velocity through the duct is

$$V_{duct} = \frac{\mathring{Q}}{\text{duct area}} = \frac{12}{60}\left(m^3/s\right) \times \frac{1}{0.1 \times 0.12\left(m^2\right)}$$

$$= 16.67 \text{ m/s}$$

and since V_1 is small,

$$-\mathring{W}_s = \mathring{m}\frac{V_{duct}^2}{2} = \frac{12}{60}\left(\frac{m^3}{s}\right) \times 1.2\left(\frac{kg}{m^3}\right) \times \frac{16.67^2}{2}\left(\frac{m^2}{s^2}\right)$$

$$= 33.3 \text{ W}$$

By Eq. (8.23), the θ-velocity at the outer periphery of the impeller is

$$V_{\theta,0} = \frac{-\mathring{W}_s}{\rho\mathring{Q}\,\omega R_0} = \frac{33.3 \text{ W}}{1.2 \times \frac{12}{60}\left(\frac{kg}{s}\right) \times \frac{720 \times 2\pi}{60}\left(\frac{1}{s}\right) \times 0.2(\text{m})}$$

$$= 9.20 \text{ m/s}$$

CHAPTER 8

From Eq. (8.29) β_0 may be obtained as

$$\cot \beta_0 = \frac{(\omega R_0 - V_{\theta,0})2\pi R_0 b}{\overset{\circ}{Q}}$$

or

$$\beta_0 = \tan^{-1} \frac{\overset{\circ}{Q}}{2\pi R_0 b(\omega R_0 - V_{\theta,0})} = 15.14°$$

Thus the inlet and outlet tip angles should be 10.62° and 15.14° respectively, and the blower consumes 33.3 W, neglecting losses.

8.6 GROUND EFFECT MACHINES—HOVERCRAFTS

A *hovercraft* is a vehicle that travels on a cushion of air maintained between its base and the terrain. Because of this cushion, a hovercraft can travel over unprepared ground, marshy land or on a water surface. There are many ways of maintaining this air cushion. One design uses an annular jet of air issuing all along the periphery of the base. This keeps the pressure inside the cushion above atmospheric (Fig. 8.10a) and supports the weight of the vehicle.

The jet of air directed inwards at an angle θ, curves back (Fig. 8.10b). If the height h of the nozzle above the ground is assumed large compared to the width t of the jet, the radius of curvature R of the jet stream can be taken as constant. By geometry it can be shown that $R = h/(1 + \cos\theta)$. Since the streamlines are curved, the pressure gradient across the jet is given by Eq. (6.28):

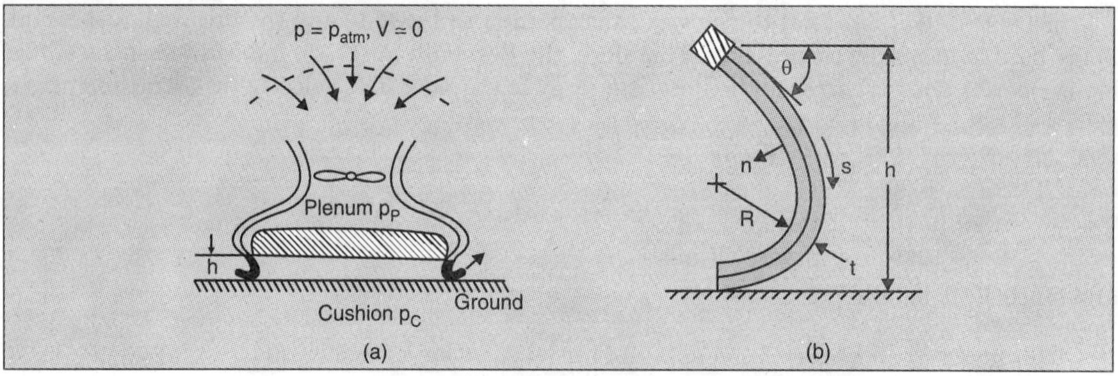

Fig. 8.10. Schematic of the air cushion in a hovercraft.

$$\frac{\partial p}{\partial n} = -\frac{\rho V^2}{R} \qquad \text{...(6.28)}$$

where n is the distance measured normal to the streamline direction and V is the velocity of air. This means that the pressure p_C on the cushion side of the jet is higher than that on the outside where it is atmospheric. Thus, a cushion of air of pressure higher than atmospheric is maintained under the vehicle by the peripheral annular jet. This pressure sustains the load F of the vehicle and

$$F = p_C A_C \quad ...(8.32)$$

where A_C is the cushion (or base) area.

The pressure p_C can be obtained by integration of Eq. (6.28) across the jet. Before doing this, the velocity of the jet should be obtained by the application of the Bernoulli equation (assuming the flow to be inviscid) between the plenum (where the pressure is p_P and the velocity is negligible) and a point in the jet. Since the pressure across the jet varies from atmospheric ($p = 0$) to the cushion pressure p_C, the velocity will vary across the jet. For an approximate analysis, assume that the average velocity can be obtained by taking the pressure in the jet to be $p_C/2$, the average pressure. Thus,

$$V_{av} = \sqrt{\frac{2\left(p_P - \frac{p_C}{2}\right)}{\rho}} \quad ...(8.33)$$

Substituting this in Eq. (6.28) one obtains

$$\frac{\partial p}{\partial n} = -2\left(p_P - \frac{p_C}{2}\right)\Big/ R$$

Integrating this equation from $n = 0$ where $p = p_C$ to $n = t$ where $p = 0$, one obtains

$$0 - p_C = -\frac{2}{R}\left(p_P - \frac{p_C}{2}\right)t$$

or

$$\frac{p_C}{p_P} = \frac{2t}{R}\Big/\left(1 + \frac{t}{R}\right)$$

Using $R = h/(1 + \cos\theta)$, one gets

$$\frac{p_C}{p_P} = \left(\frac{2(1+\cos\theta)t/h}{1+(1+\cos\theta)t/h}\right) \quad ...(8.34)$$

This equation has been obtained under the assumption that $t \ll h$ and that losses are negligible.

A more exact derivation without using the approximation $t \ll h$ and not using the average velocity in place of the varying velocity in the Bernoulli equation, gives

$$\frac{p_C}{p_P} = 1 - e^{-2(1+\cos\theta)t/h} \quad ...(8.35)$$

Figure 8.11 shows a comparison of p_C/p_P obtained from the approximate Eq. (8.34) with the exact Eq. (8.35) for $\theta = 0°$ and $\theta = 90°$. The agreement is seen to be excellent till $t/h \simeq 0.25$.

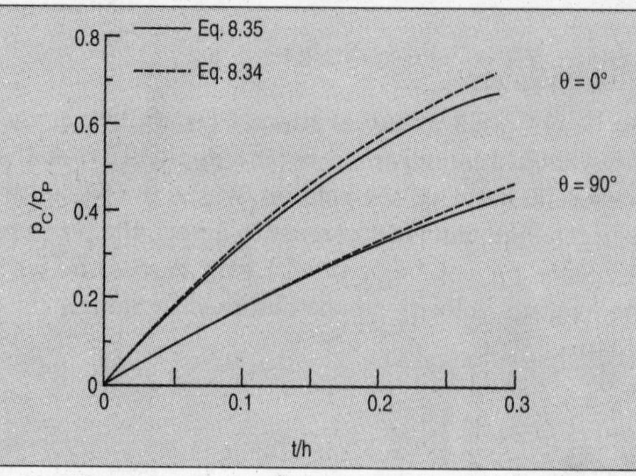

Fig. 8.11. Comparison of p_C/p_P vs. t/h from the approximate and exact equations.

The load F supported by the annular jet around a cushion of diameter D_C is

$$F = \frac{\pi D_C^2}{2} \frac{p_P (1+\cos\theta)t/h}{1+(1+\cos\theta)t/h} \quad ...(8.36)$$

The power required to support this load is equal to that supplied to the blower to maintain the jet flow. The mass flow rate of the air is

$$\mathring{m} = \rho V_{av} (\pi D_C t)$$

The blower increases the pressure of air from 0 to p_P, and so, the power required by it is

$$\text{Power} = -\mathring{W}_s = \mathring{m} \frac{p_P}{\rho} = \rho V_{av} (\pi D_C t) \frac{p_P}{\rho} \quad ...(8.37)$$

To obtain the power required to support a load F, we use Eqs. (8.32), (8.33) (8.34), (8.36) and (8.37) to get

$$\text{Power} = \frac{4}{(\pi\rho)^{1/2}} \frac{t}{D_C^2} \frac{1+(1+\cos\theta)t/h}{[(1+\cos\theta)t/h]^{3/2}} F^{3/2} \quad ...(8.38)$$

For t/h small, this reduces to

$$\text{Power} = \frac{4}{\sqrt{\pi}(1+\cos\theta)^{3/2}} \left(\frac{D_C}{t}\right)^{1/2} \left(\frac{h}{D_C}\right)^{3/2} \frac{F^{3/2}}{D_C\sqrt{\rho}} \quad ...(8.39)$$

For a constant geometry hovercraft, i.e. with D_C and t constant, the power required to hover at a constant distance h above the ground varies as $F^{3/2}$. Similarly, for a constant load, the power increases as $h^{3/2}$. For constant power, the load F times the clearance h remains constant. An idea of the numerical values of the various quantities can be had from Prob. 8.15.

8.7 FLOW MEASURING DEVICES

A few devices used to measure the flow rate or the velocity of a fluid are described here. The *exact* flow patterns through these devices are quite complex. However, a few simplifying assumptions will be made such as the absence of viscosity, one-dimensionality of flow, etc., and

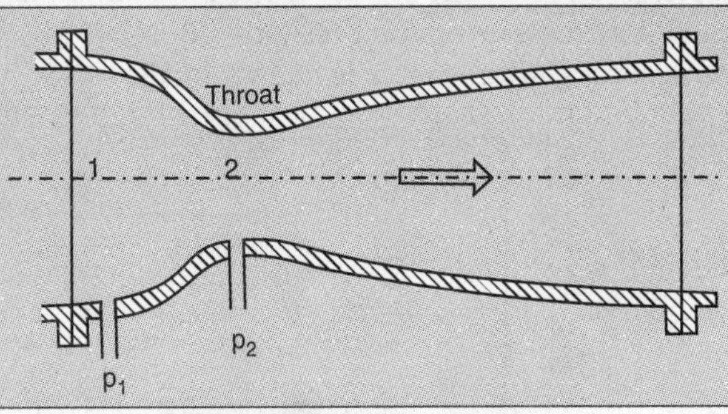

Fig. 8.12. A venturi meter.

the flow rate computed in such idealized cases using simplified models. The results so obtained are modified by experimentally determined coefficients which account for the deviations from the ideal-flow pictures. The *venturi meter* used for measuring the discharge through circular pipes is taken up first.

Figure 8.12 shows a venturi meter inserted in a circular pipe. It consists of a converging-diverging section (note that the diverging section is much longer than the converging section. Why?) with one pressure tap at the upstream straight section and another at the *throat*, the section having the minimum area. As the flow accelerates down the converging section, its pressure decreases till it is minimum at the throat. Application of the Bernoulli Eq. (7.15) between sections 1 and 2 gives

$$\frac{V_1^2}{2} + \frac{p_1}{\rho} = \frac{V_2^2}{2} + \frac{p_2}{\rho} \quad ...(8.40)$$

in the absence of viscous effects. But V_1 and V_2 are related by the continuity Eq. (4.9)

$$V_1 A_1 = V_2 A_2 \quad ...(8.41)$$

assuming that the flow is 1-*D*. From Eqs. (8.40) and (8.41), the discharge rate $\mathring{Q}_{\text{ideal}}$ is obtained as

$$\mathring{Q}_{\text{ideal}} = A_1 V_1 = \frac{A_2 \sqrt{2(p_1 - p_2)/\rho}}{\left[1 - (A_2/A_1)^2\right]^{1/2}} \quad ...(8.42)$$

The actual discharge $\mathring{Q}$ will be different from the ideal value so obtained because of the deviations from the assumptions of no viscosity and one-dimensionality. If a discharge coefficient C_d is defined as $\mathring{Q}/\mathring{Q}_{\text{ideal}}$, then

CHAPTER 8

$$\mathring{Q} = \frac{C_d A_2 \sqrt{2(p_1 - p_2)/\rho}}{\left[1-(A_2/A_1)^2\right]^{1/2}} \qquad ...(8.43)$$

For a given venturi meter, if A_2/A_1 is known and if the instrument is pre-calibrated to obtain C_d, the flow can be obtained by measuring the pressure difference $(p_1 - p_2)$ using a manometer. A typical calibration curve of a venturi meter is shown in Fig. 8.13.

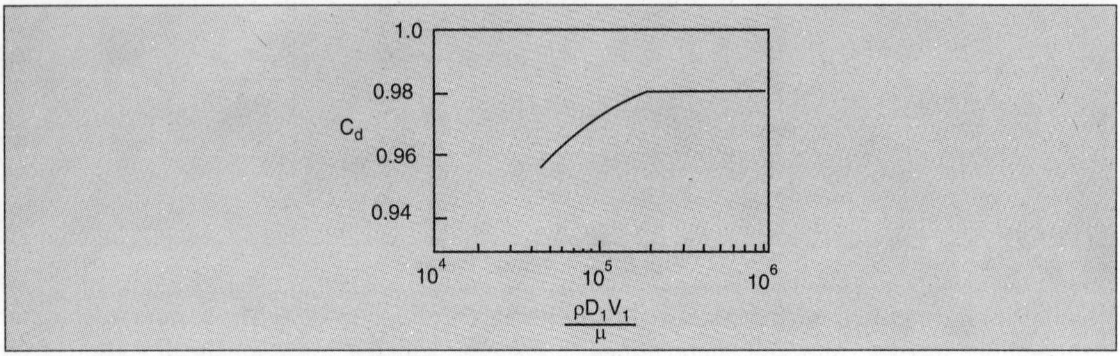

Fig. 8.13. C_d for a typical venturi meter.

Venturi meters are expensive and are usually too long for practical applications. It is often simpler to use a *flow-nozzle* (Fig. 8.14), or an *orifice meter* (Fig. 8.15) in which (Figs. 8.14 and 8.15) the fluid comes out as a jet submerged in a relatively stagnant fluid and then spreads out. It is easily seen that if the viscous action is neglected and if the flow is assumed to be 1-*D*, the ideal discharge rate is given by Eq. (8.42) as for a venturi meter. The discharge coefficients are again defined to account for deviations from the idealized behaviour and the actual flow rate is given by Eq. (8.43) where C_d is the appropriate discharge coefficient. It may be added that C_d for an orifice meter (about 0.61) is much lower than that for a venturi meter. It lies somewhere in between for a flow nozzle.

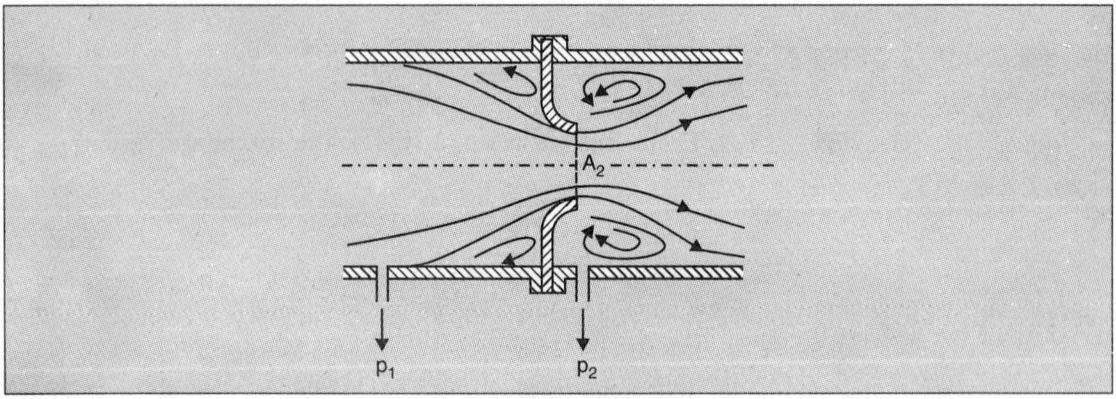

Fig. 8.14. A flow nozzle.

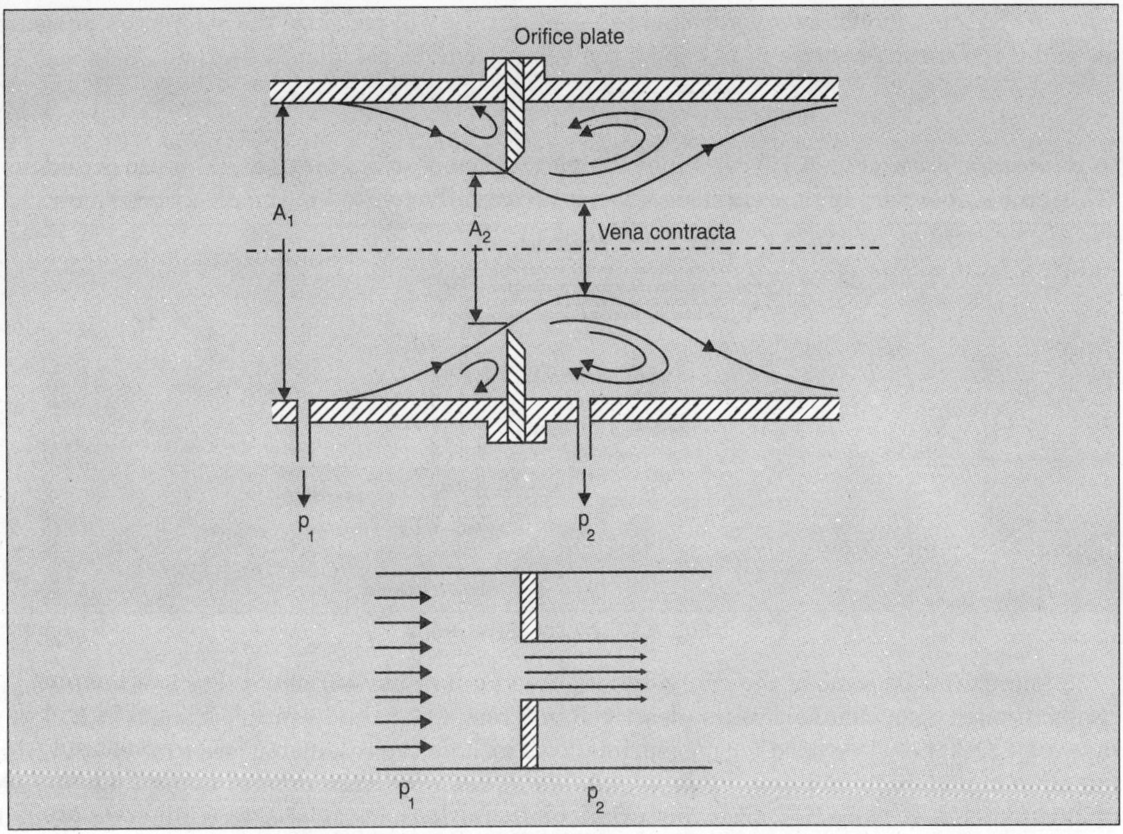

Fig. 8.15. An orifice plate meter and its idealized flow picture.

A *pitot tube* is used for measuring the detailed velocity distribution in a steady incompressible* flow field (Fig. 8.16). A fluid particle moving towards the nose (point 2) of the

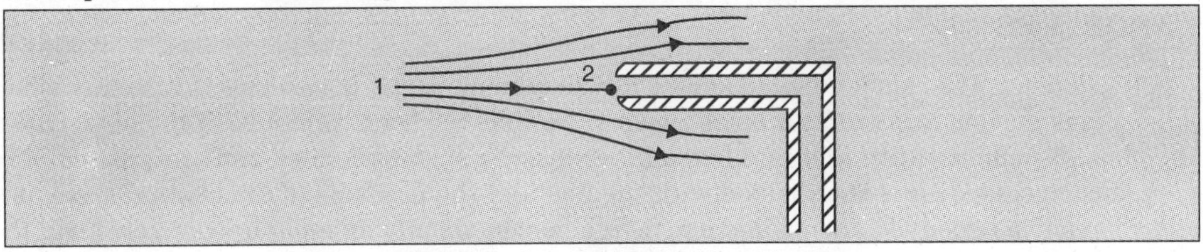

Fig. 8.16. Schematic of a pitot tube.

pitot tube is brought to rest at the nose if the tube is not "bleeding' fluid, and thus, the pressure measured at that point is given by Eq. (7.25)

$$\frac{p_2}{\rho} = \frac{p_1}{\rho} + \frac{V^2}{2} \qquad ...(8.44)$$

neglecting viscosity. The pressure p_2 at the point where the fluid is brought to rest is termed as the *stagnation* or the *total pressure* and is made up of the *static pressure* p_1 and the *dynamic*

* The use of pitot tubes for compressible flow is discussed in Examples 15.1 and 15.5.

pressure $\rho V^2/2$. A manometer connected to the pitot tube will measure the stagnation pressure p_2 and if the static pressure p_1 is known the velocity can be calculated from Eq. (8.44) as

$$V = \sqrt{2(p_2 - p_1)/\rho} \quad ...(8.45)$$

A *pitot-static probe* (Fig. 8.17) combines the measurement of stagnation and static pressures. This probe gives very high accuracies when it is carefully pointed in the direction of flow.

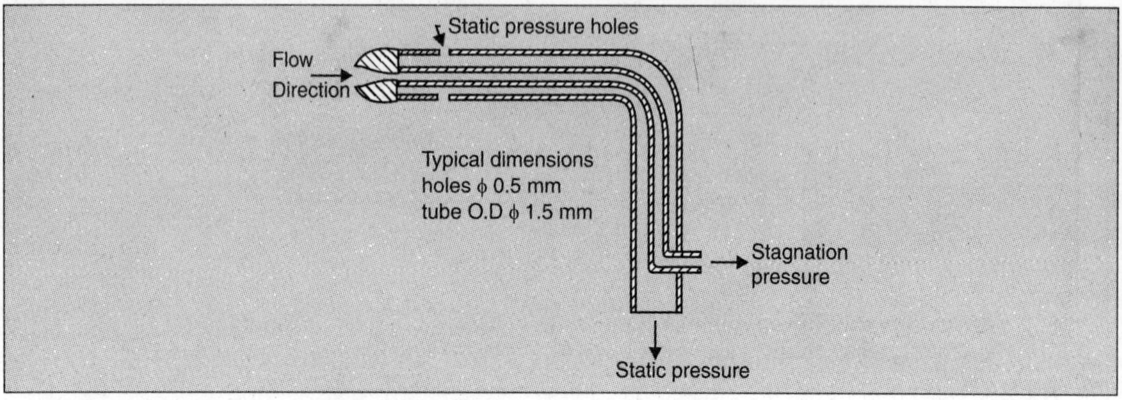

Fig. 8.17. A pitot-static probe.

These are only some of the many devices used for flow measurement. For measuring flow rates through open channels weirs (described in Example 7.8) and venturi-flumes (Prob. 7.46) are used. Other devices used for measuring flows include the rotameter (see Prob. 13.32), the rotating cup anemometer (see Prob. 13.43), etc. A hot wire anemometer is used for precise measurements of velocities, even in turbulent flows. It consists of a thin filament heated electrically. The convective heat loss from the wire is governed by the flow velocity. If this rate is known the flow velocities can readily be calculated. More details of this and other devices may be found in Bradshaw (Reading 18) and Gupta (Reading 23).

PROBLEMS

8.1 *Ram jet:* The ram jet engine shown has no moving parts. It develops thrust only when it is moving forward at a large speed V_1 so that the 'ram' pressure of the air forces a sufficient quantity of it into the engine where it is slowed down and compressed. The compressed air is then mixed with the fuel and the products of combustion leave at a relative velocity V_2 and pressure p_2 (which may be slightly different from atmospheric, the flow being supersonic). If the inlet area is A_1, exit area is A_2, and the fuel is burnt at a rate of $\mathring{m}$ kg/s, obtain the thrust developed when the engine moves at a constant velocity V_1.

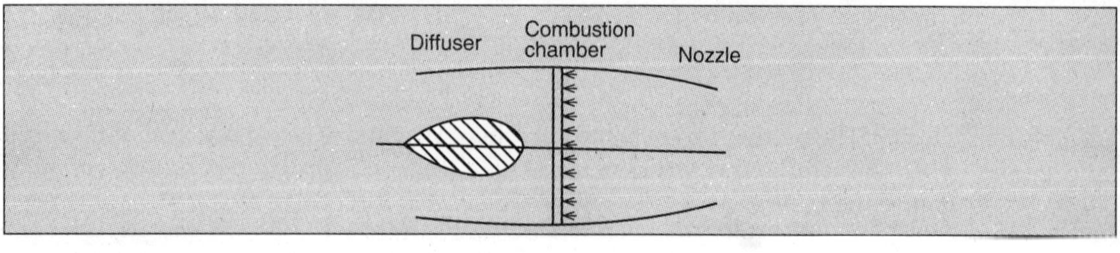

8.2 *Liquid-fuel rocket:* In a liquid-fuel rocket, the fuel and oxidizer are pumped at controlled rates into the combustion chamber C as shown, and the gases flow out of the nozzle (of exit area A_2) at a relative velocity V_2 and a pressure p_2. Since flow is usually supersonic, p_2 may be different from p_{atm}. Obtain an expression for the thrust developed in terms of the constant rate $\overset{\circ}{m}$ of combustion of fuel and oxidiser. Assume that the rocket is moving at a constant velocity.

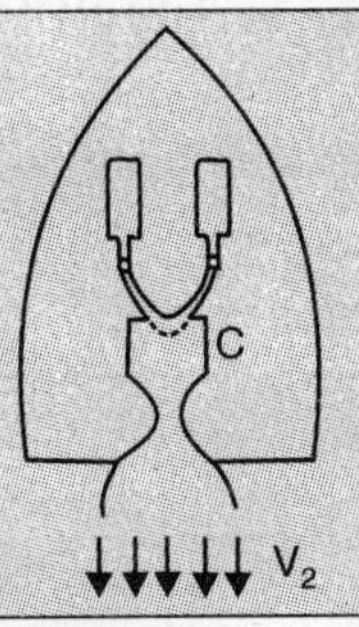

8.3 A 3 m diameter propeller is to generate a thrust of 4500 N at a speed of 100 m/s. Compute the theoretical maximum propeller efficiency and the minimum power required.

8.4 A 1400 mm diameter household ceiling fan drives air at 280 m^3/min. Find the axial force exerted by the fan. What is the pressure difference across the two sides of the 'fan disc' and the power required (neglecting the effect of the ceiling and frictional losses)? Assume that the air far upstream is almost stationary, i.e., it is being drawn from all around.

8.5 A small 2 m dia. windmill is operating at a wind speed of about 35 kmph. If the conditions are such that it is working at the maximum efficiency, obtain the power output.

8.6 A 53.4 m diameter windmill was installed in 1941 atop *Grandpa's Knob* near Rutland, Vermont (see P.S. Putnam, *Power from the Wind,* Van Nostrand, 1948). If the average wind speed at the level of the hub was 30 kmph, compute the maximum useful power generated. (The unit was operated continuously for three weeks in March 1945, generating power at an average rate of 431 kW).

8.7 Show that the highest value of η_{max} for a windmill is 59.3 per cent which occurs when $V_2 = V_1/3$.

8.8 Obtain Eq. (8.26) for the maximum value of a Pelton wheel's power output.

8.9 A jet of diameter 6 cm discharges water at 100 m/s on to a Pelton wheel of diameter 1.5 m rotating at 600 RPM. If the water is deflected through an angle of 173° compute the power developed. Also, compute the kinetic energy lost with the outgoing stream.

8.10 The *Pykara* hydroelectric scheme in South India has seven Pelton-wheel type turbines. One of these has the following specifications:

$$R = 0.925 \text{ m}$$

$$\text{RPM} = 600$$

$$V_j = 130 \text{ m/s (using a head of 855 m)}$$

$$\overset{\circ}{Q} = 1.916 \text{ m}^3/\text{s}$$

$$\beta = 15°$$

Compute the theoretical power output and compare with the actual value of 14.4 MW.

CHAPTER 8

8.11 A centrifugal pump has an impeller with

$R_i = 6.5$ cm, $R_0 = 12$ cm, $b = 2.5$ cm, $\beta_i = 30°$ and $\beta_0 = 45°$.

Compute the throughput and the power consumption at the design RPM of 1800.

8.12 *Francis turbine:* In the Francis turbine (reaction type) shown, water enters at the periphery at an absolute velocity $\mathbf{V}_i$, making an angle α_i with the tangent. (This is achieved by the use of stationary 'guide' vanes at the inlet.) If the flow rate is 0.3 m^3/s, and the turbine rotates at 100 RPM, and $R_i = 0.5$ m, $R_0 = 0.25$ m, $\beta_i = 60°$, $\beta_0 = 120°$ and $\alpha_i = 20°$, find

(a) V_i such that the water glides smoothly on to the rotating vanes,

(b) the width b of the blades, and

(c) the power output of the turbine.

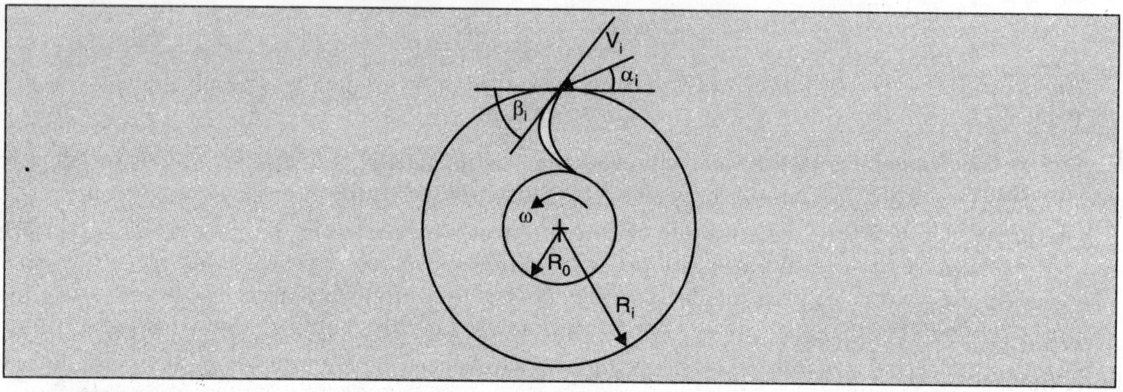

8.13 In an alternate design of the hovercraft the pressure in the plenum chamber supports the load directly as shown. Obtain an expression for the power required by the blower to support a load F. Compare this with Eq. (8.39) for a peripheral jet hovercraft and obtain the ratio of the two powers for the same F, D_c and h. Then obtain the limits on t for which the peripheral jet design is better.

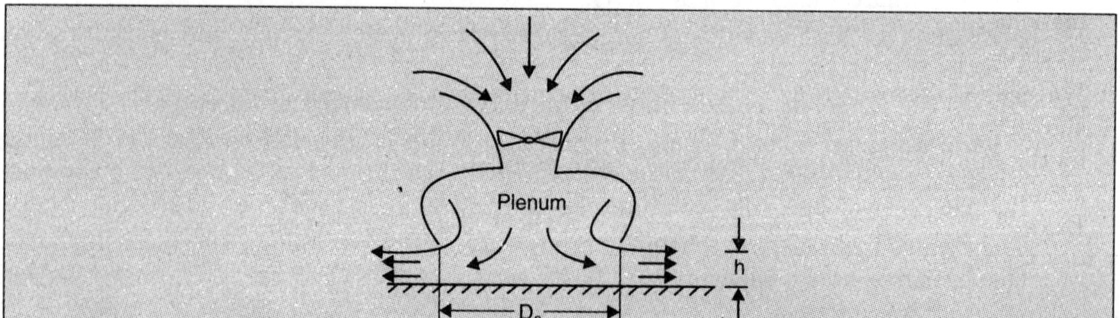

8.14 *Augmentation factor:* An augmentation factor K for a hovercraft may be defined as $K = F/F_0$, where F is the load supported and F_0 is the force exerted by an air jet directed against a flat plate as shown, and having the same $\mathring{Q}$ and the velocity V_0. Obtain the augmentation factor for both the peripheral jet and the plenum chamber (see Prob. 8.13) hovercrafts.

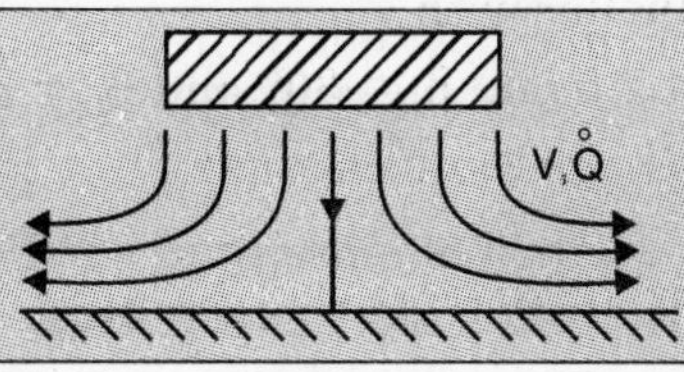

8.15 Obtain the blower power required for a hovercraft with the peripheral jet system given the following data:

weight of vehicle = 80×10^3 N
$D_c = 8$ m
$h = 2.5$ cm
$t = 1$ cm
$\theta = 45°$

Also obtain the blower power necessary if the design is of the plenum-chamber type (see Prob. 8.13) and the value of h is the same as above. Take C_d as 0.54.

8.16 Show that the frictional loss w_l per unit mass of a fluid flowing through a venturi meter can be written as

$$w_l = \frac{V_2^2}{2}\left(\frac{1}{C_d^{\,2}} - 1\right)\left(1 - \frac{D_2^4}{D_1^4}\right)$$

What is the assumption made in obtaining this result? [Hint: Use w_l in Eq. (8.40)].

8.17 A horizontal venturi meter having a throat diameter of 4 cm is installed in a 90 mm dia. pipe. Water flows through the pipe, and a water-mercury manometer reads 35 cm. What is the flow rate of water? Assume $C_d = 0.98$. Check from Fig. 8.13 if this assumption is justified. Estimate w_l using results of Prob. 8.16.

8.18 Crude oil (of $\rho = 925$ kg/m^3) flows through an 18 cm diameter pipe at a rate of 160 m^3/hr. It is desired to install an orifice meter in the pipeline to measure the flow rate. If the manometer used can read a maximum of 250 cm of water, what orifice diameter would be suitable? $C_d = 0.61$.

8.19 Water flows out of a tank with a circular orifice at the bottom. A pitot tube placed as shown measures a dynamic pressure of 110 kPa. The volume flow rate $\mathring{Q}$ as measured by the fall in the level of water in the tank is 3.2×10^{-3} m^3/s. What is the ratio of the actual velocity at the point of measurement to the theoretical value given by Eq. (b) of Example 7.4? Also compute the coefficient of discharge C_d for the orifice.

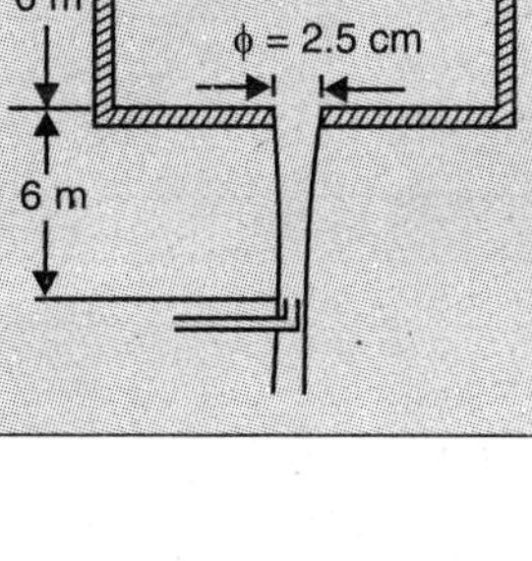

8.20 Water flows through a 6 cm dia. flow-nozzle in a 15 cm dia. pipe. Compute the reading on a water-mercury manometer if the flow rate is 0.02 m^3/s. Assume C_d to be 0.98 for these conditions.

8.21 *Venturi flume:* A common method of measuring the flow in a channel uses a section of reduced width as shown. It is observed that the level of the water *dips* by an amount h at the point of minimum cross-section. Assuming a 1-D flow upstream of the venturi flume, show that the velocity V_1 is given by

$$V_1 = \sqrt{2gh\frac{\alpha}{1-\alpha}}$$

where $\alpha = \left[\frac{W_2}{W}\left(1-\frac{h}{H}\right)\right]^2$

Show that the river does indeed dip at the contraction.

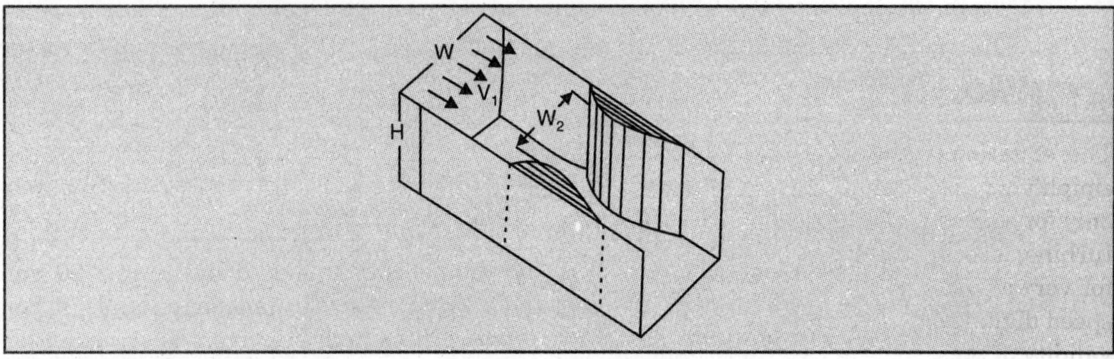

8.22 *Proportional flow weir:* A weir has a shape given by $x\sqrt{y} = k$, where x and y are as shown. The base of this weir is flush with the bottom of the channel. Show that the ideal flow rate $\mathring{Q}$ is directly proportional to the height Y. Use

$$\int_0^a \left(\frac{a}{x}-1\right)^{\frac{1}{2}} dx = \frac{\pi a}{2}.$$

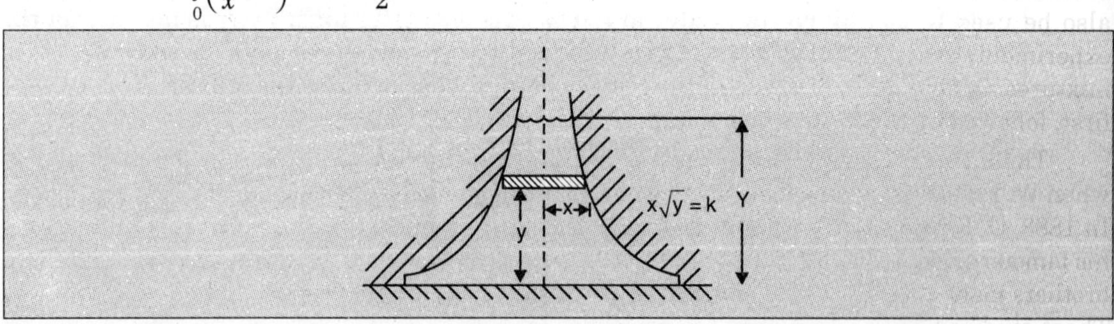

9

SIMILITUDE AND MODELLING

9.1 INTRODUCTION

The equations governing the flow of fluids are so complex even when derived under the simplifying assumptions of constant material properties that analytical solutions can be obtained only for some very simple geometries. However, an engineer designing an offshore structure, a turbine, a rocket engine, a jet airliner or a hydraulic transmission needs to calculate the flow for very complex geometries. For many such cases, ingenious simplifications and use of high-speed digital computers permit evaluation of some results, but the designers' confidence in the validity of these decreases with the amount of complexity involved. To increase his confidence, the designer resorts to the testing of scaled *models* of the design before he embarks on the construction of a *prototype*. In modern day engineering practice, any new type of machine or structure involving interaction with fluids is seldom constructed until a model test has been made. The models are frequently scaled-down versions and, therefore, can be redesigned and tested at comparatively much lower cost. Models can also be scaled-up versions in cases where the microscopic size of the prototype makes experimentation difficult. For example, the motion of small organisms in bio-fluids has been studied using scaled-up models. Scaled models can also be used to advantage to study flows that are either too fast or too slow for proper experimentation. Glacial drift has been modelled using higher speed flows of kaolin-water mixtures. In this chapter, the principles that govern the design of scaled models are discussed first, followed by the interpretation of results obtained from them.

The history of scaled-model testing in fluid engineering dates back to as early as 1869 when W. Froude first used a water basin for the design of ship hulls for the British Admiralty. In 1883, O. Reynolds (after whom the non-dimensional parameter, $\rho VL/\mu$, is named) published his famous model studies of flow through pipes. Later, at the turn of the last century, the Wright brothers made the first systematic use of wind tunnels to design the *Kitty Hawk* and the *Flyers*. These studies firmly established the principles of scale model testing.

Two of the several techniques used are presented in this book (a third technique based on the Buckingham-Pi theorem is discussed in Appendix C). The first starts with the equations of flow. The procedure used is quite general, and can easily be generalized for other systems that can be described by mathematical models. Hence, this procedure can be used in other fields of science and engineering as well. This procedure is easier to understand, in contrast to the second technique presented later in this chapter, in which some amount of physical intuition is required. However, the latter is far more powerful.

9.2 THE FIRST TECHNIQUE

9.2.1 The Dimensionless Equations of Flow

We consider in this section the modeling rules starting from the equations governing the flow, and the boundary conditions applicable to them. A few simplifying assumptions are made, e.g., steady flow, Newtonian fluid, etc., to develop the concepts, even though these are not really necessary. Consider the steady, 2-*D* flow of an incompressible fluid of density ρ and viscosity μ past an infinite solid body having an elliptical cross-section, with major and minor axes as $2L_{x0,}$ and $2L_{z0}$ (Fig. 9.1). The pressure at the *x*-axis far upstream is p_0. The equations governing the flow are

$$\nabla \cdot \mathbf{V} = 0$$

$$\rho \mathbf{V} \cdot \nabla \mathbf{V} = -\nabla p - \rho g\, \hat{\mathbf{k}} + \mu \nabla^2 \mathbf{V} \quad \text{...(9.1)}$$

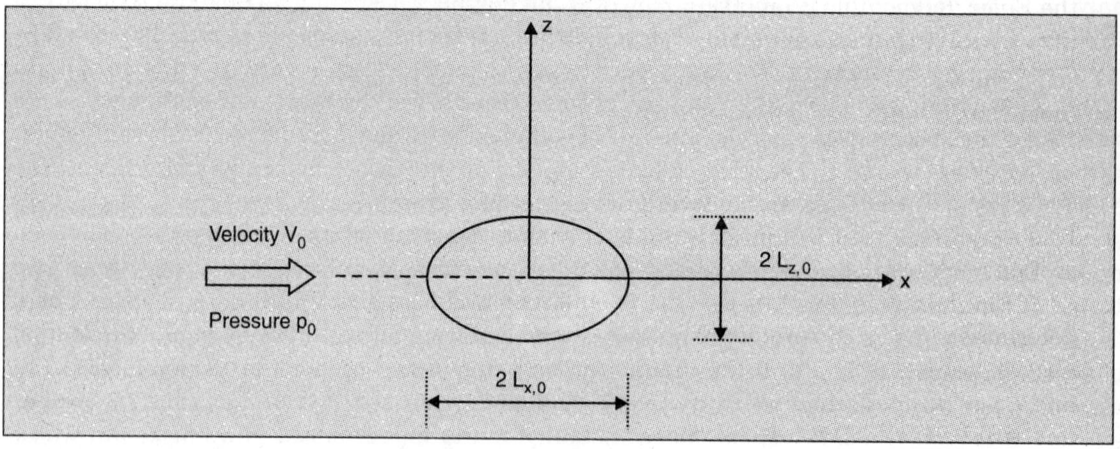

Fig. 9.1. 2-*D* flow past an infinite solid body having an elliptical cross-section.

The boundary conditions are given by

$$\mathbf{V} \to V_0 \hat{\mathbf{i}} \quad \text{as } x, z \to \pm \infty$$

$$\mathbf{V} \to 0 \quad \text{on } (x/L_{x0})^2 + (z/L_{z0})^2 = 1$$

$$p \to p_0 \quad \text{on } z = 0 \text{ as } x \to -\infty \quad \text{...(9.2)}$$

Each variable in these equations and boundary conditions is non-dimensionalized using some value characterizing the problem. The free-stream velocity, V_0 and free-stream pressure, p_0 suggest themselves as obvious choices for characterizing the velocities and pressure, respectively.

We may choose anyone of the length parameters, $L_{x,0}$ or $L_{z,0}$, for characterizing the length variables, x and z. We select $L_{x,0}$. The dimensionless variables, thus, are written as

$$x^* \equiv x/L_{x,0}\,;\; z^* \equiv z/L_{x,0} \qquad [\text{or, } \mathbf{x}^* \equiv \mathbf{x}/L_{x,0}]$$
$$V_x^* \equiv V_x/V_0\,;\; V_z^* \equiv V_z/V_0 \qquad [\text{or, } \mathbf{V}^* \equiv \mathbf{V}/V_0]$$
$$p^* \equiv p/p_0 \qquad ...(9.3)$$

On substituting these in Eqs. (9.1) and (9.2) and simplifying, one gets (see Problem 1)

$$\nabla^* \cdot \mathbf{V}^* = 0$$

$$\mathbf{V}^* \cdot \nabla^* \mathbf{V}^* = -\left(\frac{p_0}{\rho V_0^2}\right)\nabla^* p^* - \left(\frac{gL_{x,0}}{V_0^2}\right)\hat{\mathbf{k}} + \left(\frac{\mu}{\rho V_0 L_{x,0}}\right)\nabla^{*2}\mathbf{V}^* \qquad ...(9.4)$$

with the boundary conditions

$$\mathbf{V}^* \to \hat{\mathbf{i}} \qquad \text{as } x^*,\, z^* \to \pm\infty$$
$$\mathbf{V}^* \to 0 \qquad \text{on } x^{*2} + (L_{x,0}/L_{z,0})^2 z^{*2} = 1$$
$$p^* \to 1 \qquad \text{on } z^* = 0 \text{ as } x^* \to -\infty \qquad ...(9.5)$$

Here, $$\nabla^* \equiv \hat{\mathbf{i}}\frac{\partial}{\partial x^*} + \hat{\mathbf{j}}\frac{\partial}{\partial y^*} + \hat{\mathbf{k}}\frac{\partial}{\partial z^*}.$$

We define $p_0/\rho V_0^2 = 1/\text{Eu}$, $\mu/\rho V_0 L_{x,0} = 1/\text{Re}$, and $gL_{x,0}/V_0^2 = 1/\text{Fr}^2$, where Eu, Re and Fr are the Euler, Reynolds and Froude numbers, respectively (these are called pi- or Π-numbers). These involve only the (constant) characteristic values (or parameters) and *not* the variables.

The solution of Eqs. 9.4 and 9.5 can be written in the following general (functional) form (note that the *exact* solution is not required)

$$\mathbf{V}^* = \mathbf{V}^*(\mathbf{x}^*;\, \text{Eu, Re, Fr, } L_{x,0}/L_{z,0})$$
$$p^* = p^*(\mathbf{x}^*;\, \text{Eu, Re, Fr, } L_{x,0}/L_{z,0}) \qquad ...(9.6)$$

It is easy to see that 2-*D* flows of such fluids around solid bodies having more complex cross-sections will involve additional dimensionless geometric parameters.

Equation (9.6) indicates that the non-dimensional solutions, $\mathbf{V}^*(\mathbf{x}^*)$ and $p^*(\mathbf{x}^*)$, are completely determined by the values of Eu, Re and Fr, besides the geometry. Thus, the problem as mentioned above, now has only four independent parameters instead of the original six, ρ, μ, V_0, p_0, $L_{x,0}$ and $L_{z,0}$. This reduction in the number of independent variables has important consequences and will be taken up in Sec. 9.10.

The significance of the above statement can be understood by imagining two situations in which the values of the six parameters ρ, μ, V_0, p_0, $L_{x,0}$ and $L_{z,0}$ differ in the two cases, but their values are such that the values of the non-dimensional parameters Eu, Re and Fr are identical in the two cases. In such a situation the values of the *non-dimensional variables* p^* and $\mathbf{V}^*$ will be the same at the same non-dimensionalized location $\mathbf{x}^*$ and non-dimensionalized time t^*, for unsteady flows.

The equations for the dimensionless shear stresses can be obtained (see Problem 9.2) as

$$\tau^* \equiv \frac{\tau}{\rho V_0^2} = \frac{1}{\text{Re}}\left[\frac{\partial V_z^*}{\partial x^*} + \frac{\partial V_x^*}{\partial z^*}\right], \text{ etc.} \qquad ...(9.7)$$

CHAPTER 9

and so

$$\tau^* = \tau^* (\mathbf{x}^*; \text{Eu, Re, Fr}, L_{x,0}/L_{z,0}), \text{ etc.} \qquad ...(9.8)$$

Now consider the drag experienced by the solid body. The drag is the integrated x-component of the pressure and shear forces acting on the surface of the solid. The pressure drag is written as

$$\text{Pressure drag} = \oiint_{\text{Area}} p\hat{\mathbf{i}}\cdot d\mathbf{A} = \oiint_{\text{Area}} p_0 p^* \hat{\mathbf{i}}\cdot d\mathbf{A}^* L_{x,0}^2 \qquad ...(9.9)$$

where $d\mathbf{A} = dx\, dz$ and $d\mathbf{A}^* = dx^*\, dz^*$. Eq. (9.9) leads to

$$\frac{\text{Pressure drag}}{p_0 L_{x,0}^2} = \oiint_{\text{Area}} p^* \hat{\mathbf{i}}\cdot d\mathbf{A}^* = \text{fn}_1 (\text{Eu, Re, Fr}, L_{x,0}/L_{z,0}) \qquad ...(9.10)$$

or

$$\frac{\text{Pressure drag}}{\rho V_0^2 L_{x,0}^2} = \frac{p_0}{\rho V_0^2}\frac{\text{Pressure drag}}{p_0 L_{x,0}^2} = \text{fn}_1 (\text{Eu, Re, Fr}, L_{x,0}/L_{z,0})/\text{Eu}$$

$$\equiv \text{fn}_2 (\text{Eu, Re, Fr}, L_{x,0}/L_{z,0}) \qquad ...(9.11)$$

Similar results are obtained for the shear drag as well. Hence, the total drag, which is the sum of the pressure and shear components, is obtained as

$$(\text{Total Drag})/(\rho V_0^2 L_{x,0}^2) = \text{fn}_3 (\text{Eu, Re, Fr}, L_{x,0}/L_{z,0}) \qquad ...(9.12)$$

Equation 9.12 is usually expressed in terms of the drag coefficient, C_D

$$C_D \equiv \left(\frac{\textit{Total Drag}}{\frac{1}{2}\rho V_0^2\ (\textit{Characteristic Area})_0}\right) = C_D(\text{Eu, Re, Fr}, L_{x,0}/L_{z,0}) \qquad ...(9.13)$$

9.2.2 Modelling or Similarity Rules

We now develop the procedure for predicting results for a prototype using experimental results on a model. The first step is to obtain the conditions to be used for the model test. We match all the geometric ratios, e.g., $L_{x,0}$ and $L_{z,0}$ etc., for the model and prototype. This is relatively easy to achieve, since the design of the model is in our hands. This requirement is referred to as geometric similarity,* i.e., we consider two flows with geometrically similar boundaries, but with different fluids and different characteristic values of the various quantities. In addition, we carry out tests on the model (m) at the same values of Eu, Re and Fr as for the prototype (p). That is

$$\text{Eu}_m = \text{Eu}_p$$

$$\text{Re}_m = \text{Re}_p$$

$$\text{Fr}_m = \text{Fr}_p \qquad ...(9.14)$$

Equations (9.14), together with the requirement of geometric similarity, are referred to as modeling (or similarity) rules.

* Note that all angles are preserved in the model, since angles are ratios of two lengths.

9.2.3 Prediction Rules

If the values of ρ, μ, V_0, p_0, $L_{x,0}$, and $L_{z,0}$ are such that the three pi-numbers are identical for the prototype and the geometrically similar model, the non-dimensional solutions are also identical for these two. We may *then* write

$$\begin{aligned} \mathbf{V}_m^*(\mathbf{x}^*) &= \mathbf{V}_p^*(\mathbf{x}^*) && \text{(a)} \\ p_m^*(\mathbf{x}^*) &= p_p^*(\mathbf{x}^*) && \text{(b)} \\ \tau_m^*(\mathbf{x}^*) &= \tau_p^*(\mathbf{x}^*) && \text{(c)} \\ C_{D,m} &= C_{D,p} && \text{(d)} \end{aligned} \quad \text{...(9.15)}$$

We now introduce the concept of *homologous* points in the model and the prototype i.e., points having the same (*relative*) locations, $\mathbf{x}^*$. Figure 9.2 shows a pair of homologous points, *A*

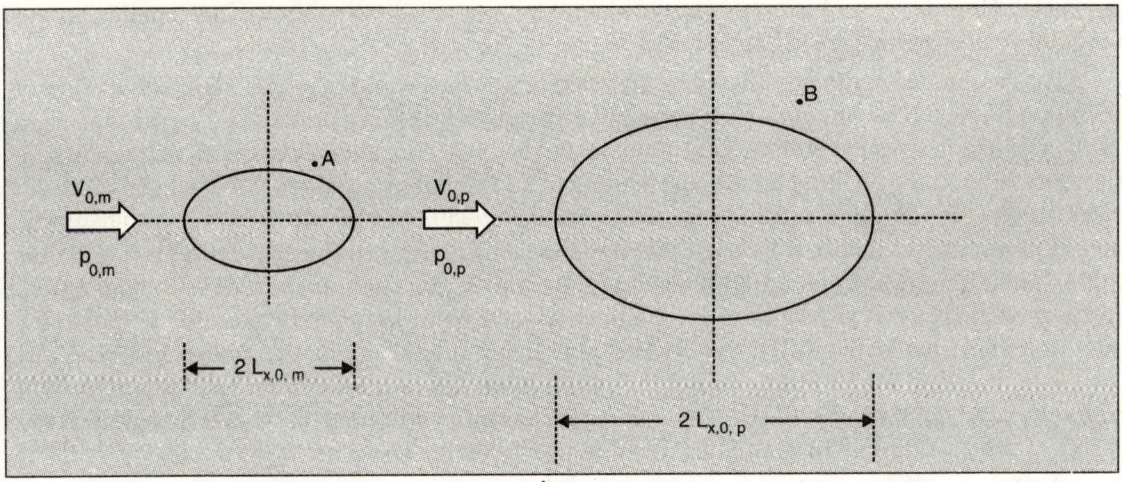

Fig. 9.2. (a) Model and (b) prototype along with a pair of homologous points, *A* and *B*.

and *B*, in the model and prototype, respectively. These have the same values of $\mathbf{x}^*$, i.e.,

$$\frac{x_A}{L_{x,0,m}} = \frac{x_B}{L_{x,0,p}}; \text{ or } x_A = x_B \frac{L_{x,0,m}}{L_{x,0,p}} \quad \text{...(9.16)}$$

Equations (9.15) suggests that the velocities, pressure and shear stresses *at these homologous points* are related by

$$\begin{aligned} V_{x,B} &= V_{x,A} \frac{V_{0,p}}{V_{0,m}} \\ V_{z,B} &= V_{z,A} \frac{V_{0,p}}{V_{0,m}} \\ p_B &= p_A \frac{p_{0,p}}{p_{0,m}} \\ \tau_{x,z,B} &= \tau_{x,z,A} \frac{\rho_p V_{0,p}^2}{\rho_m V_{0,m}^2} \end{aligned} \quad \text{...(9.17)}$$

CHAPTER 9

It is easy to see that values of **V**, p and $\tau_{x,z}$ at point B in the prototype can easily be predicted using the *experimental* values at the homologous point, A, in the more convenient model. Equation (9.17) are referred to as prediction rules.

The condition given in Eq. (9.15a), namely, $\mathbf{V}^*(B) = \mathbf{V}^*(A)$, is referred to as kinematic similarity. It suggests that for geometrically similar model and prototype flows under conditions of Eq. (9.14), the kinematic quantities are similar (at homologous points). Similarly, Eqs. (9.15 b-d) are referred to as dynamic similarity.

9.2.4 Some Comments

The characteristic lengths of the prototype will depend on the task it has to perform, e.g., the power to be developed by a turbine, the production capacity of a stirred tank used for carrying out a reaction, etc. However, no comment has been made till now about the appropriate size of the model. How small (or large) the model should be compared to the prototype depends on the availability of experimental facilities, cost, etc.

The example of scaling discussed in this section was based on the equations of steady flow of an incompressible Newtonian fluid. If there are other physical effects present, e.g., surface tension (for flow of sap in trees), unsteady effects (vibrating bodies), etc., additional terms will have to be incorporated in the modelling equations, leading to additional pi-numbers, *all* of which will need to be matched for similarity. Similarly, if a spherical solid is put slowly (with speed = 0) at the surface of a stagnant liquid, it will settle at an increasing speed (till the speed attains a constant value). Modelling equations for both the liquid as well as for the solid will need to be written, and several additional pi-numbers will be involved. Often, it may not be possible to match *all* the pi-numbers experimentally, and 'tricks' may need to be used, or the modelling rules will need to be 'relaxed' so that only some of the most important pi-numbers are matched. This will make the predictions less trustworthy and safety margins will have to be incorporated in the design of the prototype. These are discussed in the following section.

9.2.5 Reducing the Number of Pi-Numbers

As indicated in Sec. 9.2.3, the more the pi-numbers that are matched, the more are the restrictions on the choice of the independent parameters like ρ, μ, V_0, p_0, etc. At times, this may lead to physically unrealizable choices. It is, therefore, desirable to attempt to reduce the number of pi-numbers that need to be matched.

Let us examine now, in some detail, the three pi-numbers obtained in Sec. 9.2 and see if one can get around matching their values, at least in some cases. First consider $\text{Eu} = \rho V_0^2/p_0$, which uses the characteristic pressure, p_0. Note that it is not the pressure that appears in Eq. (9.1), but its gradient. Therefore, all pressures may be measured with p_0 as the datum, and we can work with a modified pressure, $(p - p_0)$. This does not change the form of Eq. (9.1), with the modified pressure replacing the pressure, but the boundary condition in Eq. (9.2) becomes

$$P = p - p_0 \to 0 \quad \text{on } z = 0 \text{ as } x \to \infty \qquad \text{...(9.18)}$$

where P signifies the modified pressure.

This modified pressure (which really is the excess pressure, i.e., the pressure *above* p_0) is no longer characterized by p_0. To characterize it, the value of excess pressure at some other point in the flow field is required. However, since no other pressure is specified in this problem, one is at liberty to choose the characteristic excess pressure arbitrarily. It is conventional to set the characteristic modified (excess) pressure* P_c as $\frac{1}{2}\rho V_0^2$. The dimensionless modified pressure can, thus, be written as

$$P^* \equiv (p - p_0)/\left(\frac{1}{2}\rho V_0^2\right) \qquad ...(9.19)$$

and we obtain

$$\nabla^* \cdot \mathbf{V}^* = 0$$

$$\mathbf{V}^* \cdot \nabla^* \mathbf{V}^* = -\left(\frac{1}{2}\right)\nabla^* P^* - \left(\frac{gL_{x,0}}{V_0^2}\right)\hat{\mathbf{k}} + \left(\frac{\mu}{\rho V_0 L_{x,0}}\right)\nabla^{*2}\mathbf{V}^*$$

$$\mathbf{V}^* \to \hat{\mathbf{i}} \qquad \text{as } x^*, z^* \to \pm\infty$$

$$\mathbf{V}^* \to 0 \qquad \text{on } x^{*2} + (L_{x,0}/L_{z,0})^2 z^{*2} = 1$$

$$P^* \to 0 \qquad \text{on } z^* = 0 \text{ as } x^* \to -\infty \qquad ...(9.20)$$

With this modification the Euler number does not appear in Eqs. (9.6), (9.8), (9.10) and (9.11). Thus, in all problems where only *one* pressure, p_0, is specified, so that a characteristic modified pressure needs to be chosen arbitrarily, the Euler number equality (in its modified sense) is automatically assured and one needs to match *only* Re and Fr.

A simplification of this kind is not always possible. Wherever two pressures are specified in the problem statement, a characteristic excess or modified pressure is defined by these two. Therefore, it can no longer be set arbitrarily at $\frac{1}{2}\rho V_0^2$. The values of the modified Euler number, then, need to be matched for the model and the prototype. One such case is the flow of liquids when cavitation (i.e., formation of vapour bubbles) occurs. Cavitation occurs whenever the pressure falls below the vapour pressure, p_v. Thus, the difference, $(p_0 - p_v)$ forms an independent characteristic modified pressure and Eu has to be defined as

$$\text{Eu} = \frac{\rho V_0^2}{p_0 - p_v}$$

and has to be matched for the model and the prototype flows.

Next consider the Froude number, Fr. It is possible in some cases to re-define the problem in such a manner that Froude number similarity can also be eliminated as a modelling requirement. A new variable, $\mathscr{P}$, is defined as

$$\mathscr{P} \equiv p + \rho g z$$

where z is the vertical co-ordinate from any arbitrary datum. This is the same $\mathscr{P}$ as introduced in Example 6.3. It may be noted that this $\mathscr{P}$ is defined so that its value throughout a stationary fluid (in which the pressure p varies due to gravity) is a constant and so $\mathscr{P}$ is called the *non-gravitational pressure.*

* This is equal to the modified pressure at the stagnation point in an inviscid flow (see Sec. 8.7). Since many flows of engineering interest can be roughly modeled by the inviscid-flow approximation (see Sec. 11.5), $\rho V_0^2/2$ is the conventional choice. However, there are situations (see Sec. 11.4) where it is not appropriate.

With the introduction of this variable $\mathscr{P}$, Eq. (9.4) is modified to

$$\nabla^* \cdot \mathbf{V}^* = 0$$

$$\mathbf{V}^* \cdot \nabla^* \mathbf{V}^* = -\frac{p_0}{\rho V_0^2} \nabla^* \mathscr{P}^* + \frac{\mu}{\rho V_0 L_0} \nabla^{*2} \mathbf{V}^* \quad \text{...(9.21)}$$

where $\mathscr{P}$ has been normalized to $\mathscr{P}^*$ using p_0. The velocity boundary conditions in Eq. (9.5) are not affected by this transformation but the pressure condition is changed to

$$\mathscr{P}^* \to 1 \text{ on } z^* = 0 \text{ as } x^* \to -\infty \quad \text{...(9.22)}$$

Thus, it is seen that Fr does not occur in the problem statement at all, and for modelling only $\text{Re} = \rho V_0 L_0/\mu$ and $\text{Eu} = \rho V_0^2/p_0$ have to be matched. If, in addition, the flow conditions are such that only one pressure p_0 is specified, one can work in terms of the modified pressure (measured above the datum, p_0) again, and use $\frac{1}{2}\rho V_0^2$ as the characteristic modified pressure. This eliminates the Euler number and there is no longer any need to match it. Thus, similarity under such conditions is attained by matching only the Reynolds number. This simplification is available in many physical problems that include flow through ducts, flow of air over bodies, flight of objects through the atmosphere, etc. In all these cases, only the Reynolds number serves as the similarity parameter. In fact, as shall be seen in Chapter 13, when the Reynolds number in flow past bluff bodies is very high, one need not match even this!

There is an important class of problems in which the simplification of dropping the matching of Fr is not possible. These problems involve a free surface, the shape of which changes with the motion. At the free surface, $z_f = f(x)$, the pressure is specified as constant (often atmospheric). Thus, the pressure boundary condition in Eq. (9.2) is replaced by

$$p = p_0 \text{ at } z = z_f$$

When the non-gravitational pressure $\mathscr{P}$ is introduced, this modifies to

$$\mathscr{P} \equiv p_0 + \rho g z_f \quad \text{at } z = z_f$$

which, on non-dimensionalization, gives

$$\mathscr{P}^* = 1 + \frac{\rho g L_{x,0} z_f^*}{p_0} \qquad \text{at } z^* = z_f^*$$

This can be recast as

$$\mathscr{P}^* = 1 + \frac{\rho V_0^2}{p_0} \frac{g L_{x,0}}{V_0^2} z_f^* \qquad \text{at } z^* = z_f^*$$

or

$$\mathscr{P}^* = 1 + (\text{Eu}/\text{Fr}^2)\, z_f^* \qquad \text{at } z^* = z_f^* \quad \text{...(9.23)}$$

Thus, Fr appears in the boundary condition though it vanishes in the equation of motion, and so needs to be matched for similarity (Eu may or may not be important depending on whether a characteristic modified pressure is defined a *priori* or can be set arbitrarily). These conditions hold for the motion of ships and other vessels close to the ocean surface. Similarly, in flows over dams, weirs and in open channels, Fr in the model must be the same as Fr in the prototype. However, for a submarine operating at large depths (compared to its size), Fr may be ignored (relaxed) as a similarity requirement.

There is an important class of problems in which only Fr similarity is required and Re similarity may be ignored (relaxed). This happens when the flow is relatively insensitive to changes in Re. It is difficult to establish a general rule for determining the applicability of this condition, but it appears that flow in harbours, rivers, estuaries, etc., are modelled well enough with only Fr similarity. For motion of ships, etc., the usual procedure calls for modelling only the Froude number and then modifying the results obtained to account for the Reynolds number differences, using 'experience'. In other gravity-dominated flows where frictional effects may be neglected, Re matching is seldom attempted. A more detailed discussion of such relaxations of modelling requirements can be found in Schuring (Reading 30) (See also Probs. 9.18, 9.27 and 9.36).

The following examples illustrate the application of the above 'rules', and also develop some more prediction rules.

Example 9.1. To estimate the power requirement of a blimp (lighter-than-air aircraft) travelling at 10 m/s (in air), it is proposed to test a one-twentieth scale model of it *in water*. What should the velocity of the model be in water and at that speed, what will be the prediction rule for the power required?

Since there is no free surface in this flow and cavitation, if any, may be neglected in the model test, only the Reynolds numbers need to be matched. Thus,

$$\left(\frac{\rho V L_{x,0}}{\mu}\right)_m = \left(\frac{\rho V L_{x,0}}{\mu}\right)_p$$

or

$$V_m = V_p \times \frac{L_{x,0,p}}{L_{x,0,m}} \times \frac{\rho_p}{\rho_m} \times \frac{\mu_m}{\mu_p}$$

$$= 10(\text{m/s}) \times 20 \times (1.226/10^3) \times (1.14 \times 10^{-3}/1.78 \times 10^{-5})$$

$$= 15.7 \text{ m/s}$$

Eq. (9.13) gives the relationship between the drag forces on the prototype and model as

$$\frac{(\textit{Drag force})_p}{(\textit{Drag force})_m} = \frac{(\rho V_0^2 L_{x,0}^2)_p}{(\rho V_0^2 L_{x,0}^2)_m}$$

The ratio of the power required is, therefore, given by

$$\frac{\textit{Power}_p}{\textit{Power}_m} = \frac{(\textit{Drag force})_p \times V_{0,p}}{(\textit{Drag force})_m \times V_{0,m}} = \frac{(\rho V_0^3 L_{x,0}^2)_p}{(\rho V_0^3 L_{x,0}^2)_m}$$

$$= (1.226/10^3) \times (10/15.7)^3 \times (20)^2 = 0.127$$

Thus, the prototype power requirement will be 12.7 per cent (i.e., lower!) of the model power requirement. The (new) prediction rule for power is given above. It may be noted that it does not matter which force and which velocity is used for the expression of the power required (even though it will, indeed, be the drag force and the far-upstream velocity, as used here), since ratios are involved, and the ratios are the same as those for the characteristic quantities.

Example 9.2. The occurrence of cavitation on the control fins of a torpedo can cause severe deflections in its trajectory. Cavitation occurs whenever the local pressure falls below the saturation vapour pressure. The higher the speed of the torpedo, the lower would be the pressure at some points on the fin and the more would be the tendency for cavitation. A torpedo is seen to be on the verge of cavitation at a speed of 30 m/s when running 10 m below the ocean surface (temperature 20°C). At what depth must it run if it is to travel at 50 m/s?

The two flows at the verge of cavitation must be similar. As 10 m is fairly large compared to the torpedo size, Fr does not need to be matched if we replace p by $\mathscr{P} = p + \rho gz$, with z measured from an appropriate reference. But since cavitation is involved, Eu (with the characteristic modified pressure now taken as $\mathscr{P}_0 - \mathscr{P}_v$) is to be matched. As the effect of viscosity is neglected, Re also is not matched, Then, Eu similarity requires that

$$[(\rho V_0^2)/(\mathscr{P}_0 - \mathscr{P}_v)]_p = [(\rho V_0^2)/(\mathscr{P}_0 - \mathscr{P}_v)]_m$$

where the 30 m/s run is (arbitrarily) labeled as the model, and the 50 m/s run as the prototype. Here,

$$\mathscr{P}_0 - \mathscr{P}_v = (p + \rho gz)_{ref} - (p + \rho gz)_{cav}$$

Taking the datum at the ocean surface where $p = p_{atm}$ and $z = 0$, we have

$$\mathscr{P}_0 - \mathscr{P}_v = p_{atm} - (p_v - \rho gh) \tag{a}$$

where h ($= -z$) is the depth of the torpedo at which cavitation occurs. Taking $p_{atm} = 101.3$ kPa, p_v at 20°C = 2.3 kPa, $\rho = 1.03 \times 10^3$ kg/m^3 for sea water,

$$(\mathscr{P}_0 - \mathscr{P}_v)_m = (101.3 - 2.3)\ (\text{kPa}) + 1.03 \times 10^3\ (\text{kg/m}^3) \times 9.81\ (\text{m/s}^2) \times 10\ (\text{m})$$
$$= 198.4 \text{ kPa}$$

Equivalence of Eu then gives (with $\rho_p = \rho_m$)

$$(\mathscr{P}_0 - \mathscr{P}_v)_p = (\mathscr{P}_0 - \mathscr{P}_v)_m\ (V_{0,p}/V_{0,m})^2$$
$$= 198.4\ (\text{kPa}) \times (50/30)^2 = 551.1 \text{ kPa}$$

Eq. (a) then gives

$$(\mathscr{P}_0 - \mathscr{P}_v)_p = p_{atm} - (p_v - \rho gh)_p$$
$$= (101.3 - 2.3)\ (\text{kPa}) + 1.03 \times 10^3\ (\text{kg/m}^3) \times 9.1 (\text{m/s}^2) \times h_p (\text{m})$$
$$= 99\ (\text{kPa}) + 10.1 \times 10^3\ h_p\ (\text{Pa})$$

Equating this to 551.1 kPa obtained above, we obtain

$$h_p = \frac{(551.1 - 99) \times 10^3}{10.1 \times 10^3} = 44.8 \text{ m}$$

Thus, the torpedo must operate at a depth of over 44.8 m if it has to run at 50 m/s without cavitation.

Example 9.3. The spillway of a hydroelectric dam passes a volume of 3×10^6 m^3/hr and is to be modeled on a one-tenth scale. What should be the volume flow rate in the model test?

In the case of such open-channel flows, Fr matching is of prime importance and Re matching will be relaxed:

$$\frac{V_{0,m}}{\sqrt{L_{x,0,m}g_m}} = \frac{V_{0,p}}{\sqrt{L_{x,0,p}g_p}}$$

or
$$\frac{V_{0,p}}{V_{0,m}} = \sqrt{\frac{L_{x,0,p}}{L_{x,0,m}}}$$

$\mathring{Q}$ can be written as the product of the characteristic V_0 and $L_{x,0}^2$ (since ratios are going to be used). Hence,

$$\frac{\mathring{Q}_m}{\mathring{Q}_p} = \frac{V_{0,m}}{V_{0,p}}\frac{L_{x,0,m}^2}{L_{x,0,p}^2} = \frac{L_{x,0,m}^{2.5}}{L_{x,0,p}^{2.5}}$$

and so

$$\mathring{Q}_m = 3\times10^6\left(\frac{\text{m}^3}{\text{hr}}\right)\times\left(\frac{1}{10}\right)^{5/2} = 0.95\times10^4\,\text{m}^3/\text{hr}$$

Example 9.4. An off-shore oil-drilling platform is expected to encounter waves of 4 m height at 0.1 Hz frequency and a steady current of 1 m/s. Determine the parameters for the model wave channel where a one-sixteenth model of the platform can be tested.

In this case, too, only Froude numbers need to be matched. Thus,

$$\frac{V_{0,m}}{\sqrt{L_{x,0,m}g_m}} = \frac{V_{0,p}}{\sqrt{L_{x,0,p}g_p}}$$

or
$$\frac{V_{0p}}{V_{0,m}} = \sqrt{\frac{L_{x,0,p}}{L_{x,0,m}}} = (16)^{1/2} = 4$$

The current in the model channel must then be given by

$$V_{0,m} = V_{0,p}/4 = 1\ (\text{m/s}) \times (1/4) = 0.25\ \text{m/s}$$

The ratio of the heights, h, of the waves, should be the same as that of $L_{x,0}$, and so

$$h_m = h_p\frac{L_{x,0,m}}{L_{x,0,p}} = 4\ (\text{m}) \times (1/16) = 0.25\ \text{m}$$

An easy way to get the scaling of the (characteristic) frequency, ω_0, is to note that

$$\frac{\omega_{0,m}}{\omega_{0,p}} = \frac{V_{0,m}}{V_{0,p}}\frac{L_{x,0,p}}{L_{x,0,m}}$$

This gives

$$\omega_{0,m} = \omega_{0,p}\frac{V_{0,m}}{V_{0,p}}\frac{L_{x,0,p}}{L_{x,0,m}} = 0.1(\text{Hz})\times\frac{1}{4}\times16 = 0.4\ \text{Hz}$$

A more formal way of obtaining the same results is to start with the unsteady version of Eq. (9.1) for incompressible Newtonian fluids

$$\nabla \cdot \mathbf{V} = 0$$

$$\rho\left(\frac{\partial}{\partial t}\mathbf{V} + \mathbf{V} \cdot \nabla \mathbf{V}\right) = -\nabla p - \rho g\, \hat{\mathbf{k}} + \mu \nabla^2 \mathbf{V} \qquad \text{(a)}$$

and using [along with Eq. (9.3)]

$$t^* \equiv \omega_0 t$$

to obtain (Problem 9.3)

$$\nabla^* \cdot \mathbf{V}^* = 0$$

$$\left(\frac{\omega_0 L_{x,0}}{V_0}\right)\frac{\partial \mathbf{V}^*}{\partial t} + \mathbf{V}^* \cdot \nabla^* \mathbf{V}^* = -\left(\frac{p_0}{\rho V_0^2}\right)\nabla^* p^* - \left(\frac{g L_{x,0}}{V_0^2}\right)\hat{\mathbf{k}} + \left(\frac{\mu}{\rho V_0 L_{x,0}}\right)\nabla^{*2}\mathbf{V}^* \qquad \text{(b)}$$

Similarity will involve the matching of the model and protoytype values of $\frac{\omega_0 L_{x,0}}{V_0}$ defined as the Strouhal number, St. This is precisely what was done intuitively above.

Scaling can similarly be carried out (almost intuitively, now) for quantities like time-interval, acceleration, mass, force, etc. Table 9.1 gives a summary of the commonly used modelling rules.

Table 9.1. Some rules of thumb for similarity requirements

No.	Examples	Should match (modelling rules)	Most important modelling rule
1.	With free surface, no cavitation, (ships, dams, harbours, off-shore platforms, etc.)	Re, Fr	Fr
2.	No free surface, no cavitation, (submarines, airplanes, satellites, enclosed flows in pipes, etc.)	Re	Re
3.	With free surface, with cavitation, (high speed ships)	Eu, Re, Fr	Fr, Eu
4.	No free surface, with cavitation, (bad pump, siphon, torpedos)	Eu, Re	Eu

At times, surface tension, compressibility effects, etc., may also be present. Obviously, Eq. (9.1) is inapplicable then. We need to develop the appropriate governing equations of these.

9.2.6 Relaxation of Rules for Exact Similarity

Let us now attempt to use two modelling rules simultaneously, and see what happens.

Example 9.5. The jump (see Problems 5.27 and 7.18) of the water level downstream of a dam (known as hydraulic jump) is to be modeled in the laboratory using a 1:12 scale model with an appropriate liquid. If the upstream velocity is estimated to be 10.5 m/s in the prototype, at what speed should the model be tested to ensure similarity? If the model shows a hydraulic jump of

8 cm, what will be the predicted height of the jump in the prototype? Obtain how the 'rate of energy-loss' will be scaled.

In order to have *complete* similarity (see Table 9.1, No. 1), we should match both Re and Fr (the latter is more important). We obtain from Re matching

$$\left(\frac{V_0 L_{x,0}}{\nu}\right)_m = \left(\frac{V_0 L_{x,0}}{\nu}\right)_p$$

or
$$\frac{\nu_p}{\nu_m} = 12\frac{V_{0,p}}{V_{0,m}}$$

Froude matching leads to

$$\frac{V_{0,p}^2}{L_{x,0,p}} = \frac{V_{0,m}^2}{L_{x,0,m}}$$

or
$$\frac{V_{0,p}}{V_{0,m}} = \left(\frac{L_{x,0,p}}{L_{x,0,m}}\right)^{1/2} = \sqrt{12}$$

Hence

$$\nu_m = \frac{\nu_p}{41.568} = 2.41 \times 10^{-8} \text{ m}^2/\text{s}$$

using the kinematic viscosity of water (ν_p) as 10^{-6} m²/s. A study of the properties of common fluids indicates that no such fluid is available. The closest is ν_{Hg} (50°C) = 1.05×10^{-7} m²/s, ν_{Hg} (200°C) = 8×10^{-8} m²/s, but we will not like to use hot mercury for tests. This is why we need to relax the requirement of Re similarity, and use only the Fr similarity (as suggested in Table 9.1) with water as the liquid in the model test. It may be noted that Re matching with water as the liquid in the model as well as the prototype, gives contradictory solutions $\left[\frac{V_{0,p}}{V_{0,m}} = \frac{L_{x,0,m}}{L_{x,0,p}} = \frac{1}{12}\right]$ as compared to that obtained with Fr matching $\left[\frac{V_{0,p}}{V_{0,m}} = \sqrt{\frac{L_{x,0,p}}{L_{x,0,m}}} = \sqrt{\frac{1}{12}}\right]$. We do not worry about matching Re (relaxation of modelling rules) and correct for the predictions using 'experience'.

The other parts of this example can easily be done (Problem 9.4).

9.3 THE SECOND TECHNIQUE

In this technique, we develop the scale factor approach and study the similarity of scaled models and prototypes. This approach is far more general and powerful than discussed in Section 9.2 since it does not require a mathematical model.

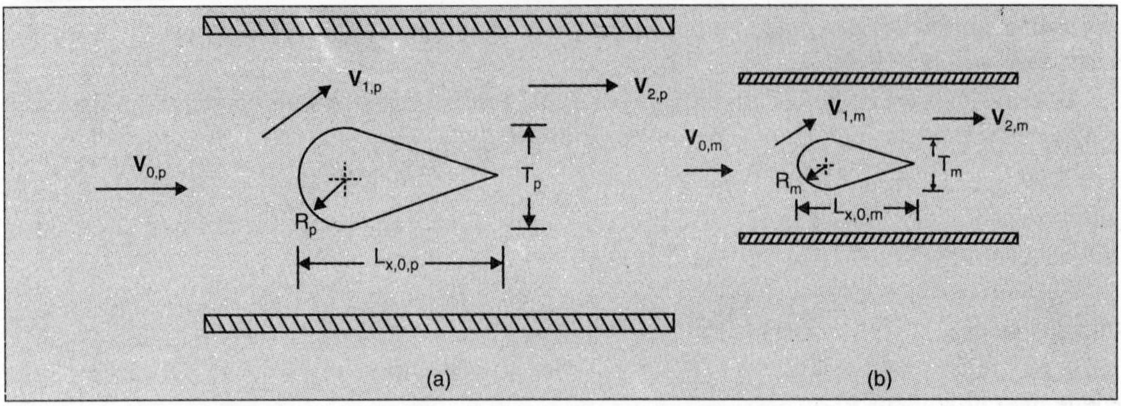

Fig. 9.3. (a) Prototype and (b) model.

9.3.1 Scale Factors

Consider a prototype and a *geometrically similar* model shown in Fig. 9.3. The solid body has a more complex geometry than that of the body shown in Fig. 9.2. Geometric similarity requires

$$\frac{R_p}{L_{x,0,p}} = \frac{R_m}{L_{x,0,m}}; \frac{T_p}{L_{x,0,p}} = \frac{T_m}{L_{x,0,m}}; \text{ etc.}$$

This can be re-written as

$$\frac{R_p}{R_m} = \frac{T_p}{T_m} = \ldots\ldots = \frac{L_{x,0,p}}{L_{x,0,m}} \equiv k_L \qquad \ldots(9.24)$$

i.e., the ratios of all *corresponding* dimensions in the model and prototype are equal. The constant ratio k_L of the corresponding length dimensions in the prototype and the model is termed the *length scale factor*. Thus $k_L > 1$ refers to models which are smaller than the prototype and $k_L < 1$ to models which are larger.

All lengths in the model, then, can be obtained from the corresponding lengths in the prototype through the use of a constant length-scale factor. Complete similarity implies that the same holds for *each* of the other quantifiable parameters (quantities, for short) such as velocity, time, stress, force, power, etc. This is to say that the value of any quantity at a point in the prototype is related to the value at the *homologous point* (i.e., points having the same *relative* locations, $\mathbf{x}^*$) in the model through the *corresponding* scale factor. For example, the velocity at a point in a prototype is related to the velocity at the homologous point in the model through the velocity scale factor k_V. If, for instance, $V_{0,p}$, $V_{1,p}$, $V_{2,p}$, ... and $V_{0,m}$, $V_{1,m}$, $V_{2,m}$, ... (see Fig. 9.3) are the velocities at homologous points in the prototype and the model, respectively, then similarity of the two requires that

$$\frac{V_{1,p}}{V_{1,m}} = \frac{V_{2,p}}{V_{2,m}} = \ldots\ldots \equiv k_V \qquad \ldots(9.25)$$

This is consistent with Eq. (9.17) where we had for pairs of homologous points

$$\frac{V_{1,m}}{V_{0,m}} = \frac{V_{1,p}}{V_{0,p}}; \frac{V_{2,m}}{V_{0,m}} = \frac{V_{2,p}}{V_{0,p}}; \text{ etc.}$$

or,
$$\frac{V_{1,p}}{V_{1,m}} = \frac{V_{2,p}}{V_{2,m}} = \frac{V_{0,p}}{V_{0,m}} \equiv k_V$$

Similarly, constant scale factors will exist for quantities like time-interval, frequency, acceleration, mass, flow rate, force, etc. Thus, the ratio of the pressures at the noses of the two aerofoils in the prototype and the model in Fig. 9.3, is equal to the ratio of the pressures at some other set of homologous points, each given by k_p, the pressure scale-factor.

We will see in Sec. 9.3.3 that the various scale factors are not all independent. The interrelationships among these scale factors will be shown to govern the design of model experiments and permit prediction of the prototype behaviour from tests on the model.

9.3.2 Requirements of Similarity

As defined earlier, two flows are said to be similar if each of the quantifiable parameters has a constant scale factor throughout the flow region. But how is this similarity to be ensured? Geometric similarity is an obvious requirement. Besides this, we have to make a proper choice of the values of the independent parameters that define the flow. These *independent* parameters include the material properties of the fluid (e.g., ρ, μ in Section 9.2.1) and the conditions at the boundaries (e.g., p_0, V_0 in Section 9.2.1). An incompressible fluid flow in the geometry described in Fig. 9.3 is completely defined by the density ρ and viscosity μ of the fluid, and by the far upstream velocity, V_0. If we were to model this system with a geometrically similar one, the only choice available would be the manipulation of the density ρ and the viscosity μ of the 'model' fluid and the far upstream velocity V_0 in the model (besides the length scale factor k_L), in order that all the quantities have constant scale factors.

Thus, *ensuring similarity consists of obtaining scaling relationships for the values of these independent quantities.* These relationships are termed as *similarity rules* or *modelling rules.*

The basic strategy in obtaining these similarity rules consists of exploiting the fact that the scale factors for various quantities are not all independent. The interrelationships between the various scale factors are converted into relations between the independent quantities for the prototype and the model. In the following sections, the relations between the scale factors are developed, first for the kinematic quantities and then for the dynamic quantities (i.e., those involving forces) and thereafter, the relations between the independent quantities are obtained.

Once the similarity between the prototype and the model is achieved by a proper choice of the independent parameters, it can be shown that the dependent quantities also have constant scale factors. The relationships between the scale factors for dependent quantities and those for the independent quantities are termed as the *prediction rules*. These are used for predicting the results for the prototype from the measured results for the model.

One example of such similarity has already been introduced in Sec. 1.6. Flows about a prototype and a model cylinder are similar (as will be shown later) when the corresponding Reynolds numbers are equal, i.e., $(\rho VD/\mu)_p = (\rho VD/\mu)_m$. Under this condition, the drag coefficients are equal, i.e., Drag/($\frac{1}{2}\rho V^2 \times$ Area) is identical in the two cases. The Reynolds number equality, then, is the similarity rule and the drag coefficient equality is a prediction rule from which we can predict the drag on the prototype cylinder from the measured drag on the model.

9.3.3 Inter-relationships between Scale Factors for Kinematic Quantities

Since the physical quantities pertaining to a system are not all independent, the various scale factors are also not so. In two geometrically similar systems, like the ones shown in Fig. 9.3, it is evident that the areas of the corresponding elements in the two systems are related by an *area scale factor* k_A that must equal k_L^2. This is because each dimension has been scaled by the factor k_L. If we consider a rectangular element,

$$dA_p = dx_p \times dy_p = k_L\, dx_m \times k_L\, dy_m$$

or

$$\frac{dA_p}{dA_m} \equiv k_A = k_L^2$$

It can be shown in the same manner that the volume scale factor $k_{\mathcal{V}}$ equals k_L^3. Therefore, we *cannot* specify the area and the volume scale factors, k_A and $k_{\mathcal{V}}$, independently of the length scale factor k_L.

The relationships between other scale factors can be found out in a similar manner. For example, the scale factor for length, time-intervals, and velocities within a system must be related through the fact that the velocity is the time derivative of the distance moved, i.e.,

$$V = dl/dt$$

where dl is the distance moved in the infinitesimal time interval dt. Similar behaviour of the model and the prototype requires that the magnitude of the velocities* at homologous points must be scaled by a constant scale factor k_V. Thus,

$$\frac{V_p}{V_m} \equiv k_V = \frac{(dl/dt)_p}{(dl/dt)_m}$$

(where the subscript on V indicating the location of the homologous points is omitted). Similarity requires that the corresponding lengths be scaled by the constant factor k_L. Similarly, time intervals must also be scaled by a constant factor k_t. Therefore, $dl_p = k_L\, dl_m$ and $dt_p = k_t\, dt_m$, so that

$$k_V = \frac{k_L dl_m}{k_t dt_m} \Big/ \frac{dl_m}{dt_m} = \frac{k_L}{k_t}$$

The velocity scale factor, then, is the ratio of the length and the time scale factors. Of these three, only two can be prescribed arbitrarily. If the aerofoils in the two systems of Fig. 9.3 were oscillating about their chords with time periods τ_p and τ_m respectively, clearly the time scale factor k_t, would be specified as τ_p/τ_m. Then, given the length scale factor k_L, the velocity scale factor k_V should equal k_L/k_t. For aerofoils where there is no time-like independent parameter, k_t is not given *a-priori*, and therefore, other relations are required for determining k_V. If the aerofoils were oscillating, the ratio of their time periods would have given a value of k_t.

The geometrically similar prototype and the model are said to have *kinematic similarity* if all the kinematic quantities such as frequency f, RPM N, angular velocity ω, acceleration a,

---

* The fact that the component velocities must be scaled by the same factor means that the velocity vectors at a pair of homologous points have the same direction.

angular acceleration α, volumetric flow rate $\mathring{Q}$, etc., have constant scale factors.* Starting from the definition of the various quantities, one can easily obtain relationships among their scale factors. As seen in Table 9.2 only two independent scale factors** are required to obtain all the kinematic scale factors. Two commonly used sets of independent scale factors, k_L, k_t and k_L, k_V have been employed in this table.

Table 9.2. Relationships between various kinematic scale factors

Scale factor	In terms of	
	k_L and k_t	k_L and k_V
k_t	k_t	k_L/k_V
k_V	k_L/k_t	k_V
k_ω, k_f, k_N	$1/k_t$	k_V/k_L
k_a	k_L/k_t^2	k_V^2/k_L
k_α	$1/k_t^2$	k_V^2/k_L^2
$k_{\mathring{Q}}$	k_L^3/k_t	$k_L^2 k_V$

9.3.4 Pi-Numbers

The relationships amongst the various scale factors as obtained in Sec. 9.3.3 are not very expressive and are conventionally transformed into more meaningful forms through the procedure illustrated below.

Take for example, the relationship $k_V = k_L/k_t$ where the factor k_V represents the ratio of velocities at any pair of homologous points, k_L is the ratio of any pair of corresponding lengths, and k_t the ratio of any pair of corresponding time intervals. Let us select *arbitrarily* a set of quantities $V_{c,p}$, $L_{c,p}$, $t_{c,p}$, and $V_{c,m}$, $L_{c,m}$, $t_{c,m}$, characterizing the similar quantities in the prototype and the model, respectively. It is understood that $V_{c,p}$ and $V_{c,m}$ are velocities at homologous points, $L_{c,p}$ and $L_{c,m}$ refer to 'similar' lengths in the prototype and the model, and $t_{c,p}$ and $t_{c,m}$ are 'similar' time intervals. For example, in the oscillating aerofoil problem of Sec. 9.3.3, the velocities (V_0) far upstream in the two cases may be taken as $V_{c,p}$ and $V_{c,m}$, chord lengths $L_{x,0,p}$ and $L_{x,0,m}$ may be taken as $L_{c,p}$ and $L_{c,m}$, and the time periods τ_p and τ_m of oscillation may be taken as $t_{c,p}$ and $t_{c,m}$. Then,

$$k_V = V_{c,p}/V_{c,m}$$
$$k_L = L_{c,p}/L_{c,m}$$
$$k_t = \tau_{c,p}/t_{c,m}$$

and the relationship $k_V = k_L/k_t$ becomes

$$\frac{V_{c,p}t_{c,p}}{L_{c,p}} = \frac{V_{c,m}t_{c,m}}{L_{c,m}} \quad ...(9.26)$$

Similarity, then, requires that the dimensionless product $(Vt/L)_c$ must have the same value in the model as in the prototype. Modelling and prediction rules are conventionally expressed in

* Kinematic similarity implies the streamline patterns are identical.

** It may be noted that this is related to the fact that all kinematic quantities have, at most, two basic dimensions.

CHAPTER 9

terms of such dimensionless products that are termed *pi-numbers* (written as Π). Thus, the condition $k_V = k_L/k_t$ requires that

$$\Pi = Vt/L$$

be identical in both the prototype and the model. The subscripts *c* in Eq. (9.26) are dropped and the characteristic nature of the values in Πs is assumed. For the oscillating aerofoils described above, similitude involves

$$\Pi = \frac{V_{0,p}\tau_p}{L_{x,0,p}} = \frac{V_{0,m}\tau_m}{L_{x,0,m}} \qquad \text{...(9.27)}$$

For given values of $V_{0,p}$, τ_p, $L_{x,0,p}$, and $V_{0,m}$, τ_m, $L_{x,0,m}$ may be chosen such that Eq. (9.27) holds. This is only a necessary condition and may not be a sufficient one.

It may be mentioned here that in this section, the inter-relationship between k_L, k_V and k_t has been used to develop a *modelling* rule.

All similar inter-relationships between scale factors lead to equality of dimensionless products of characteristic quantities. For example, the last relation in Table 9.2

$$k_{\mathring{Q}} = k_L^3/k_t = k_L^2 k_V$$

gives

$$\Pi = \left(\frac{\mathring{Q}t}{L^3}\right) \text{or} \left(\frac{\mathring{Q}}{L^2 V}\right)$$

where the characteristic nature of the quantities involved is understood. Only a few of these inter-relationships can be used as modelling rules, since there are a limited number of independent quantities to play with. In addition, there may be physical constraints present, as shown in Example 9.5.

9.3.5 Inter-relationships between Scale Factors for Dynamic Quantities

The development of the two previous sections is extended to dynamic quantities in this section. The prototype and the scaled model are said to have *dynamic similarity* if the *net* forces at homologous points in the two are related by a constant scale factor k_F. However, the force at a point in a system is related to kinematic quantities. Newton's law of motion relates the net force $\mathbf{F}$ on an element to its acceleration $\mathbf{a}$ and its inertia (i.e., the mass *m*) as $\mathbf{F} = m\mathbf{a}$. If k_F, k_m and k_a are the scale factors for force, mass and acceleration, respectively, $\mathbf{F}_p = k_F \mathbf{F}_m$, $m_p = k_m m_m$ and $\mathbf{a}_p = k_a \mathbf{a}_m$, and then, Newton's law gives

$$\mathbf{F}_m = \frac{k_m k_a}{k_F} m_m \mathbf{a}_m$$

But $\mathbf{F}_m$ must equal $m_m \mathbf{a}_m$ and, therefore,

$$\frac{k_m k_a}{k_F} = 1$$

or,

$$k_F = k_m\, k_a$$

Thus, given the scale factors k_m and k_a, the scale factor k_F for the net force can be obtained. The factor k_a has been related to k_L and k_t (*or* k_L and k_V) in Table 9.2. In fluid mechanics, the density ρ of the fluid is usually used as the quantity signifying inertia. It is easy to see that $k_m = k_\rho k_L{}^3$, and so

$$k_F = \frac{k_\rho k_L^4}{k_t^2} = k_\rho k_L^2 k_V^2$$

This result is expressed in terms of a non-dimensional Π as

$$\Pi = \frac{Ft^2}{\rho L^4} \text{ or } \frac{F}{\rho V^2 L^2}$$

where the characteristic nature of the quantities involved is understood.

Table 9.3 gives the scale factors and the corresponding Πs for some dynamic quantities of interest. All these Πs must have the same value in the model as in the prototype.

Table 9.3. Scale factors and Πs for some common dynamic quantities

Quantity	Scale factors in terms of		Π in terms of	
	k_ρ, k_L, k_t	k_ρ, k_L, k_V	ρ, L, t	ρ, L, V
Force, F	$k_\rho k_L{}^4/k_t^2$	$k_\rho k_L{}^2 k_V^2$	$Ft^2/\rho L^4$	$F/\rho L^2 V^2$
Momentum, P	$k_\rho k_L{}^4/k_t$	$k_\rho k_L{}^3 k_V$	$Pt/\rho L^4$	$P/\rho L^3 V$
Torque, T	$k_\eta k_L{}^5/k_t^2$	$k_\rho k_L{}^3 k_V^2$	$Tt^2/\rho L^5$	$T/\rho L^3 V^2$
Work, W	$k_\rho k_L{}^5/k_t^2$	$k_\rho k_L{}^3 k_V^2$	$Wt^2/\rho L^5$	$W/\rho L^3 V^2$
Power, $\mathring{W}$	$k_\rho k_L{}^5/k_t^3$	$k_\rho k_L{}^2 k_V^3$	$\mathring{W}t^3/\rho L^5$	$\mathring{W}/\rho L^2 V^3$
Pressure, p	$k_\rho k_L{}^2/k_t^2$	$k_\rho k_V{}^2$	$pt^2/\rho L^2$	$p/\rho V^2$
Shear stress, τ	$k_\eta k_L{}^2/k_t^2$	$k_\rho k_V^2$	$\tau\, t^2/\rho L^2$	$\tau/\rho V^2$

Just as the scale factors for all the kinematic quantities are made up of the power products of two scale factors, k_V and k_L (or k_L and k_t), the scale factors of all the dynamic quantities are made up of the power products of three* scale factors, k_ρ, k_L and k_t (or k_ρ, k_L and k_V). Thus, dynamic similarity implies that all the physical quantities for a prototype and its model are related to a maximum of three independent scale factors, say k_ρ, k_L and k_t.

9.3.6 Obtaining Modelling Rules

Similarity between the behaviour of a prototype and its geometrically similar model can be obtained by choosing the values of the independent parameters *properly*. These parameters influence the flow field by determining the various forces such as the viscous force, body force, surface tension force, pressure force, etc., that act on an element of fluid. Since the net force acting on a particle is the vector sum of these various forces, it stands to reason that the ratios of various component forces to the net force must be the same for both the prototype and the model. And, since the net force is scaled according to the factor k_F established earlier, each of the component forces must also be scaled according to the scale factor k_F. This will make the polygon of forces at homologous points similar (see Problem 9.7). If the force polygons are similar

* This is again related to the fact that these quantities require three independent dimensions.

CHAPTER 9

at homologous points, it is intuitively expected that the accelerations of the fluid elements at these points will be similar, and so will be the velocities (and all other kinematic quantities; Problem 9.8 demonstrates this). This forms the basis of perhaps the most convenient of the methods of establishing the modelling rules.* The method consists of the following steps:

(i) Identifying the various force components that determine the net force at a point in the flow field (this is the most intuitive of the steps and requires a lot of judgement).

(ii) Obtaining the scale factors for each of these force components using the physical laws that govern the phenomena.

(iii) Equating the scale factors so obtained to the net-force scale factor k_F obtained in Sec. 9.3.5 (since the net-force scale factor has been obtained from Newton's law of inertia, it is termed as the *inertial force factor*). This step establishes a number of relations amongst the scale factors of all the independent parameters.

(iv) Obtaining the non-dimensional Π-numbers in terms of the characteristic independent quantities from the equations established in step (iii). These give the modelling or similarity rules.

We illustrate Steps (ii) – (iv) using the viscous ($\mathbf{F}_\mu$) and inertial (net, **F**) forces. The physical laws that govern these (Step ii) are written as (for a simple 2-D problem)

$$F_{\mu,p} = \mu_p \cdot \text{area}_p \cdot (dV/dx)_p;\ F_{\mu,m} = \mu_m \cdot \text{area}_m \cdot (dV/dx)_m$$

Therefore,

$$k_{vis} = \left(\frac{F_{\mu,p}}{F_{\mu,m}}\right)_c = \frac{\mu_p \left(L^2 V/L\right)_{c,p}}{\mu_m \left(L^2 V/L\right)_{c,m}} = \frac{\mu_p}{\mu_m}\frac{L_{c,p}}{L_{c,m}}\frac{V_{c,p}}{V_{c,m}} = k_\mu k_L k_V$$

As obtained earlier,

$$k_F = k_\rho k_L^2 k_V^2 = \frac{\rho_p}{\rho_m}\frac{L_{c,p}^2}{L_{c,m}^2}\frac{V_{c,p}^2}{V_{c,m}^2}$$

Equating k_{vis} and k_F (step iii) gives

$$\left(\frac{\rho VL}{\mu}\right)_{c,p} = \left(\frac{\rho VL}{\mu}\right)_{c,m}$$

If the characteristic parameters are selected as the independent quantities, this suggests matching the Reynolds numbers.

Table 9.4 lists some of the component forces commonly encountered in fluid mechanics and the relations for their scale factors obtained by following step (ii) of the above procedure. Since the scale factors for all types of forces must be identical, each of the force scale factors is equated to the inertial scale factor k_F to obtain the relationships shown in Table 9.5. This table also lists the relevant non-dimensional Π-numbers that are formed from the independent

---

* From the arguments given here, these modelling rules are shown to be necessary conditions for similitude but sufficiency is not established. The procedures which establish sufficiency (one of which is presented in Sec. 9.2) are not as physically revealing as the one given here. We recommend this approach because of its simplicity of concept and ease in application, particularly to complex problems.

parameters. This corresponds to steps (iii) and (iv) of the above procedure. *Similarity is ensured by matching the values of these Π numbers for the prototype and the model flows.* These statements of equality comprise the modelling rules.

These numbers have a very important place in the study of fluid mechanics and have been named in honour of some of the great scientists and engineers who have made significant contributions to the subject. The numbers are denoted by the first two letters of their names. Thus, $\rho VL/\mu$, the Reynolds number, is denoted by Re, Froude number $V/\sqrt{gL}$ by Fr, and so on. Table 9.5 shows only a few of the named dimensionless numbers of which a complete list may be found in Massey (Reading 27) along with their significance.

The physical significance of the named Π-numbers can be observed by noting that each of them has been obtained from the scale factors of two kinds of forces. For example, Reynolds number involves the inertial and viscous forces. It will be shown in Chapter 11 that its value may be taken as an indication of the ratio of the magnitudes of these two forces. Thus

Re ~ inertial force/viscous force
Fr ~ inertial force/gravity force
Eu ~ inertial force/pressure force
We ~ inertial force/surface-tension force
Ca or Ma ~ inertial force/compressibility force
St ~ centrifugal force/inertial force ...(9.28)

It should be noted that not all the forces are important in all flow problems and hence, not all of the dimensionless numbers will serve as similarity rules in every problem. The following examples illustrate the use of the dimensionless numbers for obtaining modelling rules. In these examples those forces are specified which are important. Section 9.2 gave some guidelines for deciding which forces are important in a given problem and the examples in that section illustrated these choices.

Table 9.4. Component forces and their scale factor relations

		Scale factor relations		
Force	**Governing law**	**Basic**	**In terms of ks for independent parameters**	**Π**
Viscous force	$F_\mu = \mu$ (area)(vel. grad.) (Newton-Stokes relation)	$k_{F,\mu} = k_\mu k_A k_V / k_L$	$k_{F,\mu} = k_\mu k_L k_V$	$F_\mu/\mu LV$
Gravity force	$F_g = g$(mass) (Gravitation law)	$k_{F,g} = k_g k_m$	$k_{F,g} = k_g k_\rho k_L^3$	$F_g/\rho g L^3$
Pressure force	F_p = (pressure) (area)	$k_{F,p} = k_p k_A$	$k_{F,p} = k_p k_L^2$	F_p/pL^2
Surface-tension force	$F_s = \sigma$ (length)	$k_{F,s} = k_\sigma k_L$	$k_{F,s} = k_\sigma k_L$	$F_s/\sigma L$
Compressibility force	$F_c = E_s$ (area)	$k_{F,c} = k_{E_s} k_L^2$	$k_{F,c} = k_{E_s} k_L^2$	$F_c/E_s L^2$
Centrifugal force	F_ω = (mass)($\omega^2 r$)	$k_{F,\omega} = k_m k_\omega^2 k_L$	$k_{F,\omega} = k_\rho k_\omega^2 k_L$	$F_\omega/\rho\omega^2 L^4$
Inertial force	F = (mass)(acceleration) (Newton's law of motion)	$k_F = k_m k_a$	$k_F = k_\rho k_L^2 k_V^2$	$F/\rho L^2 V^2$

σ = surface tension, E_s = isentropic bulk modulus

Table 9.5. Dimensionless Π-numbers obtained by equating scale factors to k_F

Equation	Relation amongst *ks* of independent parameters	Π	Name of Π-number	Symbol
$k_{F,\mu} = k_F$	$k_\mu = k_\rho k_L k_V$	$\rho VL/\mu$	Reynolds Number	Re
$k_{F,g} = k_F$	$k_g k_L = k_V^2$	$V/\sqrt{gL}$	Froude Number	Fr
$k_{F,p} = k_F$	$k_p = k_\rho k_V^2$	$\rho V^2/p$	Euler Number	Eu
$k_{F,s} = k_F$	$k_\sigma = k_\rho k_L k_V^2$	$\rho LV^2/\sigma$	Weber Number	We
$k_{F,c} = k_F$	$k_{E_s} = k_\rho k_V^2$	$\rho V^2/E_s$	Cauchy Number	Ca
$k_{F,\omega} = k_F$	$k_\omega = k_V/k_L$	$\omega L/V$	Strouhal Number	St

For gases, $E_s/\rho = c^2$ where c is the speed of sound. Then the Cauchy number, $\rho V^2/E_s$ becomes V^2/c^2, the square root of which is termed the Mach number (Ma).

Once the similarity between the model and the prototype is established by satisfying the modelling rules obtained above, all dependent parameters will automatically be governed by their relevant scale factors. Since these scale factors can be expressed in terms of the scale factors for the independent quantities (which have already been fixed by the modelling rules), the prediction rules can be obtained directly. This is illustrated in the following examples.

Example 9.6. A seaplane is being designed for a take-off speed of 100 kmph. Model tests are to be made on a one-tenth scale model. At what speed should the model be towed to simulate the take-off condition? Similarity of inertial and gravity forces is to be ensured in this case.

Table 9.6 is constructed following the step-by-step procedure recommended in Sec. 9.3.6. In this, the first column gives the laws governing the forces whose similitudes are to be assured. The second column gives the scale factor relations in terms of the scale factors of the independent quantities. The third column gives the similarity rule obtained by equating the scale factors for the two forces, and the last column gives the similarity rule as a Π-number in terms of the characteristic independent quantities.

Table 9.6. Modelling rule for Example 9.6

Governing law	Scale factor relation	Similarity rule	Π-number
Inertia $F = ma$	$k_F = (k_\rho k_L^3)\dfrac{k_V^2}{k_L}$		
Gravitation $F_g = mg$	$k_{F,g} = (k_\rho k_L^3)k_g$	$\dfrac{k_V^2}{k_L k_g} = 1$	$\dfrac{V^2}{Lg} = (Fr^2)$

Thus, the seaplane model is governed by a single modelling rule that specifies that $(V^2/Lg)_c$ of the prototype and the model should be identical, or

$$\frac{V_m^2}{L_m g_m} = \frac{V_p^2}{L_p g_p}$$

or $$V_m = V_p \{L_m g_m / L_p g_p\}^{1/2}$$

Since $$g_m = g_p \text{ and } L_m = 0.1\ L_p$$

$$V_m = V_p\left(\frac{1}{10}\right)^{1/2} = 100(\text{km/h}) \times \sqrt{1/10} = 31.6 \text{ kmph}$$

Thus, the model should be towed at 31.6 kmph for inertial and gravity force similarity. If these are the only important forces governing the phenomenon, complete similarity is obtained by this condition.

If the force required to drag the model seaplane is 3 N, the thrust needed by the prototype at take-off can be calculated by developing the prediction rule using the force scale factor*

$$k_F = k_\rho k_L^2 k_V^2$$

Thus

$$\frac{F_p}{F_m} = \left(\frac{L_p}{L_m}\right)^2 \left(\frac{V_p}{V_m}\right)^2$$

since the density is invariant. This gives

$$\text{thrust for the prototype} = 3\ (\text{N}) \times (10)^2 \times (10) = 3000 \text{ N}.$$

Example 9.7. A one-sixth scale model of an automobile is to be tested in a wind tunnel at a speed corresponding to the prototype speed of 60 kmph. If the model drag at that speed is 510 N, what are the prototype drag and its power requirement? Only inertial and viscous forces are to be modeled.

The length scale factor k_L is given as 6, and, since the model is being tested in the same fluid, $k_\rho = 1$. First, find k_V by the modelling rule developed in Table 9.7 that specifies that the Reynolds number (the ratio of the characteristic inertial and viscous forces) must be identical in the two flows. Thus, the modelling rule is

$$\left(\frac{\rho VL}{\mu}\right)_m = \left(\frac{\rho VL}{\mu}\right)_p$$

or, since ρ and μ of the model and prototype are identical,

$$V_m = V_p \times L_p/L_m = 60 \text{ (kmph)} \times 6 = 360 \text{ kmph}$$

i.e., the model should be tested at a wind speed** of 360 kmph.

CHAPTER 9

---

* The force scale factor used is based on Newton's law of inertia. The law of gravitation could have been used as well to give the same result since $k_F = k_{F,g}$ under conditions of similarity.

** It shall be seen later in Chapter 13 that the viscous forces are not very important in cases like this and fairly good results can be got by testing the model at 'almost' any speed.

Table 9.7. Modelling rule for Example 9.7

Governing law	Scale factor relation	Similarity rule	Π-number
Inertia $F = ma$	$k_F = (k_\rho\, k_L^3)\,(k_V^2/k_L)$		
Viscous Friction	$k_{F,\mu} = k_\mu\,(k_L^2)\,(k_V/k_L)$	$k_\rho\, k_V\, k_L/\, k_\mu = 1$	$\rho VL/\mu$ (= Re)
$F_\mu = \mu$ (area)(vel. grad.)			

The prediction rule for the drag is obtained from the corresponding scale factor relation. The drag force scale factor equals the force scale factor $k_F = k_\rho\, k_L^2\, k_V^2$, or

$$k_F = 1 \times 6^2 \times \left(\frac{60}{360}\right)^2 = 1$$

Thus, the prototype drag will also be 510 N. The power requirement to overcome the wind drag can be obtained directly from the velocity and the drag as

$$\overset{\circ}{W} = 510\ (\text{N}) \times 60\ (\text{kmph}) \times 10^3\ (\text{m/km}) \times 1/3600\ (\text{hr/s}) = 8.5\ \text{kW}$$

In this case, the prediction rule for power is not relevant since the model power consumption has not been specified. However, if it were (somehow) known, the prediction rule obtained from the power scale factor $k_{\overset{\circ}{W}} = k_\rho\, k_L^2\, k_V^3$ could be used.

Example 9.8. A pump capable of lifting 5 m^3/s of water against a head of 250 m at 500 RPM is to be designed. A one-ninth scale model is constructed and tested with the same fluid against a 10 m head. At what speed should it run and what should be the volume flow rate? What would be the power scale factor? Neglect viscous forces and gravity forces within the pump.

If viscous and gravity forces are neglected, the pressure and inertial forces are to be modelled. Since this pump is a rotary machine, centrifugal forces should also be modelled. Table 9.8 shows the development of the two similarity rules from these three forces.

Table 9.8. Modelling rules for Example 9.8

Governing law	Scale factor relation	Similarity rule	Π-number
Inertia			
F = (mass) × (acc.)	$k_F = k_\rho k_L^2 k_V^2$		
Pressure			
$F_p = p \times$ *Area*	$k_{F,p} = k_p k_L^2$	$k_p = k_\rho k_V^2$	$\dfrac{p}{\rho V^2}\ \left(= \dfrac{1}{\text{Eu}}\right)$
Centrifugal			
F_ω = (mass) × $(\omega^2 L)$	$k_{F,\omega} = k_\rho k_\omega^2 k_L^4$	$k_\omega k_L = k_V$	$\dfrac{\omega L}{V}$ (= St)

Thus, Euler number, Eu, and Strouhal number, St, must be matched. Eu similarity gives.

$$\left(\frac{p}{\rho V^2}\right)_p = \left(\frac{p}{\rho V^2}\right)_m$$

or

$$k_V = \frac{V_p}{V_m} = \sqrt{\frac{p_p}{p_m} \times \frac{\rho_m}{\rho_p}}$$

or $$k_V = \sqrt{\frac{250\ (\text{m})}{10\ (\text{m})} \times 1} = 5$$

Then the St similarity gives

$$\left(\frac{\omega L}{V}\right)_m = \left(\frac{\omega L}{V}\right)_p$$

or $$\omega_m = \omega_p \frac{V_m}{V_p} \frac{L_p}{L_m} = \omega_p \times \frac{1}{5} \times 9 = \frac{9}{5}\omega_p$$

Since the RPM of the prototype is 500, the RPM of the model should be 9 × 500/5 = 900. The model pump should run at 900 RPM.

The prediction rule for the discharge is found by its scale factor relation

$$k_{\mathring{Q}} = k_L^2\, k_V = (9)^2 \times 5 = 405$$

Thus, the model pump should deliver

$$\mathring{Q}_m = \mathring{Q}_p / k_{\mathring{Q}} = 5\ (\text{m}^3/\text{s})/405 = 0.0123\ \text{m}^3/\text{s}$$

The scale factor for power consumption is

$$k_{\mathring{W}} = k_F \times k_V = k_\rho\, k_L^2\, k_V^{\,3} = 1 \times (9)^2 \times (5)^3 = 10125$$

and the prototype pump will consume 1.0125×10^4 times the power used by the model pump.

Example 9.9. The tail unit of an aircraft structure is found to vibrate violently at a frequency of 9.5 Hz as it prepares to land at 100 kmph. It is suspected that this may be due to vortex shedding from the tail unit. To investigate the phenomenon, a one-fifth scale model of the aircraft is made and tested in a *variable-density wind tunnel* ensuring similarity of inertial and viscous forces. The pressure used in the wind tunnel is 10 times the atmospheric pressure. The vortices in the model test are shed at a frequency of 23 Hz. Is the vortex shedding a probable cause of vibration?

Since the inertial and viscous forces are to be made similar, the modelling rule developed in Table 9.7 is applicable. Thus, the Reynolds number of the model and the prototype must be the same, or

$$\left(\frac{\rho V L}{\mu}\right)_m = \left(\frac{\rho V L}{\mu}\right)_p$$

From this the velocity at which the model is to be tested can be obtained as

$$V_m = V_p \times \frac{L_p}{L_m} \times \frac{\rho_p}{\rho_m} \times \frac{\mu_m}{\mu_p}$$

Since the pressure in the model is 10 times that for the prototype, $\rho_p/\rho_m = 1/10$, (by perfect gas law, the temperatures being the same). Since the effect of pressure on viscosity is negligible, $\mu_p = \mu_m$ and so

$$V_m = 100 \text{ (kmph)} \times 5 \times (1/10) \times 1 = 50 \text{ kmph}$$

Thus, the model is tested at 50 kmph.

The prediction rule for the frequency can be obtained from its scale-factor relation

$$k_f = \frac{1}{k_t} = \frac{k_V}{k_L}$$

or
$$\frac{f_p}{f_m} = \frac{V_p}{V_m} \times \frac{L_m}{L_p} = 2 \times \frac{1}{5} = 0.4$$

Thus
$$f_p = 0.4\, f_m = 0.4 \times 23 \text{ (Hz)} = 9.2 \text{ Hz}$$

Since the frequency of vibration of the tail unit is quite close to the predicted shedding frequency, the suspicion is substantiated.

Example 9.10. An aeroplane is to fly in the standard atmosphere at 360 kmph. A one-fifth scale model of it is tested in a variable density wind tunnel (at standard temperature) for similarities of inertial, viscous and compressibility forces. Find the pressure of air in the tunnel and the wind speed required. Also, find the scale factor for the pitching moment when similarity is attained.

Since similarity of three forces is desired in this case, two modelling rules result as developed in Table 9.9. Thus, the Reynolds number and Mach number must match. The speed of sound c in the model and prototype flows is the same even though the pressure is different.* Therefore, Ma similarity requires that the velocity of the two flows be identical, or that the model flow velocity should also be 360 kmph. Then Re similarity requires that

$$\left(\frac{\rho L}{\mu}\right)_m = \left(\frac{\rho L}{\mu}\right)_p$$

Table 9.9. Modelling rules for Example 9.10

Governing law	Relation between scale factors	Modelling rule	Π-number
Inertia $F = ma$	$k_F = (k_\rho\, k_L^3)(k_V^2/k_L)$		
Viscous Friction $F = \mu(\text{Area}) \times (\text{vel. grad.})$	$k_{F,\mu} = k_\mu\, (k_L^2)(k_V/k_L)$	$k_\rho\, k_V\, k_L/k_\mu = 1$	$\rho VL/\mu$ (= Re)
Compressibility $F_c = E_s(\text{Area})$ $= \rho c^2(\text{Area})$, since $c^2 = E_s/\rho$ for a perfect gas	$k_{F,c} = k_\rho\, k_c^2\, k_L^2$	$k_V/k_c = 1$	V/c (= Ma)

* Since $c = \sqrt{\gamma RT}$

But because the viscosity depends essentially on the temperature, which is constant, $\mu_m = \mu_p$, and thus

$$\rho_m = \rho_p \times \left(\frac{L_p}{L_m}\right) = 5\,\rho_p$$

To obtain a five-fold density in the model flow, the pressure in the variable density wind tunnel should be five atmospheres (it may be desirable to relax this requirement).

The scale factor for the pitching moment *M*, then, should be

$$k_M = k_L\,k_F = k_L \times (k_\rho\,k_L^2\,k_V^2) = k_\rho\,k_L^3\,k_V^2$$

Here, $k_V = 1$, $k_L = 5$, and $k_\rho = 1/5$, and so the scale factor for the pitching moment is

$$k_M = (1/5) \times (5)^3 \times 1 = 25$$

or, the pitching moment in the prototype will be twenty-five times that measured in the model study.

It may be pointed out that the occurrence of two modelling rules in this case requires the model fluid to have different properties. This difficulty is very commonly encountered in modelling studies (see Example 9.5). The more the modelling rules, the more are the constraints in the choice of the material properties and scale factors. Since the art of tailoring material properties to desired specifications is still not well developed, at times one is forced to compromise on modelling requirements. Such *relaxations* have been perfected into an art. A good description of the relaxations of modelling requirements with some interesting case studies is found in Schuring (Reading 30).

9.4 SIMPLIFICATIONS RESULTING FROM THE USE OF DIMENSIONLESS VARIABLES

It was shown in Sec. 9.2.1 that a reduction in the number of independent variables is achieved by non-dimensionalizing the equations and boundary conditions using appropriate characteristic values. For the example of flow past an infinite solid body having an elliptical cross-section, the drag coefficient C_D is a function of Re alone (in applications where Eu is not significant and where the free surface is far away):

$$C_D = C_D\,(\text{Re, geometry}) \qquad \text{...(9.29)}$$

It may be noted that the drag on a body is actually a function of ρ, V_0, μ, $L_{x,0}$ and $L_{z,0}$, i.e., of five parameters, besides additional parameters describing more complex geometries. But as shown above, the non-dimensional drag C_D is a function of the Reynolds number only (besides the geometry).* If experiments were conducted to study the effect of each of the four variables

---

* Such results can also be obtained by dimensional analysis. This technique is based on the Buckingham Pi theorem which states that if a physical process involves *n*-dimensional variables it can be reduced to a relationship between $k = n - j$ pi-numbers, where *j* equals the minimum number of independent dimensions required to describe the variables. Various techniques are available for determining the pi-numbers using dimensional analysis. A brief description of one of these procedures is included as Appendix C. We prefer the approach presented in this Chapter because it gives the pi-numbers in their most physically meaningful form.

(ρ, V_0, μ, $L_{x,0}$) on geometrically similar bodies, the original formulation would require that all the four parameters be varied independently. The non-dimensional formulation, however, suggests that studying the effect of just the Reynolds number (and geometrical ratios) will give complete information. This represents a vast saving in the experimental effort required, and therefore, such a formulation is routinely attempted before any experimental programme is undertaken (*Figure 13.16 shows experimental data on C_D* taken in this manner for cylinders). Given below are some further examples illustrating this.

Example 9.11. The time period τ of surface waves in water is found to depend on the density ρ, wavelength λ, depth h, gravity g, and surface tension σ. Rewrite this relationship in dimensionless form.

The data of the problem suggests that for surface waves, inertia, gravity and surface tension forces are relevant. The numbers that can be formed from the characteristic values of these forces are Fr and We. Thus, the two independent variables that govern the flow are $\text{Fr} = \dfrac{V}{\sqrt{gL}}$, and $\text{We} = \dfrac{\rho V^2 L}{\sigma}$, where the characteristic nature of the quantities is understood. Therefore, the non-dimensional wave-period τ/t_c is a function of $(V/\sqrt{gL})_c$, $(\rho V^2 L/\sigma)_c$ and the geometry, or

$$\frac{\tau}{t_c} = \Im\left(\frac{V_c}{\sqrt{gL_c}}, \frac{\rho V_c^2 L_c}{\sigma}, \text{ geometry}\right)$$

The geometry of the surface waves is defined by the ratio λ/h. Thus,

$$\frac{\tau}{t_c} = \Im\left(\frac{V_c}{\sqrt{gL_c}}, \frac{\rho V_c^2 L_c}{\sigma}, \frac{\lambda}{h}\right)$$

It is easy to see that either the wavelength λ or the depth h can be used as L_c. We choose (arbitrarily) λ. To obtain the estimate of V_c and t_c in terms of the given quantities, start with the relationships among their scale factors. The scale factor for velocity can be related to the scale factor for g and of length. Thus,

$$k_V = \sqrt{k_g k_L}$$

which gives $\Pi = V/\sqrt{gL}$ as invariant, with the characteristic nature of the variables understood. Since this Π is invariant, V_c may be chosen equal to $\sqrt{g\lambda}$ or some multiple of it. If $V_c = \sqrt{g\lambda}$, the Froude number becomes unity. Since Fr is now a constant, it need no longer be considered as an independent parameter. Also, the Weber number becomes $\rho g\lambda^2/\sigma$.

The characteristic time can now be obtained in terms of V_c and L_c ($= \lambda$) through the use of

$$k_V = h_L/k_t$$

which gives $\Pi = Vt/L$ as invariant. Thus, t_c can be taken as $L_c/V_c = \lambda/\sqrt{g\lambda} = \sqrt{\lambda/g}$ and the functional relation written as*

$$\frac{\tau\sqrt{g}}{\sqrt{\lambda}} = \Im\left(\frac{\rho g\lambda^2}{\sigma}, \frac{\lambda}{h}\right)$$

Example 9.12. The lift force F on an aerofoil is a function of its chord C, span S, velocity V, angle of attack α, the density ρ and viscosity μ of air, and the speed of sound c in air. Rewrite this functionality in non-dimensional terms.

The listing of variables suggests that inertial, viscous and compressibility forces are important. The relevant modelling Π-numbers are then $Re = \rho VL/\mu$ and $Ma = V/c$, where the characteristic nature of the quantities involved is understood. Thus, the non-dimensional lift force is a function of $\rho V_c L_c/\mu$, V_c/c, and the geometry. The geometry of the flow field is defined completely by the span-to-chord ratio S/C and the angle of attack α. Therefore,

$$\frac{F}{F_c} = \Im\left(\frac{\rho V_c L_c}{\mu}, \frac{V_c}{c}, \frac{S}{C}, \alpha\right)$$

V_c can be set equal to the speed of the aerofoil, and either C or S can be taken as L_c. It is conventional to choose C. To obtain the characteristic force F_c, the scale factor relation $k_F = k_\rho k_V^2 k_L^2$ is used, and it gives $\Pi = F_c/(\rho V_c^2 L_c^2)$ as invariant. Thus, one can set

$$F_c = \rho V_c^2 L_c^2 = \rho V^2 C^2$$

and obtain

$$\frac{F}{\rho V^2 C^2} = \Im\left(\frac{\rho VC}{\mu}, \frac{V}{c}, \frac{S}{C}, \alpha\right)$$

Conventionally, F_c is chosen as $\frac{1}{2}\rho V^2 CS$, which is also consistent with the scale-factor relation written above. The lift force is then

$$\frac{F}{\frac{1}{2}\rho V^2 (CS)} = \Im\left(\frac{\rho VC}{\mu}, \frac{V}{c}, \frac{S}{C}, \alpha\right)$$

Example 9.13. The power required by a pump is believed to be controlled by the pressure and inertial (including centrifugal) forces alone. Formulate, in non-dimensional terms, the problem of power required by a pump of impeller diameter D, delivering a volume flow rate $\mathring{Q}$ of a fluid of density ρ against a pressure Δp.

Since only the inertial and pressure forces are important, the non-dimensional power (in geometrically similar pumps) should be a function of the Euler number $Eu = \rho V_c^2/(\Delta p)_c$ and the Strouhal number $St = L_c/V_c$:

CHAPTER 9

---

* This result is consistent with the Buckingham-Pi theorem since the number of Π's is three (equal to the number of variables, 6, minus the number of required dimensions, 3).

$$\frac{\overset{o}{W}}{\overset{o}{W}_c} = \Im\left(\frac{\rho V_c^2}{(\Delta p)_c}, \frac{\omega_c L_c}{V_c}\right)$$

where the subscript c denotes the characteristic value of a quantity. The expressions for the characteristic quantities can be obtained in terms of the independent parameters D, $\overset{o}{Q}$, ρ and p. Clearly, $(\Delta p)_c$ can be set equal to Δp, the pressure rise. To obtain estimates of V_c and $\overset{o}{W}_c$, make use of the scale factor relations. Thus,

$$k_{\overset{o}{Q}} = k_V k_L^2$$

which gives $\Pi = \overset{o}{Q}/(VL^2)$ as an invariant. In this Π, the characteristic nature of the variables is assumed. Since Π is invariant, take $\overset{o}{Q}_c = V_c L_c^2$, or some multiple of it. We set $\overset{o}{Q}_c = V_c L_c^2$ from which $V_c = \overset{o}{Q}_c / L_c^2$. But the diameter D can be taken as a characteristic length, and the discharge $\overset{o}{Q}$ as $\overset{o}{Q}_c$. Thus, $V_c = \overset{o}{Q}/D^2$, and then $\rho V_c^2/(\Delta p)_c$ becomes $\rho \overset{o}{Q}^2/(\Delta p)D^4$. The second $\Pi = (\omega L/V)_c$ then becomes $\omega_c D^3/\overset{o}{Q}$. However, the rotational speed of a pump delivering a volume $\overset{o}{Q}$ at a head Δp cannot be specified independently and hence ω_c can be chosen as $\overset{o}{Q}/D^3$. This leaves the first Π as the independent parameter.*

Similarly, use

$$k_{\overset{o}{W}} = k_F k_V = k_\rho k_V^2 k_L^2 k_V = k_\rho \left(\frac{k_{\overset{o}{Q}}}{k_L^2}\right)^3 k_L^2 = k_\rho k_{\overset{o}{Q}}^3 k_L^{-4}$$

to obtain $\overset{o}{W}_c$. This gives $\Pi = [\overset{o}{W}/(\rho \overset{o}{Q}^3 L^{-4})]_c$ as an invariant, or that $\overset{o}{W}_c$ can be taken as $(\rho \overset{o}{Q}^3 D^{-4})$. Thus, the non-dimensional dependent variable is $\overset{o}{W}/(\rho \overset{o}{Q}^3 D^{-4})$. The functional relation, therefore, becomes

$$\frac{\overset{o}{W} D^4}{\rho \overset{o}{Q}^3} = \Im\left(\frac{\rho \overset{o}{Q}^2}{D^4(\Delta p)}\right)$$

The power requirement of geometrically similar pumps can then be expressed as a single curve (determined experimentally) between the non-dimensional variables $\overset{o}{W} D^4/\rho \overset{o}{Q}^3$ and $\rho \overset{o}{Q}^2/D^4(\Delta p)$. From this, the power requirements of any size pump under any discharge condition can be obtained.

---

* In Example 9.8, $\overset{o}{Q}$ was a dependent parameter, leaving Δp and ω as independent ones, and therefore the formulation was different. A more detailed and general discussion is given in Sec. 10.5.

PROBLEMS

9.1 Derive Eqs. (9.4) and (9.5).

9.2 Derive Eq. (9.7).

9.3 Derive Eq. (b) in Example 9.4.

9.4 Complete Example 9.5.

9.5 The velocity potential ϕ (see Chapter 12) for the flow of a non-viscous compressible fluid is given by

$$\frac{\partial^2 \phi}{\partial t^2} + \frac{\partial}{\partial t}(V_x^2 + V_y^2) + (V_x^2 - c^2)\frac{\partial^2 \phi}{\partial x^2} + (V_y^2 - c^2)\frac{\partial^2 \phi}{\partial y^2} + 2V_x V_y \frac{\partial^2 \phi}{\partial x \partial y} = 0$$

where c is the speed of sound in the gas. Non-dimensionalize this equation using characteristic quantities V_0, L_0, t_0 and c_0 (speed of sound at inlet conditions) and, thus, obtain the relevant dimensionless parameters for similarity.

9.6 *Centrifuge*: A centrifuge is a commonly used device for pollution-control. A particle-laden liquid (of density ρ and viscosity μ) flows upwards in a rotating cylindrical container as shown. Due to the centrifugal and drag forces, the solid particles move towards the wall and the relatively particle-free liquid flows out at the top. The equations of motion for a spherical particle of density ρ_s and diameter D which enters the centrifuge at $z = 0$, $r = R_i$, can be written as:

$$(\Delta\rho)\omega^2 r - \frac{18\mu}{D^2}\frac{dr}{dt} - \rho_s \frac{d^2 r}{dt^2} = 0$$

$$\frac{2\mathring{Q}(R_2^2 - r^2)}{\pi(R_2^2 - R_1^2)^2} - \frac{dz}{dt} = 0$$

with the initial condition $z = 0$ and $r = R_i$ at $t = 0$. In these equations, $\mathring{Q}$ is the flow rate and $\Delta\rho$ is the density difference, $\rho_s - \rho$. Non-dimensionalize the above equations and show that the trajectory of such a particle can be written as

$$r^* \equiv \frac{r}{R_2} = r^*\left\{\frac{z}{l}, \frac{R_i}{R_2}, \frac{R_1}{R_2}, \frac{\Delta\rho}{\rho_s}, \frac{\rho_s \omega D^2}{\mu}, \frac{\mathring{Q}}{\omega L R_2^2}\right\}$$

We define an $R_{i,cr}$ such that the particles at $R_{i,cr}$ at $z = 0$ and $t = 0$ hit R_2 at $z = L$. Then all particles with $R_i > R_{i,cr}$ are collected at the wall. For removing a given percentage of particles, $R_{i,cr}$ must be a fixed fraction of R_2. Show that the flow rate $\mathring{Q}$ that results in the required removal rate is given by the following non-dimensional relation:

$$\frac{\mathring{Q}}{\omega L R_2^2} = \Im\left\{\frac{\Delta\rho}{\rho_s}, \frac{\rho_s \omega D^2}{\mu}, \frac{R_1}{R_2}\right\}$$

An often-used expression for 50-per cent separation is

$$\frac{\mathring{Q}}{\omega L R_2^2} = \frac{\Delta\rho}{\rho_s}\frac{\rho_s \omega D^2}{\mu}\frac{2\pi(1 - R_1^2/R_2^2)}{\ln\{2/(1 - R_1^2/R_2^2)\}}$$

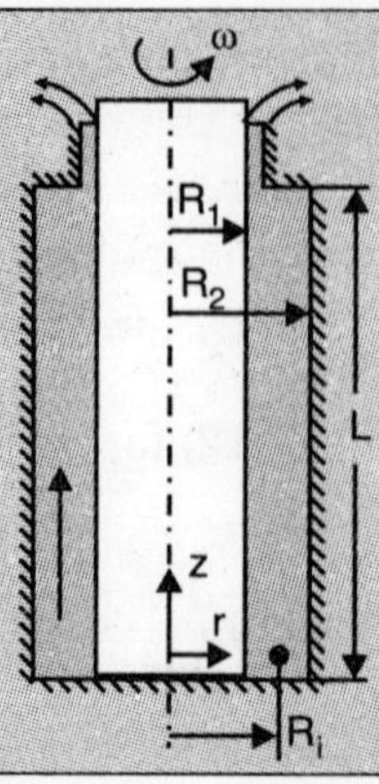

9.7 Consider the simple 2-D case where only two forces are present, the viscosity force, $\mathbf{F}_\mu$, and the gravity force, $\mathbf{F}_g$. Assume that these are perpendicular to each other. The resultant of these is the net (inertial) force, $\mathbf{F}$, acting at an angle, α, to $\mathbf{F}_\mu$. Deduce that for force polygon similarity between the model and the prototype (Hint: α must be the same in the model and the prototype):

$$k_F = k_\mu = k_g$$

9.8 Consider two geometrically similar differential fluid elements at a pair of homologous points in the model and the prototype. They are acted upon (Lagrangian approach) by net forces, $\mathbf{F}_p$ and $\mathbf{F}_m$, and experience accelerations, $\mathbf{a}_p$ and $\mathbf{a}_m$, respectively. Show that

$$\mathbf{a}_p/\mathbf{a}_m = k_F/(k_\rho\, k_L{}^3)$$

Hence deduce that if k_F, k_ρ, and k_L are the same for all pairs of homologous points, A, B, . . .

$$\left(\frac{a_p}{a_m}\right)_A = \left(\frac{a_p}{a_m}\right)_B = \ldots \equiv k_a$$

where k_a is the same for all such points. It may similarly be shown that

$$\left(\frac{V_p}{V_m}\right)_A = \left(\frac{V_p}{V_m}\right)_B = \ldots \equiv k_V$$

9.9 A 150 m long ocean liner is to be designed to travel at 25 knots (1 knot = 1.854 kmph). A 1:30 scale model of the ship is tested in a towing tank using sea water. At what speed should the model be tested assuming:

(a) wave-resistance, i.e., Fr similarity,

(b) viscous-resistance similarity,

(c) surface tension similarity?

Which of these three should normally be used? Note that these three similarities cannot be achieved simultaneously if the same fluid is used in the model as in the prototype.

9.10 A 1:25 scale model is made of the spillway of a dam. It is tested using an average water velocity of 0.5 m/s at the top and a volume flow rate of 0.075 m^3/s. Compute the velocity for the prototype at which similarity is ensured. Under these conditions, obtain the volume flow rate and the force acting on a sluice gate, if the corresponding force is measured as 2 N in the model test. If the characteristic length of the actual spillway is 15 m, obtain the Weber number for the model (σ for water = 0.07 N/m) and, thus, show that surface tension forces are insignificant even in the model test.

9.11 A propeller for a boat is to be designed using results from a 1:50 scale model test. The power generated by the model at 900 RPM is measured as 2 mW. If Froude numbers are matched (since the propeller will be operating close to the free surface), to what rotational speed of the prototype does the model test correspond? What will be the power delivered by the prototype at this speed?

9.12 An airplane is to fly at 400 kmph. An engineer suggests testing a 1:25 scale model of the plane in a wind tunnel using air under similar conditions. Is his idea feasible?

9.13 When a torpedo is fired from a submarine in addition to the drag force, it experiences a lift force as well as a moment about the *A-A* axis called the yawing moment (the yawing moment acts due to a slight asymmetry of the body). For a 1:15 scale model of a torpedo moving at 20 m/s in water, the yawing moment is measured as 10 Nm. What is the velocity of the prototype for which this test is appropriate? What will be the yawing moment on the prototype moving at that velocity? Neglect differences in the properties of water due to salinity (how is the rotation of the torpedo due to this moment avoided?). Note that the prototype velocity required for full similarity is very small. In fact, for the large Re encountered in such flows, the dependence on Re is negligible and one need not worry about Re matching. The test results are, therefore, valid for any 'high enough' prototype velocity.

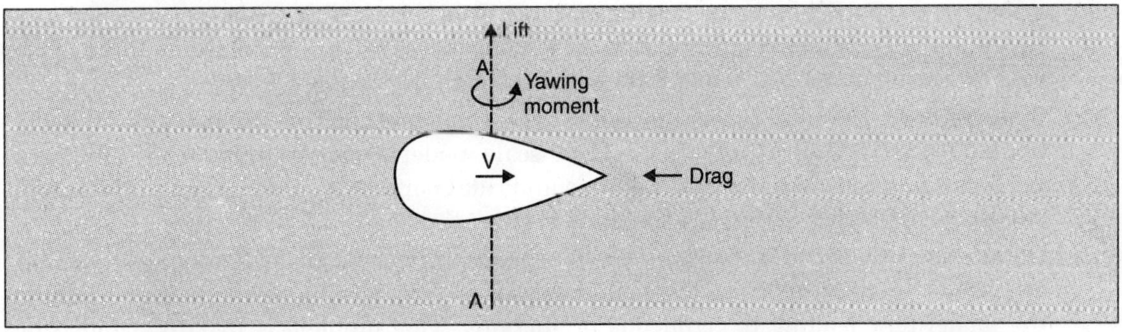

9.14 The drag experienced by a fully submerged mine because of water currents is to be estimated by tests on a 1:5 scale geometrically similar model in a wind tunnel. If the tidal current is estimated to be 4 kmph, what air velocity should be used in the model tests to ensure similarity? Under such conditions, the drag on the model is measured as 10 N. What will be the drag experienced by the mine?

9.15 In order to design a horizontal piping system for supplying water to a small plant, the pumping requirements are to be estimated by carrying out tests with air on a 1:8 scale model. The diameter of the pipe in the model is 3 cm at the supply point and the velocity there is 150 m/s. What would be the velocity in the prototype at the supply point? If the head loss in the model is measured as 1 m of mercury, obtain the head loss for the prototype in meters of water. [*Hint*: Convert the head loss into a pressure loss. Also, use the manometer equation]

9.16 The pressure drop in an orifice meter (Fig. 8.15) with oil ($\nu = 10^{-4}$ m^2/s, $\rho = 850$ kg/m^3) flowing at an upstream velocity of 1.5 m/s is to be estimated from studies on a 1:5 scale model using water. At what speed must the water flow to ensure similarity? If the pressure drop in the model is measured as 60 N/m^2, what will be the pressure drop in the prototype?

9.17 A mixing tank is to be designed using experimental data on a 1:5 scale model. Efficient mixing is observed in the model with oil (of $\nu = 10^{-3}$ m^2/s and $\rho = 850$ kg/m^3) at 2,000 RPM. At what rotational speed would you expect the prototype fluid (of $\nu = 10^{-6}$ m^2/s and $\rho = 10^3$ kg/m^3) to mix as efficiently? What will be the ratio of the power required in the two cases? Show that the mixing-time scale factor is very large and therefore this method of scaling-up is not suitable.

9.18 A solid sphere of diameter 10 cm moves in water at 5 m/s. The drag force experienced by it is 15 N. What will be the velocity of a *hollow* spherical shell of diameter 5 m moving in air in order to ensure similarity? What will be the drag experienced by it? Will your answers change if instead of a hollow shell, a solid sphere of the same size were to move in air?

9.19 Human vocal chords vibrate at about 150 Hz when exhaled air from the lungs flows past them at an average velocity of about 10 m/s. An enlarged 10:1 model of the chords is constructed and made to oscillate in water. What should be the mean velocity of water and the frequency of vibration in the model to duplicate the fluid-mechanical behaviour? (*Hint*: see Example 9.4)

9.20 The singing of a 2 cm diameter power cable is to be modelled using 0.5 cm diameter cylinders resonating in water flowing at 2 m/s. It is observed that vortices are shed at a frequency of 80 Hz for the model flow. If viscous forces are unimportant, will the cable sing at a wind speed of 1.5 m/s if its natural frequency is about 5 Hz?

9.21 In an estuary, the tidal current is found to be 1 m/s and the time period of the tides to be 12.5 hrs. In order to study this, a 1:400 scale model of the estuary is made and tides are generated in it. What current speeds and tidal periods should be used in the model to ensure similarity? (*Hint*: see Example 9.4)

Obtain the ratio of the Weber numbers for the prototype and model flows and show that even when the surface tension effects in the prototype are unimportant, they may dominate the model flow leading to erroneous predictions. This problem is usually avoided by resorting to relaxation of complete geometric similarity, something beyond the scope of this text.

9.22 A supersonic aircraft is to be designed to fly at 650 m/s at altitudes of about 10 km, where the temperature and pressure are 220 K and 25 kPa, respectively. If viscous effects are not very important, find the wind speed to be used for testing a 1:15 scale model using air under atmospheric conditions. Use the ideal gas law. Under these conditions the pitching moment is measured to be 225 Nm on the model. What will be the pitching moment on the prototype?

9.23 An aircraft travelling at speeds very close to the speed of sound, experiences severe vibrations. This is called transonic buffeting. The frequency and magnitude of the aerodynamic forces governing this phenomenon are controlled by inertial and compressibility forces. The natural structural frequencies of the aircraft are, therefore, sought to be kept away from these frequencies. In order to predict these frequencies, scale model test are undertaken. What should be the scale factors for speed and the buffeting frequencies, in terms of the scale factors for length and material properties?

If the onset of buffeting is observed at a certain angle of attack in the model, what should be the corresponding angle for the prototype?

9.24 Show that the modelling parameter for the rise of a liquid in a vertical capillary dipped in it is $\Pi = \rho g L^2/\sigma$. Note that this rule precludes working with the same fluid on scale models. However, the fact that the surface tension forces depend only on the diameter of the tube may be used and therefore, one may relax complete geometric similarity and use different length scale factors for diameter and height, k_D and k_L, respectively. Show that $k_L = 1/k_D$.

9.25 In order to estimate the drag on a parachute descending in air, a 1:80 scale model is tested in water flowing at 5 m/s. The viscous forces are not important in this flow since most of the drag is due to reduced pressure in a very large wake (see Sec. 1.7). Obtain the drag-force scale factor in terms of the velocity scale factor. If the drag on the model is measured as 2000 N, estimate the steady rate at which a parachutist (of total buoyant weight 600 N) will descend in air.

9.26 A spacecraft re-enters the earth's atmosphere and splashes in the sea. It is desired to study its behaviour using a 1:12 scale model. If the spacecraft touches down at 50 m/s, find the corresponding velocity to be used in model tests. Assume that the major effect is that associated with the splashing of water. If under these conditions, the model experiences a force of 10 N, what is the force experienced by the prototype?

9.27 A siphon operates such that cavitation occurs at its topmost point. A model test is carried out under sub-atmospheric pressures on a geometrically similar 1:15 scale model. What pressure must be used in the test to duplicate cavitation characteristics? The temperature of the water in both cases is 30°C (at which the vapour pressure = 4.24 kPa). What must be the scale factor for the velocity of the water?

9.28 Model tests of gas turbines are usually carried out assuming that inertial and pressure forces dominate the flow. If the two gases used in the prototype and model follow the ideal gas law, obtain the scale factors for the (a) RPM, (b) mass flow rate, (c) torque, and (d) specific work (i.e., rate of work per unit mass flow). Express your answers in terms of k_p, k_L, k_T and k_M, where T and M represent the temperature of the gas and its molecular weight, respectively.

9.29 The unsteady motion of a fluid over an infinite flat plate moved impulsively at velocity V_0 was considered in Prob. 6.37. The flow in this case is determined only by unsteady and viscous forces. Show using similarity arguments that the *penetration depth* must vary like $\sqrt{\nu t}$.

9.30 A rocket accelerates upwards with a constant acceleration **a**. There is considerable sloshing of the liquid fuel in the tank during take off. Care must be taken to avoid resonance between the sloshing frequency and the natural frequency of the rocket structure. If the sloshing of the fuel is determined by the *effective* gravity and inertial forces, what will be the scale factor for the sloshing frequency? Note that the strength of the effective gravity field is **a** + **g**.

CHAPTER 9

9.31 A volume $\forall$ of oil is spilt on the sea. The oil spreads out with time as a slick. The spreading action is controlled primarily by the viscous forces the seawater applies on the slick, inertial (unsteady) forces, and the buoyant weight of the slick. The surface tension forces are unimportant for the first several days. Using $\forall^{1/3}$ as a characteristic length parameter, show that the Π-numbers which determine the spread of the oil in these initial stages are $\rho_s \forall^{1/3}/g\, t^2\, \Delta\rho$ and $\rho_s \forall^{2/3}/\mu_w t$, where the subscript *s* denotes the slick, *w* the sea-water, and $\Delta\rho$ is the density difference.

9.32 When a river discharges into a tide-less sea, the seawater penetrates inland establishing a saline-wedge as shown. The extent of penetration is determined by viscous, buoyancy and inertial forces associated with the river-water. Show that the relevant modelling parameters are

$$\Pi_1 = \rho_w HV/\mu_w$$

and
$$\Pi_2 = \frac{V}{\sqrt{\dfrac{\Delta\rho}{\rho_w} gH}}.$$

The parameter Π_2 is called the *densimetric Froude number* and occurs in all situations where the buoyant weight is one of the controlling forces, such as in natural-convection heat transfer.

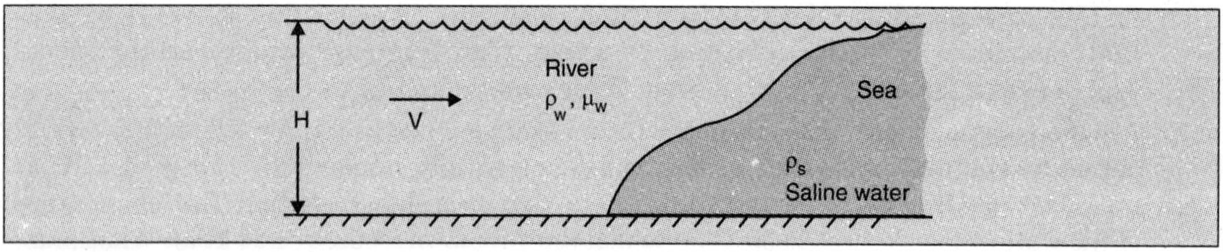

9.33 When a steel ball falls in water, or a helium balloon rises in air, a steady velocity, termed as the *terminal velocity* is attained when the drag force becomes equal to the *net* weight. In modelling this phenomenon, the following forces are of interest

(a)	gravity force net of bouyancy force	~	$(\rho_s - \rho_f)\, L^3 g$
(b)	viscous force	~	μLV
(c)	inertia of fluid (convective)	~	$\rho_f\, L^2 V^2$
(d)	inertia of fluid (unsteady)	~	$\rho_f\, L^3 V/t$
(e)	inertia of solid sphere	~	$\rho_s\, L^3 V/t$

A true similarity would require, among other things, that the density be scaled by a constant k_ρ, i.e., $\rho_{f,m}/\rho_{f,p} = \rho_{s,m}/\rho_{s,p}$. This scaling imposes a formidable restriction on the modelling experiments. If, however, we are interested in modelling just the steady portion of the motion, the fourth and fifth forces are not relevant, and a relaxation of a constant k_ρ permits modelling.

It is desired to estimate the terminal velocity with which a 10 m diameter helium-filled balloon rises in air by performing model experiments on a steel ball falling in water.

What should be the diameter of the steel ball to ensure similarity and what, then, will be the ratio of the terminal velocities? $\rho_{He} = 0.06$ kg/m^3 and $\rho_{steel} = 7.6 \times 10^3$ kg/m^3.

9.34 In Prob. 9.33, if the Reynolds number is so low that inertial forces of the fluid are negligible compared to viscous forces (this usually happens for very small spheres such as dust particles, mist droplets, etc.), and only one pi-number results. Obtain this. This means that there is no restriction on the length scale, and this pi-number serves as the prediction rule.

9.35 The pith ball supported by a jet of air described in Prob. 7.59 is to model a steel ball supported by a jet of water. What should be the scale factors for length and velocity? (*Hint*: see Prob. 9.33.) Density of pith is 300 kg/m^3.

9.36 An axial flow air compressor having a rotor diameter of 75 cm and an angular speed of 6000 RPM is to be designed to deliver a flow rate of 0.8 m^3/s. Experiments are conducted on a 1:2 scale model at 600 RPM with water. What should be the flow rate in the model? What will be the pressure developed and the power consumed in the prototype if the model values are 180 Pa and 150 W respectively?

9.37 In a centrifugal pump, pressure, inertial and centrifugal forces play a significant role. Water flows through a pump operating at 1800 RPM and generates a pressure rise of 50 kPa. If instead of water, air flows through the same pump operating at the same RPM, obtain the pressure rise. Thus see the necessity to *prime* the pump (i.e., to fill the pump casing with water before starting it). Compute the scale factor for the power consumption.

9.38 An open cylindrical tank of water is to be rotated about its vertical axis at 100 RPM. It is proposed to study the shape of the free surface of the water at steady state by tests on a one-fifth scale model using mercury. What should be the rate of rotation of the mercury tank? When the tank is rotated, the liquid in it accelerates to this steady flow due to viscous forces. What is the time required for the prototype to reach steady state if the corresponding time required by the model is 2 min? $\nu_{Hg} = 1.15 \times 10^{-7}$ m^2/s.

9.39 A 1:10 scale model of a rectangular weir has a flow rate of 0.1 m^3 when the water stands 10 cm over the crest. What will be the flow rate in the prototype when the water stands 0.5 m above its crest? Note that geometric similarity is being violated. (*Hint*: Assume the flow over the weir to be 2-*D*).

9.40 The rate of draining of a liquid from a large container can be controlled by the length of the exit tube attached to it. As the velocity V_c through the drainpipe increases, the

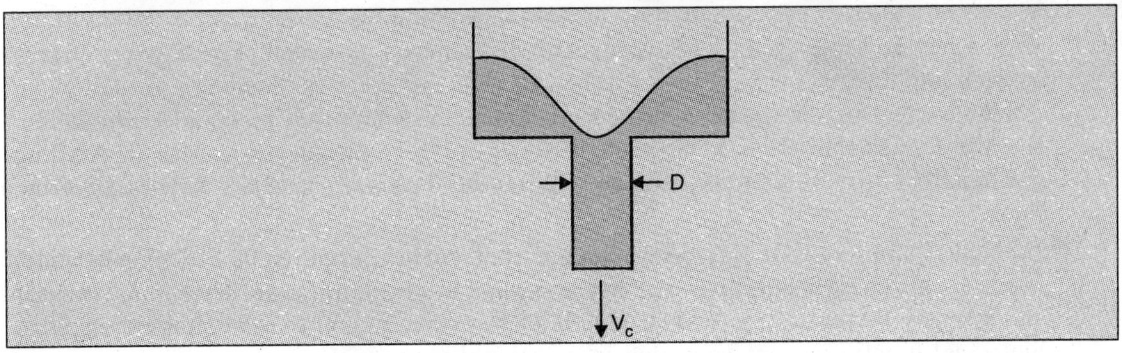

liquid level above it dips forming a cusp as shown. The maximum rate of drainage is reached when the cusp touches the drain level and any increase in velocity thereafter decreases the flow rate since the area available for water flow is decreased and the tube sucks in air. We want to conduct model tests to determine the optimum discharge velocity. What should be the velocity scale factor in terms of the length scale factor, k_L? Should the discharge tube length also be scaled by k_L? Neglect viscous effects.

9.41 Water seeps through a 30 cm long bed of sand by capillary action. At such speeds, the flow is controlled primarily by surface tension and viscous forces. This bed is modelled by a geometrically similar bed where the particle diameter is only one-fourth of the previous value. Obtain the scale factors for velocity and for the time taken by water to seep through the entire bed length. If the model bed-length is increased to 30 cm, what is the new scale factor for the time to seep through?

9.42 *Capillary inversion:* The water issuing out of an elliptical orifice has a tendency to assume a circular (i.e., a minimum perimeter) cross-section due to surface tension. In this process, however, the surface energy is transformed into kinetic energy with the result that the motion does not cease at the minimum perimeter stage but overshoots, giving rise to a series of standing waves along the surface as shown. This phenomenon is termed as the capillary inversion of a liquid jet. The jet from a small elliptic orifice is observed to undergo two inversion cycles in a given length when the average velocity is 3 m/s. If the orifice dimensions are doubled, at what velocity should the same fluid be discharged in order to obtain a geometrically similar jet? What will be the scale factor for the water level h in the tank? Note that it may not be the same as for the jet.

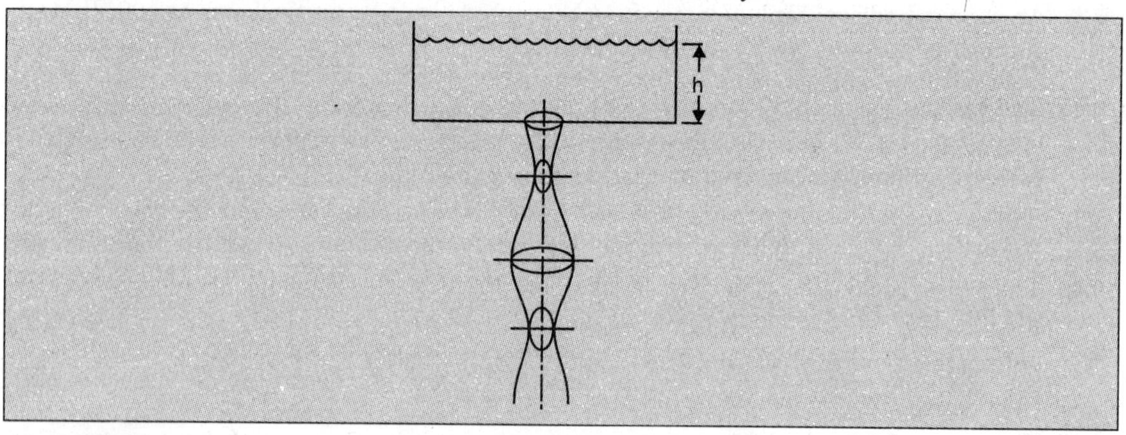

9.43 Under certain conditions, the flow through an orifice is governed primarily by surface tension and inertial forces. What will be the scale factor for the heads required by two fluids flowing through the same orifice, one having a surface tension equal to one-half of the other, such that the discharge coefficients in the two cases are identical? Assume the density of the fluids to be the same. What will be the ratio of the flow rates of the fluids?

9.44 *Biological similarity:* It has been claimed that various organisms and physiological mechanisms operate similarly and are governed by the same laws. Therefore, various species may be treated as scaled models of the same machine, and that the various biological functions must be scaled in a rational manner. Consider, for example, the metabolic rate in mammals that should depend on the rate of heat loss (proportional to the surface area). If all organs of mammals are assumed geometrically similar and made of the same material, $k_{\mathring{Q}} = k_L^2$, where $\mathring{Q}$ represents the rate of oxygen consumption. Show that $k_f = k_m^{-1/3}$, where f is the breathing frequency and m is the mass of the mammal. Similarly, if it is argued that the maximum stress level in the muscles is the same, irrespective of the scale, show that W, the weight-lifting capability of mammals is scaled by $k_W = k_m^{2/3}$. Confirm this using the following data for the 1979 world weight-lifting championships:

Weight class, kg	Total weight lifted, kg
52	242.5
56	267.5
60	282.5
67.5	332.5
75	345
82.5	370
90	380
100	385
110	410
Super heavy weight	430

What is the estimated weight of the super heavy weight champion?

9.45 The Lilliputians and the Brobdingnagians of *Gulliver's Travel* differed by a height ratio of approximately 1:100. Using scalings similar to those in Prob. 9.44 estimate their relative masses, frequencies of breathing, velocities of walking, and the maximum jumping heights. Which species would have more difficulty with surface tension of water, with viscosity of water, and with the earth's gravitational field?

9.46 The pressure drop for flow of an incompressible fluid through a rough-walled circular pipe is a function of its length L, the mean height ε of the surface irregularities, the volumetric flow rate $\mathring{Q}$, and the density ρ, and viscosity μ of the fluid. Express this functionality in dimensionless form.

9.47 The power $\mathring{W}$ required for mixing a liquid in a tank is a function of the diameter D of the stirrer, its RPM N, the density ρ, and viscosity μ of the liquid, the gravitational acceleration g, and the geometry of the tank. Non-dimensionalize this relationship.

9.48 The volumetric flow rate $\mathring{Q}$ of a liquid through a rectangular weir is related to the height H of the liquid over it, the width W of the weir, the gravitational acceleration g and the density ρ, and viscosity μ of the liquid. Non-dimensionalize this functional relationship.

9.49 The diameter d of the droplets issuing out of a spray nozzle, e.g., a flit pump, is a function of the velocity V of the jet, the diameter D of the nozzle and the density ρ, viscosity μ, and the surface tension σ of the liquid. Obtain the dimensionless parameters for this phenomenon.

9.50 The pressure drop Δp for the flow of a fluid through a porous bed of sand is a function of the average grain diameter D_p, the velocity V of the fluid, the length L of the bed, the density ρ_f and viscosity μ of the fluid, and the volumetric fraction of the void space ε (defined as the volume of free space/total volume). Obtain the non-dimensional relationship. Simplify your results using the experimental observation that Δp is directly proportional to L.

If you wish to design a large unit using experimental data from a 1:10 scale model with the same fluid, what test velocity (ratio) would you use for the model? What will be the scale factor for $\Delta p/L$?

9.51 The mass flow $\overset{\circ}{m}$ of a thin film of honey moving down a flat plate is a function of the density ρ, viscosity μ, gravitational acceleration g, width b, and thickness t of the film. Express this in dimensionless form. Simplify using the physical reasoning that $\overset{\circ}{m}$ should be directly proportional to b and g.

9.52 A cylindrical container having negligible weight and filled partially with a viscous liquid rolls down an inclined plane. After some time the container attains a steady velocity V, which is a function of the radius R of the cylinder, the viscosity μ of the liquid, the mass m of liquid per unit depth, and $g \sin\theta$, the along-the-plane component of the gravitational acceleration. Non-dimensionalize this functionality assuming that inertial forces are unimportant and show that

$$V = k\frac{mg\sin\theta}{\mu}.$$

9.53 The flow rate $\overset{\circ}{Q}$ per unit width of a river down an inclined terrain under fully developed laminar conditions is a function of the density ρ, viscosity μ, gravitational acceleration g, the angle of inclination α, and the depth h of the river. Inertial forces are unimportant for this flow. Non-dimensionalize this relationship and simplify it using the physical reasoning that $\overset{\circ}{Q}$ should vary linearly with $g \sin \alpha$.

9.54 An air bubble of volume $\forall$ rises up a vertical column of diameter D of a liquid at a constant velocity V. It is anticipated that V is a function of the density ρ, viscosity μ, and surface tension σ of the liquid, the gravitational acceleration g, the volume $\forall$ of the bubble, and the tube diameter D. Non-dimensionalize this functionality. If the surface tension forces on the bubble are very large, the bubble tends to stay at the same position. Which dimensionless group decides the importance of these forces?

10

SOME ENGINEERING APPLICATIONS - II

10.1 FLOW THROUGH PIPES

The steady, laminar, fully-developed flow through a circular pipe was analysed in Example 6.3 and the following relationship between the piezometric (or the non-gravitational) pressure gradient $\mathscr{P}'$ and the volume flow rate $\mathring{Q}$

$$\mathscr{P}' = \frac{-128\,\mu\mathring{Q}}{\pi D^4} = \frac{-32\,\mu V_{av}}{D^2} \qquad \text{...(10.1)}$$

was obtained where D is the diameter of the pipe and μ is the viscosity of the fluid.

As was noted in that example, the two conditions, namely, full development and laminar flow, make the scope of Eq. (10.1) very limited. A famous series of experiments conducted by the British engineer Osborne Reynolds in 1883 (similar to the one described in Sec. 1.8) showed that the flow remains laminar for very low velocities only. When a thin streak of dye is injected in a glass tube through which water is flowing at low velocities, there is very little spreading of the dye in the water. The dye filament flows as a thin, straight streak and small disturbances in the flow pattern die out quickly, i.e., the flow pattern is stable. As the flow rate is increased the dye filament develops a waviness before vigorous mixing takes place (see Fig. 1.26). The flow is said to be turbulent and is characterized by large scale irregular motion in all directions.

While laminar flow is dominated by viscous forces, turbulent flow is dominated by inertial forces associated with the large scale mixing (see Chapter 16). It stands to reason that in geometrically similar situations, the transition from laminar to turbulent flow should occur at a fixed ratio of inertial and viscous forces. Since the Reynolds number $(\rho VL/\mu)_c$ measures this ratio, the transition should be marked by a fixed value of the Reynolds number, $\text{Re} = \rho V_{av} D/\mu$, where V_{av} has been taken as the characteristic velocity and the diameter D as the characteristic length.

The value of the Reynolds number at which the transition from laminar to turbulent flow takes place is called the *critical Reynolds number*, Re_{cr}. If the Reynolds number is below 2,000, the flow is *always* laminar and any disturbance in the flow is quickly damped out. However, when Re is increased the flow may or may not be turbulent depending on the conditions at the inlet of the pipe and the level of vibration in the experimental set-up. If great care is taken in making the pipe walls very smooth, in suppressing all turbulence in the incoming fluid, and in isolating the experimental set-up from external vibrations, the transition can be delayed to as high a Reynolds number as 100,000. But for most engineering applications, a value of 2,300 represents a good choice for the critical Reynolds number. Thus, Eq. (10.1) should not be used for Re > 2,300. This represents a very serious limitation since an Re value of 2,300 is reached in a pipe of diameter 2 cm at water (at 20°C) velocities as low as 0.115 m/s and at air (at 20°C) velocities as low as 1.81 m/s. Most of the flow velocities encountered in engineering practice are larger than these values. In fact, optimum velocities (for minimum pumping and capital costs) in pipes range from about 1.8 to 2 m/s for liquids and from about 20 to 30 m/s for gases (see Prob. 10.23). Thus, laminar flows are relatively uncommon.

The second restriction on Eq. (10.1), namely, the flow being fully developed, is also quite difficult to meet. Near the entrance of the tube, the uniform velocity profile of the incoming flow is acted on by viscous stresses at the wall and the fluid slows down near the periphery (Fig. 10.1). This viscous effect 'penetrates' towards the axis as the flow proceeds down the pipe and the velocity profile changes with the flow direction. After a certain length a parabolic velocity profile is established and the flow is said to be *fully developed*. The length L_e over which the flow is *developing* from the uniform profile at the entrance to the parabolic profile of the fully developed laminar flow is termed as the *entrance length*. Eq. (10.1) can be applied safely only to those laminar flow situations in which the total pipe length is significantly larger than this entrance length.

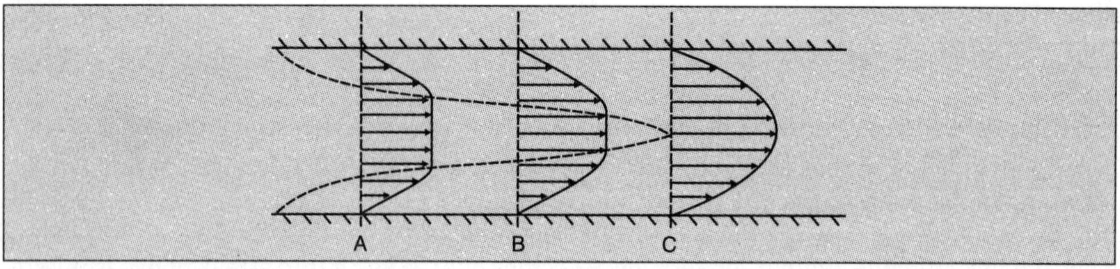

Fig. 10.1. Flow profiles in the entrance region of a pipe. Uniform velocity at entry point.

Clearly, the extent of this entrance region is controlled by the relative magnitudes of inertial and viscous forces since these two forces control the flow in this region. Therefore, the non-dimensional entrance length should be a function of the Reynolds number, or

$$\frac{L_e}{L_c} = \mathcal{F}\left(\frac{\rho V_c L_c}{\mu}\right)$$

where the subscript c denotes the characteristic value of the quantity involved. Conventionally, the diameter D is taken as L_c and the average velocity V_{av} as V_c, so that

$$\frac{L_e}{D} = \mathcal{F}\left(\frac{\rho V_{av} D}{\mu}\right) = \mathcal{F}(\text{Re}_D) \qquad ...(10.2)$$

where Re_D stands for the Reynolds number based on the diameter D.

Careful experimentation has established that in laminar flows L_e varies linearly with Re_D and the recommended correlation for laminar flow has been given by Langhaar as

$$\frac{L_e}{D} = 0.0575\,\mathrm{Re}_D \quad ...(10.3)$$

For example, when Re_D is 2,300 and the flow is kept laminar, we obtain $L_e = 138D$, which can easily be quite a significant portion of the total length of the pipe.

The recommended correlation for turbulent flow is

$$\frac{L_e}{D} = 4.4\,\mathrm{Re}_D^{0.167} \quad ...(10.4)$$

These lengths are considerably smaller. For example, when Re_D is 2,300 but the flow is turbulent, $L_e = 16.03D$ only. Thus, the entrance effect is negligible more often in turbulent flows than in laminar flows.

10.2 NON-DIMENSIONAL FORMULATION OF THE PIPE-FLOW PROBLEM

In general, the flow through a pipe is controlled by inertial, viscous and pressure forces. These three forces give two pi-numbers, typically the Reynolds number $(\rho VL/\mu)_c$ and the Euler number $(\rho V^2/\Delta p)_c$. Then, following the formulation of Sec. 9.4 it can be said that any non-dimensional dependent variable or parameter of the pipe-flow problem is a function of the Reynolds and Euler numbers and the geometry of the pipe expressed in non-dimensional form. The length L and diameter D are the obvious geometrical parameters. In addition, the French engineer M. Darcy showed experimentally in 1857 that the pressure drop in turbulent flows is also affected by the

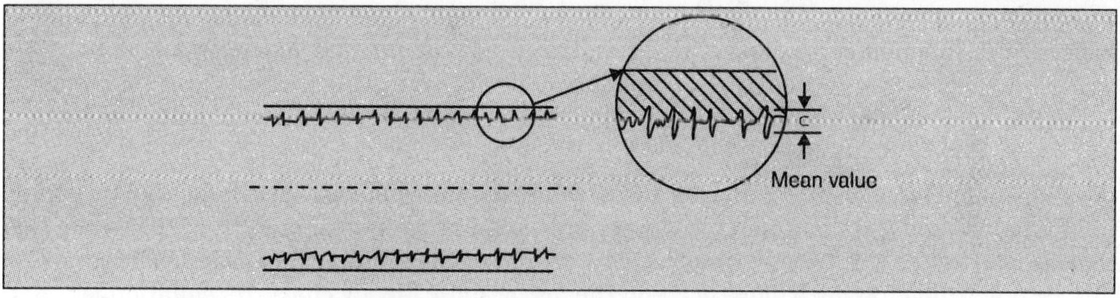

Fig. 10.2. Roughness elements at pipe walls.

roughness of the pipe walls. The wall roughness may be characterized by the mean height ε of the roughness element (Fig. 10.2). If the characteristic length is L_c, then the non-dimensional geometric parameters are L/L_c, D/L_c and ε/L_c, and any non-dimensional dependent variable is a function of

$$\mathrm{Re} = \rho V_c L_c/\mu, \mathrm{Eu} = \rho V_c^2/(\Delta p)_c, L/L_c, D/L_c \text{ and } \varepsilon/L_c$$

One of the most common problems encountered in pipe flows is to evaluate the piezometric pressure loss $\Delta\mathscr{P}$ for a given flow rate. Thus,

$$\frac{\Delta\mathscr{P}}{(\Delta p)_c} = \mathcal{F}\left(\frac{\rho V_c L_c}{\mu}, \frac{\rho V_c^2}{(\Delta p)_c}, \frac{L}{L_c}, \frac{D}{L_c}, \frac{\varepsilon}{L_c}\right) \quad ...(10.5)$$

In such problems, the diameter D is taken as the characteristic length L_c, and the average velocity V_{av} as V_c. There is no characteristic $(\Delta p)_c$ fixed by the independent parameters and its choice is arbitrary. It is conventionally set equal to $\frac{1}{2}\rho V_{av}^2$ (see Sec. 9.2.5) so that Eu becomes 2, a constant, and Eq. (10.5) simplifies to

$$\frac{\Delta \mathcal{P}}{\frac{1}{2}\rho V_{av}^2} = \mathcal{F}\left(\frac{\rho V_{av} D}{\mu}, \frac{L}{D}, \frac{\varepsilon}{D}\right) \quad ...(10.6)$$

Eq . (10.6) is valid for all flows whether laminar or turbulent, fully developed or not. If the pipe is very long compared to the entrance length L_e we may assume the flow to be fully developed along the entire length L of the pipe. Since the conditions do not change along the length of the pipe, the shear stress at the wall will be constant along the pipe length, and therefore, we expect the pressure drop per unit length to be constant. In other words, $\Delta\mathcal{P}$ should vary linearly with L. This is expressed by writing Eq. (10.6) as

$$\frac{\Delta \mathcal{P}}{\frac{1}{2}\rho V_{av}^2} = \frac{L}{D}\mathcal{F}\left(\text{Re}_D, \frac{\varepsilon}{D}\right) \quad ...(10.7)$$

The dimensionless function $\mathcal{F}(\text{Re}_D, \varepsilon/D)$ is termed as the *Darcy friction factor f* and Eq. (10.7) is usually written as

$$\Delta\mathcal{P} = f\frac{1}{2}\rho V_{av}^2 \frac{L}{D} \quad ...(10.8)$$

where f is a function of Re_D and ε/D. Thus, the head loss in a pipe of length L is

$$h_l = \frac{\Delta\mathcal{P}}{\rho g} = f\frac{V_{av}^2}{2g}\frac{L}{D} \quad ...(10.9)$$

The functional relation of f with Re_D and ε/D has been established experimentally by various investigators and the combined results are presented as a log-log graph (Fig. 10.3) called the *Moody chart* after L.F. Moody who plotted it in 1944. The laminar flow region is represented by a straight line having a slope of –1 on the logarithmic Moody chart. Its equation is

$$f = \frac{64}{\text{Re}} \quad ...(10.10)$$

It can be shown easily that this is consistent with Eq. (10.1). The shaded region in Fig. 10.3 represents the range in which the transition occurs from laminar to turbulent flow and as such, there is no reliable value of f in this range of Re_D from about 2,000 to 4,000. In the turbulent regime, the friction factor decreases with Re_D at first, and then becomes constant when the flow is *fully* turbulent. In turbulent flows, the large scale mixing of the fluid associated with turbulence results in inertial forces dominating the viscous forces everywhere except very near the wall where a thin viscous sublayer may be assumed (see Sec. 16.5). When the wall is rough, the viscous forces become small. It is in this regime that the Re_D dependence vanishes and f

TRANSPORTATION OF FLUIDS—ENERGY RELATIONS

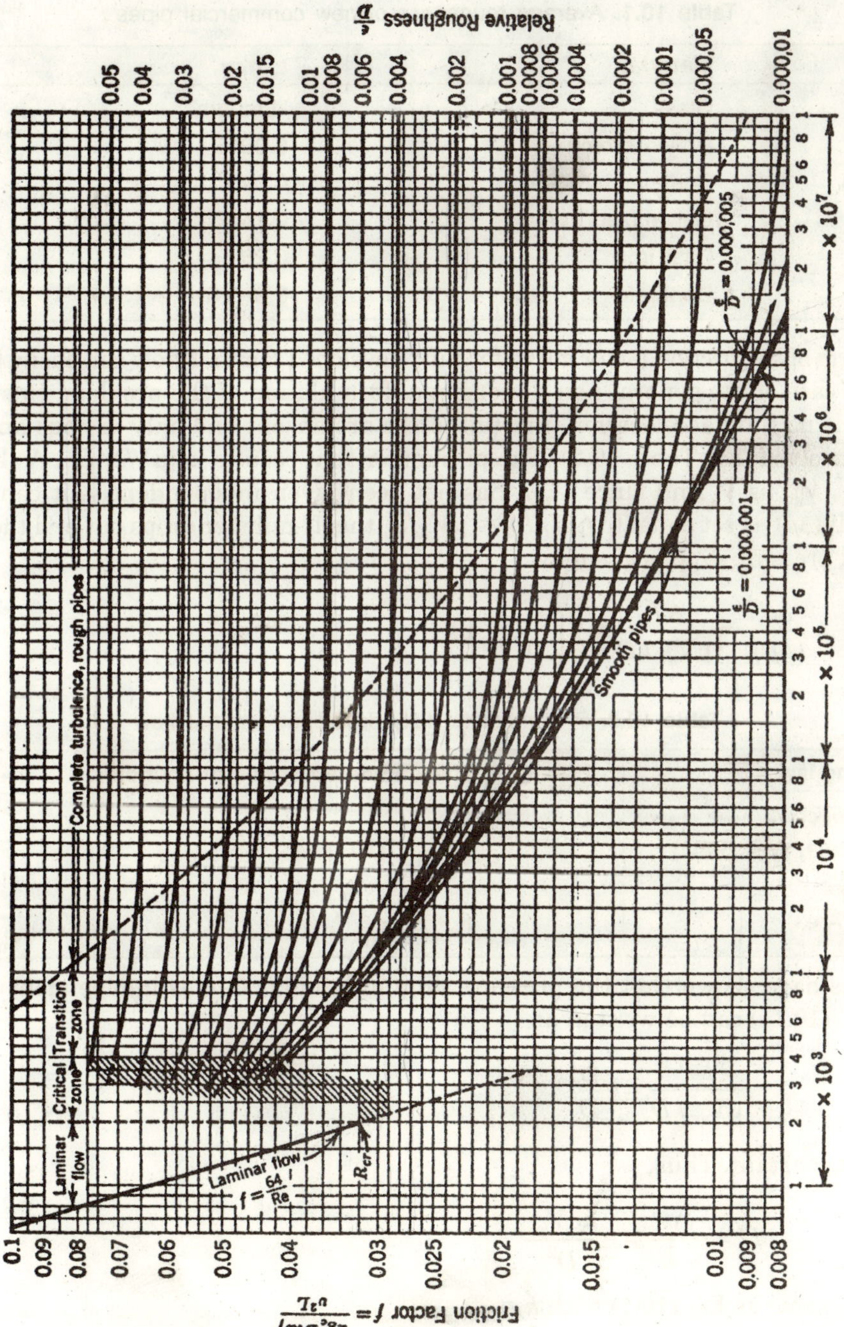

Fig. 10.3. Darcy friction factor as a function of Reynolds number with relative roughness as a parameter.

assumes a constant value depending upon ε/D alone. Table 10.1 gives the average values of the pipe-roughness factors for some commercial (new) pipes.

Table 10.1. Average roughness of new commercial pipes

Material	**ε, mm**
Glass	"Smooth"
Drawn tubing	0.0015
Commercial steel	0.046
Galvanized iron	0.15
Cast iron	0.26
Concrete	0.3–3.0

The form of the friction factor curve for fully developed laminar flow can easily be obtained as a special case of the general non-dimensional formulation of the problem. In such a flow, since there is no fluid acceleration, inertial forces are absent and only viscous and pressure forces are involved (see Table 10.2). The pi-number that results is $(\mu V/\Delta p L)_c$. As before, D is taken as L_c, V_{av} as V_c and since $(\Delta p)_c$ has not been given independently, it can be chosen arbitrarily. If $(\Delta p)_c$ is set equal to $(\mu V/L)_c = \mu V_{av}/D$, this Π-number drops out and then the non-dimensional pressure drop is given by

$$\frac{\Delta \mathcal{P}}{(\Delta p)_c} = \frac{\Delta \mathcal{P}}{\mu V_{av}/D} = \mathcal{F}\left(\frac{\varepsilon}{D}, \frac{L}{D}\right) \quad ...(10.11)$$

Table 10.2. Pi-number for fully developed laminar flow

Governing law	**Relation between characteristic quantities**	**Π-number**
Viscous Friction		
$F_\mu = \mu$ (Area) (Vel. Grad.)	$F_\mu \sim \mu L^2(V/L)$	
Pressure		
$F_p = \Delta p$ (Area)	$F_p \sim \Delta p\, L^2$	$\frac{\mu V}{(\Delta p)L}$

Experiments have shown that the pressure drop in laminar flow is independent of ε, and if $L \gg L_e$., Eq. (10.11) can be written as

$$\frac{\Delta \mathcal{P}}{\mu V_{av}/D} = k\frac{L}{D} \quad ...(10.12)$$

where k is a constant. Thus,

$$\mathcal{P}' = -\frac{\Delta \mathcal{P}}{L} = -\frac{k\mu V_{av}}{D^2} \quad ...(10.13)$$

which is the same as Eq. (10.1) with $k = 32$.

Similarly, the other limit of fully turbulent flow in rough pipes is obtained by noting that when the viscous forces become small they are dominated everywhere by inertial forces associated with the turbulent motion, and the only Π-number governing the flow is the Euler number. By choosing $(\Delta p)_c = \frac{1}{2}\rho V_c^2$ as before, and invoking full development, we get

$$\frac{\Delta \mathscr{P}}{\frac{1}{2}\rho V_{av}^2} = \frac{L}{D} f\left(\frac{\varepsilon}{D}\right) \quad ...(10.14)$$

Thus, for very large Re_D the friction factor f is constant for fixed values of ε/D, a result consistent with the Moody chart.

One consequence of the constant value of f is that the pressure drop in fully developed flow through a rough pipe varies as the square of the velocity, instead of varying directly as the velocity as in laminar flow (Fig. 10.4). In between the fully laminar and fully turbulent flows, f decreases slowly. This range of the Moody chart is correlated by the Colebrook formula:

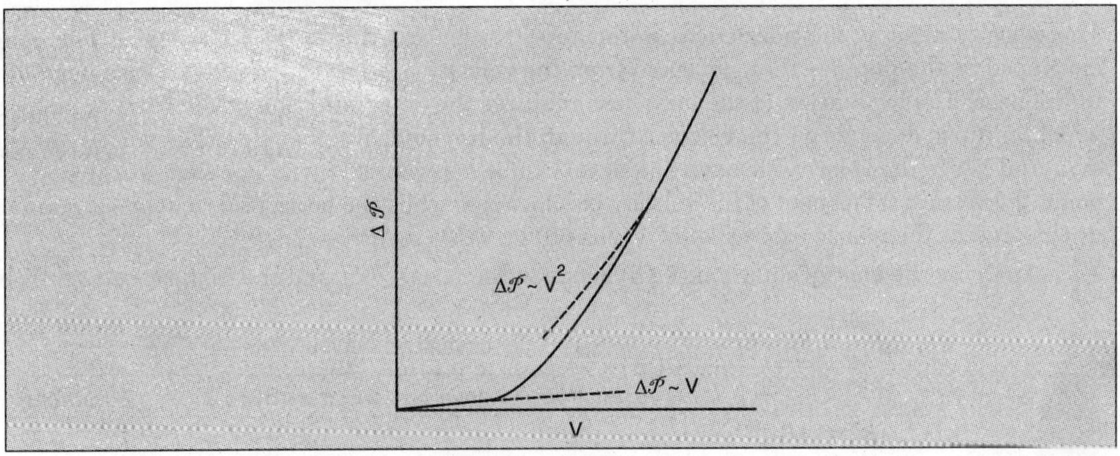

Fig. 10.4. Δ𝒫 *vs.* velocity.

$$\frac{1}{\sqrt{f}} = -2.0 \log_{10}\left[\frac{\varepsilon/D}{3.7} + \frac{2.51}{Re_D\sqrt{f}}\right] \quad ...(10.15)$$

This formula is valid for $Re_D > 4000$ and is widely used in computer calculations. For smooth pipes, another commonly used correlation is

$$f = \frac{0.184}{Re_D^{0.2}} \quad ...(10.16)$$

which is valid in the range $5 \times 10^4 < Re_D < 5 \times 10^5$.

Example 10.1. A fireman with an 8 cm dia. hose directs a jet of water at 60° to the horizontal so as to reach a fire 10 m above the ground (Fig. 10.5). If the nozzle and the hose diameters are the same and the length of the hose is 30 m, what should be the head developed by the pump? Assume the equivalent roughness of the hose to be $\varepsilon = 0.0008$ m.

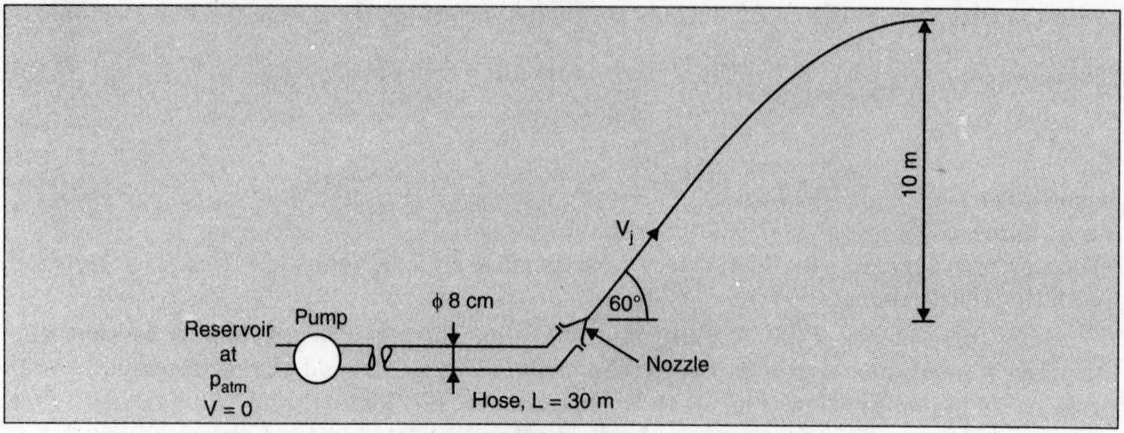

Fig. 10.5. Pump, hose and nozzle assembly for Example 10.1.

The velocity head at the nozzle exit determines the height the water jet reaches. The head developed by the pump is then obtained from the velocity head at the nozzle and the head lost in the hose. The evaluation of the head loss requires the determination of the friction factor f, which, in turn, depends on the velocity through the Reynolds number Re_D. The velocity at the exit (and hence the velocity through the hose) can be determined by noting that, at the highest point, the vertical component of the velocity becomes zero while the horizontal component remains unchanged at the nozzle exit value of $V_j \cos 60°$ or $V_j/2$ (see Prob. 7.50).

Applying the energy equation (7.13) between the nozzle exit and the highest point, we get

$$\frac{V_j^2}{2g} + \frac{p_{atm}}{\rho g} = \frac{(V_j/2)^2}{2g} + 10(\text{m}) + \frac{p_{atm}}{\rho g}$$

or $$V_j = 16.17 \text{ m/s}$$

The head loss in the pipe is given by

$$h_l = f \frac{V^2}{2g} \frac{L}{D}$$

where f is the friction factor, a function of $Re_D = \rho VD/\mu$ and the relative roughness ε/D. For the data given

$$Re_D = \frac{10^3\left(\frac{\text{kg}}{\text{m}^3}\right)\times 16.17\left(\frac{\text{m}}{\text{s}}\right)\times 0.08(\text{m})}{10^{-3}\left(\frac{\text{N s}}{\text{m}^2}\right)} = 1.294\times 10^6$$

and

$$\frac{\varepsilon}{D} = \frac{0.0008}{0.08} = 0.01$$

The corresponding friction factor f is 0.038 from Fig. 10.3. Thus, from Eq. (10.9)

$$h_l = 0.038 \times \frac{(16.17)^2 \left(\frac{\text{m}}{\text{s}}\right)^2}{2 \times 9.81 \left(\frac{\text{m}}{\text{s}^2}\right)} \times \frac{30(\text{m})}{0.08(\text{m})} = 189.9 \text{ m}$$

or a head of 189.9 m is lost within the hose. To determine the total head $-h_s$ developed by the pump, apply the energy Eq. (7,13) between the pump inlet and the nozzle exit to obtain

$$\frac{p_{\text{atm}}}{\rho g} + 0 + 0 = \frac{p_{\text{atm}}}{\rho g} + 0 + \frac{V_j^2}{2g} - h_s + h_l$$

or

$$-h_s = \frac{V_j^2}{2g} + h_l = \frac{(16.17)^2 (\text{m/s}^2)}{2 \times 9.81 (\text{m/s}^2)} + 189.9(\text{m}) = 203.2 \text{ m}$$

In other words, the pump needs to develop a head of 203.2 m, a ridiculously high value for throwing water to 10 m only. This is because the velocity in the hose is very high. To reduce this value, we must reduce the velocity through the hose, still keeping the velocity at the jet exit as 16.17 m/s so that it reaches a 10 m height. The only way of achieving this is to use a reducing nozzle at the end of the hose. If a nozzle of 2.5 cm exit diameter is used, then the velocity V_h required in the hose is

$$V_h = 16.17 \left(\frac{\text{m}}{\text{s}}\right) \times \left(\frac{2.5}{8}\right)^2 = 1.58 \text{ m/s}$$

The corresponding Re_D through the hose is then

$$\text{Re}_D = \frac{10^3 (\text{kg/m}^3) \times 1.58 (\text{m/s}) \times 0.08 (\text{m})}{10^{-3} (\text{Ns/m}^2)} = 1.264 \times 10^5$$

The value of f corresponding to this Re_D and $\varepsilon/D = 0.01$ is unchanged at 0.038 (since the flow is fully turbulent) and h_l is now

$$h_l = 0.038 \times \frac{(1.58)^2 (\text{m/s})^2}{2 \times 9.81 (\text{m/s}^2)} \times \frac{30 (\text{m})}{0.08 (\text{m})} = 1.81 \text{ m}$$

Thc total head developed by the pump is now

$$-h_s = \frac{V_j^2}{2g} + h_l = \frac{(16.17)^2 (\text{m/s})^2}{2 \times 9.81 (\text{m/s}^2)} + 1.81(\text{m}) = 15.13 \text{ m}$$

which is a much more reasonable requirement. Thus, by fitting an 8 cm to 2.5 cm reducing nozzle, the pumping head required can be cut down from 203.2 m to a reasonable 15.13 m. In both these cases, the water jet reaches up to 10 m but the rate of water delivered at that height is much lower in the second case.

It is instructive to make one more comparison, namely, the head required when the hose diameter itself is reduced to 2.5 cm, and no reducer is used. The hose velocity in this case is the same as the jet velocity, but Re_D changes to 4.04×10^5. The relative roughness ε/D is now 0.032

and the value of f is 0.059 (Fig. 10.3.) Then $h_l = 943.53$ m. Thus, the use of a large diameter hose with a reducing nozzle is indisputably indicated.

Example 10.2. In Example 7.4 the flow rate out of a tank with a long outlet tube was calculated as $\mathring{Q} = A_0\sqrt{2g(H+h)}$ neglecting the viscous losses. This result is now modified to account for these. If the tube area A_0 is much smaller than the tank area A_t, the velocities in the tank are much smaller than in the tube, and the losses in the tank are still negligible. Apply the energy Eq. (7.13) with h_l given by Eq. (10.9) between points 1 and 2 shown in Fig. 10.6 to get

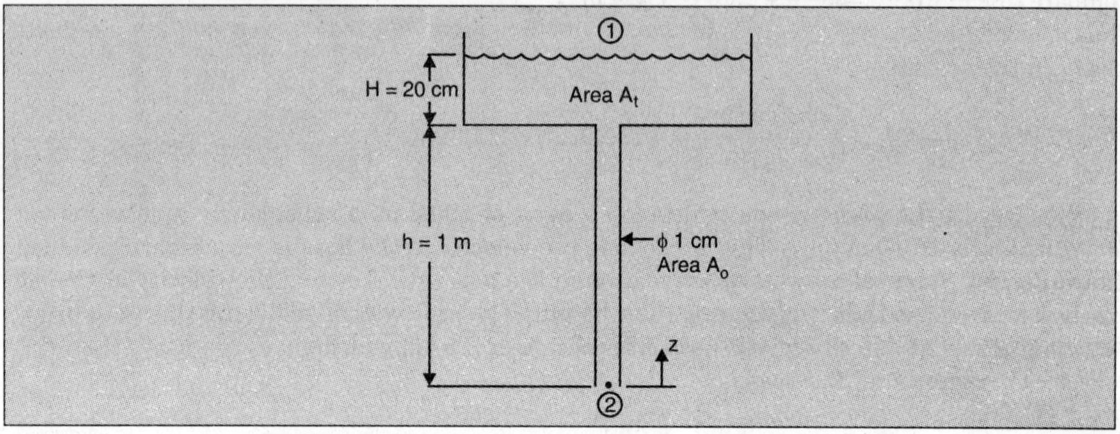

Fig. 10.6. Tank and tube assembly for Example 10.2.

$$\frac{V_2^2}{2g} + z_2 + \frac{p_1}{\rho g} = \frac{V_1^2}{2g} + z_1 + \frac{p_1}{\rho g} - f\frac{V^2}{2g}\frac{L}{D}$$

Here $V_1 \simeq 0$, $z_1 = 1.20$ m, $p_1 = p_2 = p_{atm}$, V_2 is the required velocity V, $L = 1$ m and $D = 1$ cm. Thus

$$\frac{V^2}{2g} = 1.20(\text{m}) - f\frac{V^2}{2g}\cdot\frac{1(\text{m})}{0.01(\text{m})}$$

or

$$V^2 = 23.54\left(\frac{\text{m}^2}{\text{s}^2}\right) - 100f\,V^2$$

or

$$V = \left(\frac{23.54}{1+100f}\right)^{1/2} (\text{m/s}) \quad \text{(a)}$$

This can be solved if the friction factor f is known. But f depends on Re_D, which, in turn, depends upon the unknown velocity V and so, the velocity can be obtained only by an iterative procedure. It is often convenient to assume in the first trial that the flow is fully turbulent so that f is independent of velocity. If the outlet tube is a commercial steel pipe, then $\varepsilon = 0.046$ mm from Table 10.1, or $\varepsilon/D = 0.0046$. Using the Moody chart (Fig. 10.3), we get $f = 0.029$ for a fully turbulent flow. Eq. (a) then gives $V = 2.46$ m/s. The Re_D corresponding to this velocity is

$$\text{Re}_D = \frac{10^3(\text{kg/m}^3)\times 2.46(\text{m/s})\times 0.01(\text{m})}{10^{-3}(\text{N s/m}^2)} = 2.46\times 10^4$$

This value of Re_D is not in the fully turbulent region and gives a new value of f as 0.033. Using this in Eq. (a), $V = 2.34$ m/s is obtained, which, in this second iteration, gives $Re_D = 2.34 \times 10^4$. The new value of f corresponding to this Re_D is 0.034. On carrying out this procedure through one more iteration, it is found that V converges to 2.33 m/s approximately. Thus, the volume flow rate through the pipe is

$$\mathring{Q} = A_0 V = \frac{\pi}{4}(0.01)^2 \left(\text{m}^2\right) \times 2.33 \left(\frac{\text{m}}{\text{s}}\right) = 1.83 \times 10^{-4}\,\text{m}^3/\text{s}$$

Compare this to the volume flow rate Q_{ideal} of 3.81×10^{-4} m³/s when viscous losses are neglected. Thus, the volume flow rate reduces to about half of the ideal flow rate. This is because the head loss h_l in the pipe is

$$h_l = f\frac{V^2}{2g}\frac{L}{D} = 0.035 \times \frac{(2.33)^2\,(\text{m/s})^2}{2 \times 9.81\left(\text{m/s}^2\right)} \times \frac{1\,(\text{m})}{0.01\,(\text{m})} = 0.97\ \text{m}$$

which is a significant fraction of the total head of 1.20 m.

Example 10.3. A penstock carrying water from a reservoir to a turbine was described in Example 7.2. It was shown that in the absence of viscous losses, the maximum power developed by the

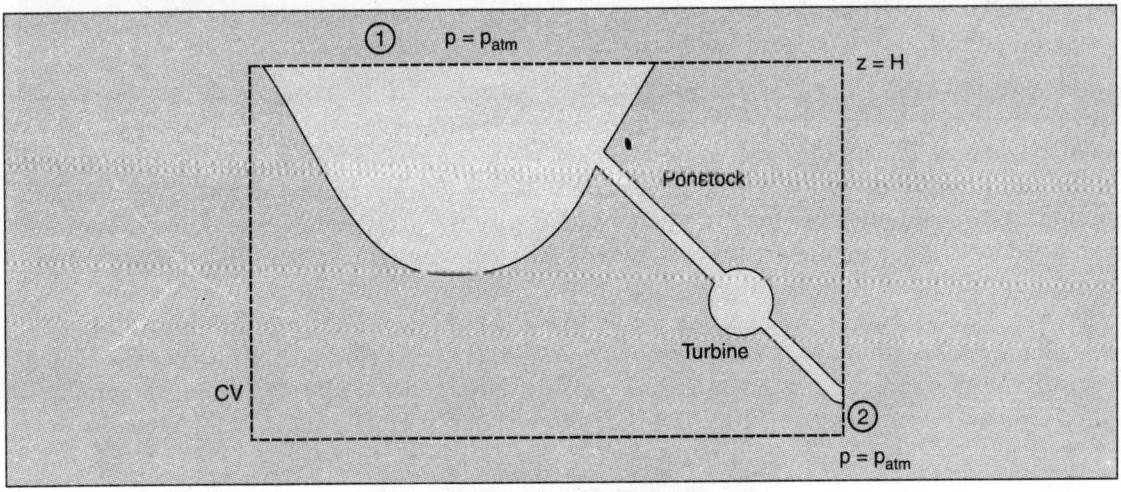

Fig. 10.7. Penstock for Example 10.3.

turbine is $\rho A_2 \left(\frac{2}{3} gH\right)^{3/2}$, where A_2 is the area of the penstock at the exit. If we include the effect of friction in the penstock of length L and average diameter D, the energy Eq. (7.12) gives

$$\frac{V_2^2}{2} + 0 + \frac{p_{atm}}{\rho} = 0 + gH + \frac{p_{atm}}{\rho} - w_s - w_l$$

The loss term w_l is given by

$$w_l = f\frac{V_2^2}{2}\frac{L}{D}$$

CHAPTER 10

so that

$$w_s = gH - \frac{V_2^2}{2}\left(1 + f\frac{L}{D}\right)$$

The power output is then

$$\overset{\circ}{W}_s = w_s\,(\rho VA)_2 = \rho V_2 A_2 \left[gH - \frac{V_2^2}{2}\left(1 + f\frac{L}{D}\right)\right]$$

If the flow is assumed to be fully turbulent, such that f is constant, the maximum power output can be determined by differentiating $\overset{\circ}{W}_s$ with respect to V_2 and setting it equal to zero. The corresponding $\overset{\circ}{W}_s$ comes out as

$$\overset{\circ}{W}_{s,\max} = \frac{\rho\, A_2\,(2gH/3)^{3/2}}{\sqrt{1+\dfrac{fL}{D}}}$$

which is $1/\sqrt{1+fL/D}$ times the $\overset{\circ}{W}_{s,\max}$ obtained when friction is assumed negligible.

For a concrete penstock of diameter 1 m, length 100 m and ε about 1 mm, the value of f for a fully developed flow is about 0.02 and $1/\sqrt{1+fL/D}$ is 0.577, i.e., the maximum power developed is only 57.7 per cent of the ideal value.

10.3 OTHER FORMS OF THE MOODY CHART

The Moody chart (Fig. 10.3) plots the variation of f with respect to Re_D and ε/D. It can be used to solve any pipe-flow problem, but is most suited for calculating viscous losses, given the flow rate and the pipe size. For other problems which require the determination of the velocity for a given head loss, or the determination of the pipe diameter to carry a given flow, Moody's chart can be used but it involves iterative solutions (Example 10.2). This is because the unknown parameter, V or D, enters the abscissa of the plot through Re_D. For such problems, the non-dimensional formulation of Eq. (10.5) is now modified. It is recognized that for flow through a pipe, any non-dimensional dependent variable is controlled by the Euler and Reynolds numbers and the geometric parameters. Thus, the non-dimensional flow velocity, V/V_c is given by

$$\frac{V}{V_c} = \mathcal{F}\left[\frac{(\Delta p)_c}{\rho V_c^2}, \frac{\rho V_c L_c}{\mu}, \frac{L}{L_c}, \frac{D}{L_c}, \frac{\varepsilon}{L_c}\right] \qquad ...(10.17)$$

Conventionally, the diameter D is chosen as L_c and the piezometric pressure drop $\Delta\mathcal{P}$ between the two ends as $(\Delta p)_c$, so that

$$\frac{V}{V_c} = \mathcal{F}\left[\frac{\Delta\mathcal{P}}{\rho V_c^2}, \frac{\rho V_c D}{\mu}, \frac{L}{D}, \frac{\varepsilon}{D}\right] \qquad ...(10.18)$$

Since no velocity has been specified as a parameter, its characteristic value can be chosen arbitrarily. Before doing this, however, it should be realized that for a fully developed flow $\Delta\mathcal{P}$ varies linearly with L, and therefore, the velocity should depend on $\Delta\mathcal{P}/L$ and not on any other combination of $\Delta\mathcal{P}$ and L. Thus, the Euler number and L/D must occur only in the combination

$$\frac{\Delta \mathscr{P}}{\rho V_c^2}\frac{D}{L}$$

and
$$\frac{V}{V_c} = \mathcal{F}\left[\frac{\Delta \mathscr{P}}{\rho V_c^2}\frac{D}{L}, \frac{\rho V_c D}{\mu}, \frac{\varepsilon}{D}\right] \quad \text{...(10.19)}$$

Since V_c can be specified arbitrarily, it may be chosen such that the first independent non-dimensional variable in Eq. (10.19) becomes a constant (= 1/2), i.e.,

$$\frac{\Delta \mathscr{P}}{\rho V_c^2}\frac{D}{L} = 1/2$$

or
$$V_c = \sqrt{\frac{2\,\Delta \mathscr{P}}{\rho}\frac{D}{L}}$$

With this characteristic velocity, then, Eq. (10.19) becomes

$$\frac{V}{\sqrt{\dfrac{2\Delta \mathscr{P}}{\rho}\dfrac{D}{L}}} = \mathcal{F}\left(\frac{\rho D}{\mu}\sqrt{\frac{2\Delta \mathscr{P}}{\rho}\frac{D}{L}}, \frac{\varepsilon}{D}\right) \quad \text{...(10.20)}$$

The non-dimensional number $\dfrac{\rho D}{\mu}\sqrt{\dfrac{2\Delta \mathscr{P}}{\rho}\dfrac{D}{L}}$ is termed as the *Karman number* Ka. (It can be seen that $\text{Ka} = \text{Re}_D\sqrt{f}$, where f is the Moody friction factor $= \dfrac{\Delta \mathscr{P}}{\frac{1}{2}\rho V^2}\dfrac{D}{L}$ and the non-dimensional velocity is $1/\sqrt{f}$).

The plot of the non-dimensional V as a function of Ka and ε/D (Fig. 10.8) is useful in calculating the flow rates directly (i.e., without iteration), given the $\Delta \mathscr{P}$ and the pipe size.

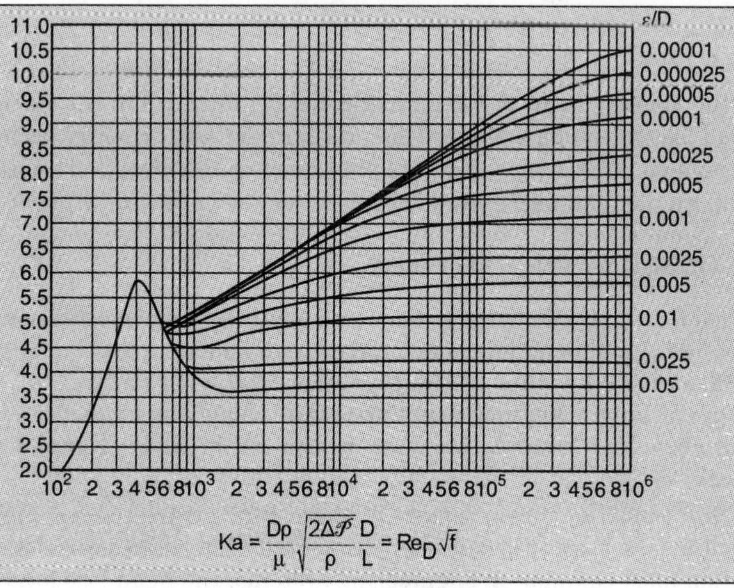

Fig. 10.8. Non-dimensional velocity as a function of Karman Number for pipe-flow problems.

In a similar manner one can set up the formulation for the diameter of the pipe required to pass a given volume flow rate $\mathring{Q}$ through a pipe of length L when a 'pressure' difference $\Delta\mathscr{P}$ is applied. (This is called the *sizing* problem). Proceeding as before, and selecting the characteristic velocity V_c as $\mathring{Q}/L_c^2$ and the characteristic length as $\mathring{Q}/\nu$, the non-dimensional diameter is obtained as

$$D^* = \frac{D}{\mathring{Q}/\nu} = \mathcal{F}\left[\frac{(\Delta\mathscr{P})\mathring{Q}^3}{\rho\, L\, \nu^5}, \frac{\varepsilon\nu}{\mathring{Q}}\right] \qquad \text{...(10.21)}$$

This functional relationship can be plotted as a graph between D^* and $\dfrac{(\Delta\mathscr{P})\mathring{Q}^3}{\rho\, L\, \nu^5}$, with $\varepsilon\,\nu/\mathring{Q}$ as a parameter. The pipe diameter can be directly obtained from such a plot when $\Delta\mathscr{P}$, $\mathring{Q}$, L and ε are given.

Example 10.4. Obtain the rate of flow of water carried by a cast iron pipe (of D = 10 cm, L = 40 m) when the applied 'head' difference across it is 2 m.

The problem is done easily using the Karman number plot. For cast iron, ε = 0.26 mm (from Table 10.1) so that ε/D = 0.26/100 = 0.0026. The Karman number Ka for this flow is

$$\text{Ka} = \frac{\rho D}{\mu}\sqrt{\frac{2\Delta\mathscr{P}}{\rho}\frac{D}{L}}$$

$$= \frac{10^3\left(\text{kg/m}^3\right)\times 0.1(\text{m})}{10^{-3}\left(\text{N s/m}^2\right)}\times\sqrt{2\times 9.81\left(\text{m/s}^2\right)\times 2\,(\text{m})\times\frac{0.1(\text{m})}{40(\text{m})}}$$

$$= 3.132\times 10^4$$

The non-dimensional velocity $V/\sqrt{2(\Delta\mathscr{P}/\rho)D/L}$ corresponding to Ka = 3.132×10^4 and ε/D = 0.0026 is 6.2 (Fig. 10.8). This gives a velocity V of 1.94 m/s.

Note that the velocity head has been neglected. In fact, $\Delta\mathscr{P}$ in the Karman number plot refers to the head lost in the pipe alone. Thus, if there are any other losses in addition to pipe losses, the head applied is not a measure of the head loss in the pipe, and unless the other losses are neglected, the Ka plot cannot be used. This is a very serious limitation of this formulation.

10.4 HEAD LOSSES IN PIPE FITTINGS

The Moody friction factor f predicts the head losses in straight pipe sections only. In any piping system there are additional head losses associated with sudden area changes in inlets, outlets and other fittings, with curvature of the pipe axis, and with valves, etc. These losses are essentially because of separation induced at area transitions and changes in the flow direction and are called *minor losses* (though they may or may not be minor compared to the straight-length pipe losses).

Since the flow patterns through most of these *fittings* are too complex to be modelled analytically, the head losses are specified as experimentally determined correlations. For turbulent flows, the condition most common in engineering systems, the head loss for a fitting is usually specified as a fraction (or multiple) K of the kinetic energy head,

$$h_l = K\frac{V^2}{2g} \qquad ...(10.22)$$

The non-dimensional factor K is termed as the *loss coefficient* and, in general, depends on the geometry of the fitting and the Reynolds number. For practical flow calculations, however, it is conventional to use the highest (asymptotic) value of K independent of Re_D, giving slightly conservative results. Table 10.3 gives the representative values of these loss coefficients for some standard pipe fittings.

Table 10.3. Loss coefficients of some standard fittings

Fitting	K	Fitting	K
Elbows (bends)		*Globe valve*	
45°	0.4	fully open	10
90°	0.9	half open	20
180°	0.9		
Unions	0.04	*Gate valve*	
Tee			
along run	0.40	fully open	0.3
along branch	1.00	half open	5

Head losses at sudden area changes are associated with trapped eddies which are formed when the boundary layer separates in the region of increasing pressure (Fig. 10.9). In Example 5.2, the rise in pressure $p_2 - p_1$ for a sudden expansion was calculated as $\frac{\mathring{m}\,V_1}{A_2}\left(1-\frac{A_1}{A_2}\right)$ where 1 and 2 refer to the smaller and larger sections respectively. Using the energy Eq. (7.13), it can be easily shown that this pressure increase corresponds to a loss in head given by

$$h_l = \left(1-\frac{A_1}{A_2}\right)^2 \frac{V_1^2}{2g} \qquad ...(10.23)$$

and, therefore, the head loss coefficient of sudden enlargements based on the velocity head through the *smaller pipe,* is $\left(1-\frac{A_1}{A_2}\right)^2$. This agrees quite closely with experimentally measured values.

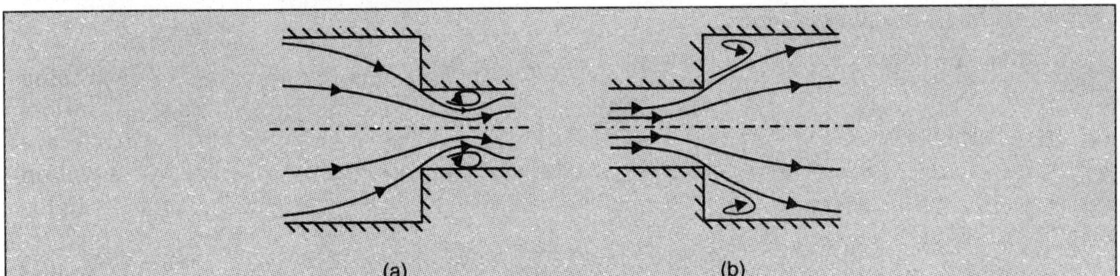

Fig. 10.9. Eddies in sudden contraction and expansion.

The head loss in a sudden contraction is much smaller, since the energy dissipating eddies are smaller. Table 10.4 gives the typical K's for some reduction ratios. The sudden contraction having

$A_1/A_2 = 0$ models a sharp edged inlet to a pipe from a large reservoir as shown in Fig. 10.10a. The loss coefficient for an inlet can be sharply reduced by rounding out the edges such that the transition is smoother. The values of K based on the *pipe* velocity head for some common entry conditions are shown in Fig. 10.10.

Table 10.4. Loss coefficients for sudden contractions and expansions (Based on velocity head in the smaller pipe)

A_1/A_2	Sudden contraction	Sudden expansion $\left[=\left(1-\frac{A_1}{A_2}\right)^2\right]$
0	0.42	1.00
0.2	0.35	0.64
0.4	0.26	0.36
0.6	0.15	0.16
0.8	0.04	0.04
1.0	0.00	0.00

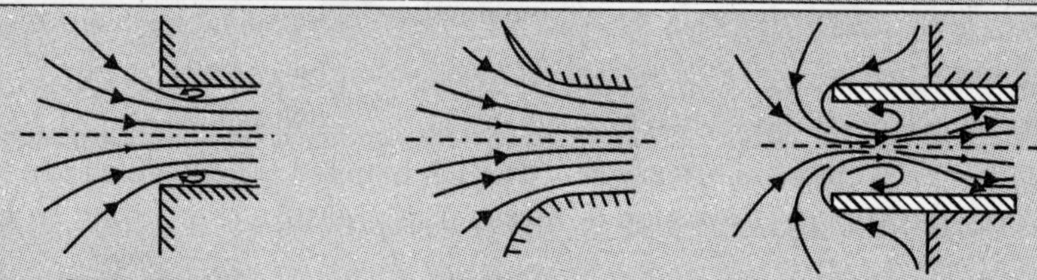

Fig. 10.10. *K*'s for some common entries to pipes.

A sharp exit from a pipe to a reservoir corresponds to an expansion with $A_1/A_2 = 0$, which gives a loss coefficient equal to 1.0. This means that the entire kinetic energy in the pipe is lost without any pressure recovery. The value of K at a pipe exit does not change from 1 even if the edges are rounded out (why?).

Example 10.5. Water flows from reservoir 1 to reservoir 2 (Fig. 10.11). Determine the volume flow rate through the pipe system.

Writing the energy Eq. (7.13) between 1 and 2, we get

$$\left(\frac{V^2}{2g}+z+\frac{p}{\rho g}\right)_2 = \left(\frac{V^2}{2g}+z+\frac{p}{\rho g}\right)_1 - h_l$$

or $$0+5(\text{m})+\frac{p_{atm}}{\rho g} = 0+40(\text{m})+\frac{p_{atm}}{\rho g} - h_l$$

or $$h_l = 35\ (\text{m})$$

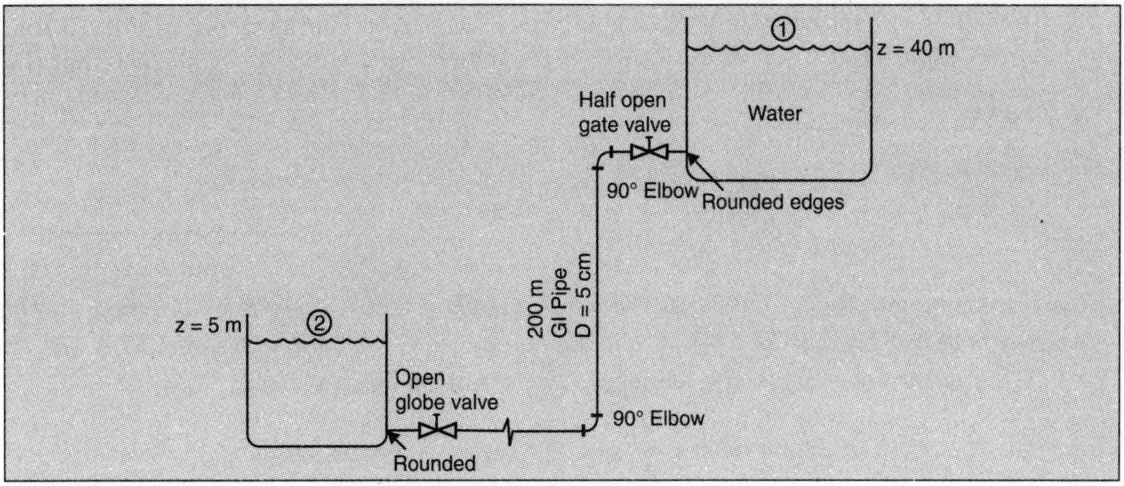

Fig. 10.11. Pipe system for Example 10.5.

This loss occurs in 200 m length of a 5 cm diameter pipe and in six fittings. Thus,

$$h_l = \sum_{i=1}^{6} K_i \frac{V_i^2}{2g} + f\frac{L}{D}\frac{V^2}{2g}$$

where the $K_i s$ represent the loss coefficients of the various fittings. For the well rounded pipe-entry, $K_1 = 0.05$; for the half open gate valve, $K_2 = 5$; for two 90° elbows, $K_3 = K_4 = 0.9$; for the fully open globe valve, $K_5 = 10$, and for an exit, $K_6 = 1.0$. Thus $\sum K_i = 17.85$, and

$$35 = \left(17.85 + f\frac{200(\text{m})}{0.05(\text{m})}\right)\frac{V^2}{2g} \quad \text{(a)}$$

where V is the velocity in the pipe. The friction factor f is unknown, being a function of Re_D and ε/D. For the GI pipe, $\varepsilon/D = 0.15 \times 10^{-3}$ (m)/5 × 10^{-2} (m) = 0.003. If the losses in the fittings are neglected as a first approximation, the head lost through the pipe is 35 m and the Karman number is

$$\text{Ka} = \frac{D\rho}{\mu}\sqrt{\frac{2\Delta\mathcal{P}}{\rho}\frac{D}{L}} = \frac{0.05(\text{m})\times 10^3\left(\text{kg/m}^3\right)}{10^{-3}\left(\text{N s/m}^2\right)} \times \sqrt{2\times 35(\text{m})\times 9.81\left(\frac{\text{m}}{\text{s}^2}\right)\times\frac{0.05}{200}}$$

$$= 2.072 \times 10^4$$

and from Fig. 10.8, the dimensionless velocity = 6.2 = $1/\sqrt{f}$. Thus $f = 0.0260$, and

$$V = 6.2\sqrt{2\frac{\Delta\mathcal{P}}{\rho}\frac{D}{L}} = 2.57 \text{ m/s}^2$$

CHAPTER 10

The right hand side of Eq. (a) is then (17.85 + 104) × 0.337 (m), and it is observed that the minor losses are indeed not negligible. A trial and error solution is then called for along with Fig. 10.3. If fully turbulent conditions are assumed, f is seen to be 0.026, for $\varepsilon/D = 0.003$, and Eq. (a) gives

$$35 = (17.85 + 0.026 \times 4000)\ \frac{V^2}{2g}$$

or $$V = 2.374 \text{ m/s}$$

This corresponds to $\text{Re}_D = 1.19 \times 10^5$ which, with $\varepsilon/D = 0.003$, gives $f = 0.027$ and another iteration is called for. Use of $f = 0.027$ in Eq. (a) gives $V = 2.34$ m/s, $\text{Re}_D = 1.17 \times 10^5$ and $f = 0.027$, and thus convergence is achieved. The resulting flow rate is

$$\mathring{Q} = 2.34 \text{ (m/s)} \times \frac{\pi}{4} \times (0.05)^2 \text{ (m}^2\text{)} = 4.6 \times 10^{-3} \text{ m}^3\text{/s}$$

10.5 PERFORMANCE CHARACTERISTICS OF TURBOMACHINERY

In Secs. 8.3 to 8.5 some simple examples of turbomachinery were analyzed. The theory developed there was a considerably simplified one as the complex 3-*D* flow fields were replaced by 1-*D* idealized velocities. With the use of modern high-speed digital computers it is possible now to relax some of these idealizations but the detailed design of turbomachinery is still an art with most of the improvements coming from experience and repeated testing. The experimental performance data needs to be organised into dimensionless characteristic curves since this reduces the number of independent parameters.

Let us first restrict our attention to pumps. The 'pressure' developed $\Delta\mathscr{P}$ by a pump running at a fixed rotational speed ω depends on the volume flow rate $\mathring{Q}$ delivered by the pump. If more $\mathring{Q}$ is drawn, the pressure developed is usually smaller. Thus, if $\Delta\mathscr{P}$ is the dependent quantity of interest, the flow rate $\mathring{Q}$ can be treated as an independent quantity. Typical pump performance characteristics at a fixed ω are shown in Fig. 10.12.

The flow through a pump is controlled by inertial, pressure and viscous forces, besides the all important geometry. Since the rotary motion of the fluid is an essential feature of turbomachines, it is convenient to divide the (convective) inertial forces into its streamwise and normal (i.e., centrifugal) components. This division is required because the two types of forces involve different parameters and one expression cannot characterize both of these. Three Π-numbers can be formed from these four forces as is shown in Table 10.5. It may be mentioned that these Π's have been formed using centrifugal forces as the reference (unlike in Tables 9.5 and 9.8, where they have been obtained using the streamwise inertial force as the reference).

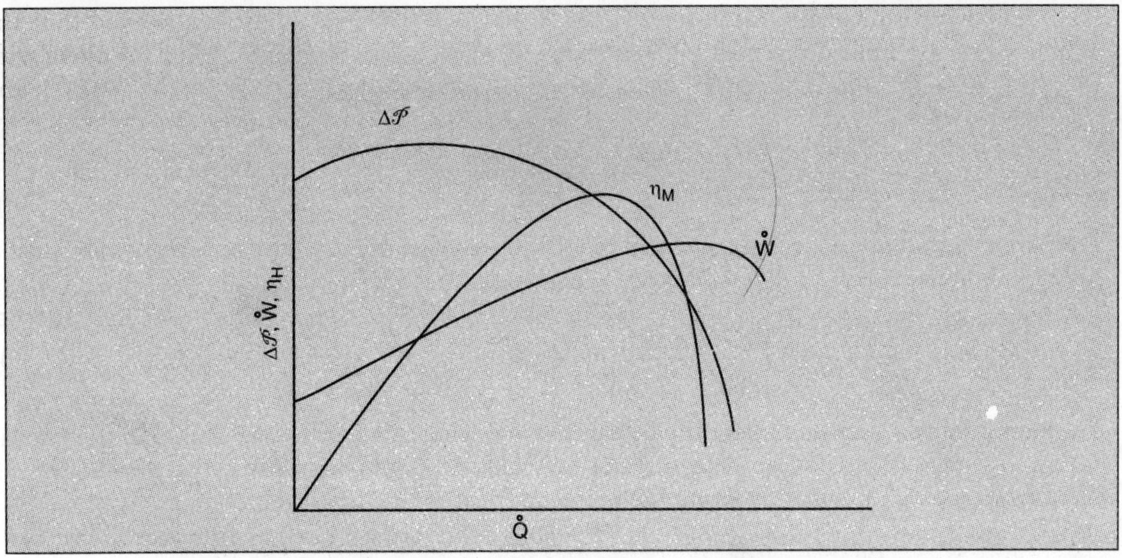

Fig. 10.12. Typical performance characteristics of centrifugal pumps at a fixed ω.

This is to conform to tradition. Thus, any non-dimensional dependent variable in *geometrically similar pumps* is a function of the three pi-numbers so formed, namely$(V/\omega L)_c$, $(p/\rho\omega^2L^2)_c$, and $\left(\frac{\rho\omega L^2}{\mu}\cdot\frac{\omega L}{V}\right)_c$. The last one is conventionally combined with $V/\omega L$ to give $(\rho\omega L^2/\mu)_c$ only. The pressure $\Delta\mathcal{P}$ developed by the pump is then given by

$$\frac{\Delta\mathcal{P}}{p_c} = \mathcal{F}\left[\left(\frac{V}{\omega L}\right)_c, \left(\frac{\rho\omega L^2}{\mu}\right)_c, \left(\frac{p}{\rho\omega^2 L^2}\right)_c\right].$$

The diameter D of the impeller is conventionally taken as the characteristic length L_c, and the angular speed n of the impeller is used as ω_c. The characteristic velocity V_c is obtained from the relation among the scale factors for $\mathring{Q}$, V and L. Thus, we may write $k_{\mathring{Q}} = k_V k_L^2$, which

Table 10.5. Similarity parameters for pumps

Governing law	Scale factor relation	Modelling rule	Π-number
Centrifugal F_ω = (mass) (w^2r)	$k_{F,\omega} = k_\rho k_\omega^2 k_L^2$		
Streamwise inertial F_i = (mass) (acc.)	$k_{F,i} = k_\rho k_V^2 k_L^2$	$\frac{k_\omega k_L}{k_V} = 1$	$\frac{\omega L}{V}$
Viscous F_μ = m (area) (vel. grad.)	$k_{F,\mu} = k_\mu k_L k_V$	$\frac{k_\rho k_\omega^2 k_L^3}{k_\mu k_V} = 1$	$\left(\frac{\rho\omega L^2}{\mu}\right)\left(\frac{\omega L}{V}\right)$
Pressure F_p = (pressure) (area)	$k_{F,p} = k_p k_L^2$	$\frac{k_\rho k_\omega^2 k_L^2}{k_p} = 1$	$\frac{\rho\omega^2 L^2}{p}$

CHAPTER 10

gives $(\overset{\circ}{Q}/VL^2)_c$ as invariant, and, therefore, we may take $V_c = \overset{\circ}{Q}_c/L_c^2$. If the actual discharge $\overset{\circ}{Q}$ is taken as $\overset{\circ}{Q}_c$, then $V_c = \overset{\circ}{Q}/D^2$. Therefore, for similar pumps,

$$\frac{\Delta\mathcal{P}}{p_c} = \mathcal{F}\left[\frac{\overset{\circ}{Q}}{nD^3}, \frac{\rho n D^2}{\mu}, \frac{p_c}{\rho n^2 D^2}\right].$$

For pump operation under a given discharge, the characteristic pressure p_c is not established *a priori* and hence it can be set arbitrarily. Choosing $p_c = \rho\, n^2D^2$, we get

$$\frac{\Delta\mathcal{P}}{\rho n^2 D^2} = \mathcal{F}\left[\frac{\overset{\circ}{Q}}{nD^3}, \frac{\rho n D^2}{\mu}\right]$$

For most pump operations (unless the fluid is very viscous), the flow is turbulent and the viscous terms are negligible compared to inertial (or centrifugal) terms. Thus, the dependence of $\Delta\mathcal{P}/\rho n^2D^2$ on $\rho nD^2/\mu$ which is like a Reynolds number, is negligible* and

$$\frac{\Delta\mathcal{P}}{\rho n^2 D^2} = \mathcal{F}\left[\frac{\overset{\circ}{Q}}{nD^3}\right]$$

Similarly, the power required $\overset{\circ}{W}$ is given by

$$\frac{\overset{\circ}{W}}{\overset{\circ}{W}_c} = \mathcal{F}\left[\frac{\overset{\circ}{Q}}{nD^3}\right]$$

An expression for $\overset{\circ}{W}_c$ can be obtained from the scale factor relation $k_{\overset{\circ}{W}} = k_\rho\, k_V^3\, k_L^2 = k_\rho\, k_\omega^3\, k_L^5$, which gives $\left(\overset{\circ}{W}/\rho\omega^3 L^5\right)_c$ as invariant. Thus, $\overset{\circ}{W}_c$ can be chosen as $(\rho\omega^3 L^5)_c$ or $\overset{\circ}{W}_c = \rho n^3 D^5$, and so

$$\frac{\overset{\circ}{W}}{\rho n^3 D^5} = \mathcal{F}\left[\frac{\overset{\circ}{Q}}{nD^3}\right]$$

The three non-dimensional parameters, namely, $\overset{\circ}{Q}/nD^3$, $\Delta\mathcal{P}/\rho n^2D^2$ and $\overset{\circ}{W}/\rho n^3D^5$ developed above, play an important role in the design and performance studies of pumps. These are termed as the discharge coefficient $C_{\overset{\circ}{Q}}$, head coefficient** C_H, and the power coefficient $C_{\overset{\circ}{W}}$, respectively

$$C_{\overset{\circ}{Q}} = \frac{\overset{\circ}{Q}}{nD^3};\, C_H = \frac{\Delta\mathcal{P}}{\rho n^2 D^2};\text{and } C_{\overset{\circ}{W}} = \frac{\overset{\circ}{W}}{\rho n^3 D^5}$$

---

*See Chapter 11.

** The head coefficient C_H is written in terms of H, the head developed as $C_H = \rho gH/\rho n^2D^2 = gH/n^2D^2$. In some books g, the acceleration due to gravity is omitted and C_H is set equal to H/n^2D^2. This is obviously not non-dimensional. In this text, we stick to the rigorous definition of C_H.

The *useful work* done by the pump is the increase in the mechanical energy of the fluid. Thus, the rate at which useful work is done by the pump is $(\Delta\mathscr{P})\ \overset{\circ}{Q}$. The hydraulic efficiency η_H of a pump is defined as

$$\eta_H = \frac{(\Delta\mathscr{P})\overset{\circ}{Q}}{\overset{\circ}{W}} = \frac{C_H C_{\overset{\circ}{Q}}}{C_{\overset{\circ}{W}}} \quad ...(10.24)$$

and this, once again, is a function of the discharge coefficient $C_{\overset{\circ}{Q}}$. The overall efficiency η of the pump is $\eta_m\eta_H$, where η_m is the mechanical efficiency of the drive-mechanism.

The non-dimensional performance curves for geometrically similar pumps are the plots of C_H, $C_{\overset{\circ}{W}}$ and η_H against $C_{\overset{\circ}{Q}}$. Typical characteristics of centrifugal pumps are shown in Fig. 10.13. The relevant non-dimensional parameters for a turbine are obtained in a similar manner, except that the hydraulic efficiency of a turbine is defined as the power extracted $\overset{\circ}{W}$ *divided* by the rate of decrease of mechanical energy of the fluid, or

$$\eta_{H,T} = \frac{\overset{\circ}{W}}{(\Delta\mathscr{P})\overset{\circ}{Q}} = \frac{C_{\overset{\circ}{W}}}{C_H C_{\overset{\circ}{Q}}}$$

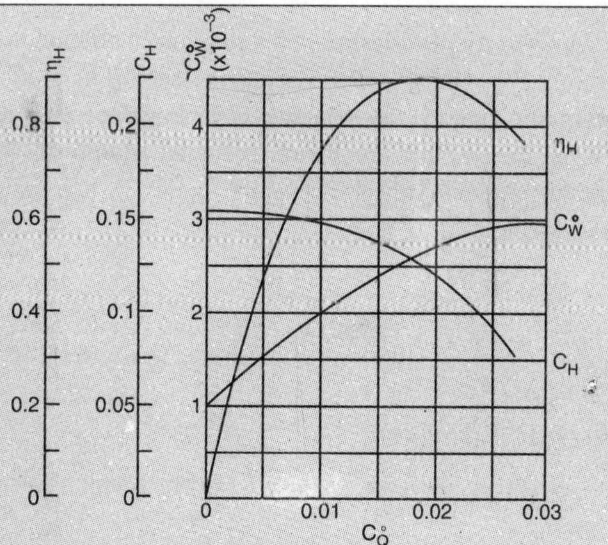

Fig. 10.13. Dimensionless pump characteristics for one design of centrifugal pumps (n in rad/s).

10.6 CLASSIFICATION OF TURBOMACHINERY

The dimensionless performance curves for one design of centrifugal pumps is shown in Fig. 10.13. Note that there is an optimum operating point corresponding to the maximum efficiency. Thus, a given family of pumps is best suited to operate at certain values, $C_{\overset{\circ}{Q}}^*$ and C_H^*, of the dimensionless flow rate and the dimensionless head irrespective of the size. For example, the centrifugal pump whose characteristics are shown in Fig. 10.13 has a $C_{\mathring{Q}}^*$ of approximately 0.019 and C_H^* of approximately 0.125. Thus, for most efficient operation,

$$\left.\begin{aligned} \mathring{Q} &= 0.019\, nD^3 && \text{(a)} \\ \text{and} \quad gH &= 0.125\, n^2D^2 && \text{(b)} \end{aligned}\right\} \qquad ...(10.25)$$

These two equations can be used to obtain the optimum values of the speed n and the size D for any given application (i.e., combination of $\mathring{Q}$ and H). However, since the pumps are usually directly coupled to induction motors, there are some constraints on n and we may be required to operate it slightly away from the optimal point. It can be verified easily (see Prob. 10.36) that for this given design of (i.e., geometrically similar) pumps we get reasonable values of D and n for high heads and low volume flow rates only. For other applications. e.g., when large flow rates are required at low heads, the D obtained is too high to be practical and the n obtained is too low for direct coupling with electric motors. A centrifugal pump of this design, therefore, is quite inefficient and so is unsuited for large flow-low head applications. It may be mentioned that a phenomenal amount of energy is being wasted in India by using centrifugal pumps for lift irrigation.

The observation that a particular design has a maximum efficiency (for all sizes of similar pumps) at fixed values, C_H^*, $C_{\mathring{Q}}^*$ and $C_{\mathring{W}}^*$ of the non-dimensional head, flow and power, permits us to develop an objective criterion for the selection of the appropriate type of turbomachine for a given application. For pumps, we use the head developed H, and the flow rate $\mathring{Q}$ to define its application. Since the non-dimensional coefficients $C_{\mathring{Q}}^*$ and C_H^* are unique for a given design, optimal operation requires that

$$\mathring{Q} = C_{\mathring{Q}}^*\, nD^3$$

and

$$gH = C_H^* n^2 D^2$$

On eliminating D from these equations

$$\frac{n\mathring{Q}^{1/2}}{(gH)^{3/4}} = \frac{C_{\mathring{Q}}^{*\,1/2}}{C_H^{*\,3/4}} \qquad ...(10.26)$$

The term on the left can be taken as the non-dimensional speed of the pump and is termed the *specific speed* C_s. Since the right hand side of Eq. (10.26) is fixed for a given design, *independent of size*, this means that C_s has an optimal value

$$C_s^* = C_{\mathring{Q}}^{*1/2} / C_H^{*3/4}$$

The speed n at which the optimal pump should run can then be obtained by setting the specific speed C_s $(= n\mathring{Q}^{1/2}/(gH)^{3/4})$ equal to its optimal value C_s^*. The value of n so obtained may or may not be suitable for direct drive from a motor and this restricts the range of $\overset{\circ}{Q}$ and H over which this design of pump (independent of size) is suitable. For the pump whose characteristics are shown in Fig. 10.13, the optimal specific speed is

$$C_s^* = \frac{C_{\overset{\circ}{Q}}^{*1/2}}{C_H^{*\,3/4}} = 0.64$$

A pump of this design when required to deliver 1 m^3/s at 85 m head will be required to run at $n = 99.31$ rad/s (or 948 RPM) corresponding to $C_s = C_s^* = 0.64$. This speed is quite suitable for an induction motor. But a $\overset{\circ}{Q}$ of 7 m^3/s at 15 m head requires an n of 10.22 rad/s (or 97.6 RPM) which is not at all suitable for direct drive. Thus, as stated earlier, this type of pump is suitable for low flow-high head applications only. This is true for all centrifugal pumps since these have fluid flowing outwards from a small mean diameter to a larger diameter (Fig. 10.14). Unless the flow rates are small, the velocity at the inlet becomes very large, giving rise to greater losses. Thus, optimum efficiency requires that the flow rate be low and correspondingly, the optimum

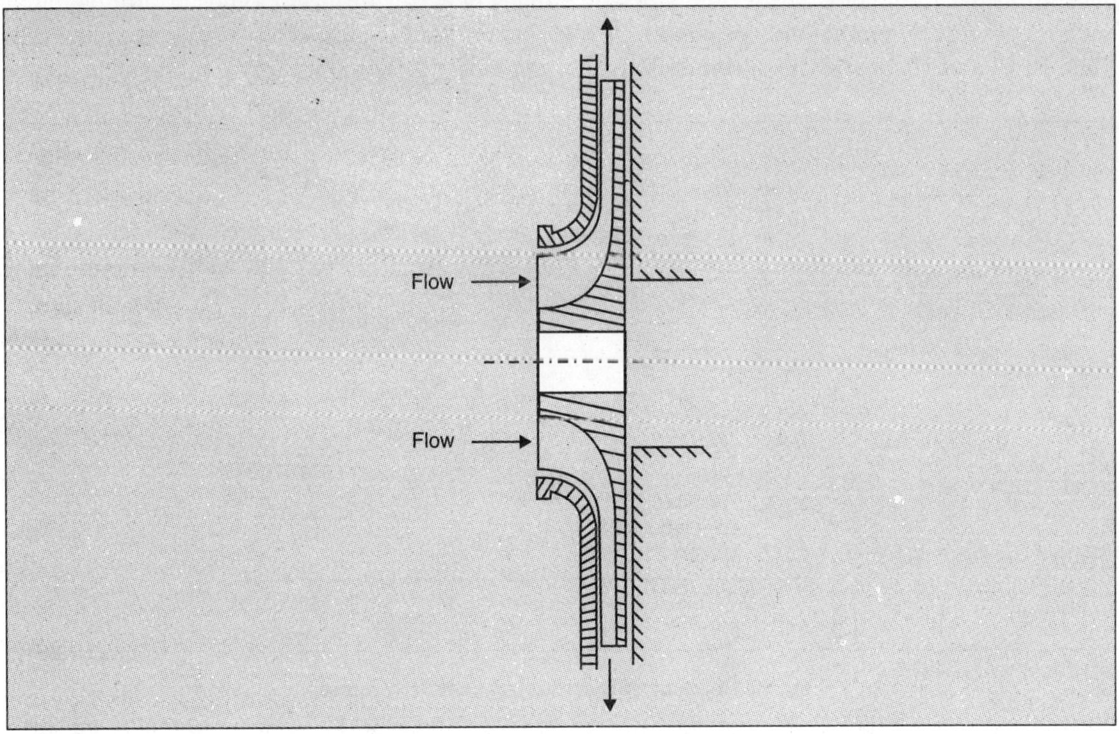

Fig. 10.14. Flow in a centrifugal pump.

specific speed is low. In an axial flow pump, on the other hand, the change in flow direction is minimal and the pump can handle large flow rates without greatly increased losses. The optimum specific speed is correspondingly higher for such machines.

For turbines, one generally specifies the head and the power developed and it can be shown easily that the corresponding definition of specific speed is

CHAPTER 10

$$C_{s,T} = \frac{n\overset{\circ}{W}^{1/2}}{\rho^{1/2}(gH)^{5/4}} = \frac{C_{\overset{\circ}{W}}^{*\ 1/2}}{C_H^{*\ 5/4}} \qquad ...(10.27)$$

This is distinguished from C_s by calling it the *power specific speed* and it can easily be seen that $C_{s,T}$ differs from C_s by a factor of the efficiency

$$C_{s,T} = \eta_{H,T}^{1/2} C_s$$

Figure 10.15 shows the plot of the *best* efficiency that can be achieved in the various types of pumps as a function of the optimal specific speed. For example, if a particular application calls for a specific speed C_s of 1.0, one should seek a design which would have $C_s^* = 1$. The figure shows that the best pump of centrifugal design which has a C_s^* of 1.0 gives an efficiency of only 0.72, while if a mixed-flow pump is chosen, the maximum efficiency achievable is 0.84. Thus, for this application, a mixed flow pump is to be preferred. Clearly, one *type* of pump is suitable over a limited range of C_s only. The suitable ranges of C_s for different types of pumps along with representative impellers are shown in Fig. 10.16. Similar plots for turbines are shown in Figs. 10.17 and 10.18 and the optimum machine can easily be found for any requirement. These

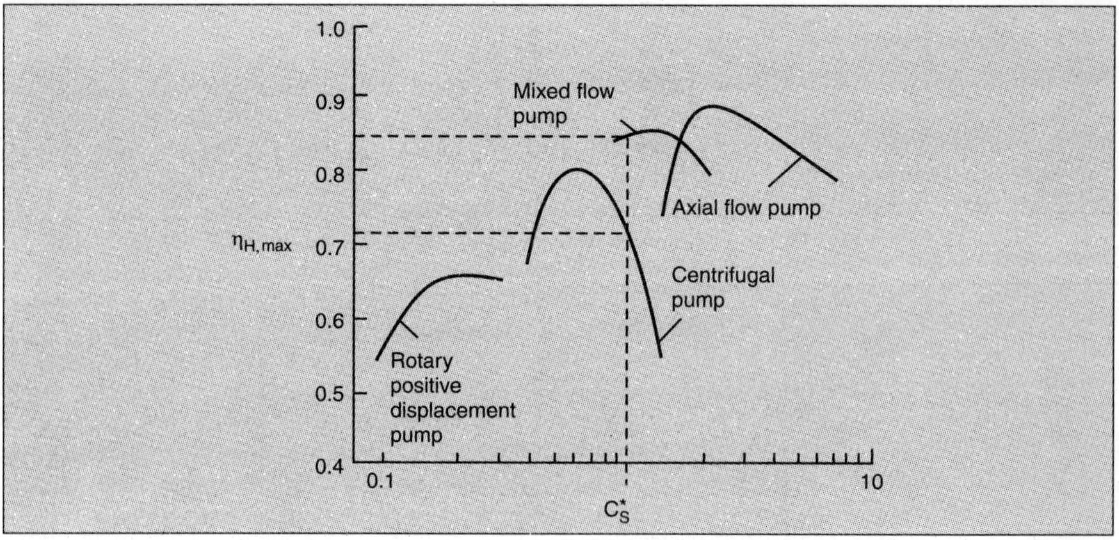

Fig. 10.15. Best efficiencies attainable in pumps.

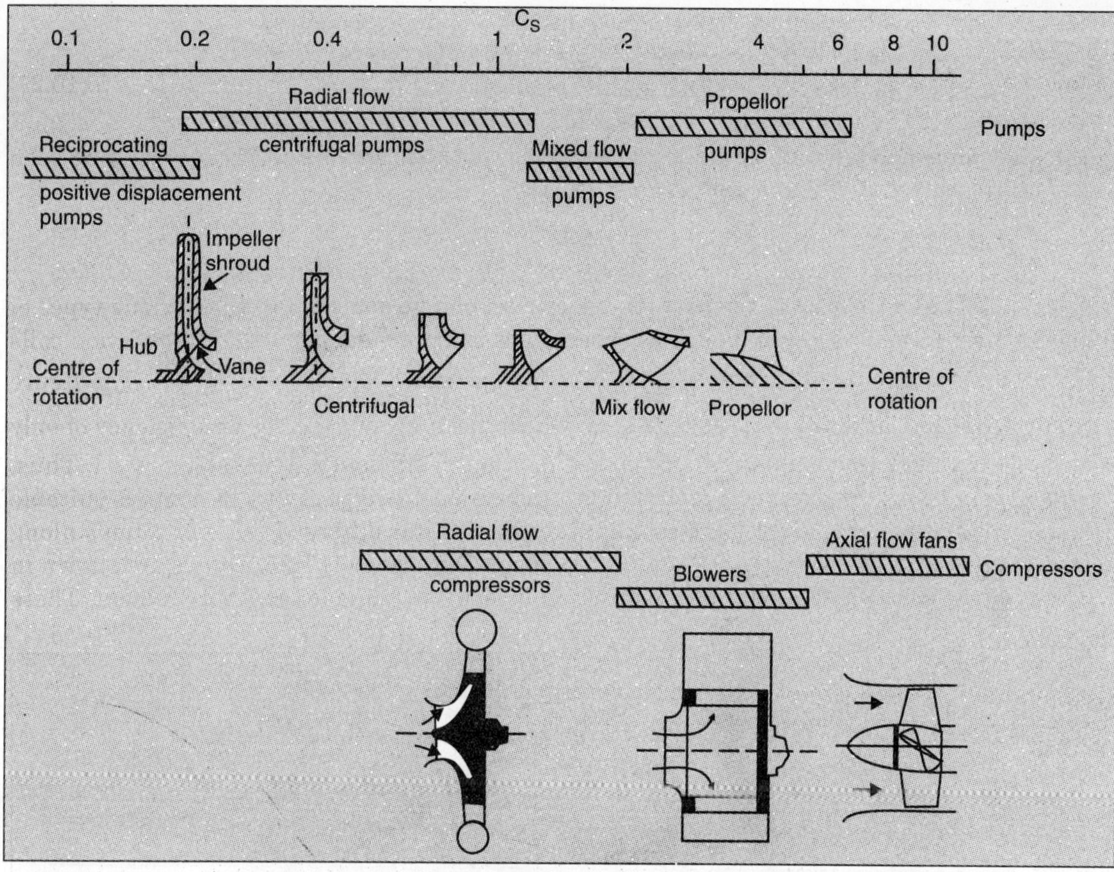

Fig. 10.16. Selection chart for pumps and compressors.

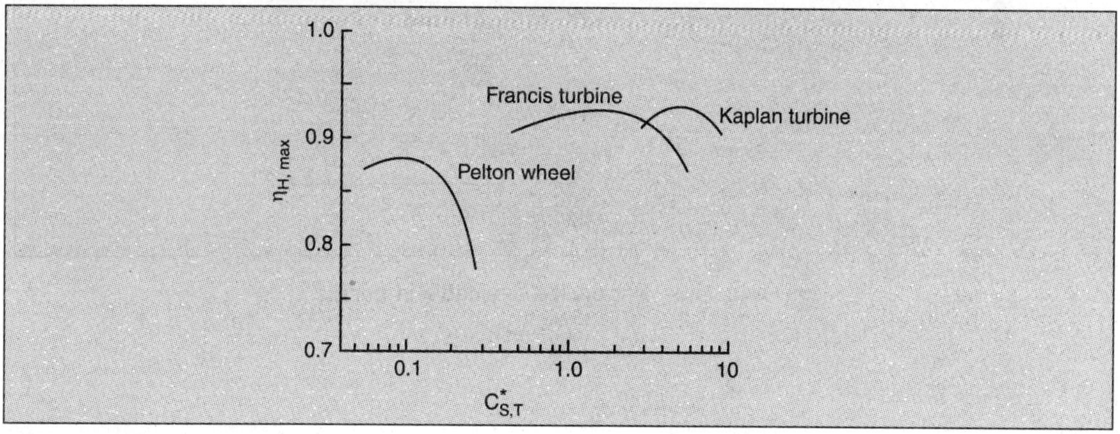

Fig. 10.17. Best efficiencies in turbines.

diagrams illustrate dramatically the power of the concepts of similitude since with minimal analysis, we have been able to provide guidelines for the choice of the basic design of equipment. Table 10.6 lists some of the largest hydraulic turbine applications in the world.

Table 10.6. Some notable hydraulic turbine applications

Location	Head	BHP	RPM	$C_{s,T}$	Type
Cimego, Italy	720 m	150,000	300	0.016	Pelton
Reisseck, Austria	1770 m	338,000	750	0.020	Pelton
Churchill Falls, Canada	316 m	618,000	200	0.614	Francis
Krasnoyarsk, Russia	95 m	685,000	93.8	1.362	Francis
John Day, USA	28.6 m	215,000	90	3.283	Kaplan

Example 10.6. A hydraulic turbine operates between two reservoirs with a level-difference of 7 m. It is required to deliver 1 MW. Find the flow rate necessary, select a suitable turbine and prescribe the rotor speed.

A well designed turbine is able to achieve hydraulic efficiencies of up to 0.9. If we further use a factor of 0.9 to account for mechanical losses, the net energy extracted from the fluid should be $1/(0.9)^2$ MW or 1.23 MW. The energy delivered by the flow is

$$\mathring{Q}\,\Delta\mathscr{P} = \mathring{Q}\,\rho g H$$

Therefore,

$$1.23\times10^6\,(W) = \mathring{Q}\,\times10^3\left(\text{kg/m}^3\right)\times9.81\left(\text{m/s}^2\right)\times7\,(\text{m})$$

or $$\mathring{Q} = 17.91\ \text{m}^3/\text{s}$$

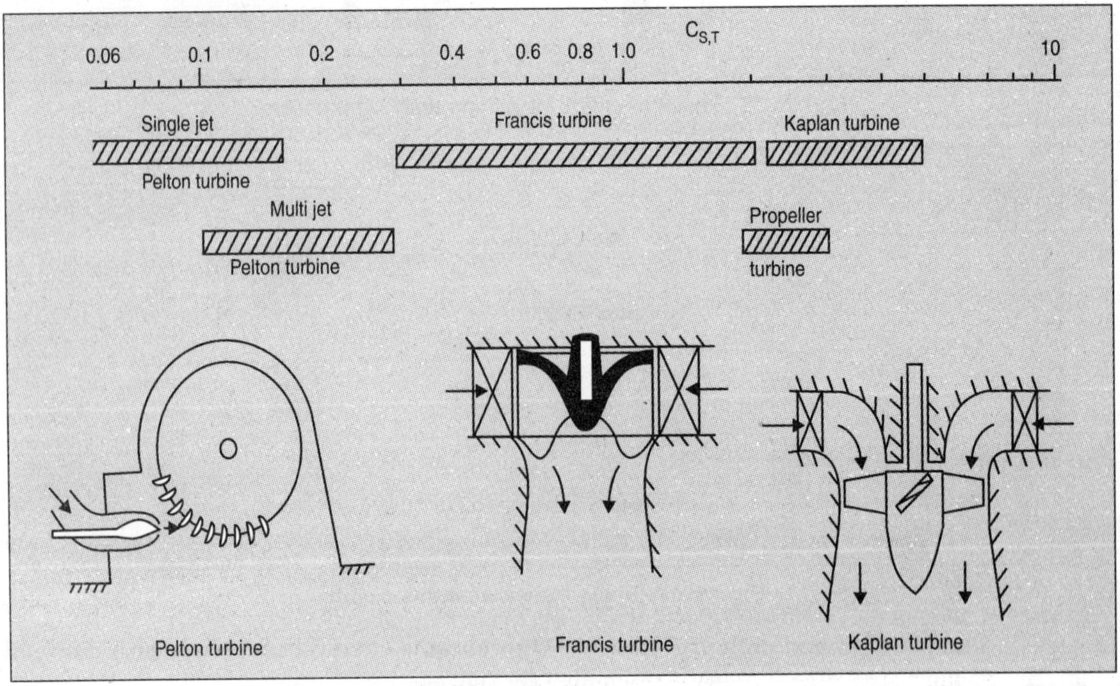

Fig. 10.18. Selection chart for turbines.

The specific speed of the turbine is then

$$C_{s,T} = \frac{n\overset{\circ}{W}^{1/2}}{\rho^{1/2}(gH)^{5/4}} = \frac{n\left(1.23\times10^6\right)^{1/2}}{\left(10^3\right)^{1/2}(9.81\times7)^{5/4}} = n\times0.18$$

Since the turbine is usually coupled directly to a 50 Hz alternator, a speed of 250 RPM or 26.18 rad/s is suitable. The corresponding specific speed $C_{s,T}$ is 4.644. It is observed from Fig. 10.17 that at this value of $C^*_{s,T}$ a Francis turbine has the best efficiency of 90 per cent, whereas a Kaplan turbine has an efficiency of 94 per cent. Thus, a Kaplan turbine should be chosen which is a propeller turbine with adjustable blades. The size of the machine can be obtained by referring to the dimensionless characteristics of the turbine selected.

PROBLEMS

10.1 Sketch the flow patterns for partly open gate and globe valves shown schematically and justify the much larger value of K for the latter. Also show that the K for a 180° standard bend will be higher than for a 90° elbow (even though both are listed as 0.9 in Table 10.3). Similarly, deduce which of the following two cases will involve a higher loss: a 90° sharp bend (case c shown), or a 'capped tee' used as a 90° elbow (case d shown).

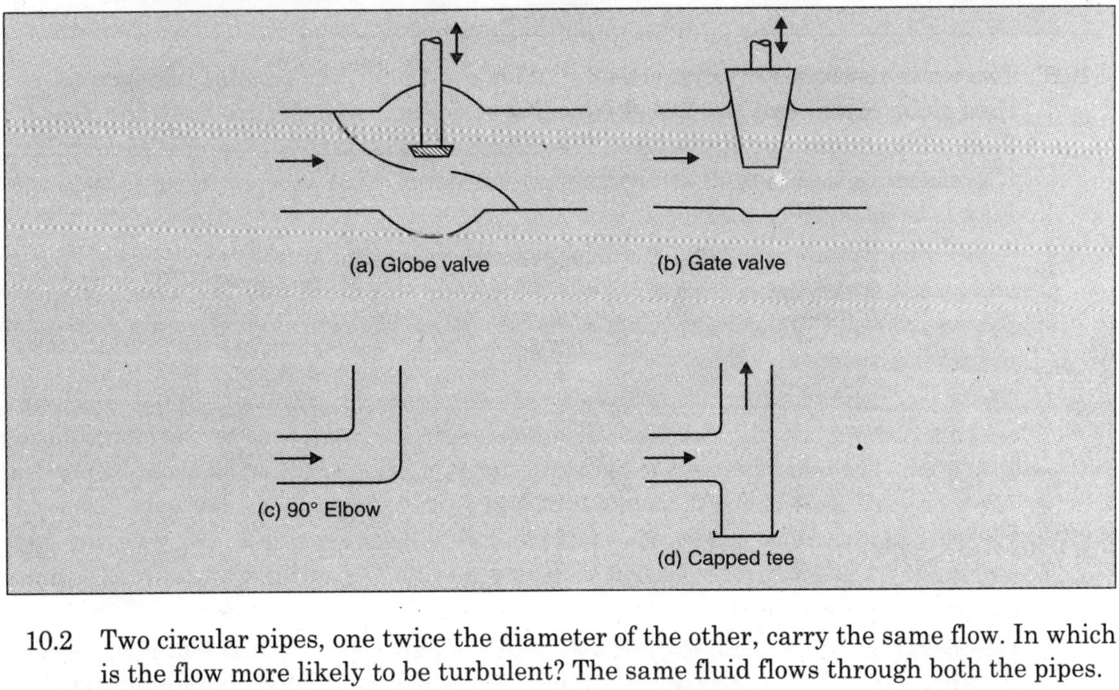

10.2 Two circular pipes, one twice the diameter of the other, carry the same flow. In which is the flow more likely to be turbulent? The same fluid flows through both the pipes.

10.3 The velocity of the water jet coming out of a rubber hose can be increased by pinching at the end. Show using the Bernoulli equation that this is possible only if the friction in the hose is not negligible. Assume the water-mains to be a constant pressure source where the velocity may be assumed negligible.

10.4 The average diameter of the blood vessels decreases by 10 per cent due to aging. Assuming that the cardiac output is unchanged and the flow in the blood vessels is always fully turbulent, estimate (*a*) V_{old}/V_{young}, (b) $\Delta p_{old}/\Delta p_{young}$, and (c) $\mathring{W}_{old}/\mathring{W}_{young}$, where Δp is the pressure generated by the heart (see Prob. 7.6) and $\mathring{W}$ is the power developed.

10.5 An engineer wishes to compute the flow rate through the pipe shown. He applies the mechanical energy equation between sections 1 and 2, and using $p_1 = p_{atm} + \rho g H$, obtains

$$V = \sqrt{\frac{2gHD}{fL}}$$

What is the fallacy in his argument and what should be the correct expression for V? At *approximately* what L/D ratios would you expect the above equation to give results within, say, one per cent of the correct value (use $f = 0.02$)?

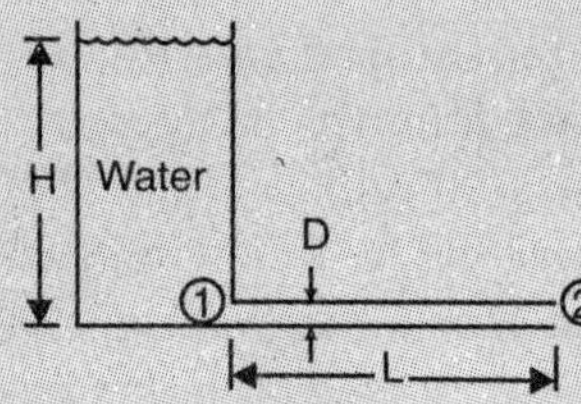

10.6 Two water reservoirs are connected by 200 m of GI pipe. The pipeline incorporates two open globe valves and two 90° elbows. The entry into and the exit from the pipe are flush, sharp-edged and submerged. The difference in heights between the tanks is 5 m. If a minimum flow rate of 360 litres/min is desired, what pipe diameter is required? Flow is to be entirely due to gravity?

10.7 A cooler pump sucks in water at atmospheric pressure and at inlet velocity close to zero and pumps it through a 1 cm dia. and 1.25 m long polythene tube to a tank 1 m above the water level. If the power consumed is 2 W, obtain the rate at which water is pumped up in litres per hour.

10.8 There are 11,000 capillaries of regenerated cellulose in an artificial kidney – each 13.5 cm long and 225 μm dia. Estimate the pressure drop experienced by the blood flowing through it. The heart pumps 5 lit/min of blood (at resting conditions), about 30 per cent of which may be assumed to be passing through the kidneys. The viscosity of blood at the body temperature is 8 cp. Check if an extra pump is required. Use pressure data for the heart as given in Prob. 7.6. Take ρ of blood as 1056 kg/m^3. Is surface tension important for these capillaries?.

10.9 One design of the Kandla-Mathura oil pipeline envisaged a 0.3 m dia. pipe carrying 2×10^4 m^3/day of oil at 10°C ($\nu = 10^{-5}$ m^2/s, $\rho = 800$ kg/m^3), with a pumping station at every 100 km. If f is 0.019 under these conditions, compute the power required at each pumping station. Neglect changes in the elevation and losses in fittings. If there are ten fully-open globe valves per 100 km, estimate the percentage error incurred in neglecting the losses in the fittings.

10.10 To control the seasonal floods in the Indo-Gangetic plain, K.L. Rao proposed a *Ganges-Cauvery link.* In the first stage, water at a rate of 1420 m^3/s is proposed to be pumped from the Ganges at Patna (height above mean sea level = 45.7 m) across the Vindhya mountains (423 m) to the Bargi reservoir (400 m) on the Narmada, about 500 km away. Compute the power requirement. Assume that the diameter of each of the 40 concrete pipes running parallel is such that the velocity is 2.44 m/s. Compare your results with the power production of 1050 MW of the entire Bhakra-Nangal hydroelectric complex. (The power requirements may be reduced by using open channel gravity flow wherever possible).

10.11 Municipal wastes from a township are to be discharged from an effluent treatment tank to a lake through a 3 m dia. cement pipe as shown. The pipe is estimated to be 2 km long and is to incorporate one gate valve. If the discharge rate is 35 m^3/s, find if a pump is required. Assume flow to be fully turbulent. Use $\varepsilon = 0.003$ m and ρ of the sludge as 1.36 times that of the lake water.

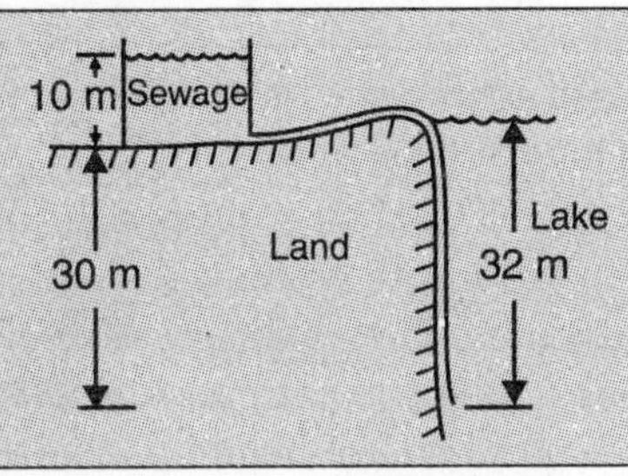

10.12 The sampling probe shown has a total length l and is used to collect water in a stream from various depths. Obtain an expression for the flow rate through the probe as a function of the depth. Assume 1-*D* flow in the river and express your result in terms of f, h_1, h_2, l, d, etc.

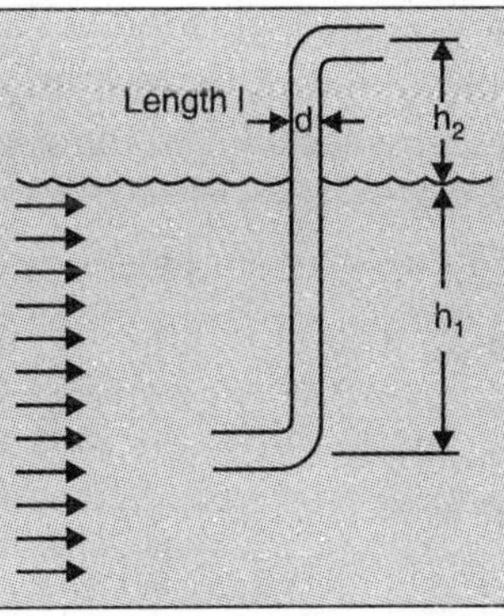

10.13 In several devices (like some designs of driers, wind tunnels, etc.), blowers are used to circulate air in closed loops. Estimate the power required for a closed-circuit wind tunnel 75 m long having an equivalent circular cross-section of approximately 2 m diameter. The maximum air velocity is to be 125 m/s. Since the pipe diameter is large, ε/D may be assumed negligible and the smooth-pipe correlation for f may be used. Also assume that the vanes used to prevent eddy-formation in the bends have an equivalent pipe length of approximately 10 times the pipe diameter.

10.14 Compute the velocity at point 5 and the pressure at point 3 in the siphon of Prob. 7.43. Assume smooth pipe of dia. 1 cm, and neglect all losses except that for the flow in the pipe.

10.15 A pump is installed, as shown, in a well which is 40 m deep. The steel pipe is 5 cm in diameter and the desired velocity of water is 2 m/s. Compute the power required assuming negligible losses in pipe-fittings.

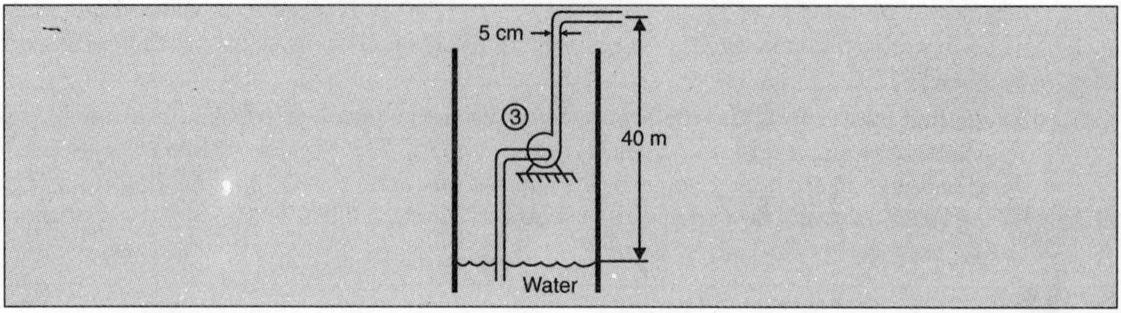

10.16 Find the minimum depth at which the pump of Prob. 10.15 must be installed to avoid cavitation at the inlet. Take p_v as 7 kPa.

10.17 *Net positive suction head (NPSH):* The actual pressure head at the pump-inlet must be higher than the vapour pressure head because the action of the impeller inside the pump further accelerates the water and lowers the pressure below the inlet value. If cavitation is to be avoided everywhere, then the net pressure *plus* velocity head at the inlet must be larger than the cavitation pressure head by a certain amount depending upon the design and the speed of the pump. This *excess* is termed the net positive suction head. If the NPSH is 2 m, what should be the proper minimum depth of the pump of Prob. 10.15.

10.18 *Priming:* If the inlet pipe of the pump of Prob. 10.15 is full of *air* when the pump is switched on, estimate the pressure at the inlet of the pump. Assume the pump develops the same *head* as when it was filled with water.

Show that this suction is insufficient to lift water to the pump and, therefore, the pump needs to be *primed* by filling it with water before switching on.

10.19 In designing the pipeline-pump system, as shown, to be used intermittently, should $(z_3 - z_1)$ or $(z_2 - z_1)$ be used in computing the minimum required power of the motor to be installed?

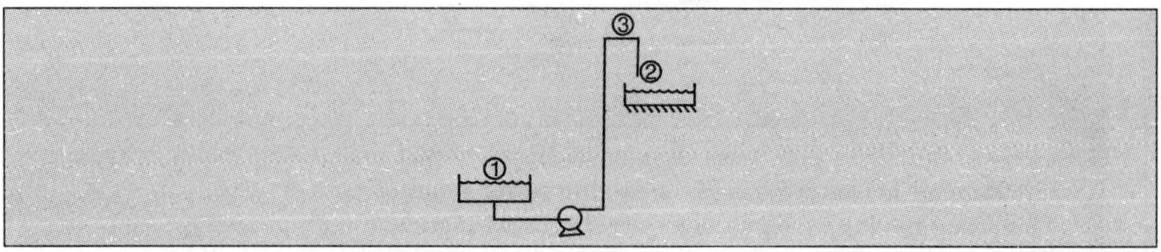

10.20 The schematic of a rear shock absorber of a scooter is shown. A piston having five symmetrically placed thin orifices of 1.5 mm dia. each, moves inside a barrel (of area 5×10^{-4} m^2) filled with oil of $\rho = 800$ kg/m^3 and $\mu = 0.1$ kg/m s on both sides. If the piston moves in at 1 m/s, estimate the damping force. Assume flow in the orifices to be fully developed. Check if this assumption is justified. If not, will the answer obtained above be an underestimate or an overestimate of the true value?

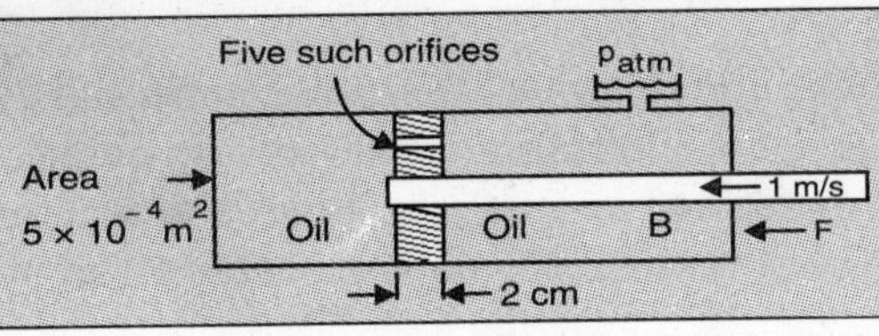

10.21 According to Prob. 7.54 the pressure p_2 is higher than the pressure p_1 for the manifold shown. A typical experimental plot of the pressure variation along x is also shown. Note that

(a) the pressure decreases linearly between adjacent ports,

(b) these *rates* of decrease of pressure become smaller towards the later ports, and

(c) that the pressure jumps are larger at the earlier ports. Explain these features.

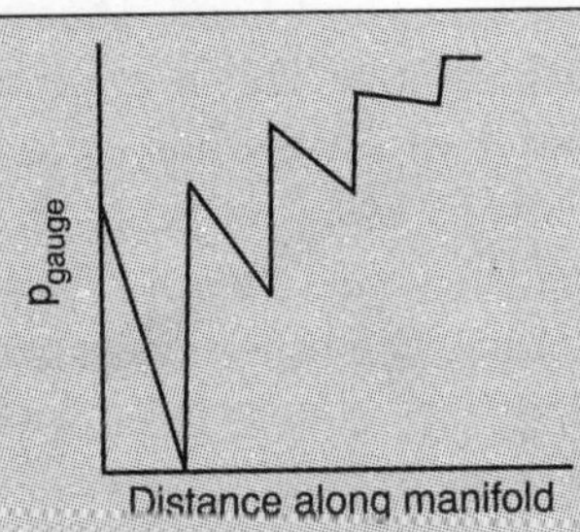

10.22 The carburettor of a scooter is shown schematically. The motion of the piston in the cylinder during the suction stroke results in flow of air down the *venturi.* The throat at 2 is the minimum pressure point and this causes the gasoline to be sucked into the tube where it is atomised. Obtain an expression for the pressure at point 2 in terms of V_3, assuming that the vaporisation of gasoline is negligible. A float valve controls the level of gasoline in the float chamber C. Obtain the level H of the gasoline required in the chamber if the rate of gasoline supply is to be $\overset{\circ}{Q}$.

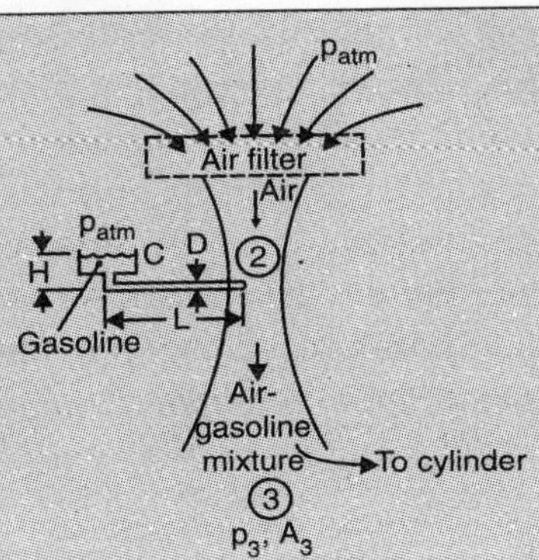

CHAPTER 10

10.23 *Economic velocity:* The annual costs for pumping a fluid per unit length of pipe can be written as

$$C_1\rho^{0.8}\mathring{Q}^{\,2.8}\mu^{0.2}D^{-4.8} + C_2$$

if the flow is turbulent and the pipe is assumed smooth. Here D is the diameter of the pipe. Justify these proportionalities qualitatively. In addition, the depreciated cost per year of the pipes is expressed as $C_3D^{1.5}$ for pipes of diameter larger than 2.5 cm. Show that there is an optimal value of the pipe diameter given by

$$D_{op} = \text{Const} \times \rho^{0.127}\mathring{Q}^{\,0.44}\mu^{0.03}$$

Then show that the optimal velocity of the fluid through the pipe varies as $\mathring{Q}^{\,0.12}$, and is therefore relatively insensitive to $\mathring{Q}$. The typical value of V_{op}, termed as the economic velocity is about 2 m/s for steel pipes.

10.24 An overhead tank containing water is connected to two faucets at the ground level through the piping system shown. D_i, L_i, f_i, and V_i are the diameter, total length, friction factor and average velocity in the ith section. This is a model for typical pipe-network problems which are solved in a fashion similar to those of electrical networks. Write down the conditions on the pressure drops and flow rates through the three sections and work out a flow chart for obtaining the total flow rate. Note that the electrical analogy cannot be carried much further because the pipe resistance is not *linear*.

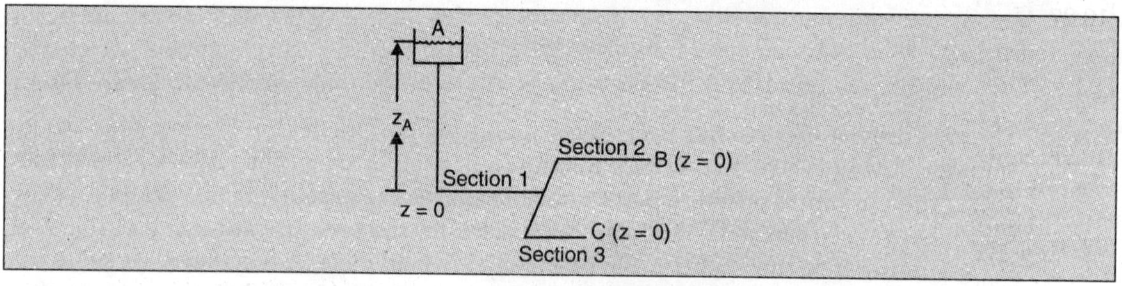

10.25 *Equivalent diameter:* The *turbulent* flow through a pipe of non-circular cross-section is often approximated by flow through an equivalent circular pipe. The diameter of such a pipe is determined from the fact that the shear stress at walls in turbulent flow depends *essentially* on the Reynolds number, independent of the cross-sectional shape (provided the cross-section is not too thin). Show by considering the balance of pressure forces and shear forces on the walls that a circular pipe having a diameter $D_e = 4A/P$ (where A and P are the area and perimeter of the cross-section of the non-circular pipe respectively) has the same pressure drop across a given length as the non-circular pipe when the shear-stress at the walls is identical in the two cases. This diameter D_e is termed the *equivalent diameter** and is used as the characteristic length in flow through

---

*A quantity more often used in the literature is *hydraulic radius* R_H defined as A/P, which is one-quarter of D_e.

non-circular pipes. The pressure drop across the equivalent circular pipe can then be obtained from Moody's chart.

Find the pressure drop across a 30 m length of a steel conduit formed as an annulus between a tube of OD = 1 cm and another of ID = 2 cm. The flow rate of water is 2.5×10^{-3} m³/s.

10.26 *Open channel flows:* The concept of equivalent diameter or hydraulic radius R_H (see Prob. 10.25) can also be used to approximate flows in open channels. Here P refers to the wetted perimeter of the cross-section and A to the flow area. For a channel of slope $S : 1$, show that the average velocity can be written as

$$V_{av} = C\sqrt{\frac{1}{f}}\sqrt{R_H S}$$

where f is the friction factor, and C is a constant. Then use the fact that Colebrook's Eq. (10.15) gives $f \sim (\varepsilon/D)^{1/3}$ in the fully turbulent rough pipe approximation, to show that the average velocity is given by

$$V_{av} = \frac{A}{\varepsilon^{1/6}} R_H^{\ 2/3}\ S^{1/2}$$

where A is a proportionality constant. This is known as the *Chezy formula* with $\varepsilon^{1/6}/A$ replaced by n which is termed as the *Manning roughness coefficient.*

10.27 Use the development of Prob. 10.26 to predict the volume flow rate through an earth drainage-ditch of rectangular cross-section (of width 1 m) and longitudinal slope of 1:50 when the level in it is 50 cm. The Manning n for earth drainage ditches in fair condition can be taken as 0.020, with R_H in m and V_{av} in m/s.

10.28 Sherlock Holmes stood contemplating at the site of crime at the edge of a circular sewer channel of 4 m diameter. The dead body had been found at 11.45 p.m. floating in the channel 7.2 km downstream. The suspect had a foolproof alibi for up to 9.15 p.m. One of the sewage plant operators had informed Sherlock Holmes that till the time the body was found, the water height had been less than a quarter of the diameter of the channel. The channel was now half-filled and a piece of wood passed by at 1 m/s. Sherlock Holmes made a quick calculation and ordered the suspect released. Can you substantiate mathematically what went on in Sherlock Holmes' mind. Assume fully turbulent flow. (Adapted from W.J. Beek and K.M.K. Muttzall, *Transport Phenomena,* Wiley, 1975. Refer to this book for more such problems). Hint: Use the results of Prob. 10.25.

10.29 *Flow through porous beds:* The flow through a bed of sand particles shown can be modelled as a flow through a bundle of n capillaries, each of diameter D_e, the equivalent diameter of the bed (as defined in Prob. 10.25). The number n is chosen so that the average velocity of the flow is the same through capillaries as in the bed. If a bed of length L_p contains N sand particles, assumed spherical of diameter D_s, show that the average perimeter is $N\pi\, D_s^2/L_p$, the void fraction ε is $1 - \dfrac{N}{A_p L_p}\dfrac{\pi}{6}D^3$, and, therefore,

$$D_e = \frac{2}{3}\left(\frac{\varepsilon}{1-\varepsilon}\right)D_s$$

The flow through the bed of spheres is usually at low enough Reynolds numbers so that the friction factor f can be replaced by k/Re, where k is a constant. Using the pipe loss Eq. (10.9) show that the head loss through the packed bed of spheres is

$$h_l = \frac{9}{4}\frac{k}{2g}\frac{\mu}{\rho}\frac{(1-\varepsilon)^2}{\varepsilon^3}\frac{V_p}{D_s^2}$$

This is known as the *Kozeny equation* and denotes the direct proportionality of V_p to the level of liquid above a porous bed (as first observed by Darcy).

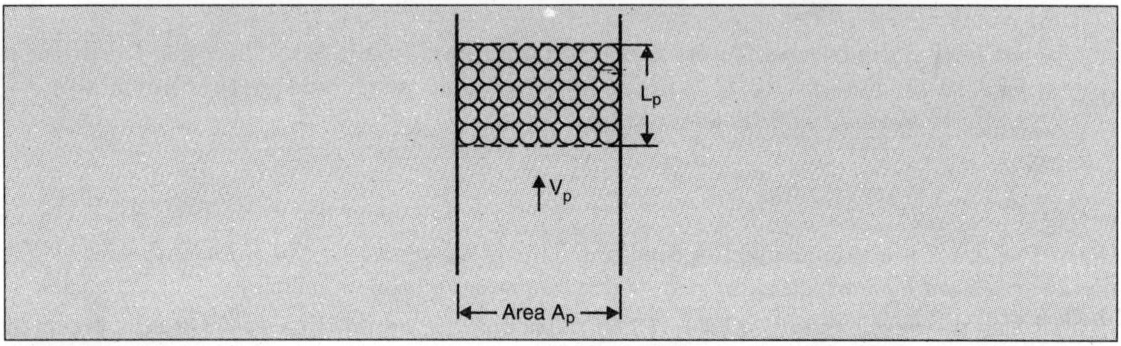

10.30 Show that the modelling and prediction rules for a pump deduced in Example 9.3 are consistent with the non-dimensional parameters developed in Sec. 10.5.

10.31 *Fan rules:* Show that a 600 mm exhaust fan delivers eight times as much volume flow rate, develops four times as much pressure, and consumes thirty two times as much power as a 300 mm geometrically similar fan running at the same RPM. If the speed is now increased by ten per cent, what is the per cent change in performance? These variations are usually condensed into what are termed as fan rules:

$$\mathring{Q} \sim D^3 n$$
$$\Delta\mathscr{P} \sim D^2 n^2$$
$$\mathring{W} \sim D^5 n^3$$

Justify.

10.32 When the voltage V supplied to the induction motor of a fan changes, the torque developed by it changes as V^2. Show using the fan rules developed in Prob. 10.31 that the speed n and the air displacement $\mathring{Q}$ vary as V.

10.33 A pump having $D = 50$ cm, $n = 157$ rad/s and whose dimensionless characteristics are given in Fig. 10.13, is used to deliver water. What are the values of $\Delta\mathscr{P}$, $\mathring{Q}$, η_H and $\mathring{W}$ if the pump is operating at the design point?

10.34 Generate dimensional $H(\mathring{Q})$, $\mathring{W}(\mathring{Q})$ and $\eta(\mathring{Q})$ curves for a centrifugal water pump whose non-dimensional characteristics are given in Fig. 10.13, if

(a) $D = 0.81$ m, n = 122.5 rad/s

(b) $D = 0.71$ m, $n = 122.5$ rad/s

10.35 Compute the value of n and D required for a pump whose non-dimensional characteristics are given in Fig. 10.13, such that it operates under near-optimal conditions when it delivers 1.26 m^3/s of water against a 125 m head.

10.36 A pumping operation requires a flow rate of 1.26 m^3/s against a head of 7.5 m. A pump from the family whose performance characteristics are shown in Fig. 10.13 is to be used. Choose the values of n and D for optimal operation. Note that this family of pumps is unsuitable for direct coupling with induction motors. If a 900 RPM motor is to be used, what type of pump would be preferable for this duty?

10.37 A 70 cm dia. centrifugal pump whose characteristics are shown in Fig. 10.13 is delivering water at 1.26 m^3/s against a 100 m head. What is its speed and power consumption? Note that the pump is not necessarily running at the design point.

10.38 Three identical pumps whose characteristics are given in Fig. 10.13 are used (a) in series, (b) in parallel. Draw the non-dimensional head *vs.* discharge curves for the two combinations.

10.39 A centrifugal pump of the type whose characteristics are shown in Fig. 10.13 rotates at 1200 RPM and has an impeller diameter of 10 cm. It is used to pump water out from the tank shown in Prob. 7.11, with $L = 1$ m and $z_i = 5$ m. Obtain the steady level, z_f.

10.40 A 100 cm dia. centrifugal pump having characteristics shown in Fig. 10.13, and rotating at 122.5 rad/s, is used to deliver water in the system shown. If minor losses are neglected and the flow is assumed fully turbulent, what is the water throughput and the power consumption? Note that the pump is not necessarily operating at its optimal point.

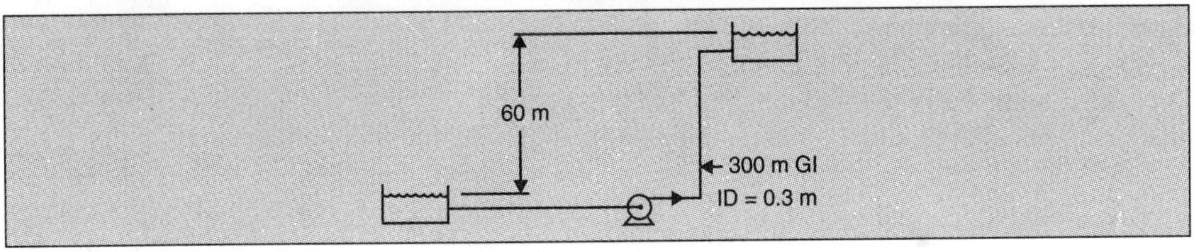

10.41 An *S.F. Fläkt* fan has the following specifications:

$\mathring{Q} = 10{,}000$ m^3/hr, $\Delta p = 200$ Pa, RPM = 980

$\eta = 83\%$ and $D = 625$ mm

What type of impeller does this employ?

10.42 The following information is available on some of the major hydroelectric installations in India:

Installation	Head (m)	$\overset{\circ}{W}$ (MW) per turbine	RPM	Installed turbine
Pykara (see Prob. 8.10)	855	14.4	600	Pelton
Koyna	475	65.2	300	Pelton
Sharavathi	442	93.25	300	Pelton
Bhakra	120	111.9	166.7	Francis
Rihand	68.5	57.44	150	Francis
Tungabhadra	20	10.30	214	Kaplan
Damodar Valley	20	1.87	250	Francis
Ganga Canal	9.62	7.2	125	Kaplan

Assuming hydraulic and mechanical efficiencies to be 0.9 each, calculate the specific speed for each application and justify the type of turbine used.

10.43 *Draft tube:* The power output of a reaction turbine is maximised by minimising the head at its outlet. For this reason one attempts to place the outlet as close to the tail water race as possible. This creates difficulties in inspection and maintenance of the turbine. For this purpose it is a common practice to locate the turbine at a convenient height and use a *draft tube* as shown. The use of the draft tube reduces the pressure level at point 1 thereby increasing the power output.

Compare the pressure at point 1 when a draft tube is used to that when it is not used. Assume that the full velocity head at point 2 is lost at the exit, and neglect friction losses in the draft tube. Show that H cannot exceed a certain critical value.

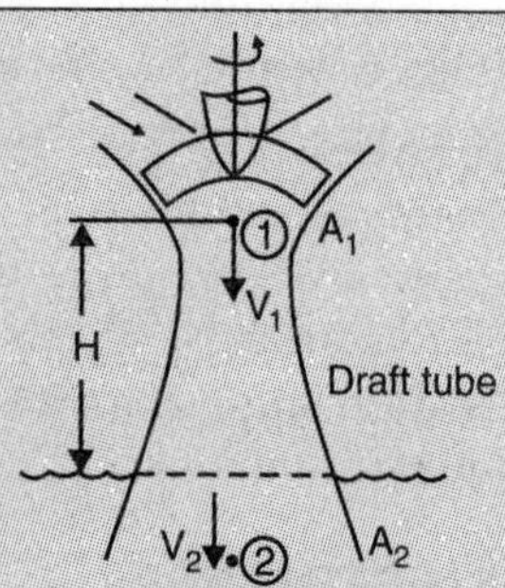

11

APPROXIMATIONS IN FLUID MECHANICS

11.1 INTRODUCTION

The equations governing the flow of fluids form a system of simultaneous, second-order, non-linear, partial differential equations which is formidably complex. Closed-form analytical solutions to these have been obtained for very simple situations only. In these cases one or more of the terms from the flow equations drop out because of such simplifying features as full-development in the flow direction (leading to inertial terms being identically zero), steady state, one-dimensionality of the flow field, etc. These were used in the few exact solutions that were obtained in Chapter 6. But such features are not encountered in all situations and one has to resort to simplifying approximations.

Many such approximations have already been encountered, for example, quasi-steady state (Example 7.4) or quasi one-dimensionality (Example 6.2), negligible viscosity, constant material properties such as density, viscosity, etc. All such approximations lead to neglecting some of the terms in the governing equations rendering them relatively easier to solve. Uptil now such approximations were made without establishing the conditions under which these were valid. In this chapter we attempt to provide a rational basis for such approximations.

11.2 ORDER OF MAGNITUDE ESTIMATES

In order to establish which terms in the governing equations are negligible, their magnitudes have to be estimated. One very powerful method of making such estimates is to first non-dimensionalize each variable by scaling it with its characteristic value. The non-dimensional variables are then expected to be of order unity, that is, except at certain isolated points in the flow field, the values of the non-dimensional variables are neither very large nor very small *compared to one*. Further, it is hoped that if the values used in the non-dimensionalization above are chosen properly, the dimensionless derivatives of the various physical quantities appearing in the equations are also of order one*. To illustrate this, consider the steady flow of

* It might be added that the identification of the *appropriate* characteristic values is a matter of skill and experience, and the beginner should not feel discouraged by being unable to spot them right away.

an incompressible fluid past a circular cylinder of radius L_0 (Fig. 11.1), the governing equations of which are

$$\nabla \cdot \mathbf{V} = 0$$

$$\rho \mathbf{V} \cdot \nabla \mathbf{V} = -\nabla p - \rho g \,\hat{\mathbf{k}} + \mu \nabla^2 \mathbf{V} \quad \text{...(11.1)}$$

The boundary equations are

$$\mathbf{V} = 0 \quad \text{on} \quad x^2 + z^2 = L_0^2$$

$$\mathbf{V} = V_0 \,\hat{\mathbf{i}} \quad \text{as} \quad x, z \to \pm \infty$$

$$p = p_0 \quad \text{at} \quad z = 0 \text{ as } x \to -\infty \quad \text{...(11.2)}$$

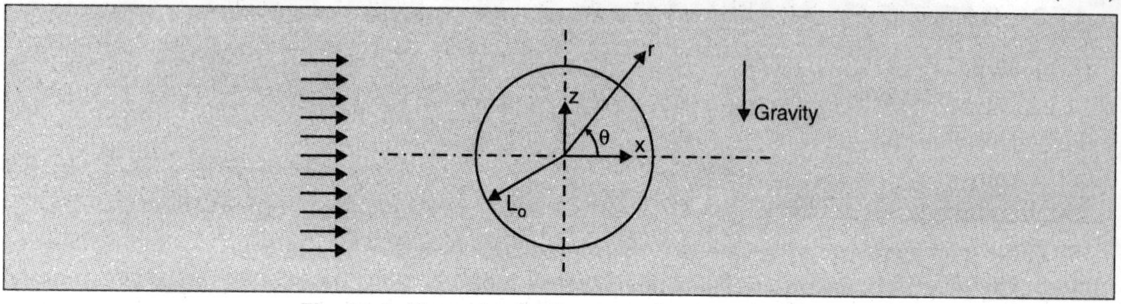

Fig. 11.1. Flow of a fluid past a circular cylinder.

It appears reasonable to take V_0, L_0 and p_0 as the values characterising the velocities, lengths and pressures respectively. If Eqs. (11.1) and (11.2) are normalized with these, the resulting equations are

$$\nabla^* \cdot \mathbf{V}^* = 0$$

$$\mathbf{V}^* \cdot \nabla^* \mathbf{V}^* = -\frac{p_0}{\rho V_0^2} \nabla^* p^* - \frac{g L_0}{V_0^2} \hat{\mathbf{k}} + \frac{\mu}{\rho V_0 L_0} \nabla^{*2} \mathbf{V}^* \quad \text{...(11.3)}$$

subject to the boundary conditions

$$\mathbf{V}^* = 0 \quad \text{on} \quad x^{*2} + z^{*2} = 1$$

$$\mathbf{V}^* \to \hat{\mathbf{i}} \quad \text{as} \quad x^*, z^* \to \pm \infty$$

$$p^* = 1 \quad \text{on} \quad z^* = 0 \text{ as } x^* \to -\infty \quad \text{...(11.4)}$$

Equations (11.3) and (11.4) are identical to Eqs. (9.4) and (9.5) respectively. If the choice of the characteristic quantities is *proper* then it is hoped that over most of the region of interest, the variables x^*, z^*, $\mathbf{V}^*$ and p^* and their derivatives $\nabla^* \cdot \mathbf{V}^*$, $\nabla^* \mathbf{V}^*$, $\nabla^* p^*$ and $\nabla^{*2} \mathbf{V}^*$ are all of order one. (This argument is to be used with caution as shall be seen shortly.) If this is so, i.e., if the non-dimensional variables and their derivatives in Eq. (11.3) are all of the same order, then the coefficients of these terms signify their relative importance. Thus, $1/\text{Eu} = p_0/\rho V_0^2$, $1/\text{Fr} = \sqrt{gL_0}/V_0$ and $1/\text{Re} = \mu/\rho\, V_0 L_0$ signify the importance of the pressure, gravity and the viscosity terms respectively, *relative to the inertial term,* the coefficient of which has been made unity. In other words, $\text{Re} = (\rho\, VL/\mu)_c$ is an estimate of the ratio of the inertial and viscous forces in the given flow field, $\text{Fr} = \left(V/\sqrt{gL}\right)_c$ is an estimate of the ratio of the inertial and gravity forces, and $\text{Eu} = (\rho V^2/p)_c$ is an estimate of the ratio of the inertial and pressure forces.* If the equations of motion had been formulated including other types of forces as well, as for example, surface tension, more pi-numbers would have resulted, each serving as an estimate of the ratio of two forces.

11.3 BASIS OF APPROXIMATIONS

If the choice of the characteristic quantities used in the non-dimensional formulation of Sec. 11.2 is indeed proper, then the magnitude of all the non-dimensional variables *and* their derivatives is expected to be of order one and, as before, the pi-numbers serve as the estimates of the ratio of the forces over most of the region of interest. This provides a reasonable basis on which approximations can be made. Thus, if the Reynolds number Re, which is an estimate of the ratio of the inertial and viscous forces in the flow field, is very small, the inertial forces are much smaller than viscous forces over most of the flow region. Hence the inertial terms may be neglected compared to the viscous terms in the equation of motion. This results in a simpler equation which probably can be integrated more easily. If, on the other hand, the value of Re is very large, inertial forces dominate viscous forces, and the latter may well be ignored. (There are important qualifications to this statement, which shall be seen shortly.) Similarly, a low value of Froude number Fr (compared to unity) implies that gravity forces dominate the flow, while large values of Fr signify that gravity forces may well be neglected.

Thus, the non-dimensional pi-numbers provide an estimate of the relative magnitudes of the various forces within the fluid and as such, form a basis of making rational approximations. In Sec. 11.4 and 11.5 two very important approximations encountered in fluid-flow analysis are outlined.

Before going further, it should be noted that the estimates of the magnitudes of the various forces can be obtained even when the applicable equations in their full forms are not available. In such cases, the estimate of each force component is obtained using the physical law governing the phenomenon. For example, the inertial forces (convective) $F_{i,c}$ are given by Newton's law as mass *times* the convective acceleration, so that

$$\begin{aligned} F_{i,c} &= (\delta m)(\mathbf{V} \cdot \nabla\, \mathbf{V}) \\ &= \left(\rho\, L_c^3 \delta m^*\right)\left(V_c^2/L_c\right)\left(\mathbf{V}^* \cdot \nabla^*\, \mathbf{V}^*\right) \\ &\sim \rho\, L_c^2\, V_c^2 \end{aligned}$$

* This was stated without proof as Eq. (9.28) in Sec. 9.3.

since the non-dimensional variables and their derivatives are expected to be of order unity. The estimate of viscous force is provided by Newton's law of viscosity as

$$F_\mu = (\delta A)\mu\left(\frac{\partial V_y}{\partial x} + \frac{\partial V_x}{\partial y}\right)$$

$$= L_c^2\left(\delta A^*\right)\mu\left(V_c/L_c\right)\left(\frac{\partial V_y^*}{\partial x^*} + \frac{\partial V_x^*}{\partial y^*}\right)$$

$$\sim \mu V_c L_c$$

The relative magnitude of inertial and viscous forces is then given by

$$F_{i,c}/F_\mu \sim \rho L_c\, V_c/\mu = \text{Re}$$

as before. Similarly an estimate of any other force component can be obtained.

Example 11.1. In Prob. 9.31, the spreading of an oil slick on sea-water was said to be controlled primarily by the viscous forces the sea-water applies on the slick, the (unsteady) inertial forces and the buoyant weight of the slick. It was mentioned that the surface tension forces are unimportant in the initial period. Obtain an estimate of this initial period during which surface tension forces may be neglected.

First obtain an estimate of the surface tension forces and then compare them with one of the other forces, say, the inertial forces. The surface tension forces are estimated as

$$F_\sigma = \sigma\,(\delta l) = \sigma\, L_c(\delta l^*) \sim \sigma\, L_c$$

where σ is the surface tension between oil and water which tends to spread the slick. The unsteady inertial forces in the slick are estimated as

$$F_i = (\delta m)(\partial V/\partial t\,) = \left(\rho_s L_c^3\ \delta m^*\right)\left(V_c/t_c \cdot \partial V^*/\partial t^*\right)$$

$$\sim \rho_s L_c^3 V_c/t_c$$

where ρ_s is the density of the slick. The surface tension forces are negligible as long as

$$F_\sigma/F_i \sim \frac{\sigma t_c}{\rho_s\ L_c^2\ V_c}$$

is small compared to one. The characteristic length L_c can be taken as $\forall^{1/3}$ where $\forall$ is the volume of oil spilt. The velocity V_c can be estimated as L_c/t_c. The actual time t can be used for t_c, and then, the relevant criterion is

$$\frac{\sigma t^2}{\rho_s \forall} \ll 1$$

or

$$t \ll \sqrt{\frac{\rho_s \forall}{\sigma}}$$

For typical values of ρ_s, σ and $\forall$, this may yield a period of several days during which surface tension forces are not significant.

11.4 LOW REYNOLDS NUMBER FLOWS

When the Reynolds number Re is much smaller than unity, the viscous forces are much larger than the inertial forces. Therefore, the inertial terms can be neglected in Eq. (11.3) to get

$$0 = -\frac{p_0}{\rho V_0^2}\nabla^* p^* - \frac{gL_0}{V_0^2}\hat{\mathbf{k}} + \frac{\mu}{\rho V_0 L_0}\nabla^{*2}\,\mathbf{V}^* \qquad ...(11.5)$$

The effect of gravity is usually combined with pressure forces by defining a non-gravitational pressure $\mathscr{P} = p + \rho g z$ [see Sec. 9.2.5 and Example 6.3] to give

$$0 = -\frac{1}{\text{Eu}}\nabla^* \mathscr{P}^* + \frac{1}{\text{Re}}\nabla^{*2}\,\mathbf{V}^* \qquad ...(11.6)$$

where $\mathscr{P}^*$ is $\mathscr{P}/p_0$. Eq. (11.6) is then the approximate equation for flows which are dominated by viscous forces. In such cases the acceleration of the fluid particles is negligible, and the viscous forces are effectively balanced by pressure forces at every point in the flow field. Changing back to dimensional form, we get

$$-\nabla\mathscr{P} + \mu\nabla^2\mathbf{V} = 0 \qquad ...(11.7)$$

This equation is a linear equation, unlike the full equation, and is evidently a much simpler one to solve. This equation is known as the *Stokes equation* after Sir G.G. Stokes, a 19th century mathematician. Fluid motions characterised by low Re can be approximated by this equation. Such motions are usually called *creeping motions*. Examples include the motion of minute dust particles, cloud droplets, etc. in air, swimming of microbes in body fluids, and the seeping flow of sub-soil water, etc. In all these situations the inertial forces are quite small compared to viscous forces. There are some other situations where, even though the fluid motion cannot be termed creeping, the inertial forces play an insignificant role since the acceleration is very small and thus the Stokes Eq. (11.7) applies. In the fully developed Couette or Poiseuille flows of Examples 6.1 and 6.3 the acceleration of fluid particles is *exactly* zero, and Eq. (11.7) applies strictly. But if the channel cross-section varies slowly enough as in the case of flow of the thin oil film in a lubricated bearing (see Example 6.2), then too the Stokes equation may be applied to obtain good approximate results. (This is equivalent to considering the flow to be quasi-fully developed as was also done in Prob. 6.42.)

The most dramatic and perhaps the most instructive application of the Stokes approximation is to the creeping motion of a sphere in a viscous fluid. The solution for this flow in a spherical polar coordinate system fixed with the centre of the sphere of radius R is given in Prob. 6.5. The streamline pattern and the pressure distribution about such a sphere is shown in Fig. 11.2. A few things are immediately noticed:

(a) The flow pattern is symmetrical, i.e., the streamlines remain the same whether the flow approaches the sphere from the left or the right.

(b) Even though the flow pattern is symmetrical, the pressure distribution is anti-symmetrical. Thus, while the pressure is maximum at the 'nose' of the sphere, it is minimum at the 'tail' even though the velocity is the same. This observation is in marked contrast to the behaviour at large Reynolds numbers (discussed in Sec. 11.4) when a symmetrical velocity distribution implies a symmetrical pressure distribution necessarily.

(c) The fluid velocity is significantly lower than the free-stream value V_0 over a *considerable* distance (compared to *R*). This means that the effect of the presence of the sphere is felt over large distances. This is also unlike that in the high Reynolds number flow discussed in Sec. 11.4.

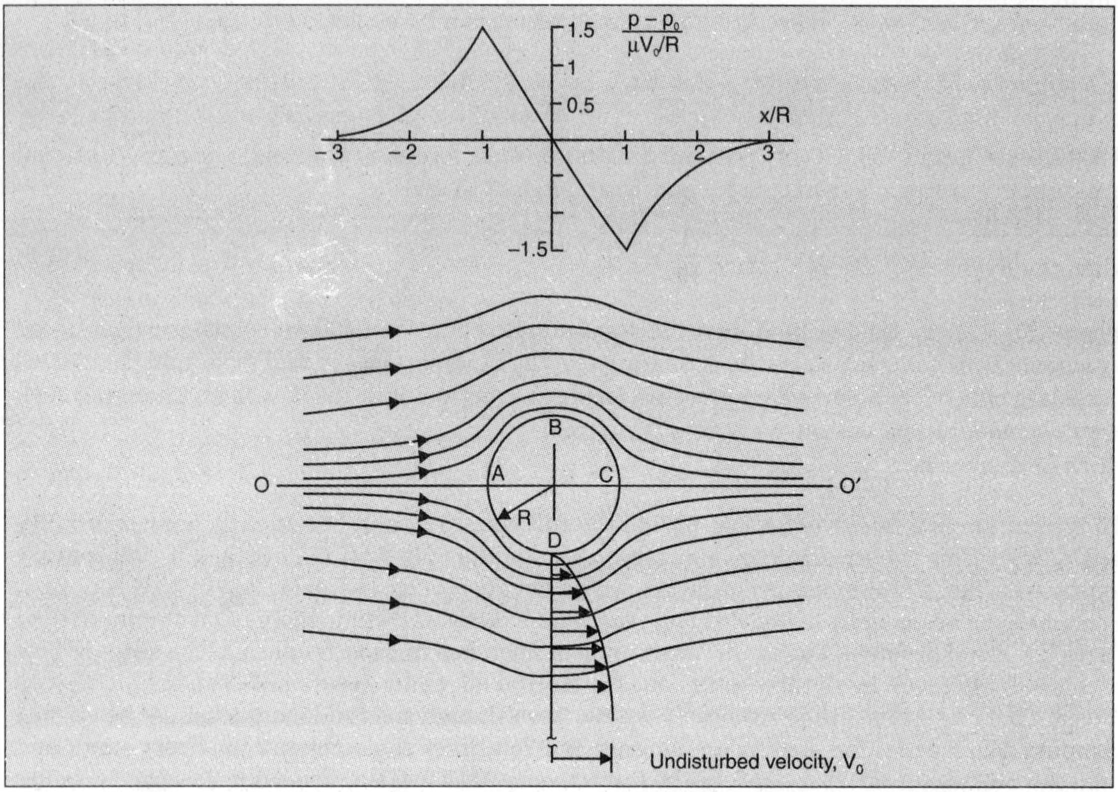

Fig. 11.2. Streamline pattern and pressure distribution on *OA* and *CO* for flow around a sphere at low Re. Values of *p* between *A* and *C* are those on the surface of the sphere.

Because of this influence of the sphere over large distances, small particles moving at low Reynolds numbers interact strongly with one another, i.e., the motion of a particle is affected by the presence of other particles even when their separation is large compared to their diameter. This is in sharp contrast to the motion of larger particles (moving at higher Reynolds numbers) which have almost no such interactions.

(d) The pressure differences near the sphere are of order *one* when scaled by $\mu V_0/R$. Thus, the characteristic pressure difference $(\Delta p)_c$ equals $\mu V_0/R$. This was to be expected, since the pressure forces balance the viscous forces and hence the two terms in Eq. (11.6) must be of the same order. This implies that Eu and Re are of the same order. Thus Eu and Re can be equated without any loss of generality. On recognizing that it is the pressure difference and not the absolute pressure which is significant in this flow (see Sec. 9.2.5), we obtain

$$\left(\frac{\Delta p}{\rho V^2}\right)_c = \left(\frac{\mu}{\rho V L}\right)_c$$

or

$$(\Delta p)_c = \left(\frac{\mu V}{L}\right)_c \qquad \text{...(11.8)}$$

This too is very different from the characteristic value $\left(= \rho V_0^2\right)$ in large Reynolds number flows.

It may be noted that most flows of engineering interest have large Reynolds numbers. Thus, for a sphere to be moving in air at a Reynolds number less than 1, when its velocity is a modest 0.1 m/s, its diameter should be less than 10^{-4} m. We shall therefore, confine our attention mostly to high Re flows, as indeed we have been doing so far.

11.5 HIGH REYNOLDS NUMBER FLOW—THE INVISCID APPROXIMATION

When the Reynolds number is much larger than one, a condition prevalent in engineering fluid flows, the viscous forces are much smaller than the inertial forces, and one may drop the last term in Eq. (11.3). This leads to (with $\mathscr{P} = p + \rho\, gz$)

$$\mathbf{V}^* \cdot \nabla^* \mathbf{V}^* = \bullet \frac{1}{\text{Eu}} \nabla^* \mathscr{P}^* \qquad \text{...(11.9)}$$

which in dimensional form gives

$$\rho(\mathbf{V} \cdot \nabla)\mathbf{V} = -\nabla \mathscr{P} \qquad \text{...(11.10)}$$

This is the *Euler equation* for steady inviscid flows, and consequently the flow is expected to behave like a fluid with zero viscosity. In such a flow, the inertial forces are balanced by the pressure forces alone.

The Euler Eq. (11.10) is a differential equation of order one and is relatively easier to solve than the full Navier-Stokes equation. But one problem arises: a first order equation needs fewer boundary conditions than does a second order one and therefore, there are too many boundary conditions. Since it is the no-slip condition at the body which is due to the action of viscosity, it is reasonable to expect that it is this condition which is to be relaxed as a consequence of negligible viscosity. The velocity boundary condition 11.4a at the body surface is then replaced simply by the condition that no fluid crosses the body surface, i.e., it is a stream surface.

Since Eq. (11.10) is still a non-linear equation, it is not that easy to solve directly. But, as will be shown in Chapter 12, some characteristics of inviscid flows permit one to obtain useful solutions to the Euler equation. This branch of fluid-mechanics which deals with the flow of inviscid fluids is one of the most mathematically developed ones and will be dealt with briefly in Chapter 12.

Figure 11.3 shows the streamline pattern for inviscid flow about a sphere. It also shows the pressure distribution at the surface of the sphere. This is the flow pattern that is expected at large Reynolds numbers, provided the approximation of neglecting viscous terms is justified. Note that, unlike in creeping flows (i.e., low Re limit), the pressure distribution is symmetric. Also, the characteristic pressure is of order ρV_0^2, unlike $\mu V_0/R$ in the creeping flows. This is expected since the two terms in Eq. (11.9) must be of the same order requiring that Eu be of

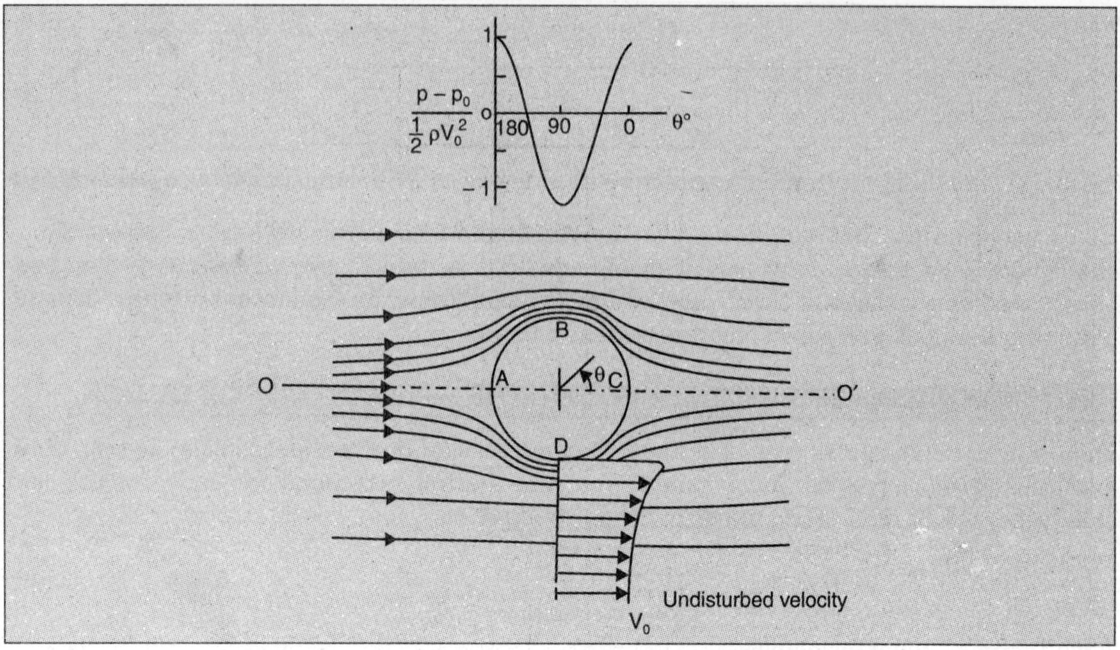

Fig. 11.3. Streamline pattern and pressure distribution along *ABC* for flow about a sphere using the inviscid flow approximation.

order unity. As before, it is the characteristic pressure difference and not the absolute pressure which is significant, and one obtains

$$(\Delta p)_c \sim \left(\rho V^2\right)_c$$

It is conventional to use a factor of one-half in $(\Delta p)_c$. So we set

$$(\Delta p)_c \sim \left(\frac{1}{2}\rho V^2\right)_c \quad ...(11.11)$$

This was the value used in Sec. 9.2.5, since large Reynolds number flows are typical of engineering fluid mechanics.

Another important thing to note is that only the streamlines very close to the sphere are displaced, the streamlines at some distance being nearly straight. The velocity is equal to the velocity V_0 of the free stream only a few diameters away from the sphere. This implies that the range of influence of a sphere in high Re flows is very small compared to that in low Re flows. Thus, the motion of large particles moving at high Re is not affected by other particles when their separation is more than a few diameters.

When the inviscid flow pattern so obtained is compared with the observed flow pattern about a sphere (Fig. 11.4), large differences are noticed immediately. A careful look, however, discloses that while the streamlines and pressures behind the sphere are quite different from those in inviscid flow, those in the front closely match it, except very near the body surface. Thus, in a vast region the inviscid flow solution indeed approximates the high Re flow. (The

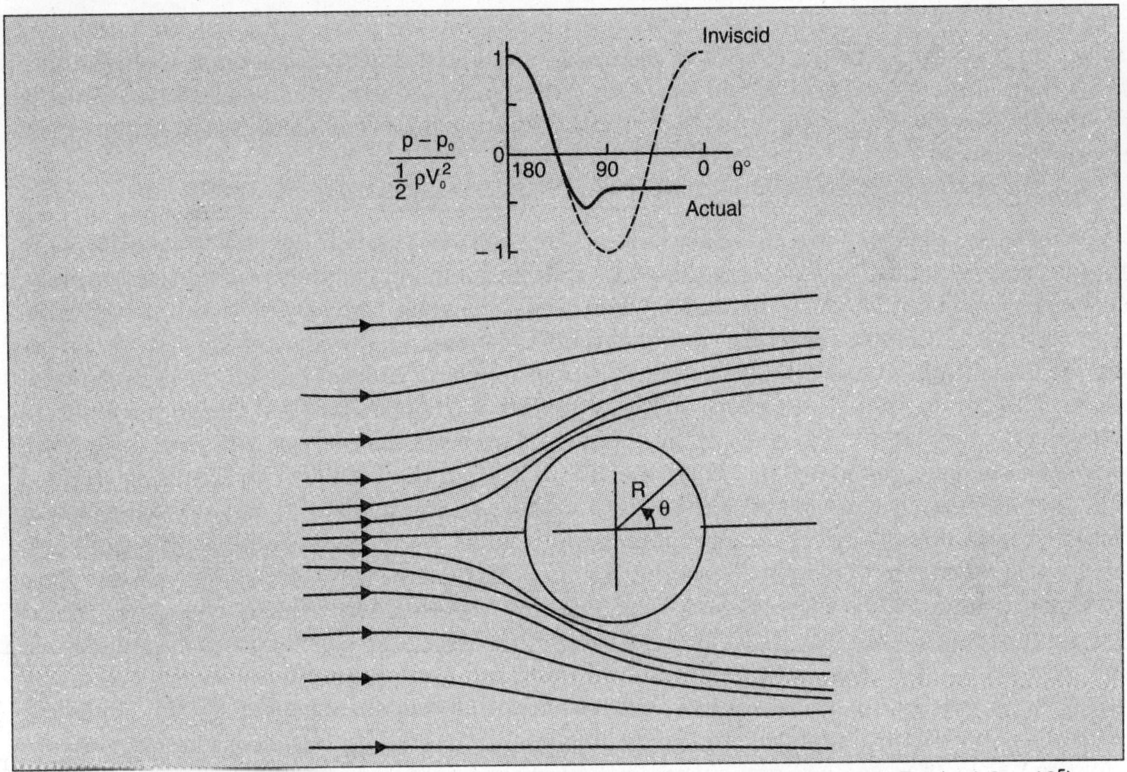

Fig. 11.4. Streamline pattern and pressure distribution about a sphere at high Re (= 1.6 × 10⁵).

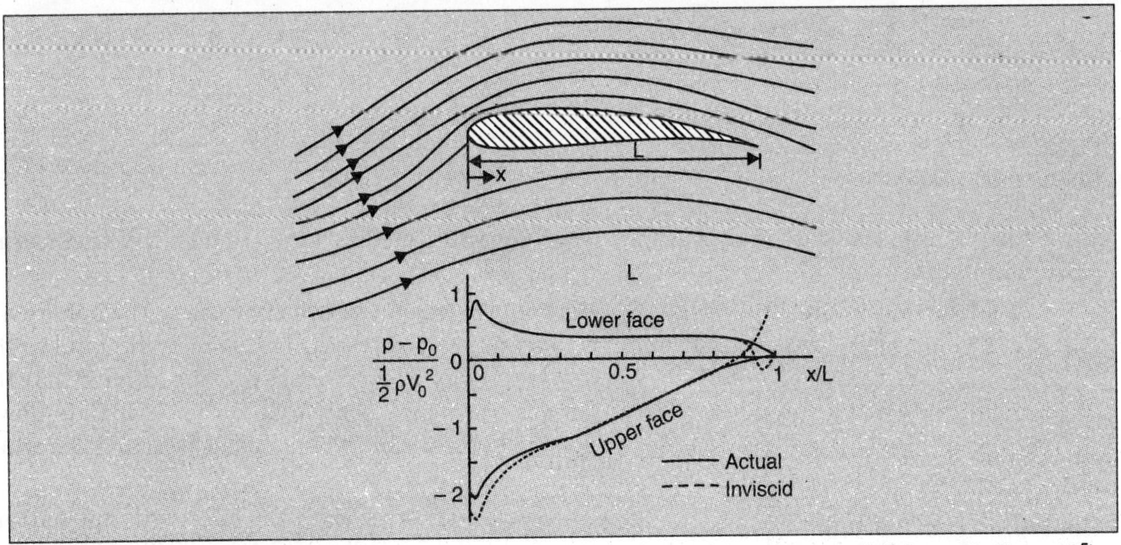

Fig. 11.5. Streamline pattern and pressure distribution for flow around an aerofoil (α = 6°, Re = 10^5).

reason for the large disagreement at the rear lies in the phenomenon termed separation — see Sec. 1.7 and 13.6). Actually, if instead of a sphere, a 'streamlined' body had been considered where separation is suppressed because of its shape, such as an aerofoil (Fig. 11.5), the inviscid flow solution would be quite close to the actual flow observed when the Reynolds number is high. The pressure in such a case is so accurately predicted that the science of aerodynamics relies, to a large extent, on inviscid flow calculations alone to predict lift forces on bodies in flight. (The prediction of drag is, however, another matter, as will be discussed in Chapter 13).

11.6 BOUNDARY LAYERS IN HIGH REYNOLDS NUMBER FLOWS

The reason for the departure of the actual flow from the inviscid flow which it was expected to approximate is not far to seek. It is due to the relaxation of one boundary condition, namely, the no-slip condition. Though it is reasonable to drop the viscous term saying that it is negligible compared to the other terms in the equation, there is no meaning to the statement that the no-slip condition applies to a negligible extent. It either applies or it does not, and the experimental evidence suggests that it continues to apply for all Reynolds numbers. Thus, the inviscid approximation which requires a slip at the solid boundary breaks down near a surface. L. Prandtl, a German scientist, postulated in 1903 in a celebrated paper the presence of a thin layer, termed as a *boundary layer,* close to the walls within which the viscous effects cannot be neglected. Outside this boundary layer, however, the viscous effects are indeed negligible and therefore, the inviscid approximation holds. Thus, Prandtl divided the flow field into two regions, one close to the body where viscous effects are not negligible, and the other away from it where the inviscid flow approximation holds (Fig. 11.6). To amplify this, consider the steady flow of an incompressible fluid parallel to a flat plate at large Reynolds numbers. If the viscous effects are neglected, one gets $V = V_0$, the far upstream velocity, everywhere. But this does not satisfy the boundary condition of no-slip at the plate. For the no-slip condition to hold, the viscous action must be important at least in some region of the flow. According to the Prandtl hypothesis, there is a

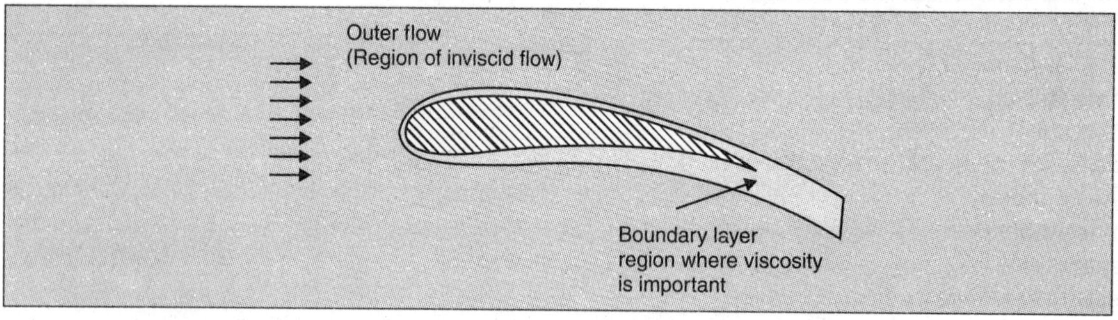

Fig. 11.6. Division of flow field into inviscid flow region and boundary layer.

small region of thickness δ near the plate over which the viscous effects are important and the velocity changes sharply from its no slip value of zero at the surface to the inviscid value V_0 at the edge of this layer. The small thickness of this layer ensures a large velocity gradient in the normal direction making the viscous forces significant.

This implies that V_0/L_0 is an incorrect estimate of the velocity gradient $\partial V_x/\partial y$ in the normal direction. To obtain a correct estimate, one must use δ, the (yet unknown) boundary layer thickness as the proper characteristic length *in the normal direction.* Thus, there are two characteristic lengths in this problem: L_0, which characterizes the lengths along the surface and δ which characterizes the lengths normal to the surface.

Therefore, the proper non-dimensionalization within the boundary layer is $x^* = x/L_0$ and $\tilde{y} = y/\delta$. The non-dimensional velocity $\mathbf{V}^*$ when differentiated with respect to these lengths x^* and $\tilde{y}$, give velocity gradients* which are of order one.

The x-component of the Navier-Stokes equation valid for flow within the boundary layer now becomes:

$$V_x^* \frac{\partial V_x^*}{\partial x^*} + \frac{L_0}{\delta} V_y^* \frac{\partial V_x^*}{\delta \tilde{y}} = -\frac{1}{2}\frac{\partial \mathscr{P}^*}{\partial x^*} + \frac{1}{\mathrm{Re}_{L_0}} \frac{\partial^2 V_x^*}{\partial x^{*2}} + \frac{1}{\mathrm{Re}_{L_0}} \frac{L_0^2}{\delta^2} \frac{\partial^2 V_x^*}{\partial \tilde{y}^2} \quad \text{...(11.12)}$$

where Eq. (11.11) has been used for Δp_c and where the subscript L_0 in Re_{L_0} refers to the fact that the characteristic length used in it is L_0. If Re_{L_0} is large, then the first of the viscous terms in Eq. (11.12) is still negligible compared to the inertial terms. But, since within this boundary layer, the viscous effects are not negligible, the second viscous term must be of the same order as the inertial terms, i.e., of order unity. Thus

$$\frac{1}{\mathrm{Re}_{L_0}} \cdot \frac{L_0^2}{\delta^2} \sim 1$$

or

$$\frac{\delta}{L} \sim \frac{1}{\sqrt{\mathrm{Re}_{L_0}}} \quad \text{...(11.13)}$$

This result is of great importance in fluid mechanics. The thickness of the boundary layer within which the viscous forces are significant shrinks as the Reynolds number increases.

In Chapter 13, the boundary layer shall be studied in greater detail. One important characteristic of a boundary layer (already referred to in Sec. 1.7) is its tendency to 'separate' from the surface whenever the flow outside it is slowing down. When this happens the viscous effects are no longer confined to a narrow region and the inviscid approximation breaks down. The boundary layer on a gently tapering 'streamlined' body is not separated and, therefore, the inviscid approximation gives very good results. But for a 'bluff' body like a sphere, the flow separates at the rear and a broad wake results in which the viscous effects predominate. These aspects shall be dealt with again in Chapter 13.

11.7 APPROXIMATIONS IN UNSTEADY FLOWS**

In this section a little more complex example of approximations in fluid flow is considered. This example illustrates, more explicitly, the meaning of the characteristic values of the physical quantities, and the origins of the boundary layer due to the necessity of using two different values to characterize a physical variable in different regions of the flow field.

Consider the problem of unsteady flow between two infinite parallel plates, a distance L_0 apart. The lower plate is held stationary and the upper one oscillates in its plane (Fig. 11.7)

---

* The y-velocity V_y within the boundary layer is not characterised by V_0, so that V_y^* *here* is not of order 1. This consideration is postponed to Chapter 13, since it is likely to divert the reader's attention here.

** This section is perhaps a little too involved for a beginner and may be omitted.

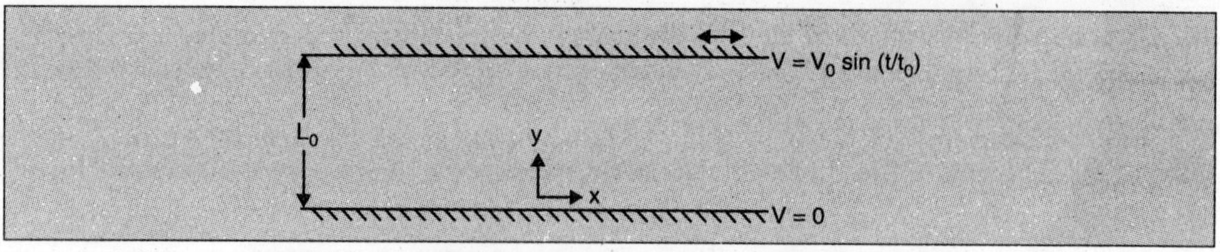

Fig. 11.7. Unsteady flow between two infinite parallel plates.

with a velocity $V_0 \sin(t/t_0)$. Due to this oscillatory motion, a fluid (of density ρ and viscosity μ) which fills the space between the plates is set in unsteady motion. Since the motion is 1-*D*, and the non-gravitational pressure $\mathscr{P}$ can be shown to be constant, the governing equation simplifies to

$$\rho \frac{\partial V_x}{\partial t} = \mu \frac{\partial^2 V_x}{\partial y^2} \qquad ...(11.14)$$

subject to the boundary conditions

$$\begin{aligned} V_x &= 0 && \text{at} \quad y = 0; \quad \text{for all } t \\ V_x &= V_0 \sin(t/t_0) && \text{at} \quad y = L_0; \quad \text{for all } t \end{aligned} \qquad ...(11.15)$$

It appears reasonable to take L_0, V_0 and t_0 as the characteristic values of length, velocity and time. This means that we are *expecting* the velocity to change from 0 to V_0 over a length L_0 and over a time period of order t_0. Introducing the non-dimensional variables

$$y^* = y/L_0, V_x^* = V_x/V_0, \text{and } t^* = t/t_0$$

The governing equation and the boundary conditions then become

$$\frac{\partial V_x^*}{\partial t^*} = \frac{\mu t_0}{\rho L_0^2} \frac{\partial^2 V_x^*}{\partial y^{*2}} \qquad ...(11.16)$$

and

$$\begin{aligned} V_x^*\left(y^* = 0, t^*\right) &= 0 \\ V_x^*\left(y^* = 1, t^*\right) &= \sin\, t^* \end{aligned} \qquad ...(11.17)$$

The non-dimensional velocity V_x^* therefore, is a function of t^* and the similarity parameter $\mu t_0/\rho L_0^2$:

$$V_x^* = \mathcal{F}\left[y^*, t^*; \frac{\mu t_0}{\rho L_0^2}\right] \qquad ...(11.18)$$

If V_0, L_0 and t_0 are indeed the *proper* characteristic values, the non-dimensional derivatives $\partial V_x/\partial t^*$ and $\partial^2 V_x^*/\partial y^{*2}$ must be of order unity, and since Eq. (11.16) is a statement of the equality of inertial (unsteady) forces and viscous forces, the Π-number $\mu t_0/\rho L_0^2$ should be of order one. If

$\mu t_0/\rho L_0^2$ is not of order one, then, clearly, one or more of the values chosen as characteristic are not really so.

First consider the case when $\mu t_0/\rho L_0^2 \gg 1$. Then from Eq. (11.16), both $\partial V_x^*/\partial t^*$ and $\partial^2 V_x^*/\partial y^{*2}$ cannot obviously be of order unity. There are two possibilities:

(a) $$\frac{\partial^2 V_x^*}{\partial y^*} \sim 1 \quad \text{and} \quad \frac{\partial V_x^*}{\partial t^*} \gg 1$$

or

(b) $$\partial V_x^*/\partial t \sim 1 \text{ and } \partial^2 V_x^*/\partial y^{*2} \ll 1$$

The first possibility, i.e., $\partial V_x^*/\partial t^* \gg 1$ implies that a wrong value has been chosen to characterize the variable t, and t^* is not of order one but much less than it. To make it of order one, then, a t_c should be used which is smaller than t_0. But a characteristic time much smaller than t_0 is not reasonable, since it implies that the velocity within the plates fluctuates much faster than the velocity at the boundary. Thus, the second possibility (b) must hold, i.e., $\partial^2 V_x^*/\partial y^{*2}$ is much less than one, and setting

$$\frac{\partial^2 V_x^*}{\partial y^{*2}} = 0 \qquad \text{...(11.19)}$$

as an approximation gives a linear velocity profile at *any fixed time*. The boundary condition applicable to this equation is the instantaneous velocity at the upper plate. Thus, the velocity at every point changes with time though the profile across the 'slab' is linear at every instant. This is the quasi steady-state approximation — the upper plate velocity changes so slowly that the flow picture at any instant resembles the corresponding steady state solution though the profile itself changes with time. This is equivalent to saying that the characteristic time t_0 of the velocity fluctuations is large compared to the characteristic penetration time L_0^2/ν (see Prob. 6.37), and hence all points in the 'slab' respond almost instantaneously to any variations in the plate velocity. (Alternately, the slab thickness L_0 is much smaller than the penetration depth $\sim \sqrt{\nu t}$).

Next consider the case $\mu t_0/\rho L_0^2 \ll 1$. Once again there are two possibilities:

(a) $\partial^2 V_x^*/\partial y^{*2} \sim 1$ and $\partial V_x^*/\partial t^* \ll 1$ and

(b) $\partial V_x^*/\partial t^* \sim 1$ but $\partial^2 V_x^*/\partial y^{*2} \gg 1$

The first possibility permits the approximation

$$\partial V_x^*/\partial t^* = 0 \qquad \text{...(11.20)}$$

which implies a constant velocity across the gap, i.e., the fluid does not respond to the fluctuations of the upper plate velocity at all. This is the case when the fluctuations are so fast (compared to the penetration time L_0^2/ν) that the fluctuating velocity does not have enough time to penetrate into the fluid.

But this simple solution does not satisfy the boundary condition at the upper plate where the velocity changes with time. This necessitates looking into possibility (b), i.e., $\partial V_x^*/\partial t^* \sim 1$ but $\partial^2 V_x^*/\partial y^{*2} \gg 1$. To make this last term of order 1 then, an L_c which is much smaller than L_0 should be used, i.e., the velocity should vary over a length much smaller than L_0. This suggests that there exists a *thin* region where the velocity varies rapidly, and in the remaining region, the earlier condition Eq. (11.20) holds. This is akin to the 'boundary layer' as discussed in Sec. 11.6.

The thickness of this layer, $\delta(=L_c)$, can be obtained by noting that when Eq. (11.14) is non-dimensionalized using this length, all the terms will be of order unity, and therefore, the non-dimensional pi which will now have the form $\mu t_0/\rho\delta^2$ should also be of order unity. Thus

$$\delta \sim \sqrt{\frac{\mu t_0}{\rho}} = \sqrt{\nu t_0}$$

which is the penetration depth for surface fluctuations. The boundary layer thickness δ is small compared to L_0 whenever $\sqrt{\nu t_0}/L_0$ or $\nu t_0/L_0^2$ is small compared to one, which is consistent with the foregoing discussion.

PROBLEMS

11.1 *Free convection:* The y-direction flow of a fluid adjacent to a vertical heated flat plate can be described by the following equation:

$$V_x \frac{\partial V_y}{\partial x} + V_y \frac{\partial V_y}{\partial y} = g\beta(T - T_0) + \frac{\mu}{\rho_0}\left(\frac{\partial^2 V_y}{\partial x^2} + \frac{\partial^2 V_y}{\partial y^2}\right)$$

where T is the temperature, β is the coefficient of thermal expansion $\left(= -\frac{1}{\rho}\left(\frac{\partial \rho}{\partial T}\right)_p\right)$ and ρ_0 is the fluid density at the ambient temperature T_0. The term $g\beta\,(T - T_0)$ represents the buoyancy effects due to the local heating of the fluid. Non-dimensionalize this equation using $x^* = x/L, y^* = y/L, V_x^* = V_x/V_0, V_y^* = V_y/V_0, \theta = \dfrac{T - T_0}{T_w - T_0}$ and show that the flow is governed by two dimensionless parameters, Re_L and the Grashof number Gr defined by

$$Gr \equiv \frac{g\beta L^3 (T_w - T_0)}{\nu^2}$$

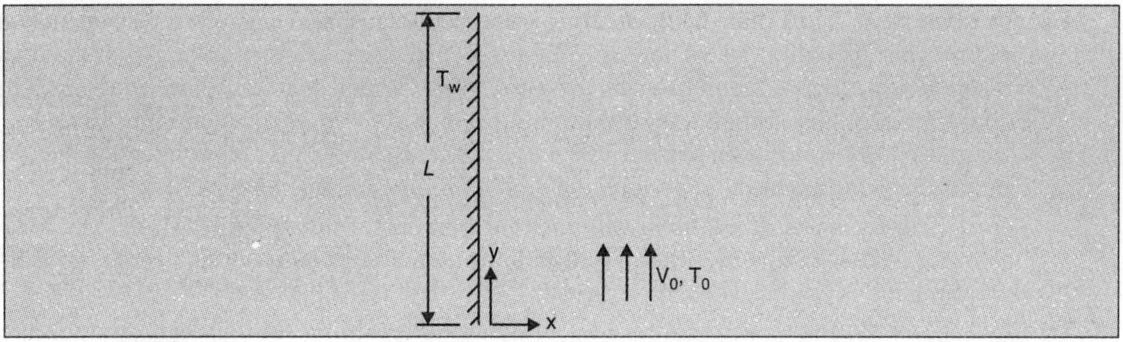

Then deduce the conditions under which the buoyancy of the fluid is unimportant, that is, the free-convection effects can be neglected. If air at 300 K flows past a 30 cm long plate at 100°C at 10 m/s, are the buoyancy terms negligible? Use $\beta = 1/T$.

For the case when $V_0 = 0$, obtain an estimate of the induced fluid velocity.

11.2 The equation of motion for solid particles in a centrifuge was written and non-dimensionalized in Prob. 9.6. Determine the conditions under which the inertial terms may be neglected. When this condition holds, the particle is in quasi-steady motion and is settling at the *local* terminal (radial) velocity.

11.3 The equation describing inviscid compressible flow was written and non-dimensionlized in Prob. 9.5. For steady flow of such fluids, obtain the conditions under which the effects of compressibility (i.e., the influence of the thermodynamic parameter *c*, the speed of sound) can be neglected. Also obtain the conditions under which the flow may be assumed to be steady.

11.4 The differential equation describing the flow of a gas undergoing a chemical reaction $A \to B$, in a circular pipe shown is given by

$$V\frac{\partial \rho_A}{\partial z} - \frac{\mathcal{D}_{AB}}{r}\frac{\partial}{\partial r}\left(r\frac{\partial \rho_A}{\partial r}\right) + k\rho_A = 0$$

where V is the velocity at the radial position r, ρ_A is the concentration of A in kg/m^3, $\mathcal{D}_{AB}$ is the radial diffusivity in m^2/s, and k is the reaction rate constant in s^{-1}. Non-dimensionalize this relationship using

$$z^* = z/L,\ r^* = r/D,\ \rho^*A = \rho_A/\rho_{A,0} \text{ and } V^* = V/V_{max}$$

Discuss the significance of the dimensionless groups obtained. Under what conditions would the first term (representing convection) dominate the second term (representing diffusion)? Under what conditions would the effect of chemical reaction be negligible?

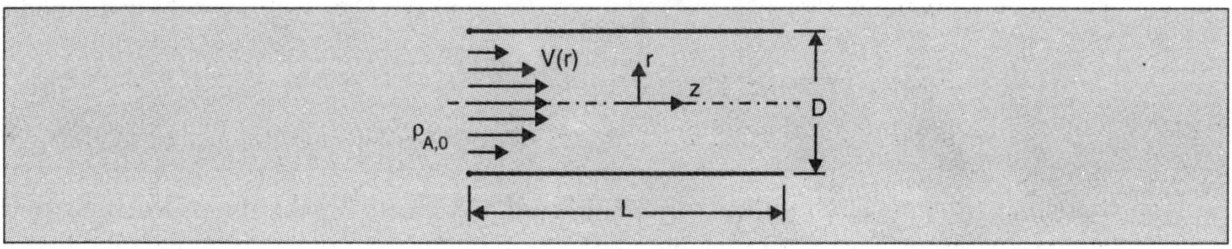

11.5 In Example 6.2 and Prob. 6.42, steady viscous flow in a channel of slowly varying cross-section was approximated by a quasi-fully-developed viscous flow solution (where inertial terms are negligible compared to pressure and viscous terms). Show that such an approximation is justified only if the magnitude of dA^*/dx^* is much smaller than Re_{L_0} where A^* = Area of flow/characteristic area, $x^* = x/L_0$, and L_0 is a characteristic length. (A similar simplification is possible for flow in a pipe having wavy walls.)

Hint: The estimate of $\partial V_x/\partial x$ in the inertial term is not provided by V_0/L_0 when the channel is slowly varying. Instead, it is obtained from the continuity equation VA = constant.

11.6 In Prob. 9.33, the various forces important in determining the unsteady motion of a solid sphere in a stationary fluid were enumerated. Show that unsteady inertial terms for the fluid are unimportant at times greater than $\rho_f D^2/\mu$ where D is the diameter of the sphere. Also show that the motion of the solid particle cannot be assumed to be steady for times less than $\rho_s D^2 \mu$. For a glass sphere of diameter 1 mm settling in water, estimate these two characteristic times. Explain the physical significance of these two times.

11.7 In modelling the flow in estuaries, harbours, etc., the surface tension effects in the model often become significant even though they are unimportant in the real system (refer Prob. 9.22). Obtain the criterion which determines if the surface tension plays a significant role in fluid flow in such models. In a model test of an estuary (using water) a characteristic dimension of 2.5 cm and a characteristic velocity of 5 cm/s are used. Determine if surface tension effects are negligible.

11.8 *Decay of vortex:* When an aircraft takes off from a runway, a swirling flow is set up in its wake. This swirling flow is termed as the 'starting vortex' and the flow within it is modelled by a free vortex (see Sec. 12.2) in which the velocity components are given by

$V_\theta = \dfrac{k}{r}$ and $V_r = 0$. The large velocity gradients near the origin set up significant viscous

forces, changing the flow pattern with time, as shown, and dissipating the energy of this vortex. The region where viscous forces are predominant and the velocity differs significantly from the free vortex field is termed the core of the vortex. Deduce how the radius of this core varies with time.

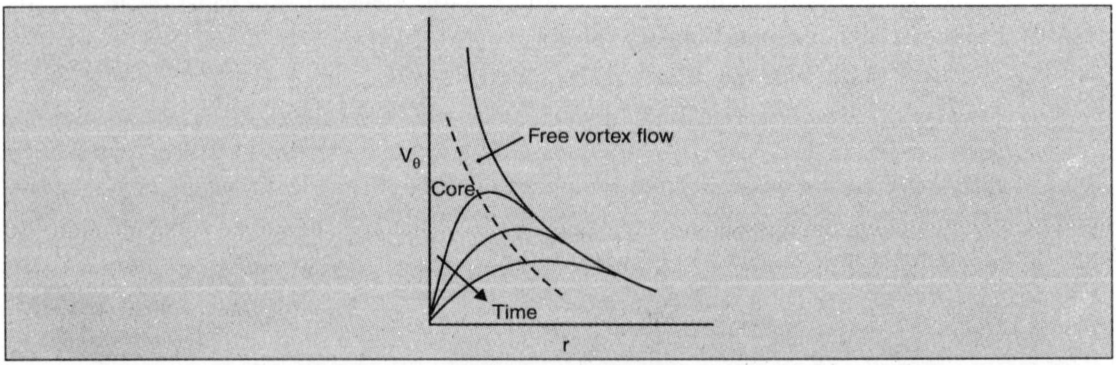

11.9 Consider the flow of a uniform stream of fluid at velocity V_0 about a pulsating sphere whose diameter can be represented as $D_0(1 + \varepsilon \sin \omega t)$. Show that the flow can be approximated by a quasi-steady flow (of a uniform stream about a sphere having the instantaneous value of the diameter) when $\varepsilon St^2 \ll 1$, where $St = \omega D_0/V_0$.

11.10 In a *uniform* flow in a horizontal channel, the pressure variations in the vertical direction are exactly hydrostatic. If, however, the flow varies slowly such that the slope of the free surface is $s \ll 1$, show that the pressure variation in the vertical direction can still be approximated as hydrostatic provided $s\,\mathrm{Fr}^2 \ll 1$.

11.11 Fluid flows through a tube whose walls are undergoing a wave motion such that the cross-sectional area at the axial location x and at time t is given by

$$A = A_0\left[1 + \varepsilon \sin\left(\omega t + \frac{x}{\lambda}\right)\right]$$

Show that the continuity equation for this flow gives

$$\frac{\partial A}{\partial t} + \frac{\partial}{\partial x}(AV_x) = 0$$

assuming that the flow is essentially 1-*D*. (This is a situation similar to the peristaltic motion in intestines described in Prob. 4.29, with $\overset{\circ}{q} = 0$). Obtain an estimate of the inertial force as $\rho V_0\,(\varepsilon\omega + V_0\varepsilon/\lambda)$ where V_0 is the characteristic velocity. Then show that the inertial terms are negligible (compared to the viscous terms) provided

$$\varepsilon\left(\frac{\rho D_0 V_0}{\mu}\right)\left(\frac{\omega D_0}{V_0}\right) \text{ and } \varepsilon\left(\frac{\rho D_0 V_0}{\mu}\right)\left(\frac{D_0}{\lambda}\right)$$

are both much smaller than unity. Thus if e and w are small enough and l is large enough such that the above two conditions are satisfied, the flow can be treated as quasi fully-developed, so that 'locally' the flow may be considered as that through a *straight tube* having the 'local' cross-sectional area.

12

INVISCID FLOWS

12.1 INTRODUCTION

The flow of fluids with zero viscosity occupies an important place in the study of fluid motions because it approximates actual flows at high Reynolds numbers. As seen in Sec. 11.5 viscous forces are negligible compared to inertial forces outside the thin boundary layer and, therefore, the flow in the region away from the walls (termed as the *outer flow*) can be taken as inviscid (Fig. 12.l). Since the boundary layer is a very thin region (of thickness of order $1/\sqrt{\mathrm{Re}}$) the validity of the inviscid flow may be extended to the wall itself as a first approximation. This simplification is possible, strictly speaking, only when the boundary layer does not separate from the wall (e.g., in flow around streamlined bodies), because when it does so, it can no longer be taken as thin. In this chapter, we look at some of the main characteristics of inviscid flows. Even though the Euler Eq. (11.10) which governs inviscid flows is a first order differential equation, it is non-linear and its solution is not very straightforward. But there is a class of flows, as shall be seen here, where only a linear equation needs to be solved and these are amenable to solution by the method of superposition. This class consists of what are termed as irrotational flows.

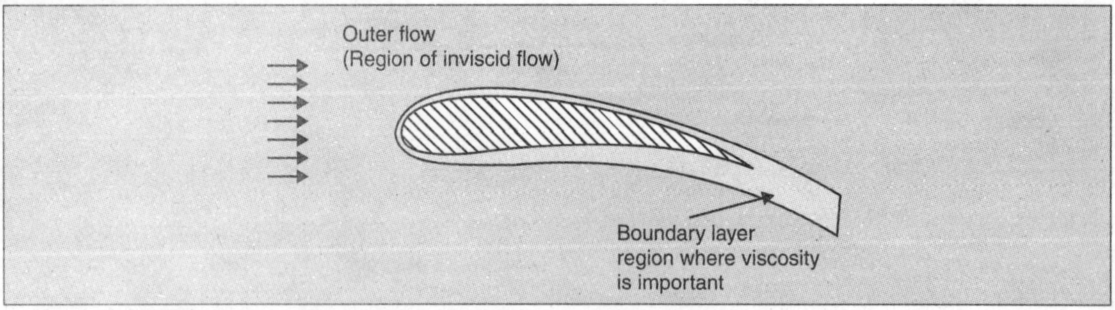

Fig. 12.1. Two regions of flow when fluid flows past a submerged body.

12.2 IRROTATIONAL FLOWS

As shown in Sec. 6.3, the motion of a fluid element can, in general, be resolved into four components—translation, dilatation, rotation and distortion. The rate of rotation of a fluid element can be measured by the average rate of rotation of two perpendicular line segments. This average rate of rotation ω_z about the z-axis, is given in terms of the gradients of the velocity components by Eq. (6.10) as

$$\omega_z = \frac{1}{2}\left(\frac{\partial V_y}{\partial x} - \frac{\partial V_x}{\partial y}\right) \quad ...(12.1)$$

Similarly the components ω_x and ω_y of the rate of rotation can be written as

$$\omega_x = \frac{1}{2}\left(\frac{\partial V_z}{\partial y} - \frac{\partial V_y}{\partial z}\right) \quad ...(12.2)$$

$$\omega_y = \frac{1}{2}\left(\frac{\partial V_x}{\partial z} - \frac{\partial V_z}{\partial x}\right) \quad ...(12.3)$$

These are readily recognized as the components of

$$\omega = \frac{1}{2}(\nabla \times \mathbf{V}) \quad ...(12.4)$$

The components of $\nabla \times \mathbf{V}$ in cylindrical and spherical polar coordinates are given in Appendix B-3.

In 2-*D* fluid motions, ω_z is the only non-zero component of the rate of rotation. If a small 'paddle wheel' is imagined (Fig. 12.2) riding on a fluid particle, the rate of spin of this paddle wheel equals ω_z. Flows in which this rate of rotation, i.e., the angular spin of the imaginary paddle wheel is zero everywhere, are termed as *irrotational* flows. Such flows, as shall be seen shortly, are intimately related to inviscid flows and play an important role in the study of flows at high Reynolds numbers.

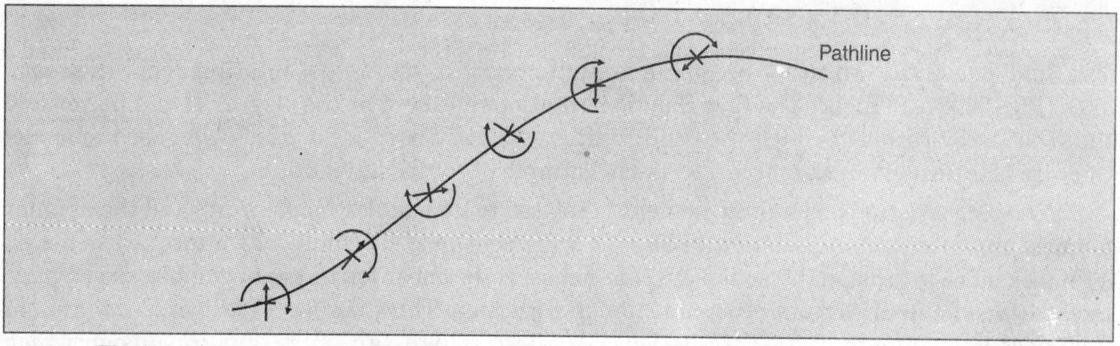

Fig. 12.2. Rotation of an imaginary paddle wheel along a pathline in a rotational flow.

It should be pointed out that the terms rotational and irrotational used here refer to the spinning of a fluid element about its *own* axis–and that too, in the sense of an *average* rotation of two perpendicular line segments. The fluid particles may or may not move in a circular arc. Thus, curved flows are not necessarily rotational and parallel flows are not necessarily irrotational. Two examples to illustrate this will suffice. Consider first the case of Couette flow in a 2-*D* channel discussed in Example 6.1. Here, although the streamlines are all straight yet the presence of a velocity gradient across the channel due to the action of viscosity makes the flow rotational (Fig. 12.3). The velocity field for the case when $dp/dx = 0$ is given by Eq. (e) of Example 6.1 as

$$V_x = \frac{V_0}{2}\left(1+\frac{y}{b}\right), V_y = 0 \qquad ...(12.5)$$

and so,

$$\omega_z = \frac{1}{2}\left(\frac{\partial V_y}{\partial x} - \frac{\partial V_x}{\partial y}\right) = -\frac{V_0}{4b} \qquad ...(12.6)$$

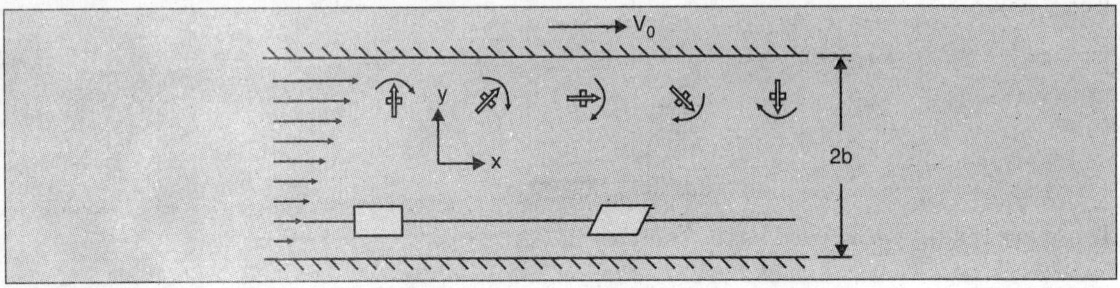

Fig. 12.3. Rotation of a paddle wheel in Couette flow.

The paddle wheel rotates in the clockwise direction [as indicated by the negative value in Eq. (12.6)] because the velocity at the upper arm is higher than that at the lower arm. Next consider the case of a *free vortex* which models the flow around the core of a hurricane. The streamlines in this case are concentric circles with the tangential velocity V_θ, the only non-zero component, decreasing with radius as

$$V_\theta = k/r \qquad ...(12.7)$$

Then,

$$\omega = \frac{1}{2}\nabla \times \mathbf{V} = 0$$

The fluid elements in this case are going around in circles, yet are not rotating about their *own* axes (Fig. 12.4), and hence the flow field is irrotational (except at the origin). It can be verified that the *forced vortex* in which the fluid rotates as a rigid body (and which was used to model the core of a tornado in Example 7.9) is an example of rotational motion.

The rate of rotation of a fluid element is related to its angular momentum, and the angular momentum must remain constant unless a torque acts on it. The only way a torque can act on a fluid element is through the action of shear forces. If the shear forces are negligible, the angular momentum of a fluid element must not change with time. Thus, *if a fluid element in an inviscid flow does not rotate to begin with, it never rotates**. This is an important conclusion which

* A rotating fluid element in an inviscid flow can, however, change its non-zero rate of rotation if it stretches—much in the same manner as a ballerina changes her speed by spreading out or drawing in her arms. Since the angular momentum is preserved in such cases, the angular velocity can change if the moment of inertia changes. However, if ω is zero to begin with, it is not possible to have a non-zero ω by this mechanism.

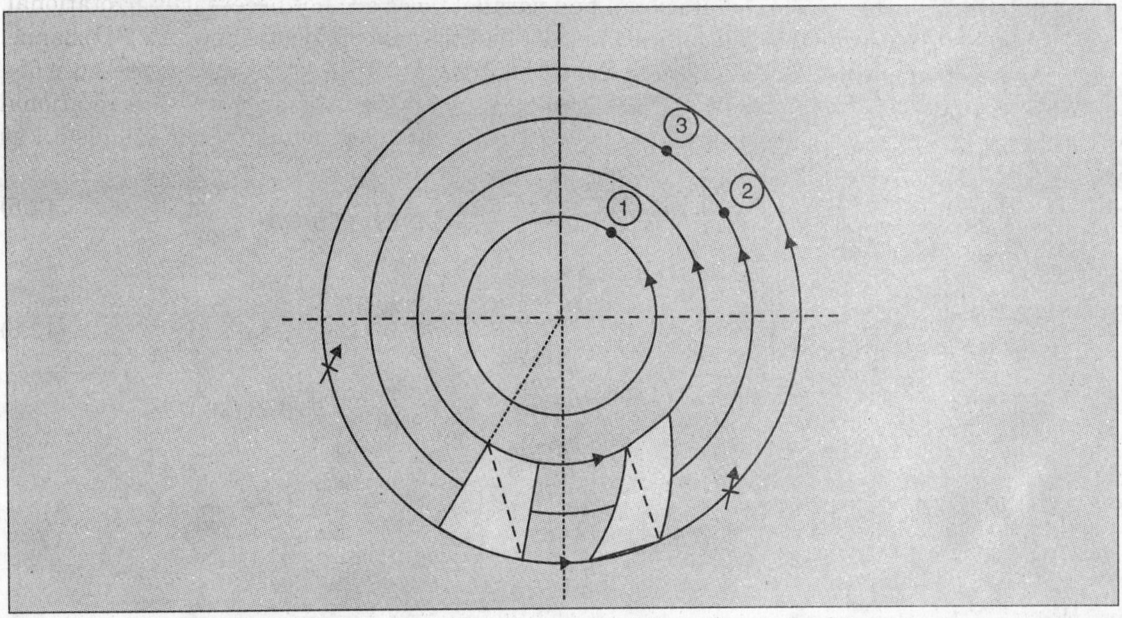

Fig. 12.4. Streamlines and fluid element deformation in the free vortex.

explains the pre-eminent position irrotational motions occupy in the flow of fluids at large Reynolds numbers. Since, in most of the fluid flow situations, either a body moves through a fluid at rest (as for example, the flight of an aeroplane in a stationary atmosphere) or a uniform stream enters a fluid machinery (as in a turbine), there usually exists a region of uniform motion (or rest). In this region of uniform motion the angular velocity of all the fluid elements is zero, as can be seen from Eq. (12.4). If the flow is inviscid, i.e., if the action of viscosity is either absent or negligible, then these fluid elements never acquire a rotation as they move downstream. Thus, inviscid flows which are initially at rest or in uniform motion are *irrotational*. Whenever a fluid is induced to flow due to gravity or pressure forces, as in rivers, channels, over weirs, in hurricanes, etc., the flow is largely irrotational as long as we keep away from the walls, or encounter large velocity gradients where viscous effects start dominating.

When a real fluid flows past a streamlined body at high Reynolds numbers, the region outside the boundary layer can be taken as inviscid. Since the velocity field is uniform upstream, every fluid element has zero rotation to begin with, and the absence of viscous action ensures that it cannot rotate as it moves downstream. The flow outside the boundary layer is, therefore, irrotational and a 'paddle wheel' riding on a fluid element will not spin (Fig. 12.5). Within the boundary layer, however, the action of viscosity rotates the fluid element (and with it, the paddle wheel).

Irrotational flows have some special properties which render the equations governing them linear and, therefore, these flows have been studied in great detail. Some of these properties are studied in the following sections. Before doing this, the concept of circulation is introduced, which plays a central role in irrotational flows.

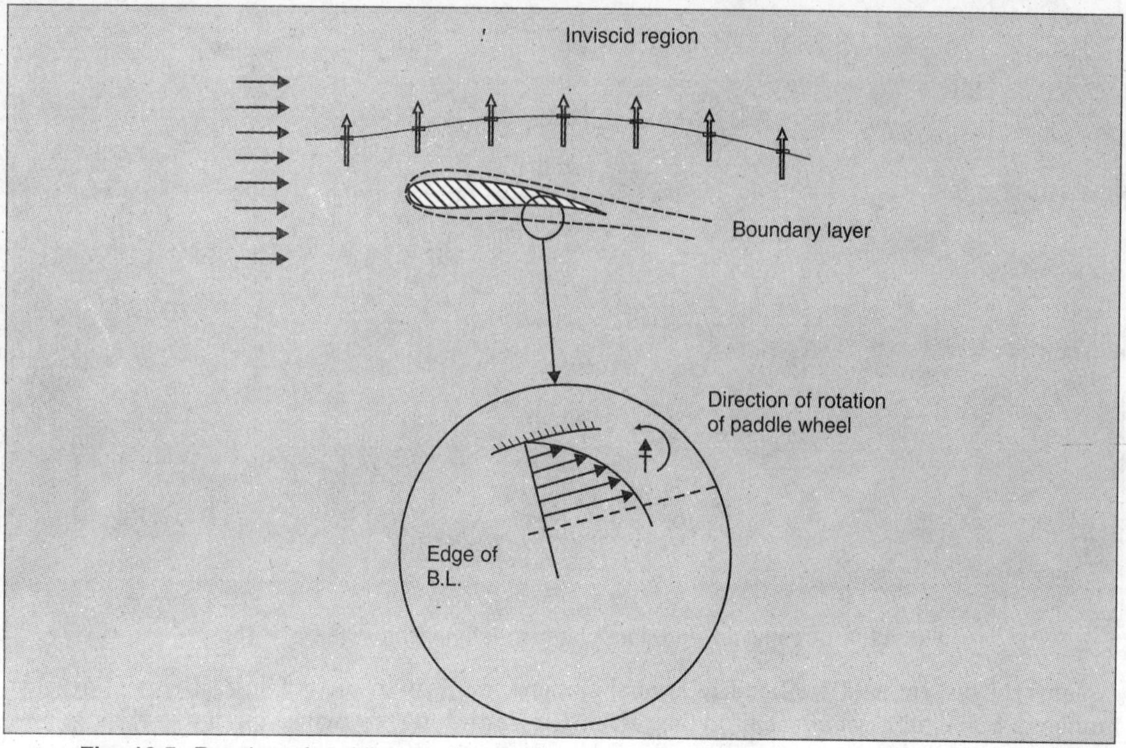

Fig. 12.5. Rotation of paddle wheel inside boundary layer and in the outer region in flow of a uniform stream about a streamlined body.

12.3 CIRCULATION

The *circulation* Γ is defined as the line integral of the tangential component of velocity about any closed contour C in a flow field:

$$\Gamma = \oint_C V_t \, ds = \oint_C \mathbf{V} \cdot d\mathbf{s} \qquad \text{...(12.8)}$$

The integration proceeds in a counter-clockwise sense along the circuit (Fig. 12.6). The circulation

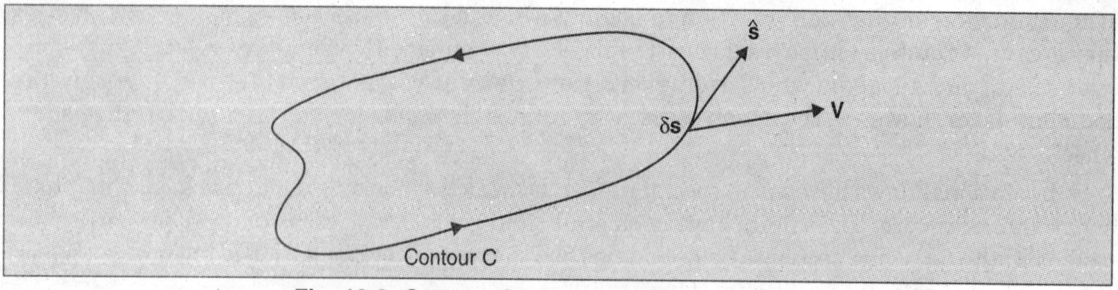

Fig. 12.6. Contour C to determine the circulation.

about a circuit is related to the angular velocity of the fluid. To demonstrate this, consider an infinitesimal circuit (Fig. 12.7). The circulation around this circuit is

$$\delta\Gamma = \oint \mathbf{V} \cdot d\mathbf{s}$$

$$= V_x \delta x + \left(V_y + \frac{\partial V_y}{\partial x}\delta x\right)\delta y - \left(V_x + \frac{\partial V_x}{\partial y}\delta y\right)\delta x - V_y \delta y$$

$$= \left(\frac{\partial V_y}{\partial x} - \frac{\partial V_x}{\partial y}\right)\delta x\, \delta y$$

or

$$\delta\Gamma = 2\omega_z\, \delta A \qquad \text{...(12.9)}$$

where δA is the area enclosed by the circuit.

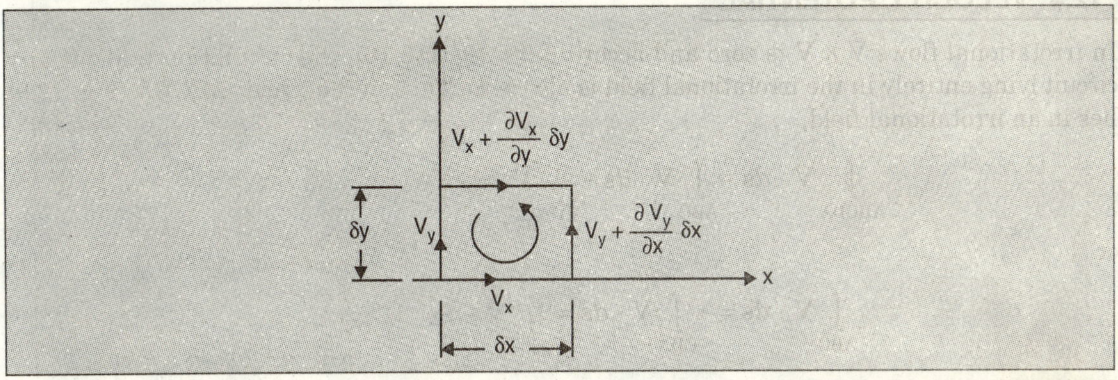

Fig. 12.7. An infinitesimal circuit associated with circulation $\delta\Gamma$.

Figure 12.8 shows that integration about a larger circuit can be replaced by the sum of integrals about the infinitesimal circuits spanning A, the area enclosed, and so

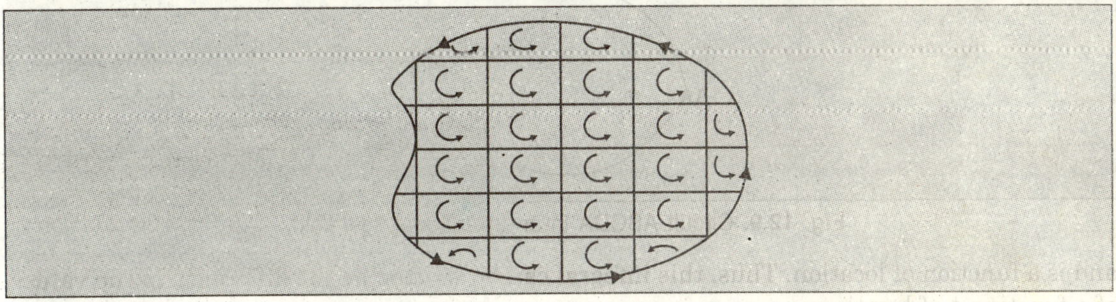

Fig. 12.8. Sum of circulations is the total circulation.

$$\Gamma = \iint_A 2\omega_z dA$$

Use of Eq. (12.1) gives

$$\Gamma = \iint_A (\nabla \times \mathbf{V})_z\, dA \qquad \text{...(12.10)}$$

If the flow in the region enclosed by the circuit is irrotational everywhere, i.e., $\nabla \times \mathbf{V}$ is zero everywhere, then Γ, the circulation, is also zero. This is an important property of irrotational

CHAPTER 12

flows. It should be noted that for Γ to be zero, the flow in the *entire* region enclosed by C should be irrotational. In the free vortex flow described in Sec. 12.2 where the streamlines are circular and $V_\theta = k/r$, the flow is irrotational everywhere except at the origin $r = 0$, and, therefore, Γ for a circuit enclosing the origin is non-zero. It can be shown by direct integration (Prob. 12.6) that the circulation about any circular path enclosing the origin is $\Gamma = 2\pi k$, independent of the radius. But for any circuit not enclosing the origin $\Gamma = 0$.

The property that Γ for any circuit enclosing an irrotational flow field is zero is used to prove the existence of a function ϕ whose gradient gives the velocity field. This is done in the next section.

12.4 VELOCITY POTENTIAL

In irrotational flows $\nabla \times \mathbf{V}$ is zero and according to Eq. (12.10), the circulation Γ about any circuit lying entirely in the irrotational field is also zero. Thus, if the circuit ABCDA (Fig. 12.9) lies in an irrotational field,

$$\oint_{ABCDA} \mathbf{V} \cdot d\mathbf{s} = \int_{ABC} \mathbf{V} \cdot d\mathbf{s} + \int_{CDA} \mathbf{V} \cdot d\mathbf{s} = 0$$

or

$$\int_{ABC} \mathbf{V} \cdot d\mathbf{s} = -\int_{CDA} \mathbf{V} \cdot d\mathbf{s} = \int_{ADC} \mathbf{V} \cdot d\mathbf{s}$$

Since ABCDA is arbitrary, this implies that the integral of $\mathbf{V} \cdot d\mathbf{s}$ is independent of the path

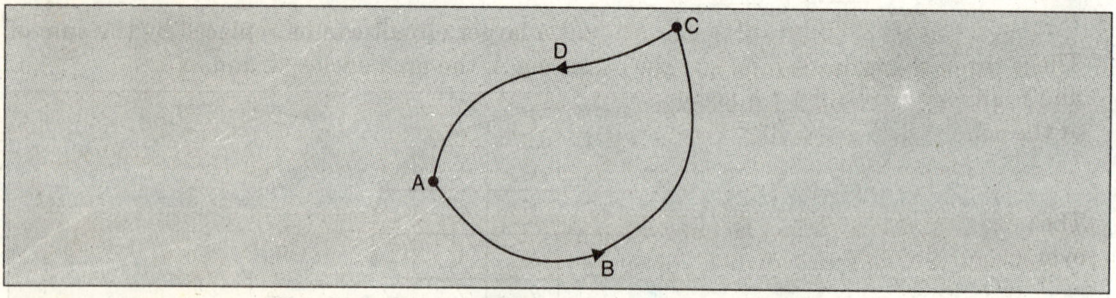

Fig. 12.9. Circuit ABCDA in an irrotational flow field.

and is a function of location. Thus, this integral can be written as the difference in the values of a function ϕ of location as

$$\int_A^C \mathbf{V} \cdot d\mathbf{s} = \phi(C) - \phi(A)$$

This is possible only if $\quad \mathbf{V} \cdot d\mathbf{s} = d\phi = \nabla \phi \cdot d\mathbf{s}$

or

$$\mathbf{V} = \nabla \phi \qquad \text{...(12.11)}$$

The function ϕ which depends on the location x and t is termed the *velocity potential.* Equation (12.11) states that whenever the flow field is irrotational, the velocity can be written as the gradient of a scalar function ϕ termed as the velocity potential. For this reason, irrotational flows are also termed as *potential flows.*

The velocity components in cartesian coordinates are then written as

$$V_x = \frac{\partial \phi}{\partial x}, V_y = \frac{\partial \phi}{\partial y}, V_z = \frac{\partial \phi}{\partial z} \quad \text{...(12.12)}$$

with corresponding equations for cylindrical and spherical polar coordinates given in Appendix B-1. This permits replacing the three dependent variables, V_x, V_y and V_z by a *single* scalar function ϕ and results in a dramatic simplification of the equations governing the flow, as shown in the next section.

CHAPTER 12

12.5 EQUATIONS GOVERNING POTENTIAL FLOWS

In general, the equations governing 2-*D* incompressible and inviscid flows are:

$$\text{Continuity equation: } \frac{\partial V_x}{\partial x} + \frac{\partial V_y}{\partial y} = 0 \quad \text{...(12.13)}$$

Two components of the Euler Eq. (11.10):

$$\frac{\partial V_x}{\partial t} + V_x \frac{\partial V_x}{\partial x} + V_y \frac{\partial V_x}{\partial y} = -\frac{1}{\rho}\frac{\partial p}{\partial x} + f_x \quad \text{...(12.14)}$$

$$\frac{\partial V_y}{\partial t} + V_x \frac{\partial V_y}{\partial x} + V_y \frac{\partial V_y}{\partial y} = -\frac{1}{\rho}\frac{\partial p}{\partial y} + f_y \quad \text{...(12.15)}$$

These three equations form a system of *simultaneous* equations for the three variables V_x, V_y and *p,* and are to be solved subject to the condition that the normal velocity component vanishes at the solid stationary walls:

$$V_n = \mathbf{V} \cdot \hat{\mathbf{n}} = 0 \quad \text{(at walls)} \quad \text{...(12.16)}$$

The no-slip condition cannot be imposed at such boundaries since otherwise the problem becomes overspecified as has already been discussed in Sec. 11.5.

If, in addition to being incompressible and inviscid, the flow is irrotational as well, there exists a scalar velocity potential ϕ, such that

$$\mathbf{V} = \frac{\partial \phi}{\partial x}\hat{\mathbf{i}} + \frac{\partial \phi}{\partial y}\hat{\mathbf{j}}$$

according to the development in Sec. 12.4. Under such conditions, the two velocity components are obtained from a single scalar function ϕ. Use of the definition of ϕ in the continuity Eq. (12.13), gives

$$\frac{\partial^2 \phi}{\partial x^2} + \frac{\partial^2 \phi}{\partial y^2} = 0$$

In vector notation, this is

$$\nabla^2 \phi = 0 \quad \text{...(12.17)}$$

where $\nabla^2 = \nabla \cdot \nabla$ is the Laplacian operator. In Cartesian coordinates, it is given by

$$\nabla^2 = \frac{\partial^2}{\partial x^2} + \frac{\partial^2}{\partial y^2} \qquad \text{...(12.18)}$$

Equation (12.17) applies to 3-D irrotational flows as well. The form of the Laplacian in various coordinate systems is given in Appendix B-4.

Before going to the next equation, notice an interesting property of potential flows. Since there is only one dependent variable ϕ in Eq. (12.17), it is no longer coupled to the Euler equation and can be solved independently. Thus, the velocity field for incompressible irrotational flows is obtained directly as a solution of the Laplace Eq. (12.17) subject to the boundary condition Eq. (12.16), which, in terms of ϕ becomes

$$\frac{\partial \phi}{\partial n} = 0 \qquad \text{(at walls)} \qquad \text{...(12.19)}$$

The Laplace Eq. (12.17) subject to boundary conditions of the type given in Eq.(12.19) is encountered in diverse fields such as elasticity, electromagnetic field theory, etc. Well developed methods for solution of this equation are available. Instead of going into these, only a few interesting cases will be *presented* here. Readers are referred to Krishnamurty (Reading 25) and Milne-Thomson (Reading 28) for methods of obtaining solutions for potential flows.

Let us point out some interesting consequences of the fact that the velocity field is determined fully by the Laplace equation. First Eqs. (12.17) and (12.19) are valid for *both steady and unsteady flows* even though the time variation does not enter the equations explicitly. But since the instantaneous boundary conditions are applied, if the boundary conditions are changing with time, the solution also changes with time. The rate of change of the boundary condition does not affect the solution at a given instant. This means that the solution depends on the *current* boundary conditions and not on the *history* of flow and is, in effect, *quasi-steady* with every point in the flow field responding immediately to any changes in the boundary conditions. This is a consequence of the assumption of incompressible flow which requires that the speed at which disturbances are propagated in a medium is infinite (see Sec. 15.2).

The second interesting consequence is that the velocity field can be obtained independently of the pressure field, but not *vice versa*. Thus, the pressure gradient adjusts itself depending upon the inertia of the fluid (measured by the density ρ) to give the desired velocity gradients.

Thirdly, the Laplace equation and the relevant boundary conditions are linear and, therefore, it is possible to build-up solutions to complex flow fields using superposition of simple flows. This is illustrated in Sec. 12.6.

Now integrate the x- and y-components of the Euler Eqs. (12.14 and 12.15). Attention shall now be restricted to flows where gravity is the only body force. Then, $f_x = -g\partial h/\partial x$ and $f_y = -g\,\partial h/\partial y$, where h is the coordinate measured vertically upwards. First use the irrotationality condition

$$\frac{\partial V_y}{\partial x} - \frac{\partial V_x}{\partial y} = 0$$

to substitute for $\partial V_x/\partial y$ and $\partial V_y/\partial x$ in Eqs. (12.14) and (12.15) respectively to obtain

$$\frac{\partial V_x}{\partial t}+\frac{\partial}{\partial x}\left(\frac{V_x^2+V_y^2}{2}\right)=-\frac{1}{\rho}\frac{\partial p}{\partial x}-g\frac{\partial h}{\partial x} \quad \text{...(12.20)}$$

$$\frac{\partial V_y}{\partial t}+\frac{\partial}{\partial y}\left(\frac{V_x^2+V_y^2}{2}\right)=-\frac{1}{\rho}\frac{\partial p}{\partial y}-g\frac{\partial h}{\partial y} \quad \text{...(12.21)}$$

Then, since $V_x^2+V_y^2=V^2$ and $V_x=\partial\phi/\partial x$, $V_y=\partial\phi/\partial y$, Eqs. (12.20) and (12.21) become

$$\frac{\partial}{\partial x}\left[\frac{\partial\phi}{\partial t}+\frac{V^2}{2}+\frac{p}{\rho}+gh\right]=0 \quad \text{...(12.22)}$$

and
$$\frac{\partial}{\partial y}\left[\frac{\partial\phi}{\partial t}+\frac{V^2}{2}+\frac{p}{\rho}+gh\right]=0 \quad \text{...(12.23)}$$

Since x and y can be chosen arbitrarily, one gets

$$\frac{\partial\phi}{\partial t}+\frac{V^2}{2}+\frac{p}{\rho}+gh = \text{constant, throughout the flow field}$$

Conventionally, the distance h in the vertical direction is represented by z. Then

$$\frac{\partial\phi}{\partial t}+\frac{V^2}{2}+\frac{p}{\rho}+gz = \text{constant, throughout flow field} \quad \text{...(12.24)}$$

This equation is also valid for 3-*D* irrotational flows.

For steady flow fields, this equation has the same form as the Bernoulli Eq. (7.15). However, Eq. (7.15) as derived, was valid *along a streamline only* for both rotational as well as irrotational *inviscid* flows. Thus, the equation

$$\frac{V^2}{2}+\frac{p}{\rho}+gz = \text{constant}$$

applies to *steady, inviscid* flows

(a) throughout the flow field if the flow is irrotational, and

(b) along any streamline only, if the flow is rotational.

Example 12.1. Check the extension of the Bernoulli equation to irrotational flows by applying it between points 1 and 2 of the irrotational free vortex of Fig. 12.4.

Application of Eq. (12.24) gives

$$\frac{V_1^2}{2}+\frac{p_1}{\rho}=\frac{V_2^2}{2}+\frac{p_2}{\rho}$$

if the two points are at the same vertical level z. Then,

$$p_2-p_1=\frac{1}{2}\rho\left(V_1^2-V_2^2\right)=\frac{\rho}{2}k^2\left(\frac{1}{r_1^2}-\frac{1}{r_2^2}\right)$$

CHAPTER 12

The same result is obtained on using Eq. (7.17) which gives the pressure variation in a direction normal to streamlines. Thus, for points 1 and 3 on the same radial line,

$$\frac{\partial p}{\partial n} = -\frac{\rho V^2}{r}$$

Here, $V = k/r$, $n = -r$, and so

$$\frac{\partial p}{\partial r} = \frac{\rho k^2}{r^3}$$

or

$$p = -\frac{1}{2}\frac{\rho k^2}{r^2} + \text{constant}$$

or

$$p_3 - p_1 = \frac{\rho k^2}{2}\left(\frac{1}{r_1^2} - \frac{1}{r_3^2}\right)$$

Then the use of the Bernoulli equation between points 2 and 3 which lie on the same streamline gives the same result for $p_2 - p_1$ as before.

12.6 SOME SIMPLE 2-D POTENTIAL FLOWS

In this section a few simple potential functions which describe some commonly encountered flows are *presented*. Many complex flows can be written as superpositions of these and, therefore, these elementary functions represent the basic building blocks of potential flow theory.

First consider the flow given by the potential function

$$\phi = Ax + By$$

This obviously is a solution of Eq. (12.17). The velocity components are obtained using Eq. (12.12) as

$$V_x = A,\ V_y = B$$

both constant. This potential function, thus, represents a *uniform* flow with velocity vector $A\hat{\mathbf{i}} + B\hat{\mathbf{j}}$. The streamlines are shown in Fig. 12.10. Thus, a fluid flowing uniformly at a velocity V_0 at an angle θ to the x-axis has the potential function

$$\phi = (V_0 \cos\theta)\, x + (V_0 \sin\theta)\, y \qquad \text{...(12.25)}$$

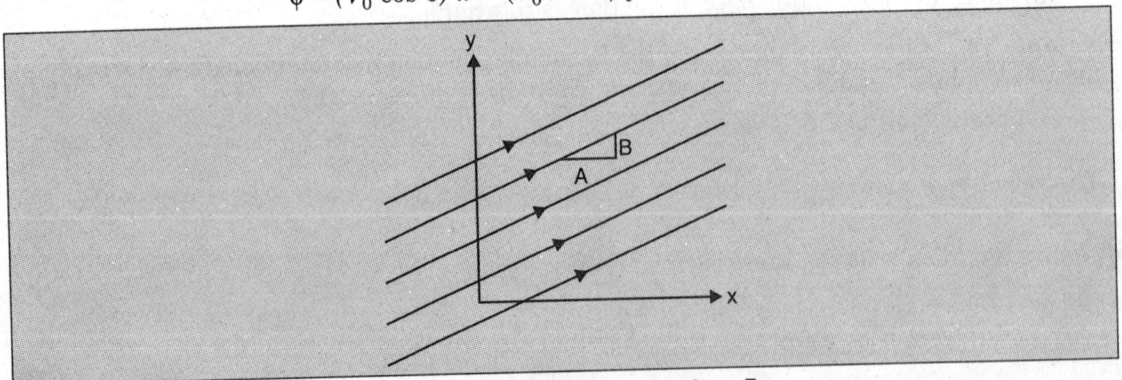

Fig. 12.10. Streamlines for $\phi = Ax + By$.

Note that the equation $\nabla^2 \phi = 0$ can be solved only up to a constant and so an arbitrary constant can always be added to ϕ.

Next consider the flow described by

$$\phi = A \ln r$$

It can be confirmed that this satisfies the Laplace equation, and the velocity components in the r- and θ-directions are

$$V_r = \frac{\partial \phi}{\partial r} = \frac{A}{r}$$

$$V_\theta = \frac{1}{r}\frac{\partial \phi}{\partial \theta} = 0$$

This represents a flow directed away from the origin (Fig. 12.11), with velocity decreasing linearly with r. This is termed as a simple *2-D source.*

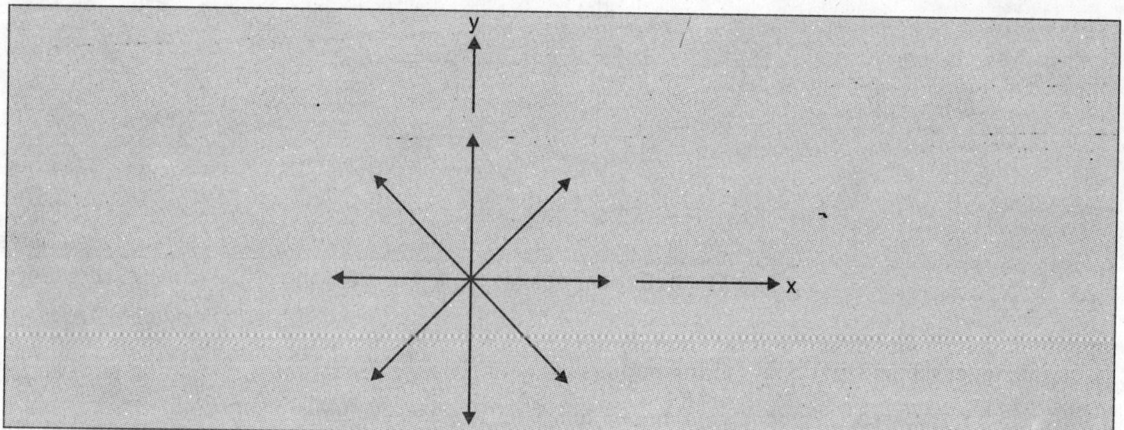

Fig. 12.11. Streamlines for $\phi = A \ln r$, $A > 0$.

Consider a circular surface of radius r and unit depth (in the z-direction).

The volumetric flow rate crossing it is $V_r \cdot 2\pi r \cdot 1 = (A/r)(2\pi r) = 2\pi A$, which is independent of r. This is defined as the *strength* of a 2-D source and is denoted by q. Thus, the velocity potential of a 2-D source of strength q is

$$\phi = \frac{q}{2\pi} \ln r \qquad ...(12.26)$$

A negative q represents a 2-D sink.

Lastly, consider the flow characterized by

$$\phi = k\theta$$

The velocity components of this are

$$V_r = \frac{\partial \phi}{\partial r} = 0$$

$$V_\theta = \frac{1}{r}\cdot\frac{\partial \phi}{\partial \theta} = \frac{k}{r}$$

The streamlines of this flow are concentric circles (Fig. 12.12). This is the *free vortex* flow introduced in Sec. 12.2. In Sec. 12.3, it was shown that the circulation for this vortex about any circular path with the origin as centre is $\Gamma = 2\,\pi k$. This can be taken as a measure of the strength of the free vortex. Thus, the velocity potential of a free vortex in terms of Γ is

$$\phi = \frac{\Gamma\theta}{2\pi} \qquad ...(12.27)$$

Table 12.1 lists the potential functions of these three elementary rows for easy reference. Also listed are their stream functions ψ which can be

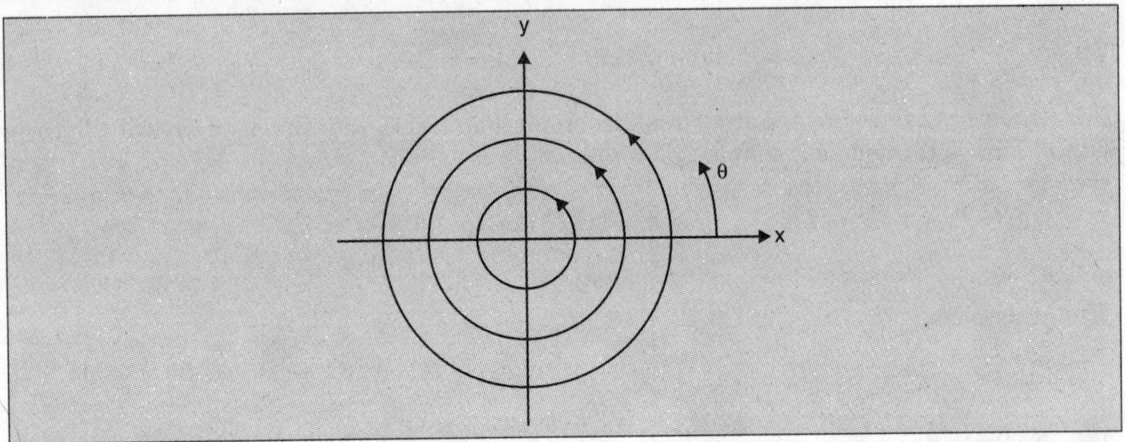

Fig. 12.12. Streamlines for $\phi = k\,\theta$, $k > 0$

obtained by straightforward integration of the velocity components according to Eqs. (4.13) and (4.14), or their counterparts in cylindrical coordinates (see Prob. 4.17).

Table 12.1. Potential and stream functions of some elementary 2-*D* flows

Flow	Potential function ϕ	Stream function ψ
Uniform flow	$(V_0 \cos\theta)\,x + (V_0 \sin\theta)\,y$	$(V_0 \cos\theta)\,y - (V_0 \sin\theta)\,x$
Source	$\frac{q}{2\pi}\ln r$	$\frac{q}{2\pi}\theta$
Free vortex	$\frac{\Gamma}{2\pi}\theta$	$-\frac{\Gamma}{2\pi}\ln r$

It can be verified using the irrotationality condition $\omega_z = 0$ that the stream function ψ of potential flows* also satisfies the Laplace equation

$$\nabla^2\psi = 0$$

Thus, just as the potential function ϕ can be obtained by the superposition of the ϕs of elementary flows, so also can the stream function be obtained by the superposition of the ψs of elementary flows.

---

* Note that stream functions are defined only for 2-*D* flows (or for axi-symmetric flows), though ϕ is defined for 3-D flows as well.

12.7 SOME POTENTIAL FLOW SOLUTIONS BY SUPERPOSITION

Given in this section are a few examples of how to construct potential and stream functions for relatively complex flows by superposition.

(a) *Rankine half-body:* Consider first the superposition of a source of strength q at the origin and a uniform flow of velocity V_0 along the x-axis from the left. Then the potential and stream functions of the combined flow are given by

$$\phi = V_0 x + \frac{q}{2\pi} \ln r \quad ...(12.28)$$

$$\psi = V_0 y + \frac{q\theta}{2\pi} \quad ...(12.29)$$

The velocity components associated with these ϕ and ψ functions are

$$V_x = \frac{\partial \phi}{\partial x} = \frac{\partial \psi}{\partial y} = V_0 + \frac{q}{2\pi}\frac{x}{x^2 + y^2} \quad ...(12.30)$$

$$V_y = \frac{\partial \phi}{\partial y} = -\frac{\partial \psi}{\partial x} = -\frac{q}{2\pi}\frac{y}{x^2 + y^2} \quad ...(12.31)$$

The streamline pattern is shown in Fig. 12.13. Any streamline of this flow pattern can be replaced by a solid wall and then Eqs. (12.28) and (12.29) represent the potential and stream functions, respectively, of the inviscid flow about that wall. This is possible because there is no flow

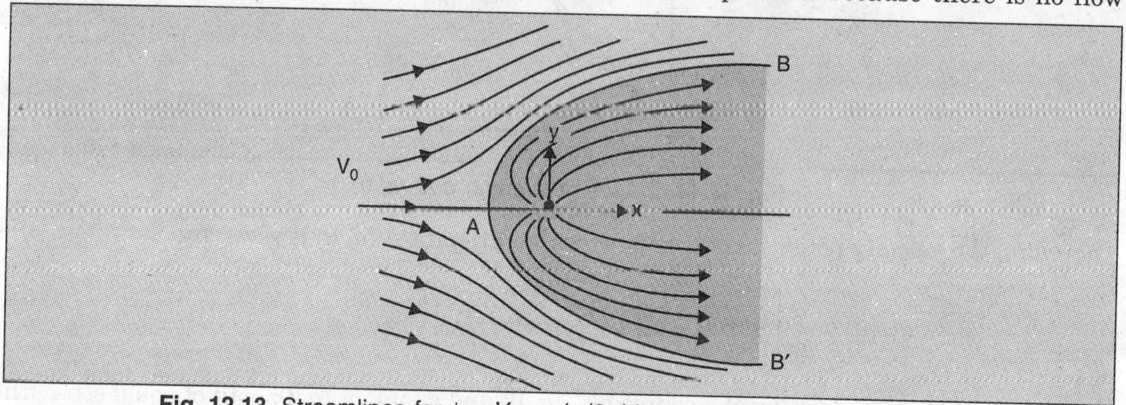

Fig. 12.13. Streamlines for $\phi = V_0 x + (q/2\pi) \ln r$ and the Rankine half-body. Point A is at $x = -q/(2\pi V_0)$.

across a streamline, and thus, a streamline is akin to an impervious wall (see Example 4.5). The streamline BAB' is one streamline of interest which defines a symmetric body extending to infinity. Thus, Eqs. (12.28) and (12.29) define the potential flow about the body shown shaded. This is known as the Rankine half-body. The point A where the streamline bifurcates must necessarily be a stagnation point. This is so because a streamline is tangential to the velocity vector and this is violated if the velocity at A is not zero. Therefore, from Eqs. (12.30) and (12.31), the coordinates of point A are ($x = -q/2\pi V_0$, $y = 0$). The streamline passing through A is then given by

$$\psi = V_0 y + \frac{q\theta}{2\pi} = q/2$$

or

$$V_0 r \sin\theta + \frac{q\theta}{2\pi} = q/2 \qquad ...(12.32)$$

which represents the boundary of the Rankine half-body. Prob. 12.12 describes a physical situation where the flow can be modelled as one about such a body.

(b) *Doublet:* A doublet consists of a source and a sink of equal strength placed very close together. Consider first a source of strength q placed at $x = a$ on the x-axis (Fig. 12.14). The velocity potential at an arbitrary point $P(r, \theta)$ is

$$\phi(r,\theta) = \frac{q}{2\pi}\ln r_1$$

where

$$r_1 = \sqrt{r^2\sin^2\theta + (r\cos\theta - a)^2}$$

or

$$r_1 = r\left(1 + \frac{a^2}{r^2} - 2\frac{a}{r}\cos\theta\right)^{1/2}$$

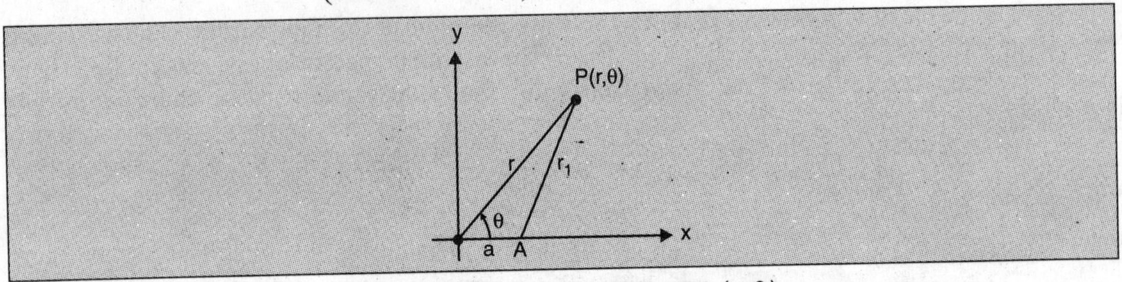

Fig. 12.14. Source of strength q at $(a, 0)$.

Therefore, the velocity potential of a source located at point $A(a, 0)$ is given by

$$\phi(r,\theta) = \frac{q}{2\pi}\left[\ln r + \frac{1}{2}\ln\left(1 - \frac{2a}{r}\cos\theta + \frac{a^2}{r^2}\right)\right]$$

Next consider a pair consisting of a source at $(-a, 0)$ and a sink at $(a, 0)$ both of equal strength (Fig. 12.15). Then, by superposition the ϕ of the combination is

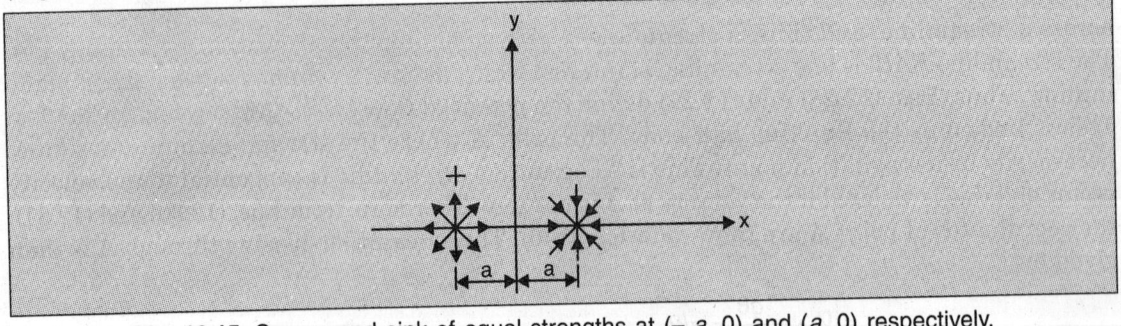

Fig. 12.15. Source and sink of equal strengths at $(-a, 0)$ and $(a, 0)$ respectively.

$$\phi = \frac{q}{2\pi}\left[\ln r + \frac{1}{2}\ln\left(1+\frac{2a}{r}\cos\theta + \frac{a^2}{r^2}\right)\right]$$

$$-\frac{q}{2\pi}\left[\ln r + \frac{1}{2}\ln\left(1-\frac{2a}{r}\cos\theta + \frac{a^2}{r^2}\right)\right]$$

$$= \frac{q}{4\pi}\left[\ln\left(1+\frac{2a}{r}\cos\theta + \frac{a^2}{r^2}\right) - \ln\left(1-\frac{2a}{r}\cos\theta + \frac{a^2}{r^2}\right)\right]$$

Now, bringing the source and the sink close together such that $a/r \ll 1$, the term $(a/r)^2$ drops out in comparison with the a/r term and the use of the Taylor expansion $\ln(1 \pm \varepsilon) = \pm\varepsilon + \ldots$ leads to

$$\phi = \frac{qa}{\pi r}\cos\theta + \text{ terms of order } q\left(\frac{a}{r}\right)^2$$

An interesting case arises when a approaches zero with q increasing such that qa remains constant at K. Then

$$\phi \to \frac{K}{\pi r}\cos\theta$$

This is called a doublet of strength $m \equiv K/\pi$. The potential ϕ of such a doublet is given by

$$\phi = \frac{m}{r}\cos\theta \qquad \text{...(12.33)}$$

The velocity components of this flow are

$$V_r = \frac{\partial\phi}{\partial r} = -\frac{m}{r^2}\cos\theta$$

$$V_\theta = \frac{1}{r}\frac{\partial\phi}{\partial\theta} = -\frac{m}{r^2}\sin\theta$$

Integrate these to obtain the stream function ψ as

$$\psi = -\frac{m}{r}\sin\theta \qquad \text{...(12.34)}$$

The plot of these streamlines is shown in Fig. 12.16.

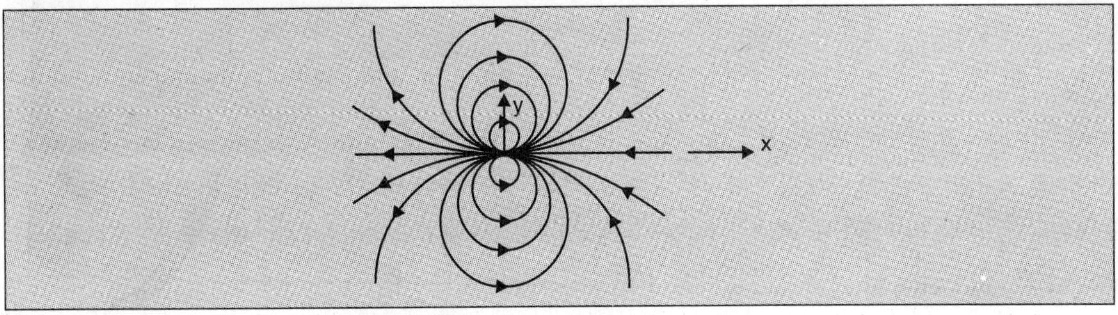

Fig. 12.16. Streamlines for a doublet.

(c) *Flow past a circular cylinder:* Next consider the superposition of a doublet of strength m and a uniform stream of velocity V_0 in the x-direction. The resultant potential and stream functions are

$$\phi = V_0 x + \frac{m}{r}\cos\theta$$

and

$$\psi = V_0 y - \frac{m}{r}\sin\theta$$

Replacing x and y by $r\cos\theta$ and $r\sin\theta$ respectively,

$$\phi = V_0\, r\left(1+\frac{m}{V_0 r^2}\right)\cos\theta \quad ...(12.35)$$

and

$$\psi = V_0\, r\left(1-\frac{m}{V_0 r^2}\right)\sin\theta \quad ...(12.36)$$

The corresponding streamline pattern is shown in Fig. 12.17. The locations of the stagnation points A and B can be found by setting the velocity components equal to zero. This gives the coordinates of point A as $\left(r=\sqrt{m/V_0},\ \theta=-\pi\right)$ and of point B as $\left(r=\sqrt{m/V_0},\ \theta=0\right)$. It can be shown that the streamline passing through points A and B is given by

$$\psi = V_0 r\left(1-\frac{m}{V_0 r^2}\right)\sin\theta = 0 \quad ...(12.37)$$

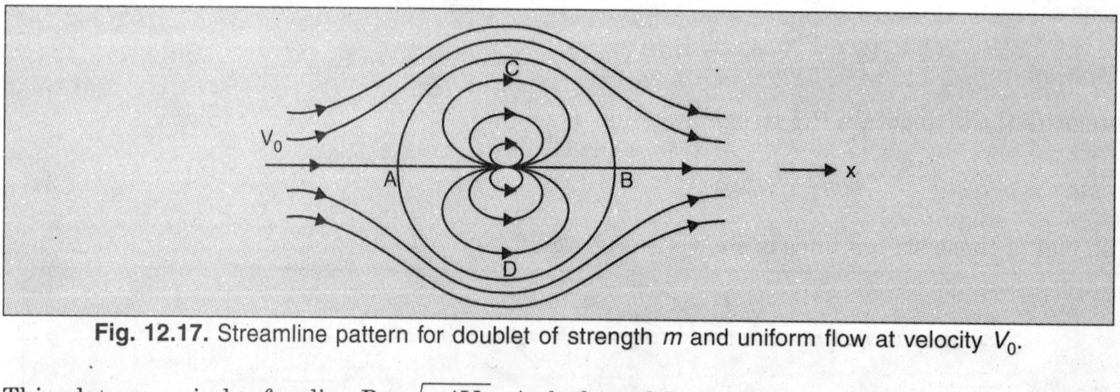

Fig. 12.17. Streamline pattern for doublet of strength m and uniform flow at velocity V_0.

This plots as a circle of radius $R=\sqrt{m/V_0}$. As before, this streamline can be considered as the boundary of a solid body. Thus, Eqs. (12.35) and (12.36) represent the inviscid flow of a uniform stream past a circular cylinder of radius $R=\sqrt{m/V_0}$. It is convenient to rewrite Eqs. (12.35) and (12.36) in terms of $R=\sqrt{m/V_0}$ as

$$\phi = V_0 r\left(1+\frac{R^2}{r^2}\right)\cos\theta \qquad ...(12.38)$$

and,
$$\psi = V_0 r\left(1-\frac{R^2}{r^2}\right)\sin\theta \qquad ...(12.39)$$

The pressure at any point in the flow field can now be obtained by using Eq. (12.24) with $\partial\phi/\partial t = 0$ between that point and a point far upstream where the velocity is V_0 and the pressure is p_0. The pressure at any point at the *cylinder surface* can then be obtained by setting $r = R$. Carrying out these operations, one gets

at $\theta = 90° \;:\; V_\theta = -2V_0$

$\theta = -90° \;:\; V_\theta = 2V_0$

$$p_{r=R} - p_0 = \frac{\rho V_0^2}{2}\left(1-4\sin^2\theta\right) \qquad ...(12.40)$$

the same result as obtained in Example 7.7. This is a symmetrical pressure distribution (see Fig. 7.21) and therefore, the net force on the cylinder is zero.

12.8 ROBINS-MAGNUS EFFECT

Consider the flow of a uniform stream at velocity $V_0\,\hat{\mathbf{i}}$ past a circular cylinder of radius R. The potential function for this can be obtained by the superposition of a uniform stream and a doublet of strength $m = V_0R^2$ according to Eq. (12.38):

$$\phi = V_0 r\left(1+\frac{R^2}{r^2}\right)\cos\theta$$

If a vortex of strength Γ is further superimposed on this, the potential and stream functions become

$$\phi = V_0 r\left(1+\frac{R^2}{r^2}\right)\cos\theta + \frac{\Gamma\theta}{2\pi} \qquad ...(12.41)$$

and
$$\psi = V_0 r\left(1-\frac{R^2}{r^2}\right)\sin\theta - \frac{\Gamma}{2\pi}\ln r \qquad ...(12.42)$$

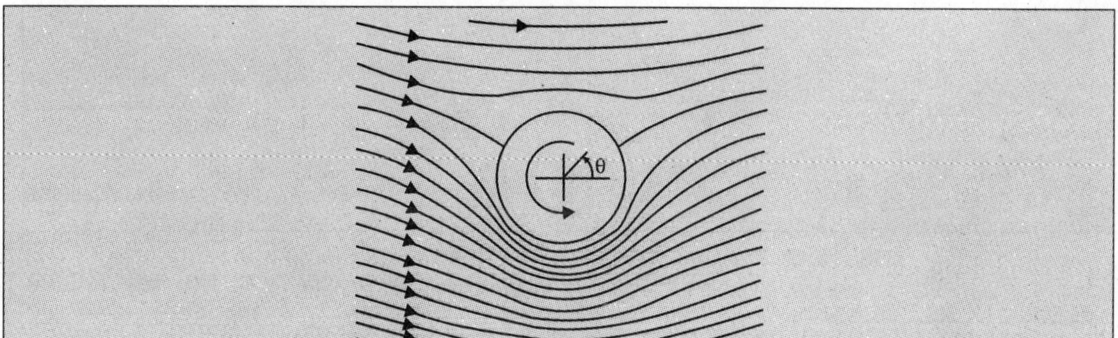

Fig. 12.18. Streamlines for $\phi = V_0 r\,(1+R^2/r^2)\cos\theta + \Gamma\theta/2\pi$ with $\Gamma = 2\pi V_0 R$.

The streamlines for this flow are shown in Fig. 12.18. The circle of radius R is still a streamline so that Eqs. (12.41) and (12.42) still represent the flow past a circular cylinder, but now the stagnation points are shifted away from the 'nose' and 'tail' of the cylinder. It can be shown that the stagnation points occur at $r = R$, and $\theta = \sin^{-1}(\Gamma/4\,\pi R V_0)$. The flow velocities at the top are now *less* than those at the bottom, and the flow picture looks like as if a swirl has been added to the flow about the cylinder shown in Fig. 12.17. Therefore, Eqs. (12.41) and (12.42) are used to model the flow about a cylinder spinning anticlockwise about its own axis*.

The flow picture is still symmetrical about the y-axis but the symmetry about the x-axis is destroyed. Calculating the pressure at a point on the *cylinder surface* using Eq. (12.24), the pressure distribution is obtained as

$$p_{r=R} - p_0 = \frac{1}{2}\rho V_0^2\left[1 - 4\sin^2\theta + \frac{2\Gamma\sin\theta}{\pi V_0 R} - \frac{\Gamma^2}{4\pi^2\, V_0^2\, R^2}\right] \quad ...(12.43)$$

This pressure distribution (Fig. 12.19) still has the fore and aft symmetry leading to the prediction of zero drag. But since the velocities at the lower half-surface are larger than at the upper half-surface, the pressures there are lower. This leads to a vertically downward force acting on the cylinder. Conventionally, a force normal to the mainstream and directed upwards is termed a *lift* F_L. The force on this cylinder, then, is a negative lift. The total force can be calculated by integrating the vertical component of the pressure force. This gives

$$F_L = \int_0^{2\pi} -(p_{r=R} - p_0) R\, d\theta\, \sin\theta = -\rho V_0 \Gamma \quad ...(12.44)$$

It can be shown easily that Γ is also the circulation about any closed circuit enclosing the cylinder (see Prob. 12.7).

Equation (12.44) which gives the lift force in terms of the density, the free stream velocity and the circulation has been obtained here for a circular cylinder. It can, however, be shown (by arguments beyond the level of this text) that this equation also gives the lift force about any

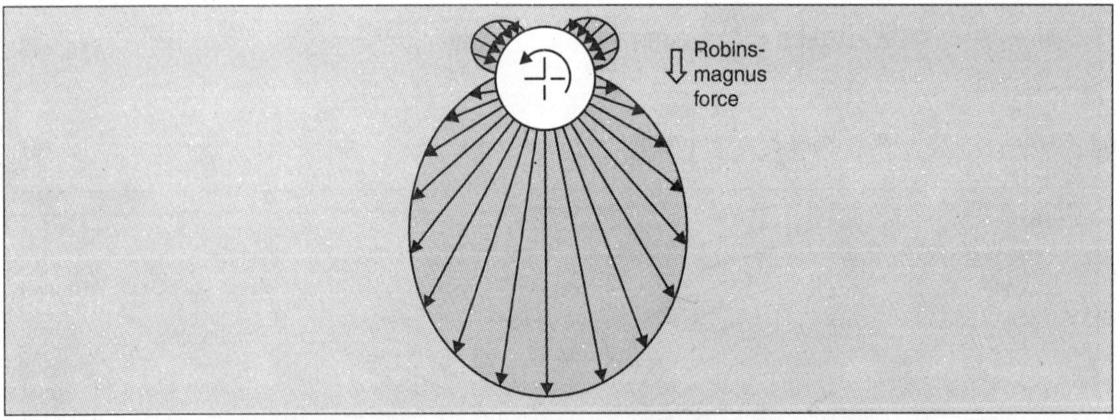

Fig. 12.19. Pressure distribution for a rotating cylinder in a uniform flow with $\Gamma = 2\pi V_0 R$. Flow is from left to right.

* The spin of the cylinder will not affect the flow if the fluid is *truly* inviscid. But the presence of viscosity, however small, requires 'no slip' at the wall, thus spinning the fluid along with the rotating cylinder. The above formulation is, therefore, not valid for a truly inviscid flow and requires the presence of viscosity for setting up the swirl.

body, as long as the body is two-dimensional and the flow is incompressible. In non-circular bodies, obviously the circulation Γ is not due to the spinning of the body about its axis, but rather due to the asymmetry of the top and bottom halves of the surfaces. Thus the lift on an asymmetric 2-D aerofoil (of Fig. 7.22) also equals $-\rho V_0 \Gamma$ where Γ is the circulation about the aerofoil. The problem of lift on an aerofoil will be discussed again in Chapter 14.

The generation of lift on a translating 2-D or 3-D body due to the circulation around it is termed as the *Robins-Magnus effect.* The origin of lift on a sphere can be explained in essentially the same manner as on a cylinder, except that the mathematics is far more complex.

The lift force F_L is generally non-dimensionalized in the same manner as the drag force, giving a lift coefficient C_L as

$$C_L = \frac{F_L}{\frac{1}{2}\rho V_0^2 A_c}$$

where A_c is the characteristic area. The Robins-Magnus effect gives

$$C_L = \frac{\rho V_0 \Gamma}{\frac{1}{2}\rho V_0^2 A_c} = \frac{2\Gamma}{V_0 A_c} \qquad \text{... (12.45)}$$

For cylinders and spheres, A_c is taken as the frontal area. The measured values of C_L and C_D for spinning cylinders and spheres are shown in Fig. 12.20.

The swerving and curving balls in such games as tennis, table tennis, soccer, golf and baseball are due to the Robins-Magnus effect (Fig. 12.21). The swing of a cricket ball, however, has an altogether different explanation as discussed in Sec. 1.7. See also M.M. Sivaramakrishnan, *Science Today,* 16 (3), 51 (1982).

The drag acting on a circular cylinder has been shown to be zero in the above discussion. It can also be shown (by arguments beyond the level of this text) that the drag on any 2-D body in an inviscid flow is zero.

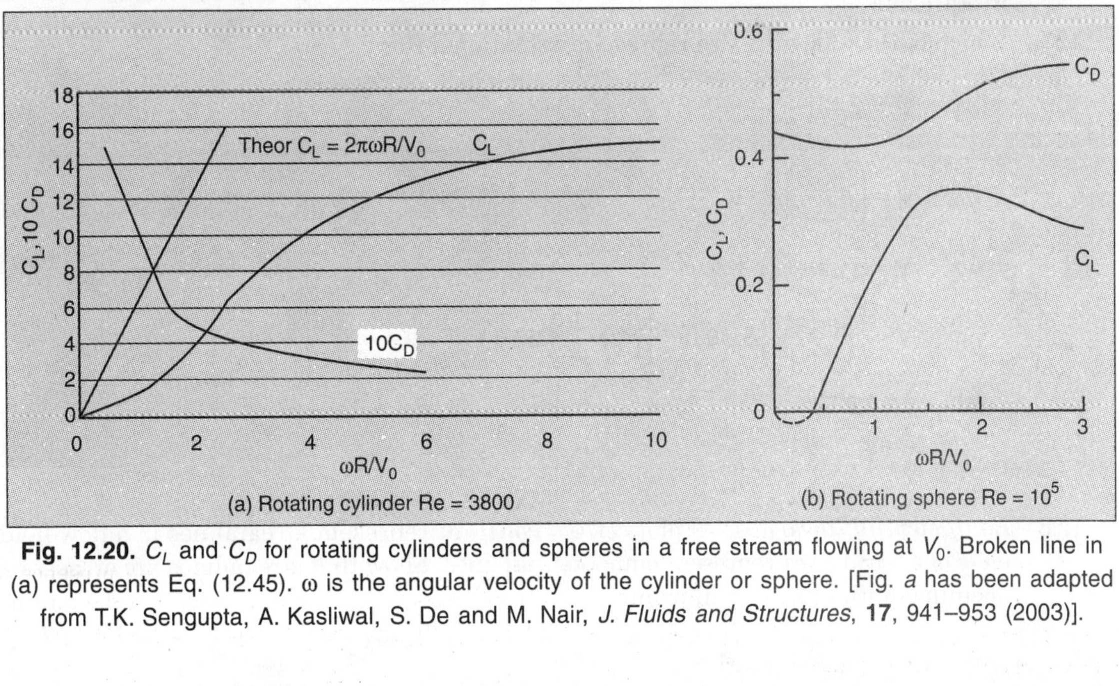

Fig. 12.20. C_L and C_D for rotating cylinders and spheres in a free stream flowing at V_0. Broken line in (a) represents Eq. (12.45). ω is the angular velocity of the cylinder or sphere. [Fig. *a* has been adapted from T.K. Sengupta, A. Kasliwal, S. De and M. Nair, *J. Fluids and Structures*, **17**, 941–953 (2003)].

Obviously, this does not conform to the actual observation of a finite drag on bodies moving at large Reynolds numbers, when the viscous effects are supposedly negligible. This becomes particularly intriguing since the lift predicted by inviscid flow theory matches the experimental results fairly well. The discrepancy between the experimental and the predicted values of the

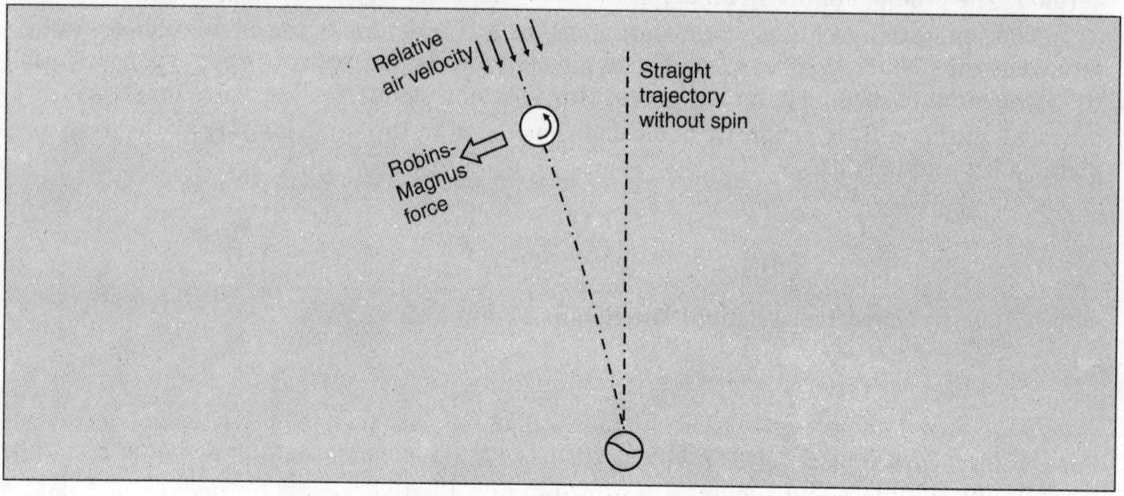

Fig. 12.21. Robins-Magnus effect on a spinning ball.

drag is known as the *d'Alembert paradox.* This discrepancy can be explained by the separation of the boundary layer which forms on the body surface. This is discussed in Sec. 13.6.

PROBLEMS

12.1 Show that the stream function ψ satisfies the Laplace equation when the flow is irrotational.

12.2 Which of the following can represent irrotational flow?

(a) $\psi = 2 \ln\left(\dfrac{x}{2y}\right)$

(b) $\psi = \left(r - \dfrac{2}{r}\right)\sin\theta$

(c) $\mathbf{V} = 2y\,\hat{\mathbf{i}} + (2x + 3y)\,\hat{\mathbf{j}}$

(d) $\mathbf{V} = \left[\left(1 - \dfrac{1}{r^2}\right)\cos\theta\right]\hat{\mathbf{r}} - \left[\left(1 + \dfrac{1}{r^2}\right)\sin\theta + \dfrac{1}{2\pi r}\right]\hat{\theta}$

(e) $\phi = xy$

(f) $\phi = \left(r + \dfrac{2}{r}\right)\cos\theta - 5\theta$

12.3 *Equipotential lines:* Just as plots of ψ = constant represent streamlines in a flow field, plots of ϕ = constant represent equipotential lines. Show that streamlines are orthogonal to equipotential lines at all points.

12.4 *Graphical superposition of streamlines:* Two potential flows A and B are superposed to generate another potential flow. If $\psi_A = 1, 2, 3$ etc. (in arbitrary units) represent the family of streamlines for flow A and $\psi_B = 1, 2, 3$, etc. (again in arbitrary units) represent the family of streamlines for flow B, show that the principle of superposition requires that the dashed curves $\psi = 2, 3, 4, 5$, etc. shown represent the family of streamlines for the combined flow. Use this method to plot the streamlines for a Rankine half-body.

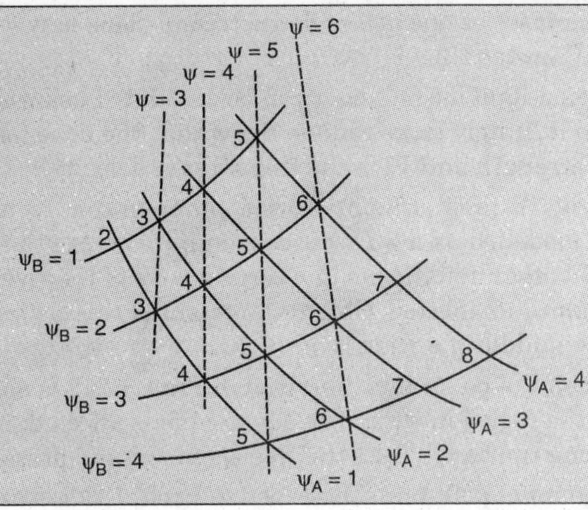

12.5 Obtain the stream function ψ for flow described by $\phi = (x^2 - y^2)/2$. What flow does this represent physically?

12.6 Show that if the flow field around a solid body is irrotational, the circulation around any circuit C enclosing the body is the same. (*Hint*: Take the circuit PQRSTUP shown.) Consider now the free vortex with $V_0 = k/r$. Show that the circulation of any path enclosing the origin is $2\pi k$.

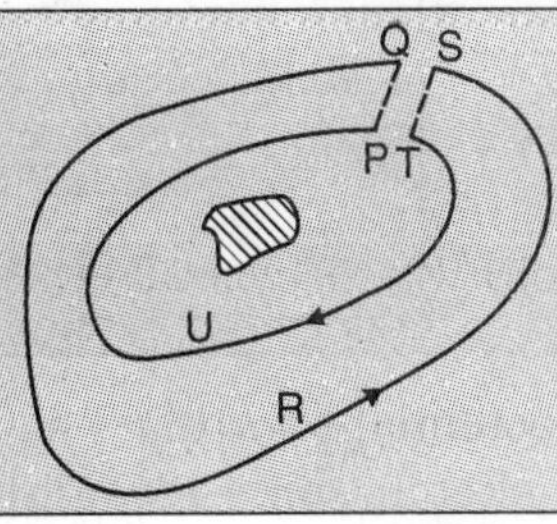

12.7 Show by direct integration that the circulation about any circuit enclosing the cylinder in Fig. 12.18 is $-\Gamma$.

12.8 Obtain the pressure difference $\nabla\mathscr{P}$ between points A (r_1, θ_1) and B (r_2, θ_2) for the following flow fields:

(a) the uniform flow given by Eq. (12.25)

(b) the 2-D source given by Eq. (12.26)

(c) the free vortex given by Eq. (12.27)

Also obtain the stream functions for each of these flows.

12.9 Show that the 'breadth' of a Rankine half-body (described in Sec. 12.7) far downstream approaches asymptotically the value q/V_0. Also, find the breadth of the body at the location of the source.

12.10 Obtain an expression for the pressure p_s on the surface of a Rankine half-body in terms of p_0 and V_0. Plot $(p_s - p_0)/\frac{1}{2}\rho V_0^2$ as a function of θ and see if there is a region where the pressure increases as one moves downstream. (This may lead to flow separation as will be seen in Chapter 13).

12.11 The drag force on a Rankine half-body can be calculated using a control volume bounded by a circle of an infinitely large radius. Show that the drag force equals $\rho q V_0/2$ where q is the source strength and V_0 is the free-stream velocity.

12.12 The discharge of 50 m^3/s effluent from a plant into a 10 m deep river flowing at 0.2 m/s can be modelled as a 2-*D* source spanning across the river depth. It is found that the fish and other aquatic life in a certain zone of the river die out, whereas those outside it are *almost* unaffected. Obtain the extent of this critical region if the discharge is located in the middle of a wide river.

12.13 The discharge point of pollutants described in Prob. 12.12 is now located at one bank of the river. Find the region in which the aquatic life is affected. Also, sketch the pressure distribution along the bank. Take the free-stream static pressure far away to be p_0.

12.14 The discharge point of pollutants described in Prob. 12.12 is now shifted to the outside corner of a 90° bend in a wide river as shown. Carefully sketch the streamlines and indicate the region in which you now expect the effects of pollutants to be confined? What is the extent of this region along the bank on the upstream side of the bend? (The potential function for flow in a 90° corner in absence of a source is $\phi = A\,(x^2 - y^2) = Ar^2 \cos 2\theta$ (See Prob. 12.5). Also, sketch the pressure distribution along the bank, upstream and downstream from the corner.

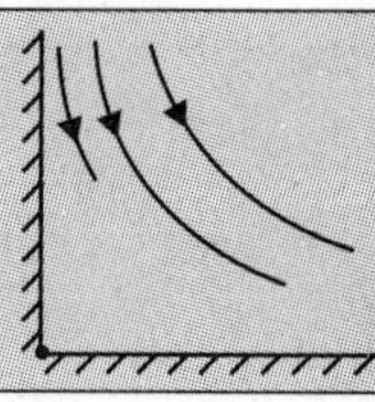

12.15 *Spiral vortex:* Consider the potential $\phi = k_1 \ln r + k_2\theta$ with k_1 and k_2 constants. Determine the stream function and sketch some streamlines. (You may use the graphical superposition method of Prob. 12.4). This flow is the superposition of a vortex and a source/sink and represents such diverse flows as that in a kitchen sink, tornado and a centrifugal pump. Obtain the vortex strength and source/sink strength in terms of k_1 and k_2.

12.16 *Tornado:* The spiral vortex of Prob. 12.15 is a better model for the flow around the core of a stationary tornado or a hurricane. For such flows k_1 is negative, representing a sink. The flow inside the core of radius R has radial, axial and swirl components, but the swirl component usually dominates and the flow within the core is often idealized as a forced vortex with fluid undergoing a solid body-type rotation.

Obtain

(a) the potential function of the outer flow in terms of the circulation Γ and q, the influx per unit height,

(b) V_r and V_θ for the outer flow region $r \geqslant R$,

(c) the variation of pressure for the outer region,

(d) V_θ for the core region $r \leqslant R$, and

(e) the pressure variation for the core region, if V_r and V_z there are much smaller than V_θ.

12.17 In a typical stationary tornado the pressure is found to be 720 Pa below atmospheric at a radial distance of 220 m from its centre, and the influx of air is 25×10^3 m^3/s per metre height. Compute

(a) the sink and the vortex strengths,

(b) V_r at 220 m from the centre,

(c) the velocity at that point, and

(d) the pressure at 300 m from the centre.

12.18 In July 1977, Delhi was hit by a freak tornado and the Miranda House library walls collapsed. Which way do you think the debris of walls fell: into the library or outwards? Give reasons.

12.19 A whirlpool in a tank from which water is draining out as shown, can be described essentially by a free vortex over most of its flow field. The tangential velocity at $r = 40$ cm is 7.5 cm/s. Find the shape $z(r)$ of the free surface and compute the draw down Δz at $r = 40$ cm and at $r = 5$ cm. Do you expect the free vortex flow to be valid at small values of r as well?

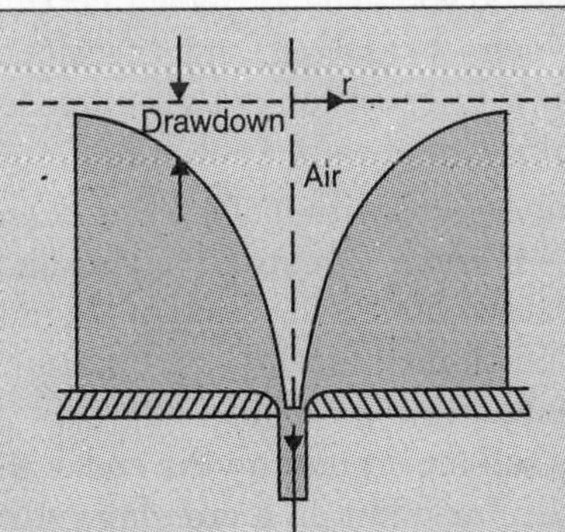

12.20 Consider a uniform flow $\phi = V_0 x$ past a vortex of strength Γ situated at the origin. Write the stream function and plot the streamlines using the graphical method of Prob. 12.4. Obtain the velocity components V_r and V_θ of the flow, and write down the expression for the pressure at any point (r, θ) in the flow field.

12.21 Consider a pair of sources, each of strength q located along the y-axis at $(0, +a)$ and $(0, -a)$ respectively. Sketch the streamline pattern using the graphical technique of Prob. 12.4. Write down the stream and the potential functions of this flow. Find the velocity distribution $V(x)$ and the pressure distribution $p(x)$ along the x-axis.

CHAPTER 12

12.22 *Rankine oval:* Consider the superposition of a free-stream $\mathbf{V} = V_0\hat{\mathbf{i}}$, a source of strength m at $(-a, 0)$, and a sink of strength m at $(a, 0)$. Write the stream function for this combination. The plot of these streamlines gives a closed cylindrical body called a *Rankine oval* as shown. Determine the location of stagnation points A and B. Also obtain the thickness $2h$ of the body and the velocity at points C and D.

$$\left(\text{Use } \tan^{-1}\alpha - \tan^{-1}\beta = \tan^{-1}\frac{\alpha-\beta}{1+\alpha\beta}\right).$$

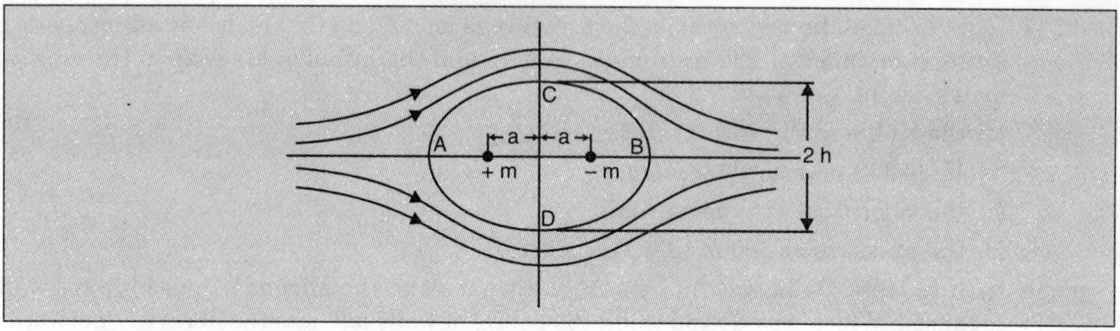

12.23 Show that the stream function of the combination of a uniform flow with velocity V_0 along the x-axis, a source of strength q at $(+a, 0)$, and a sink of the same strength at $(-a, 0)$ is given by

$$\psi = V_0 y + \frac{q}{2\pi}\tan^{-1}\frac{2ay}{x^2+y^2-a^2}$$

What streamline pattern do you expect? Show that this flow does not have a stagnation point on the x-axis between $-\infty$ and $-a$, and, therefore, it cannot represent a flow about a closed body. Note that if the position of the source and sink are interchanged, we get the flow about the Rankine oval as in Prob. 12.22.

12.24 *Kelvin oval:* Consider the superposition of a free stream and a vortex pair at $(0, +a)$ and $(0, -a)$ of circulations $-\Gamma$ and $+\Gamma$ respectively as shown. Obtain the stream function for this combination. Determine the location of stagnation points along the x-axis and, thus, determine the condition when this combination may represent the flow about a closed body. Plot the streamline $\psi = 0$ for $\Gamma/2\pi V_0 a = 1$. This represents a Kelvin oval. Find the velocity at the top most point on the surface of this oval, and note that this is larger than that at a similar point on a circular cylinder.

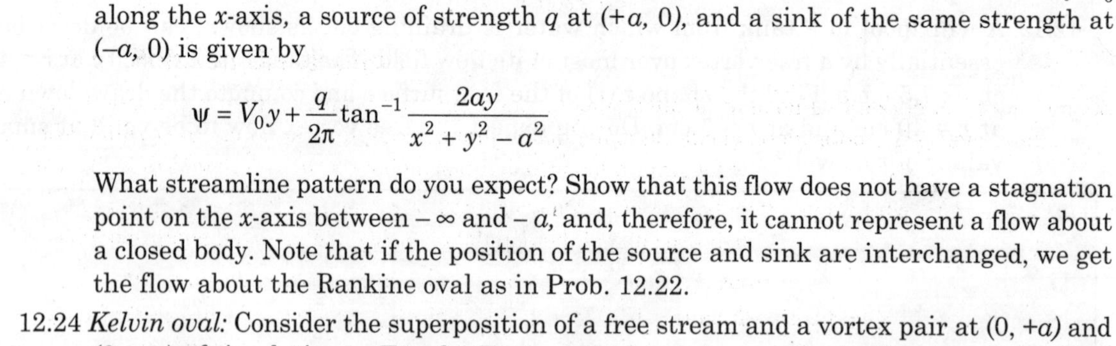

12.25 A uniform stream $\psi = V_0 y$ is superposed on the pair of sources of Prob. 12.21. What are the stream and potential functions of the flow? Obtain the condition under which one or more stagnation point is located on the x-axis. Can this combination represent a flow about a closed body or about a half-body (i.e., a semi-infinite body)? This is at times used as a crude approximation for the separating flow normal to a flat plate. By locating more sources along $(-a, +a)$ on the y-axis, a better approximation can be obtained. The flow pattern for 101 sources located along the dashed line is shown.

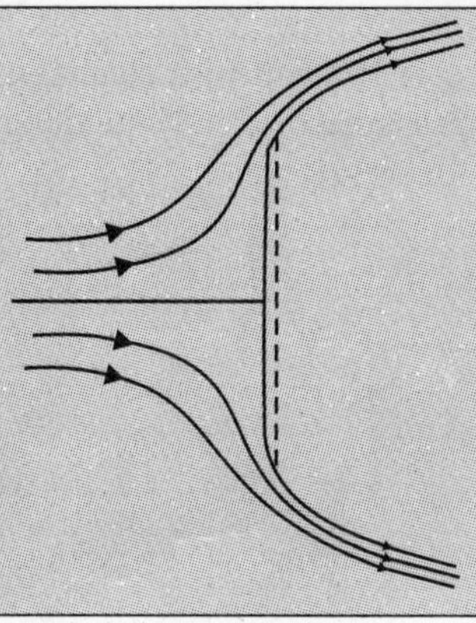

12.26 *Method of images:* In many physical situations, the potential flow around a submerged body with walls nearby is of interest. An example is a free vortex of circulation Γ placed near a wall as shown, with a uniform flow from the left. A common technique to solve such problems is to invoke symmetry and consider the case of a uniform flow with two vortices of strengths Γ and $-\Gamma$ at $(0, +a)$ and $(0, -a)$ respectively. A straight streamline at $y = 0$ can then be interpreted as a wall. This is known as the method of images. Obtain the pressure distribution at the wall for the case shown. Also compare this pressure distribution with that at the corresponding location in the absence of the wall.

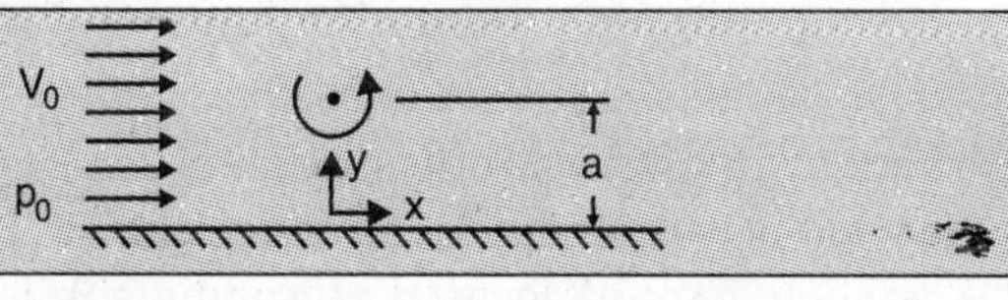

12.27 A source of strength q is located at a distance a from a wall as shown. Use the method of images (Prob. 12.26) to show that the flow picture is the same as obtained in Prob. 12.25. Obtain the velocity at the wall and locate the points at which the velocity is maximum. Also obtain the pressure distribution along the wall and the region where the pressure gradient is positive (and so flow separation is likely to occur as discussed in Chapter 13).

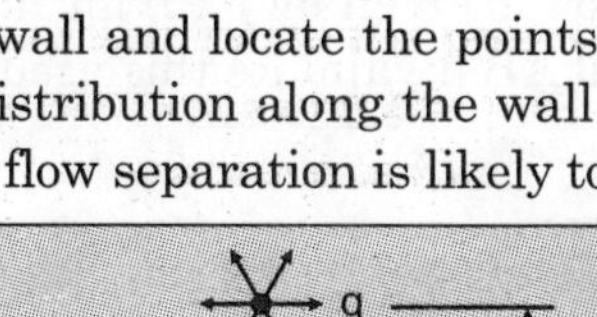

12.28 A vortex of strength $-\Gamma$ is located at a distance a from an infinite wall. If a uniform flow is superposed on this, show that the stream function for the flow is the same as for the Kelvin oval of Prob. 12.24. Find the velocity and pressure distribution along the wall.

12.29 If a source of strength q is located between two walls as shown, what system of images would be required to model the flow using the method of images of Prob. 12.26?

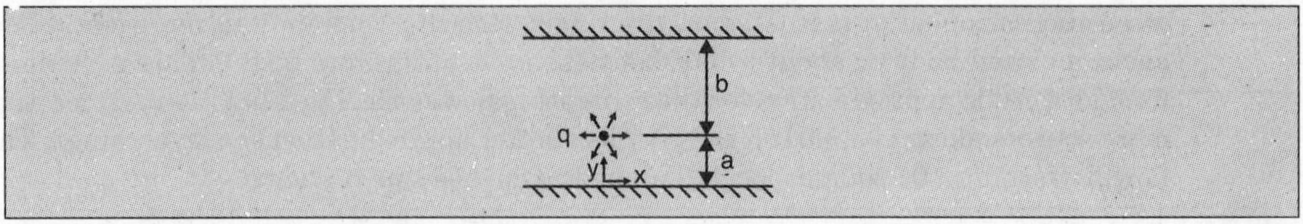

12.30 A 2-D source of strength q is located in a corner as shown. Use the method of images (Prob. 12.26) to obtain the pressure distribution along the walls. (Hint: Superpose velocities of the individual sources directly).

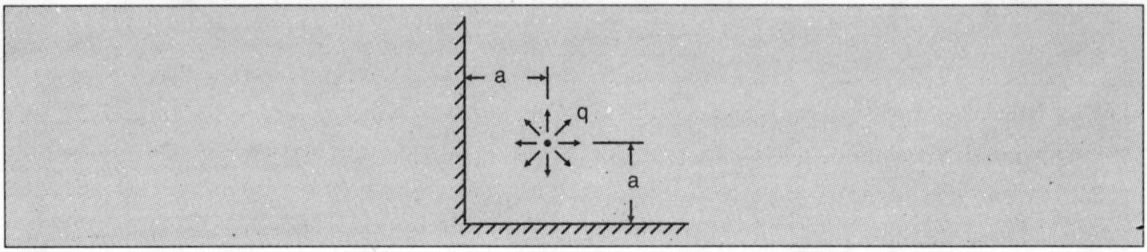

12.31 A tall cylindrical chimney of outside diameter D is subjected to a wind of velocity V as shown. If the flow about it can be assumed potential (a highly unlikely situation), find the *net* forces (per unit length of the chimney), at sections PR and QS. Assume pressure inside the chimney as atmospheric.

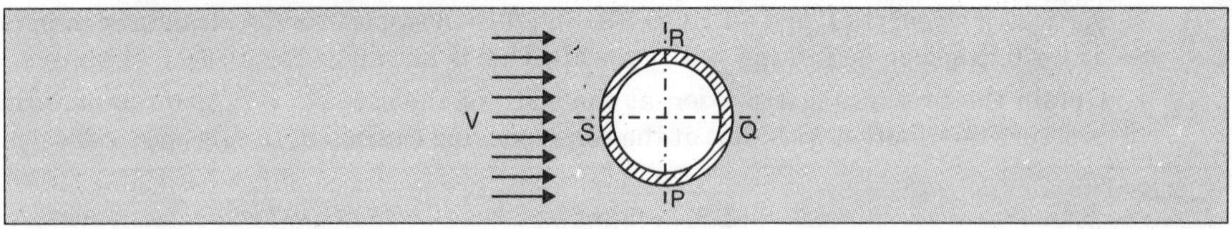

12.32 An aircraft hangar in an Antarctic station is semi-cylindrical in shape and is oriented such that the wind is usually in the direction shown. It is normally well sealed to prevent heat loss. A window pane W located at an angle θ_1 is broken. Obtain the net lift acting on the structure. At what θ_1 should W be located so that this force is zero? Assume that the pressure on the outside is given by $p(\theta)$ for inviscid flow around a cylinder. Does the presence of the ground AB invalidate this assumption?

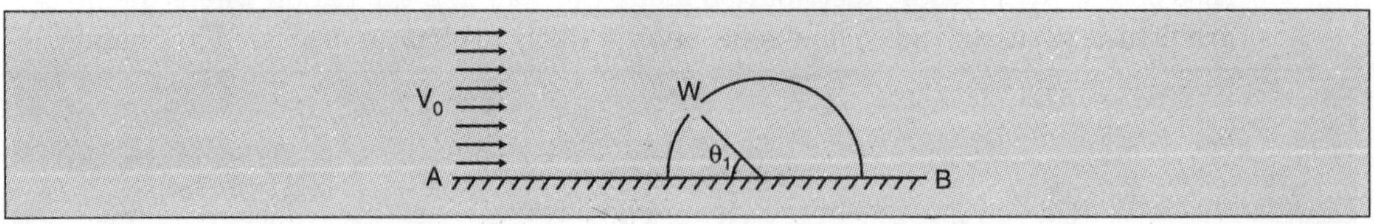

12.33 *Yaw meter:* A cylindrical probe with three radial holes (connected to manometers) is sometimes used for measuring the yaw angle between the flight path and the longitudinal axis of a flight vehicle as shown. Whenever the pressure on the side holes is equal, the centre hole points in the direction of flow and the yaw angle is zero. Obtain a relation between the yaw angle γ and the measured pressure difference $p_A - p_B$. Then show that the maximum sensitivity of the instrument is attained when $\theta_1 = 45°$ and when the yaw angle is zero.

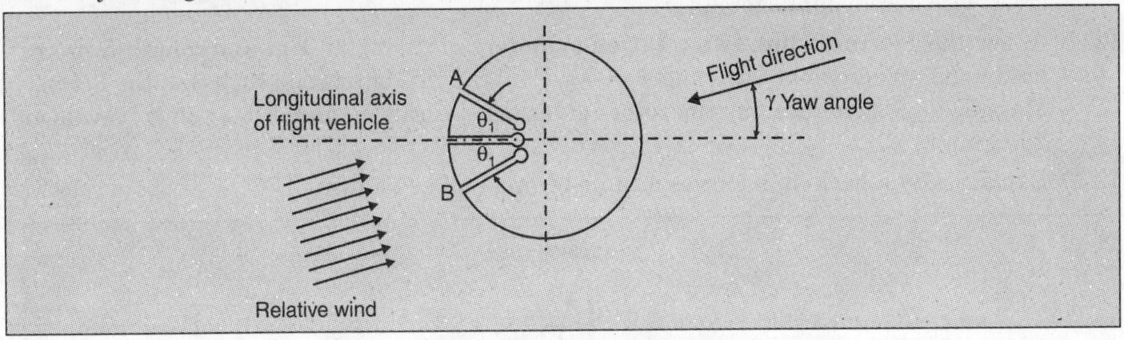

12.34 What should be the location of the holes A and B in the yaw meter of Prob. 12.33 such that the pressure measured at A and B is the free stream pressure when γ is zero?

12.35 Most of the drag force acting on bodies moving through fluids arises because the separation of the boundary layer modifies drastically the velocity and pressure distributions on the rear of the body. One estimate of the drag acting on a circular cylinder can be obtained by assuming that on the front half of the cylinder ($\pi/2 \leqslant \theta \leqslant 3\pi/2$) the pressure distribution is the same as in the inviscid flow, but on the rear half, the boundary layer separates from the surface and the pressure does not recover from the minimum value it acquires at the shoulder ($\theta = \pm\, \pi/2$). Obtain the drag coefficient C_D (based on frontal area) given by such a model. Note that the measured value of C_D for laminar boundary layers at high Re is about 1.2.

12.36 A 1 m dia. cylinder rotates at 1500 RPM in an air flow of 20 m/s. Compute the Robins-Magnus force acting on it per unit length.

12.37 The *Flettner rotor ship* built in Germany in 1924, used two rotating vertical cylinders instead of sails. Each cylinder had a diameter of 3 m and was 15 m long, rotating at 720 RPM as shown. Find the magnitude and direction of the Robins-Magnus force

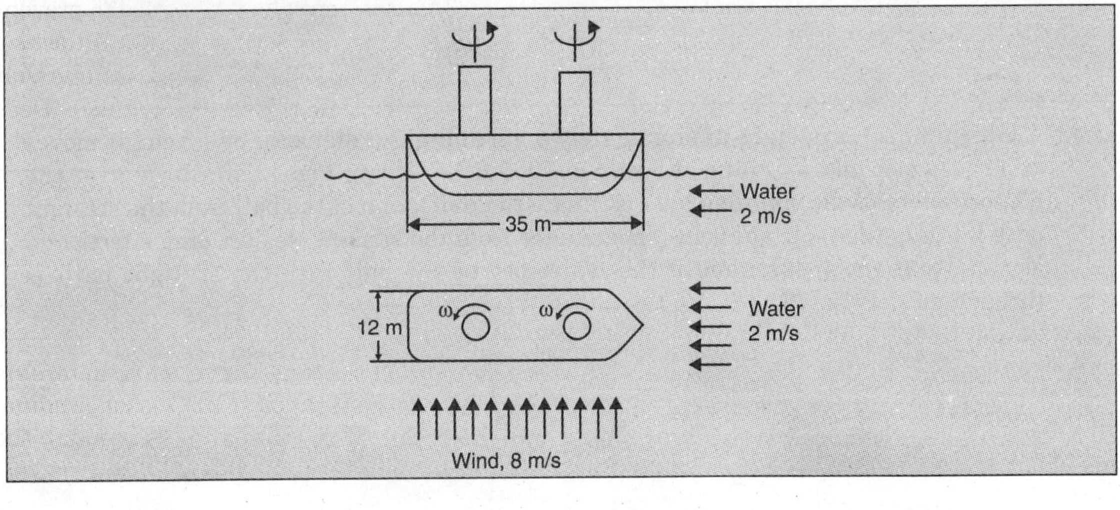

developed by the rotors in a cross breeze of 8 m/s. If 1.9 m of the ship (weight 800 tons) were below the water level, find the velocity with which the ship moved forward in a water current of 2 m/s in the direction shown. Assume that the drag coefficient based on frontal area is 0.1. Also assume that the keel area is sufficient to counteract the cross-wise aerodynamic force.

12.38 As for the Flettner ship, two rotating cylinders of diameter 1 m and length 5 m are used to lift an aeroplane weighing 4000 kg. Find the rate of rotation such that the Robins-Magnus force generated by the rotors gives the aircraft enough lift when it travels at 250 kmph.

12.39 Explain why a back-spin curves a table-tennis ball as shown.

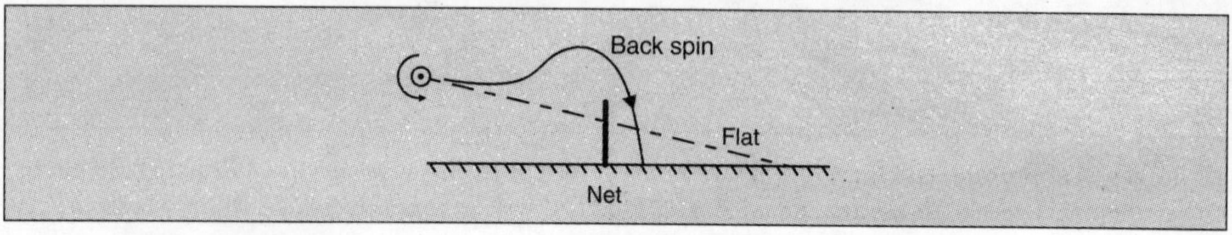

12.40 Explain why at high altitudes (e.g., Mexico city), baseball pitchers are not as effective in curving the ball as at sea level. Also explain why, when the wind is blowing from centre field towards the home plate, the curve-ball pitcher is not as effective.

12.41 A light cardboard cylinder *C* rolls down the ramp BA and takes the trajectory as shown. Explain the departure from the parabolic trajectory of a cardboard box *D* which slides down the same ramp.

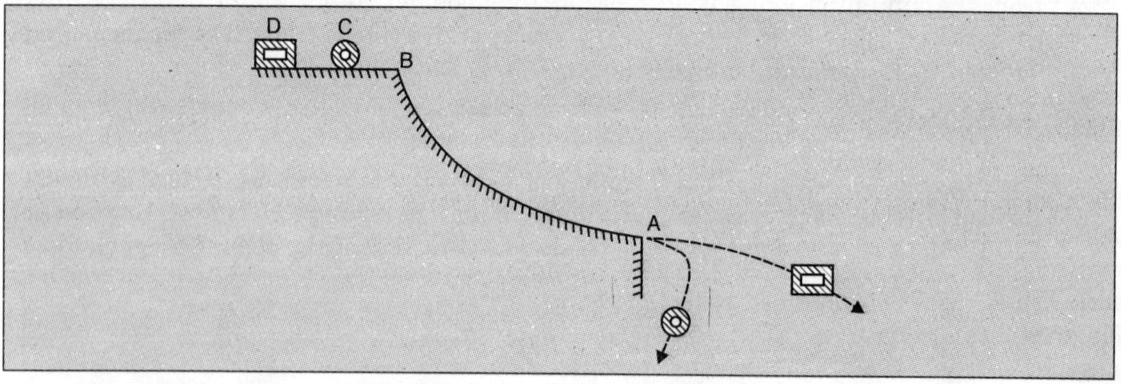

12.42 A table-tennis ball weighs approximately 3 gm and has a diameter of 3.5 cm. It moves at a speed of 3 m/s, spinning at a rate of 3000 RPM. Using Fig. 12.20 which gives the lift coefficient of the spinning ball, estimate the deflection of the ball from the straight path if it hits the table at about 3 m distance from the striker. Neglect drag effects and assume that the inclination of the trajectory of the ball with the straight path is throughout very small.

13

BOUNDARY LAYERS

13.1 INTRODUCTION

As seen earlier in Sec. 11.5 and 12.1 when a fluid flows at high Reynolds numbers past a body, the viscous effects may be neglected everywhere except in a thin region in the vicinity of the walls (Fig. 13.1). This region, termed as the *boundary layer* by L. Prandtl who first postulated its existence in 1903, is necessitated by the requirement of the no-slip condition at the walls, a condition which cannot be satisfied by inviscid flows. Even though the boundary layer is comparatively very thin, the dynamics of this layer, in particular, its tendency to *separate* from the walls under certain conditions, affects the entire flow picture in a rather dramatic manner. A thorough understanding of the boundary layer is, therefore, very crucial for realistic modelling of many flow situations.

Since the viscous forces are not negligible within the boundary layer, the Euler equation (the x- and y-components of which are given by Eqs. (12.14) and (12.15)) is no longer applicable and we have to revert to the Navier-Stokes Eq. (6.24). But the fact that the boundary layer is very thin permits some simplification of the governing equations. The resulting equations are termed the *Prandtl boundary-layer equations* and are generally easier to solve than the full Navier-Stokes equation. In this chapter, first these equations are obtained and then some important properties of boundary layers are studied.

13.2 PRANDTL BOUNDARY-LAYER EQUATIONS

First consider a steady 2-*D* flow of an incompressible fluid past a flat wall (Fig. 13.2). Let the *inviscid* velocity $V_w(x)$ at the wall* be prescribed as a function of x. This would have been the velocity at the wall if the flow was truly inviscid, but the presence of viscosity, however small, requires *no slip* at the wall and a boundary layer develops along it. At any location x the

* Note that V_w on a flat wall can be a function of x, as for example, when the wall forms part of a converging or a diverging channel, as also in the developing flow region in a straight channel (Fig. 13.1b).

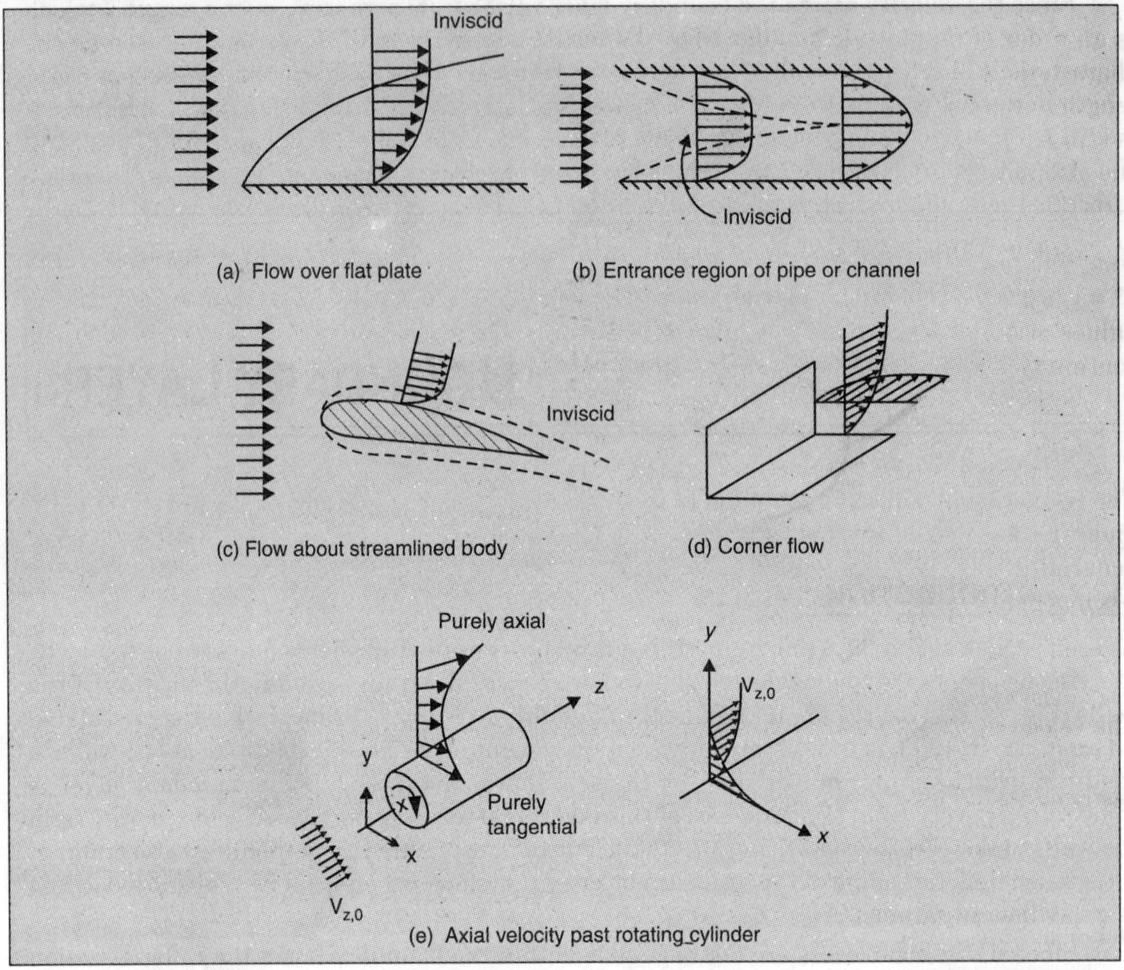

Fig. 13.1. Some boundary layer flows. Thickness of boundary layers has been shown enlarged for clarity. Figure on right in (e) shows the resolution into two boundary layers.

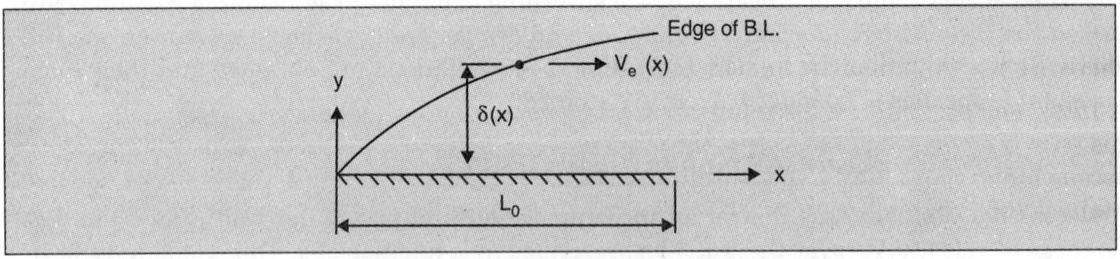

Fig. 13.2. Boundary layer on a flat plate.

velocity component V_x varies across the boundary layer from zero at the wall to $V_e(x)$ at its edge. Since the boundary layer is thin one can, as a first approximation, set $V_e(x)$, the velocity at the outer edge of the boundary layer equal to $V_w(x)$.

Since the velocity across the boundary layer varies from zero to V_w over a length δ which is an order of magnitude smaller than the length along the wall, L_0 cannot be used as the characteristic length in the y-direction across the boundary layer. The appropriate characteristic length in the y-direction, therefore, is δ, as was also seen in Sec. 11.6. But in the x-direction, it is still L_0. Thus, boundary-layer problems require two different characteristic lengths, one in the streamwise and the other in the normal direction. The velocity components in these two directions may also be expected to have different characteristic values. Denoting these by $V_{x,0}$ and $V_{y,0}$ then, the non-dimensional variables $V_x^* = V_x/V_{x,0}$, $V_y^* = V_y/V_{y,0}$, $x^* = x/L_0$ and $y^* = y/\delta$ are expected to be of order unity. The value $V_{x,0}$ is of the same order as $V_w(x)$, but the values of δ and $V_{y,0}$ are as yet unknown. If these dimensionless variables are used in the continuity Eq. (4.20) for steady, 2-D incompressible flows, we get

$$\frac{\partial V_x^*}{\partial x^*} + \frac{V_{y,0}}{V_{x,0}} \frac{L_0}{\delta} \frac{\partial V_y^*}{\partial y^*} = 0 \qquad \text{...(13.1)}$$

The two terms in this equation must be of the same order, otherwise one of them will drop out making the equation degenerate. Thus $V_{y,0}\, L_0/V_{x,0}\, \delta$ must be of order one. Without loss of generality, it can be set equal to one and an estimate of the characteristic value of the y-component of velocity can be got as

$$V_{y,0} = V_{x,0} \frac{\delta}{L_0} \qquad \text{...(13.2)}$$

The velocities in the y-direction, therefore, are much smaller than those in the x-direction.

Next non-dimensionalize the x- and y-components of the Navier-Stokes equation. It can easily be verified that the resulting equations are

$$V_x^* \frac{\partial V_x^*}{\partial x^*} + V_y^* \frac{\partial V_x^*}{\partial y^*} = -\frac{\Delta \mathscr{P}_0}{\rho V_{x,0}^2} \frac{\partial \mathscr{P}^*}{\partial x^*} + \frac{\mu}{\rho V_{x,0} L_0} \frac{\partial^2 V_x^*}{\partial x^{*2}} + \frac{\mu L_0}{\rho V_{x,0} \delta^2} \frac{\partial^2 V_x^*}{\partial y^{*2}} \qquad \text{...(13.3)}$$

and

$$V_x^* \frac{\partial V_y^*}{\partial x^*} + V_y^* \frac{\partial V_y^*}{\partial y^*} = -\frac{\Delta \mathscr{P}_0}{\rho V_{x,0}^2} \frac{L_0^2}{\delta^2} \frac{\partial \mathscr{P}^*}{\partial y^*} + \frac{\mu}{\rho V_{x,0} L_0} \frac{\partial^2 V_y^*}{\partial x^{*2}} + \frac{\mu}{\rho V_{x,0} L_0} \frac{L_0^2}{\delta^2} \frac{\partial^2 V_y^*}{\partial y^{*2}} \qquad \text{...(13.4)}$$

where use has been made of Eq. (13.2). In Eq. (13.3) the coefficient of $\partial^2 V_x^* / \partial x^{*2}$ is recognized as $1/\mathrm{Re}_{L_0}$ and since Re_{L_0} is large (an essential condition for the boundary-layer approximation) this term is small compared to one, the order of the inertial terms, and can be neglected. Since viscous forces must play a role within the boundary layer, the second viscous term *must* be retained and thus

$$\frac{\mu L_0}{\rho V_{x,0} \delta^2} = 1$$

or

$$\frac{\delta}{L_0} = \sqrt{\frac{\mu}{\rho V_{x,0} L_0}} = \sqrt{\frac{1}{\mathrm{Re}_{L_0}}} \qquad \text{...(13.5)}$$

This is the same result as obtained in Sec. 11.6 and shows that the boundary layer is indeed very thin.

It is easy to see from Eq. (13.3) that the characteristic pressure variation $\Delta\mathscr{P}_0$ along the x-direction should be of order $\rho V_{x,0}^2$, the same as in inviscid flow. Next consider Eq. (13.4). Using the estimate of δ/L_0 obtained in Eq. (13.5), it is noticed that the coefficient of the pressure term dominates, being equal to Re_{L_0}, if $\Delta\mathscr{P}_0$ along the y-direction is taken as $\rho V_{x,0}^2$, the same as that along the x-direction. Clearly, $\rho V_{x,0}^2$ is too high an estimate of the pressure variation *across* the boundary layer. The pressure variation in the y-direction may thus be neglected in comparison with the variation of pressure in the streamwise direction (which is of the same order as the pressure variation in the outer inviscid flow). Therefore, it can be concluded that to a first approximation, the pressure does not change across the thickness of the boundary layer. This means that the inviscid pressure is "impressed" on the boundary layer and the pressure distribution at the edge of the boundary layer approximates that at the wall. This is why the lift force on an aerofoil calculated using the inviscid-flow pressure distribution estimates the actual lift quite accurately (see Fig. 11.5).

Using the above simplifications and reverting back to dimensional variables, the following equations governing the flow within the boundary layer, are obtained

$$\frac{\partial V_x}{\partial x} + \frac{\partial V_y}{\partial y} = 0 \qquad \text{...(13.6)}$$

and

$$V_x \frac{\partial V_x}{\partial x} + V_y \frac{\partial V_x}{\partial y} = -\frac{1}{\rho}\frac{dp_w}{dx} + \frac{\mu}{\rho}\frac{\partial^2 V_x}{\partial y^2} \qquad \text{...(13.7)}$$

where p_w is the pressure at the wall calculated using the inviscid approximation.* These equations are due to Prandtl and are applicable within the boundary layer.

Before the boundary conditions applicable to these equations are listed it should be pointed out that the boundary layer thickness δ used in the above formulation is not a definite distance within which the viscous effects are predominant and above which they are *totally* absent. In fact, these effects are present everywhere but decrease asymptotically as we move away from the wall. Thus, δ represents an arbitrary distance beyond which it may be assumed that the inviscid velocity approximates the actual velocity closely. It is conventional to set this distance at the point where the difference between the two is only one per cent. Thus, at $y = \delta$, $V_x(x, \delta) = V_e(x) = 0.99\ V_i(x, \delta)$, where V_i represents the inviscid velocity field (Fig. 13.3). If δ is small, $V_i(x, \delta)$ can be approximated by $V_i(x, 0)$, that is, the inviscid velocity $V_w(x)$ at the wall. Thus, the Prandtl boundary-layer Eqs. (13.6) and (13.7) are to be solved subject to the boundary conditions

$$V_x = V_y = 0 \quad \text{at} \quad y = 0 \qquad \text{...(13.8)}$$

and

$$V_x = 0.99\ V_w(x) \quad \text{at} \quad y = \delta \qquad \text{...(13.9)}$$

* The information provided by Eq. (13.4) that the pressure variation across the bundary layer is negligible compared to that along it has been used in Eq. (13.7) by changing $\partial p/\partial x$ to dp_w/dx. Eq. (13.4) provides no further information at this first order of approximation, and has been dropped.

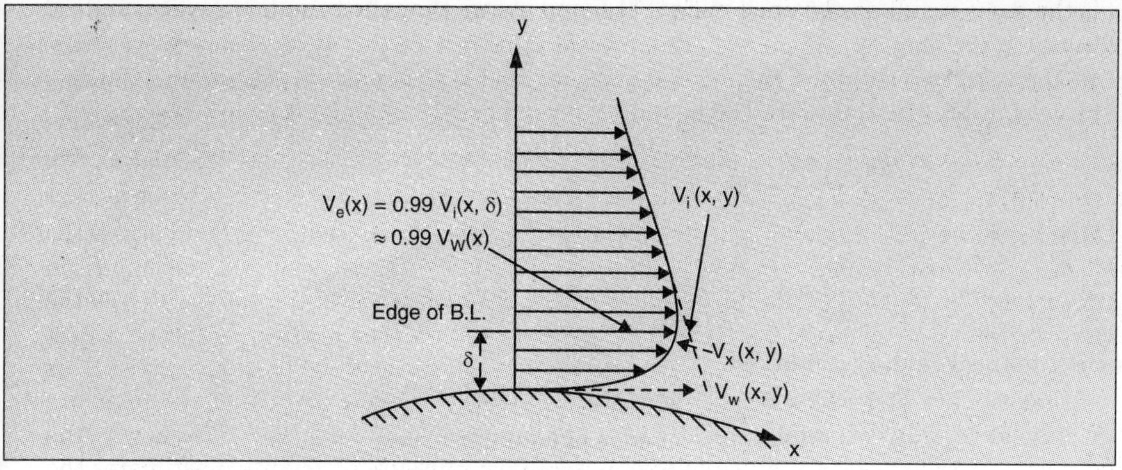

Fig. 13.3. Boundary condition at edge of boundary layer, *y*-scale is exaggerated.

These equations have been obtained for a flat wall subject to an arbitrary pressure gradient, but are also valid for a curved wall provided the radius of curvature of the surface is large compared to the boundary layer thickness δ everywhere. On curved walls x is the streamwise coordinate (measured along the wall) and y is normal to it.

It must be stated that the above simplifications are valid only as long as the boundary layer thickness δ, the region in which viscous effects are important, is small compared to L_0. This condition breaks down when the boundary layer separates from the wall.

Equations (13.6) and (13.7) are relatively easier to solve* than the full Navier-Stokes equation and have been extensively studied in the last eighty years. The development of fluid mechanics is roughly divided into two eras—before and after Prandtl. Before Prandtl, even though the basic equations of flow were available, very few realistic solutions were obtained. The science of fluid mechanics was sharply divided between mathematicians who constructed an elaborate mathematical edifice based on the Euler equation, but having little relation to practical flows, and the engineers who manipulated a large amount of empirical data without any clear understanding of what was involved. After Prandtl, a deeper understanding of a wide variety of flow phenomena has been achieved and it has become possible to compute many flow fields, at least up to the separation point. Extensive numerical methods using high-speed digital computers have been developed for solving the boundary layer equations both for laminar and turbulent flows.

13.3 BOUNDARY LAYER ON A FLAT PLATE

When an infinite expanse of fluid flows parallel to a flat plate, the action of viscosity brings the fluid in immediate contact with it, to rest. This fluid, in turn, slows down the adjoining layers, thereby giving rise to a boundary layer. The boundary layer thickness δ grows in the downstream

* Those familiar with the theory of partial differential equations will note that the Navier-Stokes equation is an elliptic equation and, therefore, it is essential that it be solved simultaneously over the entire flow field. On the other hand, Prandtl boundary-layer equations are parabolic and can be solved by beginning at the leading edge and 'marching' downstream, resulting in a considerable saving of computational effort.

direction because more and more fluid is brought under the influence of viscosity as the flow proceeds. If the plate is thin enough, the inviscid velocity $V_i(x, y) = V_0$, is constant everywhere. Since the velocity is constant, the inviscid pressure field is also constant everywhere and dp_w/dx to be used in Eq. (13.7) is zero. The boundary layer Eqs. (13.6) and (13.7) now become

$$\frac{\partial V_x}{\partial x} + \frac{\partial V_y}{\partial y} = 0$$

$$V_x \frac{\partial V_x}{\partial x} + V_y \frac{\partial V_x}{\partial y} = \frac{\mu}{\rho} \frac{\partial^2 V_x}{\partial y^2} \quad \text{...(13.10)}$$

subject to the boundary conditions

$$V_x = V_y = 0 \qquad \text{at } y = 0$$

$$V_x \to 0.99\ V_0 \qquad \text{at edge of boundary layer } y = \delta \quad \text{...(13.11)}$$

Since δ is unknown, it is more convenient to replace the outer boundary condition by

$$V_x \to V_0 \qquad \text{as } y \to \infty$$

These equations were first solved by H. Blasius using numerical techniques. He showed that the velocity at any location x varies with y such that at $y = 5\sqrt{\mu x/\rho V_0}$ from the wall, V_x equals $0.992\ V_0$. Therefore, the boundary layer thickness can be taken as $\delta = 5\sqrt{\mu x/\rho V_0}$, or in non-dimensional form

$$\frac{\delta}{x} = \frac{5}{\sqrt{\text{Re}_x}} \quad \text{...(13.12)}$$

where $\text{Re}_x = \rho V_0 x/\mu$, the Reynolds number based on the *local* x as the characteristic length. This result is consistent with the boundary-layer assumption that the region in which viscous forces are predominant, is thin. It is also consistent with the order of magnitude estimates obtained in Sec. 11.6 and 13.2.

Another important fact established by Blasius concerns the shape of the velocity profile across the boundary-layer. He showed that on a flat plate, the velocity profiles at various positions along the plate are similar to one another in shape. This property is termed as *self-similarity* and requires that the 'non-dimensional shape' of the velocity profile be independent of the streamwise coordinate x. Thus, if we non-dimensionalize the velocity V_x by the *local* inviscid velocity $V_w(x)$ and the distance y by the *local* boundary-layer thickness $\delta(x)$, then $V_x/V_w(x)$ is a function of $y/\delta(x)$, independent of x. For the case of the flat plate $V_w(x)$ is, of course, a constant V_0 and so

$$\frac{V_x}{V_0} = \mathcal{F}(\eta)$$

where

$$\eta = \frac{y}{\delta(x)} = \frac{y}{5\sqrt{\dfrac{\mu x}{\rho V_0}}} = \frac{1}{5}\frac{y}{x}\sqrt{\text{Re}_x}$$

It is conventional to drop the factor of 5 from this expression and write V_x/V_0 as a function of ξ where

$$\xi = \frac{y}{x}\sqrt{\mathrm{Re}_x} \qquad \text{...(13.13)}$$

The resulting velocity profile V_x/V_0 as a function of ξ is shown in Fig. 13.4. Since the velocity profiles are self-similar, it is possible to obtain the actual profile at any x from the single non-dimensional profile shown in this figure. It is seen that the velocity V_x approaches its asymptotic value V_0 very rapidly and at a value of ξ of about 5, V_x/V_0 is 0.992. The slope at $\xi = 5$ can be seen to be almost zero, signifying the absence of shear stresses at the edge of the boundary layer, thus justifying the inviscid assumption beyond this point.

The stress at the plate depends upon the velocity gradient at the wall, i.e., on $\partial V_x/\partial y$ at $y = 0$. The value of this can be obtained from the slope of the non-dimensional velocity profile at $\xi = 0$.

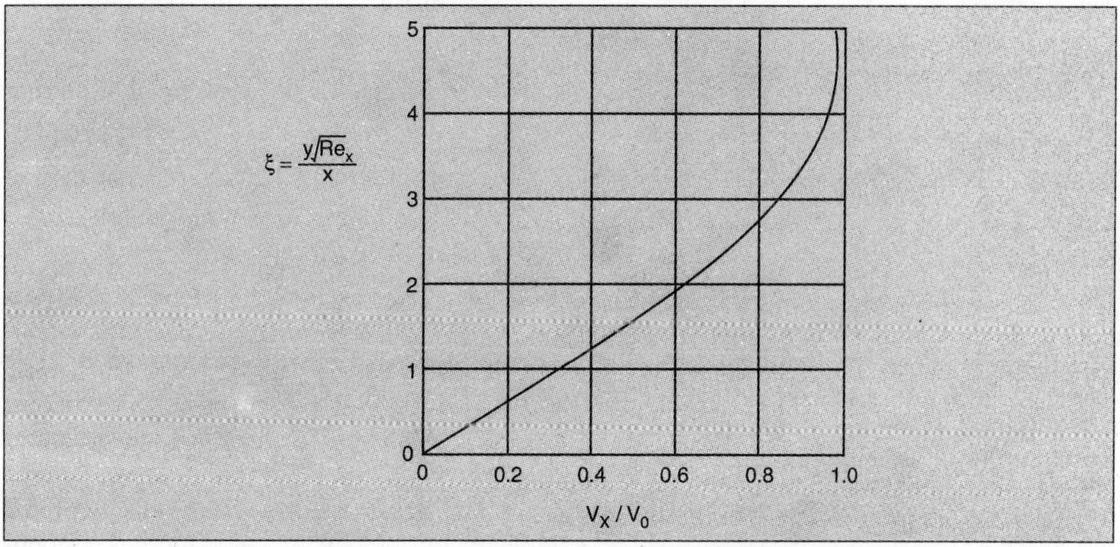

Fig. 13.4. The non-dimensional velocity profile for flow over a flat plate.

The computed value of $\partial(V_x/V_0)/\partial\xi$ is 0.332 at $\xi = 0$. Therefore, the shear stress at the wall is given by

$$\tau_w = \mu\left(\frac{\partial V_x}{\partial y}\right)_{y=0} = \mu V_0\left[\frac{\partial(V_x/V_0)}{\partial\xi}\frac{\partial\xi}{\partial y}\right]_{y=0}$$

$$= \mu V_0\sqrt{\frac{\rho V_0}{\mu x}}\left(\frac{\partial(V_x/V_0)}{\partial\xi}\right)_{\xi=0}$$

Using the value of 0.332 for the slope of the non-dimensional velocity profile, the local wall-shear-stress is

$$\tau_w = 0.332\frac{\rho V_0^2}{\sqrt{\mathrm{Re}_x}} \qquad \text{...(13.14)}$$

CHAPTER 13

For high Reynolds number flows, it is conventional to take $\rho V_0^2/2$ as the characteristic value of the stress ($\rho V_0^2/2$ was shown in Chapter 11 to be the characteristic pressure for such flows). The non-dimensional wall shear stress, termed the *skin-friction coefficient* C_f is then given by

$$C_f = \frac{\tau_w}{\frac{1}{2}\rho V_0^2} = \frac{0.664}{\sqrt{\text{Re}_x}} \quad \text{...(13.15)}$$

For high Reynolds numbers, the value of C_f is much smaller than one, signifying that the shear stresses are much smaller than the normal stresses (that is, the pressure forces), a result consistent with the fact that large Reynolds numbers imply small viscous effects.

The drag (on one side only) on a flat plate of length L (and width b) can be obtained by integrating the shear stress

$$F_D = b\int_0^L \tau_w(x)dx = \frac{1.328}{\sqrt{\text{Re}_L}}\left(\frac{1}{2}\rho\, V_0^2\right)(bL)$$

Introducing the *drag coefficient*

$$C_D = \frac{F_D}{\frac{1}{2}\rho\, V_0^2 A}$$

where A is the area bL of the plate, we get

$$C_D = \frac{1.328}{\sqrt{\text{Re}_L}} \quad \text{...(13.16)}$$

For Re_L values between 10^3 and 5×10^5, the agreement of Eq. (13.16) with the measured values of C_D on flat plates is very good (Fig. 13.5). The departure at lower values of Re_L is explained by the fact that at these values, the boundary layer is no longer very thin and thus the thin boundary-layer assumption breaks down.

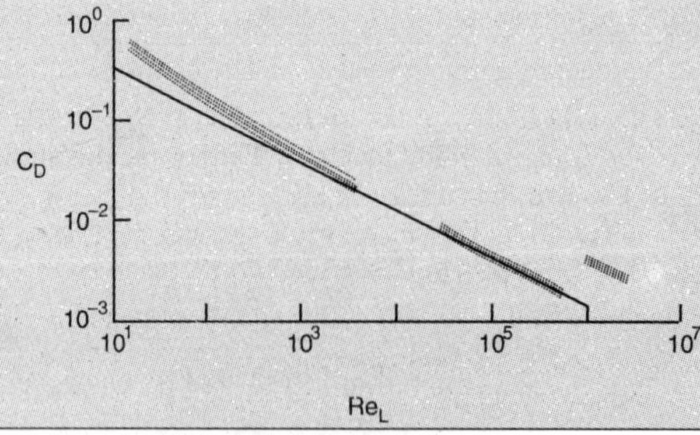

Fig. 13.5. C_D *vs.* Re_L for flow over a flate plate. Shaded region shows range of experimental data.

The rather dramatic departure of experimental results from the Blasius solution at values of Re higher than about 5×10^5 is due to the fact that the above formulation is valid for laminar flows only. As Re_x, the *local* Reynolds number increases, the flow *within* the boundary layer becomes unstable and at some distance downstream from the leading edge, a transition from laminar to turbulent flow takes place. The shear forces at the wall, when the boundary-layer is turbulent are considerably larger than those in laminar flows and, therefore, C_D is also larger. The turbulent boundary-layers will be discussed in Sec. 13.5.

13.4 APPROXIMATE SOLUTION OF BOUNDARY-LAYER EQUATIONS — INTEGRAL METHOD

In this section is presented an approximation technique, first introduced by Th. von Karman (after whom it is known) for solving the boundary-layer equations. In this approach, instead of solving the boundary-layer equations at every point of the flow field, one uses a control volume of infinitesimal length *dx* in the flow direction but which has a finite height extending from the wall to the edge of the boundary layer (Fig. 13.6). Figure 13.6 (a) shows the mass balance for such a *CV* for a boundary layer on a flat plate (refer Prob. 4.9). Clearly, there must be an entrainment ($-d\mathring{m}$) of fluid at the top edge o1f the boundary layer. It can be seen that

$$d\mathring{m} = -\frac{d}{dx}\left[\int_0^{\delta} \rho\, V_x dy\right] dx \qquad \text{...(13.17)}$$

where δ is a function of x.

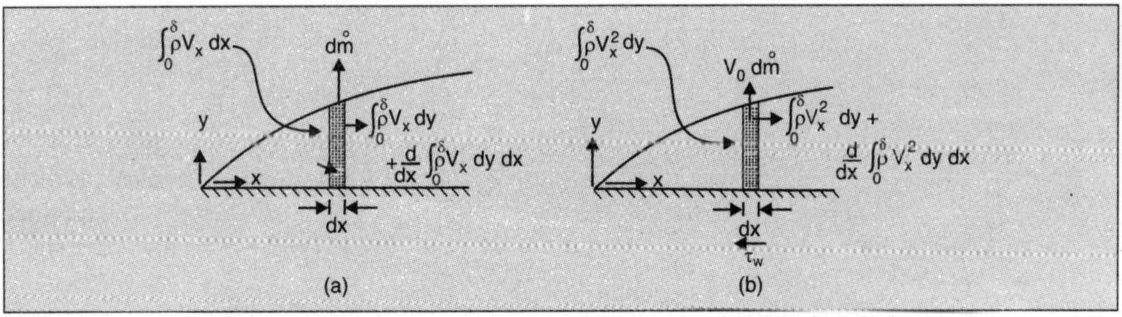

Fig. 13.6. Mass and momentum flows across the control surface in a *CV* of unit width in flow over a flate plate.

The momentum fluxes for the same *CV* are shown in Fig. 13.6(b). The x-velocity at the edge of the boundary layer is the inviscid velocity V_0. The pressure variations in the y-direction are negligible according to the thin boundary-layer assumption, and since $dp_w/dx = 0$, the only external force acting on the *CV* is the shear stress at the wall. If $\tau_w(x)$ represents this shear stress, the momentum Eq. (5.3) for the *CV* gives

$$-\tau_w d_x \cdot 1 = \frac{d}{dx}\left[\int_0^{\delta} \rho\, V_x^2 dy \cdot 1\right] dx + V_0\, d\mathring{m}$$

or, on using Eq. (13.17),

$$-\tau_w = \frac{d}{dx}\left[\int_0^{\delta} \rho\, V_x^2 dy\right] - V_0 \frac{d}{dx}\left[\int_0^{\delta} \rho\, V_x dy\right]$$

CHAPTER 13

On noting that V_0 is a constant, this equation can be written as

$$\frac{d}{dx}\int_0^\delta \frac{V_x}{V_0}\left(1-\frac{V_x}{V_0}\right)dy = \frac{\tau_w}{\rho V_0^2}$$

If y is non-dimensionalized with respect to δ, the local boundary-layer thickness, one gets

$$\frac{d}{dx}\left[\delta(x)\int_0^1 \frac{V_x}{V_0}\left(1-\frac{V_x}{V_0}\right)d\left(\frac{y}{\delta}\right)\right] = \frac{\tau_w}{\rho\, V_0^2} \quad \text{...(13.18)}$$

This is the momentum integral equation for a boundary layer over a flat plate — a special case of the von Karman momentum integral equation. The solution of this equation depends on one important observation: that if the boundary layer profile is self-similar, i.e., if the non-dimensional velocity V_x/V_0 is a function of the non-dimensional coordinate $\eta = y/\delta$ alone, the integral on the left in Eq. (13.18) is a constant, independent of x. Then Eq. (13.18) reduces to an ordinary differential equation for δ. For laminar boundary layers, $\tau_w\ (x)$ can be found from the velocity profile using Newton's viscosity relation $\tau = \mu\, dV_x/dy$. Thus, if $V_x/V_0 = f(\eta)$ with $\eta = y/\delta$, then

$$\tau_w = \mu\left[\frac{dV_x}{dy}\right]_{y=0} = \mu\frac{V_0}{\delta}\left[\frac{df}{d\eta}\right]_{\eta=0}$$

and Eq. (13.18) becomes

$$\alpha\frac{d\delta}{dx} = \frac{\mu}{\rho\, V_0\delta}\beta \quad \text{...(13.19)}$$

where

$$\alpha = \int_0^1 f(1-f)\,d\eta \quad \text{...(13.20)}$$

and

$$\beta = \left[\frac{df}{d\eta}\right]_{\eta=0} \quad \text{...(13.21)}$$

The solution of Eq. (13.19) can readily be obtained as

$$\delta = \sqrt{\frac{2\,\beta\,\mu\,x}{\rho\, V_0\alpha} + C}$$

where C is the constant of integration. At the leading edge $x = 0$ of the plate, the boundary-layer thickness is zero, so that $C = 0$, and

$$\frac{\delta}{x} = \sqrt{\frac{2\,\mu}{\rho\, V_0\, x}\frac{\beta}{\alpha}} = \frac{1}{\sqrt{\text{Re}_x}}\sqrt{\frac{2\,\beta}{\alpha}} \quad \text{...(13.22)}$$

where $\sqrt{2\,\beta/\alpha}$ is a function of the shape of the boundary layer velocity profile. One can similarly deduce that $C_D = \sqrt{8\,\alpha\,\beta/\text{Re}_L}$.

Equation (13.22) [with α, β defined by Eqs. (13.20) and (13.21)] represents a considerable progress from the original set of equations but it is not enough since α and β depend on the shape of the velocity profile — and that is not known. von Karman showed that it is possible to obtain *good approximate solutions* by using any reasonable shape of the velocity profile. This means that the estimate of δ (and consequently of C_f and C_D) is *relatively* insensitive to the shape of the self-similar velocity profile assumed.

To illustrate the insensitivity of the results to the velocity profile chosen, let us begin with a linear velocity profile:

$$f = a\eta + b$$

Obviously $f = V_x/V_0$ must be zero at $\eta = 0$, and equal to one at $\eta = 1$. Thus, $b = 0$ and $a = 1$, and so the self-similar linear velocity profile to be used in Eqs. (13.20), (13.21) and (13.22) is

$$f = \eta$$

This gives

$$\alpha = \int_0^1 \eta(1-\eta)\,d\eta = 1/6$$

$$\beta = \left[\frac{df}{d\eta}\right]_{\eta=0} = 1$$

and so,

$$\frac{\delta}{x} = \frac{3.46}{\sqrt{\text{Re}_x}} \qquad \text{...(13.23)}$$

Comparison of this with the 'exact' solution Eq. (13.12) shows that Eq. (13.23) is not such a bad result considering the amount of effort saved. The drag coefficient using this profile comes out as $C_D = 1{\cdot}155/\sqrt{\text{Re}_L}$ which also compares well with Eq. (13.16).

We next try a parabolic profile

$$f = a + b\eta + c\eta^2$$

We need three boundary conditions: $f(0) = 0$ and $f(1) = 1$ are obvious. The third boundary condition is obtained from the fact that the shear stress at the edge of the boundary layer is negligible, or $f'(1) = 0$. These give

$$f = 2\eta - \eta^2 \qquad \text{...(13.24)}$$

as the parabolic self-similar profile. With this, Eqs. (13.20), (13.21) and (13.22) give

$$\alpha = \frac{2}{15}$$

and

$$\beta = 2$$

so that

$$\frac{\delta}{x} = \frac{5.48}{\sqrt{\text{Re}_x}} \qquad \text{...(13.25)}$$

which is very close to the 'exact' result of Eq. (13.12). The drag coefficient comes out as $C_D = 1.463\ \sqrt{\text{Re}_L}$. Figure 13.7 shows the comparison of the linear, parabolic and the Blasius velocity profiles.

This method can be generalized for obtaining the boundary-layer results in the presence of a pressure gradient as well, where they are found to be extremely useful.

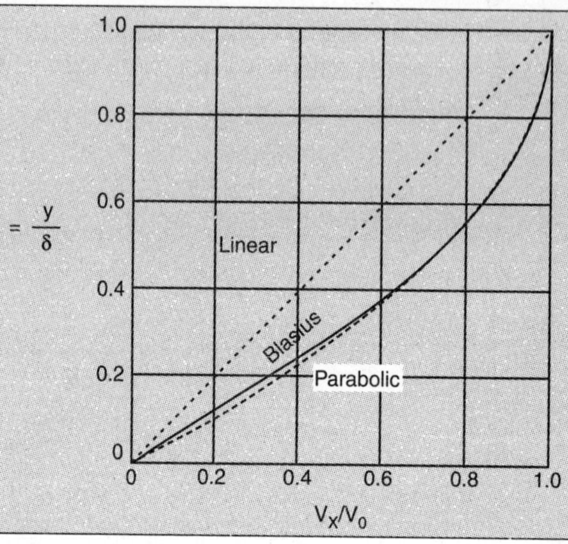

Fig. 13.7. Comparison of linear, parabolic and Blasius profiles for flow over a flat plate.

13.5 TURBULENT BOUNDARY LAYERS

The flow within the thin viscous layer that develops adjacent to a boundary remains laminar only as long as the Reynolds number Re_δ (= $\rho\ V_0\ \delta/\mu$) based on the boundary layer thickness δ remains less than about 1800. As the flow proceeds down the surface, the boundary layer thickness δ increases and the flow becomes unstable. Disturbances in the flow tend to grow and a transition to turbulence, characterized by permanent and random fluctuations in the velocities, takes place. The mean velocities still exhibit the characteristics of a boundary layer, though the thickness is now much more. For flow over a flat plate, an Re_δ of 1800 corresponds (through Eq. 13.12) to an Re_x of about 1.3×10^5, and represents the *minimum* value of the Reynolds number at which turbulence first appears. The transition from laminar to turbulent boundary layer on a flat plate can be delayed to an Re_x as high as 4×10^6, if the level of turbulence in the external stream is kept very low and if the plate is isolated from all vibrations. A value 5×10^5 is a reasonable mean value of the *critical Reynolds number* Re_{crit} representing laminar to turbulent transition in boundary layers on flat plates.

When the Reynolds number Re_L equa ls Re_{crit}, the turbulent boundary layer first appears at the trailing edge of the plate. As Re_L increases, the transition point moves upstream, being always located at the point where Re_x equals Re_{crit}. Upstream of the transition point the boundary layer is laminar while downstream, it is turbulent. The turbulent boundary layer is relatively thicker (Fig. 13.8).

A look at the velocity profile across the turbulent boundary layer in Fig. 13.8 reveals the same pattern as seen in the turbulent velocity profile in circular tubes (see Fig. 1.28): the profile

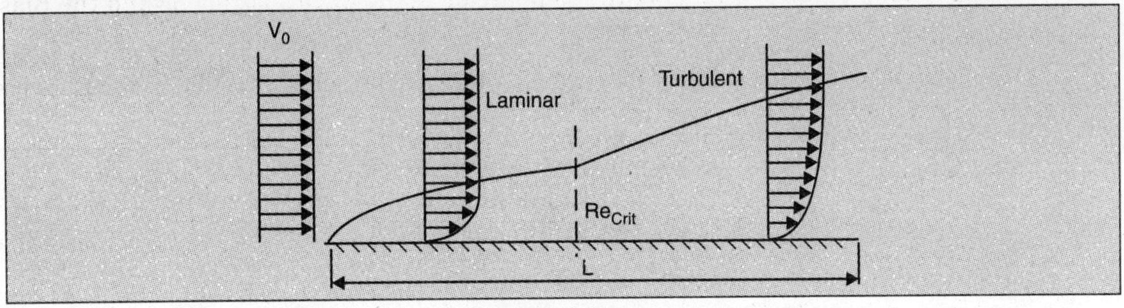

Fig. 13.8. Laminar and turbulent boundary layers on a flat plate when $Re_L > Re_{crit}$.

is sharper at the plate representing a much higher level of shear stress at the wall. The velocity profile can be approximated by a one-seventh power law relationship

$$\frac{V_x}{V_0} = \left[\frac{y}{\delta}\right]^{1/7} \quad ...(13.26)$$

An 'exact' analysis for turbulent boundary layers is not yet available since the phenomenon of turbulence is still incompletely understood. One can, however, make use of the integral method developed in Sec. 13.4 to obtain approximate results. But here also one runs into a problem since the wall shear stress τ_w to be used in Eq. (13.18) can no longer be evaluated from the velocity profile 13.26 using the Newton's law of viscosity. The shear stress in turbulent flow depends on the nature of mixing caused by the eddies in specific flow situations. It is, therefore, necessary to use empirical information for τ_w. One such correlation expressed in terms of the non-dimensional skin-friction coefficient C_f is given by

$$C_f = \frac{\tau_w}{\frac{1}{2}\rho V_0^2} = \frac{0.045}{(Re_\delta)^{1/4}} \quad ...(13.27)$$

where Re_δ is the Reynolds number based on the local boundary layer thickness δ.

Use of Eqs. (13.26) and (13.27) in the integral momentum Eq. (13.18) gives the differential equation for the turbulent boundary-layer thickness δ as

$$\frac{7}{72}\frac{d\delta}{dx} = 0.0225\left[\frac{\mu}{\rho V_0 \delta}\right]^{1/4} \quad ...(13.28)$$

This equation is valid from the point of transition to the trailing edge of the plate. If we assume, *for simplicity,* that the transition occurs at the leading edge of the plate and that the boundary layer is turbulent everywhere, we obtain

$$\frac{\delta}{x} = 0.37(Re_x)^{-1/5} \quad ...(13.29)$$

CHAPTER 13

and

$$C_f = \frac{\tau_w(x)}{\frac{1}{2}\rho V_0^2} = 0.0576(\text{Re}_x)^{-1/5} \quad ...(13.30)$$

The drag coefficient then is obtained as

$$C_D = \frac{1}{\frac{1}{2}\rho V_0^2 \cdot L \cdot 1}\int_0^L \tau_w(x)\, dx \cdot 1 = 0.072(\text{Re}_L)^{-1/5} \quad ...(13.31)$$

This equation has been plotted in Fig. 13.9 where it is seen that there is good agreement with experimental data in the Re_L range of 5×10^5 to 10^7. Beyond $\text{Re}_L = 10^7$, Eq. (13.31) increasingly underestimates the total drag, and the following is the recommended correlation

$$C_D = 0.455\,(\log_{10} \text{Re}_L)^{-2.58} \quad ...(13.32)$$

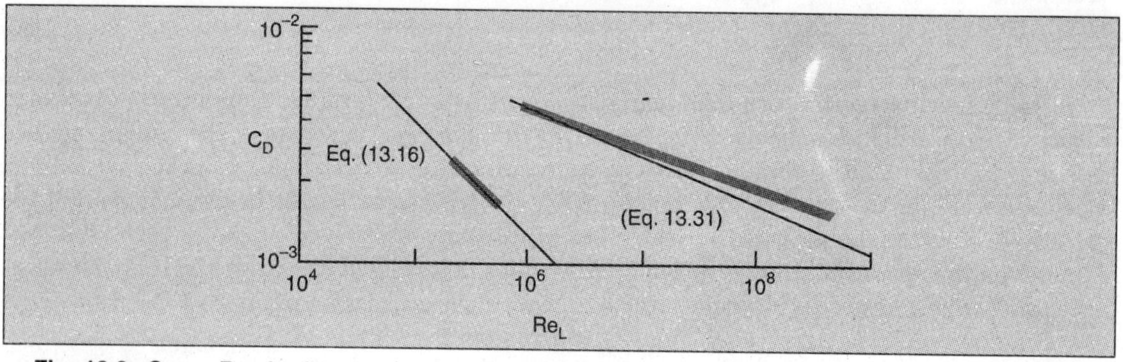

Fig. 13.9. C_D *vs.* Re_L for flow over a flat plate. Shaded region shows range of experimental points.

Equation (13.31) has been obtained by assuming that the turbulent layer starts at the leading edge itself. But all boundary layers start out as laminar, and become turbulent at $x = x_{\text{crit}}$ where the local Reynolds number reaches the critical value Re_{crit}. Thus, in the integral of Eq. (13.31), the value of τ_w for x between 0 and x_{crit} should be the laminar one (given by Eq. 13.15), and between x_{crit} and L, it should be the turbulent one (given by Eq. 13.30). The drag equation then becomes

$$C_D = 0.072\,(\text{Re}_L)^{-1/5} - \frac{A}{\text{Re}_L} \quad ...(13.33)$$

where the value of A depends on the value of the critical Reynolds number. For $\text{Re}_{\text{crit}} = 5 \times 10^5$, A takes the value of 1670 (refer Prob. 13.4).

The drag formulae 13.31 to 13.33 are valid only when the plate surface can be taken as smooth. When the surface is rough, the drag behaviour (see Fig. 13.10) is similar to that in the case of flow through a pipe. The drag is higher than that on a 'smooth' plate, increasing with the surface roughness measured by $\in/L$, where $\in$ is the mean roughness height. As Re_L increases, C_D acquires a constant value depending upon the value of $\in/L$. As before, the flow is said to be fully turbulent since the viscous effects are truly negligible in this region.

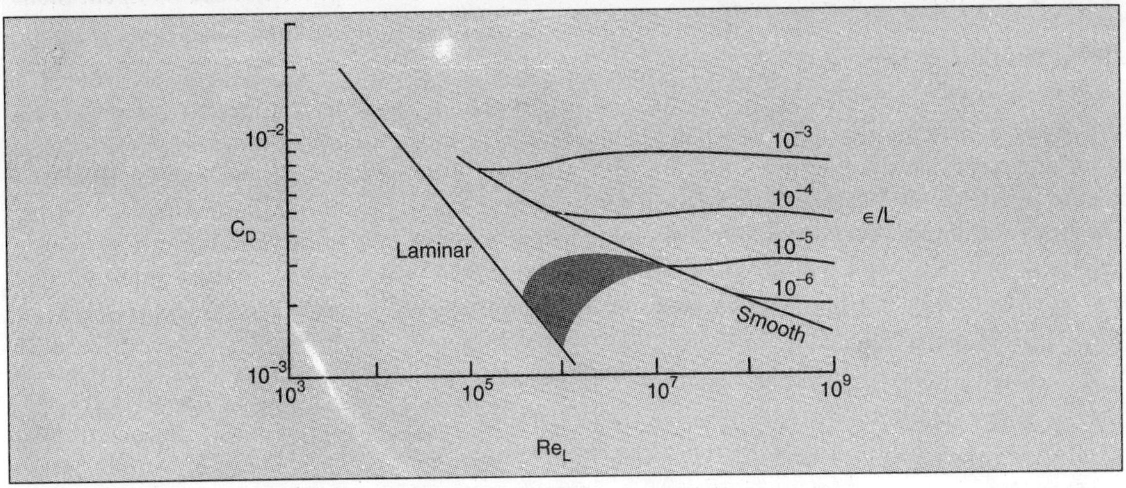

Fig. 13.10. C_D vs. Re_L for smooth and rough flat plates.

13.6 BOUNDARY-LAYER SEPARATION

The most important phenomenon connected with the development of boundary layers, in flows at high Reynolds numbers, is that the flow in the immediate neighbourhood of a wall is reversed under certain conditions, resulting in a back flow (Fig. 13.11). Under these conditions, it can

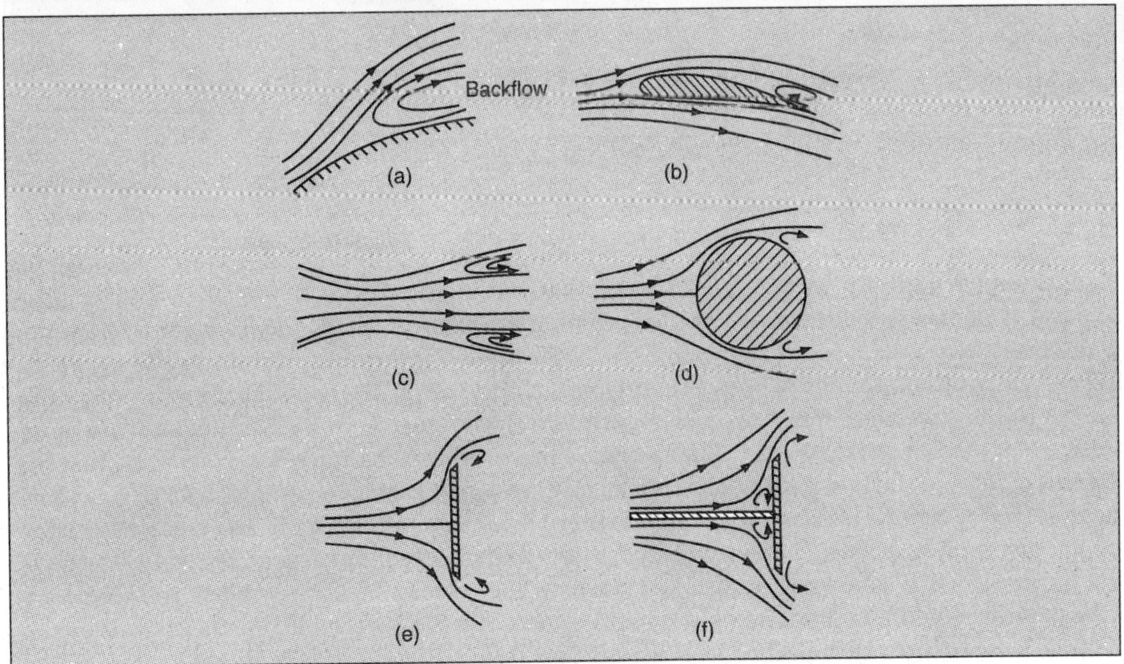

Fig. 13.11. Some examples of boundary layer separation.

no longer be assumed that the viscous effects are confined to within a thin region near the wall, and the boundary layer is said to have 'separated'. Once the boundary layer separates from the wall, there is a drastic change in the flow pattern and the inviscid flow is no longer a good approximation of the actual flow picture.

In order to understand the conditions under which the very important phenomenon of boundary-layer separation takes place, consider the flow of a steady, uniform high Re stream of an incompressible fluid past a 2-*D* circular cylinder. The inviscid flow pattern for this case is shown in Fig. 12.17 and the corresponding pressure distribution is plotted in Fig. 13.12. The fluid accelerates on either side of the stagnation point *A*, acquiring the maximum velocity ($= 2V_0$) at points *C* and *D*. Thereafter, the fluid decelerates to the rear stagnation point *B*. The pressure decreases on the upstream half of the cylinder but increases on the downstream half.

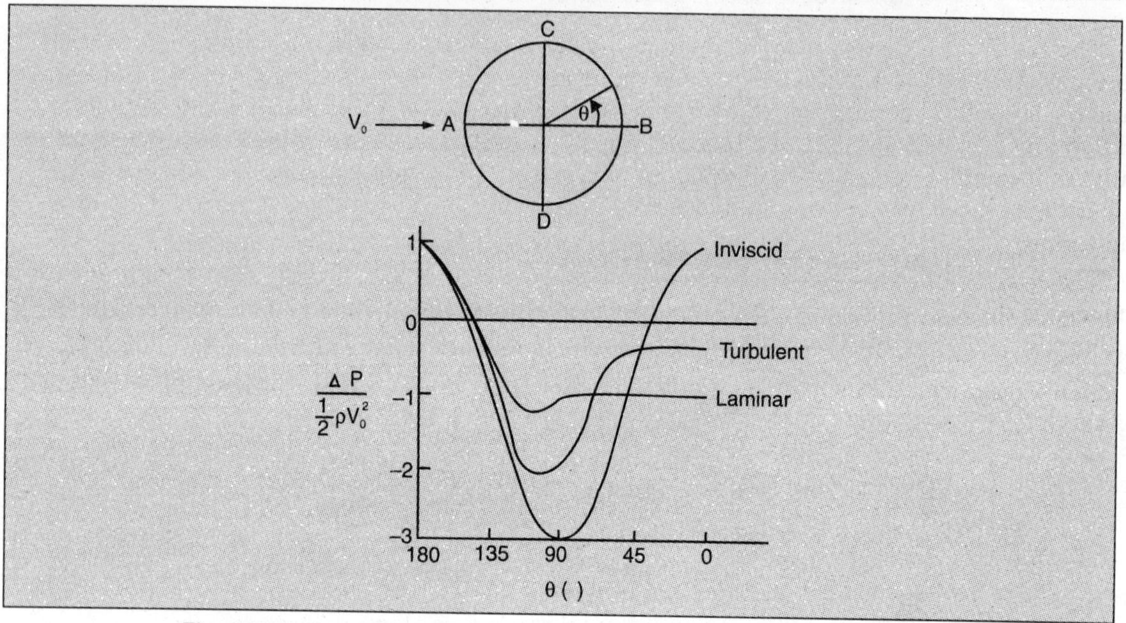

Fig. 13.12. Pressure distribution in flow around a 2-*D* circular cylinder.

It has been shown in Sec. 13.2 that *across* the boundary layer the pressure remains constant. This means that the streamwise pressure gradient in the boundary layer is the same as the pressure gradient along the wall in the outer (inviscid) flow, i.e., the external pressure is 'impressed' upon the boundary layer. Therefore, the flow within the boundary layer is subjected to a decreasing pressure (i.e., a negative gradient) in the upstream half and to an increasing pressure in the downstream half. The two pressure gradients balance out exactly, so that an element of fluid in the inviscid region starting from rest at *A* comes to rest again at *B*. However, the presence of viscous action within the boundary layer dissipates energy and a particle *inside* the boundary layer is slowed down by viscosity as well as by the pressure gradient between *C* and *B*. Therefore, it comes to rest at a point where the velocity V_e at the edge of the boundary layer (nearly equal to the inviscid velocity at the wall) is still appreciable. Further downstream, the fluid close to the wall is pushed in a direction opposite to the main flow, establishing a region of back flow resulting in a sequence of velocity profiles shown in Fig. 13.13. Since the slope $\partial V_x/\partial n$ changes from a positive value to a negative one at separation, the point of separation *S* corresponds to the case $\partial V_x/\partial n = 0$. Therefore, the shear stress vanishes at this point.

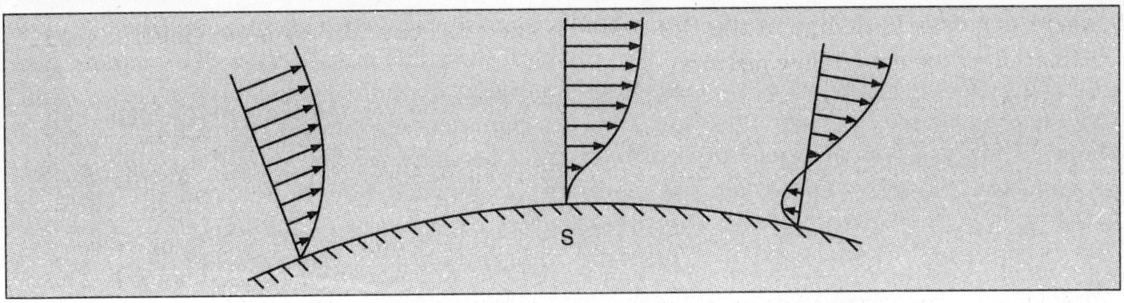

Fig. 13.13. Velocity profiles near S, the point of separation.

The same phenomenon also occurs in a divergent channel. The inviscid flow requires the velocity to decrease with pressure increasing downstream in the divergent portion. But the boundary-layer fluid cannot overcome the increasing pressure and comes to rest, setting up a backflow and eddies (Fig. 13.11 (c)). If the divergent channel is used as a diffuser, the separation results in incomplete pressure recovery. Therefore, divergent diffusers used for slowing down fluid streams are made to taper very gently (for example in a venturi-meter, Fig. 8.12, or a wind tunnel, Fig. 7.16).

The *necessary* condition for boundary layer separation that emerges from the above discussion is the presence of a positive streamwise pressure gradient. Since separation is normally (but not always) a condition to be avoided, a positive $\partial p/\partial s$ is termed as an 'adverse' pressure

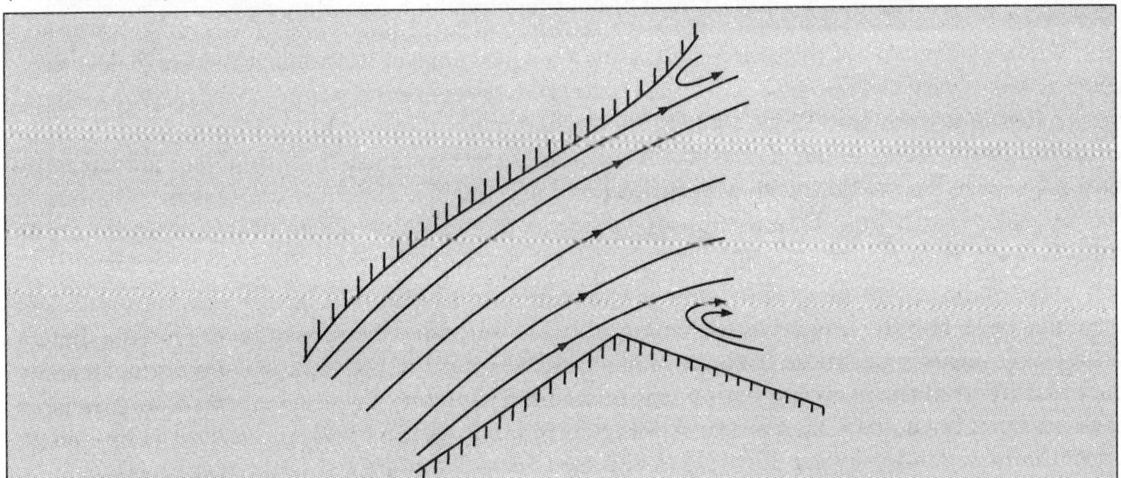

Fig. 13.14. Separation occurring on smoothly and sharply curving walls.

gradient (as opposed to a favourable one). If a wall has a sharp corner, separation will normally occur there (Fig. 13.14), but if it is curving smoothly the exact point of separation point is quite sensitive to small changes in the shape of the wall, especially when these small changes in the wall geometry produce large changes in the pressure distribution. Once the separation takes place, the eddies formed in the region of back flow alter the flow pattern, *even upstream of the separation point.* In some cases, the effect is confined to a local region with the flow reattaching itself a little downstream of the point of separation, as in the pipe entry flow (Fig. 13.15). But more often, the effects are much more extensive, modifying the flow pattern drastically, as for example in the case of cylinders and spheres. In such cases, the viscous flow is not confined to a thin region close to the boundaries and therefore the inviscid solution about the body can no longer be taken as a first approximation to the high Reynolds number flow. This can be seen from the fact that the point of separation on a circular cylinder is observed to occur at 81° from the front stagnation point, upstream from the point of maximum velocity. This is the region where the inviscid flow theory predicts a 'favourable' pressure gradient, and thus separation is not expected. Clearly, separation has modified the inviscid flow region and the pressure distribution significantly, bringing the adverse pressure gradient region upstream of $\theta = 90°$. This is confirmed by the measured pressure distribution on a circular cylinder shown in Fig. 13.12.

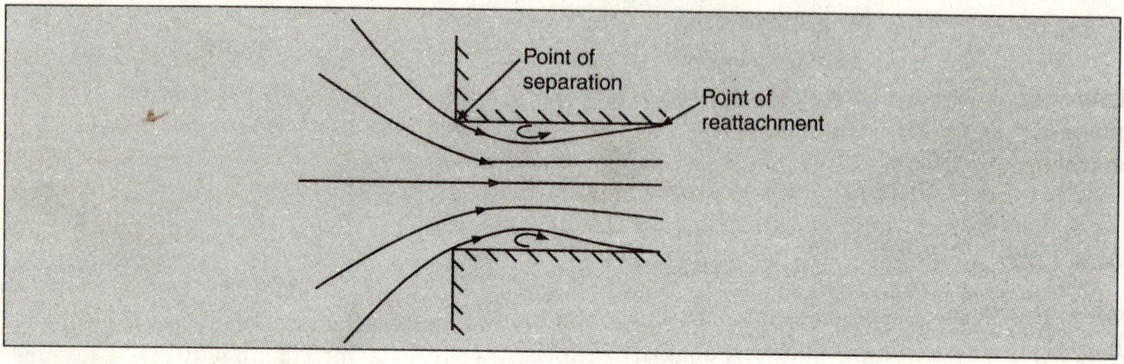

Fig. 13.15. Reattachment of boundary layer in a pipe-entry flow.

An interesting point to observe is that over a wide range of Reynolds numbers (below Re_{crit} when the boundary layer becomes turbulent), the point of separation is quite insensitive to changes in the Reynolds number. Thus, as long as the Reynolds number is high enough so that the boundary layer assumption is a reasonable one (and is low enough so that transition to turbulence does not occur), the location of the separation point and the non-dimensional pressure distribution do not change with Re_D. This results in a fairly constant value of the drag coefficient C_D as shall be seen in Sec. 13.7.

Separation is not restricted to laminar boundary layers alone, and can take place in turbulent layers as well. But the large scale mixing within the turbulent boundary layers makes them a little less prone to separation. In the case of a circular cylinder, the separation point occurs way back at 120° from the *front* stagnation point when the boundary layer is turbulent, as compared to 81° when it is laminar. This property was referred to in Sec. 1.7 and is exploited in the design of golf-balls and in the swing of a cricket ball (see Sec. 1.7 and Probs. 13.40 and 13.41).

13.7 DRAG ON BODIES MOVING THROUGH FLUIDS

When a body moves through a fluid at rest, it experiences a force whose component, opposing the motion, is termed the drag. This force is the resultant of the streamwise components of the shear stresses and the normal stresses acting on the surface of the body.* In a truly inviscid flow shear stresses are zero, and as has been pointed out in Chapter 12, the resultant of the streamwise components of the pressure forces (i.e., the normal stress contribution) is exactly zero, so that the net drag predicted in inviscid flow is zero. It is, therefore, apparent that drag

* The effect of gravity on the pressure distribution is conventionally *not* included in, the drag force.

is a phenomenon related to viscosity. It is conventional to express the drag in terms of a *drag coefficient* C_D defined as Drag/$\left(\frac{1}{2}\rho V_0^2\right) A_c$, where A_c is some characteristic area of the body. A steady fluid flow is controlled by inertial and viscous forces and, therefore, the drag coefficient C_D is a function of the Reynolds number Re.

For low values of Re the viscous forces dominate the inertial forces. The pressure forces are then of the order of viscous forces ($\sim \mu V_0/R$ for a sphere, as seen in Sec. 11.4) so that the drag forces must be of the order of $(\mu V_0/L_0)$ (Area), or, of the order of $\mu V_0 L_0$. The drag relation for very low Re flows over spheres has been given in Prob. 6.5 as

$$\text{Drag} = 6\,\pi\mu V_0 R \qquad \text{...(13.34)}$$

This is known as the *Stokes law* and bears out the validity of our estimates. Thus, for a sphere moving at low Reynolds numbers, the drag coefficient C_D based on the *frontal* area πR^2 as the characteristic area A_c, is obtained as

$$C_D = \frac{6\pi\mu V_0 R}{\frac{1}{2}\rho V_0^2 \cdot \pi R^2} = \frac{24}{\text{Re}_D} \qquad \text{...(13.35)}$$

where Re_D is the Reynolds number based on the diameter of the sphere. For low values of Re, then, C_D is of order larger than one, as is to be expected since C_D is formulated with the relatively smaller inertial force $\frac{1}{2}\rho V_0^2 A_c$ as the characteristic force.

As Re_D increases, C_D should decrease since the relative magnitude of the viscous forces decreases. In the limit of high Reynolds numbers, it was seen earlier that the flow increasingly resembles the inviscid flow except within a thin layer at the boundary, and the pressure 'impressed' on this boundary layer is the pressure distribution obtained in the outer inviscid flow. But, as mentioned earlier, the net contribution of the inviscid pressure distribution to the drag is zero, so that the viscous stresses are the only *expected* contributors to the drag. The drag force, therefore, should be of the order of the viscous stresses *times* the area. But viscous stresses are no longer of the order of $\mu V_0/L_0$. Instead, since the velocity variations are confined to within the boundary layer thickness δ, the correct order of viscous stresses is $\mu V_0/\delta$ and so

$$C_D \sim \frac{(\mu V_0/\delta)\, L_0^2}{\frac{1}{2}\rho V_0^2\, L_0^2}$$

which, with $\delta/L_0 \sim 1/\sqrt{\text{Re}_{L_0}}$, gives $C_D \sim \sqrt{\text{Re}_{L_0}}$. This was the order of magnitude of C_D obtained for a flat plate aligned with the flow. For Re_{L_0} beyond the critical values for transition, the correct $\delta/L_0 \sim 1/\text{Re}_{L_0}^{0.2}$, and, therefore, $C_D \sim 1/\text{Re}_{L_0}^{0.2}$. The measured drag coefficients for a flat plate agree quite well with these estimates as can be seen from Figs. 13.5 and 13.9. The drag coefficient for large values of Re_L is significantly smaller than one, confirming that only viscous stresses are contributing to the drag, and these are small compared to $\rho V_0^2/2$.

The behaviour of the drag coefficient for a flat plate, however, cannot be extended to bodies of other shapes. If, instead, we consider the variation of C_D with Re_D for a sphere, as shown in Fig. 13.16, we notice some drastic differences. While the curve coincides quite well with *Stokes*

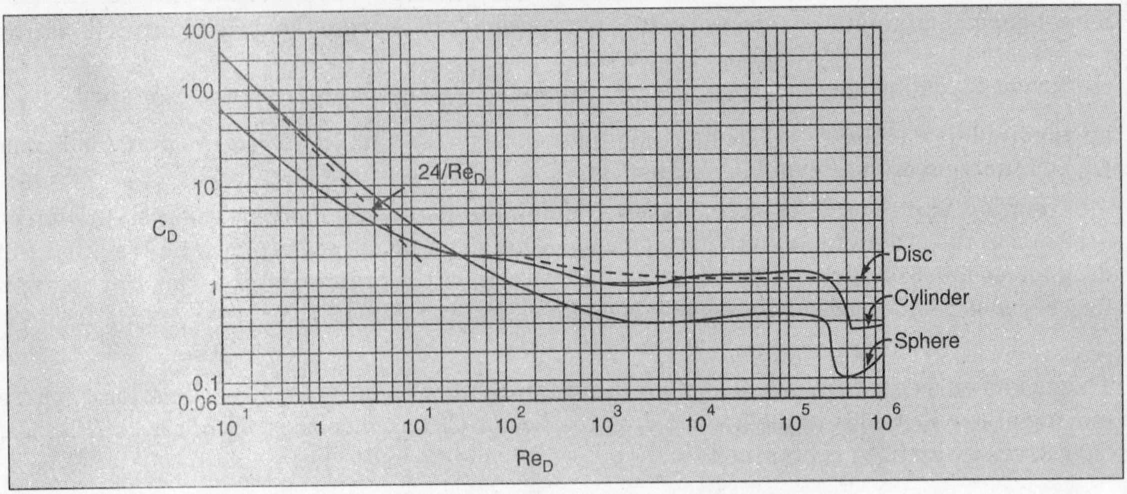

Fig. 13.16. C_D vs. Re_D for sphere, cylinder and disc.

law at very low Reynolds numbers (< 0.5) and C_D decreases with Re_D, it no longer does so in the high Re_D range. In fact, in the Re_D range of 10^3 to 2×10^5, C_D varies little and has a value of between 0.4 and 0.5. Similarly, for a 2-*D* circular cylinder C_D has a value of about one within the same range of Re_D.

These C_D values of order one suggest that the prime contributors to the drag on such bodies are pressure differences, which are of the order of $\rho V_0^2/2$ in high Re flows (see Sec. 11.5). This implies that the streamwise components of the pressure forces around the body do not integrate out to zero as was expected on the basis of the inviscid flow theory. The reason for this departure is the drastic change produced in the flow pattern due to 'separation'. When a boundary layer separates due to an adverse pressure gradient on the leeward side of a body, the fluid no longer 'follows' the body contour and the pressure does not 'recover' to the stagnation value. This causes a difference in the fore-and-aft pressures, giving rise to what is called the *pressure drag,* as opposed to the drag due to the surface shear stresses, termed as *skin-friction drag.* The distributions of pressure on the surface of a sphere (see Fig. 1.20) and a circular cylinder (see Fig. 13.12) confirm this hypothesis. In fact, the drag calculated from the measured distribution on a circular cylinder accounts for almost 97 per cent of its total drag.

Since the pressure drag arises because of separation, it is very sensitive to changes in the shape of the body. Slight modifications of the body contour can, at times, advance or delay the separation, changing the drag drastically. It is for this reason that pressure drag is sometimes referred to as *form drag.* Figure 13.17 shows graphically that a slight rounding of the leading edge of a cylinder (with axis along the flow direction) produces a delayed separation, a narrower wake, and a significantly lower C_D. Also, as was seen in Sec. 13.6, transition to turbulence delays separation and one, therefore, observes a sudden decrease in the drag on a sphere (Figs. 13.16 and 13.18). The reader is referred to Shapiro (Reading 1) for a very lucid discussion of many curious observations regarding drag on immersed bodies which can be explained by the nature of boundary-layer separation.

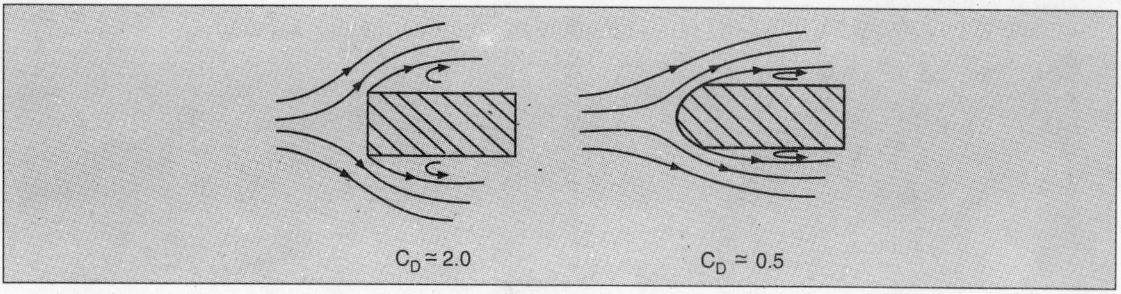

Fig. 13.17. Reduction in C_D by rounding the corners of a cylinder with axis aligned to flow.

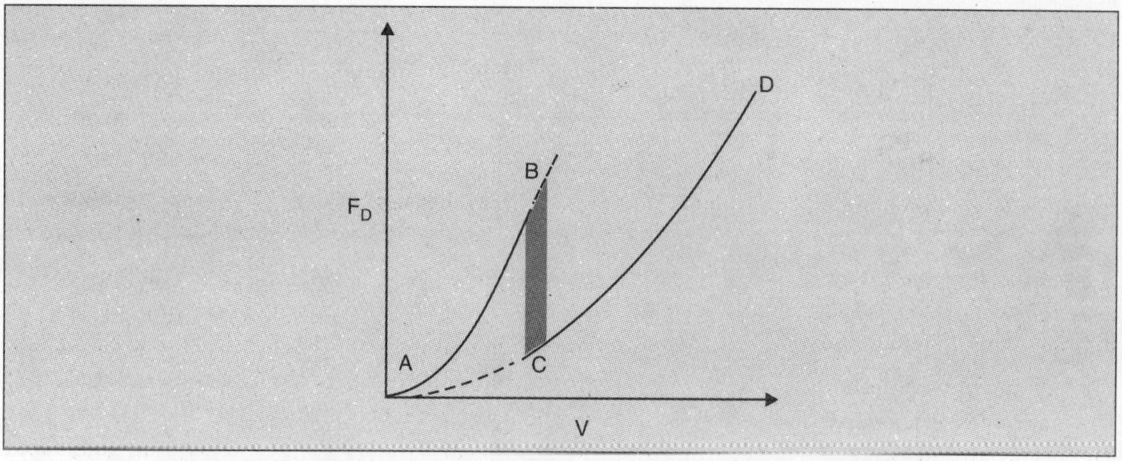

Fig. 13.18. Drag force as a function of speed for a sphere moving in a fluid. Parabolic shapes of *AB* and *CD* signify that C_D is constant in the two zones.

The fact that C_D for a sphere or a cylinder is essentially constant in a large Re_D range from 10^3 to 2×10^5 is because the point of separation is independent of Re_D, provided boundary-layer transition does not occur. It is, therefore, conventional to approximate C_D by two constant values, one when the flow is laminar, and the other (a lower value) when it is turbulent. For bodies with sharp edges, separation is forced at that point irrespective of whether the boundary layer is laminar or turbulent, and a single value of C_D characterizes the flow. Table 13.1 lists these typical values of C_D for some common geometries. It is seen very clearly that the shapes producing broader wakes have larger values of C_D, the pressure drag being the dominant component of drag.

Such bodies, where the pressure drag predominates are referred to as *bluff bodies,* and since in these the width of the wake is critical, the frontal area is generally used as the characteristic area. In other bodies, namely *streamlined bodies* (see Sec. 13.8) where viscous forces are the prime contributors, the total area of the body on which the shear stresses act, is important. Therefore, C_D for such bodies is based on the wetted area. This is the case for hulls of ships and barges. For aerofoils, however, it is conventional to choose half the wetted area, which is approximated by the planform area.

Table 13.1. Values* of C_D for some common geometries at high Re

Shape		C_D	
2-*D*			
Circular cylinder		1.2	(laminar)
		0.33	(turbulent)
Half-circular solid cylinder		1.16	
Half-circular convex cup	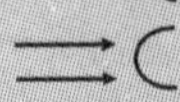	1.2	
Half-circular concave cup	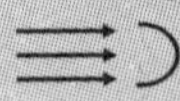	2.3	
Plate normal to flow		2.0	
Elliptic (infinite) cylinder			
Slenderness** ratio 1/2		0.6	(laminar)
		0.20	(turbulent)
Slenderness ratio 1/8		0.28	(laminar)
		0.10	(turbulent)
3-*D*			
Sphere		0.44	(laminar)
		0.20	(turbulent)
Solid hemisphere	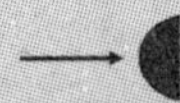	0.38	
Hollow half-cup, convex	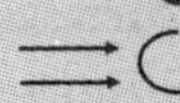	0.38	
Hollow half-cup, concave	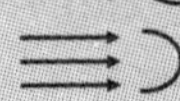	1.42	
Disc		1.17	

Example 13.1. Find the steady state velocities of steel spheres of diameter 1 m, 10 cm, 1 cm, 1 mm, 0.1 mm and 0.01 mm falling in water at 20°C (when $\nu = 1.003 \times 10^{-6}$ m²/s).

As a sphere of diameter D falls through water, starting from rest, it is accelerated by gravity and opposed by buoyancy and the drag. The first two forces are independent of velocity but the drag is not and, therefore, the sphere accelerates till the drag is sufficient to balance the *net* weight.

The limiting velocity is termed the *terminal* or *settling velocity* V_t. Under these conditions,

$$C_D\left(\frac{1}{2}\rho_w V_t^2\right)\frac{\pi D^2}{4} = (\rho_s - \rho_w)\frac{\pi D^3}{6} g \qquad \text{(a)}$$

where ρ_w and ρ_s (= 7800 kg/m³) are the densities of water and steel, respectively. Therefore,

---

* All C_D's based on frontal areas, flow from left to right.

** Slenderness ratio refers to minor/major axis of the cross-section.

$$V_t = \sqrt{\frac{4g\,D}{3\,C_D}\left(\frac{\rho_s}{\rho_w}-1\right)} = 9.43\left(\frac{m^{1/2}}{s}\right)\sqrt{\frac{D}{C_D}} \qquad \text{(b)}$$

where C_D is a function of Reynolds number (see Fig. 13.16) and so of V_t. Thus, a trial-and-error solution is called for.

Start with the largest sphere and assume that the flow is turbulent. For such a flow, C_D is approximated as constant at 0.2.

The velocity V_t is then

$$V_t = 9.43\sqrt{\frac{1}{0.2}} = 21.1\text{m/s}$$

The corresponding Reynolds number is $9.97 \times 10^5 \left(\frac{s}{m^2}\right) DV_t = 2.1 \times 10^7$, well within the turbulent zone. Hence, the 1 m dia. sphere travels at a velocity of 21.1 m/s.

For the next smaller sphere with D =10 cm, the flow may again be turbulent and computations may be started using $C_D = 0.2$. Equation (b) gives $V_t = 9.43\ \sqrt{0.1/0.2} = 6.67$ m/s. The corresponding Reynolds number is 6.65×10^5, again being in the turbulent range. Hence the 10 cm dia. sphere has a terminal velocity of 6.67 m/s.

For the 1 cm dia. sphere, using $C_D = 0.2$ gives $V_t = 9.43\ \sqrt{0.01/0.2} = 2.11$ m/s, which gives $Re_D = 2.10 \times 10^4$. This is too low for turbulent flow and one should use $C_D = 0.44$ which gives $V_t = 9.43\ \sqrt{0.01/0.44} = 1.42$ m/s, and $Re_D = 1.42 \times 10^4$. $C_D = 0.44$ is a reasonable value for this Re_D. Thus, the 1 cm sphere travels at 1.42 m/s.

Using $C_D = 0.44$, the 1 mm dia. sphere travels at $V_t = 9.43\ \sqrt{10^{-3}/0.44} = 0.45$ m/s, giving $Re_D = 448$. This value of Re_D is too low for C_D to be taken as 0.44. A couple of iterations gives $V_t = 0.38$ m/s corresponding to $Re_D = 378$ and $C_D = 0.62$.

The sphere of dia. 10^{-4} m is going to have a very low Reynolds number. Assume, as a first approximation, that $Re_D < 0.5$, so that Stokes law is valid. Since in this range, an analytical expression for C_D is available (Eq. 13.34), Eq. (a) may be reformulated to write

$$3\pi\mu\ DV_t = (\rho_s - \rho_w)\frac{\pi\,D^3}{6}\ g$$

which gives

$$V_t = \frac{(\rho_s - \rho_w)g\ D^2}{18\mu} = 3.695 \times 10^6\left(\frac{1}{m^3 s}\right)D^2 \qquad \text{(c)}$$

For $D = 10^{-4}$ m, this gives $V_t = 0.037$ m/s. The computed Re_D is 3.7 which is not low enough for Stokes law to be applicable. Once again, the formulation of Eq. (a) has to be used. A couple of iterations give $V_t = 9.43 \times \sqrt{10^{-4}/10.1} = 0.0297$ m/s, with $Re_D = 2.96$ and C_D from Fig. 13.16 as 10.1.

It is clear that spheres of diameters smaller than 10^{-4} m are likely to have C_D in the Stokes law region. For $D = 10^{-5}$ m, Eq. (c) predicts $V_t = 3.7 \times 10^{-4}$ m/s. The corresponding Re_D is 3.7×10^{-3}, which is much smaller than one.

Figure 13.19 shows the calculated variation of the terminal velocity with the diameter of the sphere. One notices from Eq. (b) that for large D, when C_D is nearly constant, $V_t \sim D^{1/2}$, and from Eq. (c) that for small D, $V_t \sim D^2$.

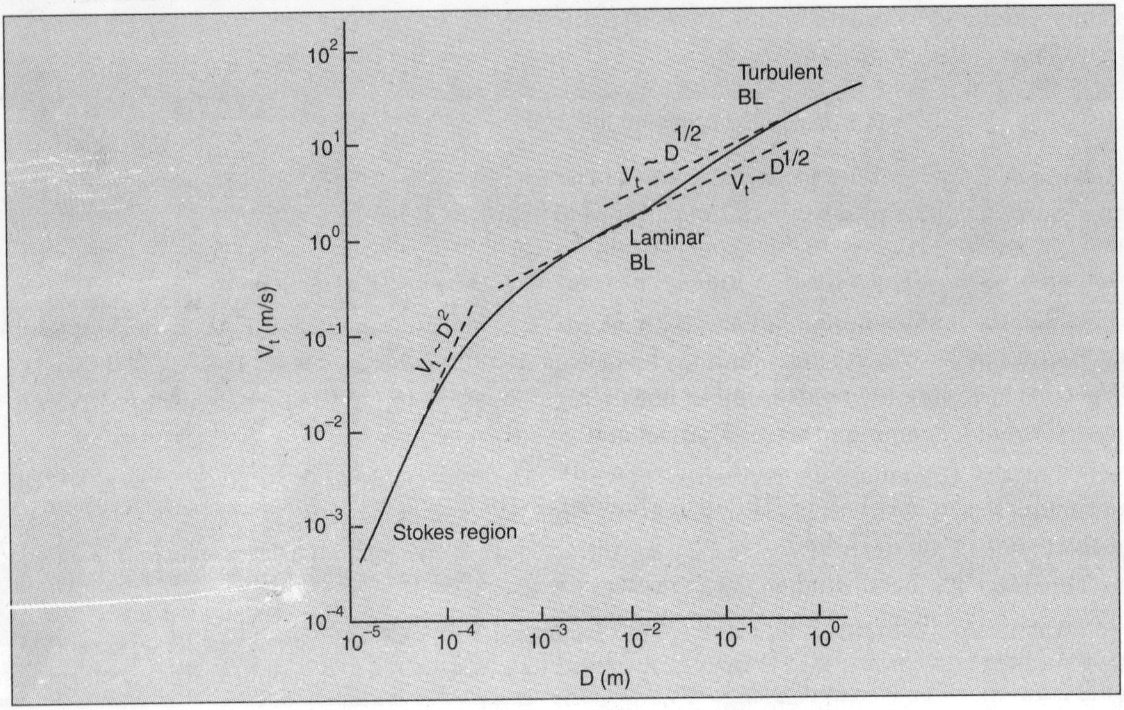

Fig. 13.19. Terminal velocities for steel spheres in water.

Example 13.2. A cricket ball leaves the bat at a velocity of 35 m/s and an angle of 45°. Calculate the range in air at 20°C if the ball is not spinning and the air is stationary. Also calculate the range if the ball is hit (a) into, (b) with, and (c) across a 5 m/s wind. The diameter and mass of a cricket ball are 7 cm and 156 gm respectively.

A ball in flight in still air is subjected to two forces, the weight mg and the drag $F_D = C_D\,(\rho V^2/2)\,(\pi D^2/4)$ acting opposite to the *relative* motion. The equations of motion in the horizontal and vertical directions can easily be written with reference to Fig. 13.20 as

$$m\,\frac{d}{dt}(V\,\sin\theta) = -mg - F_D \sin\theta \tag{a}$$

and

$$m\,\frac{d}{dt}(V\,\cos\theta) = -F_D \cos\theta \tag{b}$$

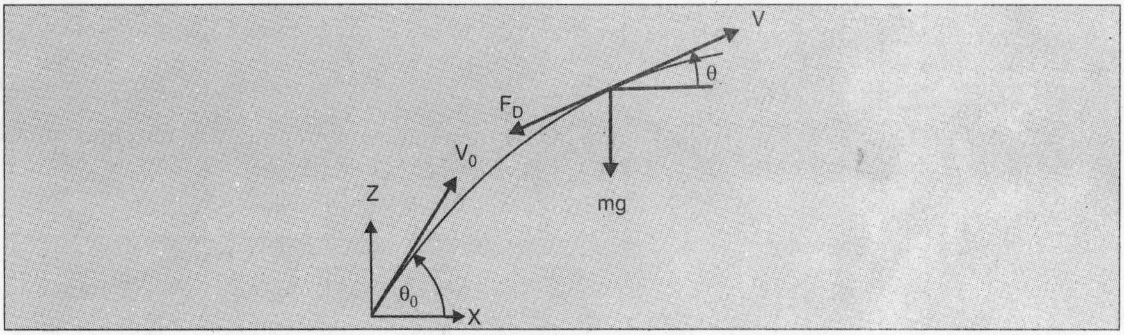

Fig. 13.20. Forces acting on the cricket ball of Example 13.2.

The value of F_D to be used in these equations can be obtained from the drag coefficient C_D at the flight Reynolds number. For the cricket ball the maximum value of Re_D = 1.22 (kg/m^3) × 35 (m/s) × 0.07 (m)/1.8 × 10^{-5} (N s/m^2) = 1.66 × 10^5 which is below the critical value. Hence, C_D = 0.44 is used throughout the range.

Since the force F_D in Eqs. (a) and (b) depends on the variable V, a direct solution of these equations is not possible. These equations are solved numerically by breaking up the flight time into a number of small time intervals of duration δt each. If δt is small enough, it is reasonable to assume that within an interval, $F_D \cos\theta$ and $F_D \sin\theta$ are constant. Following this step-by-step procedure, it is possible to obtain quite accurate trajectories. The numerically calculated trajectory of the cricket ball in still air is shown as curve I (Fig. 13.21). The range X is 78.8 m and the time of flight is 4.4 s.

The effect of a head wind on the trajectory can be obtained by computing the flight of the ball *relative to the moving air* and then allowing for the movement of air relative to the ground. Figure 13.22(a) shows the construction to obtain the *relative initial velocity* $V_{0,r}$ of the ball and the *relative initial angle* of projection $\theta_{0,r}$. Clearly

$$V_{0,r} = \sqrt{V_0^2 + 2V_0V_a\cos\theta_0 + V_a^2} = 38.7 \text{ m/s}$$

and

$$\theta_{0,r} = \tan^{-1}\frac{V_0 \sin\theta_0}{V_0\cos\theta_0 + V_a} = 39.8°$$

The range X_r with respect to the wind and the time of flight T obtained using these initial values in the numerical solutions of Eqs. (a) and (b) are 91.2 m and 4.4 s. In time T, the air moves a distance $V_a T$ = 22 m and thus, with respect to the ground, the range $X = X_r - V_aT$ = 69.2 m. The corresponding trajectory (in a frame of reference fixed with ground) is also shown in Fig. 13.21.

CHAPTER 13

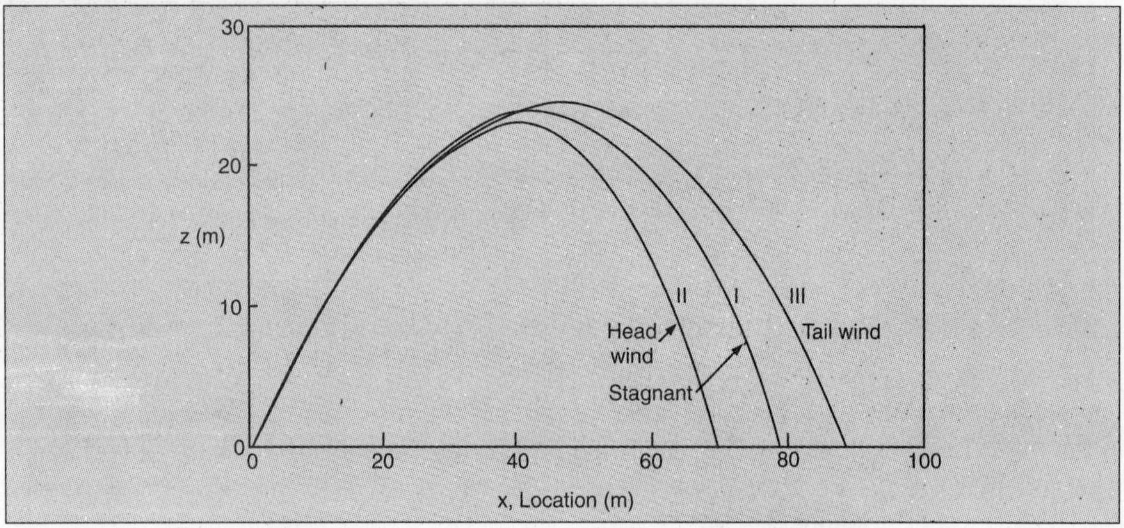

Fig. 13.21. Trajectory of the cricket ball of Example 13.2.

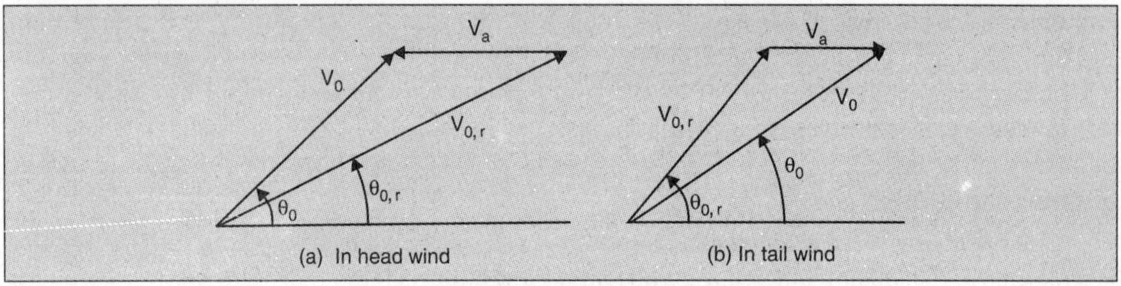

Fig. 13.22. Construction required for computing initial relative velocity and angle of cricket ball.

The relative initial velocity for the tail wind is seen from Fig. 13.22 (b) to be

$$V_{0,r} = \sqrt{V_0^2 - 2V_0 V_a \cos\theta_0 + V_a^2} = 31.7 \text{ m/s}$$

and the initial inclination

$$\theta_{0,r} = \tan^{-1} \frac{V_0 \sin\theta_0}{V_0 \cos\theta_0 - V_a} = 51.4°$$

The range X_r with respect to the wind and time of flight T with these initial values come out as 66.54 m and 4.5 s. The corrected range with respect to the ground with a tail wind then is $X = X_r + V_a T = 89.0$ m. The corresponding trajectory is shown as curve III in Fig. 13.21.

A wind blowing across the direction of flight has little effect on the range but produces a transverse deviation. Working with the relative velocities, it is found that the vertical component of the initial ball velocity is unchanged at $V_0 \sin \theta_0$, but the *relative* horizontal component is now the resultant of $V_0 \cos \theta_0$ and V_a (Fig. 13.23).

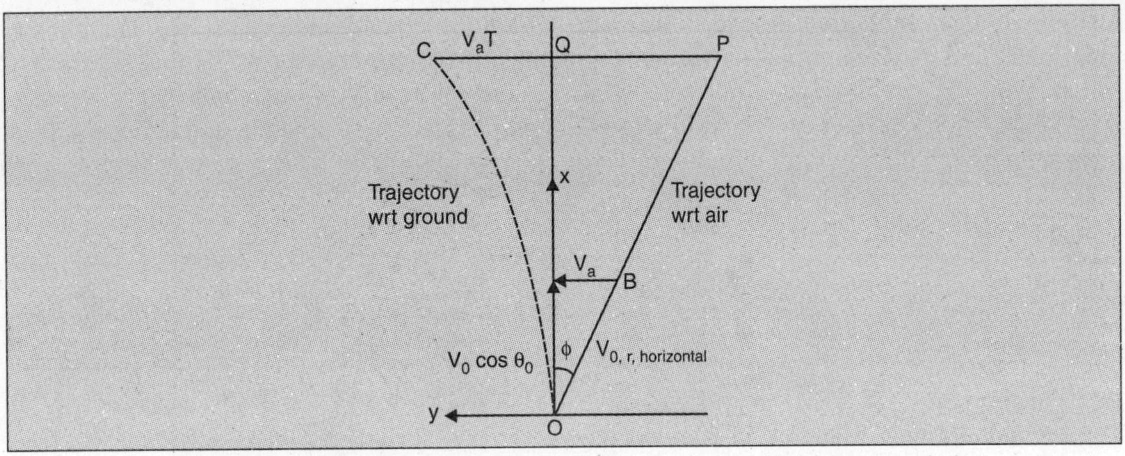

Fig. 13.23. Initial horizontal relative velocity of cricket ball in a cross wind.

Thus, $V_{0,r,\text{ horizontal}} = \sqrt{V_0^2 \cos^2 \theta_0 + V_a^2} = 25.25$ m/s

and,

$$V_{0,r} = \sqrt{V_{0,r,\text{ horizontal}}^2 + V_{0,r,\text{ vertical}}^2}$$

$$= \sqrt{V_0^2 \cos^2 \theta_0 + V_a^2 + V_0^2 \sin^2 \theta_0}$$

$$= \sqrt{V_0^2 + V_a^2} = 35.4 \text{ m/s.}$$

The relative angle of inclination with the horizontal is

$$\theta_{0,r} = \tan^{-1} (V_{0,r,\text{ vertical}} \,/\, V_{0,r,\text{ horizontal}})$$

$$= \tan^{-1} \frac{V_0 \sin \theta_0}{\left(V_0^2 \cos^2 \theta_0 + V_a^2\right)^{1/2}} = 44.4°$$

Based on these, the numerical calculations give a range relative to the wind as 82.02 m and the time of flight as 4.4 s. This range is in a frame of reference moving with the air, i.e., along the direction *OB* in Fig. 13.23, of the relative horizontal velocity, inclined at an angle $\phi = \tan^{-1} (V_a/V_0 \cos \theta_0) = 11.4°$ to the direction *OQ*. The actual trajectory *OC* (with respect to the ground) can be obtained by correcting for the effect of the wind. In time *T*, the wind carries the ball a distance $PC = V_aT = 22$ m. Thus the actual range in the direction of throw *OQ* is equal to (82.02) cos ϕ = 80.4 m, and the transverse deviation is $PC - PQ = V_aT - 82.02 \sin \phi = 5.8$ m.

13.8 STREAMLINING

The fact that most of the drag acting on a bluff body moving through a fluid at high Re is due to the pressure distribution caused by separation is made use of in the reduction of drag. If the separation can be delayed, or completely eliminated, large reductions in drag can be achieved.

As seen earlier, a positive pressure gradient (which is termed 'adverse') is responsible for separation, and so, if the adverse pressure gradient can be reduced to the minimum, separation can, hopefully, be delayed or eliminated. This is achieved by shapes with long, tapering tails (Fig. 13.24) which decelerate the fluid slowly enough. An aerofoil, a bomb, and a dolphin (see

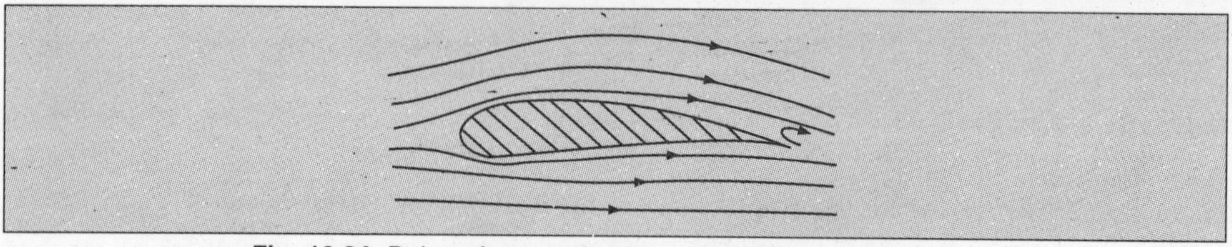

Fig. 13.24. Delay of separation by use of a long tapering tail.

Fig. 1.23a) have such shapes. As can be seen from the pressure distribution on an aerofoil (see Fig. 11.5), the positive pressure gradients on its tail are relatively small, such that the separation is delayed till very near the trailing edge. This results in a great reduction in the pressure drag, so much so, that in many cases the skin-friction drag (which formed barely 3 per cent of the total drag on a circular cylinder) dominates. Such shapes are termed *streamlined shapes.* Nearly all objects (including fish and birds) intended to move at high speeds through fluids have streamlined shapes so as to reduce the drag. The greatly reduced drag in streamlined bodies has been highlighted in Fig. 13.25 which shows relative sizes of a streamlined body and a circular cylinder having the same drag.

Fig. 13.25. Streamlined body and a cylinder having the same drag.

Since most of the drag on streamlined bodies is due to the viscous stress on the wall, the total drag depends on the wetted area of the body and, therefore, it is conventional to use the wetted area or the planform area as the characteristic area. It is to be noted that in streamlined bodies, C_D has a significant dependence on Re, unlike that in bluff bodies where C_D is largely independent of it. This is because the shear stress at the wall is a function of Re, while the (non-dimensional) pressure distribution depends basically on the location of the separation point which does not change much as long as flow transition does not occur. Also, unlike in (smoothly contoured) bluff bodies where C_D decreases on transition, the C_D of streamlined bodies generally increases when the boundary layer becomes turbulent. That is why high-performance gliders tend to have wings which are carefully contoured to keep the flow laminar upto as far back as possible.

Example 13.3. An aircraft model is mounted in a wind tunnel for drag determination. To decrease the drag contribution of the mounting system, the 1 m long column is streamlined into an aerofoil shape with a chord length of 10 cm (Fig. 13.26). At a flow speed of 80 m/s in the wind tunnel,

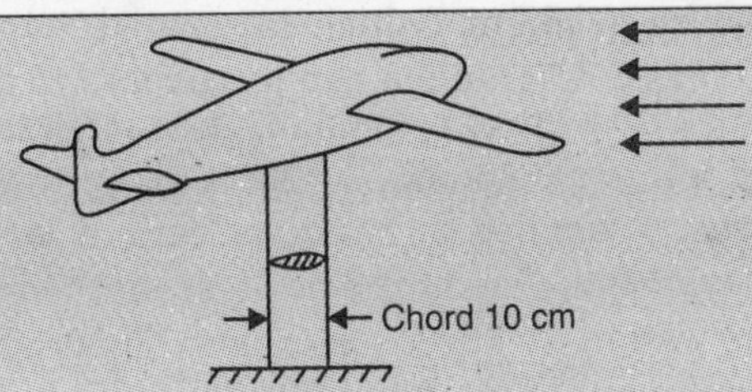

Fig. 13.26. Mounted aircraft in a wind tunnel.

estimate the error due to the drag on the mounting column, if the variation of C_D of the streamlined column with Re is given by Table 13.2.

Table 13.2. C_D *vs* Re for mounting column

Re	C_D
3×10^5	0.0045
6×10^5	0.0055
1×10^6	0.0060
3×10^6	0.0065

The drag coefficient and the Reynolds number for 2-*D* streamlined shapes are based on the planform area and the chord length *C*, respectively. Thus,

$$\text{Re} = \frac{\rho VC}{\mu} = \frac{1.22\left(\text{kg/m}^3\right)\times 80\left(\text{m/s}\right)\times 0.10\left(\text{m}\right)}{1.80\times 10^{-5}\left(\text{N s/m}^2\right)}$$

$$= 5.42 \times 10^5$$

The C_D expected at this Re is 0.0053 from Table 13.2. The drag force on the column then is

$$F_D = C_D \frac{1}{2}\rho V^2 A_c$$

$$= 0.0053 \times \frac{1}{2}\times 1.22\left(\text{kg/m}^3\right)\times 80^2\left(\text{m/s}\right)^2 \times \left(0.10\times 1\right)\left(\text{m}^2\right)$$

$$= 2.1 \text{ N}$$

The error due to the drag on the column is, therefore, 2.1 N, which is usually a very small fraction of the total drag.

If the mounting had been approximated as a flat plate, the drag coefficient would be given by Eq. (13.32) since $\text{Re}_L = 5.4 \times 10^5$, as

$$C_D = 0.072\ \text{Re}_L^{-0.2} - 1670/\text{Re}_L = 0.0021$$

This is for one side of the flat plate only. For both sides, the equivalent C_D is 0.0042. Thus, 75 per cent of the total drag on the streamlined body can be accounted for by the flat plate drag.

CHAPTER 13

13.9 BOUNDARY-LAYER CONTROL

The control of boundary layers entails suppression of its tendency to separate from the walls. Giving a long, slowly-tapering tail to a body moving through a fluid, i.e., streamlining it, is one method of control. But this cannot be applied everywhere. Various artificial methods of controlling boundary layers have been developed. One of the earliest methods was suggested by Prandtl himself. It involves moving the wall in the direction of flow. This reduces the velocity difference, thereby suppressing the development of the boundary layer and consequently its separation. But this method is quite difficult to achieve physically. One simple application, where it may be used, is in flow past a cylinder. If the cylinder rotates counterclockwise as shown in Fig. 12.18, the wall on the underside moves along with the flow. If the rate of rotation is large enough, separation on this side may be entirely prevented. Moreover, on the upper side the separation is so slight that the streamline pattern is remarkably similar to the one obtained by the inviscid approximation. It is because of this reason that the lift on a rotating cylinder can be predicted by the Robins-Magnus effect formula, Eq. (12.44).

Another effective method is *boundary layer suction* in which the slower moving fluid elements near the wall are sucked through slits in the wall before they are brought to rest by the adverse pressure gradient. These slower particles are then replaced by swifter particles from

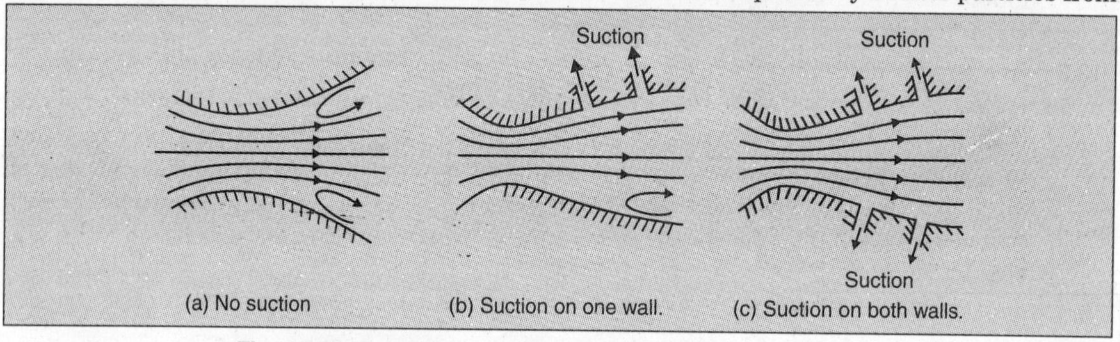

Fig. 13.27. Prevention of separation by suction in a diffuser.

farther away from the walls, thereby delaying separation. Figure 13.27 shows how in the diverging section of a diffuser, the flow can be made to resemble an inviscid one by applying a strong enough suction at the walls. The pressure recovery is then enhanced and is given by the Bernoulli equation. The energy used in suction is comparatively much smaller than that saved because of the prevention of separation.

Yet another method of preventing separation is *blowing,* in which faster moving particles are injected at the wall, upstream of the separation point, thereby energizing the boundary layer and preventing it from separating. The wings of almost all modern aeroplanes use this to some extent. The higher pressures on the underside (Fig. 13.28) force air through the slots into the slowing boundary layer on the upper surface. This accelerates the boundary-layer flow, thereby suppressing separation. This conventional explanation, attributed to Prandtl, is now being contested (see E.L. Houghton and P.W. Carpenter, Aerodynamics for Engineering Students, 5th Ed., Butterworth Heinemann, London, 2006). The explanation is that the 'slat' in the front creates a circulation and energizes the fluid, which leads to a reduction of the severity of the adverse pressure gradient on the following part of the aerofoil.

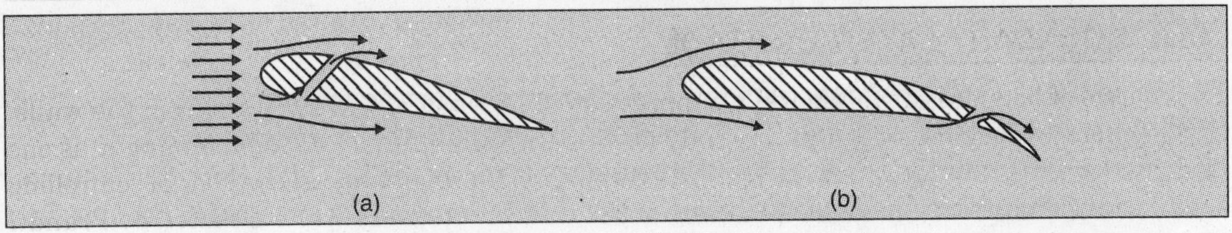

Fig. 13.28. Suppression of separation by blowing (a) nose slat, (b) slotted wing.

PROBLEMS

13.1 Use the velocity profile $V_x = a \sin(by)$ inside a laminar boundary layer on a flat plate to obtain expressions for the boundary-layer thickness and the drag coefficient.

13.2 Determine the constants C_0, C_1, C_2 and C_3 in the cubic velocity profile

$$\frac{V_x}{V_0} = C_0 + C_1 \frac{y}{\delta} + C_2 \left(\frac{y}{\delta}\right)^2 + C_3 \left(\frac{y}{\delta}\right)^3$$

used to approximate the boundary layer velocity distribution on a flat plate. (Hint: The fourth condition is obtained from the boundary-layer Eq. (13.7) applied at $y = 0$, the wall). Use this profile to obtain estimates of δ, C_f and C_D.

13.3 In a tray drier, relatively dry air is passed over moist solids spread in trays as shown. If there are 10 trays to be dried (each 1 m × 1m) in a unit, and the air velocity above the trays is to be kept at 3 m/s (to give high drying rates), estimate the power required by a blower whose efficiency is 60 per cent. Assume that the properties of the air do not alter significantly with the moisture pick-up. Note that a substantially higher power will be required due to the energy lost because of boundary-layer separation at bends, etc.

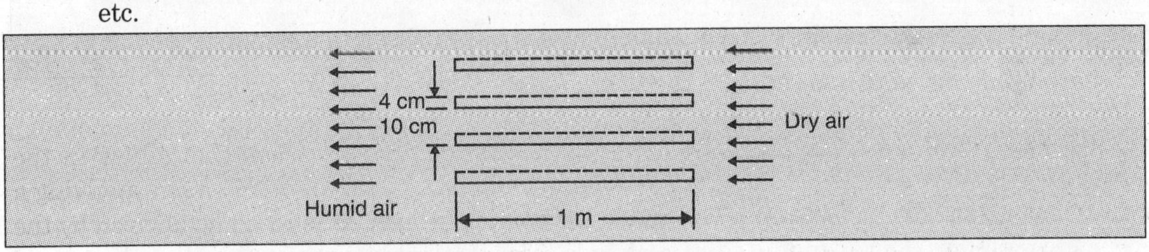

13.4 Assuming that the boundary layer over a flat plate becomes turbulent at a value of x where $Re_x = Re_{crit} = 5 \times 10^5$ (instead of being turbulent over the entire plate), show that

$$C_D = 0.072\,(Re_L)^{-1/5} - \frac{1670}{Re_L} \qquad \text{(a)}$$

and at $x = L$,

$$\delta = \left[\frac{2.73 \times 10^4}{(V_0/\nu)^{1.25}} + \frac{0.289}{(V_0/\nu)^{0.25}}\left(L - \frac{5 \times 10^5}{V_0/\nu}\right)\right]^{0.8} \qquad \text{(b)}$$

What assumptions are involved in obtaining Eq. (a)?

CHAPTER 13

13.5 The ocean linear *Queen Elizabeth* is approximately 310 m long and 36 m wide. Its normal draft (i.e., immersed depth) is 12 m. Assuming the base and sides to be flat plates of uniform width, estimate the total skin-friction drag acting when the ship cruises at its normal speed of 28.5 knot (= 53 kmph). Also compute the power required to overcome this drag and the thickness of the boundary layer at the tail-end. Note that a further contribution to the total force will arise due to form and wave drags. Assume fluid properties to be the same as those for fresh water.

13.6 Estimate the skin-friction drag acting on a rocket of diameter 5 m and length 40 m, travelling at 2000 kmph at an altitude of 8000 m in standard atmosphere (see Fig. 2.6). The viscosity varies like $\sqrt{T}$ and is almost independent of pressure.

13.7 The entrance length in a pipe can be estimated very crudely as the distance taken by the laminar boundary layer on a flat plate to grow to a thickness $D/2$. Show that this gives $L_e/D = 0.01\ \mathrm{Re}_D$ (cf. Eq. 10.3). What are the possible sources of error?

13.8 It is said that one should lie down on a beach to avoid the wind. Check this by computing the velocity of air at a height of 15 cm from the ground and at a height of 1.5 m. Assume that at the beach the boundary layer is 75 m thick and the velocity V_e outside the layer is 10 m/s. Assume also that the boundary layer at the beach is turbulent.

13.9 *Displacement thickness:* Since the velocity within the boundary layer approaches the inviscid velocity asymptotically, the boundary layer thickness is not exactly defined. The use of $V_x(\delta) = 0.99\ V_i$ is one way of overcoming the problem, but at times a more precise definition is needed. The displacement thickness is one such concept. It is defined as the distance by which the wall will have to be displaced if the entire flow were moving at the inviscid velocity. Alternately, it represents the normal displacement of the streamlines in the outer flow region, as shown. Show that the actual volume of fluid that passes up to $y = h$ (which is slightly larger than δ) would have passed in a distance $y' = \frac{1}{V_e}\int_0^h V_x\, dy$ if the flow were inviscid, and therefore, the displacement of streamlines is given by

$$\delta^* = h - y' = \int_0^h \left(1 - \frac{V_x}{V_e}\right) dy$$

Since $(1 - V_x/V_e)$ is almost zero outside the boundary layer, it is possible (and convenient) to replace the limit of integration h by ∞. The displacement thickness, thus, represents the amount by which the walls of the body should be displaced outwards for the purposes of calculating the outer flow. Next obtain δ^* for the parabolic velocity profile of Eq. (13.24).

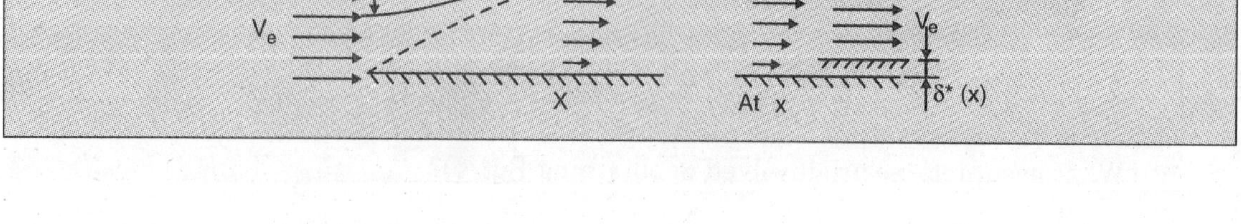

13.10 *Momentum thickness:* The momentum thickness of a boundary layer is defined as the thickness of the layer of fluid of velocity V_e for which the momentum flux rate equals the difference between the momentum-flux in the boundary layer at velocity V_e and the actual momentum flux. Show that

$$\theta = \int_0^\infty \frac{V_x}{V_e}\left[1 - \frac{V_x}{V_e}\right] dy$$

Obtain the momentum thickness θ for the flat plate boundary layer with a parabolic profile. Momentum thickness θ is usually more useful in engineering fluid mechanics where one is concerned with drag representing momentum losses.

13.11 *Shape factor:* While both δ^* and θ, the displacement and momentum thicknesses (see Probs. 13.9 and 13.10) are functions of the streamwise coordinate x, show that their ratio $H = \delta^*/\theta$ is independent of x in the case of self-similar boundary-layer flows. Thus, in such flows H is purely a function of the 'shape' of the velocity profile and is termed the *shape factor*. Since the separation point is marked by a specific shape of the velocity profile (dV_x/dy at the wall being zero), the location of the separation point at the wall in numerical calculations is indicated by H acquiring a specific value H_0 (usually 3.5 for laminar flows and 2.4 for turbulent flows).

13.12 The development of a boundary layer over a flat plate may be modelled as the 'penetration' of the effect of the wall into the fluid stream. Thus, the boundary-layer thickness at x is the penetration depth at that location. Using the result of Prob. 6.37, show that the boundary-layer thickness is given by

$$\delta/x = 3.6/\sqrt{\mathrm{Re}_x}$$

13.13 It is seen that a person can float easily in the *Dead Sea* (which has a salt concentration of over 26 per cent). Is it easier to swim in it too?

13.14 With the background of Chapter 13, redraw the flow patterns for the various geometrics of Probs. 1.22, 1.23, 1.24 and 1.25.

13.15 Draw schematically the flow pattern around the two rotating cylinders shown. How does the flow pattern change when the cylinders (a) do not rotate, and (b) rotate in the opposite direction to that shown?

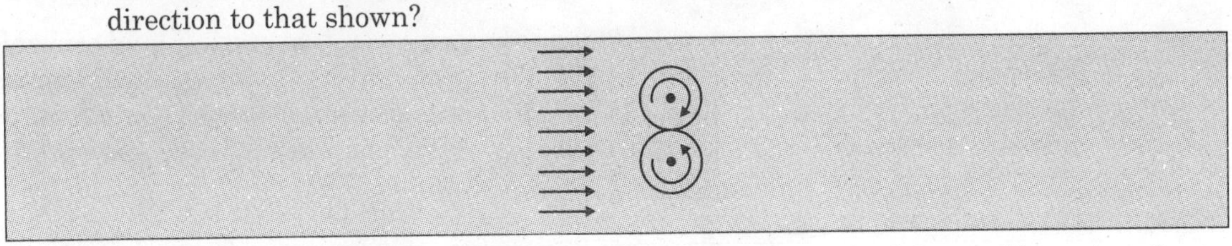

13.16 If we take two streamlined bodies A and B as shown, which of these will require a larger force to move in air, and why?

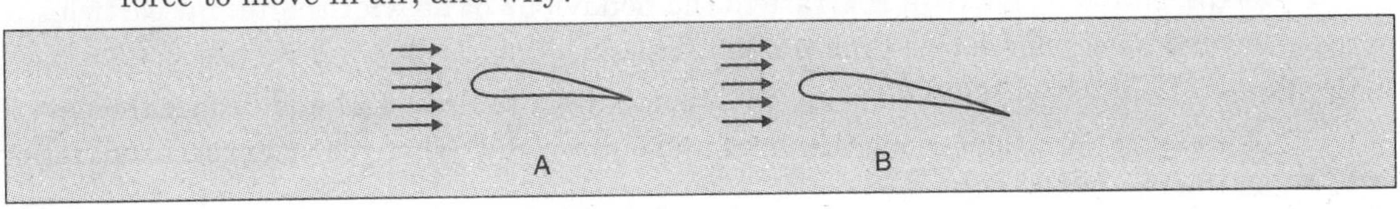

CHAPTER 13

13.17 Draw the streamlines for flow about two identical streamline bodies shown, one with a thin trip-wire and one without it. Which of these would have a higher drag force?

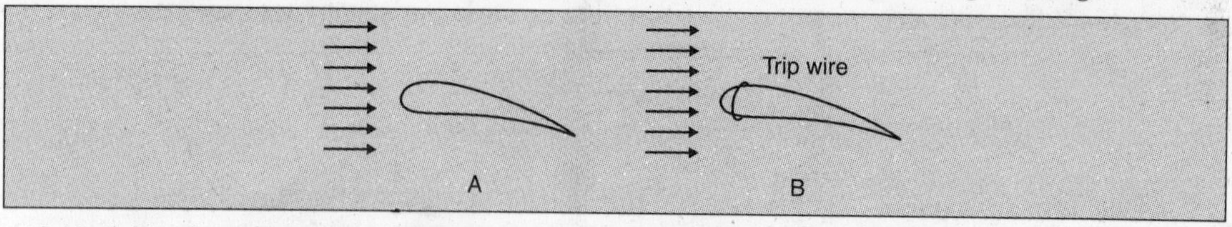

13.18 A bomb and a sphere having the same frontal area are dropped together, first in air and then in glycerine. The sphere travels slower than the bomb in air, but faster in the much more viscous glycerine. Explain.

13.19 A submarine periscope is a cylinder of 0.3 m dia. It is enclosed in a streamlined strut having the NACA 4412 profile (whose characteristics are given as Fig. 14.4) in order to reduce the drag force acting on it. Estimate the minimum drag acting on it per unit length when the submarine is travelling in a still ocean at 15 m/s. Use density and viscosity of sea water as 1.025×10^3 kg/m^3 and 1.09×10^{-3} Pa s respectively.

13.20 The drag coefficient of a cylinder decreases as its *aspect ratio* (defined as the length-to-diameter ratio) decreases. For an infinite cylinder $C_D = 1.2$, while for one of $L/D = 1$, it is 0.6. Explain why.

13.21 It is found that a bubble of air rising up a column of a viscous liquid is subjected to negligible form drag. Explain.

13.22 Car-manufacturers use the following relationship for the power required by an automobile

$$\overset{\circ}{W} = (k_1\, mg + k_2\, AV^2)\, V$$

where A is the projected area. Justify this relationship.

13.23 *Secondary flow:* Consider the forced vortex (i.e., solid-body type motion) set up in water contained in a stationary beaker. By considering the radial forces on the two fluid elements, A in the main body, and B within the boundary layer at the bottom, show that a recirculating motion results as shown. Such motions are known as secondary flows.

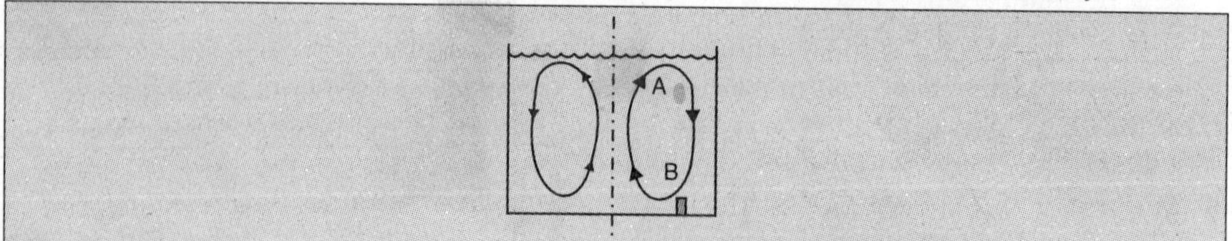

13.24 On stirring sugar in a tea cup one notices that the sugar grains pile up near the axis at the bottom. This is in contrast to the behaviour of heavier-than-fluid particles in devices like cyclones (see Prob. 13.56). Explain.

13.25 Consider the flow of a fluid along a 90° bend. Refer to Prob. 7.57 and deduce the nature of pressure variations along the inner and outer walls. Where do you expect separation on the two walls?

13.26 If the inviscid pressure distribution is used to compute the lift on the aerofoil at a large angle of attack as shown, will a good estimate be obtained or not? Why?

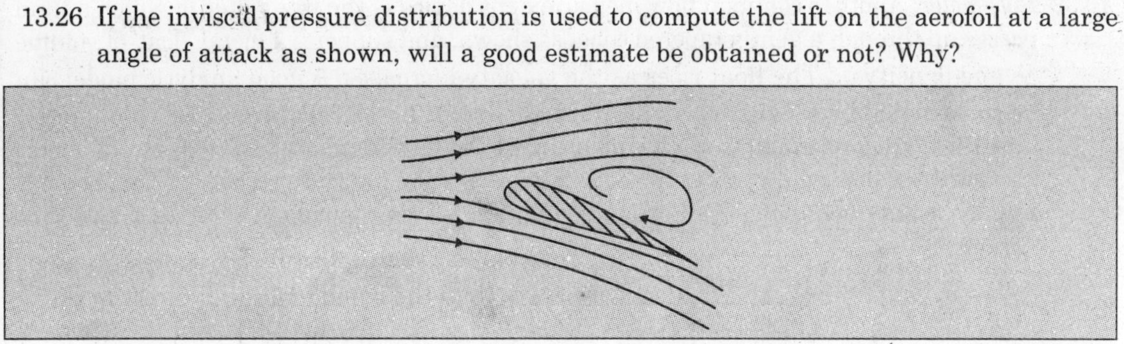

13.27 The loss coefficient K (with V_1 as the characteristic velocity) of a divergent section in a pipeline (for a given area-ratio A_1/A_2) is shown as a function of the divergence angle. Explain why K has a minimum, i.e., why is K large when α is either too small or too large?

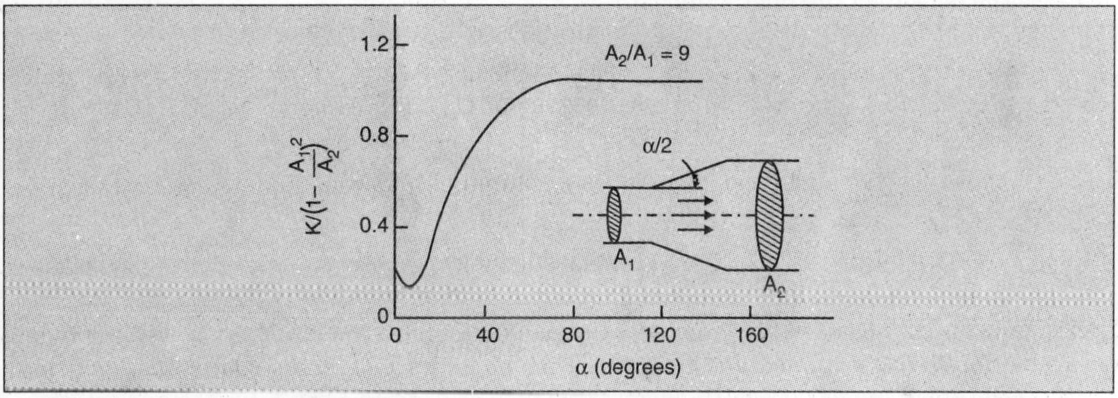

13.28 We experience more drag when we move in water than in air. Is this primarily because of the higher density or the higher viscosity of water?

13.29 The potential-flow pattern around a rotating cylinder was discussed in Sec. 12.8. At low rates of counterclockwise rotation of the cylinder, the boundary layer at the bottom surface may remain laminar while at the top surface it may become turbulent. An asymmetry in the flow pattern may, therefore, develop due to the delayed separation at the top. Show that the lift-force acting on the cylinder due to this asymmetry is in a direction opposite to that of the Robins-Magnus force.

13.30 The wall shear stress on the flat plate of Fig. 13.2 was non-dimensionalized using $\frac{1}{2}\rho V_0^2$. What do you think is a more appropriate characteristic shear-stress with which τ_w should be non-dimensionalized? Does the new dimensionless skin-friction coefficient become of order one?

13.31 A *drag-force flowmeter* indicates the flow velocity by measuring the drag experienced by a hollow sphere roughened on the outside. Does this instrument have a linear scale (i.e., is the meter reading, which is proportional to the drag, also proportional to velocity)? What is the use of roughening the outside surface of the ball? Such an instrument has a high dynamic response and can measure fluctuations of frequencies around 70 to 200 Hz.

13.32 *Rotameter:* Another common flow measurement device is the rotameter in which a fluid passes up through a gently tapered tube, as shown, and supports a metal 'float' of volume $\forall$ and density ρ_s. The float rises as the velocity increases. A good analytic model can be constructed by idealizing the pressure distribution as constant equal to p_1 on the under-surface, and equal to p_2 (the pressure at the minimum area section) on the upper surface since the separation at the edge suppresses the pressure recovery. The pressure p_2 may be obtained using the Bernoulli equation assuming negligible losses. Show that

$$V_1 = C_r A_2 \sqrt{\frac{2(\rho_s - \rho_f)\forall g}{\rho_f A_3 (A_1 + A_2)(A_1 - A_2)}} \simeq C_r\, A_2 \sqrt{\frac{2(\rho_s - \rho_f)\forall g}{(\rho_f A_1 A_3^2)}}$$

where C_r is an empirically determined *rotameter coefficient*. Thus, observe the direct proportionality of the flow rate to the area A_2. It is observed that in a rotameter the calibration curves for both water and alcohol (of density 790 kg/m^3) are identical. What is the density of the float material?

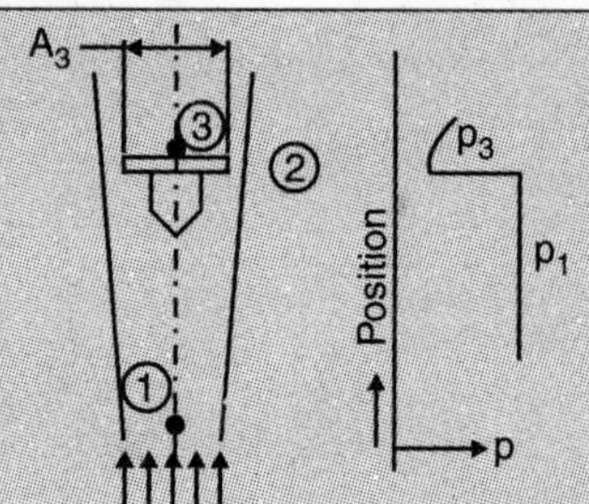

13.33 Towards the end of World War I, the Germans astounded the Parisians by shelling them by *Big Bertha* stationed 110 km away. They had, by accident, discovered that by firing shells at 52° (angles larger than what was then considered to be optimal for range) they could hit targets much farther away. Can you think of a reason why? (*Hint*: Larger angles increased the altitude attained by the shells to about 40 km.)

13.34 A railway locomotive has a C_D of 1.9. By streamlining the frontal shape of the locomotive using a well rounded bullet nose, its C_D can be lowered to 0.5 [see M.M. Sivaramakrishnan, *Science Today,* **14**(8),48 (1980)]. If the average speed of 80 kmph of the locomotive is kept unchanged, how much percentage savings in power (and so of fuel) does this represent? Assume that rolling friction is negligible compared to drag. Also, how much fractional reduction in the total fuel bill is anticipated if, in addition, the average speed of the locomotive is reduced to 70 kmph?

13.35 Air at 20 m/s blows across a chimney of 1 m dia. and 30 m length. Compute the drag force acting on the chimney. What is the bending moment at its base?

13.36 The two main planes of the early aeroplanes were braced by a number of thin cylindrical wires. This design was later on replaced by monoplane design which did away with the wires. One of the reasons for the change was the inordinate amount of drag introduced by these wires. Estimate the power required to overcome this drag if the total length of 2 mm dia bracing wire in one design of aeroplane is 200 m, and the aeroplane speed is 180 kmph. Compare this with the total thrust horse power of 200.

13.37 Estimate the drag force acting on a 0.25 m × 0.25 m sieve when clean water flows through it at an average velocity of 2 m/s. The sieve consists of 0.08 cm dia. wires weaved in a 1 cm × 1 cm square grid. Assume that the flow pattern across any wire is not influenced by the neighbouring wires. Do you expect the force to be larger or smaller than the computed value, if the interaction between the flows around neighbouring wires are accounted for?

13.38 The rowing force on a boat is generated by the drag on the oar moving backwards in water. The oar may be modelled as a flat disc of radius R with its centre a distance l below the oar lock. Write an equation to obtain the steady velocity V_b of the boat if the oars may be assumed to be moving in a 90° arc at an average rotational speed ω, and that the average thrust acting on the boat is one half of the thrust when the oars are at the position shown. The drag coefficient of the boat is $C_{D,B}$ based on area A_B.

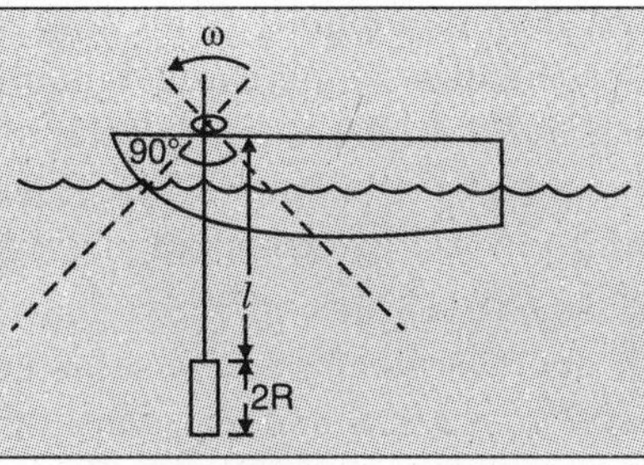

13.39 A bumblebee can hover at the same position by flapping its wings (of span 1.7 cm and chord 0.73 cm). The drag force acting on the wings during their downward motion balances its mass of 0.88 g. To obtain a first approximation, replace the motion of a wing in a 90° arc by a parallel motion of a flat plate such that the amplitude of motion is the same as the amplitude of the wing tips. The plate moves down at a constant velocity V_w. Assuming that the upstroke is drag-free and takes half the time as the downstroke, and that the time-averaged drag force is equal to one-half the maximum force during the downstroke, estimate the rate of flapping for hovering. Compare with the actual value of approximately 195 Hz.

13.40 *Swing of cricket ball:* The laminar-to-turbulent transition on a new cricket ball (of diameter 7.2 cm) occurs at an Re_D of about 1.4×10^5 if the flow does not encounter the seam. But it can be triggered by the seam (when it is at 30° to the airflow) at an Re_D as low as 9.5×10^4. Obtain the corresponding critical velocities corresponding to these transition points. What happens when a seam bowler bowls the ball at a speed which is

(a) higher than the upper critical value

(b) below the lower critical value.

[For more information on the swing of a cricket ball, see N.G. Barton, *Proc. Roy. Soc. Lon.*, **A 379**, 109 (1982)].

13.41 According to wind-tunnel experiments, the lateral force on a new cricket ball bowled at about 105 kmph with seam at 10–15° to the air-flow (which is found to give the best swing) is about 40 per cent of its weight. Estimate the total lateral displacement of the ball (if the length of the pitch = 20 m) assuming that the slowing down of the ball is negligible and that the angle that the trajectory makes with the straight path is small throughout.

13.42 Cricket commentators often talk of late swings referring to balls that swing unpredictably late in flight. Explain this phenomenon. Consider a cricket ball of mass 0.156 kg and diameter 7.2 cm, being bowled at a speed of V_0 (> upper critical speed of Prob. 13.40). If the drag coefficient above the upper critical speed is constant at 0.15, estimate the velocity at which the ball must be bowled so that it starts to late-swing at a distance of about 15 m from the bowling end. From your results, do you feel that a swing bowler can plan his delivery for a late swing or whether its delivery is just a matter of chance?

13.43 *Half-cup anemometer:* The anemometer of Prob. 1.16 has hemispherical cups of diameter 4 cm each at 12.5 cm from the central axis. Find the torque acting when a wind of 12 m/s velocity is blowing and the anemometer is jammed with one arm oriented normal to the wind. Next consider the case when the anemometer is rotating such that the frictional torque acting on it is 8.5×10^{-3} N m. Find the rotational speed ω if the average wing torque acting on it can be approximated by the value when one pair of cups is oriented normal to the wind. Assume quasi-steady flow.

13.44 The density of air in Mexico City (whose altitude above mean sea level = 2200 m) is 0.95 kg/m^3. If an Olympic sprinter develops the same power as at sea level, estimate the time he takes to run 100 m at Mexico, if he covers this distance in 10 s at sea level. Neglect the time required for initial acceleration and assume that the major amount of effort required goes in overcoming the air drag. (Actually the sprinter takes about the same time. What could be the possible causes?)

13.45 *Trajectory of a rotating ball in air:* Consider the flight in stagnant air of a spherical ball (of density ρ_s) rotating about a horizontal axis as shown. Write the x- and z-direction

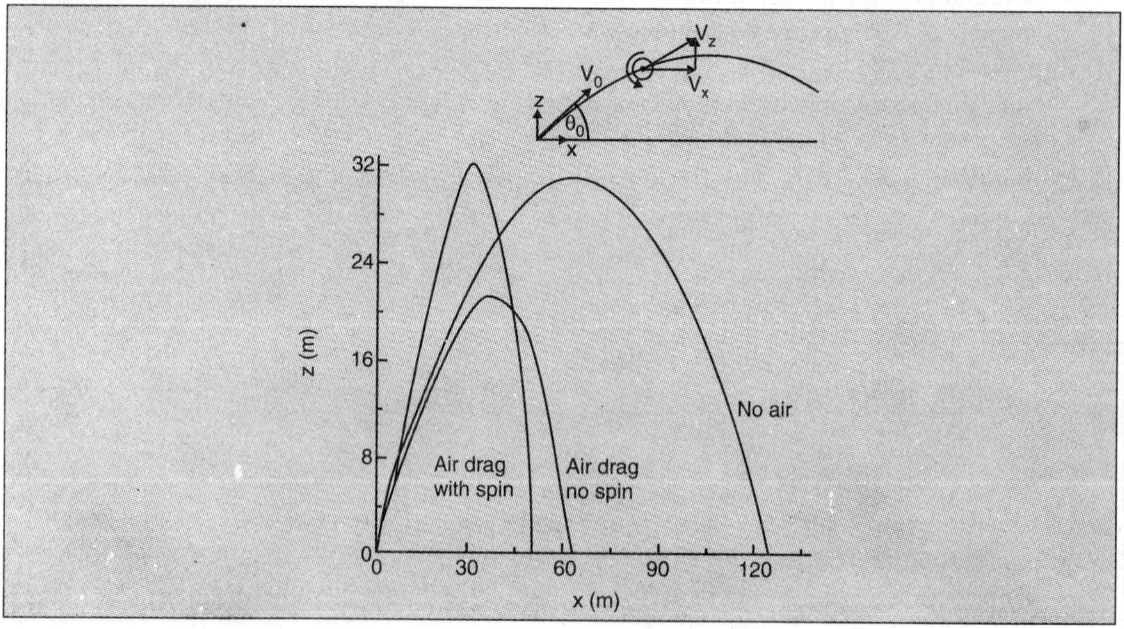

equations of motion in terms of C_L and C_D. Do your equations reduce to that for a projectile in the absence of air? These coupled equations can be integrated numerically to give the trajectory of a ball. For a soccer ball (of dia. 22.1 cm, weight 0.43 kg) kicked with 35 m/s at 45° to the horizontal and with a 'back spin' of 1500 RPM, the trajectory obtained on integration is as shown.

13.46 One notices that all vehicles moving at large speeds—be they racing cars, aeroplanes, rockets, or space shuttles—are highly streamlined. But the Lunar Module (LEM) of Apollo fame was an ugly looking structure with all kinds of sharp corners, protrusions, etc., though, it also travelled at very high speeds. Defend NASA.

13.47 In the rotating cylinder of Sec. 12.8 it is observed that at large rotational speeds there is very little separation of the boundary layer. Explain.

13.48 In the use of parachutes, a reasonable vertical speed of descent is 7 m/s. If a certain design of hemispherical parachutes gives a C_D of 1.17, what should be the canopy diameter if it weighs 5 kg and is to be used for a payload of 65 kg?

13.49 Compute the terminal velocity for a glass bead of diameter 53×10^{-6} m (i.e., 270 mesh size) settling in water. Density of glass = 2600 kg/m^3. Would a particle much larger than this be settling such that Re is in the Stokes region?

13.50 In the *initial* period of fall of a glass bead in water, the drag coefficient depends on the Reynolds number $(\rho DV/\mu)_c$ as well as the Strouhal number $(D/Vt)_c$. Show that whenever $\rho_f D^2/\mu t_c \ll 1$ one can neglect unsteady effects and assume the flow to be quasi-steady.

13.51 A small solid sphere starts to fall in a liquid with zero velocity. Assuming that Re < 1 throughout its trajectory and the flow to be quasi-steady, obtain the position of the sphere and its velocity as a function of time. Then estimate the time taken and the distance travelled by the glass sphere of Prob. 13.49 to attain 99 per cent of its terminal velocity. Find out if your results are consistent with the conditions deduced in Prob. 11.6 for neglecting the unsteady portion of the trajectory.

13.52 A raindrop falling in air breaks up when the viscous forces acting on it exceed the surface tension forces that tend to keep it together. The viscous forces are estimated by the drag force $C_D \frac{1}{2} \rho V^2 \frac{\pi D^2}{4}$ and the surface tension forces by $\pi\sigma D$. Find an estimate of the maximum size of rain droplets.

13.53 *Elutriation:* A common method of separating solid particles according to their size is to drop them in a column through which a fluid is flowing upwards as shown. Solids

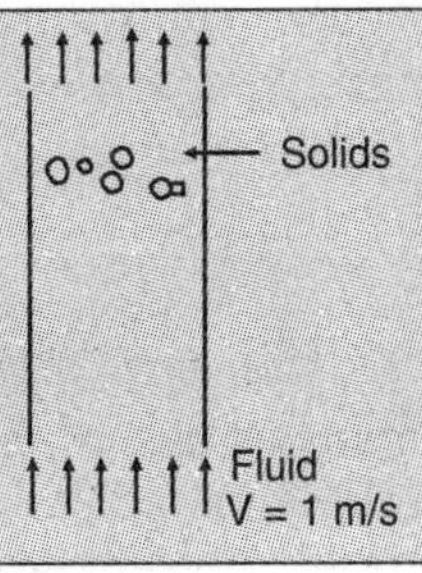

CHAPTER 13

above a certain critical size will fall to the bottom, while those below this size will get carried along with the fluid. This is called elutriation. Find out the critical size for a solid of density 2230 kg/m^3 being elutriated with water flowing at 1 m/s. Note that flow is not necessarily in the Stokes-law region.

13.54 A sphere of salt (of density ρ_s) is dropped gently in a cylinder filled with water. The salt dissolves at such a rate that its diameter D varies according to the equation

$$\frac{-dD}{dt} = a + bV$$

where V is the instantaneous velocity. Obtain an expression for the initial diameter D_0 such that the salt is completely dissolved by the time it reaches the bottom of the cylinder of height H. Assume Stokes law to be valid and the motion of the salt sphere to be quasi-steady. Note that a similar analysis is applicable when a soluble gas bubble rises up in a liquid as in aeration in fermentation processes.

13.55 *Settling tank:* A settling tank is a common pollution control device used for separating solids from fluids. A solid-laden liquid (liquid density ρ_f and viscosity μ) enters the tank as shown. The flow within it is essentially 1-D

$$V_x = V_0 = \mathring{Q}/ZY$$

$$V_y = V_z = 0$$

The solid particles settle down relative to the liquid. It is assumed that only those particles which reach the bottom are removed, the rest flowing out with the fluid. Obtain the limiting diameter D_{crit} of the particles of density ρ_s which are removed in this manner. Assume that the time for acceleration of the solid particles is small compared to the total time in residence, and that the particles are small enough for $\text{Re}_D < 1$. What particle trajectories do you expect?

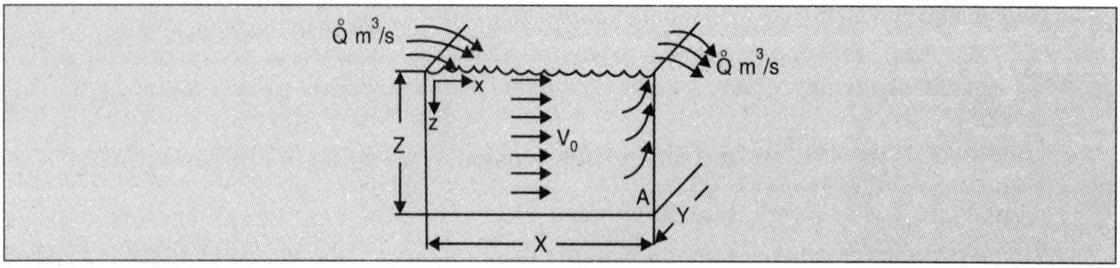

13.56 *Cyclone dust collector:* Dust-laden gases enter a 'cyclone' tangentially at high speed, swirl around and pass out through a central port as shown. The radial and tangential (i.e., swirl) components of the gas at any radius r are given by

$$V_\theta = V_1\sqrt{\frac{R_1}{r}}, \text{ and } V_r = -\,\mathring{Q}/2\pi rL$$

A dust particle is acted upon by the centrifugal force directed outwards, centrifugal buoyancy (see Prob. 6.43) acting inwards, and the drag in the radial direction. A particle moves radially till these force balance out and thereafter, it rotates at a constant r depending upon its size D. It is seen that all particles with r greater than half the core-

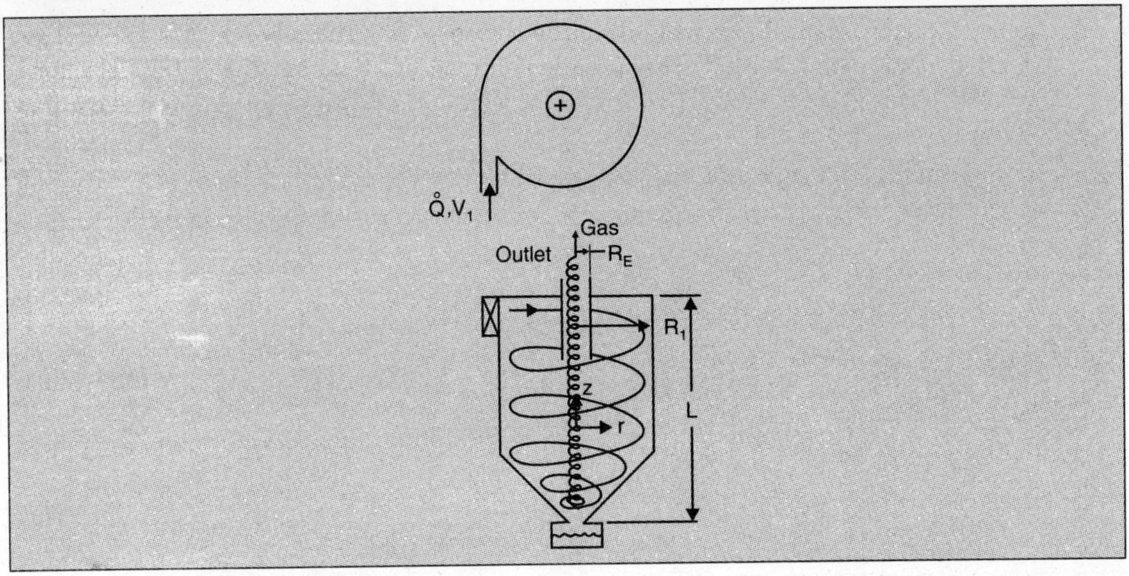

tube radius R_e are retained in the cyclone. Obtain a relationship between r and D in terms of the solid density ρ_s and the other parameters. Assume all particles are small enough for Stokes law to be valid.

13.57 *Centrifuge:* Using the results of Prob. 6.43 for the centrifugal buoyancy force acting on a solid particle immersed in a rotating liquid, show that the r- and z-components of the equation of motion of a small solid sphere of diameter D in a centrifuge given in Prob. 9.6 are correct. Assume that the z-velocity of the liquid is parabolic and that the particle is moving at the same velocity as the liquid in this direction. Note that since the centrifugal field increases with r, the particle does not attain a constant radial velocity, in contrast to the behaviour in the settling tank of Prob. 13.55.

13.58 *Inertial impaction:* The collection of solid particles by inertial impaction on a cylinder was described in Prob. 6.46. Show that the force responsible for 'catch' is estimated by $\Delta\rho D^3 V_0^2$ and the viscous drag opposing it is estimated by $\mu D V_{rel}$, where V_0 is the stream velocity, and V_{rel} is the velocity of a particle of diameter D, *relative* to the fluid. Thus, show that V_{rel} varies as D^2, thereby indicating that small particles are caught less efficiently.

13.59 Even though the drag exerted by a steady irrotational flow of a fluid across a body is zero, this is not true of unsteady flows. Why?

14

SOME ENGINEERING APPLICATIONS - III

14.1 LIFTING SURFACES

It has been known for a long time that a flat plate inclined at a small angle to the direction of flow has a force component normal to the flow direction (Fig. 14.1). This force component is termed as the *lift,* because it can sustain the weight of bodies in horizontal flight. This is how a kite flies and an eagle glides. Heavier-than air flight vehicles also generate their lift in a similar manner. There is also a component of force in the direction of air flow. This component is the *drag* which opposes the motion of the object.

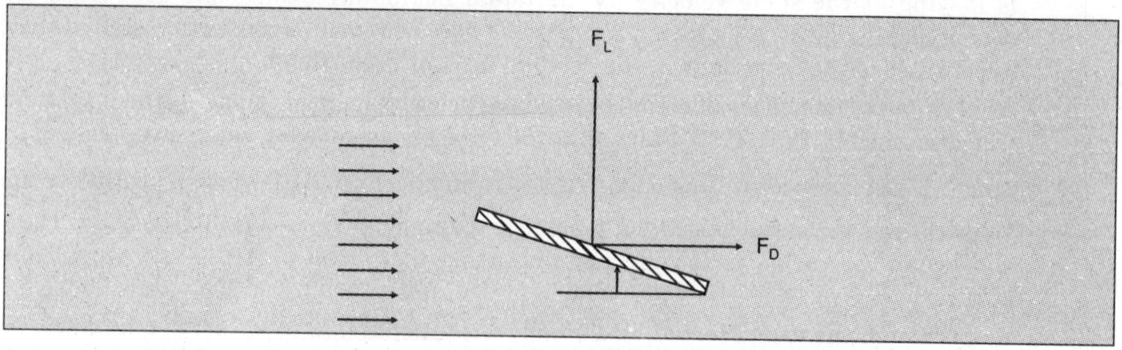

Fig. 14.1. Drag and lift forces on a flat plate inclined at an angle α to the fluid flow.

The lift F_L required to sustain an object in steady flight must equal its weight mg, while the thrust developed must equal the drag force F_D. The drag-to-lift ratio F_D/F_L is the thrust required per unit weight of the body and, therefore, its reciprocal F_L/F_D can be taken as a measure of the efficiency of the lifting surface. In the case of a flat plate the maximum F_L/F_D is obtained when its *angle of attack* α, which is the angle the plate makes with the flow direction, is small (around 4°). The lift force then is considerably larger than the drag force and, therefore, the thrust required to sustain flight is much less than the weight of the body.

As seen in Chapter 13, the drag on bodies moving through fluids has two contributions: the form drag arising because of the incomplete pressure recovery behind the object due to separation, and the skin friction. The sharp leading edge of the plate at an angle of attack induces flow separation on the upper surface* of the plate leading to a relatively large drag (Fig. 14.2a). One way of reducing the drag is to delay this separation by guiding the flow smoothly on the upper surface of the plate. This results in a narrower wake and can be achieved by giving the plate a curvature (Fig. 14.2b). This curvature is termed as *camber*. Also the lift of cambered plates is larger than that of flat plates and this results in much higher values of F_L/F_D ratios.

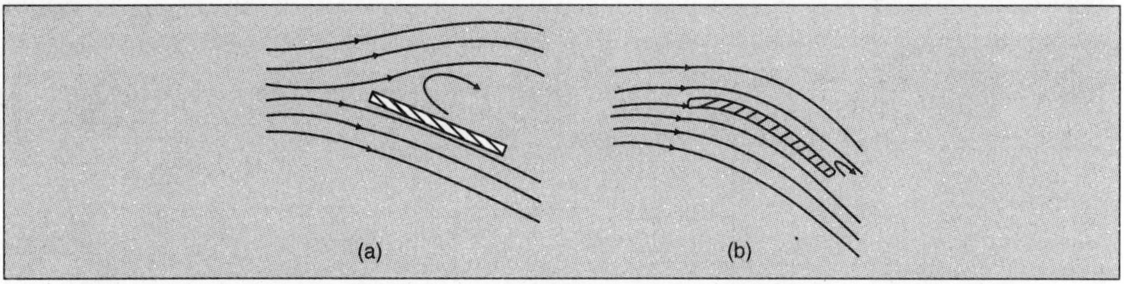

Fig. 14.2. (a) Separation on top surface of a flat plate, and (b) flow on a cambered plate.

Still higher F_L/F_D ratios can be obtained by using *aerofoil shapes* which are thin, streamlined, cambered bodies with rounded noses and long, slowly tapering tails (Fig. 14.3). Separation on such a body can be delayed to very near the sharp trailing edge, thereby giving very low drag for even fairly high angles of attack α. The slowly tapering tail keeps the adverse pressure gradient low enough for the flow to remain attached all along the surface except very near the trailing edge.

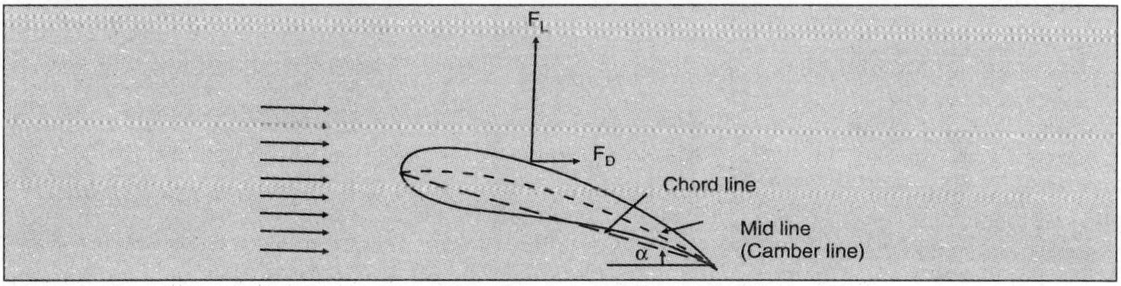

Fig. 14.3. An aerofoil shape.

All modern low speed aircrafts use wings which have aerofoil-shaped cross-sections.** The F_L/F_D ratios of high performance glider wings can be as high as 100.

The lift and drag forces on wings are usually expressed in non-dimensional terms using the lift and drag coefficients $C_L = F_L/\frac{1}{2}\rho\, V_0^2\, A$, and $C_D = F_D/\frac{1}{2}\rho\, V_0^2\, A$ respectively. Here, V_0 is the flow velocity and A is the characteristic area, usually taken as the planform area. For

* Separation is not a problem on the under-surface because the pressure gradient there is favourable.

** High speed flight in contrast requires sharp edges instead of rounded leading edges, because of the problems of compressibility associated with high speed (compared to the speed of sound) flows.

CHAPTER 14

incompressible flows, the lift and drag forces depend on the inertial and viscous forces and on the geometry. Therefore, in non-dimensional terms

$$C_L, C_D = \mathcal{F}(\text{Re, geometry})$$

The geometry can be specified by the shape of the profile and the angle of attack, and for a fixed profile

$$C_L, C_D = \mathcal{F}(\text{Re}, \alpha) \quad ...(14.1)$$

Since the Reynolds numbers for most flight vehicles is very large (> 10^5), its effect is very small and usually C_L and C_D are expressed as a function of the angle of attack only. Figure 14.4 shows the curves of C_L, C_D and their ratio C_L/C_D for the aerofoil shape designated as NACA 4412.

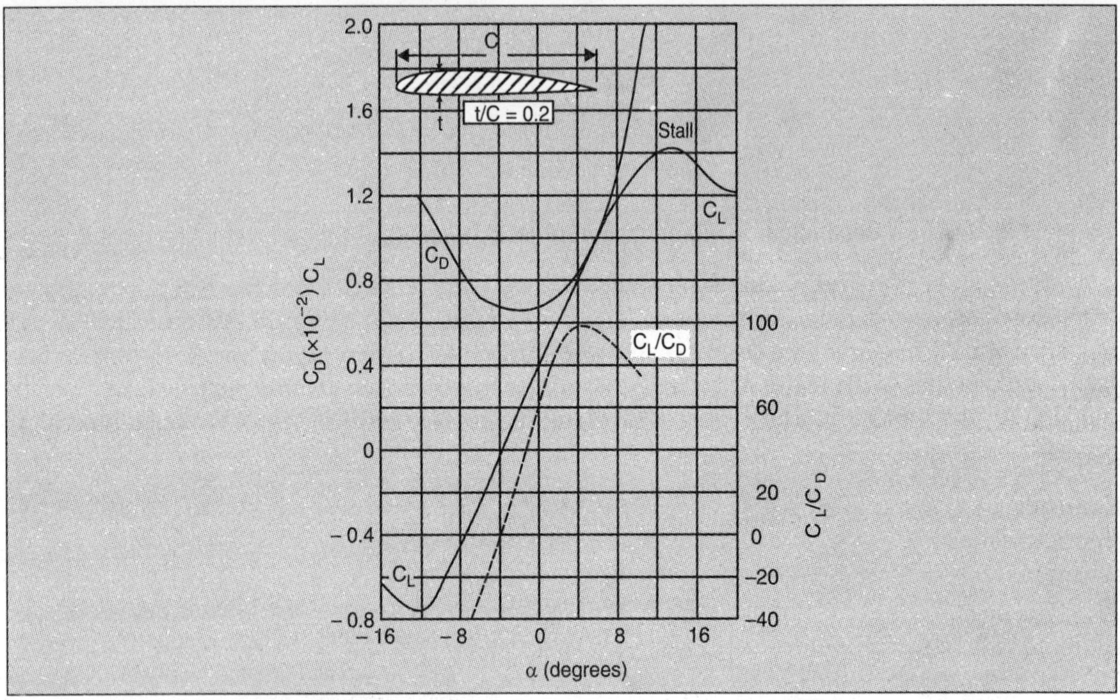

Fig. 14.4. C_L,C_D and their ratio for NACA 4412 aerofoil.

These variations are typical of most aerofoils at large Re. The lift coefficient C_L increases linearly with α upto a point after which it decreases rather suddenly. The drag coefficient C_D also shows a sudden increase at this point and the aerofoil is said to be *stalled.* Stalling is explained by the fact that at large α, the flow separates very close to the leading edge and therefore does not follow the contour of the aerofoil resulting in a loss of lift. The resulting wide wake (Fig. 14.5) is responsible for the simultaneous increase in the drag coefficient. In the range where C_L varies linearly with α, the lift developed by the wings can be written as

$$F_L = k\, \alpha\, V^2 \quad ...(14.2)$$

where k is a constant of proportionality. Since the total lift F_L of an aerofoil must equal the weight mg supported throughout the flight, an aeroplane having a fixed wing geometry has the maximum angle of attack α when its velocity is the lowest, i.e., at take-off and landing. Equation (14.2) shows that an aircraft carrying more load requires a higher take-off and landing velocity and thus requires longer take-off and landing runs. An aeroplane in flight must not slow down below a certain speed termed the *stall speed,* because then even the maximum permissible α will not give enough lift, and increasing α any further will mean a sudden loss of lift due to the stalling of the aerofoil.

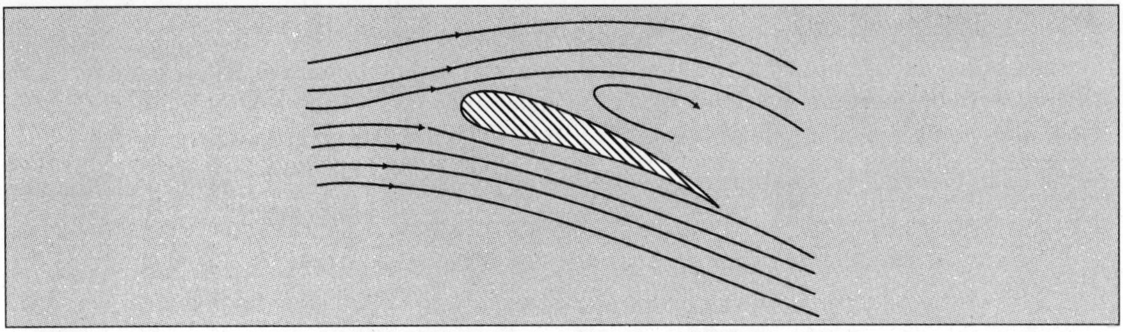

Fig. 14.5. Flow beyond the stall point.

The nomenclature of the force component normal to the direction of flow as 'lift' should not be misconstrued to imply that the force is always in the vertical direction. The 'lift' force acting on a vertical sail of a yacht is horizontal, normal to the sail. The larger value of F_L compared to F_D on a well designed sail can make the boat go at an acute angle to the wind direction, as seen in Fig. 1.24. The horizontal lift force acting on the blades of an aeroplane or ship propeller generates a thrust in the axial direction, as discussed in Sec. 14.3. Similarly, the design of fan blades and the blades of axial-flow pumps and turbines relies on the large F_L/F_D ratios of aerofoil shapes.

Example 14.1. A transport plane weighs 100,000 kg and has wings of NACA 4412 section whose characteristics are given in Fig. 14.4. If the wing span is 50 m and the chord length is 6 m, estimate the minimum take-off speed of the aeroplane at sea level.

The stall angle α for the given section is 14° (Fig. 14.4). If the maximum permissible α is set at 12° allowing for a margin of safety, this gives a C_L of 1.45. The total weight lifted mg is 10^5 (kg) × 9.81(m/s²) = 9.81×10^5 N. A force balance in the vertical direction gives

$$mg = F_L = C_L \frac{1}{2} \rho V_0^2 \text{ (span} \times \text{chord)}$$

$$9.81 \times 10^5 \text{ (N)} = 1.45 \times \frac{1}{2} \times 1.22 \text{ (kg/m}^3) \times V_0^2 \times 50 \text{ (m)} \times 6 \text{ (m)}$$

or

$$V_0 = 60.8 \text{ m/s}$$

This is the minimum take-off speed. The acceleration of the aircraft under its own power gives the minimum length of the run-way required. The actual length needed will be considerably larger than this because safety regulations require that it should be possible to bring the aircraft to rest on the run-way if the take off is aborted just before reaching the speed V_0 obtained above.

If the same aircraft is flying at its cruise speed of 120 m/s, the C_L required will be given by

$$9.81 \times 10^5 \text{ (N)} = C_L \times \frac{1}{2} \times 1.22 \text{ (kg/m}^3) \times (120)^2 \text{ (m/s)}^2 \times 50 \text{ (m)} \times 6 \text{ (m)}$$

CHAPTER 14

or

$$C_L = 0.372$$

which corresponds to an angle of attack* of about 0°.

14.2 ORIGINS OF LIFT

As mentioned in Example 7.7 the lift force on a streamlined aerofoil arises because of the differences in the pressure distributions on the upper and lower surfaces of the aerofoil. Consider a 2-*D* aerofoil at a small angle of attack (Fig. 14.6). The pressure p at any point of the aerofoil surface can be found by the Bernoulli equation as $p - p_0 = \rho(V_0^2 - V^2)/2$, where V is the local

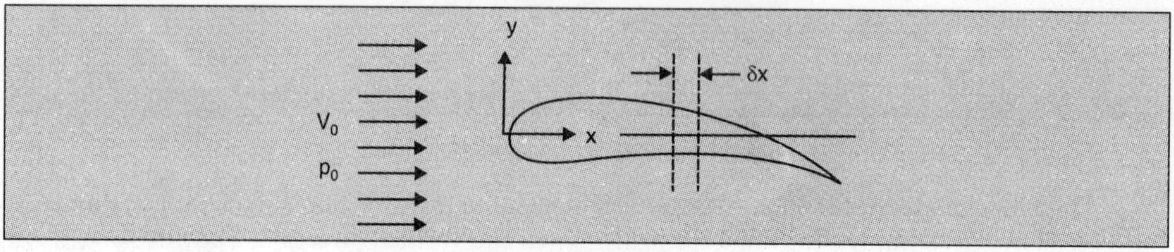

Fig. 14.6. Aerofoil at small angle of attack.

velocity. Consider a small element of the aerofoil of length δx along the x-axis. The lift contribution of the upper and lower aerofoil surfaces of this element per unit span (in the y-direction) is seen as

$$\delta F_L = (-p_u + p_l)\ \delta x$$

δx being the projected area of both the upper and the lower surfaces. The pressure difference $p_l - p_u$ can be found by the Bernoulli equation as $\rho\left(V_u^2 - V_l^2\right)/2$, and therefore,

$$\delta F_L = \frac{1}{2}\rho\left(V_u^2 - V_l^2\right)\delta x = \rho\ \frac{(V_u + V_l)(V_u - V_l)}{2}\ \delta x$$

For thin aerofoils, V_l and V_u are only slightly different from V_0 and therefore,

$$\delta F_L \simeq \rho V_0 (V_u - V_l)\delta x$$

The total lift per unit span is then

$$F_L = \rho V \int_0^C (V_u \bullet V_l)\ dx$$

where C is the chord-length.

The circulation Γ around the contour defined by the body shape is approximated by

$$\Gamma = \int_0^C V_l dx + \int_C^0 V_u dx = \int_0^C (V_l \bullet V_u) dx$$

* Note that for cambered aerofoils, small negative α also give positive value of F_L.

Since the flow outside may be assumed irrotational, the circulation about any closed contour enclosing the body is Γ (see Prob. 12.6). Thus, the lift per unit span of the aerofoil is given by

$$F_L = -\rho V_0 \Gamma \qquad \text{...(14.3)}$$

where Γ is the circulation about any circuit enclosing the body. This is known as the *Kutta Joukowski theorem.** This is the same result as the Robins-Magnus law Eq. (12.44) which was obtained for a spinning cylinder in irrotational flow. The circulation here, of course, is not established by the spin of the aerofoil.

Though the above result has been obtained for inviscid flows, no lift is generated in a *truly* inviscid flow. This is because the setting up of the circulation Γ depends on the presence of

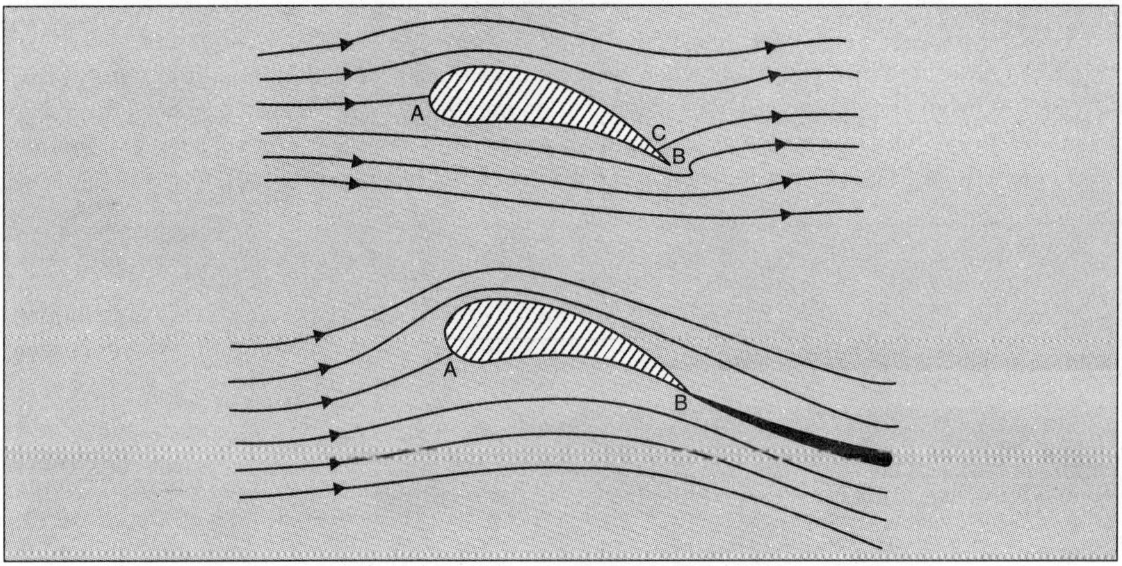

Fig. 14.7. Flow around an aerofoil in (a) truly inviscid fluid, and (b) a real fluid. Shaded region represents velocity discontinuity because of separation.

viscosity, howsoever small. If an aerofoil is set in motion in a truly inviscid fluid, the flow pattern around it is as shown in Fig. 14.7(a), with stagnation points at *A* and C. There is no separation and the net lift is zero. But in any real fluid, at howsoever large a Reynolds number, a thin boundary layer would form at the surface and would separate at the sharp trailing edge at *B*. A recirculating vortex is set up between *B* and *C*, which washes downstream, and the stagnation point moves down to the separation point at *B*. Thus, the flow pattern is modified to the one shown in Fig. 14.7(b). This flow pattern, unlike that in Fig. 14.7(a), has a circulation, giving larger velocities at the upper surface than on the lower one. This explains the generation of lift on a 2-*D* aerofoil.

As can be seen clearly, the drag on an aerofoil is principally the skin-friction drag, with pressure drag playing a very minor role, if any. Therefore, the drag coefficient is higher when

CHAPTER 14

* The Kutta Joukowski theorem is valid for all shapes (not necessarily thin) in inviscid flow and can be obtained exactly using the mathematical theory of irrotational flows.

the boundary layer flow is turbulent than when it is laminar. This is in contrast to the case of round-nosed bluff bodies where a turbulent boundary layer means a narrower wake with a dramatic reduction in the pressure, and hence the total, drag. It is for this reason that high performance glider wings are designed to keep the boundary layer flow laminar till as far downstream as possible.

So far only 2-*D* aerofoils have been considered. When the span of the aerofoil is finite, the lift generated per unit span decreases while the drag increases. Discussion of these effects is beyond the scope of this text and the reader is referred to Kuethe and Chow (Reading 26).

14.3 PROPELLERS

The thrust generated by a propeller, whether of an aeroplane or of a ship, has its origins in the lift produced by an aerofoil-shaped body moving at an angle of attack. Figure 14.8 shows a typical two-bladed propeller as it rotates at an angular speed ω and advances at a velocity V_0 through a stationary fluid.* The tangential section *AA* at any radial location r has an aerofoil shape as

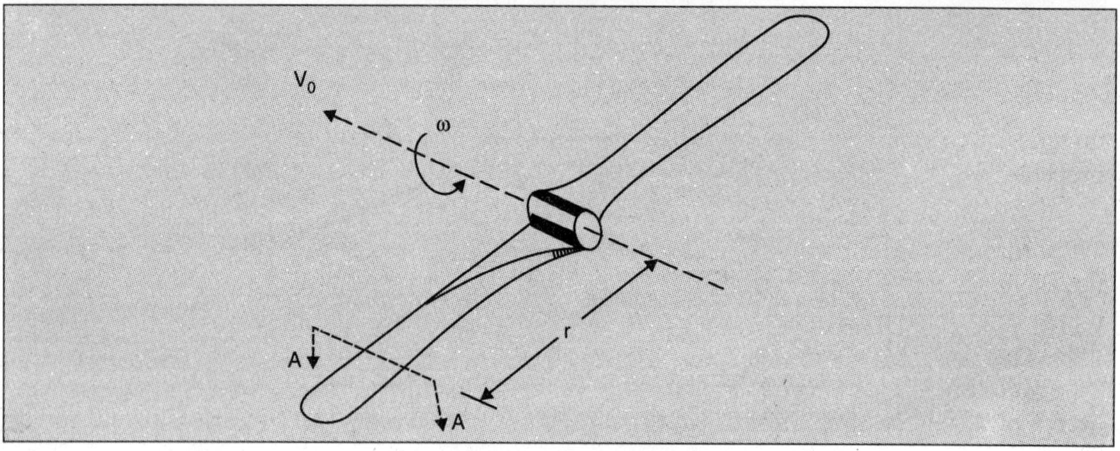

Fig. 14.8. Two-bladed propeller.

seen in Fig. 14.9, which shows the section of a blade when it is at the horizontal position. At this point, the *blade element* has an axial velocity V_0 and a tangential velocity ωr, so that the resultant velocity of the blade element is inclined at an angle $\phi = \tan^{-1} (V_0 / \omega r)$ to the tangential (vertical) direction. The relative velocity of the flow over this blade element is then in the opposite direction. If the *pitch angle* of the blade element is θ (measured from the tangential direction), the angle of attack of the aerofoil is $\alpha = \theta - \phi$. The lift and drag forces, δF_L and δF_D, on the blade element are in the direction shown, one normal to, and the other along the relative flow

* Note that V_0 is the velocity of the fluid far away relative to the propeller.

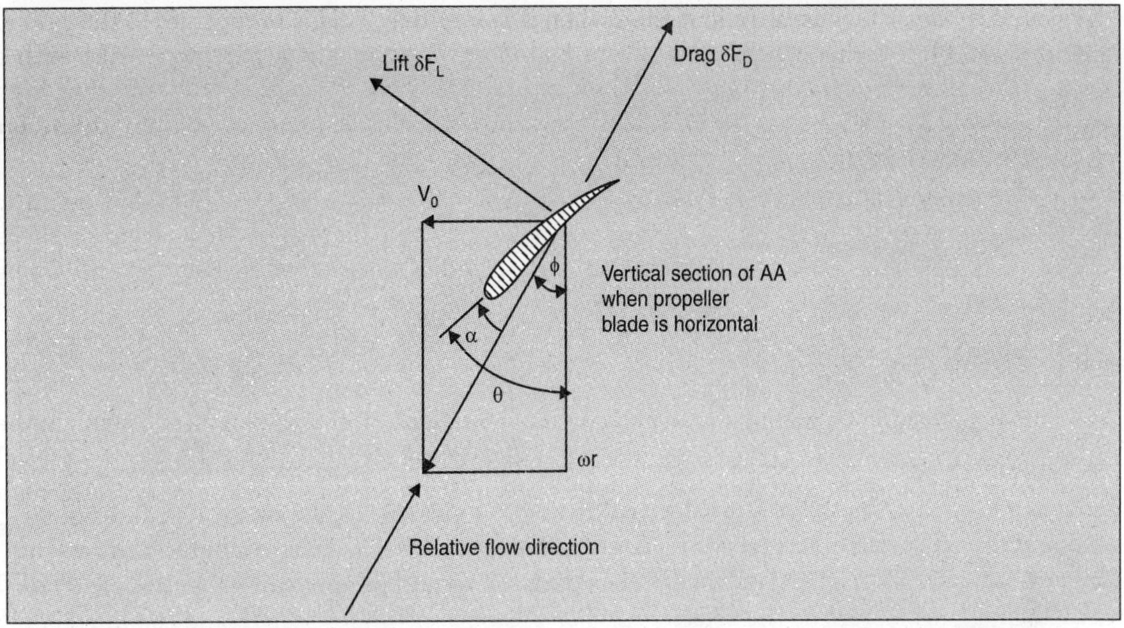

Fig. 14.9. Velocities and forces on the aerofoil-shaped section *AA* of the propeller of Fig. 14.8.

direction. The axial components of these two forces contribute to the thrust δF_T produced by the propeller

$$\delta F_T = \delta F_L \cos\phi - \delta F_D \sin\phi.$$

Thus, a significant thrust is obtained when $\delta F_L/\delta F_D$ is large. The tangential components of the forces, $\delta F_L \sin\phi + \delta F_D \cos\phi$ contribute to the torque acting on the shaft and must be overcome by the engine. The design of a propeller is aimed at maximizing the ratio of the axial thrust developed to the tangential torque required.

Note that as the velocity of advance V_0 changes, it affects the angle ϕ as also the angle of attack α. The relative magnitudes of the lift and drag forces, as also their orientations, change. Thus, the effectiveness of a fixed pitch (i.e., constant θ) propeller changes with the velocity of advance. It is, therefore, conventional to have mechanisms incorporated in the hub of the propeller which change the pitch of the blades so as to optimize the performance of the propeller under different operating conditions.

The angle ϕ, which the relative velocity vector makes with the plane of the propeller changes as r changes, being less near the tip than near the hub. This means that the angle of attack α of a *straight propeller blade* (one in which θ is constant with r) increases from the hub to the tip. Since the maximum aerofoil efficiency (measured by the maximum ratio of C_L/C_D) occurs at a fixed value of α, it is conventional to vary the pitch of the blade along its length by twisting it such that θ is larger near the hub than near the tip (see Fig. 14.8).

The characteristics of propellers are usually presented in terms of non-dimensional performance curves. Since the flow on a propeller is governed by inertial (both streamwise and centrifugal) and viscous forces (in the absence of compressibility effects, which is a valid assumption to make if the tip speed is well below the speed of sound), the governing parameters are seen from Table 14.1 to be $(V/\omega L)_c$ and $(\mu V/\rho\omega^2 L^3)_c$. It is conventional to take the velocity

Table 14.1. Similarity parameters for propellers

Governing Law	Scale Factor Relation	Similarity Rule	Π-number
Intertial (centrifugal) F_ω = (mass) $(\omega^2 L)$	$k_{F,\omega} = k_\rho k_\omega^2 k_L^4$		
Intertial (streamwise) F = (mass) $(\mathbf{V} \cdot \nabla \mathbf{V})$	$k_F = k_\rho k_L^2 k_V^2$	$\dfrac{k_V}{k_\omega k_L} = 1$	$\dfrac{V}{\omega L}$
Viscous $F_\mu = \mu$ (area) (Vel. grad.)	$k_{F,\mu} = k_\mu k_L k_V$	$\dfrac{k_\mu k_V}{k_\rho k_\omega^2 k_L^3} = 1$	$\dfrac{\mu V}{\rho \omega^2 L^3}$

of advance V_0 as the characteristic velocity, the angular speed ω as the characteristic angular speed and the tip-circle diameter D of the propeller as the characteristic length. The two pi-numbers formed are $V_0/\omega D$ and $\mu V_0/\rho\omega^2 D^3$. Dividing the second pi-number by the first gives the two modelling parameters as the *velocity ratio* or *advance ratio* $\text{Ve} = V_0/\omega D$, and the Reynolds number $\text{Re} = \rho\omega D^2/\mu$.

Any non-dimensional dependent parameter then is a function of these two parameters. The *thrust coefficient* C_T is defined as

$$C_T = \frac{F_T}{F_{T,c}} = \mathcal{F}(\text{Ve, Re})$$

To obtain the characteristic thrust $F_{T,c}$ use the scale factor relation $k_F = k_\rho\ k_L^2\ k_V^2$, which gives $F/\rho L^2 V^2$ or $F/\rho\omega^2 L^4$ as invariant. Therefore $F_{T,c}$ can be set equal to $(\rho\omega^2 L^4)_c = \rho\ \omega^2\ D^4$. Thus, the thrust coefficient is

$$C_T = \frac{F_T}{\rho\ \omega^2\ D^4} = \mathcal{F}(\text{Ve, Re}) \qquad \text{...(14.4)}$$

Similarly, the non-dimensional *power coefficient* $C_{\mathring{W}}$ is given by

$$C_{\mathring{W}} = \frac{\mathring{W}}{\rho\ \omega^3\ D^5} = \mathcal{F}(\text{Ve, Re}) \qquad \text{...(14.5)}$$

where $\mathring{W}$ is the power input to the propeller. The efficiency of a propeller is defined as the ratio of the useful power output $F_T V_0$ and the power input $\mathring{W}$

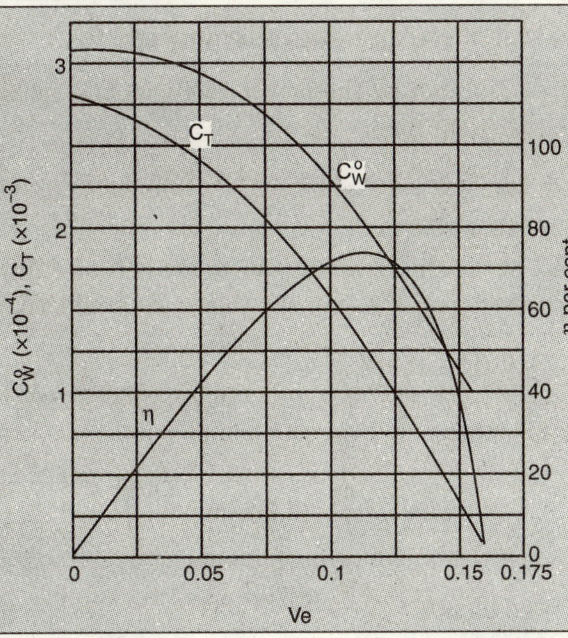

Fig. 14.10. Characteristics of a typical aircraft propeller.

$$\eta = \frac{F_T V_0}{\mathring{W}} = \frac{C_T \cdot \rho\, \omega^2\, D^4 \cdot V_0}{C_{\mathring{W}}\, \rho\, \omega^3\, D^5} = \frac{C_T \text{Ve}}{C_{\mathring{W}}} \quad \text{...(14.6)}$$

Therefore, η is also a function of Ve and Re. For very large Reynolds numbers, the viscous effects are negligible and C_T, $C_{\mathring{W}}$ and η are then functions of the velocity ratio Ve alone. The characteristics of a typical aircraft propeller are shown in Fig. 14.10. It should be observed that the thrust coefficient is maximum when the aircraft is stationary and decreases as Ve increases. The efficiency η is seen to be zero when the aircraft is stationary because no useful work is being done. The efficiency increases as the velocity of advance increases till it reaches a maximum. Thereafter, η decreases sharply and becomes zero when the propeller is advancing so fast that the net thrust (and C_T) is zero.

Example 14.2. A helicopter rotor is much like a propeller spinning in a horizontal plane so that the thrust produced supports the weight of the helicopter. A helicopter weighing 1500 kg has a rotor whose dimensionless characteristics are given in Fig. 14.10. If the rotor diameter is 6.5 m find the rotor speed, the power required and the slip-stream velocity far down-stream when the helicopter is in hovering-flight.

The total lift on the helicopter and thus the thrust required is 1500 (kg) × 9.81 (m/s²) = 1.47×10^4 N. Since the helicopter is not climbing, its velocity ratio Ve is zero and, therefore, $C_T = 2.8 \times 10^{-3}$. This gives

$$1.47 \times 10^4 \,(N) = C_T\, \rho\, \omega^2\, D^4$$

$$= 2.8 \times 10^{-3} \times 1.22 \,(\text{kg/m}^3) \times \omega^2 \times (6.5)^4 \,(\text{m})^4$$

CHAPTER 14

or

$$\omega = 49.1 \text{ rad/s} = 470 \text{ RPM}$$

The input power required is found by the power coefficient $C_{\mathring{W}}$, which equals 3.1×10^{-4} at Ve = 0. Therefore, the power required is

$$\mathring{W} = C_{\mathring{W}} \, \rho \, \omega^3 D^5 = 3.1 \times 10^{-4} \times 1.22 \text{ (kg/m}^3) \times (49.1)^3 \text{ (s}^{-3}) \times (6.5)^5 \text{ (m)}^5$$

$$= 519.4 \text{ kW}$$

To determine the velocity of the slip stream, refer to the actuator-disc analysis of a propeller given in Sec. 8.2. In the notation of Fig. 8.4, the thrust F_T produced is given by Eq. (8.13) as

$$F_T = \rho A_3 (V_2 - V_1) \frac{V_2 + V_1}{2}$$

Here $V_1 = 0$ and so

$$1.47 \times 10^4 \text{ (N)} = 1.22 \text{ (kg/m}^3) \times \frac{\pi}{4} \times (6.5)^2 \text{ (m)}^2 \times \frac{V_2^2}{2}$$

or

$$V_2 = 26.95 \text{ m/s}$$

This is the steady slip stream velocity underneath the hovering rotor. The rate of conversion of engine power to kinetic energy of the fluid stream is given by

$$\mathring{W}_{KE} = \mathring{m} V_2^2/2 = (\rho \, V_3 \, A_3) V_2^2/2 = \rho \, A_3 \, V_2^3/4$$

since V_3, the velocity at the rotor is given by Eq. (8.14) as $(V_1 + V_2)/2 = V_2/2$. Thus,

$$\mathring{W}_{KE} = \frac{1}{4} \times 1.22 \left(\text{kg/m}^3\right) \times \frac{\pi}{4} \times (6.5)^2 \left(\text{m}\right)^2 \times (26.95)^3 \left(\text{m/s}\right)^3$$

$$= 198.1 \text{ kW}$$

Since no useful work is being done, the helicopter being stationary in hovering flight, the total input power of 519.4 kW is transferred to the fluid. But not all of it appears as the kinetic energy, the rest (519.4 – 198.1 = 321.3 kW) being dissipated as various losses.

14.4 HYDROFOILS

A lifting surface moving submerged in water is termed as a *hydrofoil.* The lifting wings of hydrocrafts, blades of ship-propellers and Kaplan turbines, and vanes on the impellers of water pumps are all examples of hydrofoils. The analysis of hydrofoils proceeds along the same lines as that of aerofoils, as long as the pressure at any point on the hydrofoil surface does not fall below the vapour pressure at the local temperature.

As water flows past a hydrofoil, the pressure on the 'top' surface decreases. It may go below the vapour pressure quite close to the leading edge. Whenever this happens, and it is not at all uncommon, water vaporizes locally, forming vapour bubbles. This phenomenon is termed as *cavitation.* As the flow proceeds downstream, the vapour bubbles grow very rapidly and then collapse suddenly as they are washed down into the higher pressure region. At higher hydrofoil speeds the cavitation may be so extensive that the whole body may appear to be surrounded by a vapour cloud.

Cavitation is highly undesirable from two points of view. First, the vaporization of water modifies the pressure distribution on the top surface of the hydrofoil, and prevents it from falling below the vapour pressure resulting in a loss of lift. Secondly, the repeated implosive collapse of vapour bubble erodes and pits the hydrofoil surface. This condition is quite serious in marine propellers, pump impellers and many hydraulic structures and has to be avoided.

The pressure difference in flows at large Re (typical for flow past hydrofoils) is of the order of $\rho V_0^2/2$ as shown in Sec. 11.5. Therefore, the potential for cavitation is measured by the similarity number $(p_0 - p_v)/\frac{1}{2}\rho V_0^2$, where p_v is the vapour pressure. This parameter is known as the *Thoma number* Th. Cavitation takes place whenever Th is below a critical value Th_{crit}. The value of Th_{crit} depends on the geometry of the hydrofoil. For a given hydrofoil, the larger the angle of attack α, the lesser are the expected pressures on the 'top' surface, the more is the tendency for cavitation, and therefore Th_{crit} is higher.

The drag and lift coefficients of hydrofoils are functions of the Thoma number, besides the usual Re, the angle of attack α, and the geometry. Whenever Th is greater than Th_{crit}, there is no tendency to cavitate, and C_L and C_D are independent of Th. But for Th below Th_{crit} (which increases with α) the lift coefficient C_L decreases and the drag coefficient C_D increases as Th decreases. Typical variations of C_L and C_D with Th for two different values of the angle of attack are shown in Fig. 14.11. Note the surprising feature that as Th becomes very low, the drag coefficient starts decreasing with decreasing α.

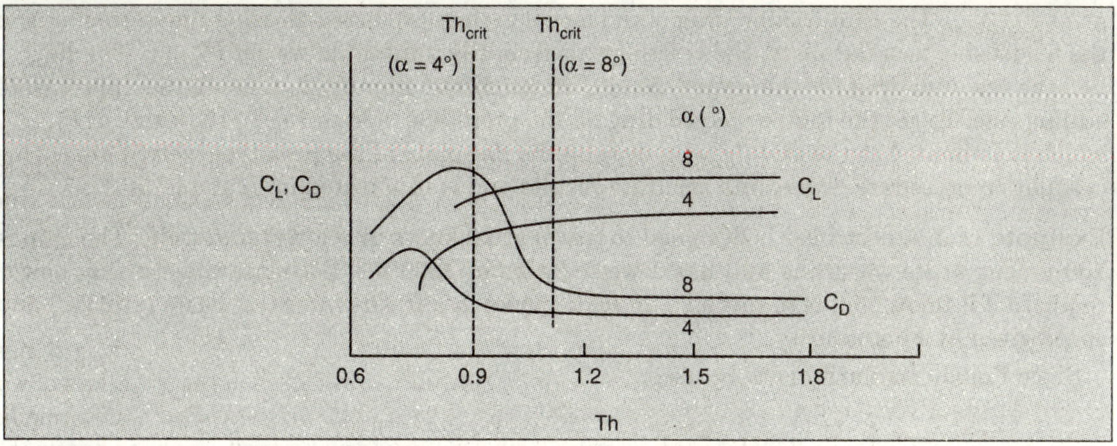

Fig. 14.11. C_L and C_D as a function of Th.

14.5 MODELLING OF DRAG ON SHIPS

When an object moves in water at the interface with air or very close to it, as for example a ship, it produces travelling waves on the surface. Since water is brought to rest (in a frame fixed with the ship) at the leading edge and the pressure at the surface of water is fixed at atmospheric, the elevation of the water surface increases locally, thereby converting the kinetic

CHAPTER 14

energy to potential energy. As the ship moves on, the water elevation decreases again, setting up a system of waves that travel away from the ship, carrying away energy and resulting in increased drag. The actual drag on the ship is still the integrated value of the streamwise components of the shear and pressure forces, but now the pressure distribution is modified because of the travelling waves.

In Sec. 9.2.5 it was seen that whereas it is possible to model the flow about bodies moving far away from the liquid-gas interface by Reynolds number similarity alone, Froude similarity is required as well in the case of ships and other bodies moving close to the free surface. This is so because while the introduction of the non-gravitational pressure $\mathscr{P} = p + \rho gz$ in the first case results in only the Re occurring in the equation, in the case of flows with free surfaces this only shifts the Froude number to the boundary condition as in Eq. (9.23). Thus, the drag on ship models can be used for predicting the drag on prototype ships only if Fr and Re are both matched simultaneously. But as was seen earlier, if water is used in model experiments also, it is impossible to match both Fr and Re as long as k_L is different from one.

This difficulty is usually overcome by realizing that the drag on a ship hull (or any other object for that matter) is made up of two contributions: skin-friction drag and pressure drag. Whereas the skin-friction drag is essentially a function of Re, the pressure drag is independent of it. In the case of ships, the wave and form drags are the two components of the pressure drag and are determined essentially by Fr. Thus, the total drag of the ship can be broken up into two components

$$F_{D,\text{total}} = F_{D,\text{skin}} + F_{D,\text{res}}$$

The *residual drag* $F_{D,\text{res}}$ includes the wave drag. It is conventional to conduct model tests matching the Froude numbers (the Reynolds number of the model flow under this condition is several orders of magnitude lower than the Re for the prototype) and measuring the drag. This gives $F_{D,\text{total},m}$. The skin-friction drag of the model is then predicted treating the wetted area of the model ship as a flat plate. Subtracting this from the total drag, we get $F_{D,\text{res},m}$. The $F_{D,\text{res},p}$, i.e., the residual drag for the prototype is then predicted from $F_{D,\text{res},m}$ using the appropriate scaling rule. To get the total expected drag on the prototype, add to $F_{D,\text{res},p}$ the value of $F_{D,\text{skin},p}$ which is estimated as the skin-friction drag on the flat plate of the prototype wetted area. This calculation procedure gives quite accurate results and is illustrated in Example 14.3.

Example 14.3. A destroyer is designed to travel at 20 knots in sea-water at 20°C. The ship is 100 m long at its waterline and has a wetted area of 2500 m^2. Estimate the cruising power required if it takes 65.2 N to tow a 1 : 20 scale model in a fresh-water test basin (at 20°C) at a speed given by Fr similarity.

Since Froude similarity is to be used,

$$\left(\frac{V}{\sqrt{gL}}\right)_p = \left(\frac{V}{\sqrt{gL}}\right)_m$$

or

$$\frac{V_p}{V_m} = k_V = \sqrt{\frac{L_p}{L_m}} = k_L^{1/2} = \sqrt{20} = 4.47$$

which means V_m = 20 (knots) × 0.515 (m/s knot)/ 4.47 = 2.3 m/s.

First estimate the skin-friction drag on the model and subtract it from the total drag to obtain the residual drag for the model. The length of the 1 : 20 model is 5 m and therefore, its Reynolds number in fresh water is

$$\mathrm{Re}_m = \frac{998.3\,(\mathrm{kg/m^3}) \times 2.3\,(\mathrm{m/s}) \times 5\,(\mathrm{m})}{1.002 \times 10^{-3}\,(\mathrm{Pa\ s})} = 1.146 \times 10^7$$

The skin-friction coefficient at this Re is estimated by Eq. (13.32), the boundary layer being turbulent. Thus

$$C_{D,m} = 0.455\,(\log_{10} \mathrm{Re}_L)^{-2.58}$$
$$= 2.94 \times 10^{-3}$$

and the skin-friction drag on a wetted area of 2500 (m^2) × $(1/20)^2$ = 6.25 m^2 is

$$F_{D,\mathrm{skin},m} = (C_D \times \rho V^2 \times A_{\mathrm{wetted}}/2)_m$$
$$= 2.94 \times 10^{-3} \times 9.983 \times 10^2\,(\mathrm{kg/m^3}) \times (2.3)^2\,(\mathrm{m/s})^2 \times 6.25\,(\mathrm{m^2})/2$$
$$= 48.5\ \mathrm{N}$$

The residual drag on the model is, therefore (65.2 – 48.5) N or 16.7 N.

This residual drag is now scaled-up for the prototype using the prediction rule obtained from the scale factor for forces

$$k_F = k_\rho\, k_L^2\, k_V^2 = k_\rho\, k_L^3 = \left(\frac{1.025}{0.9983}\right) \times 20^3 = 8214$$

and, therefore,

$$F_{D,\mathrm{res},p} = 8214 \times F_{D,\mathrm{res},m} = 8214 \times 16.7\,(\mathrm{N}) = 1.37 \times 10^5\ \mathrm{N}$$

The theoretically estimated skin-friction drag must be added to this to give the total drag on the prototype. The Reynolds number for the prototype is

$$\mathrm{Re}_p = \frac{1.025 \times 10^3\,(\mathrm{kg/m^3}) \times (20 \times 0.514)(\mathrm{m/s}) \times 100\,(\mathrm{m})}{1.09 \times 10^{-3}\,(\mathrm{Pa\ s})}$$
$$= 9.667 \times 10^8$$

The skin-friction drag coefficient is given by Eq. (13.32) as

$$C_{D,p} = 0.455\,(\log_{10} 9.667 \times 10^8)^{-2.58} = 1.577 \times 10^{-3}$$

The skin-friction drag is then

$$F_{D,\mathrm{skin},p} = (C_D \cdot \rho V^2 \cdot A_{\mathrm{wetted}}/2)_p$$
$$= 1.577 \times 10^{-3} \times \frac{1}{2} \times 1.025 \times 10^3\,(\mathrm{kg/m^3}) \times (10.28)^2\,(\mathrm{m/s})^2 \times 2500\,(\mathrm{m^2})$$
$$= 2.13 \times 10^5\ \mathrm{N}$$

The total drag in the prototype is then ($2.13 \times 10^5 + 1.37 \times 10^5$) N or 3.5×10^5 N, and the power required to propel the ship is

$$\mathring{W} = F_{D,\mathrm{total},p} \times V_p = 3.5 \times 10^5\,(\mathrm{N}) \times 10.28\,(\mathrm{m/s}) = 3.6\ \mathrm{MW}$$

14.6 FLUIDICS

Fluidics is a relatively new technology employing fluid flow phenomenon in specially designed 'no-moving-parts' devices to perform sensing, logic and control functions. These devices were introduced in 1960 and serve purposes similar to those normally served by electronic devices. The devices are much more economical, faster and smaller than conventional hydraulic control elements employing moving parts such as valves, etc., and can operate in conditions where electronic devices are unsatisfactory, such as at high temperatures and humidity, in presence of severe vibration, in high fire-risk environments or where ionizing radiations are present.

The earliest of fluidic devices employed certain properties of boundary layers and fluid jets. Suppose a fluid jet emerges from an orifice into a region bounded by a side wall (Fig. 14.12). It is observed that as the jet emerges, it attaches itself to the nearby wall and follows it. This

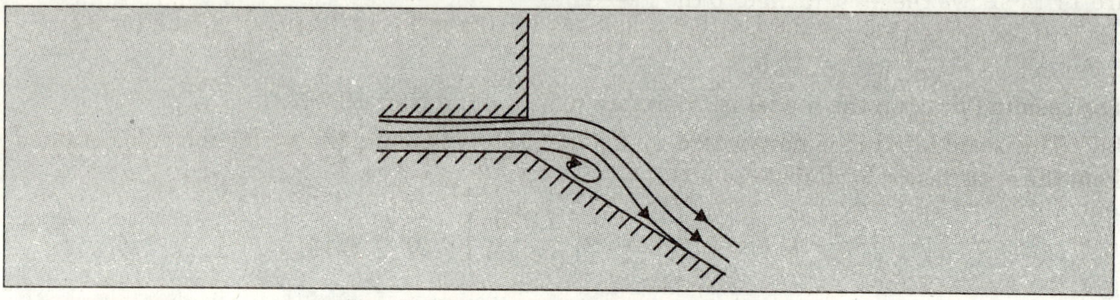

Fig. 14.12. Attachment of a jet to a wall.

phenomenon is known as the *Coanda effect* after H. Coanda who first reported it in 1932. It is because of this effect that a jet striking a sphere curves around it (see Prob. 5.32) and a fluid draining from a cup runs around the lip, (Fig. 14.13).

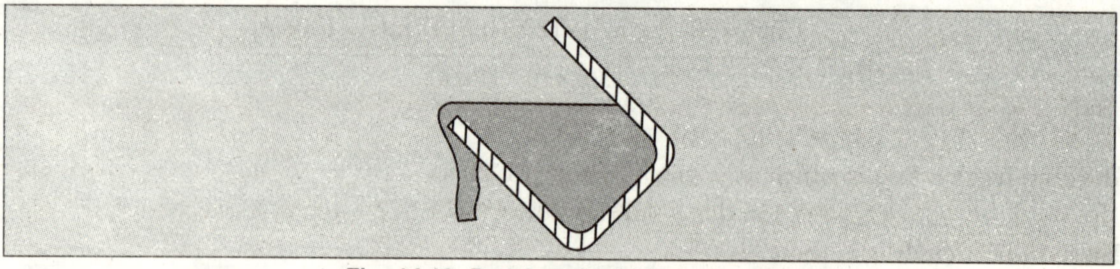

Fig. 14.13. Draining of fluid from a cup.

The explanation of this effect invokes the *entrainment* of fluid by a 2-*D* jet (see Prob. 4.11). As a jet issues from an orifice, it drags the surrounding fluid along from the sides due to viscous action, thereby 'entraining' fluid mass. The presence of a wall does not allow the jet to draw fluid in from that side and it is drawn down on to the wall. The curving fluid stream ensures that the fluid pressure in the trapped eddy is lower than that on the other side and the jet remains attached to the wall. The Coanda effect occurs more readily in a turbulent jet than in a laminar one because the rate of entrainment is higher in the former. Also, the effect is more pronounced in 2-*D* jets than in 3-*D* jets because in the latter, the fluid can leak in from the sides.

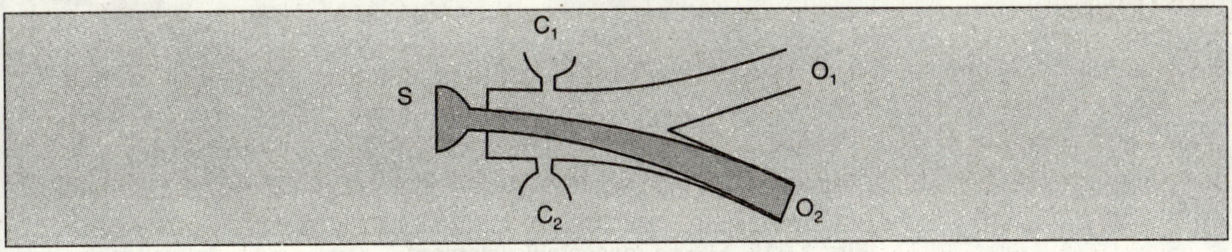

Fig. 14.14. Fluidic bi-stable switch.

The schematic layout of a wall-attachment fluidic amplifier is shown in Fig. 14.14. It consists of a supply port S, two output ports O_1 and O_2, and two control ports C_1 and C_2. The Coanda effect is used to force the power stream issuing from S as a jet to attach to one or the other side of the wall causing the entire stream to exit through the corresponding output port. Figure 14.14 shows the jet attached to the lower wall. When no pressure is applied at C_1 and C_2, the jet remains attached to the lower wall and the output is continuously received at O_2. Now, if a control signal is applied at C_2 in the form of a small flow rate, sufficient to supply the entrainment needs, the jet is pushed to the other wall and the output is received at O_1. The jet remains attached to this wall even when C_2 is withdrawn. To 'switch' back the jet to O_2, one has to apply a signal at C_1. The various possible states of inputs (i.e., control) and outputs are presented as what is known as a *truth-table* in Table 14.2.

Table 14.2. Truth-table for a fluidic bi-stable switch

Control		Output	
C_1	C_2	O_1	O_2
√	0	0	√
0	0	0	√
0	√	√	0
0	0	√	0

Since the output flow rates received at O_1 and O_2 are much larger than (normally 8 to as high as 30 times) the flow rates required at C_1 and C_2, a great deal of amplification is obtained. The device works as a *bi-stable switch* working on one mode until triggered to the other mode. The digital nature of this device permits development of logic circuits for a wide variety of applications. In comparison with other digital amplifiers, the device is characterized with high response speeds—of the order of 10^{-4} s, with pressure recoveries of as high as 80 to 90 per cent of the supply pressure in the power stream.

The basic wall-attachment amplifier can be modified to give various other digital logic devices. By applying a bias signal to control port C_1 (Fig. 14.15), the device is made into a simple switch, that is, a *monostable element*. The device in its stable operation gives a signal at port O_2. The output is switched to O_1 whenever the signal at C_2 exceeds the bias signal at C_1, but reverts to O_2 when C_2 is withdrawn. Its truth table is given as Table 14.3.

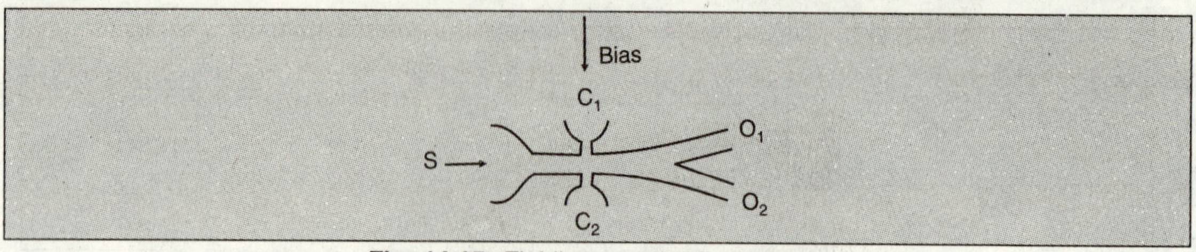

Fig. 14.15. Fluidic monostable element.

Table 14.3. Truth-table for a fluidic mono-stable element

Control		Output	
C_1	C_2	O_1	O_2
√	0	0	√
√	√	√	0
√	0	0	√

A wall attachment *diode* which appears in many fluidic assemblies is shown in Fig. 14.16. When the fluid flows from P_2 to P_1, the fluid stream attaches itself to the walls and enters the reentrant loops, causing interference with the main flow. This results in a much higher resistance to flow when fluid flows from P_2 to P_1 than when it does from P_1 to P_2. Thus, the device works as a diode.

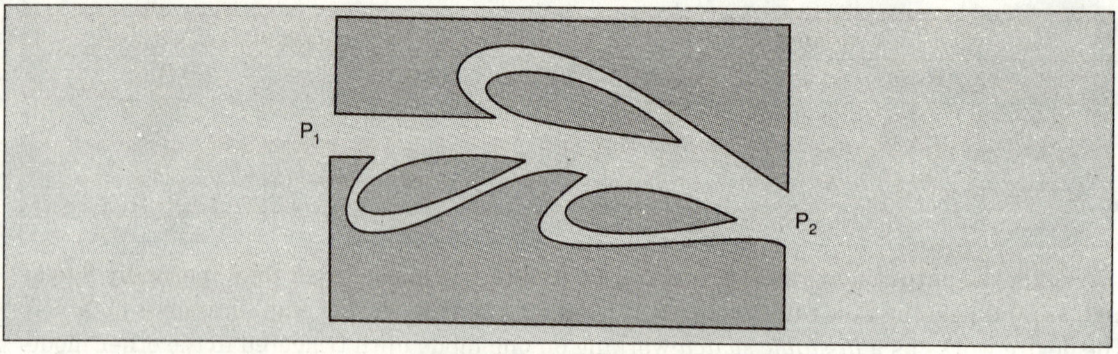

Fig. 14.16. A fluidic diode.

A simple application of a monostable switch to a process liquid-level controller is shown in Fig. 14.17. The asymmetrical construction of the wall attachment device makes the flow attach to the right wall, discharging the liquid into the tank. The open end of a tube attached to the left control-port (the only one in this device) is placed at the desired level of fluid in the tank. As soon as the liquid in the tank reaches this level, the entrainment of atmospheric air through the control tube ceases and the liquid jet switches to the left wall, getting diverted from the tank.

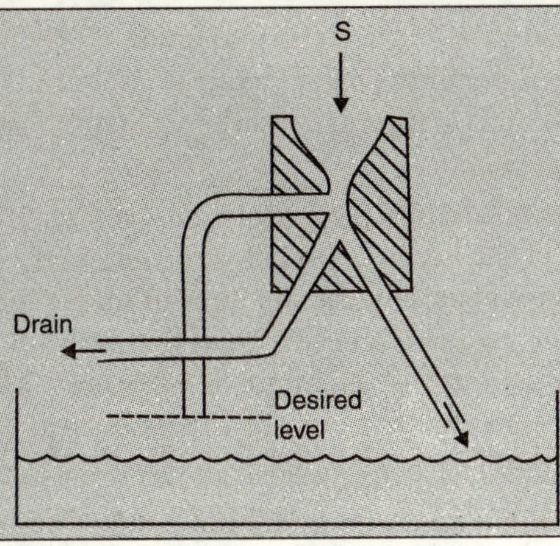

Fig. 14.17. A fluidic level controller.

There are various other digital and proportional fluidic devices using wall attachment or some other fluid-dynamic phenomena. A very cogent description of these is given in Humphrey and Tarumoto (Reading 24).

PROBLEMS

14.1 What happens when an aircraft moving slightly above the stall speed meets (a) a tail wind, and (b) a head wind?

14.2 Discuss why the payload capacity of aircrafts is decreased when operating at high altitudes.

14.3 If the bumble-bee (see Prob. 13.39) can glide at 10 m/s, without flapping its wings, what is the effective lift coefficient of its wings? Is this value typical of lifting surfaces?

14.4 A kite having a chord of 0.5 m, span of 1 m, and mass of 0.2 kg, flies in a wind of 30 kmph under the most efficient flying conditions [$\alpha = 9.5°$, $C_L = 0.51$ and $C_D = 0.08$. See *Science Today.* 16(1), 44 (1982)]. Obtain the tension in the chord and the angle it makes with the vertical.

14.5 A Boeing 727 has a wing span of 45 m and a chord length of 6 m. It weighs 150,000 kg, and is designed to cruise at 260 m/s at an altitude where $\rho_{air} = 0.3$ kg/m³. Compute C_L and the angle of attack, if the characteristics of its wings are as given below

$\alpha°$	C_L	C_D
–4	0	0.014
–2	0.2	0.011
0	0.4	0.013
2	0.6	0.018
4	0.8	0.028
6	0.98	0.040
8	1.13	0.056
10	1.21	0.077
12	1.26	0.103 (stall)

Is this aircraft designed to operate at the maximum C_L/C_D point? Estimate the drag on the wings and the corresponding power required. Note that the actual power required will be much higher because of the various other contributions to the drag. Also estimate the minimum landing speed.

14.6 While turning, an aircraft banks (i.e., tilts about its longitudinal axis) such that the horizontal component of the lift force provides the necessary centripetal force. Write the appropriate force balance equations for the aircraft under such conditions. Compute the angle of attack and the radius of curvature of the path taken by the Boeing 727 (see Prob. 14.5) cruising at 260 m/s if the angle of bank is (a) 10° and (b) 20°. What is the radius of the sharpest turn that the aircraft can execute at this speed without stalling?

14.7 A glider weighing 200 kg has wings whose characteristics are given by Fig. 14.4. Its chord is 1.2 m and its span is 7 m. At what angle to the horizontal should it glide down so that its range (i.e., the horizontal distance covered) is the maximum? What is the speed under such conditions?

14.8 A 4.5 m dia. propeller having the NACA 4412 aerofoil section (with characteristics of Fig. 14.4) is designed to operate at 400 kmph and 188.5 rad/s (i.e., 1800 RPM). Obtain the pitch angle θ at $r = 1$ m and 2 m if $\alpha = 4°$ (giving maximum C_L/C_D) at all sections. Also compute the axial thrust and the torque per unit length at these positions. Assume blade chord is 15 cm at either location.

14.9 A 3.5 m dia. propeller whose characteristics are given by Fig. 14.10 rotates at 1320 RPM. Determine the air-speed, thrust, and power required at the maximum efficiency point.

14.10 An experimental vehicle is driven by a 2 m dia. airscrew having characteristics shown in Fig. 14.10. The thrust required to start the vehicle is computed to be 300 N. At what speed should the airscrew rotate?

14.11 A 2 m dia. propeller whose characteristics are shown in Fig. 14.10 is designed to withstand a maximum thrust of 2000 N. Find the maximum speed of rotation if it is used to drive a vehicle in (a) air, and (b) water. What is the power of the motor required to drive the propeller?

14.12 A propeller rotating at speed ω_1 gives enough thrust to a *blimp* (see Example 9.1) so that it cruises at a steady speed of V_1. If the speed of rotation of the propeller is doubled, it is seen that the cruising speed also doubles. What conclusions can you draw about the drag characteristics of the blimp? How does the power requirement change?

14.13 If the windmill described in Prob. 8.6 had characteristics as given in Fig. 14.10, and if it were operated under conditions such that $V_2 = V_1/3$ (see Prob. 8.6), compute the rate of rotation of the propeller. Can the power output be found using the C_W° curve?

14.14 To reduce the drag encountered in high speed boats, hydrofoils are often used to support the boat clear off water. A boat travelling at 40 kmph uses a completely immersed hydrofoil of the NACA 4412 section (see Fig. 14.4), with platform area of 2 m^2. What is the weight of boat if the hydrofoil is operating at maximum C_L/C_D point? What power is required to overcome the drag encountered by the hydrofoil? If the hydrofoil is at a depth of 1 m below the surface, will cavitation pose any problem at 30°C? Use Fig. 14.11 to estimate Th_{crit}.

14.15 The nose-cap of a torpedo travelling 10 m below the surface of the sea may be assumed to be a hemisphere of 1 m dia. It is assumed that the pressure distribution on the nose can be well approximated by the actual pressure distribution on a sphere. Compute the velocity at which cavitation is likely to occur at 30°C. Use fresh water properties. (The minimum value of the pressure coefficient for a laminar boundary layer is – 0.55 and for a turbulent boundary layer is – 1.05, see Fig. 1.20.)

14.16 The shaft horse-power generated by the engines of the ocean-liner *Queen-Elizabeth* at normal cruising speed (see Prob. 13.5) is 74.3 MW and the efficiency of its propulsion system is 70 per cent. What is the residual drag acting on it?

14.17 A prototype ship having a submerged area of 400 m^2 and a length of 50 m is to be designed after conducting tests on a geometrically similar 1 : 25 model in fresh water. The force required to tow the model at 2 m/s is 20 N. Obtain the drag on the prototype and the velocity to which the model test corresponds.

14.18 By constructing the *truth tables* for the two fluidic devices shown, confirm that they are 'and' and 'or' logic elements.

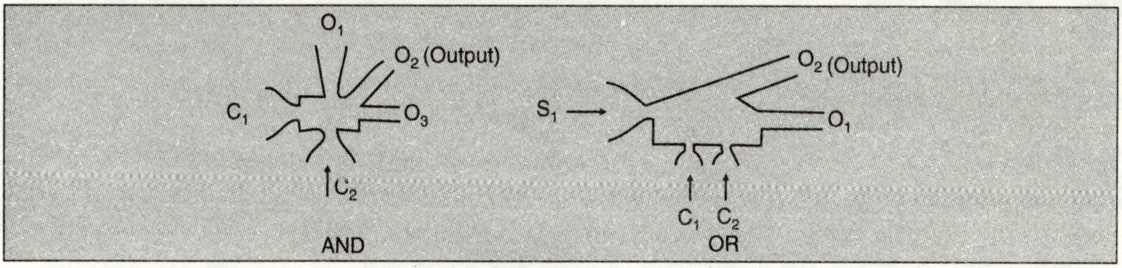

14.19 *Jet deflection proportional amplifier:* The deflection 0 of the main stream of a jet deflection proportional amplifier shown depends on the longitudinal momentum of the main jet and the transverse differential momentum introduced by the two control jets. The deflection θ is reflected in a difference in the mass flow rates at ports O_1 and O_2. Show that for small mass flow rates, $\mathring{m}_{C,1}$ and $\mathring{m}_{C,2}$ the response $\mathring{m}_{O,2} - \mathring{m}_{O,1}$ varies linearly with $\mathring{m}_{C,1} - \mathring{m}_{C,2}$. An actual characteristic is shown. Explain the *saturation*, i.e., the limiting value of the response.

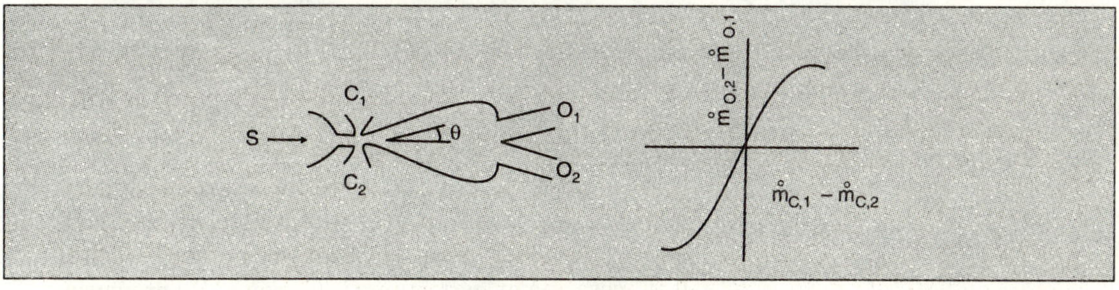

14.20 An application of the monostable switch (see Fig. 14.15) is in an artificial heart as shown. Explain the working of the device [See *Scientific American*, **211**(6), 81 (1964)].

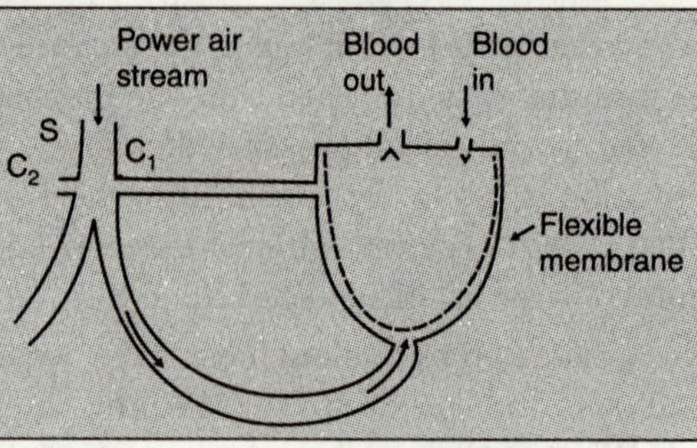

15

EFFECTS OF COMPRESSIBILITY

15.1 INTRODUCTION

So far attention was restricted only to those *flow fields* where the variations $\Delta\rho$ in the density are absent. These are termed as incompressible *flows.* The flow of incompressible *fluids,* i.e., those in which the density does not change with pressure at all, obviously fulfils this condition. But all real fluids are compressible to some extent. Thus, the flow of any real fluid is truly 'incompressible' only in very restrictive circumstances when the pressure field is uniform throughout. Such a flow field is obviously not of much interest. But there exist a wide range of conditions under which $\Delta\rho/\rho$, the fractional change in density is very small compared to unity, and the *flow* may be approximated as incompressible even though the *fluid* itself may be a highly compressible gas. The conditions under which the effects of compressibility are negligible shall be explored now.

Consider the steady flow of an *inviscid* fluid. Let *s* be the coordinate in the streamline direction. The dominant terms of the continuity Eq. (4.18) are then $\rho\partial V/\partial s$ and $V\partial\rho/\partial s$. If the variations in the density are to be negligible,

$$|V\,\partial\rho/\partial s| \ll |\rho\,\partial V/\partial s|$$

or

$$\left|\frac{1}{\rho}\frac{\partial\rho}{\partial s}\right| \ll \left|\frac{1}{V}\frac{\partial V}{\partial s}\right| \qquad \text{...(15.1)}$$

An estimate of $\partial V/\partial s$ is obtained from the Euler Eq. (6.27) for steady inviscid flow. On neglecting the gravity term we get,

$$\rho V\frac{\partial V}{\partial s} = -\frac{\partial p}{\partial s}$$

so that

$$\frac{1}{V}\frac{\partial V}{\partial s} = -\frac{1}{\rho V^2}\frac{\partial p}{\partial s} \qquad ...(15.2)$$

Combining Eqs. (15.1) and (15.2), the following condition is got for the validity of the incompressibility assumption:

$$\left|\frac{1}{\rho}\frac{\partial \rho}{\partial s}\right| << \left|\frac{1}{\rho V^2}\frac{\partial p}{\partial s}\right|$$

or

$$\left|\frac{\Delta\rho}{\rho}\right| << \frac{1}{\rho V^2}|\Delta p| \qquad ...(15.3)$$

The density and pressure variations in a fluid are related through its elasticity as

$$E = \rho\frac{\partial p}{\partial \rho} \qquad ...(15.4)$$

and, therefore, condition 15.3 becomes

$$\frac{1}{E}|\Delta p| << \frac{1}{\rho V^2}|\Delta p|$$

or

$$\frac{V^2}{E/\rho} << 1 \qquad ...(15.5)$$

It will be shown in Sec. 15.2 that $\sqrt{E/\rho}$ measures the speed c of weak waves (e.g., sound waves) in the fluid and, therefore, the incompressibility condition becomes

$$\frac{V^2}{c^2} << 1 \qquad ...(15.6)$$

The ratio of the fluid velocity to the speed of sound is termed the Mach number Ma after Ernst Mach, an Austrian physicist. The value of the speed of sound c in air is around 340 m/s so that a velocity as high as 100 m/s gives $Ma^2 = 0.087$, an order of magnitude smaller than unity. Hence, the flow of air at a velocity as high as 100 m/s (which is higher than the economic velocity for flow of air in pipes) can be treated as incompressible. However, at speeds higher than this, the effects of compressibility have to be accounted for.

15.2 VELOCITY OF WEAK PRESSURE WAVES

Consider a long tube filled with a stationary fluid. Imagine a piston at one end which starts moving at time $t = 0$ with a small velocity δV. If the fluid is truly incompressible, the continuity equation will require that every fluid particle is set in motion instantaneously at time $t = 0$. However, the presence of compressibility, howsoever small, results in a slight compression of fluid in the immediate neighbourhood of the piston, raising its pressure by, say, δp. The increased pressure then acts on the adjacent fluid particles, thereby, accelerating them. These particles get compressed a little. The consequent increase in their pressure acts, in turn, on further particles. This results in a pressure pulse or 'wave' moving down the tube. The fluid particles at a given point are set into motion only when the wave reaches that location.

Let us assume that this wave front travels with a velocity c. The particles 'ahead' of it are stationary and 'behind' it are all moving with the velocity δV of the piston (Fig. 15.1a). One can transform the flow pattern into a steady one by attaching the frame of reference to the wavefront itself. Then the fluid approaches the wavefront at a velocity c from the right and leaves it at a velocity $c - \delta V$, and at an increased pressure $p_0 + \delta p$ (Fig. 15.1 b). Let the densities ahead of and behind the wavefront be ρ_0 and $\rho_0 + \delta\rho$ respectively. Consider a *CV* of infinitesimal length enclosing the wavefront. The continuity and momentum equations applied to this *CV* per unit cross-sectional area give

$$(\rho_0 + \delta\rho)(c - \delta V) - \rho_0 c = 0 \quad ...(15.7)$$

and

$$\delta p = -(\rho_0 + \delta\rho)(c - \delta V)^2 + \rho_0 c^2 \quad ...(15.8)$$

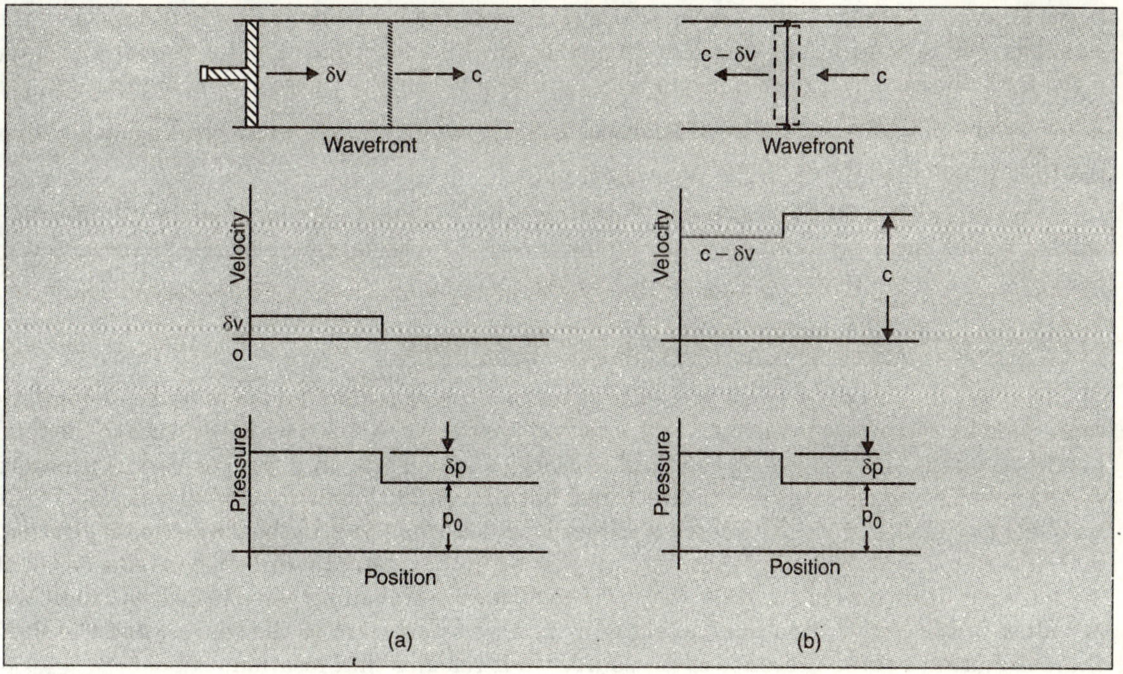

Fig. 15.1. (a) Pressure pulse travelling down a tube. (b) Frame of reference attached to wavefront.

For a weak wave δp, $\delta\rho$ and δV are small and on retaining terms of the first order only

$$c\,\delta\rho - \rho_0\,\delta V = 0$$

and

$$\delta p = 2\rho_0\, c\,\delta V - c^2\,\delta\rho$$

CHAPTER 15

Eliminating δV from these equations

$$\delta p = c^2 \delta \rho$$

or

$$c^2 = \frac{\delta p}{\delta \rho} \qquad ...(15.9)$$

For waves of infinitesimal strength, the compression process is essentially adiabatic and reversible, so that $\delta p/\delta \rho$ is replaced by E_s/ρ, where E_s is the *isentropic compressibility* or *bulk modulus* defined as $\rho(\partial p/\partial \rho)_s$. For ideal gases, it can be shown* that $E_s = \gamma p$, where γ is C_p/C_v, the ratio of the specific heats at constant pressure and constant volume respectively. Thus, the speed of the pressure wave in an ideal gas is given by

$$c^2 = \left(\frac{\partial p}{\partial \rho}\right)_s = \frac{E_s}{\rho} = \frac{\gamma p}{\rho} = \frac{\gamma R^* T}{M} \qquad ...(15.10)$$

where R^* is the universal gas constant (= 8.317 kJ/kg-mol K), and M is the molecular weight of the gas. For air at 20°C, this gives c as 343 m/s.

It should be noted that c is the speed at which a pressure wave travels and is not related to the local flow velocity. Also, the speed c is relative to the local flow, i.e., it is the speed of the signal as seen by a frame of reference attached to the fluid downstream of the wavefront. Thus, if the fluid ahead of the pulse is moving at a uniform velocity V (instead of being at rest), the wave speed in a stationary frame of reference is $V \pm c$, where $c = \sqrt{E_s/\rho}$, depending on whether the fluid is moving into the pulse or away from it.

The expression for the wave speed obtained above is valid only for weak or infinitesimal waves. Waves carrying finite pressure jumps travel at much larger velocities as discussed in Sec. 15.6.

15.3 CONSEQUENCES OF FINITE WAVE SPEED

The fact that the speed c of weak pressure waves in a compressible fluid is finite has important consequences. As mentioned in Sec. 15.2, c represents the wave speed with reference to the fluid particles, and is superposed on the fluid velocity, if any. Thus, in a fluid at rest, a pressure wave travels at equal rates in all directions, but in a fluid that is moving at velocity V, the wave velocity upstream from a source is reduced to $c - V$ while that in the downstream direction is increased to $c + V$. As V increases, the wave speed in the upstream direction is reduced till at $V = c$ (i.e., at a Mach number Ma = V/c = 1) the upstream travelling wave vanishes. Therefore, at values of Ma larger than one, no signal is received upstream of the source, and the fluid there is unaware of the presence of the source till it reaches that location.

To elucidate, imagine a point source of pressure disturbance (such as a small bullet) travelling with velocity V through a medium where the speed of sound is c. The passage of this source tends to push the fluid particles out of the way and creates instantaneous disturbances all along the path it traverses. These disturbances propagate as spherical wavefronts away from

---

* The reader is referred to any text on thermodynamics, e.g., Reynolds, W.C. and Perkins, H.C., *Engineering Thermodynamics*, McGraw-Hill, 1977, 2nd edn.

the point source at velocity c. Let O in Fig. 15.2 represent the instantaneous location of the source, and –1, –2, –3, etc., its location at 1, 2, 3, ... seconds earlier. Pressure pulses are generated at each point and spread out spherically. The pressure pulse generated at point –1 has travelled a distance equal to c in 1 second and therefore at $t = 0$, the corresponding wavefront is a sphere of radius c centred at point –1. Similarly, the pressure pulse generated at point –2 is at a sphere of radius $2c$ centred at point –2. The flow behaviour is markedly different

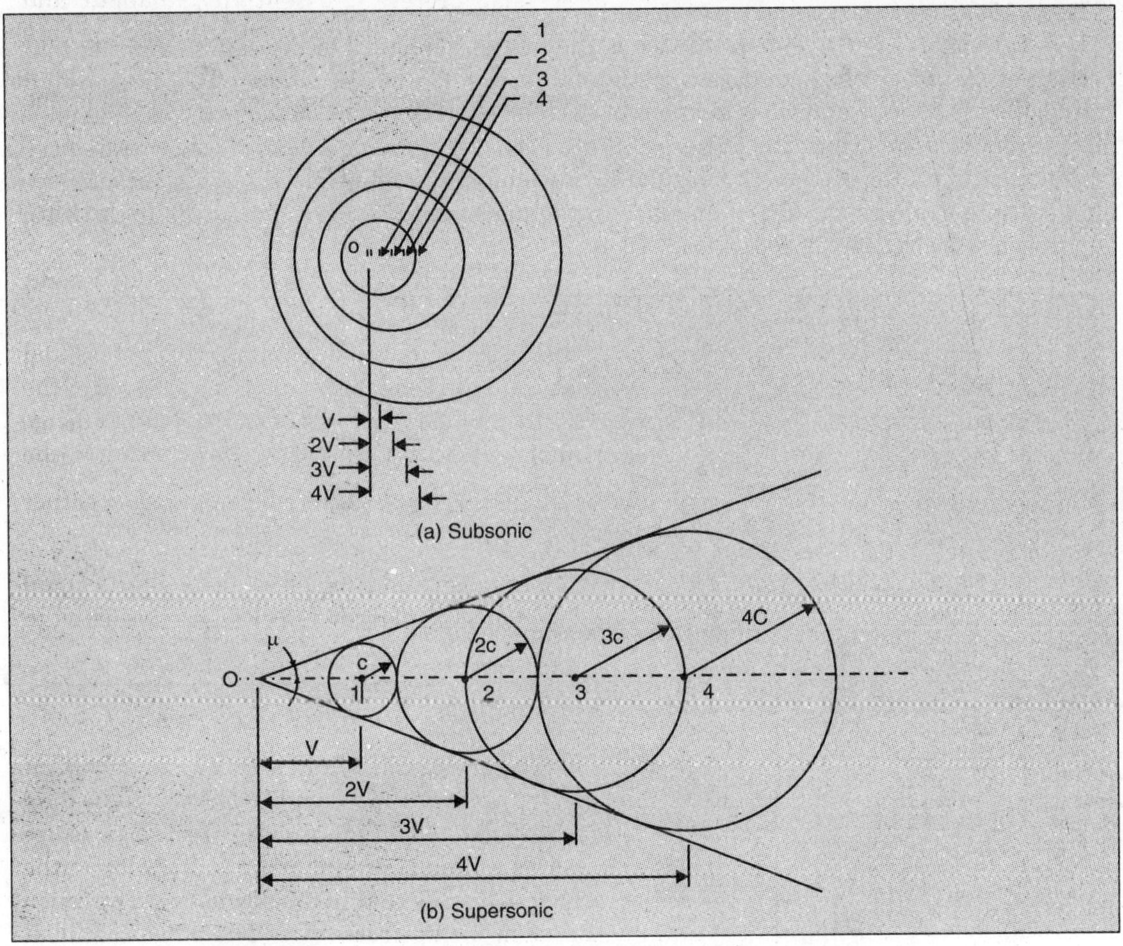

Fig. 15.2. Pressure waves for (a) subsonic, and (b) supersonic cases.

depending on whether the source speed V is smaller or larger than the wave speed c, i.e., depending on whether the Mach number $Ma = V/c$ for the source is lesser or greater than 1. Figures 15.2 (a) and (b) show the typical patterns for these two cases. The presence of the source is felt at all points of the flow field in the subsonic case of $Ma < 1$. This is not so for the supersonic case when $Ma > 1$ and the presence of the source is felt only in a limited region. At $t = 0$, then, the pressure waves are confined in a conical region (Fig. 15.2b). All points outside this region have not received any pulse from the source and are, in effect, unaware of its presence. As such, the conical region where the pulses are confined is termed as the *zone of action,* while the region outside it is termed as the *zone of silence.* This is the reason why we cannot hear a supersonic aircraft till it is well past overhead and we are within its zone of action. The zone of action is also termed as the *Mach cone*. It should be clear from the construction of Fig. 15.2(b) that the half-vertex angle μ of the Mach cone which is termed as the *Mach angle,* is given by

$$\mu = \sin^{-1}\left(\frac{1}{\text{Ma}}\right)$$

Figure 15.3 shows the special case when the source is moving at the speed of sound, i.e., when Ma = 1. In this case the zone of silence is the half-space ahead of the source and the zone of action or the Mach cone comprises of the half space behind the source. The pressure pulse originating from *all* the previous locations of the source intersect now on the plane passing through its current location. Thus, the pulses reinforce each other, giving a zone of concentrated action at the location of the source. This results in much higher pressures at the source location when Ma = 1 and explains the sharp increase in the measured pressures when the flight velocity approaches the velocity of sound (Fig. 15.4).

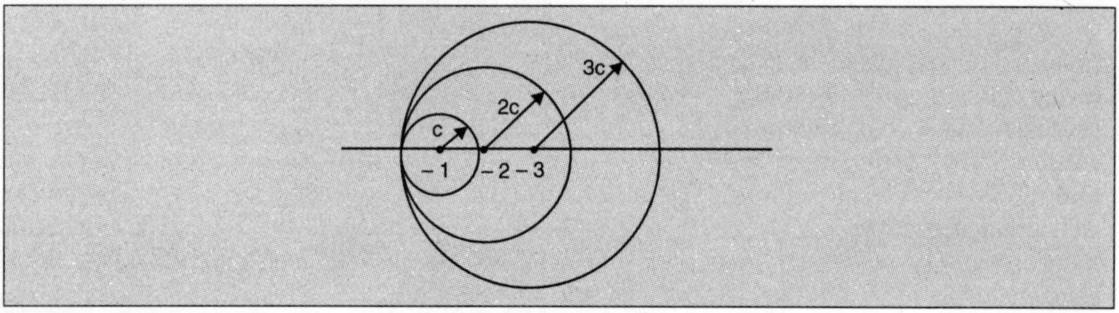

Fig. 15.3. Zone of action for Ma = 1.

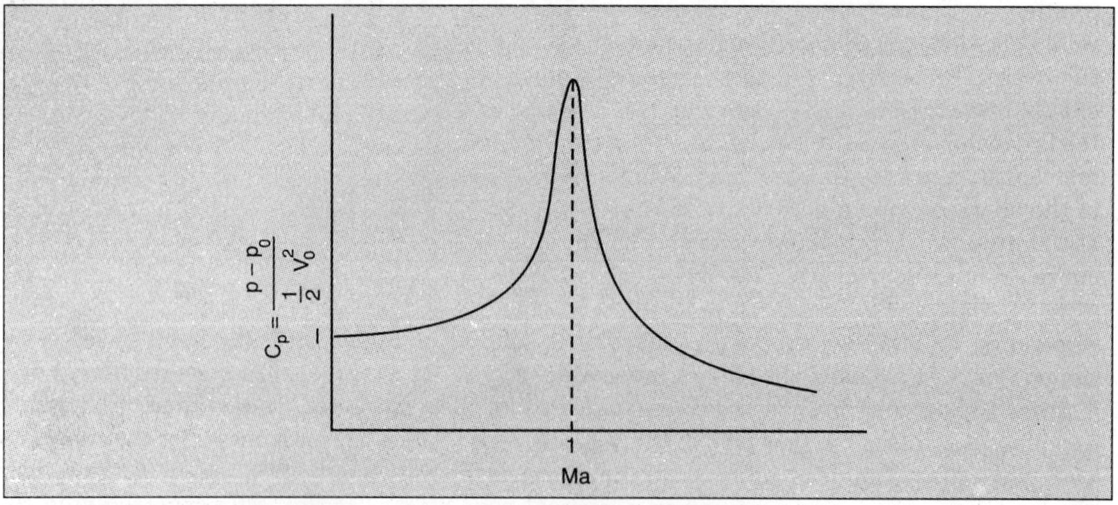

Fig. 15.4. Measured pressure at source location.

The zones of action and silence from a frame of reference attached to the source are shown in Fig. 15.5. The fluid now approaches the source at a supersonic velocity (i.e., at Ma > 1) and, therefore, the wave fronts are 'washed' downstream, giving a Mach cone. The fluid outside of this zone of action is unaware of the presence of the source.

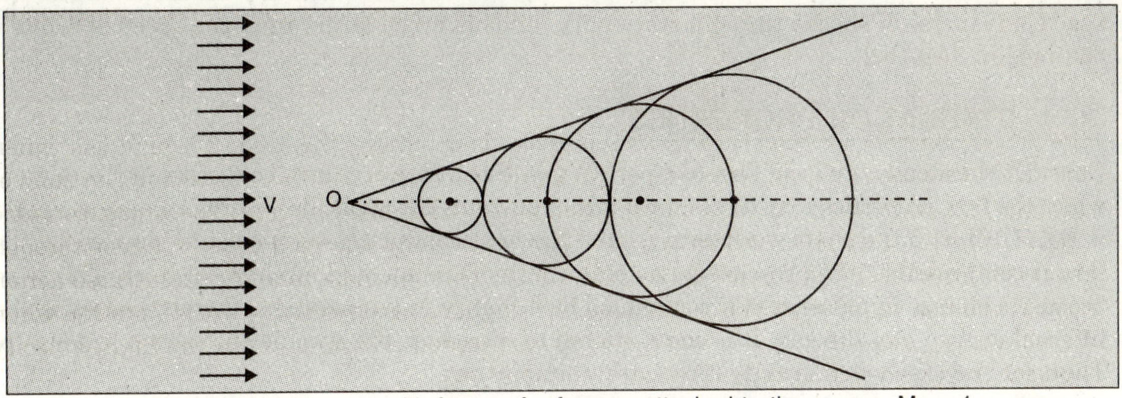

Fig. 15.5. Mach cone with frame of reference attached to the source, Ma > 1.

Even though the above concepts have been developed for motions past an infinitesimal pressure-pulse source, these have practical significance for flow past finite bodies. As the flow meets the body, the pressure at the body surface is different from the ambient pressure and, therefore, the body surface acts as a pressure source (though of finite strength). In the case of subsonic flows, the presence of the body is signalled upstream by the forward travelling waves and the flow adjusts itself for the passage of the body as signified by the curved streamlines well ahead of it (see Fig. 1.7). For supersonic flows, on the other hand, the fluid upstream cannot know of the presence of the body and, therefore, the streamlines are straight upto the Mach cone enclosing it. This creates a complication. The flow velocities at the nose of the body (Fig. 15.6) must be zero since the fluid is brought to rest there. But, since the fluid ahead of the body is supposed to be unaware of its presence, the fluid velocity just upstream of it should be unaffected. This suggests that the fluid is *suddenly* brought to rest at the nose. This solution would not allow the continuity equation to be satisfied. Since the streamlines need to curve around the nose of the body, the fluid there must know of its presence and therefore, the flow there should be subsonic. This is achieved through the presence of a sudden *discontinuity* ahead of the body where the fluid velocity changes abruptly from supersonic to subsonic, while the pressure and density across it increase. Such a discontinuity is termed as a *shock*. Since the flow upstream of the shock is supersonic, the fluid there is unaware of the presence of the body and all the streamlines are straight upto the shock. The streamlines downstream of the shock, however, curve around the body, the flow there being subsonic.

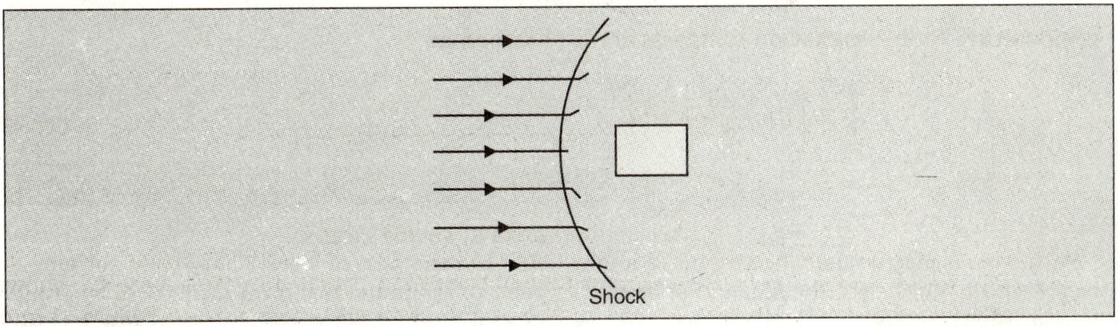

Fig. 15.6. Flow pattern for a finite body for Ma > 1. Reference frame attached to the body.

Shocks are an essential feature of supersonic flow past bodies. An elementary analysis of normal shocks is presented in Sec. 15.6. Since shocks represent finite pressure changes, the speed of traverse of shocks through stationary fluids is much larger than the speed of sound c obtained in Sec. 15.2.

15.4 STAGNATION PROPERTIES

Consider the steady, inviscid flow of a perfect gas. For a *thin* streamtube (across any section of which the flow properties may be assumed fairly constant) the applicable energy balance equation is Eq. (7.10) with the shaft-work term w_s and the heat-conduction term q set as zero. Since the flow is compressible, the pressure p has both thermodynamic and mechanical roles. This is so because a change in pressure is accompanied by a change in temperature, thereby changing the internal energy u of the gas. It is conventional to introduce the specific enthalpy $h = u + p/\rho$. Thus, for the case when gravity terms are unimportant,

$$h + \frac{V^2}{2} = \text{constant along a streamline} \qquad \text{...(15.11)}$$

From thermodynamics, it is known that the enthalpy h is related to the temperature T of a perfect gas by

$$h = C_p T = \frac{\gamma}{\gamma - 1} \frac{R^*}{M} T$$

and, therefore,

$$\frac{\gamma}{\gamma - 1} \frac{R^*}{M} T + \frac{V^2}{2} = \text{constant along a streamline} \qquad \text{...(15.12)}$$

The constant for a streamline can be evaluated in terms of the properties at a stagnation point, i.e., in terms of the properties at a point where the velocity is zero.* If the temperature at such a point, termed as the stagnation temperature, is T_0, then

$$\frac{\gamma}{\gamma - 1} \frac{R^*}{M} T + \frac{V^2}{2} = \frac{\gamma}{\gamma - 1} \frac{R^*}{M} T_0$$

On dividing through by $\frac{\gamma}{\gamma - 1} \frac{R^*}{M} T$ and replacing $\gamma R^* T/M$ by c^2, the ratio of the gas temperature to its stagnation temperature is obtained as

$$\frac{T}{T_0} = \left[1 + \frac{\gamma - 1}{2} \text{Ma}^2\right]^{-1} \qquad \text{...(15.13)}$$

---

* Such a point may actually be present in a flow field (e.g., the nose of a stationary body), or may not be present in which case the stagnation properties refer to the properties if the fluid could be brought to rest isentropically.

where Ma = V/c is the *local* Mach number. The ratios of the density and the pressure at any point to their respective stagnation-point values are then obtained from Eq. (15.13) and the isentropic relations as

$$\frac{p}{p_0} = \left(\frac{T}{T_0}\right)^{\gamma/(\gamma-1)} = \left[1 + \frac{\gamma - 1}{2}\,\text{Ma}^2\right]^{-\gamma/(\gamma-1)} \quad ...(15.14)$$

and

$$\frac{\rho}{\rho_0} = \left(\frac{T}{T_0}\right)^{1/(\gamma-1)} = \left[1 + \frac{\gamma - 1}{2}\,\text{Ma}^2\right]^{-1/(\gamma-1)} \quad ...(15.15)$$

These relations are plotted in Fig. 15.7.

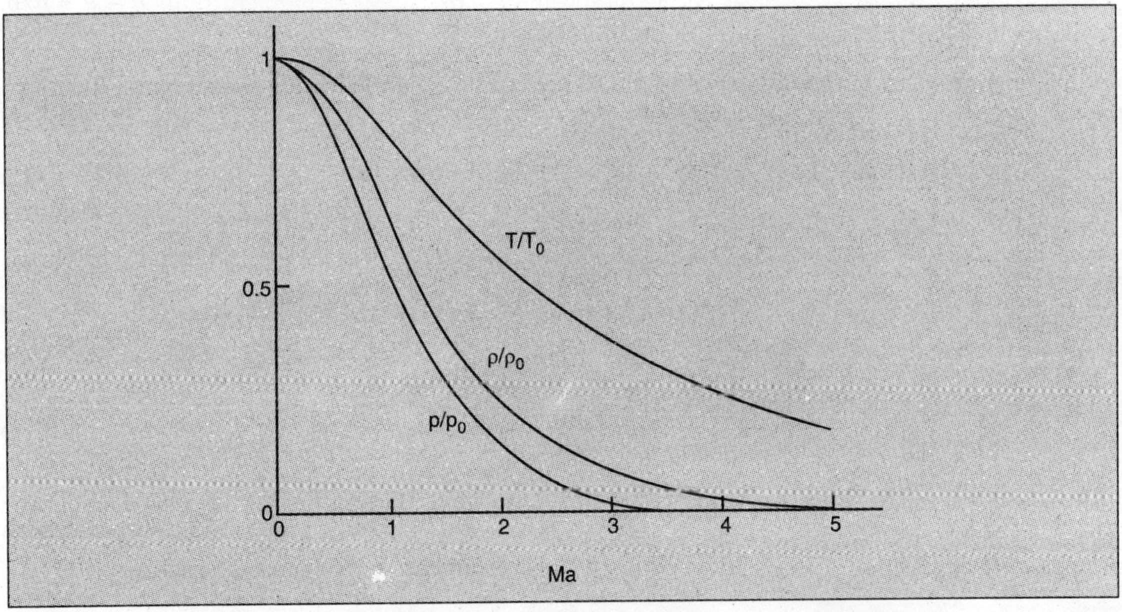

Fig. 15.7. T/T_0, p/p_0 and ρ/ρ_0 as a function of Ma for a perfect gas (γ = 1.4).

Example 15.1. Flow velocity measurements using pitot tubes were discussed in Sec. 8.7 where the Bernoulli equation applicable for incompressible flows was used. Airspeed indicators used in aircrafts are calibrated according to Eq. (8.45) treating the airflow to be incompressible. At higher speeds, compressibility effects are brought into play and the indicated air speed is in error. Obtain the correction factor required when the aircraft Mach number is *less* than one.

The air at the tip of the pitot tube is brought to rest so that the pitot probe measures the stagnation pressure. To make the flow steady, fix the frame of reference with the probe. Air then flows past the probe (Fig. 15.8) at a Mach number Ma and at a pressure equal to the atmospheric pressure p. The stagnation pressure of this air is p_0, the pressure measured by the probe. Thus, Eq. (15.14), with γ = 1.4, gives

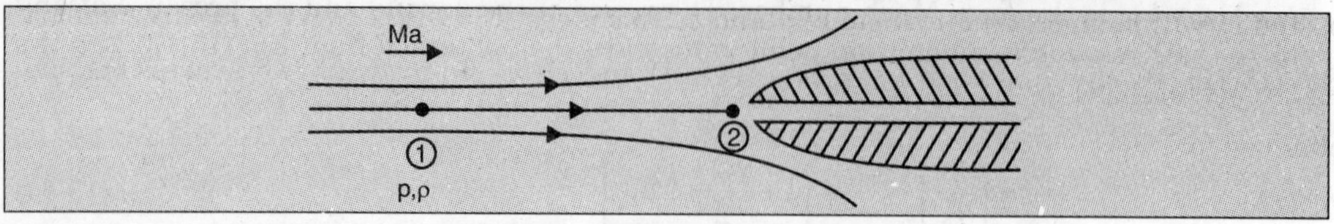

Fig. 15.8. A pitot tube. Frame of reference fixed with the probe.

$$\frac{p}{p_0} = \left[1 + \frac{\gamma - 1}{2}\text{ Ma}^2\right]^{-\gamma/(\gamma-1)} = (1 + 0.2\text{ Ma}^2)^{-3.5}$$

or

$$\frac{p_0 - p}{p} = \left(1 + 0.2\text{ Ma}^2\right)^{3.5} - 1 \qquad \text{(a)}$$

For low enough Mach numbers, the right hand side can be expanded using the binomial theorem to get

$$\frac{p_0 - p}{p} = 0.7\text{ Ma}^2 + 0.175\text{ Ma}^4 + ...$$

where Ma is based on the speed of sound at point 1, i.e., at the pressure p and density ρ. Thus, $\text{Ma}^2 = V^2/c^2 = \rho V^2/\gamma p$.

This gives

$$\frac{p_0 - p}{\frac{1}{2}\rho V^2} = 1 + \frac{1}{4}\text{ Ma}^2 + ...$$

If the indicator is calibrated according to the incompressible Bernoulli equation based on ρ, the atmospheric density, $p_0 - p = \frac{1}{2}\rho V_i^2$, where V_i is the indicated air speed. Therefore,

$$\left(\frac{V_i}{V}\right)^2 = 1 + \frac{1}{4}\text{ Ma}^2 + ...$$

or

$$\frac{V_i}{V} = 1 + \frac{1}{8}\text{ Ma}^2 + ... \qquad \text{(b)}$$

The true air speed is, therefore, less than the indicated air speed but the error introduced is quite small. Thus, for Ma = 0.2, V_i is only 0.5 per cent larger than V.

An interesting point to note here is that Eq. (b) is applicable only for flight speed measurements, since the conditions in the atmosphere are supposed to be known, and these are used for the purpose of defining the Mach number and the density to be used for calibration. If the flow speed in a duct were to be measured where ρ and T (and hence c) are not known, the measurement of these quantities will usually involve bringing the fluid to rest and, thus,

ρ_0 and c_0 will be measured instead of ρ and c. To convert Ma = V/c to V/c_0, use the energy Eq. (15.12). By dividing each term by $c_0^2(\gamma - 1)$ and $c^2/(\gamma - 1)$, in turn, two expressions are obtained for c^2/c_0^2 one in terms of Ma = V/c and the other in terms of V/c_0. Equating these two, we get

$$1+\frac{(\gamma-1)}{2}\mathrm{Ma}^2=\left[1-\frac{\gamma-1}{2}\left(\frac{V}{c_0}\right)^2\right]^{-1}$$

and then Eq. (15.14) gives

$$\frac{p}{p_0}=\left[1-\frac{\gamma-1}{2}\left(\frac{V}{c_0}\right)^2\right]^{\gamma/(\gamma-1)}=\left[1-0.2\left(\frac{V}{c_0}\right)^2\right]^{3.5} \quad \text{for } \gamma = 1.4$$

Proceeding as before, we obtain

$$\frac{p_0-p}{\frac{1}{2}\rho_0 V^2}=1-\frac{1}{8}\left(\frac{V}{c_0}\right)^2+\ldots$$

which states that the true speed is *more* than the speed indicated by the incompressible formula.

15.5 STEADY INVISCID COMPRESSIBLE FLOW IN A CHANNEL OF SLOWLY VARYING CROSS-SECTION

Consider the steady inviscid flow of fluid in a channel whose cross-sectional area varies slowly enough for the flow to be modelled as 1-*D*. The application of the mass balance Eq. (4.5) gives

$$\rho VA = \text{constant} \qquad \ldots(15.16)$$

along the channel. The applicable momentum equation is the Euler Eq. (12.14) which on neglecting the gravity terms gives

$$\rho V\frac{dV}{dx}=-\frac{dp}{dx}$$

or

$$\left(\frac{1}{2}\rho V^2+p\right)=\text{constant} \qquad \ldots(15.17)$$

It is usually more instructive to cast these equations in differential form. By logarithmically differentiating Eq. (15.16), we get

$$\frac{d\rho}{\rho}+\frac{dV}{V}+\frac{dA}{A}=0 \qquad \ldots(15.18)$$

Equation (15.17) gives

$$dp+\rho V\,dV=0 \qquad \ldots(15.19)$$

On eliminating dV one obtains

$$\frac{d\rho}{\rho} - \frac{dp}{\rho V^2} + \frac{dA}{A} = 0$$

Since the flow is inviscid and adiabatic, it is isentropic, and dp and $d\rho$ are related through Eq. (15.9) to c^2. A simple manipulation gives

$$\frac{dp}{\rho V^2} = \frac{1}{1 - \text{Ma}^2}\frac{dA}{A} \qquad \text{...(15.20)}$$

Similarly, one can obtain

$$\frac{dV}{V} = -\frac{1}{1 - \text{Ma}^2}\frac{dA}{A} \qquad \text{...(15.21)}$$

and

$$\frac{d\rho}{\rho} = \frac{\text{Ma}^2}{1 - \text{Ma}^2}\frac{dA}{A} \qquad \text{...(15.22)}$$

The corresponding equations for incompressible flow are obtained from the Bernoulli Eq. (7.15) as

$$\frac{dp}{\rho V^2} = \frac{dA}{A}, \text{ and } \frac{dV}{V} = -\frac{dA}{A}$$

indicating that the pressure increases and the velocity decreases as the area increases.

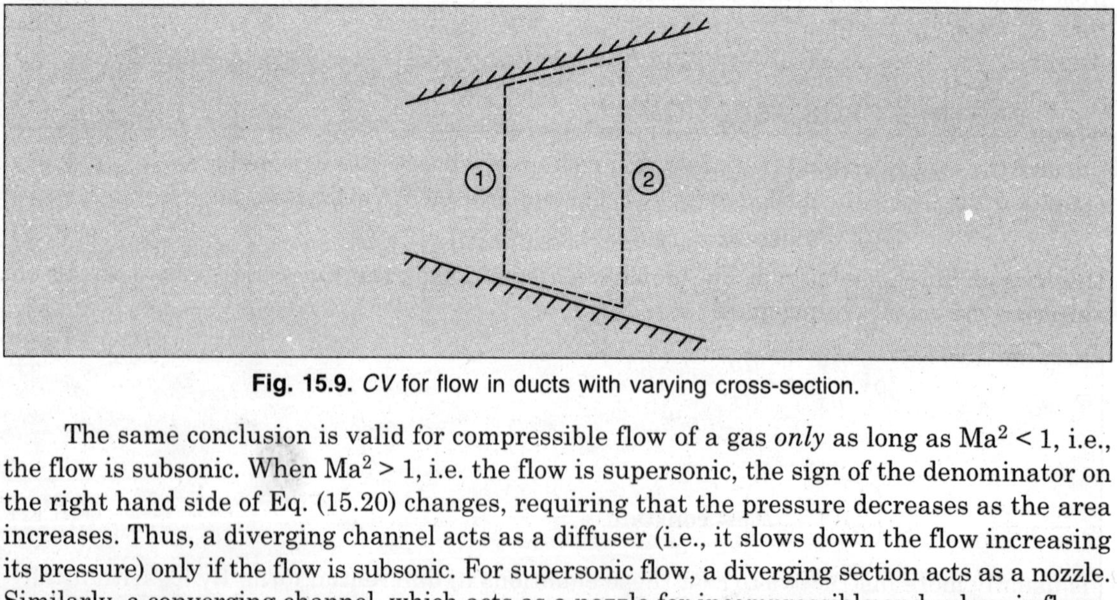

Fig. 15.9. *CV* for flow in ducts with varying cross-section.

The same conclusion is valid for compressible flow of a gas *only* as long as $\text{Ma}^2 < 1$, i.e., the flow is subsonic. When $\text{Ma}^2 > 1$, i.e. the flow is supersonic, the sign of the denominator on the right hand side of Eq. (15.20) changes, requiring that the pressure decreases as the area increases. Thus, a diverging channel acts as a diffuser (i.e., it slows down the flow increasing its pressure) only if the flow is subsonic. For supersonic flow, a diverging section acts as a nozzle. Similarly, a converging channel, which acts as a nozzle for incompressible and subsonic flows, acts as a diffuser for supersonic flows.

It can also be shown quite easily (see Prob. 15.5) that the local Mach number of the flow changes with area according to the following relation

$$\frac{d\text{Ma}}{\text{Ma}} = -\frac{1 + \frac{\gamma - 1}{2}\text{Ma}^2}{1 - \text{Ma}^2}\frac{dA}{A} \qquad \text{...(15.23)}$$

so that for subsonic flows, the Mach number decreases as the area increases while for supersonic flows, it increases with the area. These results are summarized in Table 15.1.

The dramatic difference in the behaviour of subsonic and supersonic flows has important consequences. Some of these are summarized below:

Table 15.1. Changes of p, V, ρ and Ma with area changes

Converging Section $dA < 0$	Diverging Section $dA > 0$
Subsonic flow (Ma < 1)	
$dV > 0$	$dV < 0$
$dp < 0$	$dp > 0$
$d\rho < 0$	$d\rho > 0$
$d\text{Ma} > 0$	$d\text{Ma} < 0$
Supersonic flow (Ma > 1)	
$dV < 0$	$dV > 0$
$dp > 0$	$dp < 0$
$d\rho > 0$	$d\rho < 0$
$d\text{Ma} < 0$	$d\text{Ma} > 0$

(a) As the flow proceeds down a converging channel, its Mach number approaches unity *from below* in the case of subsonic flow, and approaches unity *from above* in the case of supersonic flow. Conversely, for flow down a diverging channel, the Mach number moves away from one, decreasing in the case of subsonic flow and increasing in the case of supersonic flow.

(b) A subsonic flow remains subsonic, and a supersonic flow remains supersonic in a channel whose cross-sectional area is either increasing or decreasing monotonically.

(c) To change a subsonic flow to a supersonic flow or vice-versa, a converging-diverging channel has to be used as shown in curves (c) and (b) in Fig. 15.10, respectively.

(d) Sonic conditions (i.e., Ma = 1) can occur only at the section of minimum area, i.e., at the *throat*. This is seen from Eq. (15.23) in which the denominator on the right hand side vanishes at Ma = 1, requiring that dA/dx be zero. Note that the minimum area alone does not

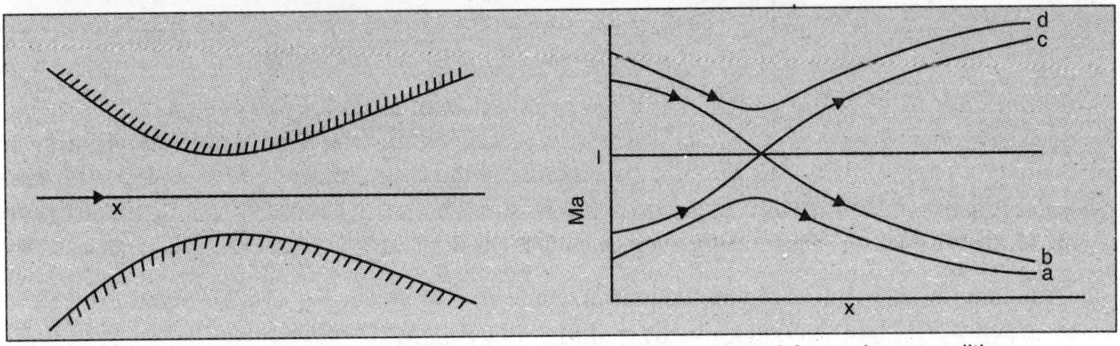

Fig. 15.10. Ma variations along a converging-diverging channel for various conditions.

ensure sonic flow. If the flow upstream of it is subsonic, it may well be the same at the minimum area section as for curve (a) in Fig. 15.10. Eqs. (15.20)–(15.22) then require that dp/dx, $d\rho/dx$ and dV/dx be zero. The downstream section then acts as a diffuser with Ma decreasing from

CHAPTER 15

the already subsonic value. Similarly, if in the converging portion the flow is supersonic, the Ma at the throat may still be larger than unity but then the diverging portion acts like a nozzle, accelerating the flow further into the supersonic regime as for curve (d) in Fig. (15.10).

(e) Consider a converging duct as shown in Fig. 15.11 which accelerates the flow from rest where its density, pressure and temperature are ρ_0, p_0 and T_0 respectively. As the back pressure p_b at the downstream end is decreased, the velocity through the duct increases, thus increasing the mass flow rate. At $p_b/p_0 = 1$ there is no flow through the duct. As p_b/p_0 decreases, the flow rate through the duct increases. The exit pressure p_e equals p_b as for curve (a) in Fig. 15.11, and the Mach number at the exit can be obtained using Eq. (15.14) as

$$\text{Ma}_e = \sqrt{\frac{2}{\gamma-1}\left[\left(\frac{p_b}{p_0}\right)^{-(\gamma-1)/\gamma} - 1\right]} \quad \text{...(15.24)}$$

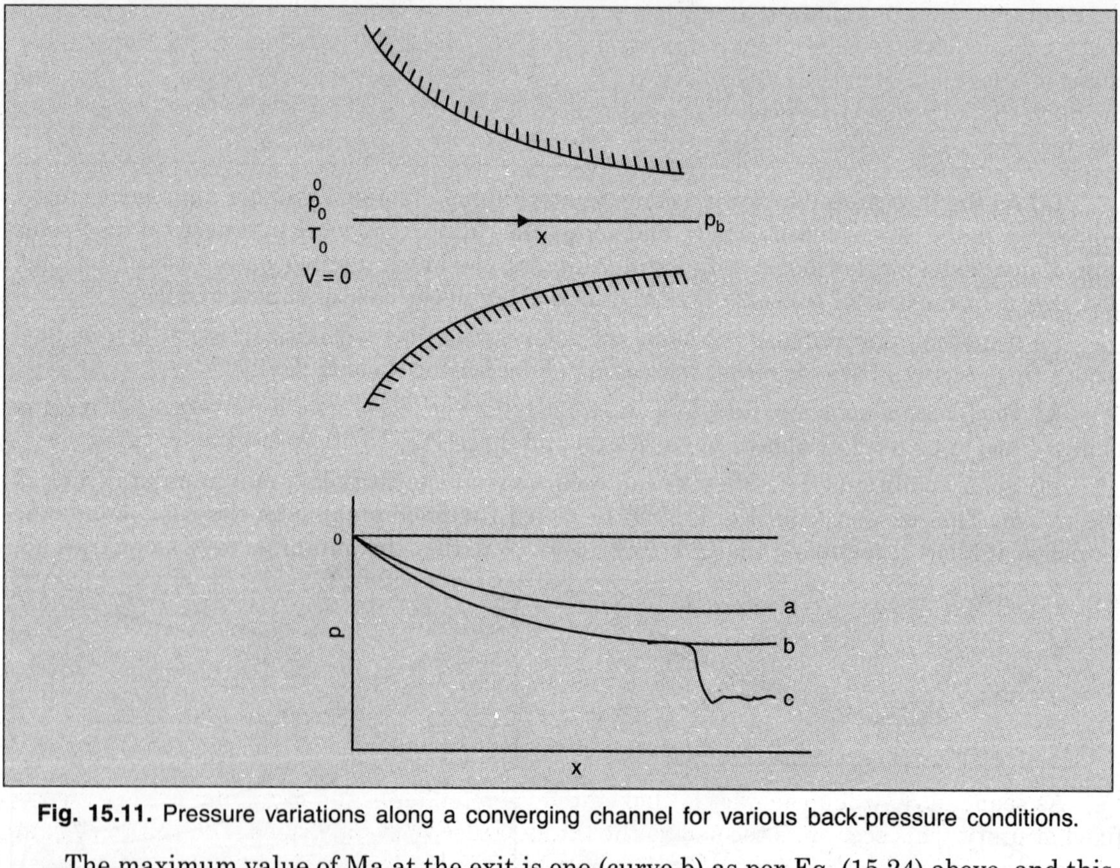

Fig. 15.11. Pressure variations along a converging channel for various back-pressure conditions.

The maximum value of Ma at the exit is one (curve b) as per Eq. (15.24) above, and this occurs when p_b/p_0 has decreased to $[(\gamma + 1)/2]^{-\gamma/(\gamma-1)}$. Any further decrease in the pressure downstream of the exit cannot increase the Ma there beyond one. Thus, for all values of p_b/p_0 less than $[(\gamma + 1)/2]^{-\gamma/(\gamma-1)}$ the values of the Mach number Ma_e and the pressure p_e at the exit are fixed at unity and $p_0\,[(\gamma + 1)/2]^{-\gamma/(\gamma-1)}$ respectively, the decrease of pressure from this value to p_b taking place outside the duct as for curve (c) in Fig. 15.11. Since the flow conditions inside the duct no longer change, the flow carried by it is constant for all values of p_b less than $p_0\ [(\gamma+1)/2]^{-\gamma/(\gamma-1)}$. The convergent duct is said to be *choked* under this condition. Any increase in the flow rate for the given stagnation conditions, can be achieved only by increasing the throat area.

The phenomenon of choking can also be explained in terms of the signal speed. In supersonic flows (including the limiting case of sonic flows) waves cannot travel upstream and so the upstream flow cannot sense any change in the downstream conditions (Sec. 15.3). Thus, once the sonic conditions are attained at the exit of the duct any further changes in the back pressure cannot be propagated upstream and the flow through the duct is unaffected.

A similar phenomenon is observed in the case of a convergent-divergent duct (Fig. 15.12). For a moderate drop in pressure, $p_0 - p_b$, the flow accelerates through the converging portion and decelerates through the diverging portion, remaining subsonic everywhere (see curve (a) in Fig. 15.10). Curves (a) and (b) of Fig. 15.12 show such an operation. The pressure p_e at the exit equals the back pressure p_b. As p_b is lowered further, the flow through the duct increases till the conditions at the minimum area point become sonic. The flow conditions in the convergent portion do not change thereafter. The flow in the divergent portion follows either curve (c) or curve (d), depending on the back pressure. In case (c), the flow remains subsonic throughout the duct, while in case (d), it becomes supersonic in the divergent portion. The flow rate through the duct remains the same in either case. The back pressure conditions between (c) and (d), however, do not permit any isentropic solution. In such a case, a shock is established in the divergent portion of the duct, changing the flow from supersonic to subsonic behind the shock. The curve (f) is an example of such a flow. The reader is referred to specialized books on compressible flow, for example, Shapiro (Reading 31) or Thompson (Reading 34) for further discussion.

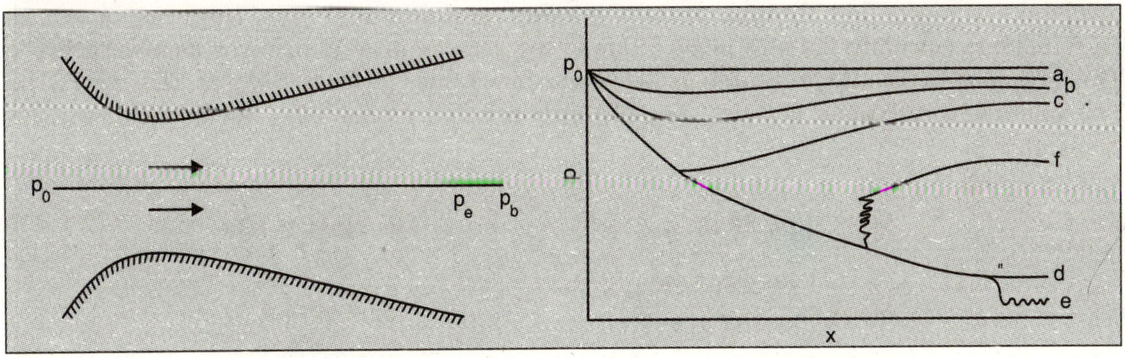

Fig. 15.12. Pressure variation in a converging-diverging nozzle under various back-pressure conditions.

As seen above, when the duct is choked the value of Ma at the minimum-area section is fixed at unity. This section is termed as the *throat* under such conditions. The value of Ma at any other section of area A is then determined completely by the ratio A/A^* where the superscript* denotes the value at the 'throat'. To obtain this relationship, begin with Eq. (15.16) to get

$$\frac{A}{A^*} = \frac{\rho^* V^*}{\rho V}$$

To express ρ^*/ρ and V^*/V as functions of Ma, write

$$\frac{\rho^*}{\rho}=\frac{\rho^*}{\rho_0}\frac{\rho_0}{\rho}$$

Equation (15.15) gives ρ/ρ_0 as well as ρ^*/ρ_0 (when Ma is set equal to 1). Thus,

$$\frac{\rho^*}{\rho}=\left[\frac{2}{\gamma+1}\left\{1+\frac{\gamma-1}{2}\mathrm{Ma}^2\right\}\right]^{1/(\gamma-1)}$$

Similarly,

$$\frac{V^*}{V}=\frac{V^*/c^*}{V/c}\frac{c^*}{c}=\frac{1}{\mathrm{Ma}}\sqrt{\frac{T^*}{T}}=\frac{1}{\mathrm{Ma}}\sqrt{\frac{T^*}{T_0}\frac{T_0}{T}}$$

On using Eq. (15.13) to obtain T^*/T_0 and T/T_0, we get

$$\frac{V^*}{V}=\frac{1}{\mathrm{Ma}}\left[\frac{2}{\gamma+1}\left\{1+\frac{\gamma-1}{2}\mathrm{Ma}^2\right\}\right]^{1/2}$$

and finally

$$\frac{A}{A^*}=\frac{1}{\mathrm{Ma}}\left[\frac{2}{\gamma+1}\left\{1+\frac{\gamma-1}{2}\mathrm{Ma}^2\right\}\right]^{(\gamma+1)/2\,(\gamma-1)} \quad ...(15.25)$$

This equation, along with Eqs. (15.13–15.15) for T/T_0, p/p_0 and ρ/ρ_0, allow us to solve any problem involving isentropic flow through ducts of varying cross-sectional area. However, inverting Eq. (15.25) to calculate Ma at a given value of the area ratio A/A^* involves messy algebra*. This inversion is usually presented as isentropic flow tables [refer to Shapiro, (Reading 31)]. The following correlations, however, provide fairly accurate estimates of Ma for the flow of air ($\gamma = 1.4$):

Subsonic flow

$$\mathrm{Ma} = 1 - 0.88\,(\ln A/A^*)^{0.45} \qquad \text{for } A/A^* < 1.34$$

$$\mathrm{Ma}=\frac{1+0.27(A/A^*)^{-2}}{1.73(A/A^*)} \qquad \text{for } A/A^* > 1.34$$

Supersonic flow

$$\mathrm{Ma} = 1 + 1.2(A/A^* - 1)^{1/2} \qquad \text{for } A/A^* < 2.9$$

$$\mathrm{Ma} = [216\, A/A^* - 254\,(A/A^*)^{2/3}]^{1/5} \qquad \text{for } A/A^* > 2.9 \quad ...(15.26)$$

Equations (15.13–15.15) simplify for air (i.e., $\gamma = 1.4$) to

$$\frac{T}{T_0}=\frac{1}{1+0.2\,\mathrm{Ma}^2} \quad ...(15.27)$$

---

* It may be noted that two values of Ma, one subsonic and the other supersonic, are possible for any value of A/A^*.

$$\frac{p}{p_0} = \frac{1}{\left(1+0.2\ \text{Ma}^2\right)^{3.5}} \qquad ...(15.28)$$

and

$$\frac{\rho}{\rho_0} = \frac{1}{\left(1+0.2\ \text{Ma}^2\right)^{2.5}} \qquad ...(15.29)$$

or, inversely,

$$\text{Ma}^2 = 5\left[\frac{T_0}{T} - 1\right] = 5\left[\left(\frac{p_0}{p}\right)^{2/7} - 1\right] = 5\left[\left(\frac{\rho_0}{\rho}\right)^{2/5} - 1\right] \qquad ...(15.30)$$

When the duct is not choked, i.e., the flow rate is less than the maximum possible through the given duct (for the given stagnation conditions) the minimum-area section is not termed as the throat. However, it is possible to *imagine* a throat section where the sonic conditions will be achieved if the duct were extended to it. The conditions at this hypothetical throat can be used as the reference conditions in Eqs. (15.25) and (15.26).

The following examples illustrate the use of Eqs. (15.26–15.30) for some simple problems of isentropic flow of air through ducts.

Example 15.2. A supersonic wind-tunnel uses a convergent-divergent nozzle to accelerate air contained in pressure tanks at a stagnation pressure and temperature of 8×10^5 Pa and 303 K, to a Mach number of 2. If the test-section area is 0.1 m^2, find the throat area required. Also determine the flow rate through the nozzle.

Assuming the flow to be isentropic, the throat area can be obtained from Eq. (15.25). For $A = 0.1$ m^2, Ma = 2, and $\gamma = 1.4$, it gives

$$\frac{0.1\left(\text{m}^2\right)}{A^*} = \frac{1}{2}\left[\frac{2}{2.4}\left\{1+0.2\times 2^2\right\}\right]^{\frac{2.4}{2\times 0.4}}$$

or

$$A^* = 0.059\ \text{m}^2$$

To calculate the flow rate, first determine the conditions at the throat. The pressure p^* and temperature T^* at the throat are obtained from Eq. (15.14) and Eq. (15.13) respectively, as

$$p^* = 8 \times 10^5\ (\text{Pa}) \times (1 + 0.2 \times 1)^{-3.5} = 4.23 \times 10^5\ \text{Pa}$$

and

$$T^* = 300\ (\text{K}) \times (1 + 0.2 \times 1)^{-1} = 250\ \text{K}$$

The corresponding ρ^* and V^* are then

$$\rho^* = \frac{p^* M}{R^* T} = \frac{4.23\times 10^5\,(\text{Pa})\times 28.98(\text{kg/kg-mol})}{8314.3\left(\dfrac{\text{J}}{\text{kg-mol K}}\right)\times 250(\text{K})}$$

$$= 5.898\ \text{kg/m}^3$$

and

$$V^* = c^* = \sqrt{\frac{\gamma R^* T^*}{M}} = \sqrt{\frac{1.4 \times 8314.3 \left(\frac{\text{J}}{\text{kg-mol K}}\right) \times 250(\text{K})}{28.98 \left(\frac{\text{kg}}{\text{kg-mol}}\right)}}$$

$$= 316.9 \text{ m/s}$$

Therefore,

$$\overset{\circ}{m} = \rho^* V^* A^* = 5.898 \left(\frac{\text{kg}}{\text{m}^3}\right) \times 316.9 \left(\frac{\text{m}}{\text{s}}\right) \times 0.059 \ (\text{m}^2)$$

$$= 110.3 \text{ kg/s}$$

Example 15.3. Find the thrust developed by the solid propellant rocket shown in Fig. 15.13. The pressure and temperature in the combustion chamber are given as p_0 and T_0 respectively.

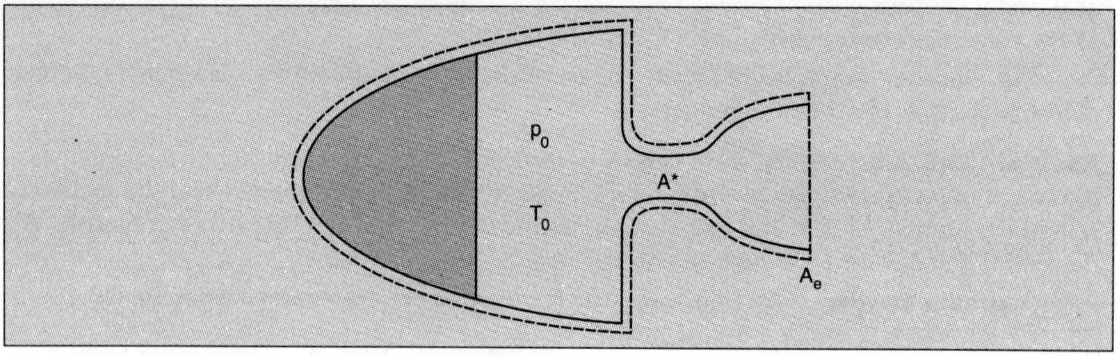

Fig. 15.13. A solid propellant rocket and the *CV* to be used in Example 15.3.

Application of the momentum theorem Eq. (5.2) to the *CV* shown gives the net thrust F_t developed by a stationary rocket, or by one moving at a constant velocity, as

$$F_t = -\overset{\circ}{m}\ V_e - (p_e - p_a)\ A_e$$

where $\overset{\circ}{m}$ is the mass flux, V_e and p_e are the velocity and pressure respectively at the exit plane of the nozzle, and p_a is the ambient pressure. Note the possibility of p_e being different from p_a. Since $\overset{\circ}{m} = \rho_e V_e A_e$ and $\rho_e = M p_e / R^* T_e = \gamma p_e / c_e^2$, $\overset{\circ}{m}\ V_e$ becomes $\gamma p_e \ \text{Ma}_e^2 A_e$, so that

$$|F_t| = A_e p_e \left[\gamma \ \text{Ma}_e^2 + 1 - \frac{p_a}{p_e}\right]$$

For a rocket motor operating in vacuum, $p_a = 0$. In addition, for ordinary rockets, $\gamma \text{Ma}_e^2 >> 1$, so that

$$|F_t| = A_e p_e \gamma \ \text{Ma}_e^2$$

It is conventional to write this result in terms of the throat area A^* and the combustion chamber pressure p_0, which is the stagnation pressure. Using Eqs. (15.25) and (15.14) to obtain A_e/A^* and p_e/p_0, we get

$$|F_t| = kA^* p_0$$

where

$$k = \frac{\gamma}{\left(\dfrac{\gamma+1}{2}\right)^{(\gamma+1)/(2\gamma-2)}} \mathrm{Ma}_e \left[1+\frac{\gamma-1}{2}\mathrm{Ma}_e^2\right]^{-1/2}$$

Ma_e is a function of the ratio A_e/A^*, and so k is a function of A_e/A^* and of γ of the products of combustion.

It is conventional to rate rocket motors by their *specific impulse I* defined as the thrust produced per unit rate of the fuel consumed. Then, for negligible exit pressure,

$$I = V_e$$

The velocity V_e is related to Ma_e through the speed of sound at the exit which in turn, is related to T_e. The exit temperature is determined by T_0 and Ma_e through Eq. (15.13), and so

$$I = \sqrt{\frac{\gamma R^* T_0}{M}} \frac{\mathrm{Ma}_e}{\left(1+\dfrac{\gamma-1}{2}\mathrm{Ma}_e^2\right)^{1/2}}$$

Since Ma_e depends solely on the ratio A_e/A^* (see Eq. 15.25), I depends solely on the combustion chamber temperature T_0.

15.6 NORMAL SHOCK

The fact that pressure waves cannot travel upstream in supersonic flow precludes the possibility of a continuous flow in the case of supersonic flights of objects (Sec. 15.3). Instead, a discontinuity occurs in the flow field where the flow suddenly 'jumps' from supersonic to subsonic conditions. Such a discontinuity is termed as a *shock*. Similarly, it was seen in Sec. 15.5 that in the case of flow through a convergent-divergent duct, a continuous isentropic flow cannot occur for certain back-pressure conditions and a shock is established in the diverging portion.

Physically, a shock is a thin region (of thickness of the order of the mean free-path) where the velocity gradients are very large, leading to the presence of significant viscous effects. Thus, the flow is no longer isentropic and Eqs. (15.13 – 15.16) do not apply *across* a shock. However, since the flow on either side of the shock may be assumed isentropic, these equations are valid if attention is restricted to *any one side* of the shock.

In this section, the simplest type of shock, namely, a stationary normal shock is analysed. A moving shock can be made stationary by attaching the frame of reference to it. Consider a thin *CV* enclosing the shock shown in Fig. 15.14. Let 1 refer to the upstream side of the flow and 2 the downstream side. The continuity equation for the *CV* gives

$$\rho_1 V_1 = \rho_2 V_2 \qquad ...(15.31)$$

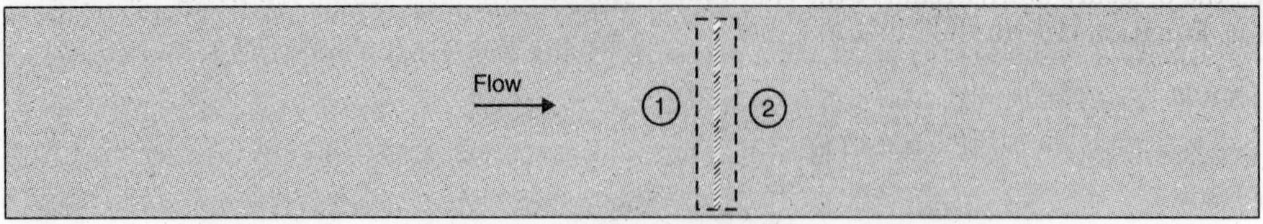

Fig. 15.14. *CV* enclosing a normal shock.

Since the shock is very thin, the application of the momentum and energy equations gives

$$p_1 + \rho_1 V_1^2 = p_2 + \rho_2 V_2^2 \quad \text{...(15.32)}$$

and

$$\frac{\gamma}{\gamma - 1} \frac{R^*}{M} T_1 + \frac{V_1^2}{2} = \frac{\gamma}{\gamma - 1} \frac{R^*}{M} T_2 + \frac{V_2^2}{2} \quad \text{...(15.33)}$$

or

$$C_p T_1 + \frac{V_1^2}{2} = C_p T_2 + \frac{V_2^2}{2} \quad \text{...(15.34)}$$

since $C_p = R^*\gamma/(\gamma - 1) M$. Assume that the conditions on side 1 are known while those on side 2 are to be determined. Since there are four unknowns, one more relation is needed and this is provided by the perfect gas law

$$\frac{p_1}{\rho_1 T_1} = \frac{p_2}{\rho_2 T_2} \quad \text{...(15.35)}$$

On eliminating ρ_2, V_2, and the temperatures T_1 and T_2 from these equations

$$\frac{p_2}{p_1} = \frac{1}{\gamma + 1}\left[2\gamma \text{ Ma}_1^2 - (\gamma - 1)\right] \quad \text{...(15.36)}$$

where $\text{Ma}_1 = V_1/(\gamma R^* T_1/M)^{1/2}$. Similarly,

$$\frac{\rho_2}{\rho_1} = \frac{V_1}{V_2} = \frac{(\gamma + 1) \text{ Ma}_1^2}{(\gamma - 1) \text{ Ma}_1^2 + 2} \quad \text{...(15.37)}$$

The pressure ratio p_2/p_1 can be taken as a measure of the shock strength and, as seen from Eq. (15.36), it is uniquely determined by Ma_1 and γ.

To obtain the Mach number Ma_2 behind the shock, note that Eqs. (15.31–15.35) are symmetrical with respect to points 1 and 2 and, therefore, Eq. (15.36) which is obtained from these, should also be symmetrical. Thus

$$\frac{p_1}{p_2} = \frac{1}{\gamma + 1}\left[2 \gamma \text{ Ma}_2^2 - (\gamma - 1)\right]$$

Equating this ratio of p_1/p_2 with that from Eq. (15.36) the relation between Ma_1 and Ma_2 is obtained as

$$\text{Ma}_2^2 = \frac{(\gamma - 1) \text{ Ma}_1^2 + 2}{2\gamma \text{ Ma}_1^2 - (\gamma - 1)} \quad \text{...(15.38)}$$

Since a shock represents a region of concentrated viscous action, the entropy across it must increase. Further, the second law of thermodynamics implies that p_2 should be greater than p_1, i.e., the pressure downstream of the shock must *always* be greater than the pressure upstream of it. Equation (15.36) then requires that

$$\frac{2\gamma\, \text{Ma}_1^2 - (\gamma - 1)}{\gamma + 1} > 1$$

or

$$\text{Ma}_1 > 1$$

Equation (15.38) then gives $\text{Ma}_2 < 1$.

To summarize, then,

(a) A shock represents a sudden increase in pressure.

(b) The flow upstream of a *normal* shock is always supersonic, while that downstream is always subsonic. Thus, a shock slows down the fluid, increasing its pressure and density.

(c) The shock strength is determined uniquely by the upstream Mach number Ma_1. The larger the upstream supersonic Mach number Ma_1, the greater is the shock strength and the lower is the downstream subsonic Mach number Ma_2.

(d) Even though the total energy ahead of a shock and behind it are constant, a shock, representing concentrated viscous action, converts mechanical energy into thermal energy. Thus, while the stagnation temperature does not change across the shock (the total energy being the same), the stagnation pressure decreases (see Prob. 15.10).

(e) If a frame of reference is attached with the upstream flow, then the shock moves into the stationary fluid at a Mach number equal to Ma_1 which is greater than one. This implies that the speed of a shock wave, which is a *finite* pressure pulse, is larger than *c*, the speed of a weak or infinitesimal pressure pulse. Moreover, the stronger the pulse (i.e., the larger the value of p_2/p_1) the greater is Ma_1 and the higher is its speed.

Shocks are not always normal to the flow direction. When a shock is inclined to the incoming stream, it is termed as an *oblique shock*. It can be shown that in such a case it is only the normal component of velocity whose value jumps across the shock, the tangential component remaining unchanged (Fig. 15.15). The jump in the normal velocity component from $V_{1,n}$ to $V_{2,n}$ is governed by the normal–shock relations, so that Eqs. (15.36–15.38) all apply with Ma_1 changed to $\text{Ma}_{1,n}$ defined as $V_{1,n}/c_1$ and $\text{Ma}_{2,n} = V_{2,n}/c_2$. Conclusions (*a*)-(*d*) above still apply, except

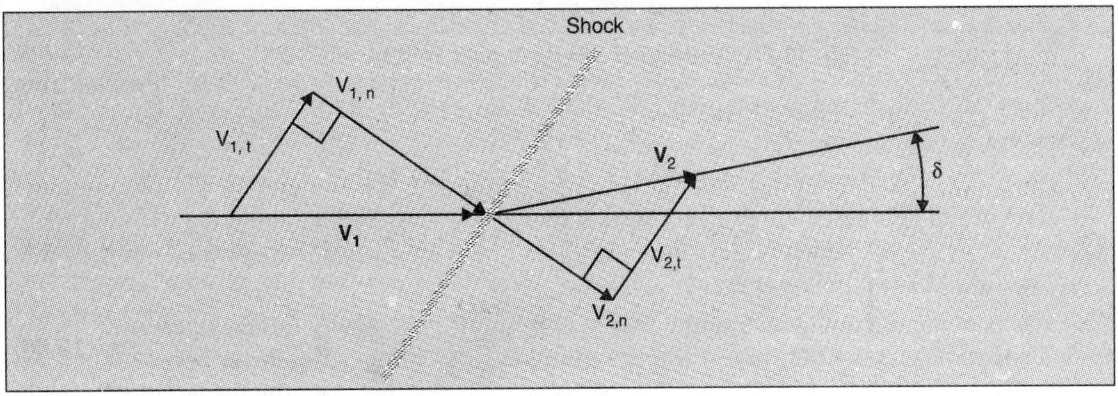

Fig. 15.15. Velocity components in an oblique shock.

that $Ma_{2,n}$ instead of Ma_2 is now required to be subsonic. Thus, $V_{2,n}$ is less than c_2 but since $V_{2,t}$ is unchanged, V_2 itself may or may not be less than c_2 and the downstream flow may still be supersonic. When Ma_2 is subsonic, the shock is termed as a *strong shock* but when Ma_2 is supersonic (note that Ma_1 is always supersonic) the shock is termed as a *weak shock*.

It may be noted that the velocity vector $\mathbf{V}_2$ is at an angle δ to $\mathbf{V}_1$ (Fig. 15.15). Thus, the flow deflects on passing through an oblique shock. Conversely, whenever a supersonic flow has to turn suddenly through a small (positive) angle δ, an oblique shock is established at the corner (Fig. 15.16).

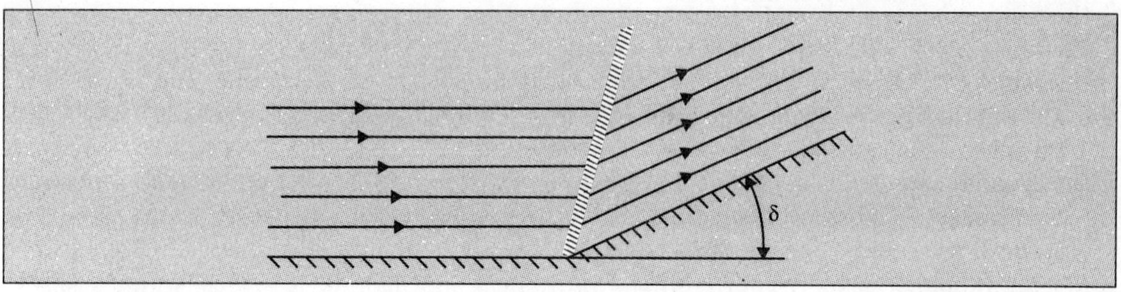

Fig. 15.16. Formation of an oblique shock when a supersonic flow turns at a sharp corner.

Since shock waves represent a conversion of mechanical energy into thermal energy, one should always endeavour to avoid them or to make them as weak as possible. Thus, oblique shocks are preferred over normal shocks since they are weaker.

Example 15.4. A convergent-divergent nozzle (Fig. 15.17) of a supersonic wind tunnel has $A_t/A^* = 2$, where A_t is the test-section area. What Mach number is it designed for? In one application, the downstream conditions are not properly adjusted so that a shock stands in the nozzle where $A/A^* = 1.5$. What is the Mach number in the test section under these circumstances?

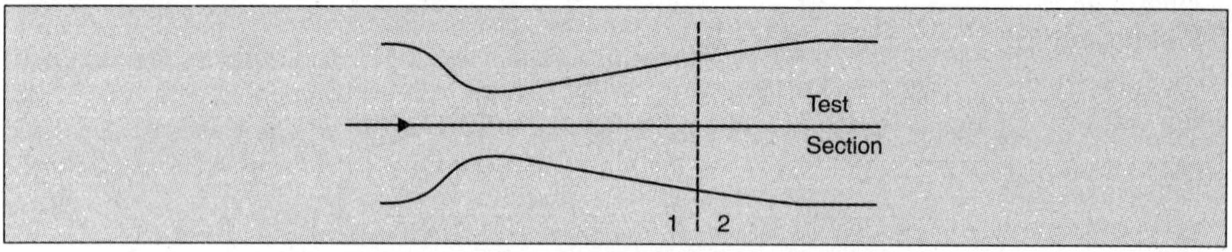

Fig. 15.17. Convergent-divergent duct for Example 15.4.

Since air is used, the approximate relations of Eq. (15.26) can be employed. For $A_t/A^* = 2$, we obtain

$$Ma_t = 1 + 1.2\,(2-1)^{1/2} = 2.2$$

The accuracy of this estimate can be checked by substituting this value in Eq. (15.25), which yields $A_t/A^* = 2.0050$, which is close enough. Thus, the wind tunnel test section is designed for a test-section Mach number of 2.2.

When a shock stands at a point where $A/A^* = 1.5$, the Mach number across it 'jumps' from a supersonic value Ma_1 to a subsonic value Ma_2, the value of the latter being determined by Ma_1. The value of Ma_1 is determined by the isentropic flow between the throat and that point (which corresponds to an area ratio $A_1/A^* = 1.5$). Equation (15.26) gives

$$\text{Ma}_1 = 1 + 1.2\,(1.5 - 1)^{1/2} = 1.85$$

The value of Ma_2 is then obtained from Eq. (15.38) as

$$\text{Ma}_2^2 = \frac{(1.4-1)(1.85)^2 + 2}{2\times 1.4\times (1.85)^2 - 0.4} = 0.367$$

or $\text{Ma}_2 = 0.60$, which is the Mach number just behind the shock. The flow then slows down through the divergent passage isentropically. Before calculating Ma_t' the new Mach number in the test section, note that Ma = 0.60 corresponds to an A/A^* of 1.19 only as given by Eq. (15.26). This means that if the flow behind the shock is passed through a convergent section isentropically, the sonic or throat conditions will be arrived at an area which is 1/1.19 of the area at the location of the shock. This is much larger than the minimum area in the actual nozzle (which is only 1/1.5 of the area at the shock location). The flow after the shock, therefore, cannot pass through the original throat.

The ratio of the test-section area to the required new (hypothetical) throat area is thus seen to be 1.19 × 2/1.5 or 1.59. When this ratio is used in Eq. (15.26) it yields a new test-section Mach number of

$$\text{Ma}_t' = \frac{1 + 0.27\times(1.59)^{-2}}{1.73\times 1.59} = 0.40$$

The presence of a shock at the given location in the divergent section reduces the test-section Mach number drastically from the design value of 2.2 to 0.4.

Example 15.5. The use of a pitot tube in subsonic flow was discussed in Example 15.2. The formulation of that example cannot be used for supersonic flows because a shock must stand between the supersonic region and the mouth of the tube where the flow is brought to rest. Obtain an expression for the Mach number of the flow in terms of the measured pitot pressure and the static pressure of the free stream.

The required expression can be obtained if it is assumed that the stagnation streamline is normal to the shock (Fig. 15.18), and that the fluid is brought to rest isentropically behind

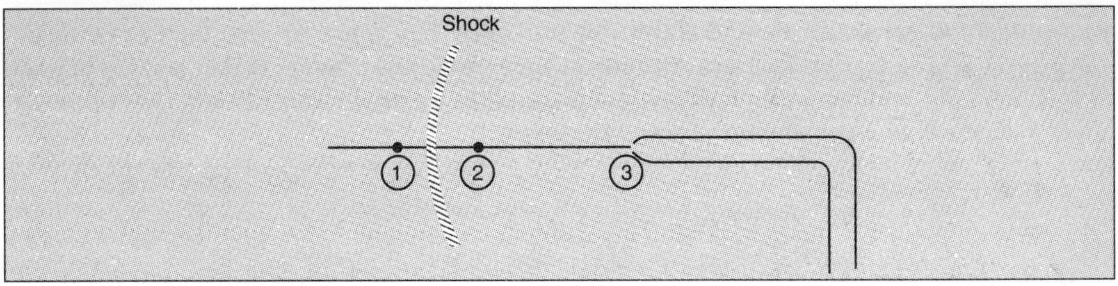

Fig. 15.18. Shock ahead of a pitot tube in a supersonic flow.

the shock from point 2 to point 3. Thus, the pressure measured by the pitot tube is $p_3 = p_{0,2}$, the stagnation pressure behind the shock. Therefore,

$$\frac{p_{0,2}}{p_1} = \frac{p_{0,2}}{p_2}\,\frac{p_2}{p_1}$$

The ratio p_2/p_1 is related to Ma_1 by Eq. (15.36). The ratio $p_2/p_{0,2}$ can be obtained in terms of Ma_1 by first determining Ma_2 using Eq. (15.14) and then using Eq. (15.38). Carrying out these operations gives

$$\frac{p_{0,2}}{p_1} = \frac{\left[\dfrac{\gamma+1}{2}\mathrm{Ma}_1^2\right]^{\gamma/(\gamma-1)}}{\left[\dfrac{2\gamma}{\gamma+1}\mathrm{Ma}_1^2 - \dfrac{\gamma-1}{\gamma+1}\right]^{1/(\gamma-1)}}$$

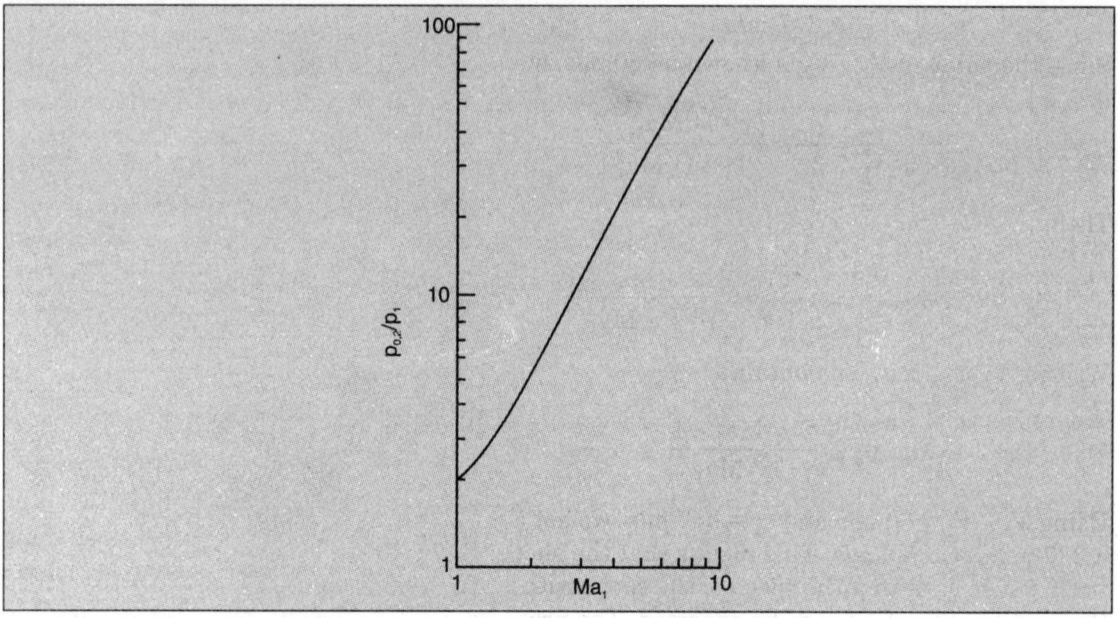

Fig. 15.19. $p_{0,2}/p_1$ *vs.* Ma_1 for supersonic flow for a pitot tube.

Figure 15.19 gives a plot of this equation for $\gamma = 1.4$. Thus, the flow Mach number Ma_1 can be obtained directly if $p_{0,2}/p_1$ is measured.

Example 15.6. Air at 10^5 Pa and 300 K flows steadily in a constant area duct at 60 m/s. A valve at the end of the duct is closed abruptly at time $t = 0$. Since the air at that point is brought to rest suddenly, a shock wave develops and propagates upstream (Fig. 15.20). Determine the shock velocity V_s and the pressure behind the shock.

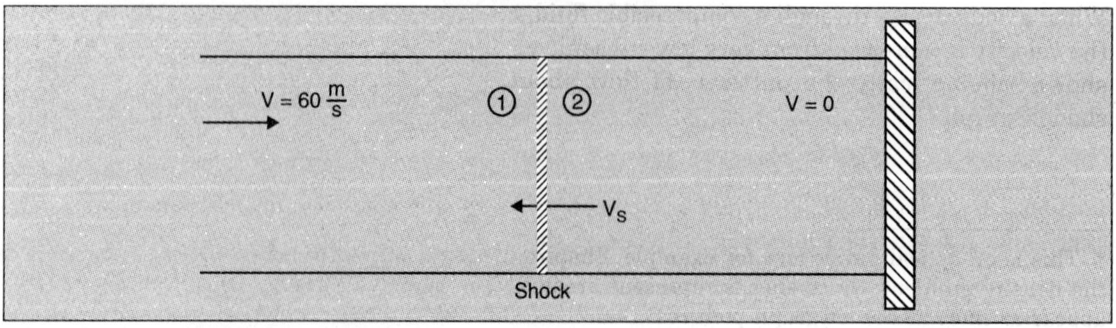

Fig. 15.20. Shock in a pipe with end closed abruptly.

In a frame of reference fixed with the shock, the upstream velocity is $V_1 = V_s + 60$ m/s, while the downstream velocity is $V_2 = V_s$. The speed of sound at point 1 is as

$$c_1 = \sqrt{\frac{\gamma R^* T_1}{M}} = \left(\frac{1.4 \times 8314.3 \left(\frac{\text{J}}{\text{kg-mol K}} \right) \times 300(\text{K})}{28.98\ (\text{kg/kg-mol})} \right)^{1/2}$$

$$= 347 \text{ m/s}$$

Since the value of $V_1 - V_2$ is known as 60 m/s, start with Eq. (15.37)

$$\frac{V_2}{V_1} = \frac{\rho_1}{\rho_2} = \frac{(\gamma - 1)\ \text{Ma}_1^2 + 2}{(\gamma + 1)\ \text{Ma}_1^2}$$

Then,

$$\frac{V_1 - V_2}{V_1} = 1 - \frac{V_2}{V_1} = \frac{2}{\gamma + 1} \frac{\text{Ma}_1^2 - 1}{\text{Ma}_1^2}$$

Writing $V_1 = c_1\ \text{Ma}_1$, one obtains

$$V_1 - V_2 = \frac{2\ c_1}{\gamma + 1} \frac{\text{Ma}_1^2 - 1}{\text{Ma}_1}$$

Using $V_1 - V_2 = 60$ m/s and $c_1 = 347$ m/s, we get (for $\gamma = 1.4$) $\text{Ma}_1 = 1.11$. Therefore, $V_1 = 1.11 \times 347$ (m/s) = 385.2 m/s. This means that the shock travels to the left at a rate of (385.2 – 60) = 325.2 m/s. Note that the speed of the shock with respect to the upstream fluid is greater than the speed of sound, as is to be expected. The pressure behind the shock, i.e., in the fluid that has come to rest, is obtained from Eq. (15.36) as 1.271×10^5 Pa, which represents a pressure jump of 2.71×10^4 Pa.

It may be added that the phenomenon of *water hammer,* observed when a valve is suddenly closed in a pipeline, is the same as above. The pressure pulses are reflected repeatedly at the two ends of the pipe, creating a hammering effect.

15.7 FLIGHT OF BODIES THROUGH A COMPRESSIBLE FLUID

When a body moves through a compressible fluid, a whole series of phenomena unfolds itself as the velocity is increased from very low subsonic values to high supersonic values. Figure 15.21 shows schematically the patterns of flow about an aerofoil as seen by a device called a shadow-graph.*

* This is an optical device (see for example, Shapiro, Reading 31) which brings out the variations in the density gradient. Since shocks represent areas of very rapidly changing density gradients, they appear as lines in shadowgraph pictures.

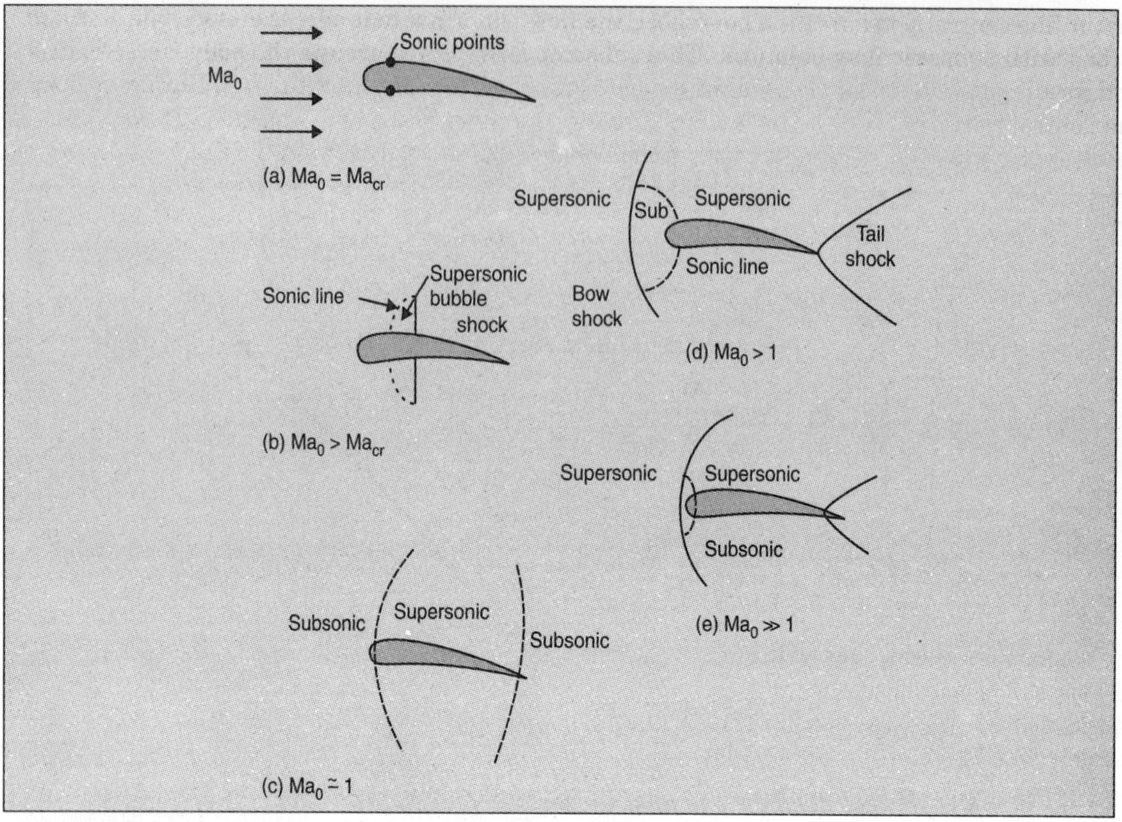

Fig. 15.21. Supersonic and subsonic regions at different Ma_0 for flow around an aerofoil.

At very low speeds, the flow may be treated as incompressible. In a frame of reference attached with the aerofoil, the fluid accelerates as it moves over the aerofoil surface. Consequently, the pressure there decreases. As the free-stream velocity V_0 (equal to the flight velocity) increases, compressibility effects come into play but till a certain value of the *free stream* Mach number Ma_0 (still below unity), the flow picture does not change significantly. At this value of Ma_0 the flow velocity on the surface of the aerofoil reaches the sonic value (Fig. 15.21a). The value of Ma_0 at which the flow first becomes sonic on the body surface is termed as the *critical Mach number* $Ma_{0,cr}$. It depends on the geometry of the body and is always less than one. Any increase in Ma_0 beyond $Ma_{0,cr}$ makes the flow supersonic at least on some portions of the aerofoil. As the flow decelerates down the surface, it becomes subsonic again, but a gradual isentropic change back to subsonic flow is difficult to achieve. The transition usually takes place across a shock, as shown in Fig. 15.21 (b). The region where the *local* Mach number $Ma > 1$ is known as the *supersonic bubble* which terminates in a shock. In a very carefully designed aerofoil (e.g., in *Whitcomb supercritical aerofoil)* it is possible to make the shock very weak over a significant range of Ma_0. As Ma_0 increases still further, the size of the supersonic bubble increases till for Ma_0 slightly below one it may be spread over nearly the entire aerofoil surface (Fig. 15.21c).

When the free-stream Mach number Ma_0 becomes supersonic, the upstream fluid cannot be aware of the presence of the body and as discussed in Sec. 15.3, this results in a bow-shaped shock standing a little distance upstream from the rounded leading edge (Fig. 15.21d). Note that in the central portion of the bow shock the flow is normal to it and, therefore, the shock is strong with subsonic flow behind it. This subsonic flow curves around the body, accelerating and forming supersonic zones. These zones terminate in shocks near the tail. But away from the central portion the bow shock is inclined to the flow. Thus, the conditions of an oblique shock apply, and even though the normal component of the velocity behind the shock is subsonic,

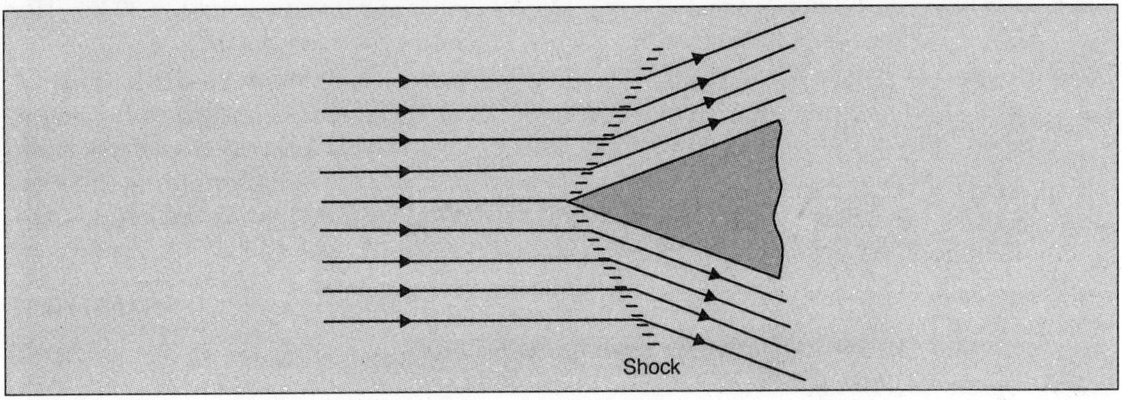

Fig. 15.22. Attached oblique shocks on a sharp-edged body.

the net velocity may be supersonic. The subsonic flow behind a bow shock is, therefore, restricted to a small region near the leading rounded edge as shown. As Ma_0 is further increased, the bow shock is swept back further and the region of subsonic flow shrinks towards the nose of the aerofoil (Fig. 15.21e).

If the nose of the aerofoil is made sharp instead of rounded, subsonic flow is not required near the tip and instead of a bow shock, attached oblique shocks result (Fig. 15.22). Since the oblique shocks are weaker, they represent less loss of mechanical energy (and, therefore, less drag) than bow shocks. It is for this reason that supersonic aerofoils have sharp leading edges rather than rounded ones. The preferred shapes are the half diamond or diamond (Fig. 15.23). This is in sharp contrast to the well rounded leading edges of subsonic aerofoils required to avoid separation.

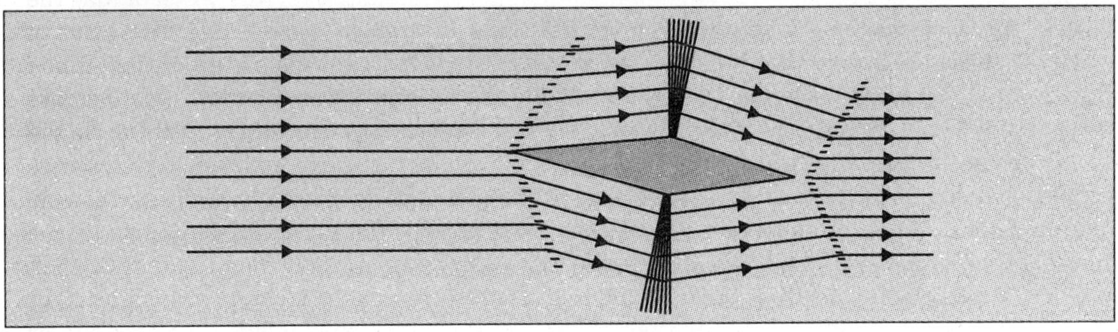

Fig. 15.23. Diamond-edged aerofoils for supersonic aircrafts.

PROBLEMS

15.1 A sonar signal bounces off a submarine and returns in 15 s. If the average temperature of the ocean water is 10°C, and its isentropic compressibility E_s is 2.07 × 10^9 Pa, at what distance is the submarine? Assume $\rho = 10^3$ kg/m^3.

15.2 A supersonic *Concorde* passes by at a Mach number of 2.0 and a height of 20 km. How far ahead must one try to look for it when one hears the sonic boom?

15.3 Show that a right-moving compression wave imparts a rightward velocity to the fluid as it passes by, irrespective of whether the fluid ahead of it is moving to the left or right. Also, if a wave travelling to the right in a pipe meets a rigidly closed end, show that the wave must get reflected and start travelling to the left after hitting this end, and that the pressure near the right end will be higher than that on the left, i.e., the reflected wave is a compression wave.

15.4 Show that Eq. (15.14) reduces to $p_0 = p + \frac{1}{2}\rho V^2$ as Ma → 0, and is, thus, consistent with the Bernoulli equation for incompressible flows.

15.5 Derive Eq. (15.23).

15.6 Show that in a channel with monotonically decreasing or increasing area, the flow remains either subsonic or supersonic throughout. Thus, prove that the flow through the bulge shown cannot be sonic at any point.

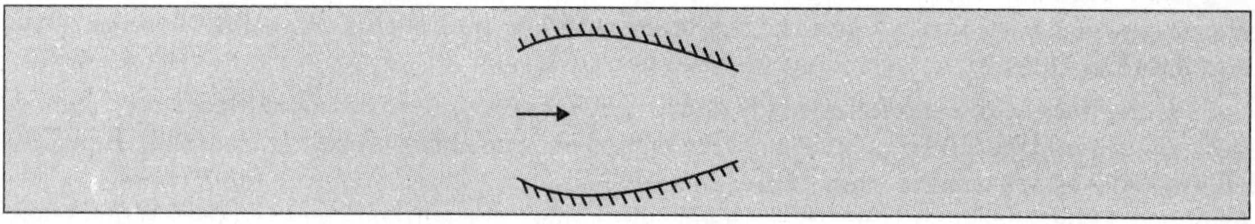

15.7 Consider a 2-*D* source-type compressible flow. Simplify Eqs. (15.20–15.23) for this situation and show that such a source flow cannot exist inside a core of radius r^* at which Ma = 1. If the flow is supersonic for $r > r^*$, show that the pressure decreases with increasing radius.

15.8 Air at a stagnation temperature of 300 K is flowing in a duct. A pitot-static tube is used to measure the velocity. The static tube measures the pressure of the air as 0.3 × 10^5 Pa gauge (when $p_{atm} = 10^5$ Pa), while the differential pressure is recorded as 0.7 × 10^5 Pa. Compute the velocity of the air and compare your results with those obtained assuming air to be incompressible.

15.9 Air at 2 × 10^5 Pa and 27°C flows out to the atmosphere from a large tank through a converging nozzle having a minimum area of 60 cm^2. Compute the mass flow rate, the Mach number and the temperature at the exit of the nozzle.

If the volume of the tank is 5 m^3, and if the flow can be assumed to be quasi-steady, how would you obtain the time taken for the pressure of the tank to decrease to, say, 1.5 × 10^5 Pa? Assume that the air in the tank remains at constant temperature.

15.10 Show mathematically that the stagnation temperature is unchanged across a shock while the stagnation pressure decreases.

15.11 A convergent-divergent duct having a minimum area of 2.5×10^{-3} m^2 and an exit area of 7.5×10^{-3} m^2 draws in air at a pressure of 9×10^5 Pa and temperature of 400 K. Compute the design value of the pressure, temperature and Mach number at the exit. Also compute the mass flow rate under these conditions. Find out the range of exit pressures when a shock exists in the divergent section.

15.12 A rocket having nozzle with $A_e/A^* = 5$ is designed to operate at a height of 20 km, where the temperature and pressure are 217 K and 4×10^3 Pa respectively. If the products of combustion are generated at 30×10^5 Pa and have a γ of approximately 1.4, estimate the thrust generated per unit throat area.

15.13 In Fig. 15.16, the upstream Mach number, pressure and temperature are 2.5, 10^5 Pa and 300 K respectively. The angle of deflection is 5°. Compute the downstream temperature, pressure, and Mach number and the angle θ which the shock makes with $\mathbf{V}_1$. Note that the pressure downstream depends on the value of θ (or δ) and so, by having a diamond aerofoil with different values of δ at the top and bottom surfaces, one can easily generate a lift force.

15.14 A spherical shock wave is created by an explosion in air. At an instant, the pressure just inside the shock is 12×10^5 Pa. Obtain the rate at which the shock is expanding and the velocity of the gas just behind the wave front. Take the atmosphere to be at 300 K and 10^5 Pa.

15.15 Determine the maximum temperature encountered by the skin of a *Concorde* when flying at Ma = 2 at 20 km altitude where the ambient temperature is – 56°C.

15.16 A supersonic tunnel has a convergent-divergent *nozzle* to accelerate the flow. The test section where Ma = 2 is followed by a convergent-divergent *diffuser* to recover the kinetic energy of the flow so that the pressure difference required across the tunnel is minimized. The complexities of the flow do not usually permit a smooth flow through the circuit and shock(s) are inevitable. Optimum design calls for restricting a shock to the diverging portion of the diffuser as shown. If the test-section area is 0.1 m^2 obtain the following:

(a) the throat area A_N^* for the nozzle,

(b) the pressure and temperature at the test section (with $p_0 = 10^5$ Pa and $T_0 = 300$ K),

(c) the Mach numbers at locations D, the minimum-area section of the diffuser, and S, the point where the shock stands, given that $A_S = 1.2\, A_N^*$ and $A_D = 1.1\, A_N^*$,

(d) the pressure and temperature just upstream of the shock, and

(e) the Mach number, pressure and temperature just behind the shock.

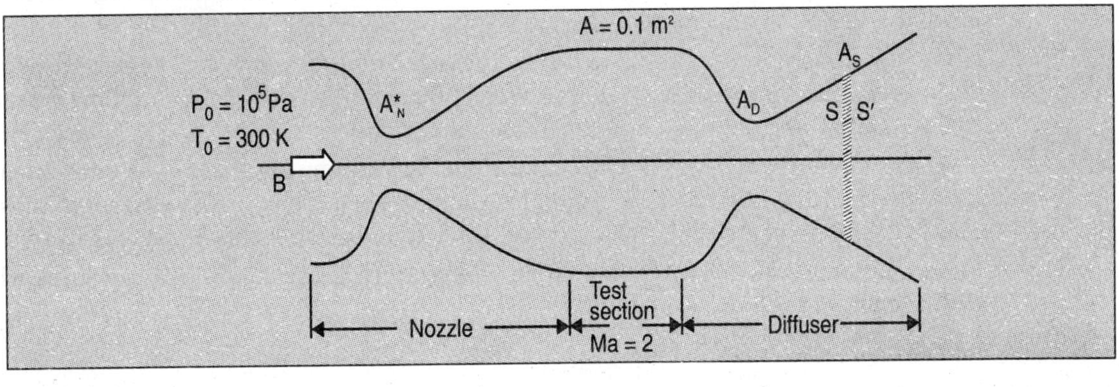

CHAPTER 15

15.17 Consider the design of the supersonic wind-tunnel of Prob. 15.16. If the minimum area A_D of the diffuser equals A_{N^*} , the throat area of the nozzle, we expect that for the isentropic flow, the flow is sonic at D and subsonic thereafter. But the presence of viscosity, howsoever small, decreases the stagnation pressure of the flow, while keeping the stagnation temperature unchanged. Show that this requires that the area A_D must be larger than A_{N^*} .

16

INTRODUCTION TO TURBULENT FLOWS

16.1 NATURE OF TURBULENCE

The turbulent motion of fluids has been referred to frequently in the previous chapters where it was mentioned that most flows of engineering interest as also those occurring in nature are turbulent. The detailed nature of such flows has not been discussed beyond noting that they are marked by highly irregular and rapid fluctuations of velocities, both with respect to time and location (see Sec. 1.7). It was also mentioned that turbulent motion involves rapid mixing of the fluid resulting in higher rates of heat transfer and of dispersal of species in multi-component flows. Turbulent flows are, therefore, used extensively in industry.

To get an understanding of the nature of turbulent flows, consider the flow of a fluid in the wake of a circular cylinder. As the Reynolds number increases, a series of phenomena unfolds itself as described in Sec. 1.7. For values of Re_D below about 30 the flow in the wake is steady and is entirely predictable. But as Re_D is increased, the flow in the wake becomes periodic, and at values of Re_D of about 60, one can see eddies being shed alternately from the top and bottom of the cylinder. Figure 1.17 shows the fluctuations of velocity at a fixed point in the wake for a series of Reynolds numbers. Notice that as Re_D increases the periodic fluctuations in the wake degenerate into chaotic or random variations. Such flows are termed as turbulent. Since randomness is an essential characteristic of turbulent flows, it is essential to use a statistical description for it. Instead of a single value for the velocity $\mathbf{V}$ at a point, a probability distribution is now expected for it. It is found that the probability distribution of the velocity $\mathbf{V}$ is a *normal* or *Gaussian* distribution. One important parameter used to specify a Gaussian distribution is its mean value.

To understand the significance of the mean value of velocity distribution consider, for example, the flow of a liquid through a pipe under a constant head. If the flow is turbulent, the velocity varies not only with the location, but also with time. A single measurement of velocity at a given point will be meaningless because it will not tell us what the value will be at the same point at a later instant, or at a geometrically similar point even at the same instant. However, if we average the velocity over time, the mean value obtained is expected to be the same at geometrically similar points, because we expect the same probability distribution at such points. For a pair of geometrically dissimilar points, the distributions, and hence the mean values are expected to be different. Thus, the mean values provide a basis for studying the spatial variations of velocity.

The instantaneous velocity is then written as

$$\mathbf{V} = \overline{\mathbf{V}} + \mathbf{V}' \qquad \text{...(16.1)}$$

where $\overline{\mathbf{V}}$ is the average velocity and $\mathbf{V}'$ is termed as the turbulent fluctuation. The fluctuations are fairly small, typically up to about 10 per cent of $\overline{\mathbf{V}}$. But these small fluctuations produce drastic changes in shear stresses so much so that the pressure drop across a pipe can be up to a hundred times larger than if the flow was kept laminar.

The definition of the mean velocity as the time-averaged velocity is possible only in cases where the boundary conditions are constant with time, as for example, when the head producing the flow is constant. Such a flow is termed statistically *stationary*. If, on the other hand, the head is steadily decreasing with time, one has to distinguish between the long-term trend and the short-term fluctuations. Figure 16.1 shows such a situation. Clearly, a time-averaged velocity has no meaning in such cases. Such flows are termed as *non-stationary* flows. In these flows, the time-average is replaced by what is termed as the *ensemble average*. This involves repeating the experiments a large number of times and measuring the instantaneous velocity in each attempt at precisely the same relative position and time. The large number of values so obtained constitute an ensemble and this represents the probability distribution of velocity at that point and time. The ensemble average then replaces the mean velocity. In the following discussion

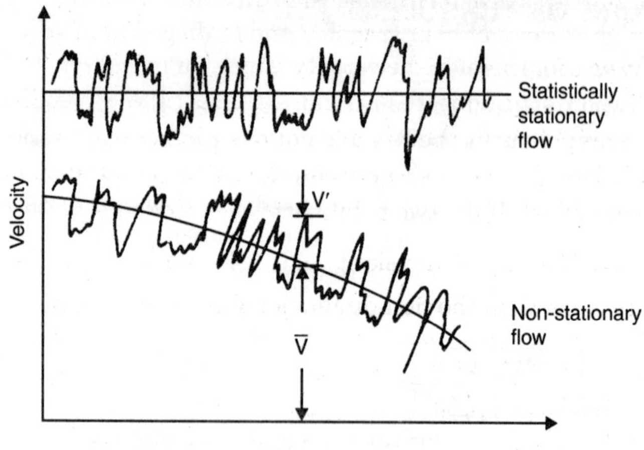

Fig. 16.1. Velocity fluctuations in stationary and non-stationary flows.

our attention shall be restricted to stationary flows only and, therefore, the mean velocity is used throughout.

From the definition of Eq. (16.1) it is easily seen that the mean value of the fluctuations $\overline{\mathbf{V}'}$ is zero, i.e. $\overline{\mathbf{V}'} = \mathbf{0}$ and, therefore, the level of turbulence is not indicated by the mean value $\overline{\mathbf{V}'}$.

Such an information is provided by a statistical measure called *variance*, which is defined as the mean of the squared value of the velocity fluctuations, namely, $\overline{\mathbf{V}'\cdot\mathbf{V}'}$. The variance of velocity fluctuations can also be interpreted in terms of the kinetic energy of turbulent motion. Consider the kinetic energy (per unit mass) associated with the instantaneous x-component of velocity

$$\frac{1}{2}\left(\overline{V}_x + V'_x\right)^2 = \frac{1}{2}\left(\overline{V}_x^2 + 2\,\overline{V}_x\, V'_x + V'^2_x\right)$$

The average value of this is obtained by taking the time average of each term. Since $\overline{V}_x$ is constant as far as this averaging process is concerned and since $\overline{V'_x}$ is zero, the contribution of the second term is zero and the average kinetic energy per unit mass associated with the x-velocity is $\frac{1}{2}\left(\overline{V}_x^2 + \overline{V'^2_x}\right)$. In this expression, $\frac{1}{2}\overline{V}_x^2$ is the kinetic energy associated with the mean motion and, therefore, $\frac{1}{2}\overline{V'^2_x}$ is the energy associated with the turbulent fluctuations. The total kinetic energy associated with turbulent motion is $\frac{1}{2}\left(\overline{V'^2_x} + \overline{V'^2_y} + \overline{V'^2_z}\right) = \frac{1}{2}\overline{\mathbf{V}'.\mathbf{V}'}$. The average $\overline{\mathbf{V}'\cdot\mathbf{V}'}$ is, therefore, taken as a measure of the level of turbulent fluctuations. This is often expressed as a percentage of the square of the mean velocity $\overline{V}^2$.

It should be noted that an essential characteristic of turbulent motion is the presence of the components of $\mathbf{V}'$ in all directions. Thus, turbulence is a three-dimensional phenomena even when the *mean* motion is unidirectional.

16.2 STRUCTURE OF TURBULENT FLOWS

Though the fluctuating component of the velocity $\mathbf{V}'$ at every point of the flow field has a Gaussian probability distribution signifying that it can be treated as a random process, the velocity fluctuations at two neighbouring points are not completely independent. This dependence is measured by the statistical covariance between the same fluctuating quantity measured at two points a little distance apart. If the two points have coordinates **x** and **x + r**, then the covariance of the V'_x components of the turbulent velocity of these two points is defined as $\overline{V'_x(\mathbf{x})V'_x(\mathbf{x}+\mathbf{r})}$. It is conventional to work with the non-dimensional form of the covariance, namely

$$R_{xx}(\mathbf{r}) = \frac{\overline{V'_x(\mathbf{x})V'_x(\mathbf{x}+\mathbf{r})}}{\left\{\overline{V'^2_x(\mathbf{x})}\;\overline{V'^2_x(\mathbf{x}+\mathbf{r})}\right\}^{1/2}} \quad ...(16.2)$$

A value of R_{xx} (**r**) of unity signifies a perfect correlation of the two quantities involved and means that the two points are always moving together, i.e., their motion is in phase. A negative value denotes, on the other hand, that the velocities at the two points are mostly out of phase. Figure 16.2 shows typical variations of the correlation R with increasing separation **r**.

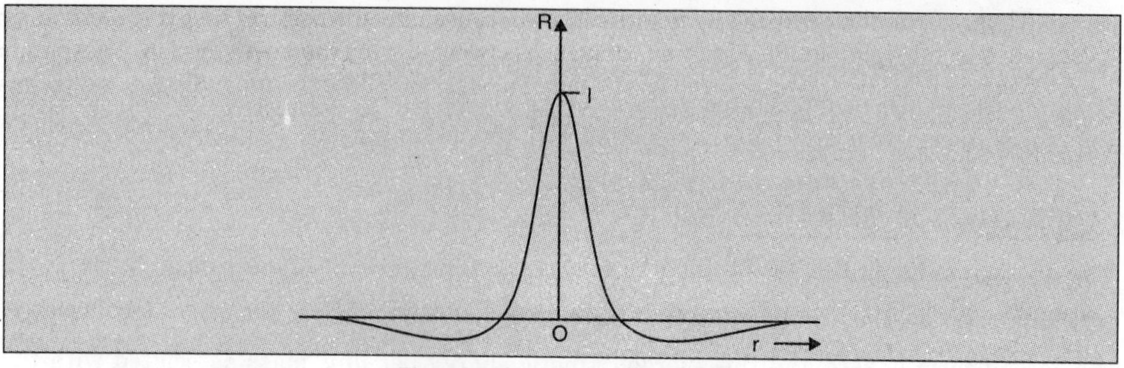

Fig. 16.2. Variation of R with **r**.

The existence of a positive correlation indicates that the fluid can be modelled as travelling more or less in lumps. Since rotational motion is an essential characteristic of turbulent motion, these lumps are visualized as *eddies* of various sizes. Physically, the correlation $R(\mathbf{r})$ is a measure of the strength of eddies whose size is larger than the magnitude of **r**. This is because the velocities at two points are correlated only when they lie on the same eddy and, therefore, eddies of size smaller than **r** do not contribute to the correlation $R(\mathbf{r})$.

It should be noted that the eddies or lumps of fluid that are seen to be in quasi-rigid-body like coherent motion are of a scale much larger than the molecular scale. Thus, we are talking here of *macroscopic* fluid balls of varying sizes. These eddies or lumps of fluid are clearly visible in shadowgraphs.* The size of these eddies changes continuously through agglomeration and disintegration.

The eddying motion of these lumps of fluid is superposed on the mean motion so that these lumps are convected with the flow. One obvious way of studying this convection is to compare the correlation of the velocity fluctuations at two points a distance **r** apart, with and without a time delay τ. If the eddies were being convected as truly rigid balls of fluid with the mean flow, the correlation $R(\mathbf{r})$ without a time delay should be identical with the correlation $R(\mathbf{r} + \overline{\mathbf{V}}\,\tau)$ with a time delay of τ. Figure 16.3 shows typical results. It is seen that the peaks of the two

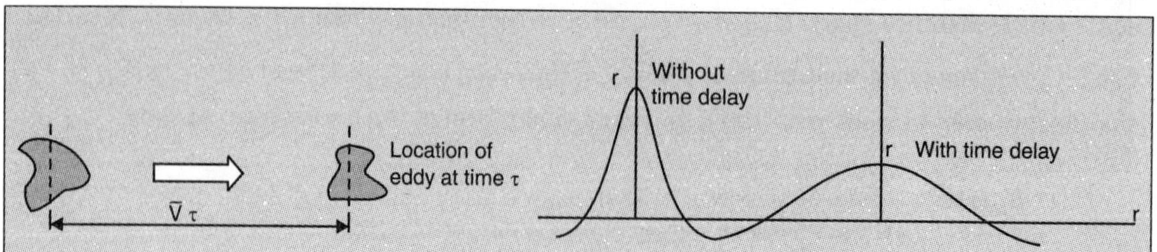

Fig. 16.3. Comparison of correlations with and without time delay, τ.

correlations are indeed separated by a distance equal to $\overline{\mathbf{V}}\,\tau$, though the delayed correlation is weaker. This is because the eddies are changing all the time and their convection can at best be treated as semi-rigid.

* Shadowgraphy is an optical technique of flow visualization which makes use of the variation of the refractive index of the fluid. It gives a pattern of bright and dark regions related to the second derivative of the density variations produced by the flow structure.

Thus, correlation studies and flow visualization experiments confirm that turbulent motion consists of semi-rigid motion of eddies which are convected with the mean motion. These eddies are of varying sizes — from as large as the dimensions controlling the mean flow, to quite small values. But the smallest of eddies are much larger than the molecular mean-free paths, so that turbulent motion belongs to the realm of continuum.

16.3 ORIGIN OF TURBULENCE

In flows which are originally laminar, turbulence arises from the instability of the laminar flows at high Reynolds numbers. There are various kinds of instabilities associated with fluid motions. Described here is the shear layer instability, the kind that is most often responsible for turbulence. A shear layer is a region with sharp velocity gradients such as that which occurs downstream from a point of separation. The velocities on the two sides of the surface of separation are quite different, giving rise to a thin region in which the velocity changes sharply. Figure 16.4 shows some such situations.

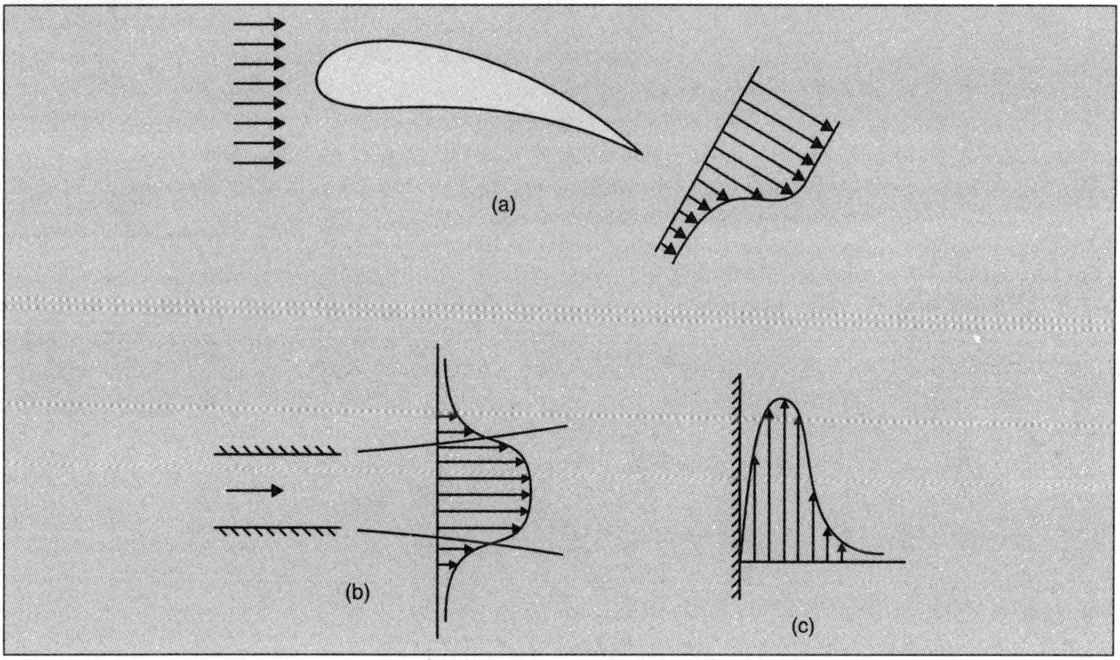

Fig. 16.4. Some examples of shear layers. (a) At trailing edge of an aerofoil, (b) a jet, and (c) in natural convection.

At large Reynolds numbers the shear layers may be idealized as surfaces of velocity discontinuities (Fig. 16.5 a). Consider such a discontinuity in a frame of reference in which the velocities on the two sides of the surface of discontinuity are equal and opposite (Fig. 16.5 b). In such a frame of reference, let the surface of discontinuity undergo a small disturbance, namely, develop a slight waviness as shown in Fig. 16.5 (c). This will increase slightly the velocity of the fluid on the convex parts of the surface, e.g., on A_u, B_l, and C_u, while it will decrease slightly the velocity on the concave parts, e.g., on A_l, B_u and C_l. If the flow is considered

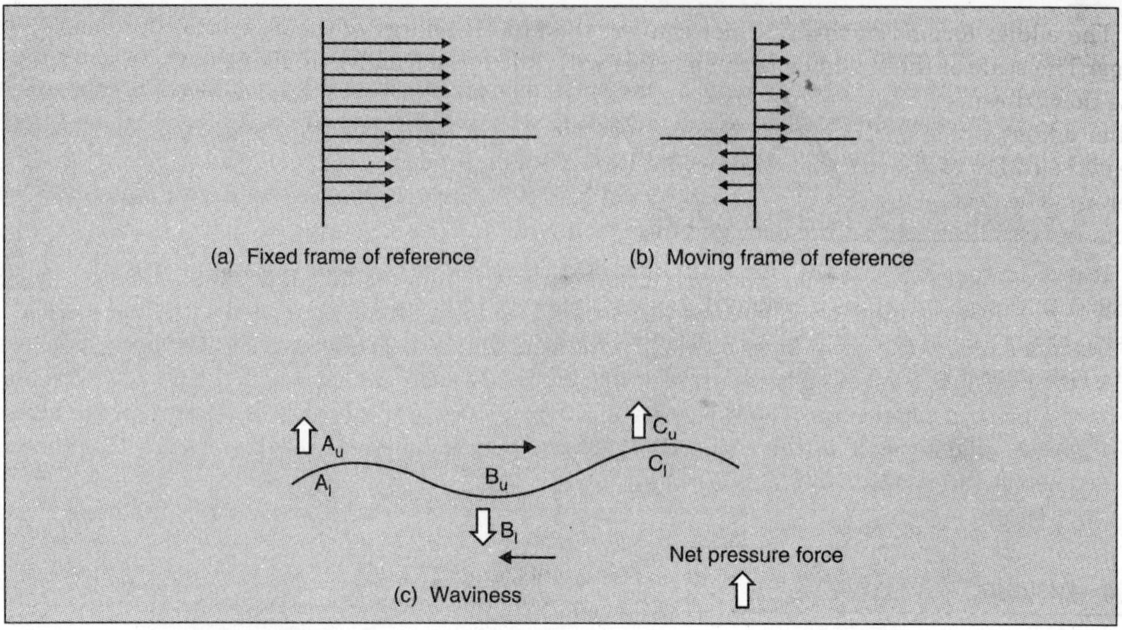

Fig. 16.5. (a) and (b) Idealization of a shear layer at high Re, and (c) the conditions when a slight waviness appears on the surface of discontinuity.

as steady, the application of the Bernoulli equation will indicate that a net pressure force acts so as to amplify the disturbances. Thus, a shear layer is unstable and tends to curl up. The wave becomes more and more distorted with time and finally coils up into vortices (Fig. 16.6).

Fig. 16.6. Curling up of a shear layer.

This process of eddy formation appears only at high Reynolds numbers when viscous effects are negligible. At low Reynolds numbers, the action of viscosity tends to dissipate or stabilize the velocity discontinuity, giving rise to laminar flow. Thus, laminar pipe-flow becomes turbulent at a Reynolds number (based on average velocity and diameter) of around 2300 unless small disturbances which may develop into eddies are very carefully avoided. Similarly boundary layers become unstable when the Reynolds number (based on the displacement thickness, see Prob. 13.9) exceeds 600 or so.

The eddies formed in this manner convert the kinetic energy of the flow into turbulent energy. The scale of these eddies is comparable with, though normally smaller than, the scale of the flow. However, these eddies are usually unstable themselves and tend to break up into smaller eddies which derive their energies from the larger ones. This process goes on till the scale of the eddies is so small that the relevant Reynolds number is not high enough for instability to persist. The decreasing size of eddies gives rise to increasing velocity gradients, which finally results in viscous effects taking over and dissipating the turbulent energy as thermal energy.

It is clear that since the Reynolds number of smallest eddies is determined by the limit of stability, in flows with the same length scales, the larger the Reynolds number, the lower is the size of the smallest eddies. This 'cascade' of turbulent energy from the larger to smaller eddies and then finally to viscous dissipation has been summarized by L.F.G. Richardson* as

Big whirls make little whirls,
Which feed on their velocity.
Little whirls make lesser whirls,
And so on to viscosity.

16.4 REYNOLDS STRESSES

In this section, an explanation is sought of how velocity fluctuations which are quite small compared to the mean velocities produce large changes in stresses. For this start with the momentum equation obtained in Chapter 6 which is also valid for turbulent flows provided the instantaneous values of all the quantities are taken. Starting with the x-component of the momentum Eq. (6.21) modified for 3-D flows:

$$\rho\left[\frac{\partial V_x}{\partial t}+V_x\frac{\partial V_x}{\partial x}+V_y\frac{\partial V_x}{\partial y}+V_z\frac{\partial V_x}{\partial z}\right]=\rho f_x+\frac{\partial \sigma_{xx}}{\partial x}+\frac{\partial \tau_{yx}}{\partial y}+\frac{\partial \tau_{zx}}{\partial z} \quad \text{...(16.3)}$$

where σ_{xx} and τ_{yx} are given by Newton-Stokes relations like Eqs. (6.12) and (6.14).

Writing the instantaneous values of all quantities (i.e., the time-averaged values *plus* the turbulent fluctuations), one gets

$$\rho\left[\frac{\partial}{\partial t}(\bar{V}_x+V_x')+(\bar{V}_x+V_x')\frac{\partial(\bar{V}_x+V_x')}{\partial x}+(\bar{V}_y+V_y')\frac{\partial(\bar{V}_x+V_x')}{\partial y}+(\bar{V}_z+V_z')\frac{\partial(\bar{V}_x+V_x')}{\partial z}\right]$$

$$=\rho f_x+\frac{\partial(\overline{\sigma_{xx}}+\sigma'_{xx})}{\partial x}+\frac{\partial(\overline{\tau_{yx}}+\tau'_{yx})}{\partial y}+\frac{\partial(\overline{\tau_{zx}}+\tau'_{zx})}{\partial z} \quad \text{...(16.4)}$$

Next take the time average of the entire equation. For this, the following obvious rules should be noted: If $F = \bar{F} + F'$, $G = \bar{G} + G'$ and $H = \bar{H} + H'$ are three fluctuating quantities with $\bar{F}$, $\bar{G}$ and $\bar{H}$ as the respective mean values, then

* As quoted by Shapiro (in Reading 2).

$$\overline{F'} = \overline{G'} = \overline{H'} = 0$$

$$\overline{F+G} = \overline{F} + \overline{G}$$

$$\overline{H(F+G)} = \overline{HF} + \overline{HG}$$

$$\overline{\text{Const.}F} = \text{Const. } \overline{F}$$

and

$$\overline{\partial F/\partial x} = \partial\overline{F}/\partial x, \text{ etc.}$$

From these it follows that

$$\overline{\frac{G\partial F}{\partial x}} = \overline{(\overline{G}+G')\frac{\partial F}{\partial x}} = \overline{G}\frac{\partial \overline{F}}{\partial x} + \overline{G'\frac{\partial(\overline{F}+F')}{\partial x}}$$

$$= \overline{G}\frac{\partial \overline{F}}{\partial x} + \overline{G'\frac{\partial F'}{\partial x}}$$

$$= \overline{G}\frac{\partial \overline{F}}{\partial x} + \frac{\partial\overline{(G'F')}}{\partial x} - \overline{F'\frac{\partial G'}{\partial x}}$$

With these averaging rules, it can be seen quite easily that the LHS of Eq. (16.4) averages out to

$$\rho\left[\frac{\partial \overline{V}_x}{\partial t} + \overline{V}_x\frac{\partial \overline{V}_x}{\partial x} + \frac{\partial\overline{(V_x')^2}}{\partial x} - \overline{V_x'\frac{\partial V_x'}{\partial x}} + \overline{V}_y\frac{\partial \overline{V}_x}{\partial y} + \frac{\partial\overline{(V_x' V_y')}}{\partial y}\right.$$

$$\left. - \overline{V_x'\frac{\partial V_y'}{\partial y}} + \overline{V}_z\frac{\partial \overline{V}_x}{\partial z} + \frac{\partial\overline{(V_x' V_z')}}{\partial z} - \overline{V_x'\frac{\partial V_z'}{\partial z}}\right]$$

$$= \rho\left[\frac{\partial \overline{V}_x}{\partial t} + \overline{V}_x\frac{\partial \overline{V}_x}{\partial x} + \overline{V}_y\frac{\partial \overline{V}_x}{\partial y} + \overline{V}_z\frac{\partial \overline{V}_x}{\partial z}\right]$$

$$+ \rho\left[\frac{\partial}{\partial x}\overline{(V_x')^2} + \frac{\partial}{\partial y}\overline{(V_x' V_y')} + \frac{\partial}{\partial z}\overline{(V_x' V_z')}\right]$$

$$- \overline{\rho V_x'\left[\frac{\partial V_x'}{\partial x} + \frac{\partial V_y'}{\partial y} + \frac{\partial V_z'}{\partial z}\right]}$$

Starting with the continuity equation $\nabla \cdot \mathbf{V} = 0$, and replacing the instantaneous velocities by their mean values and the fluctuating components, it can easily be shown by averaging that $\nabla \cdot \overline{\mathbf{V}} = 0$. On substracting $\nabla \cdot \overline{\mathbf{V}} = 0$ from $\nabla \cdot \mathbf{V} = 0$ one obtains $\nabla \cdot \mathbf{V}' = 0$, i.e., the continuity equation for both the fluctuating components and the mean velocities have the same form. The expression within the brackets in the last term of the above expression, therefore, drops out. The RHS of Eq. (16.4) simplifies to

$$\rho f_x + \frac{\partial \bar{\sigma}_{xx}}{\partial x} + \frac{\partial \bar{\tau}_{yx}}{\partial y} + \frac{\partial \bar{\tau}_{zx}}{\partial z}$$

Thus, on averaging, Eq. (16.4) becomes

$$\rho\left[\frac{\partial \bar{V}_x}{\partial t} + \bar{V}_x \frac{\partial \bar{V}_x}{\partial x} + \bar{V}_y \frac{\partial \bar{V}_x}{\partial y} + \bar{V}_z \frac{\partial \bar{V}_x}{\partial z}\right]$$

$$+ \rho\left[\frac{\partial}{\partial x}\overline{(V_x')^2} + \frac{\partial}{\partial y}\overline{(V_x' V_y')} + \frac{\partial}{\partial z}\overline{(V_x' V_z')}\right]$$

$$= \rho f_x + \frac{\partial \bar{\sigma}_{xx}}{\partial x} + \frac{\partial \bar{\tau}_{yx}}{\partial y} + \frac{\partial \bar{\tau}_{zx}}{\partial z} \quad \text{...(16.5)}$$

or

$$\rho\left[\frac{\partial \bar{V}_x}{\partial t} + \bar{V}_x \frac{\partial \bar{V}_x}{\partial x} + \bar{V}_y \frac{\partial \bar{V}_x}{\partial y} + \bar{V}_z \frac{\partial \bar{V}_x}{\partial z}\right]$$

$$= \rho f_x + \frac{\partial}{\partial x}\left(\bar{\sigma}_{xx} - \rho \overline{V_x'^2}\right) + \frac{\partial}{\partial y}\left(\bar{\tau}_{yx} - \rho \overline{V_x' V_y'}\right) + \frac{\partial}{\partial z}\left(\bar{\tau}_{zx} - \rho \overline{V_x' V_z'}\right) \quad \text{...(16.6)}$$

with similar equations for the *y*- and *z*-components of momentum. These equations which govern the 'mean values' in turbulent flows have exactly the same form as the laminar-flow equations except that the stresses are now modified. These additional stresses represent the action of velocity fluctuations on the mean flow arising because of the non-linearity in the convective acceleration terms.

The additional stresses $-\rho\overline{V_x'^2}$, $-\rho\overline{V_x' V_y'}$, $-\rho\overline{V_x' V_z'}$, etc., are known as the *Reynolds stresses* and are often large compared to the corresponding viscous stresses, so that, at times, viscous effects may even be dropped. That is why *fully* turbulent flows have been modelled to be independent of viscous forces as in Sec. 10.2. It is because of these large Reynolds stresses that the mean velocity or pressure distributions in turbulent flows are very different from the laminar ones.

Each Reynolds stress represents the covariance of two fluctuating components of velocity at a given point and is non-zero whenever these fluctuations are not independent. The Reynolds normal stresses $-\rho\overline{V_x'^2}$, $-\rho\overline{V_y'^2}$ and $-\rho\overline{V_z'^2}$ are clearly non-zero and are related to the mean turbulent kinetic energy. It shall be presently shown that it is possible to have dependence of the fluctuating components in two different directions and hence non-zero values of covariances like $\overline{V_x' V_y'}$. Consider, for example, a shear flow with mean velocity $\bar{V}_x(y)$ increasing with *y* and with $\bar{V}_y = \bar{V}_z = 0$ everywhere (Fig. 16.7). The stress component $-\rho\overline{V_x' V_y'}$ can be interpreted as an averaged transport of *x*-momentum across a surface with its normal in the *y*-direction. It is seen that a particle which travels upwards across the surface at $y = y^*$ has a positive V_y'.

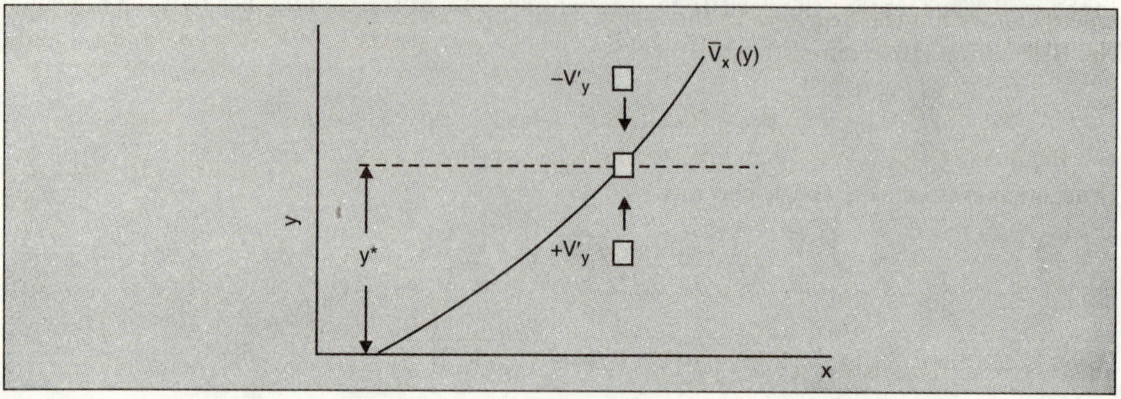

Fig. 16.7. Transfer of x-momentum across $y = y^*$ because of motion of eddies.

Since it is coming from a region where $\overline{V}_x$ is lower than $\overline{V}_x(y^*)$ the $V_x'(y^*)$, i.e., the velocity fluctuation measured at y^* is more likely to be negative. Thus, the product $V_x'V_y'$ is likely to be negative. On the other hand, a particle crossing $y = y^*$ from above has a negative value of V_y' and is likely to result in a positive indication of V_x'. Thus, this also makes a negative contribution to $V_x'V_y'$ and, therefore, the Reynolds stress $-\rho\overline{V_x'V_y'}$ is likely to be positive, the same sign as the viscous stress in this flow.

The development above explains how the Reynolds stresses are related to the detailed velocity fluctuations. In most engineering problems one is interested only in calculating the mean velocity distributions and, therefore, prefers to write Reynolds stresses in terms of the mean velocities and their gradients. Various models have been developed which do just this, but all are far from satisfactory. One reason for this state of affairs is that since the Reynolds stresses depend on the structure of the turbulent eddies, dimensions of the largest of which may be of the same order as that of the main flow field itself, it is too much to expect that Reynolds stresses can be cast as functions of local values of the mean velocities or even their gradients. Better models are continually being developed, but most of the early information about the turbulent velocity profiles is at best semi-empirical in nature.

16.5 TURBULENT FLOW NEAR A WALL

Turbulent flows in the neighbourhood of solid walls show certain characteristic features which are demonstrated both in internal (i.e., confined flows as such as Couette or Poiseuille flows) as well as in external flows (i.e., boundary layers around obstacles). At such surfaces, the applicable boundary condition is the no-slip condition. This implies that at the walls both the mean velocity and the fluctuating component of velocity are zero. Thus, for a 2-D flow along a wall at $y = 0$

$$\overline{V}_x\ (y = 0) = 0, \text{ and } V_x'\ (y = 0) = 0 \text{ at all times.}$$

Also, the requirement of an impervious wall gives

$$\overline{V}_y\ (y = 0) = 0, \text{ and } V_y'\ (y = 0) = 0 \text{ at all times.}$$

Since $\partial V_x'/\partial x$ and $\partial V_z'/\partial z$ are zero at the wall, the continuity equation for the fluctuating components gives $\partial V_y'/\partial y = 0$, and therefore V_y' must vary at least as rapidly as y^2 very close to the wall. The fact that the velocity fluctuations are zero at the wall implies that the Reynolds stresses there vanish. In particular, the Reynolds shear stress $-\rho\overline{V_x'V_y'}$ is zero, and the only shear stress exerted directly at the wall is the viscous stress.

As we move away from the wall, the fluctuating velocity components increase rapidly and the Reynolds stresses are much larger than the viscous stresses. The total shear stress which is the sum of the viscous stress and the Reynolds stress

$$\overline{\tau}_{xy} = \mu\partial\overline{V}_x/\partial y - \rho\overline{V_x'V_y'}$$

cannot change very rapidly with y, otherwise, it will give very large mean accelerations. In fact, it can be shown that in Couette flow (i.e., flow between two parallel infinite plates) the shear stress is constant with y. Consequently the viscous stress $\mu\partial\overline{V}_x/\partial y$, at the wall is of the same order as the Reynolds stress $-\rho\overline{V_x'V_y'}$ further out, and though the total stress does not vary, each of the two components shows rapid variations. Figure 16.8 shows typical variations of the Reynolds stress and viscous stress across a turbulent boundary-layer on a flat plate. The

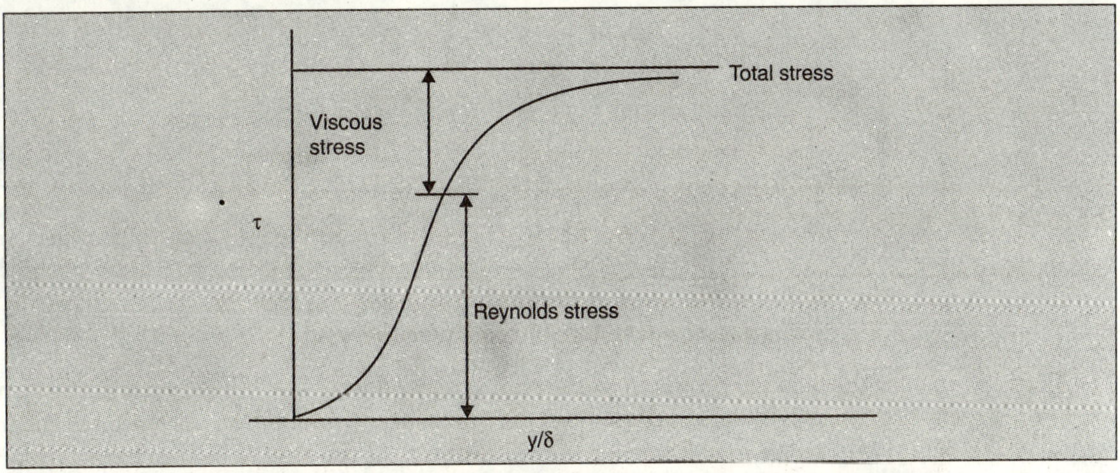

Fig. 16.8. Variation of viscous and Reynolds stresses near a flat plate.

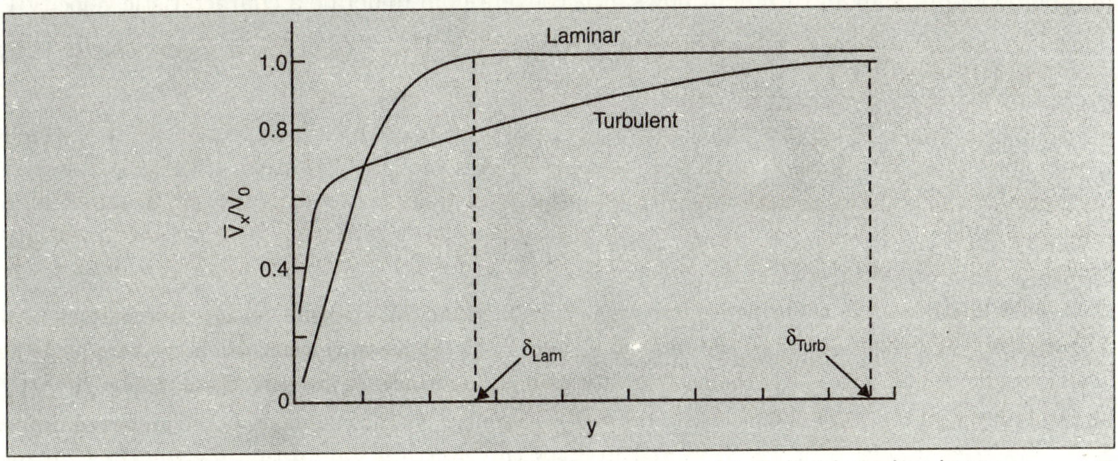

Fig. 16.9. Comparison of velocity profiles in laminar and turbulent boundary layers.

region where viscous effects dominate the Reynolds stresses is only a very small fraction of the total boundary layer thickness. This region is termed the *viscous sublayer** and is around one per cent of the total boundary layer thickness δ. The given distribution of viscous stresses across this layer requires that the mean velocity profile must rise sharply at the wall, and then become relatively flat, well within the overall boundary layer thickness. Figure 16.9 compares a typical turbulent boundary layer profile with a laminar one. The same effect is observed in internal flows as well. Figure 16.10(a) shows the fully-developed laminar and turbulent profiles across a circular pipe discharging the same flow rates. Figure 16.10(b) shows the profiles when the pressure gradients in the laminar and turbulent cases are identical. Note that in this case the shear stress at the walls must be identical, and since the shear stress at the wall is viscous in either case, the velocity profiles have identical slopes there. But the presence of Reynolds stresses forces a much earlier levelling off of the velocity profiles resulting in a marked decrease of flow rate.

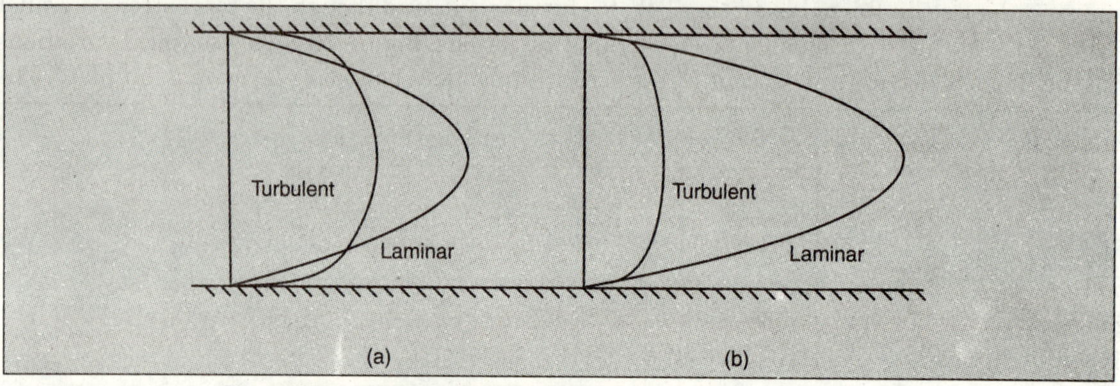

Fig. 16.10. Laminar and turbulent velocity profiles in a pipe for (a) same flow rate and (b) same pressure drag.

Let us consider the velocity profile near the wall in some detail. We will base our discussion on arguments based essentially on similarity and dimensional considerations. As the total shear stress varies only slowly across a wall, we can take its value τ_w at the wall to be the characteristic of the shear stress. Also, since we expect the internal processes (through Reynolds stresses) to play a fundamental role in all turbulent flows, it is conventional to define a characteristic velocity V_τ based on the equivalence of τ_w and internal stresses ρV_τ^2. Thus, we define a *shear velocity* V_τ by

$$V_\tau = \sqrt{\frac{\tau_w}{\rho}} \qquad \text{...(16.7)}$$

and use this as a characteristic velocity all across the flow.

---

* In older literature this sublayer is often termed as 'laminar' because the velocity fluctuations near the wall are very small. But since the ratio of V_x' and V_y' to the mean velocity $\overline{V}_x$ is not zero near the wall (even though V_x' and V_y' are themselves very small), the turbulence intensity measured by $\sqrt{V_x'^{\,2}}/\overline{V}_x$ is rather large (of the order of 0.3 close to the wall). Therefore, viscous sublayer is the preferred usage.

In the viscous sublayer the dominant stress is the viscous stress $\mu\, \partial \overline{V}_x/\partial y$ and, therefore, the similarity parameter is obtained from the ratio of $\mu \partial \overline{V}_x/\partial y$ to the wall stress τ_w. Since the characteristic value y_c of y is not fixed *a-priori* we obtain y_c by equating the characteristic value of the viscous stresses, i.e., $\mu V_\tau/y_c$ to τ_w. This gives

$$y_c = \frac{\mu V_\tau}{\tau_w} = \frac{\mu V_\tau}{\rho V_\tau^2} = \frac{\mu}{\rho V_\tau} = \frac{\nu}{V_\tau} \qquad \text{...(16.8)}$$

Thus the non-dimensional velocity across the viscous sublayer is given by the following functional relationship

$$\frac{\overline{V}_x}{V_\tau} = \mathcal{F}\left(\frac{yV_\tau}{\nu}\right)$$

This is known as the *law of the wall.* It is found experimentally that this is a universal law in the sense that a single functional relationship describes the velocity profile very near the wall in diverse flow situations. This law is valid for values of yV_τ/ν up to about 40. Within this region, for values of yV_τ/ν up to about 8, the velocity profile may be approximated as linear

$$\frac{\overline{V}_x}{V_\tau} = \left(\frac{yV_\tau}{\nu}\right) \text{; for } \left(\frac{yV_\tau}{\nu}\right) < 8 \qquad \text{...(16.9)}$$

Beyond $yV_\tau/\nu = 40$, the viscous stresses are dominated by the Reynolds stresses, and can therefore be ignored. It can be shown, by arguments slightly beyond the scope of this text, that the velocity profile beyond $yV_\tau/\nu = 40$, but still close to the wall, follows a logarithmic law

$$\frac{\overline{V}_x}{V_\tau} = A \log \frac{V_\tau y}{\nu} + B \qquad \text{...(16.10)}$$

where A and B are universal constants (equal to about 5.6 and 4.9 respectively). Figure 16.11 shows a typical velocity profile close to a wall. It is seen that the profile within the viscous sublayer and in the *log-law region* is independent of the nature of the outer flow. Thus, Eqs. (16.9) and (16.10) are universal velocity profiles valid for all flows close to the wall. Typically, the log-law region extends to about 20 per cent of the turbulent boundary layer thickness δ. Within this region the characteristic length and velocity of the main flow do not enter the expression for the velocity. Once τ_w is determined, the structure of this inner layer close to the wall is determined completely by the dynamics of the turbulent eddies close to the wall which is seen to be independent of the nature of the outer flow. The upper limit on the logarithmic profile occurs where the dynamics of the overall boundary layer starts playing a significant role.

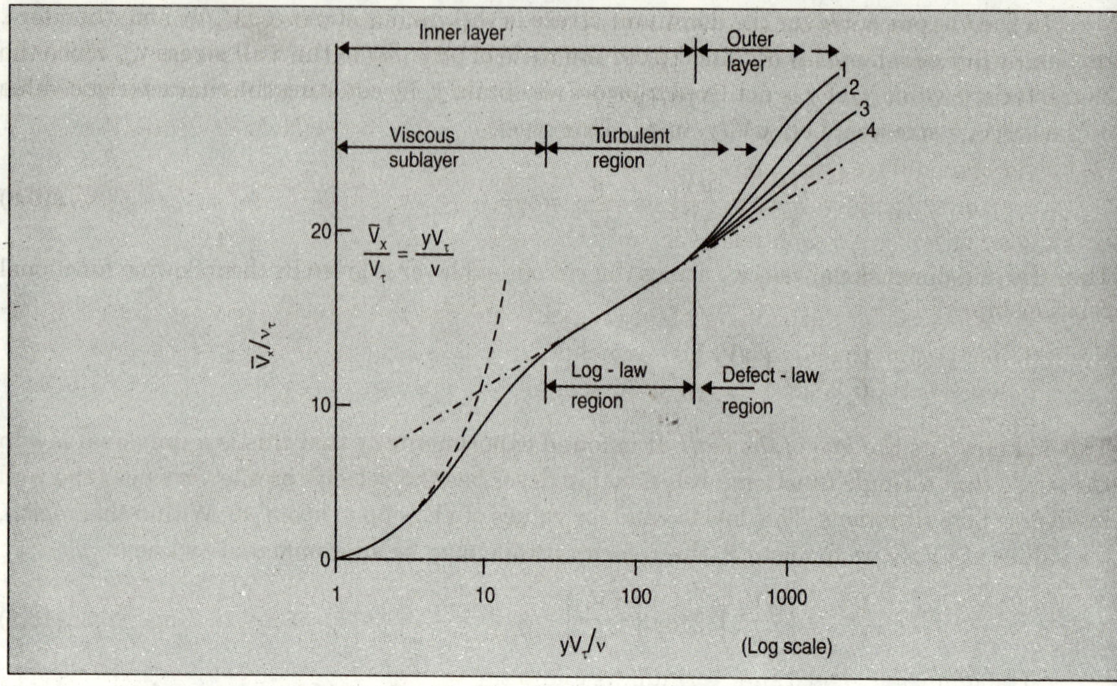

Fig. 16.11. Velocity variations close to the wall. (1) for strong increasing pressure; (2) flat-plate flow; (3) pipe flow; (4) strong decreasing pressure.

16.6 TURBULENT BOUNDARY LAYERS

As noted in Sec. 13.5, the flow within the thin boundary layer that develops close to a solid boundary remains laminar only as long as the Reynolds number Re_δ $(= \rho V_0\, \delta/\mu)$ based on the boundary layer thickness remains less than about 1800. As the flow proceeds down the surface, the boundary layer thickness δ increases and the flow becomes unstable. These instabilities develop into turbulent motions within the boundary layer and the boundary layer is said to be turbulent. Extensive experiments have established that the character of the flow within the turbulent boundary layer changes with the distance from the wall and it is convenient to model the thin boundary layer to be made up of still thinner layers. Very close to the wall the fluctuating velocities and, therefore, the Reynolds stresses tend to be negligible while the total shear stress remains nearly constant. Thus, the viscous shear stresses $\mu\partial \bar{V}_x/\partial y$ dominate and the region is termed as the viscous sublayer as in Sec. 16.5. As shown there, the characteristic velocity and length within this layer are the shear velocity V_τ $(= \sqrt{\tau_w/\rho}\,)$ and ν/V_τ respectively. This region extends from the wall up to $y = 40\,\nu/V_\tau$, and the velocity is given by the functional relationship $\bar{V}_x/V_\tau = \mathcal{F}(yV_\tau/\nu)$. Beyond $y = 40\,\nu/V_\tau$, the viscous shear stresses are dominated by the Reynolds stresses, and as has been mentioned in Sec. 16.5, the velocity follows the logarithmic law

$$\frac{\bar{V}_x}{V_\tau} = A \log \frac{V_\tau y}{\nu} + B$$

where A and B are constants. A large body of experimental data has established that A and B are universal constants, i.e., their values are independent of the nature of the outer flow. Therefore, all the formulations of these regions are valid for internal flows as well as for boundary layers, provided the flow does not change rapidly in the x-direction. The log-law region and the viscous sublayer are, thus, said to form an *inner layer* where the mechanics of the flow are controlled completely by the shear stress at the wall, the outer flow having no effect. The inner layer occupies a region relatively very close to the wall, say, within 20 per cent of the turbulent boundary layer thickness.

The rest of the boundary layer region is termed as the *outer layer* which constitutes about 80 per cent of the boundary layer thickness. In this region, we expect the velocity profile to be independent of the direct effects of viscosity, and the only reasonable characteristic length is δ, the boundary layer thickness. Since the Reynolds stresses are the only relevant shear stresses, V_τ is still a characteristic velocity, but the velocity at the edge of the boundary layer (= V_0 in the case of flow over a flat plate) also enters the formulation. The velocities in this layer are given by what is termed as the *velocity-defect law*

$$\frac{\bar{V}_x - V_0}{V_\tau} = \mathcal{F}\left(\frac{y}{\delta}\right)$$

The following logarithmic formula is found to be a reasonably satisfactory correlation for the velocity in this region

$$\frac{\bar{V}_x - V_0}{V_\tau} = 8.6 \log\left(\frac{y}{\delta}\right) \qquad \text{...(16.11)}$$

At the outer edge of the boundary layer, i.e., as the free stream is approached, the flow is no longer fully turbulent. Experimental observations with smoke-filled boundary layers indicate that the flow at a given point becomes turbulent intermittently. Figure 16.12 shows the irregularly fluctuating character of the edge of the boundary layer. Viscous stresses again start dominating the shear at this outer edge. It is, therefore, convenient to hypothesize a *viscous super-layer* at the outer edge of the boundary layer, the thickness of which is of the order of the smallest dissipating eddies. This viscous super-layer acts as a buffer in which viscous stresses transfer fluctuating vorticity to the outer irrotational flow. If ε represents the rate of transfer of turbulent kinetic energy per unit mass down the 'cascade' to the smallest dissipating eddies, then using dimensional analysis it can be shown that the relevant characteristic length and velocity are $(\nu^3/\varepsilon)^{1/4}$ and $(\nu\varepsilon)^{1/4}$ respectively. Therefore, the thickness of the viscous super-layer is of the order of $(\nu^3/\varepsilon)^{1/4}$ which is usually slightly more than the sublayer thickness of V_τ/ν.

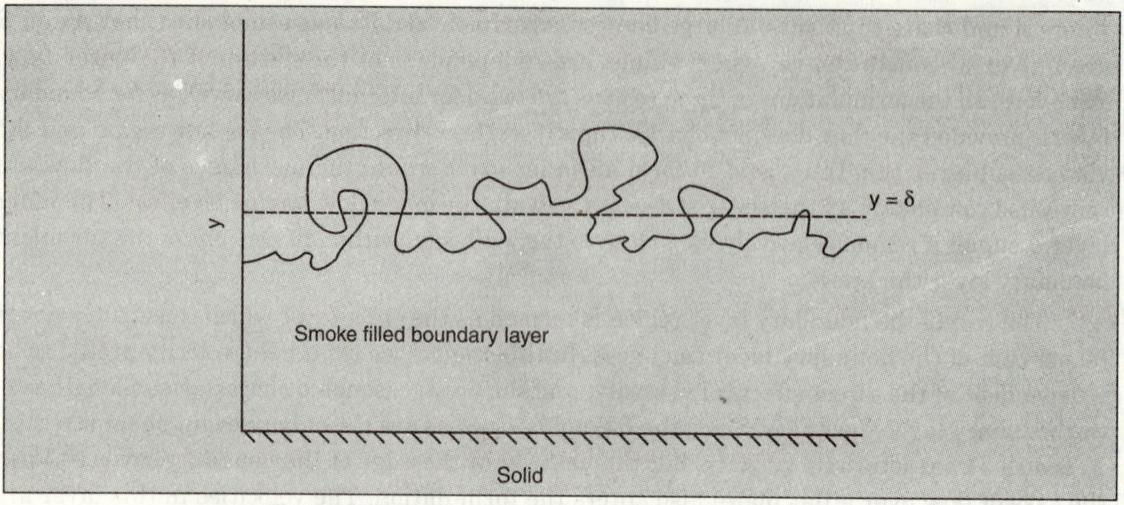

Fig. 16.12. Fluctuations at the edge of boundary layer.

Table 16.1 summarizes the multi-layered model of the turbulent boundary layer discussed above.

Table 16.1. Various layers in the turbulent boundary layer

	Inner Layer		Outer Layer	
	Viscous sublayer	Log-law region	Outer-law region	Viscous superlayer
Range	$y < 40 \dfrac{\nu}{V_\tau}$	$\dfrac{40\,\nu}{V_\tau} < y < 0.2\delta$	$y > 0.2\,\delta$	Near $y = \delta$
Characteristic length	ν/V_τ	ν/V_τ	δ	$(\nu^3/\varepsilon)^{1/4}$
Characteristic velocity	$V_\tau = \sqrt{\dfrac{\tau_w}{\rho}}$	V_τ	V_0, V_τ	$(\nu\,\varepsilon)^{1/4}$
Dominating stress	Viscous	Turbulent	Turbulent	Viscous
Nature of flow	Intermittent turbulent	Fully turbulent	Fully turbulent	Intermittent turbulent
Type of velocity law	$\dfrac{\bar{V}_x}{V_\tau} = f\left(\dfrac{yV_\tau}{\nu}\right)$ Linear for $y < 8\nu/V_\tau$	Log law $\dfrac{\bar{V}_x}{V_\tau} = A \log \dfrac{yV_\tau}{\nu} + B$	Velocity defect law $\dfrac{\bar{V}_x - V_0}{V_\tau} = f\left(\dfrac{y}{\delta}\right)$	

The logarithmic relationships presented above are valid over almost the entire range of Reynolds numbers for which experimental data is available. But these are relatively inconvenient to work with. Power law velocity profiles similar to those presented in Sec. 13.5 are simpler to use and give explicit relationships for C_f, but are applicable over restricted Reynolds number ranges. Thus, for the range of $3000 < Re_\delta < 70{,}000$, where Re_δ is the Reynolds number based on the boundary layer thickness δ, it is possible to use a *one-seventh* power law to approximate the velocity profile over almost the entire boundary layer thickness

$$\frac{\bar{V}_x}{V_\tau} = 8.74 \left(\frac{yV_\tau}{\nu}\right)^{1/7}$$

Using the condition that $\bar{V}_x = V_0$ at $y = \delta$, we get

$$V_\tau = \left(\frac{V_0}{8.74}\right)^{7/8} \left(\frac{\nu}{\delta}\right)^{1/8} \qquad \text{...(16.12)}$$

Then Eq. (16.7) gives

$$C_f = \frac{\tau_w}{\frac{1}{2}\rho V_0^2} = 0.045\,(Re_\delta)^{-1/4},$$

which is the same expression as Eq. (13.27). Then, as shown in Sec. 13.5, we can use the integral momentum Eq. (13.18) to obtain the boundary layer thickness as

$$\frac{\delta}{x} = 0.37\,(Re_x)^{-1/5}$$

which is the same as Eq. (13.25).

This expression is valid for Re_δ range from 3,000 to 70,000 which corresponds roughly to an Re_x range from 5×10^5 to 10^7.

EPILOGUE

As we have seen so far, liquids and gases in motion behave in very complicated ways, giving rise to a wide variety of complex flow phenomena. We review here in this epilogue some of the varied phenomena dealt with in this text, classified according to the kinds of forces that control them.

A fluid particle within the flow field can be subjected to either body or surface forces. The gravitational force is the only body force considered here. The surface forces can either be normal or shear forces. It is an essential property of fluids that shear forces occur only if the fluid is flowing and is being *continually* strained due to the relative motion of fluid particles. Thus, a fluid at rest or one which is moving as a rigid body (i.e., in which there is no relative motion of particles) is free of shear stresses. Such flows were discussed in Chapter 2 and in Prob. 6.13.

While analysing the motion of fluids, it is convenient to hypothesize the presence of an inertial force equal to mass *times* the acceleration and acting in the direction of the fluid acceleration. The problem of *dynamics* is then reduced to one of force balance with the inertial force balancing out all other forces acting on the fluid element.

The simplest flows to analyse are obviously those where the inertial forces are zero, i.e., the fluid acceleration is absent. In Chapter 2 on fluid statics, the case when the fluid is at rest was considered. Under these conditions, the gravity forces are balanced by the pressure forces, the viscous forces being zero. This results in a pressure distribution given by Eq. (2.8) which, for a fluid of constant density gives the linear hydrostatic pressure distribution. Note that one obtains the same variation of pressure in the vertical direction even if the fluid is moving but has no component of acceleration (and thus of inertial force) in the vertical direction. The case of a liquid in a bucket rotating about a vertical axis (Example 6.4) is an example.

Another class of flows where inertial forces are absent are the so called fully developed flows like the Couette flow between infinite parallel plates or cylinders (Example 6.1 and Prob. 6.18) and Poiseuille flow in long tubes (Example 6.3). In these examples, the surface and the body forces are in complete equilibrium. If we neglect gravity forces (or absorb them in the non-gravitational pressure $\mathscr{P}$ defined in Example the pressure forces are balanced by the viscous forces exactly. The pressure drop in Poiseuille flow, therefore, is determined by the viscous stresses at the walls. In the case of Couette flow where the pressure gradient is zero the net viscous force on the fluid element vanishes, the stresses on opposite faces being equal and opposite.

The expression for the inertial forces in fluids is a bit more complex than in solids, because in the study of fluid mechanics, the use of field description is preferred to the particle description. As shown in Chapter 3, the true acceleration of a fluid particle in the field description can be obtained as a combination of the local (i.e., unsteady) term and the convective term. Even when the flow is steady, i.e., the conditions at a fixed point do not change with time, for example, in a flow of an inviscid fluid about a cylinder (see Fig. 12.17), a fluid particle experiences acceleration as it moves downstream. Thus, a particle in steady flow decelerates while approaching the cylinder, accelerates as it moves from the stagnation point up to the shoulder, decelerates on the downstream half up to the rear stagnation point, and then accelerates once again to the free stream velocity far downstream. The flow over a flat plate set impulsively in motion in its own plane (Prob. 6.37) is one case where the convective inertial force is zero and we obtain a balance of unsteady inertial forces and viscous stresses. These contribute to a diffusion type of phenomena, with the effect of the plate motion penetrating into the fluid with time.

From the estimates of the unsteady and convective inertial forces obtained in Chapter 11, it is easily seen that whenever the Strouhal number is small compared to unity, the unsteady effects may be neglected and the convective effects dominate. Most fluid motions of engineering interest are dominated by convective effects (and are, therefore, quite difficult to analyse mathematically since the convective terms are non-linear).

Whenever the viscous forces are absent or negligible, the inertial force balances the pressure and the gravity forces and the motion is controlled by the Euler Eq. (11.10). As before, the gravity force may be combined with the pressure force. Since the net pressure force acting on a fluid particle is the negative of the pressure gradient, the equilibrium of the pressure and inertial forces suggests that the pressure gradient is in the same direction as the (negative of) fluid acceleration. Thus, in inviscid fluid motions, the pressure is higher where the velocity is smaller and vice versa. This same fact is stated by the Bernoulli Eq. (7.15). Figures 7.16, 7.21 and 12.19 show typical pressure variations in accelerated flows.

When the streamlines are curved, the convective inertial force includes the centripetal force as well and, therefore, the pressure gradient in inviscid flows has a component normal to the streamlines also. Examples include the flow through a pipe bend (see Fig. 6.19) or through a converging channel (see Fig. 7.27). A simple but significant example is a liquid in a rotating bucket such that the motion is like that of a rotating rigid body. Since centripetal acceleration is the only acceleration, the pressure gradient $\Delta\mathscr{P}$ is along the radial direction. This results in a parabolic free surface as seen in Example 6.4 and gives rise to a centrifugal buoyancy as discussed in Prob. 6.43. This is exploited in the design of several pollution control devices as in Probs. 13.56 and 13.57.

The picture is further complicated when the flow is viewed in a rotating reference, as for example, the motion in the Earth's atmosphere. In this case, the Coriolis component of acceleration dominates and the air currents are along the isobars rather than across them, as can be verified from any weather map. Such flows have not been discussed in this text but the reader is referred to a very lucid presentation in Scorer (Reading 29).

In flows where the free surface of a liquid (i.e., its interface with a gas) is not involved, for example, in the flow of gases, in flows completely confined by solid walls, or in the motion of deep sea submarines, the contribution of the gravity force can be visualized as limited to providing a buoyancy force (according to the Archimides' principle Eq. 2.14), and the fluid motion can be treated as being essentially under the influence of the modified pressure field $\mathscr{P}$. The net force on an object then is the integrated non-gravitational pressure force *plus* the buoyancy. In problems where the buoyancy is small compared to other forces (for example, the weight of the object), the consideration of gravity forces may be altogether eliminated as is done in the field of aerodynamics. In flows where the free surface of a liquid is central to the problem, the gravity effects cannot be isolated to the buoyancy force alone. The location of the free surface, determined by the condition that the pressure there equals the gas pressure everywhere, is itself governed by the dynamics of flow and, therefore, gravity affects the flow in a more complex manner. For this reason, we cannot introduce $\mathscr{P}$ and have to include the gravity effects in the analysis. The variation of the liquid level behind a dam (see Fig. 5.11), the nappe or sheet of liquid falling over a weir (see Fig. 7.24a), the geometry of a liquid jet issuing out of an orifice (see Fig. 7.11) are some situations where the gravity force is of central importance. In all of these cases, the Bernoulli equation 7.15 in its full form (i.e., without introduction of $\mathscr{P} = p + \rho gz$) is called for. In these cases, it is the pressure head *plus* the gravity head that reflects the amount of acceleration that the fluid has undergone (Example 7.8) and we cannot neglect the gravity head.

The gravity force also controls open channel flows and the drag on ships where the surface waves carry energy away from the ship and contribute to the drag (Sec. 14.5).

Another type of force is the surface tension force which has been mentioned only briefly. It is applicable only to a free surface. The capillary action (Sec. 2.9) is based on this force. Also, since the surface tension force tends to minimize the free-surface area, it is responsible for setting up waves similar to gravity waves. These surface tension waves are known as capillary waves and are also responsible for increasing the drag on ships (though by a negligibly small amount for full-sized ships). One example of capillary waves is given in Prob. 9.42 where the mechanism of their formation is also described. These effects are important only when the Weber number is sufficiently small compared to unity. In most hydraulic structures, this number is usually much larger than one, and hence capillary effects are generally confined to small-scale models only. For example, while the full scale bridge piers exhibit only gravity waves, in their small scale models the capillary waves are also present, superposed on the gravity waves.

The viscous stresses in a fluid depend on the severity of the rate of distortion produced. As explained in Sec. 6.3 the rate of distortion depends on the velocity gradients. It has been shown in Sec. 11.2 that the Reynolds number based on an appropriate characteristic length measures the relative magnitude of the inertial and viscous forces and that the viscous forces are significant only when Re is small. Great care, of course, needs to be exercised in selecting the proper characteristic length. When $\text{Re} \ll 1$, we may neglect the inertial forces altogether and expect the pressure forces to be balanced by the viscous forces. This is the Stokes flow regime discussed in Sec. 11.4 and is marked by large values of drag coefficients. The creeping flows as, for example, the motion of dust particles in air or the seepage flow of ground water and the quasi-fully-developed flows as in lubricated bearings (Example 6.2) are all characterized by negligible inertia. The characteristics of such flows are summarized in Sec. 11.4.

The other limit when $\text{Re} \gg 1$ is of greater interest physically since most of the engineering flows fall in this regime. Here one expects the viscous effects to be negligible suggesting that the flow patterns should be similar to those for inviscid flow as discussed in Sec. 11.5. In such a flow field, the net pressure force on a closed surface integrates out to zero, predicting the absence of drag. This result is, of course, ridiculous and, as was shown in Sec. 11.6 and Chapter 13, the presence of viscosity, howsoever small, results in a thin region next to the solid walls where the shear effects must be taken into account. In this region, termed as the boundary layer, the viscous forces are of the same order as the inertial forces. The solution of the flow problem at high Reynolds numbers is generally obtained by patching two solutions — one for the thin viscous boundary layer and the other for the remaining outer region where the flow may be assumed inviscid. Even though the boundary layer involves three types of forces, inertial, viscous and pressure, the last of these can be taken as known *a-priori*—from the inviscid outer flow, the solution of which is obtained first. Such an analysis gives a skin-friction drag coefficient of the order of $Re^{-1/2}$. While the measured drag coefficients for streamlined bodies agree with this estimate, those for bluff bodies are much larger. The reason for this departure lies in the dynamics of the boundary layer flow as discussed in Sec. 13.6. If the pressure increases in the downstream direction, the boundary layer tends to separate from the wall, changing the entire flow pattern drastically. Thus, the shear effects which are normally confined to a narrow region near the walls and, therefore, let the outer flow field be approximately the same as predicted by inviscid flow theory, spread much farther into the flow field on separation. The pressures at the rear of

Some fluid flow phenomena involving two types of forces

Types of forces	Pressure p or $\mathscr{P}$	Gravity	Surface tension	Viscous
Inertial	Inviscid flows (Ch.12)	Waves on surface	Capillary waves (Prob. 9.42)	Impulsive motion of plate (unsteady convective inertia, Prob. 6.37) Steady boundary layer on flat plate (Sec. 13.3)
	Outer flows (Sec.11.15) Fluid moving as rigid body: accelerating tanker (Prob.6.13) rotating bucket (Example 6.4)	Liquid jets (Example 7.4) Flow over weir (Example 7.8)		
Pressure p or $\mathscr{P}$		Fluid statics (Ch. 2)	Soap bubbles (Prob. 2.28)	Fully developed Poiseuille flow (Example 6.1) Stokes flow (Sec.11.4) Lubricated bearings (Example 6.2)
Gravity			Capillarity (Sec. 2.9)	

a bluff body are then no longer predicted by the Bernoulli equation and are lower than those at the front. This results in a net pressure drag or form drag, which, for bluff bodies, is orders of magnitude larger than the skin-friction drag.

We have outlined above only a few of the multitude of fluid-flow phenomena that are encountered. The table on page 465 summarizes some of the situations from those discussed in the text which involve a balance of two types of forces. We have restricted ourselves only to laminar and incompressible flow problems, the consideration of which forms the bulk of this text. There is still a wide variety of phenomena which we have not touched upon or have barely done so. Some of the books listed in further readings dwell upon these in greater detail. The structure presented in this text should equip the reader to undertake a profitable study of these.

FURTHER READING

The following books give a good introduction to fluid-flow phenomena. These have been written as notes for a series of excellent cine-films (which are distributed in India by National Education and Information Films Ltd., Apollo Bunder, Mumbai).

1. Shapiro, A.H., *Shape and Flow,* Heinemann, 1961.
2. National Committee of Fluid Mechanics Films, *Illustrated Experiments in Fluid Mechanics,* MIT Press, 1972.

   The following contains an excellent collection of actual flow photographs:
3. Van Dyke, M., *An Album of Fluid Motion,* Parabolic Press, 1982.

   The following are some outstanding texts written for engineers at about the same level as this book:
4. Li, W.H. and Lam, S.H., *Principles of Fluid Mechanics,* Addison Wesley, 1964.
5. Lu, P.C., *Fluid Mechanics,* Iowa State Univ. Press, 1979.
6. Massey, B.S., *Mechanics of Fluids,* Van Nostrand Reinhold, 1979.
7. Rouse, H., *Elementary Mechanics of Fluids,* Wiley, 1979.
8. Sabersky, R.H., Acosta, A.J. Hauptmann, E.G. and Gates, E.M., *Fluid Flow—A First Course in Fluid Mechanics,* 4th Edn., Collier-Macmillan, 1998.
9. White, F.M., *Fluid Mechanics,* McGraw-Hill, 6th Edn., 2006.

   The following are some general texts at a higher level:
10. Batchelor, G.K. and Batchelor, G.K., *An Introduction to Fluid Dynamics,* Cambridge, 2000.
11. Landau, L.D. and Lifshitz, E.M., *Fluid Mechanics,* Pergamon, 1987.
12. Prandtl, L. and Tietjens, O.G., *Fundamentals of Hydro- and Aeromechanics,* Dover, 1957.
13. Prandtl, L. and Tietjens, O.G., *Applied Hydro– and Aeromechanics,* Dover, 1957.
14. Rosenhead, L. (Ed.), *Laminar Boundary Layer,* Dover, 1988.
15. Schlichting, H., *Boundary Layer Theory,* McGraw-Hill, 2000.
16. Tritton, D.J., *Physical Fluid Dynamics,* Van Nostrand Reinhold, 1988.
17. White, F.M., *Viscous Fluid Flow,* McGraw-Hill, 2005.

   The following are books on particular branches of fluid mechanics:
18. Bradshaw, P., *Experimental Fluid Mechanics,* Pergamon, 1970, 2nd Edn.
19. Bradshaw, P., *An Introduction to Turbulence and its Measurement,* Pergamon, 1971.

20. Daish, C.B., *The Physics of Ball Games,* English Universities Press, 1972.
21. Golding, E.W., *The Generation of Electricity by Wind Power,* E. and F.N. Spon, 1976, 2nd Edn.
22. Goody, R.M. and Walker, J.C.G., *Atmospheres,* Prentice-Hall, 1972.
23. Gupta, S.K., *Momentum Transfer Operations,* Tata McGraw-Hill, 1979.
24. Humphrey, E.F. and Tarumoto, D.M. (Eds.), ***Fluidics,*** Fluid Amplifiers Associates, 1965.
25. Krishnamurty, K., *Principles of Ideal Fluid-Aerodynamics,* Wiley, 1980.
26. Kuethe, A.M. and Chow, C.Y., *Foundations of Aerodynamics,* Wiley, 1997.
27. Massey, B.S., *Units, Dimensional Analysis, and Physical Similarity,* Van Nostrand Reinhold, 1971.
28. Milne-Thomson, L.M., *Theoretical Hydrodynamics,* Macmillan, 1996.
29. Scorer, R.S., *Natural Aerodynamics,* Pergamon, 1958.
30. Schuring, D.J., *Scale Models in Engineering—Fundamentals and Applications,* Pergamon, 1977.
31. Shapiro, A.H., *The Dynamics and Thermodynamics of Compressible Fluid Flows,* 2 Vols., Ronald, 1958.
32. Shepherd, D.G., *Principles of Turbomachinery,* Macmillan, 1971.
33. Tennekes, H. and Lumley, J.L., *A First Course in Turbulence,* MIT Press, 1972.
34. Thompson, P.A., *Compressible-Fluid Dynamics,* McGraw-Hill, 1972.

APPENDIX - A

UNITS AND DIMENSIONS

All quantitative studies of physical phenomena involve the measurement and specification of the magnitudes of such quantities as length, time, velocity, force, pressure and energy. The measurements are made in terms of certain defined standards termed as *units*. Thus, a metre is a defined standard of length and to say that a certain length is 7 m implies a comparison with that standard. In principle, an independent unit could be defined for each type of physical quantity, for example, independent units of length, time, velocity, acceleration, etc. But this would be too cumbersome – for velocity is related to length and time in a certain manner and instead of using an independent unit for velocity, a *derived unit* m/s could be used which would suffice. These derived units are obtained through the use of physical laws relating various quantities.

In the Système International d'Unités (k) the use of which is recommended universally, the units of mass, length and time are the basic units in terms of which all other units (besides thermal, magnetic, and electrical) are obtained. These units are kilogram, metre and second respectively. All other units are obtained as products of powers of these units. The dimensions of a physical quantity are the powers of the basic units contained in its derived unit. If M represents the mass, L, the length, and T, the time dimension, then the dimensions of velocity are LT^{-1} and of acceleration, LT^{-2}.

Since the SI units use mass, length and time as the basic dimensions, it is termed as an *MLT* system. Another *MLT* system based on units of pound, foot and second (the FPS system) is also prevalent.

As pointed out earlier, the derived units are obtained through the use of physical laws relating various quantities. These physical laws are in the nature of statements of proportionality. Thus, the Newton second Law requires that

$$\text{force} \propto (\text{mass}) \times (\text{acceleration}) \tag{A.1}$$

or

$$\text{force} = k \times (\text{mass}) \times (\text{acceleration}) \tag{A.2}$$

where k is a universal constant, i.e., independent of the physical situation, though not necessarily of the system of units.

In *MLT* systems, the value of k is set as unity to obtain the unit of force as that magnitude which produces a unit acceleration on a unit mass. For example, in SI units, a unit force accelerates a 1 kg mass through 1 m/s^2. Thus, the derived unit is kg m/s^2. It is termed as a *Newton* (N). Similarly, in the FPS system, k is unity and the unit of force is a *poundal,* which produces an acceleration of 1 ft/sec^2 on a mass of 1 lb.

It is not essential to have only three basic units and dimensions, namely, *M, L* and *T.* In fact, the engineering system of units in use till recently, and in terms of which a large body of literature still exists, uses force also as a primary quantity and, thus, the system is an *FMLT* system. In the British Engineering system, besides the units of mass, length and time as pound, foot and second respectively, another basic unit is used, the pound-force (lbf) which is defined as the force of gravity acting on one pound mass. In this system, then, the value of k in Eq. (A.2) is not unity and also, is not dimensionless. Using the fact that the standard acceleration due to gravity is 32.2 ft/sec^2, we get

$$1 \text{ lbf} = k \times 1 \text{ (lb)} \times 32.2 \text{ (ft/s}^2) \tag{A.3}$$

or

$$k = \frac{1}{32.2} \; \frac{\text{lbf}}{\text{lb}\left(\text{ft/s}^2\right)} \tag{A.4}$$

Newton's second law should thus read in the *FMLT* British engineering system as*

$$\text{force} = \frac{1}{32.2} \times \text{(mass)} \times \text{(acceleration)} \tag{A.5}$$

Similarly, in the MKKS (metre-kilogram-kg force-second) system, the basic unit of force is the kilogram-force (kgf) which is the gravity force on a 1 kilogram mass (kg). Thus,

$$k = \frac{1}{9.81} \frac{\text{kgf}}{\text{kg}\left(\text{m/s}^2\right)} \tag{A.6}$$

and

$$\text{force} = \frac{1}{9.81} \times \text{(mass)} \times \text{(acceleration)} \tag{A.7}$$

A British *FLT* (force-length-time) system has also been used in engineering practice. This uses the basic units of pound-force, foot and second, and the value of k is put as unity. The corresponding *derived* unit of mass is termed as the *slug.* Thus, Eq. (A.2) can be written in this system as

$$\text{force} = 1 \times \text{(mass)} \times \text{(acceleration)} \tag{A. 8}$$

Table A.1 gives the units (with conversion factors) of several important variables in some commonly used systems.

---

* In some books, k is written as $\frac{1}{g_c}$ or $\frac{1}{g_o}$.

Table A.1. Some common units and conversion factors

Quantity	SI Units MLT	MKS Units FMLT	British Engineering Units FMLT	British Engineering Units FLT	Other Units
length	metre(m)		foot (ft = 0.3048 m)		mile (mi = 1.6 km) inch (in = 1/12 ft)
mass	kilogram (kg)		pound (lb = 0.4536 kg)	slug (= 32.2 lb) (= 14.6 kg)	
time	second (s)		second (sec)		hour (hr = 3600 sec)
force	kg m/s^2 ≡ Newton (N)	kilogram force (kgf = 9.8 N)	pound force (lbf = 4.45 N)		
volume	m^3		ft^3 (= 2.83 × 10^{-2} m^3)		gallon imperial (gal = 4.73 × 10^{-3} m^3) litre (L = 10^{-3} m^3) gallon U.S. (= 3.78 × 10^{-3} m^3)
volume flux	m^3/s		ft^3/sec (= 2.83 × 10^{-3} m^3/s)		ft^3/min (cfm = 0.17 m^3/s) gal/min (= 7.89 × 10^{-2} m^3/s) gal U.S./min (= 6.31 × 10^{-2} m^3/s)
density	kg/m^3		lb/ft^3 (= 16.02 kg/m^3)	slug/ft^3 (= 515.4 kg/m^3)	
velocity	m/s		ft/sec (= 0.3048 m/s)		knot (= 1.688 ft/sec) (= 0.514 m/s) miles/hr (= 0.447 m/s)
acceleration	m/s^2		ft/sec^2 (= 0.3048 m/s^2)		
pressure	N/m^2 ≡ Pascal (Pa)	kgf/m^2 (= 9.81 Pa)	lbf/ft^2 (psf = 47.88 Pa)		bar (=10^5 Pa) atmosphere (= 1.013 × 10^5 Pa) pound per in^2 (psi = 6.89 × 10^3 Pa)
energy or work	N/m ≡ Joule (J)	kgf/m (= 9.81 J)	ft-lbf (= 1.36 J)		Btu (= 1.055 × 10^3 J) calorie (= 4.187 J)
power	J/s ≡ Watt (W)	kgf m/s (= 9.81 W)	ft-lbf/sec (= 1.36 W)		horse power (HP = 7.46 × 10^2 W)
specific energy	J/kg	kgf m/kg (= 9.81 J/kg)	ft-lbf/lb (= 0.167 J/kg)	ft-lbf/slug (= 19.86 J/kg)	Btu/lb (= 2.326 × 10^3 J/kg)
dynamic viscosity	Pa s (= N s/m^2) (= kg/ms)	kg/m s (= Pa s)	lbf-sec/ft^2 (= 47.88 Pa s)	slug/ft-sec (= 47.88 Pa s)	Poise (p = 0.1 Pa s) centipoise (cp = 10^{-3} Pa s) lb/ft-sec (= 1.488 Pa s)
kinematic viscosity		m^2/s	ft^2/sec (= 9.29 × 10^{-2} m^2/s)		Stoke (St = 2.58 × 10^{-5} m^2/s)

APPENDIX - B

SOME USEFUL FORMULAE

B.1. Gradient of a Scalar, $\nabla\eta$

Cartesian

$$\nabla\eta = \frac{\partial\eta}{\partial x}\hat{\mathbf{i}} + \frac{\partial\eta}{\partial y}\hat{\mathbf{j}} + \frac{\partial\eta}{\partial z}\hat{\mathbf{k}}$$

Cylindrical

$$\nabla\eta = \frac{\partial\eta}{\partial r}\hat{\mathbf{r}} + \frac{1}{r}\frac{\partial\eta}{\partial\theta}\hat{\boldsymbol{\theta}} + \frac{\partial\eta}{\partial z}\hat{\mathbf{k}}$$

Spherical

$$\nabla\eta = \frac{\partial\eta}{\partial r}\hat{\mathbf{r}} + \frac{1}{r}\frac{\partial\eta}{\partial\theta}\hat{\boldsymbol{\theta}} + \frac{1}{r\sin\theta}\frac{\partial\eta}{\partial\phi}\hat{\phi}$$

B.2. Divergence of a Vector, $\nabla \cdot \mathbf{A}$

Cartesian

$$\nabla \cdot \mathbf{A} = \frac{\partial A_x}{\partial x} + \frac{\partial A_y}{\partial y} + \frac{\partial A_z}{\partial z}$$

Cylindrical

$$\nabla \cdot \mathbf{A} = \frac{1}{r}\frac{\partial}{\partial r}(rA_r) + \frac{1}{r}\frac{\partial A_\theta}{\partial\theta} + \frac{\partial A_z}{\partial z}$$

$$= \frac{\partial A_r}{\partial r} + \frac{A_r}{r} + \frac{1}{r}\frac{\partial A_\theta}{\partial\theta} + \frac{\partial A_z}{\partial z}$$

Spherical

$$\nabla \cdot \mathbf{A} = \frac{1}{r^2}\frac{\partial}{\partial r}\left(r^2 A_r\right) + \frac{1}{r\sin\theta}\frac{\partial\left(A_\theta \sin\theta\right)}{\partial\theta} + \frac{1}{r\sin\theta}\frac{\partial A_\phi}{\partial\phi}$$

$$= \frac{\partial A_r}{\partial r} + \frac{2A_r}{r} + \frac{1}{r}\frac{\partial A_\theta}{\partial\theta} + \frac{A_\theta\cot\theta}{r} + \frac{1}{r\sin\theta}\frac{\partial A_\phi}{\partial\phi}$$

B.3. Curl of a Vector, ∇ × A

Cartesian

$$\nabla \times \mathbf{A} = \begin{vmatrix} \hat{\mathbf{i}} & \hat{\mathbf{j}} & \hat{\mathbf{k}} \\ \dfrac{\partial}{\partial x} & \dfrac{\partial}{\partial y} & \dfrac{\partial}{\partial z} \\ A_x & A_y & A_z \end{vmatrix}$$

Cylindrical

$$\nabla \times \mathbf{A} = \begin{vmatrix} \dfrac{1}{r}\hat{\mathbf{r}} & \hat{\boldsymbol{\theta}} & \dfrac{1}{r}\hat{\mathbf{k}} \\ \dfrac{\partial}{\partial r} & \dfrac{\partial}{\partial \theta} & \dfrac{\partial}{\partial z} \\ A_r & rA_\theta & A_z \end{vmatrix}$$

Spherical

$$\nabla \times \mathbf{A} = \begin{vmatrix} \dfrac{1}{r^2 \sin\theta}\hat{\mathbf{r}} & \dfrac{1}{r\sin\theta}\hat{\boldsymbol{\theta}} & \dfrac{1}{r}\hat{\boldsymbol{\phi}} \\ \dfrac{\partial}{\partial r} & \dfrac{\partial}{\partial \theta} & \dfrac{\partial}{\partial \phi} \\ A_r & rA_0 & r\sin\theta\, A_\phi \end{vmatrix}$$

B.4. Laplace Operator, $\nabla^2 \eta$

When η is a scalar, **Φ**

Cartesian $$\nabla^2 \Phi = \frac{\partial^2 \Phi}{\partial x^2} + \frac{\partial^2 \Phi}{\partial y^2} + \frac{\partial^2 \Phi}{\partial z^2}$$

Cylindrical $$\nabla^2 \Phi = \frac{1}{r}\frac{\partial}{\partial r}\left(\frac{r\partial \Phi}{\partial r}\right) + \frac{1}{r^2}\frac{\partial^2 \Phi}{\partial \theta^2} + \frac{\partial^2 \Phi}{\partial z^2}$$

Spherical $$\nabla^2 \Phi = \frac{1}{r^2}\frac{\partial}{\partial r}\left(\frac{r^2\partial \Phi}{\partial r}\right) + \frac{1}{r^2 \sin\theta}\frac{\partial}{\partial \theta}\left(\sin\theta \frac{\partial \Phi}{\partial \theta}\right) + \frac{1}{r^2 \sin^2\theta}\frac{\partial^2 \Phi}{\partial \phi^2}$$

When η is a vector **V**

The relevant form is the one contained within the curly brackets on the RHS of Navier-Stokes equation, Sec. B.9.

B.5. Material Rate of Change

$$\frac{D\eta}{Dt} = \frac{\partial \eta}{\partial t} + (\mathbf{V} \cdot \nabla)\eta$$

When η is a scalar Φ

Cartesian $$\frac{D\Phi}{Dt} = \frac{\partial \Phi}{\partial t} + V_x \frac{\partial \Phi}{\partial x} + V_y \frac{\partial \Phi}{\partial y} + V_z \frac{\partial \Phi}{\partial z}$$

Cylindrical $$\frac{D\Phi}{Dt} = \frac{\partial \Phi}{\partial t} + V_r \frac{\partial \Phi}{\partial r} + \frac{V_\theta}{r} \frac{\partial \Phi}{\partial \theta} + V_z \frac{\partial \Phi}{\partial z}$$

Spherical $$\frac{D\Phi}{Dt} = \frac{\partial \Phi}{\partial t} + V_r \frac{\partial \Phi}{\partial r} + \frac{V_\theta}{r} \frac{\partial \Phi}{\partial \theta} + \frac{V_\phi}{r \sin\theta} \frac{\partial \Phi}{\partial \phi}$$

When η is a vector **V**

The relevant form is the one contained within the curly brackets on the LHS of the Navier-Stokes equation, Sec. B.9.

B.6. Rates of Deformations

Cartesian

$$\mathring{\epsilon}_x = \partial V_x/\partial x \qquad \mathring{\epsilon}_y = \partial V_y/\partial y \qquad \mathring{\epsilon}_z = \partial V_z/\partial z$$

$$\omega = \frac{1}{2}(\nabla \times \mathbf{V}) \qquad \text{(Sec. B.3)}$$

$$\mathring{\gamma}_{xy} = \mathring{\gamma}_{yx} = \frac{\partial V_y}{\partial x} + \frac{\partial V_x}{\partial y}$$

$$\mathring{\gamma}_{yz} = \mathring{\gamma}_{zy} = \frac{\partial V_z}{\partial y} + \frac{\partial V_y}{\partial z}$$

$$\mathring{\gamma}_{zx} = \mathring{\gamma}_{xz} = \frac{\partial V_x}{\partial z} + \frac{\partial V_z}{\partial x}$$

Cylindrical

$$\mathring{\epsilon}_r = \partial V_r/\partial r \qquad \mathring{\epsilon}_\theta = \frac{1}{r}\frac{\partial V_\theta}{\partial \theta} + \frac{V_r}{r} \qquad \mathring{\epsilon}_z = \partial V_z/\partial z$$

$$\omega = \frac{1}{2}(\nabla \times \mathbf{V}) \qquad \text{(Sec. B.3)}$$

$$\mathring{\gamma}_{r\theta} = \mathring{\gamma}_{\theta r} = \frac{\partial V_\theta}{\partial r} - \frac{V_\theta}{r} + \frac{1}{r}\frac{\partial V_r}{\partial_\theta}$$

$$\mathring{\gamma}_{\theta z} = \mathring{\gamma}_{z\theta} = \frac{1}{r}\frac{\partial V_z}{\partial \theta} + \frac{\partial V_\theta}{\partial z}$$

$$\mathring{\gamma}_{zr} = \mathring{\gamma}_{rz} = \frac{\partial V_r}{\partial z} + \frac{\partial V_z}{\partial r}$$

Spherical

$$\mathring{\epsilon}_r = \frac{\partial V_r}{\partial r} \quad \mathring{\epsilon}_\phi = \frac{1}{r\sin\theta}\frac{\partial V_\phi}{\partial \phi} + \frac{V_r}{r} + \frac{V_\theta \cot\theta}{r} \quad \mathring{\epsilon}_\theta = \frac{1}{r}\frac{\partial V_\theta}{\partial \theta} + \frac{V_r}{r}$$

$$\omega = \frac{1}{2}(\nabla \times \mathbf{V}) \qquad \text{(Sec. B.3)}$$

$$\mathring{\gamma}_{r\theta} = \mathring{\gamma}_{\theta r} = \frac{r\partial(V_\theta/r)}{\partial r} + \frac{1}{r}\frac{\partial V_r}{\partial \theta}$$

$$\mathring{\gamma}_{\phi\theta} = \mathring{\gamma}_{\theta\phi} = \frac{\sin\theta}{r}\frac{\partial}{\partial \theta}\left(\frac{V_\phi}{\sin\theta}\right) + \frac{1}{r\sin\theta}\frac{\partial V_\theta}{\partial \phi}$$

$$\mathring{\gamma}_{\phi r} = \mathring{\gamma}_{r\phi} = \frac{1}{r\sin\theta}\frac{\partial V_r}{\partial \phi} + r\frac{\partial}{\partial r}\left(\frac{V_\phi}{r}\right)$$

B.7. Newton-Stokes Law (see Sec. B.6. also)

Cartesian

$$\sigma_{xx} = -p + 2\mu\mathring{\epsilon}_x - \frac{2}{3}\mu\left(\frac{\partial V_x}{\partial x} + \frac{\partial V_y}{\partial y} + \frac{\partial V_z}{\partial z}\right)$$

$$\sigma_{yy} = -p + 2\mu\mathring{\epsilon}_y - \frac{2}{3}\mu\left(\frac{\partial V_x}{\partial x} + \frac{\partial V_y}{\partial y} + \frac{\partial V_z}{\partial z}\right)$$

$$\sigma_{zz} = -p + 2\mu\mathring{\epsilon}_z - \frac{2}{3}\mu\left(\frac{\partial V_x}{\partial x} + \frac{\partial V_y}{\partial y} + \frac{\partial V_z}{\partial z}\right)$$

$$\tau_{xy} = \tau_{yx} = \mu\mathring{\gamma}_{xy}$$

$$\tau_{yz} = \tau_{zy} = \mu\mathring{\gamma}_{yz}$$

$$\tau_{zx} = \tau_{xz} = \mu\mathring{\gamma}_{zx}$$

Cylindrical

$$\sigma_{rr} = -p + 2\mu \overset{\circ}{\epsilon}_r - \frac{2}{3}\mu \left(\frac{\partial V_r}{\partial r} + \frac{1}{r}\frac{\partial V_\theta}{\partial \theta} + \frac{\partial V_z}{\partial z} + \frac{V_r}{r} \right)$$

$$\sigma_{\theta\theta} = -p + 2\mu \overset{\circ}{\epsilon}_\theta - \frac{2}{3}\mu \left(\frac{\partial V_r}{\partial r} + \frac{1}{r}\frac{\partial V_\theta}{\partial \theta} + \frac{\partial V_z}{\partial z} + \frac{V_r}{r} \right)$$

$$\sigma_{zz} = -p + 2\mu \overset{\circ}{\epsilon}_z - \frac{2}{3}\mu \left(\frac{\partial V_r}{\partial r} + \frac{1}{r}\frac{\partial V_\theta}{\partial \theta} + \frac{\partial V_z}{\partial z} + \frac{V_r}{r} \right)$$

$$\tau_{r\theta} = \tau_{\theta r} = \mu \overset{\circ}{\gamma}_{r\theta};\ \tau_{\theta z} = \tau_{z\theta} = \mu \overset{\circ}{\gamma}_{\theta z};\ \tau_{zr} = \tau_{rz} = \mu \overset{\circ}{\gamma}_{zr}$$

Spherical

$$\sigma_{rr} = -p + 2\mu \overset{\circ}{\epsilon}_r - \frac{2}{3}\mu \left[\frac{1}{r^2}\frac{\partial}{\partial r}\left(r^2 V_r\right) + \frac{1}{r\sin\theta}\frac{\partial}{\partial \theta}(V_\theta \sin\theta) + \frac{1}{r\sin\theta}\frac{\partial V_\phi}{\partial \phi} \right]$$

$$\equiv -p + 2\mu \overset{\circ}{\epsilon}_r - \frac{2}{3}\mu\,(\nabla\cdot\mathbf{V})$$

$$\sigma_{\phi\phi} = -p + 2\mu \overset{\circ}{\epsilon}_\phi - \frac{2}{3}\mu\,(\nabla\cdot\mathbf{V})$$

$$\sigma_{\theta\theta} = -p + 2\mu \overset{\circ}{\epsilon}_\theta - \frac{2}{3}\mu\,(\nabla\cdot\mathbf{V})$$

$$\tau_{r\theta} = \tau_{\theta r} = \mu \overset{\circ}{\gamma}_{\theta r};\ \tau_{r\phi} = \tau_{\phi r} = \mu \overset{\circ}{\gamma}_{r\phi};\ \tau_{\theta\phi} = \tau_{\phi\theta} = \mu \overset{\circ}{\gamma}_{\phi\theta}$$

B.8. Continuity Equation

$$\boxed{\frac{\partial \rho}{\partial t} + \{\nabla.(\rho\mathbf{V})\} = 0}$$

Cartesian $$\frac{\partial \rho}{\partial t} + \left\{ \frac{\partial}{\partial x}(\rho V_x) + \frac{\partial}{\partial y}(\rho V_y) + \frac{\partial}{\partial z}(\rho V_z) \right\} = 0$$

Cylindrical $$\frac{\partial \rho}{\partial t} + \left\{ \frac{1}{r}\frac{\partial}{\partial r}(r\rho V_r) + \frac{1}{r}\frac{\partial}{\partial \theta}(\rho V_\theta) + \frac{\partial}{\partial z}(\rho V_z) \right\} = 0$$

Spherical $$\frac{\partial \rho}{\partial t} + \left\{ \frac{1}{r^2}\frac{\partial}{\partial r}\left(r^2\rho V_r\right) + \frac{1}{r\sin\theta}\frac{\partial}{\partial \theta}(\rho V_\theta \sin\theta) + \frac{1}{r\sin\theta}\frac{\partial}{\partial \phi}(\rho V_\phi) \right\} = 0$$

B.9. Navier-Stokes Equation

$$\boxed{\rho\left\{\frac{D\mathbf{V}}{Dt}\right\} \equiv \rho\left\{\frac{\partial \mathbf{V}}{\partial t} + (\mathbf{V}\cdot\nabla)\mathbf{V}\right\} = \rho\mathbf{f}\,\nabla p + \mu\left\{\nabla^2\mathbf{V}\right\}}$$

Cartesian

$$\rho\left\{\frac{\partial V_x}{\partial t} + V_x\frac{\partial V_x}{\partial x} + V_y\frac{\partial V_x}{\partial y} + V_z\frac{\partial V_x}{\partial z}\right\}$$

$$= \rho f_x - \frac{\partial p}{\partial x} + \mu\left\{\frac{\partial^2 V_x}{\partial x^2} + \frac{\partial^2 V_x}{\partial y^2} + \frac{\partial^2 V_x}{\partial z^2}\right\}$$

$$\rho\left\{\frac{\partial V_y}{\partial t} + V_x\frac{\partial V_y}{\partial x} + V_y\frac{\partial V_y}{\partial y} + V_z\frac{\partial V_y}{\partial z}\right\}$$

$$= \rho f_y - \frac{\partial p}{\partial y} + \mu\left\{\frac{\partial^2 V_y}{\partial x^2} + \frac{\partial^2 V_y}{\partial y^2} + \frac{\partial^2 V_y}{\partial z^2}\right\}$$

$$\rho\left\{\frac{\partial V_z}{\partial t} + V_x\frac{\partial V_z}{\partial x} + V_y\frac{\partial V_z}{\partial y} + V_z\frac{\partial V_z}{\partial z}\right\}$$

$$= \rho f_z - \frac{\partial p}{\partial z} + \mu\left\{\frac{\partial^2 V_z}{\partial x^2} + \frac{\partial^2 V_z}{\partial y^2} + \frac{\partial^2 V_z}{\partial z^2}\right\}$$

Cylindrical

$$\rho\left\{\frac{\partial V_r}{\partial t} + V_r\frac{\partial V_r}{\partial r} + \frac{V_\theta}{r}\frac{\partial V_r}{\partial \theta} - \frac{V_\theta^2}{r} + V_z\frac{\partial V_r}{\partial z}\right\}$$

$$= \rho f_r - \frac{\partial p}{\partial r} + \mu\left\{\frac{\partial}{\partial r}\left(\frac{1}{r}\frac{\partial}{\partial r}(rV_r)\right) + \frac{1}{r^2}\frac{\partial^2 V_r}{\partial \theta^2} - \frac{2}{r^2}\frac{\partial V_\theta}{\partial \theta} + \frac{\partial^2 V_r}{\partial z^2}\right\}$$

$$\rho\left\{\frac{\partial V_\theta}{\partial t} + V_r\frac{\partial V_\theta}{\partial r} + \frac{V_\theta}{r}\frac{\partial V_\theta}{\partial \theta} + \frac{V_r V_\theta}{r} + V_z\frac{\partial V_\theta}{\partial z}\right\}$$

$$= \rho f_\theta - \frac{1}{r}\frac{\partial p}{\partial \theta} + \mu\left\{\frac{\partial}{\partial r}\left(\frac{1}{r}\frac{\partial}{\partial r}(rV_\theta)\right) + \frac{1}{r^2}\frac{\partial^2 V_\theta}{\partial \theta^2} + \frac{2}{r^2}\frac{\partial V_r}{\partial \theta} + \frac{\partial^2 V_\theta}{\partial z^2}\right\}$$

$$\rho\left\{\frac{\partial V_z}{\partial t} + V_r\frac{\partial V_z}{\partial r} + \frac{V_\theta}{r}\frac{\partial V_z}{\partial \theta} + V_z\frac{\partial V_z}{\partial z}\right\}$$

$$= \rho f_z - \frac{\partial p}{\partial z} + \mu\left\{\frac{1}{r}\frac{\partial}{\partial r}\left(r\frac{\partial V_z}{\partial r}\right) + \frac{1}{r^2}\frac{\partial^2 V_z}{\partial \theta^2} + \frac{\partial^2 V_z}{\partial z^2}\right\}$$

Spherical

$$\rho\left\{\frac{\partial V_r}{\partial t}+V_r\frac{\partial V_r}{\partial r}+\frac{V_\theta}{r}\frac{\partial V_r}{\partial \theta}+\frac{V_\phi}{r\sin\theta}\frac{\partial V_r}{\partial \phi}-\frac{V_\theta^2+V_\phi^2}{r}\right\}$$

$$=\rho f_r-\frac{\partial p}{\partial r}+\mu\left\{\nabla^2 V_r-\frac{2\,V_r}{r^2}-\frac{2}{r^2}\frac{\partial V_\theta}{\partial \theta}-\frac{2}{r^2}V_\theta\cot\theta-\frac{2\left(\partial V_\phi/\partial\phi\right)}{r^2\sin\theta}\right\}$$

$$\rho\left\{\frac{\partial V_\theta}{\partial t}+V_r\frac{\partial V_\theta}{\partial r}+\frac{V_\theta}{r}\frac{\partial V_\theta}{\partial \theta}+\frac{V_\phi}{r\sin\theta}\frac{\partial V_\theta}{\partial \phi}+\frac{V_r V_\theta}{r}-\frac{V_\phi^2\cot\theta}{r}\right\}$$

$$=\rho f_\theta-\frac{1}{r}\frac{\partial p}{\partial \theta}+\mu\left\{\nabla^2 V_\theta+\frac{2}{r^2}\frac{\partial V_r}{\partial \theta}-\frac{V_\theta}{r^2\sin^2\theta}-\frac{2\cos\theta}{r^2\sin^2\theta}\frac{\partial V_\phi}{\partial \phi}\right\}$$

$$\rho\left\{\frac{\partial V_\phi}{\partial t}+V_r\frac{\partial V_\phi}{\partial r}+\frac{V_\theta}{r}\frac{\partial V_\phi}{\partial \theta}+\frac{V_\phi}{r\sin\theta}\frac{\partial V_\phi}{\partial \phi}+\frac{V_\phi V_r}{r}+\frac{V_\theta V_\phi}{r}\cot\theta\right\}$$

$$=\rho f_\phi-\frac{1}{r\sin\theta}\frac{\partial p}{\partial \phi}+\mu\left\{\nabla^2 V_\phi-\frac{V_\phi}{r^2\sin^2\theta}+\frac{2}{r^2\sin\theta}\frac{\partial V_r}{\partial \phi}+\frac{2\cos\theta}{r^2\sin^2\theta}\frac{\partial V_\theta}{\partial \phi}\right\}$$

where $$\nabla^2=\frac{1}{r^2}\frac{\partial}{\partial r}\left(\frac{r^2\partial}{\partial r}\right)+\frac{1}{r^2\sin\theta}\frac{\partial}{\partial \theta}\left(\sin\theta\frac{\partial}{\partial \theta}\right)+\frac{1}{r^2\sin^2\theta}\left(\frac{\partial^2}{\partial \phi^2}\right)$$

APPENDIX - C

DIMENSIONAL ANALYSIS

As seen in Sec. 9.4, the non-dimensional pi-numbers that result from scale-factor considerations can be used as dimensionless variables and allow a significant reduction in experimental and/or analytical effort required. Dimensional analysis is an *alternative approach* to the same problem and provides procedural techniques whereby the variables that are assumed to be controlling the physics of a problem can be formed into dimensionless groups.

Outlined below are the principal steps involved in determining the non-dimensional Π's.

(a) List all the physical variables or parameters involved in the problem, e.g. X_1, X_2, X_3, ... , X_n, a total of *n* quantities.

(b) Choose a system of dimensions*; *MLT*, *FLT* or *FMLT* (see Appendix A).

(c) Write the dimensions of all the relevant variables in the chosen system. Table C.l gives dimensions of some physical quantities in *MLT, FLT* and *FMLT* systems. Determine the minimum number *r* of dimensions required to express these variables. For *most* problems in fluid mechanics, this number equals 3.

(d) Select a 'basic' group of *r independent* variables or parameters. The selection is made such that all the dimensions are involved in these, and that non-dimensional parameters cannot be formed from the products of powers of these chosen variables. It is usually convenient to select one geometric, one kinematic and one dynamic variable or parameter. Let these be X_1, X_2, X_3.

(e) It is now possible to form $(n-r)$ independent Π*'s* out of all the variables.

$$\Pi_{i-3} = X_i \; X_1^a \; X_2^b \; X_3^c$$

for $i = 4, \ldots, n$.

---

* A beginner may use any system, but there are some sophisticated considerations by which the power of dimensional analysis can be increased by a proper choice. The reader is referred to a very fine book: Taylor, E.S., *Dimensional Analysis for Engineers,* Clarendon, 1974 which is an essential reading for all those who want to use dimensional analysis for new problems.

Table C.1. Dimensions of some physical quantities

Quantities	Dimension		
	MLT	*FLT*	*FMLT*
Length	L	L	L
Time	T	T	T
Mass	M	$FL^{-1}T^2$	M
Force	MLT^{-2}	F	F
Velocity	LT^{-1}	LT^{-1}	LT^{-1}
Acceleration	LT^{-2}	LT^{-2}	LT^{-2}
Newton Law Constant, k	None	None	$FM^{-1}L^{-1}T^2$
Momentum	MLT^{-1}	FT	MLT^{-1}
Work	ML^2T^{-2}	FL	FL
Pressure	$ML^{-1}T^{-2}$	FL^{-2}	FL^{-2}
Density	ML^{-3}	$FL^{-4}T^2$	ML^{-3}
Viscosity	$ML^{-1}T^{-1}$	$FL^{-2}T$	$FL^{-2}T$
Kinematic Viscosity	L^2T^{-1}	L^2T^{-1}	L^2T^{-1}
Surface Tension	MT^{-2}	FL^{-1}	FL^{-1}
Stress	$ML^{-1}T^{-2}$	FL^{-2}	FL^{-2}
Mass Flow Rate	MT^{-1}	$FL^{-1}T$	MT^{-1}
Power	ML^2T^{-3}	FLT^{-1}	FLT^{-1}

(f) Determine the values of a, b and c such that the Π's are dimensionless. A fundamental theorem known as the Buckingham Pi theorem states that *the total number of independent parameters which can be formed by combining the n physical variables and parameters of a problem is equal to (n–r), where r is the number of primary dimensions required to express the dimensional formulae of the n physical quantities.**

Illustrated below is the use of this procedure to determine the non-dimensional parameters involved in turbomachinery such as pumps. Consider a pump handling a volume rate of flow $\overset{\circ}{Q}$, density ρ, viscosity μ, bulk modulus E_s, working against a pressure difference $\Delta\mathscr{P}$ and consuming power at a rate $\overset{\circ}{W}$. Let the rotational speed be n and the characteristic dimension D. Thus, there are $n = 8$ parameters involved. Working in the *MLT* system of dimensions, 3 basic variables are required so that $r = 3$, and, thus, one can form 5 (= 8 – 3) non-dimensional Π's. Table C.2 shows one way of organizing the information. In this, D, n and ρ are selected as the three quantities forming the basic group. The number under 'values of exponents' are the values of a, b and c required to make the respective groups non-dimensional. These non-dimensional Π's are similar to those obtained in Chapter 10 using the method of Sec. 9.4. The

* A more rigorous statement of the theorem requires r to be the *rank of the matrix* formed by the dimensional exponents of all the physical quantities involved. See Langhaar, H.L., *Dimensional Analysis and Theory of Models,* Wiley, 1951, for a complete discussion including proof.

additional parameter Π_4 arises because of the effects of compressibility which was not included in Chapter 10.

Table C.2. Calculation of Π's for turbomachinery

	Variables							
	Basic group			**Others**				
	D	n	ρ	$\mathring{Q}$	$\Delta\mathscr{P}$	μ	E_s	$\mathring{W}$
Dimensions	L	T^{-1}	ML^{-3}	L^3T^{-1}	$ML^{-1}T^{-2}$	$ML^{-1}T^{-1}$	$ML^{-1}T^{-1}$	ML^2T^{-3}
	D^a	n^b	ρ^c					
Non-dimensional parameters	Values of exponents							
$\Pi_1 = C_{\mathring{Q}}$	−3	−1	0	$\mathring{Q}/nD^3$				
$\Pi_2 = C_H$	−2	−2	−1		$\Delta\mathscr{P}/\rho n^2D^2$			
$\Pi_3 = 1/\text{Re}$	−2	−1	−1			$\mu/\rho n^2D^2$		
$\Pi_4 = 1/\text{Ma}$	−2	−2	−1				$E_s/\rho n^2D^2$	
$\Pi_5 = C_{\mathring{W}}$	−5	−3	−1					$\mathring{W}/\rho n^3D^5$

APPENDIX - D

PROPERTIES OF FLUIDS

Table D.1. Density, viscosity and surface tension (in contact with air) of some common liquids at 20°C

	ρ (kg/m^3)	μ, cp (10^{-3} Pa s)	σ (10^{-3} N/m)
Benzene	881.3	0.651	28.90
Blood* (human)	1056.0	8.0	—
Ethanol	788.6	1.201	22.77
Gasoline	680.3	0.292	—
Glycerin	1262.7	1488.0	63.34
Kerosene	804.0	1.914	27.73
Mercury	13555.0	1.555	513.7
SAE 10 oil	917.4	81.34	36.49
SAE 30 oil	917.4	440.2	35.025
Water, fresh	998.3	1.002	72.8
Water, sea	1025.6	1.09	73.0

*At body temperature.

Table D.2. Properties of water

T (°C)	ρ (kg/m^3)	ν (10^{-6} m^2/s)	Vapour pressure (Pa)
0	1000	1.788	—
0.01	—	—	611
10	1000	1.307	1227
20	998.3	1.004	2340

Contd...

30	996	0.802	4240
40	992	0.662	7380
50	988	0.555	12340
60	983	0.475	19920
70	978	0.414	31160
80	972	0.365	47360
90	965	0.327	70110
100	958	0.295	101300

Table D.3. Properties of air (at 1 atm pressure)

T (K)	ρ (kg/m³)	ν (10^{-6} m²/s)
250	1.413	9.49
293	1.220	14.75
300	1.177	15.68
350	0.998	20.76
500	0.705	37.90
1000	0.352	117.8

ANSWERS TO PROBLEMS

CHAPTER 1

1.2: 0.2 N/m^2 in $-x$ direction

1.3: 1.945×10^{-3} Pa s

1.11: 0.75

CHAPTER 2

2.1: $p_A - p_B = (\rho_2 - \rho_1)\, gh + \rho_2\, gh_1$;
$p_C - p_A = -(\rho_1 gh_2 + \rho_2 gh_1)$

2.2: 0

2.3: No

2.4: No

2.5: $p_1 = p_{atm} + (\rho_2 - \rho_1)gh_1 - \rho_1 gh_2$; No

2.8: $h^2 - (H+L)h + HL - p_{atm}h/\rho g$

2.10: 2.275×10^5 Pa; 2.226×10^5 N

2.11: 3.3×10^4 N

2.12: 3270 N

2.13: 0

2.14: $\sqrt{3}\ R$

2.15: 811.4 N on top bolts; 1671.6 N on bottom bolts

2.16: $\rho_{gas}gh$; 7.6 litres

2.17: $h = h_1\,(2\rho_{Hg} - \rho_{gas})/\rho_{gas}$

2.18: 27.82 m

2.19: 2.48 m; 1.08 m from top

2.20: 220 kg

2.21: 7.4 kg; 29 kg

2.22: 4:3 cm above interface

2.23: $\rho = W/8(L_1A + ha)$; $\Delta h/\Delta\rho = -W/ag\rho^2$

2.24: 26.1 cm; 19.5 cm

2.25: $2\sigma\cos\theta/\rho gd$

2.26: $R^3 p_{atm} + 2\sigma R^2 = p_{atm}\left(R_1^3 + R_2^3\right) + 2\sigma\left(R_1^2 + R_2^2\right)$

2.27: 80 Pa

2.28: 0.015 cm^3

2.30: $\rho g \mathring{Q}; \rho g \mathring{Q}$

2.31: 0.02 m

CHAPTER 3

3.1: $V_r = 0$; $V_\theta = \omega r$; $a_r = -\omega^2 r$; $a_\theta = 0$

3.2: 205 units

3.3: 0; Yes

3.4: Yes; $(\mathring{Q}/A_0)^2\, e^{-x}/(1 + e^{-x})^3$

3.5: $6\hat{\mathbf{i}} + 2\hat{\mathbf{j}}$; 4.99; 3.88

3.6: 0.1 °C/s; – 0.4 °C/s

3.7: Rs. 300/day

3.9: 5.9 km/hr

3.11: Streamlines

3.12: Streaklines

3.13: $\partial P/\partial t = (B-D)P/1000 - E + I$

3.14: $\rho \forall \dfrac{\partial m_A}{\partial t} = -R\forall + (m_{A,0} - m_A)\,\mathring{m}$; $m_A = m_{A,0} - R\forall/\mathring{m}$

CHAPTER 4

4.1: 26.3 m^2/s

4.2: 40 m/s

4.3: 24 km/hr

4.5: 0.426 m/s

4.6: $V_p(t)D/4h$

4.7: $(V_P A_P + V_A A_A + V_B A_B)/A_0$; $(\rho_A V_A A_A + \rho_B V_B A_B)/(V_A A_A + V_B A_B)$

4.8: 14.76 m/s at 88.35° from normal

4.9: $\dfrac{3}{8} V_0 d\delta/dx$

4.10: $-\dfrac{V_1}{a} + \sqrt{\left(\dfrac{V_1}{a}\right)^2 + \dfrac{2A_2 H}{A_1 a}}$

4.12: 18.35 cm; 0.98 cm

4.19: $V_r = V_0 \cos\theta\,(1 - R^2/r^2)$; $V_\theta = -V_0 \sin\theta\,(1 + R^2/r^2)$

4.20: $\overset{\circ}{Q}/2\pi r$; $\Gamma/2\pi r$

4.21: – 64

4.23: $e^{-x}\sinh y$; $e^{-x}\sinh y + y + \text{const}$; $V_0(y^2/ax^2 - 2y^3/3a^2x^3)$; $V_0(y^2/ax - y^3/3a^2x^2) + \text{const.}$

4.25: $f(z)/r$

4.26: $\partial\delta/\partial t + \partial(\delta V_x)/\partial x = 0$

4.27: $r\,\partial\delta/\partial t + \partial(rV\delta)/\partial r = 0$

4.28: $dV/dx - V/(L-x) = \omega/\tan\theta$ with $\theta = \theta_0 - \int_0^t \omega\, dt$

4.29: $\cos^2(\omega t - kx)(\partial V_x/\partial x) + (kV_x - \omega)\sin 2(\omega t - kx) = -4\overset{\circ}{q}/\pi A^2$

CHAPTER 5

5.3: No

5.4: 2.9×10^5 N

5.5: Compression

5.6: No

5.7: $\rho V_1^2 A$ downwards

5.8: $(9.7\,\hat{\mathbf{i}} + 17.8\,\hat{\mathbf{j}}) \times 10^3$ N

5.9: 62.5 N in tension

5.10: $\overset{\circ}{Q}_2 = 3\overset{\circ}{Q}_1/4$; $\overset{\circ}{Q}_3 = \overset{\circ}{Q}_1/4$; $F_n = \rho V_1^2 A_1 \sin 60°$

5.11: 6.2×10^5 N to the right

5.12: cannot be calculated

5.14: $\{\rho_f V(A_c - A_p)(V_p + V) + \frac{1}{2}\rho_f V_p^2 A_c\} / \frac{(A_c - A_p)}{A_p}$

5.15: 6.36 N; 3.18 N; 6.36 N; 0.98 N

5.17: $-58.9t^{-1/2} + 49(2 - \sqrt{t})^2$ N

5.19: $F_D / \frac{1}{2}\rho V_0^2\, LW \equiv C_D = \frac{1}{3}(\delta/L)$

5.21: $-\rho V_1^2 \cos^2\alpha_1 (\tan\alpha_1 + \tan\alpha_2)\, L$ (Width)

5.22: $V_0 \sin\beta_1/\sin(\beta_1 - \alpha_1)$; $V_0 \sin\beta_2/\sin(\alpha_2 - \beta_2)$ at angle given by $\sin\alpha_2/\sin(\alpha_2 - \beta_2)$ $= \sin\alpha_1/\sin(\beta_1 - \alpha_1)$; $\overset{\circ}{m}V_0 \sin\alpha_1(\cos\beta_1 + \cos\beta_2)/\sin(\beta_1 - \alpha_1)$

5.23: $1.5\,\rho V_1^2 A_1$; 4.39 m/s; 0.099 m

5.24: $h\left[1 + \frac{2V^2}{g}\left(\frac{1}{h} - \frac{1}{h_0}\right)\right]^{1/2}$

5:25: $(p_1 - p_2)(R^2 - r^2)/4\mu L$

5.26: $\tau = -(p_1 - p_2)\, r/2L$

5.27: $-h_1/2 \pm \sqrt{(h_1/2)^2 + 2h_1 V_1^2/g};\ h_1$

5.31: $\partial(V_x\delta)/\partial t + \partial(V_x^2\,\delta)/\delta x = -\delta g\, d\delta/dx$

5.32: $\rho V^2 A \sin\theta\, \hat{\mathbf{i}} - \rho V^2 A\,(1 - \cos\theta)\, \hat{\mathbf{j}}$

5.33: $(p_2 - p_1 + \frac{1}{3}\rho V_0^2)\pi R^2$

5.34: $\frac{5}{9}\rho V_1^2$

5.35: 44.18×10^{-3} N m

5.37: 1.92 N

5.38: –237.5 N m

5.39: $\rho V^2 RA;\ \mathring{m}/\rho RA$

5.40: $\sqrt{3}\, t_1/2$ from centre, upwards

CHAPTER 6

6.1: Yes

6.4: $-(C_1C_2/\mu^2 z^2)\,\xi/\left(1+\frac{\xi^2}{4}\right)^3;\ -(C_1C_2/2\,\mu^2 z^2)\,\xi/\left(1+\frac{\xi^2}{4}\right)^3$

6.6 : $V_x = 0,\ V_y = -\mathring{Q}/A$

6.10: $-\frac{3}{4}\left(\frac{V_0}{\delta}\right)\left(1-\frac{y^2}{\delta^2}\right);\ \frac{3}{10}\sqrt{\mu V_0^3\,\rho/x}\ ;\ 1.2\ \sqrt{\mu/\rho\ V_0\ L}$

6.11: $p_0 + p'x;\ p'y;\ -p'y/2\mu$

6.12: $1/\sqrt{29}\ (56\ \hat{\mathbf{i}} + 11\ \hat{\mathbf{j}})$ Pa

6.13: Straight line, slope $-a/g$

6.14: 5.98×10^4 Pa

6.15: 6.54 m/s^2; – 3.92×10^4 Pa; 19.6 m/s^2

6.16: 3.8 m/s^2, downwards

6.17: $\theta = 29.2°$

6.18: $T\,(b^2 - a^2)/4\,\pi\,\Omega\,L\,a^2\,b^2$

6.19: $T = 4\,\pi\,a^2\,L\,\mu\,\Omega$

6.21: $V_x = \rho\, g \sin\theta\,(h - y/2)\, y/\mu;\ \mathring{Q}/W = \rho g h^3 \sin\theta/3\,\mu$

6.22: $W(V_0 h - \rho g h^3/3\mu)$

6.24: $V_{y,w} = \rho_w g\,(xh - x^2/2)/\mu_w + \rho_0 gh\, x/\mu_w$;

$$V_{y,0} = \rho_0 g\,(2xh - x^2/2)/\mu_0 + \rho_0 gh^2 \left(\frac{1}{2}\mu_0\rho_w/\mu_w\rho_0 + \mu_0/\mu_w - \frac{3}{2}\right)/\mu_0$$

6.25: $V_{x,A} = \left(\dfrac{\mu_B}{\mu_A + \mu_B}\right)\dfrac{2V_0 y}{h}$; $V_{x,B} = \left(\dfrac{\mu_A}{\mu_A + \mu_B}\right)\dfrac{2V_0(y-h)}{h} + V_0$

6.26: $(\rho g/2\mu)\,[(R-h)^2 \ln R/r - (R^2 - r^2)/2]$

6.27: $\mathcal{P}_1 + \rho V_1^2 (1 - R_1^2/r^2)/2$

6.28: $(V_0 \ln r/R_2)/\ln R_1/R_2$; $2\mu\pi V_0^2 L/\ln R_1/R_2$

6.29: $\delta = 2\left[\frac{1}{2}R_1^2 \ln R_2/R_1 - \left(R_2^2 - R_1^2\right)/4\right]/(2R_1+\delta)\ln(R_1/R_2)$

6.30: $(\rho g R^2/4\mu)\,[(1 - r^2/R^2) + 2(R_1/R)^2 \ln(r/R)]$

6.31: $W(V_0 b - 2p'b^3/3\mu)$; $W(\mu V_0/2b - p'b)$

6.32: $3\mu V_0 (l/2b)^2$

6.33: $W[3\mu V_0 (L/2b)^2 + \rho g L^2 (\sin\theta)/2]$

6.34: No

6.35: $(24\pi R\,\mu/\delta^3)(\delta\omega R/2 - \mathring{Q}/W)$; $4\pi\mu\omega (R/\delta)^2 (2W\delta\omega R - 3\mathring{Q})$

6.36: 15.7 N s/m^2

6.39: $(\mu\omega/\cot\theta_1)\, 2\pi R^3/3$, 1.6 Pa s

6.40: $4h_0 T/\omega\pi R^4$

6.41: $\pi(p_0 - p_{atm})(R^2 - R_0^2)/2\ln(R/R_0)$; $(4\pi h^3/3\mu)(p_0 - p_{atm})/\ln(R/R_0)$

6.43: 0.02 N

6.44: $\sin^{-1} g/\omega^2 L$; $m\omega^2 L$; $\sin^{-1} g/\omega^2 L$; $m\omega^2 L(1 - \rho_f/\rho_s)$

6.48: $[\mathring{q}/(R_2 - R_1)]^2 \rho \ln(R_2/R_1)$

CHAPTER 7

7.3: $\mu\omega^2 d^2/\rho C_p (D-d)^2$

7.6: –1.156 W; – 6.9 W

7.7: from 1 to 2

7.8: turbine; 3 to 4

7.9: 250 W

7.10: – 4.09 × 10^4 Pa gauge; 4.8 m/s

7.11: $(p_{atm}/\rho g)(z_i - z_f)/(L + z_i - z_f) - h_s^*$

7.12: 722.12 Pa gauge

7.13: 2

7.14: $R_j = 0.84\,R$

.15: 1.69×10^5 Pa; 0.26 m; 6×10^{-4} °C

.19: No

.20: 7.9 m/s; 8×10^{-4}

.21: $R = 0.631\ y^{1/4}$

.22: 9.3 s

.24: 6.26 m/s; 8 m from bottom

.25: 88.86 m/s

.26: 7.93×10^4 N

.27: 118 gm

.28: 860 kg/m^3

.30: 104.1 N tension

.33 3.6 kW

.34: $\rho A_j(2w_s - 3gh_j)\left(\frac{3}{2} - \frac{A_j}{A_1}\right)$; $\left(\frac{A_j}{A_1}\right)(2w_s - 2gh_j)$; $2(w_s - gh_j)$

.35: 0.694; 0.538

.36: $A_p^2\ (V_1^3 A_3 + V_j^3 A_j) - (V_j A_j + V_1 A_3)^2 = 2A_j A_3\ (V_j A_j + V_1 A_3)\ (V_1 - V_j)^2$ with $A_3 = A_p - A_j$

.37: 2.08×10^5 Pa; 1.81×10^5 Pa; –451 N; 259.6 N

.38: $p_{atm} + \rho\ (\mathring{Q}/8\pi^2\delta^2)\ (1/R_c^2 - 1/R_B^2)$; $p_{atm} - \rho\ V_A^2/2$

.39: 61 W; 0.62 cm; 6.1 N

.41: $p_{atm} + \rho\ (V_2^2 - V_1^2)/2$; 3.09 cm

.42: 2.3 m/s

.43: 3.13 m/s; -9.5×10^4 Pa; -9.02×10^4 Pa; -9.5×10^4 Pa

.44: $\frac{\pi}{4} D^2 \sqrt{V_1^2 - 2gh_2}$

.45: 83.6 kW

.46: 0.16m, 0.45m

.47: $\frac{8}{15} C_d \sqrt{2gh^5}\ \tan\frac{\alpha}{2}$; 0.025 m^3/s

.48: –0.17 m; 0.22 m; 0.78 m; –168.5 N/m

.49: –0.48 m; 0.55 m; 3.89 m

.51: $p_0 + \rho V^2/2$

.57: k/r; $p_0 + \rho k^2/2r^2$

CHAPTER 8

.1: $p_{2,g} A_2 + \rho_1 V_1 A_1\ (V_2 - A_1) + \mathring{m} V_2$

.2: $p_{2,g} A_2 + \mathring{m} V_2$

.3: 97.5%; 461 kW

8.4: 34.5 N; 22.4 Pa; 104.5 W
8.5: 1.56 kW
8.6: 468.5 kW
8.9: 1.4 MW; 9.9 kW
8.10: 15.7 MW
8.11: 0.07 m^3/s; 30.7 kW
8.12: 7.05 m/s; 3.96 cm; 10.54 kW
8.13: $-\overset{\circ}{W} = C_d \pi D_c h \sqrt{2/\rho}\,(4F/\pi D_c^2)^{3/2}$
8.14: $D_c/8h$; $(1 + \cos\theta)D_c/4h$
8.15: 21.54 kW; 27.58 kW
8.17: 0.012 m^3/s; 1.72 m^2/s^2
8.18: 16.31 cm
8.19: 0.967; 0.601
8.20: 20.5 cm

CHAPTER 9

9.5: $V_0 t_0/L$; V_0/c_0
9.9: 4.56 knot; 750 knot; 136.9 knot; Froude
9.10: 2.5 m/s; 234.4 m^3/s; 31.25 kN; 2143
9.11: 127.3 RPM; 1.77 kW
9.12: $V_m = 10^4$ kmph, No (compressibility important)
9.13: 1.33 m/s; 150 N m
9.14: 300 km/hr; 36.4 N
9.15: 1.25 m/s; 77.4 cm
9.16: 0.075 m/s; 2×10^4 Pa
9.17: 0.08 RPM; 2.35×10^{-10}
9.18: 1.5 m/s; 4.12 N; No
9.19: 0.067 m/s; 0.1 Hz
9.20: No
9.21: 0.05 m/s; 0.625 hr; 1.6×10^5
9.22: 759 m/s; 1.88×10^5 N m
9.23: $\sqrt{k_{E_s}/k_\rho}$; $(1/k_L)\sqrt{k_{E_s}/k_\rho}$; $k_\theta = 1$
9.25: $k_\rho\ k_V^2 k_L^2$; 0.98 m/s
9.26: 14.43 m/s; 17.28 kN
9.27: 1.07×10^4 Pa; $\sqrt{15}$
9.28: $k_L^{-1}\sqrt{k_T/k_M}$; $k_p k_L^2 \sqrt{k_M/k_T}$; $k_p k_L^3$; k_T/k_M

9.30: $\sqrt{k_{a+g}/k_L}$

9.33: 86.2 cm; 1.293

9.34: $\Delta\rho L^2 g/\mu V$

9.35: 0.546; 0.122

9.36: 0.01 m^3/s; 87.8 Pa; 5.86 kW

9.37: 61 Pa; 820

9.38: 223.6 RPM; 5.75 min

9.39: 11.2 m^3/s

9.40: $\sqrt{k_L}$; Yes

9.41: 1; 4; 1

9.42: 2.12 m/s; 0.5

9.43: 1/2; $1/\sqrt{2}$

9.44: 111.9 kg

9.45: 10^{-6}; 100; 1; 1 Lilliputians; Lilliputians; Brobdingnagians

9.46: $\Delta p\, D^4/\rho \mathring{Q}^2 = \mathcal{F}(\rho\mathring{Q}/\mu D, L/D, \in/D)$

9.47: $\mathring{W}/\rho N^3 D^5 = \mathcal{F}(\rho N D^2/\mu, N^2 D/g, \text{geometry})$

9.48: $\mathring{Q}/\sqrt{gH^5} = \mathcal{F}(\rho\sqrt{gH^3}/\mu, W/H)$

9.49: $d/D = \mathcal{F}(\rho VD/\mu, \rho DV^2/\sigma)$

9.50: $\Delta p/\rho_f V^2 = \mathcal{F}(\rho VD_p/\mu, L/D_p, \in)$; 0.1; 10^{-3}

9.51: $\mathring{m}/\rho\sqrt{gt^5} = \mathcal{F}(\rho\sqrt{gt^3}/\mu, b/t)$; $\mathring{m} \sim \rho^2 g b t^3/\mu$

9.53: $\mathring{Q}\mu/\rho g h^3 = \mathcal{F}(\alpha)$

9.54: $V/\sqrt{gD} = \mathcal{F}(\rho\sqrt{gD^3}/\mu, \rho g D^2/\sigma, \forall/D^3)$; $\dfrac{\rho g D^2}{\sigma}$

CHAPTER 10

10.4: 1.235; 1.694; 1.694

10.5: $L/D > 3730$

10.6: 8.5 cm

10.7: 466.5 litre/hr

10.8: Yes; No

10.9: 6.28 MW; 1.6%

10.10: 1.4×10^4 MW

10.11: Yes, pump needed

10.12: $(\pi d^2/4)\sqrt{(V^2 - 2gh_2)/(1 + f\ L/d)}$

10.13: 1.9 MW

10.14: 1.4 m/s; $p_{atm} - 7.25 \times 10^3$ Pa

10.15: 1.68 kW

10.16: 31.36 m

10.17: 33.02 m

10.18: 522 Pa

10.20: 805 N; No; underestimate

10.22: $p_{atm} - \rho_{air} V_3^2 (A_3/A_2)^2/2$;

$(-\rho_{air}/\rho_g)(A_3/A_2)^2(V_{3,air}^2/2g) + (\mathring{Q}^2/A_0^2)(1 + fL/D + \Sigma K_i)/2g$

10.25: 5.2×10^6 Pa

10.27: 1.4 m^3/s

10.33: 8×10^5 Pa; 0.373 m^3/s; 0.9; 331.4 kW

10.35: 117.7 rad/s; 0.83 m

10.36: 136 RPM; 1.67 m

10.37: 145 rad/s; 1.48 MW

10.39: 3.58 m

10.40: 0.98 m^3/s; 3.125 MW

CHAPTER 11

11.1: $Gr/Re_L^2 << 1$; Yes; $\rho_0 g\beta (\Delta T)\ L_c^2/\mu$

11.2: $\rho_s \omega D^2/18\mu << 1$

11.3: $L/c_0t_0 << 1$, $V_0/c_0 << 1$; $L/c_0t_0 << 1$

11.4: $(D_{AB}/V_{max}\ D)L/D << 1$; $kL/V_{max} << 1$

11.6: 1 s; 2.2 s

11.7: $\sigma/\rho V_0^2 L_0 << 1$ imples negligible surface tension effects

11.8: $\sim\sqrt{\nu t}$

CHAPTER 12

12.2: *b*, *c*, *d*, *e* and *f*

12.5: *xy* + const.

12.8: 0; $(\rho q^2/8\pi^2)(1/r_1^2 - 1/r_2^2)$; $(\rho\Gamma^2/8\pi^2)(1/r_1^2 - 1/r_2^2)$

12.9: $q/2V_0$

12.10: $p_0 - \rho V_0^2\ (\xi^2 + 2\xi\cos\theta)/2$ with $\xi = q/2\pi r V_0$

12.12: width far away = 25 m

12.14: $\sqrt{q/\pi A}$

12.15: $k_1\theta - k_2 \ln r$; $2\pi k_1$; $2\pi k_2$

12.16: $-(q/2\pi)\ln r + \Gamma\theta/2\pi$; $-q/2\pi r$, $\Gamma/2\pi r$;
$p_{atm} - \rho(q^2 + \Gamma^2)/8\pi^2 r^2$; $\Gamma/2\pi R^2$;
$p_{atm} + \rho\Gamma^2 r^2/8\pi^2 R^4 - \rho(q^2 + 2\Gamma^2)/8\pi^2 R^2$

12.17: $q = 2.5 \times 10^4$ m²/s; $\Gamma = 4.04 \times 10^4$ m²/s; 18.09 m/s;
34.36 m/s; $p_{atm} - 387.2$ Pa

12.19: $\Delta z = 45.87/r^2$, z and r in cm; 0.03 cm; 1.83 cm

12.20: $p = p_0 - (\rho\Gamma/2\pi r)(\Gamma/4\pi r - V_0 \sin\theta)$

12.21: $p(x) = p_0 - \rho(qx/\pi)^2/2(x^2 + a^2)^2$

12.22: $x = \pm\sqrt{a^2 + ma/\pi V_0}$; thickness $2h$ from

$$\tan\left[\left(\frac{h}{a}\right)\left(\frac{2\pi V_0 a}{m}\right)\right] = 2\left(\frac{h}{a}\right)/\left[\left(\frac{h}{a}\right)^2 - 1\right]$$

12.24: $x = \pm\sqrt{\Gamma a/\pi V_0 - a^2}$; $2.45V_0$

12.25: $q/2\pi V_0 a > 1$; halfbody

12.26: $p_0 - \rho\,[\Gamma a/\pi(x^2 + a^2)][2V_0 + \Gamma a/(x^2 + a^2)\pi]/2$
$p_0 - \rho\,[\Gamma a/\pi(x^2 + a^2)][V_0 + \Gamma/4\pi a]/2$

12.27: $p_0 - \rho\,(q^2/\pi^2)(x^2/2)/(x^2 + a^2)^2$

12.28: $V_{x,w} = V_0 - \Gamma a/\pi(x^2 + a^2)$;
$p_w = p_0 + \rho\,[\Gamma a/\pi(x^2 + a^2)]\,[2V_0 - \Gamma a/\pi(x^2 + a^2)]/2$

12.30: $p_0 - \rho\left[\frac{q}{\pi}\left\{\frac{x-a}{(x-a)^2 + a^2} + \frac{x+a}{(x+a)^2 + a^2}\right\}\right]^2/2$

12.31: $-\rho V_0^2 R/6$; $-5\rho V_0^2 R/6$

12.32: $[2(p_{in} - p_0) + \frac{5}{3}\rho + V_0^2]R \times$ width; 54.76°

12.33: $p_A - p_B = 2\rho V_0^2 \sin(2\theta_1)\sin(2\gamma)$

12.34: 30°

12.35: 8/3

12.36: 6.02 kN

12.37: 3.12×10^5 N; 14.55 m/s

12.38: 281.6 RPM

12.42: 2.3 m

CHAPTER 13

13.1: $4.8x/\sqrt{\mathrm{Re}_x}$; $1.31/\sqrt{\mathrm{Re}_L}$

13.2: 0; 3/2; 0; –1/2; $4.64x/\sqrt{\mathrm{Re}_x}$; $0.6465/\sqrt{\mathrm{Re}_x}$; $1.29/\sqrt{\mathrm{Re}_L}$

13.3: 1.6 W

13.5: 2.64×10^6 N; 38.86 MW; 1.342 m

13.6: 7.66×10^4 N

13.9: $1.83x/\sqrt{\mathrm{Re}_x}$

13.10: $0.73x/\sqrt{\mathrm{Re}_x}$

13.16: B

13.19: 1.38×10^4 N

13.30: $\mu V_0/\delta$; Yes

13.31: No

13.32: 1790 kg/m^3

13.34: 73.7%; 79.9%

13.35: 4392 N; 6.6×10^4 N m

13.36: 30.5 kW

13.37: 20 N

13.38: $C_{D,B}A_B\left(\frac{1}{2}\rho V_b^2\right) = C_{D,0}\,\frac{1}{2}\rho\,(\omega l - V_b)^2\,\pi R^2$

13.39: 533.3 Hz

13.40: 104 KMPH; 70 KMPH

13.41: 0.92 m

13.42: $1.036\ V_{\text{upper critical}}$

13.43: – 0.0143 N m, 112.3 RPM

13.44: 9.2 s

13.45: $F_L \sin\theta + F_D \cos\theta = -m(dV_x/dt)$;

$$\sqrt{V_x^2 + V_z^2}\ (C_L V_x - C_D V_z) - \frac{4}{3} D \frac{\rho_s}{\rho_f} g = \frac{4}{3}\frac{\rho_s}{\rho_f} D\,(dV_z/dt)$$

13.48: 5 m

13.49: 2.5×10^{-3} m/s

13.51: $x = (\Delta\rho\, gD^2/18\mu)\ [t + (\rho_s D^2/18\mu)\ \{\exp(-18\mu t/\rho_s D^2) - 1\}]$;
$V = (\Delta\rho\, gD^2/18\mu)\ [1 - \exp(-18\mu t/\rho_s D^2)]$; 1.87×10^{-3} s; 3.6×10^{-6} m

13.52: 6.54 mm

13.53: 2.73 cm

13.54: $bH + A\tan^{-1}(D_0/A)$ with $A = \sqrt{a\Delta\rho g/18\mu\, b}$

13.55: $(18\mu V_0 z/\Delta\rho\, gX)^{1/2}$; straight lines

13.56: $r = \pi\,(\rho_s - \rho_f)\, D^2 V_1^2\; R_1 L/9\mu\mathring{Q}$

CHAPTER 14

14.3: 1.14

14.4: 8.99 N ; 10.8°

14.5: 0.537; 1.4°; 4.5×10^4 N; 11.7 MW; 173 m/s

14.6: 1.45°, 39.1 km; 1.72°, 18.9 km; 3.2 km

14.7: 0.57°; 21.88 m/s

14.8: 34.5°; 773.8 N/m; 466.5 N m/m; 20.4°; 864.1 N/m; 528.1 N m/m

14.9: 55.6 m/s; 4.37 kN; 338.5 kW

14.10: 707.5 RPM

14.11: 1826.7 RPM; 63.8 RPM; 84.7 kW; 2.96 kW

14.12: $\mathring{W}_2 = 8\mathring{W}_1$

14.13: 16.55 RPM

14.14: 10^4 kg; 10.97 kW; No

14.15: 19.3 m/s

14.16: 8.93×10^5 N

14.17: 2.84×10^5 N; 10 m/s

CHAPTER 15

15.1: 10.8 km

15.2: 34.6 km

15.8: 263.9 m/s; 304.5 m/s

15.9: 2.8 kg/s; 1; 249.9 K

15.11: 4.48×10^4 Pa; 169.8 K; 2.6; 4.7 kg/s; 4.48×10^4 to 8.76×10^5 Pa

15.12: 44.5×10^5 N/m^2

15.13: 329.8 K; 1.39×10^5 Pa; 1.9; 27.5°

15.14: 1120.8 m/s; 844.5 m/s

15.15: 117.6 °C

15.16: 0.06 m^2; 1.28×10^4 Pa; 166.7 K; 1.38; 1.54; 2.58×10^4 Pa; 203.7 K; 0.69; 6.68×10^4 Pa; 274 K.

INDEX

T

U